Malaysian Technical University Conference on Engineering and Technology (MUCET 2017)

IOP Conference Series: Materials Science and Engineering
Volume 318

Penang, Malaysia
6 – 7 December 2017

ISBN: 978-1-5108-6036-0
ISSN: 1757-8981

Printed from e-media with permission by:

Curran Associates, Inc.
57 Morehouse Lane
Red Hook, NY 12571

Some format issues inherent in the e-media version may also appear in this print version.

Printed by Curran Associates, Inc. (2018)

For permission requests, please contact the Institute of Physics
at the address below.

Institute of Physics
Dirac House, Temple Back
Bristol BS1 6BE UK

Phone: 44 1 17 929 7481
Fax: 44 1 17 920 0979

techtracking@iop.org

Additional copies of this publication are available from:

Curran Associates, Inc.
57 Morehouse Lane
Red Hook, NY 12571 USA
Phone: 845-758-0400
Fax: 845-758-2633
Email: curran@proceedings.com
Web: www.proceedings.com

Table of contents

Volume 318

Malaysian Technical Universities Conference on Engineering and Technology 2017 (MUCET 2017)

6–7 December 2017, Penang, Malaysia

Accepted papers received: 16 February 2018
Published online: 19 March 2018

Preface

Papers

IOP Conf. Series: Materials Science and Engineering **318** (2018) 011001 doi:10.1088/1757-899X/318/1/011001

Malaysian Technical Universities Conference on Engineering and Technology
(MUCET 2017)
6th – 7th December 2017, Park Royal Hotel, Penang, Malaysia

Preface

MUCET 2017 aims to provide a platform for academics and researchers to share knowledge and ideas, widen networking, present their research findings and explore any other potential benefits that might be contribute to the better future.

This conference highlights the contributions of innovative research in science, engineering and technology. Topics presented includes of Electronic and Electrical Engineering, Civil Engineering, Chemical Engineering & Natural Resources, Mechanical & Manufacturing Engineering, Education, Social Science & Technology Management, Science, and Information & Communication Technology.

Research fields of the presented papers including:

- Electronical and Electronic Engineering (EE)
- Civil Engineering (CE)
- Chemical Engineering & Natural Resources (CHEM)
- Mechanical & Manufacturing Engineering (MME)
- Education, Social Sciences & Technology Management (ESTM)
- Information & Communication Technology (ICT), Science (SCI) & Mathematics (SM)

IOP Conf. Series: Materials Science and Engineering **318** (2018) 011002 doi:10.1088/1757-899X/318/1/011002

Malaysian Technical Universities Conference on Engineering and Technology (MUCET 2017)
6th – 7th December 2017, Park Royal Hotel, Penang, Malaysia

Welcoming Note

CHAIRMAN OF MALAYSIAN TECHNICAL UNIVERSITIES CONFERENCE ON ENGINEERING AND TECHNOLOGY (MUCET 2017)

Assalamualaikum w.b.t and Salam 1Malaysia.

It is great pleasure to welcome all delegates to the Malaysian Technical Universities Conference on Engineering and Technology (MUCET 2017).

Competitive advantage in today's globalized economy lies principally in fostering the skills, knowledge, creativity and competencies of people and these cuts across nations, regions, sectors and organizations. This is very significant as it is explicit recognition that individual human development is fundamental to our collective development as societies. As the success in the 21st century is driven by excellence in engineering and technology innovation, by pushing the boundaries of knowledge and braving through discoveries of new, innovative and improved product and processes, we, as technical universities must play key role in fostering these skills and provides the best effort to be the pillar for these success. While we face significant competition for developing and instilling the engineering and technology skills within our region, other nations are also putting in place measures to boost and strengthen their capacity. Hence we must seize every opportunity to stand at par with other countries, while making sure sustainability and dynamic growth of Malaysia economy.

The theme of this year's MUCET, "Leading Innovation in Engineering, Science and Technology" is indeed in tandem with the government's strategic initiatives at advancing, in a more effective manner on innovative human capital towards economic growth by utilizing every aspects of engineering, science and technology. Innovation, in deed is the key to global competitiveness that be able to provide new and better jobs, a resilient economy and strategic key in gaining the national goals. As a developing nation, Malaysia relies on its human capital and universities have a pivotal role in nurturing and producing this much-needed pool of competent and highly skilled labour force at every levels in our endeavour towards national transformation 2050 (TN50).

I sincerely wish that this conference will provide a useful platform for participants, especially between technical universities academia, to interact, brain storming, and come up with concrete resolutions towards more effective collaboration in the area of science, engineering and technology. It is hoped that from this conference, participants can grab the opportunities to emerge stronger and more dynamic in research to stimulate the development of the new products, processes and so that they will match with the latest global industrial trends.

On behalf of the Research, Management and Innovation Centre, UniMAP, I would like to record my deepest gratitude and thanks to all MUCET 2017 committee, Universiti Malaysia Perlis (UniMAP), other MTUN research and management offices and as well as to all parties, for your dedication in ensuring the success of this conference.

Thank you.

PROFESSOR DR SYED ALWEE ALJUNID SYED JUNID
Chairman
Malaysian Technical Universities Conference on Engineering and Technology (MUCET 2017)

IOP Conf. Series: Materials Science and Engineering **318** (2018) 011003 doi:10.1088/1757-899X/318/1/011003

Peer review statement

All papers published in this volume of *IOP Conference Series: Materials Science and Engineering* have been peer reviewed through processes administered by the proceedings Editors. Reviews were conducted by expert referees to the professional and scientific standards expected of a proceedings journal published by IOP Publishing.

Practicability of Lignocellulosic Waste Composite in Controlling Air Pollution from Leaves Litter through Bioethanol Production

Mahadevan Tarrsini[1], Yi Peng Teoh[1], Qi Hwa Ng[1, *], Balakrishnan Kunasundari[1], Zhong Xian Ooi,[2] Hoong Siew Shuit[1] and Peng Yong Hoo[1]
[1]Department of Chemical Engineering Technology, Faculty of Engineering Technology, Universiti Malaysia Perlis (UniMAP), 02100, Sg. Chuchuh, Padang Besar, Perlis Malaysia
[2]Department of Chemical Science, Faculty of Science, Universiti Tunku Abdul Rahman, Jalan Universiti, Bandar Barat, 31900, Kampar, Perak, Malaysia.

E-mail: qhng@unimap.edu.my

Abstract. Environmental degradation through greenhouse emission have spurred nation's interest on feedstock-based fuel. Yet, development of this clean biofuel is obstructed by the expensive feedstock which takes up most of the production cost. Therefore, as an alternative, utilization of widely available lignocellulosic residues with relatively no commercial significance has been considered. This present work emphasizes on mango (*Mangifera indica*) leaves one of the most abundant lignocellulosic waste in Malaysia. Through implementation of this biomass for bioethanol production, continuous allowance of air pollution with a deleterious impact to the country's environment could be reduced. The high concentration of sugar (16-18%w/v) in the form of cellulose and hemicellulose is ultimately the reason behind the selection of these leaves as a substrate for bioethanol production. Hence, in this study, a comparison of biomass composition in Harum Manis, Sunshine and Chokanan mango leaves were conducted to detect the most suitable substrate source to produce biofuel. At the end of the biomass evaluation, Harum Manis mango leaves turned out to be the most competitive bioethanol crop as these leaves reported to be made up of 34.71% cellulose and 44.02% hemicellulose which summed up to give highest fermentable sugar source with a lignin content of 19.45%.

1. Introduction

Bioethanol has received much attention in recent years due to its potential as a clean sustainable petroleum based fuel that offers vital economic values [1]. Besides its contribution as an alternative fuels source, bioethanol have also become an ultimate priority of chemical industry in terms of solvent, disinfectant and sterilizing agent [2]. Initially, bioethanol is obtained from conversion of reducing sugar stored in sugarcane, corn and maize through fermentation process [3]. However, as the demand for bioethanol increased tremendously, utilization of human feedstock for reducing sugar supply has been less practicable [4]. Hence, to ensure a long sustainability of bioethanol, biotechnologists considered non edible lignocellulose material (LCM) as a promising option to produce bioethanol [5].

There has been extensive research regarding the use of LCM as a source of fermentable sugars for bioethanol production because these substrates are inexpensive and highly available. Thus, in terms of raw material, there will be a great reduction in cost [6]. Yet, there is not much recognition given to agricultural leafy residues, specifically leaves of mango (*Mangifera indica*), a LCM source with a potential to produce bioethanol. *Mangifera indica* of Anacardiaceae family is well distributed in tropical regions, especially in Malaysia, India, Thailand, Philippines, China and Srilanka [7]. In Malaysia, the equatorial climate facilitates a vastly growth of mangoes [8]. Harum manis, Chokanan and Sunshine are known to be among the popular mango varieties in Malaysia [9].

Although this tropical fruit have recorded an overall production of 42 million tons worldwide for its juice and rich taste, the remaining part of the plant remains as waste [10]. The leaves are generally the most widely available crop residue apart of the flowers, stems and barks. These mango leaves usually overcrowds the mango plantations during tree pruning and harvesting season. This scenario leads to an alarming disposal issue in various countries [11]. In some countries, the open burning of collected leaves waste is practiced as a step to manage the leaves waste. However, this action led to air pollution whereby 85% particle of the leaf smoke could cause severe chemical effect to the lung when inhaled [12]. Eventually, these airborne particles will aggravate to chronic respiratory illness and shorten life span of humans. Furthermore, when the burning goes out of control, the nation will be economically challenged to overcome the cost of property damages, personal injuries and fire station services [13].

Therefore, using the mango leaves as LCM for producing bioethanol could be a strategic goal to reduce air pollution. According to Das and co-workers [14], cellulose (26.3%) and hemicellulose (54.4%) content of leafy biomass could be promptly hydrolysed into fermentable sugar that subsequently is bio transformed into ethanol. However, there is no any clear data on the variety of mango leaves that could aid in bioethanol production. Hence, this paper presents a comparison between mango leaves varieties with regards to biomass composition through gravimetric analysis, as this is a crucial step for determination of the most viable substrate in bioethanol production.

2. Materials and Methods

2.1 Chemicals

The extraction solvent, acetone with a Molecular Weight (MW)= 58.08 g/mol was purchased from Fisher Scientific. Sodium hydroxide (MW = 40 g/mol) was supplied by MacRon. Sulphuric acid (Purity= 96%: MW=98.079 g/mol) was received from Fisher Scientific. Distilled water was used as a diluting agent and for washing the biomass.

2.2 Substrate Preparation

The dried leaves of Harum manis, Chokanan and Sunshine mango (*Mangifera indica*) were collected from Institute of Sustainable Agrotechnology (INSAT), Perlis, Malaysia. The collection was carried out to accumulate 500 grams of leaf sample for each mango variety. The leaves were washed with running tap water thrice to ensure removal of dirt and dust particles and then cut into small pieces. Next, the leaf samples were dried overnight in oven (Memmert, Germany) at 80 °C. For an efficient fermentation process, the leaves were then ground using laboratory blender (GM 200, Retsch, Germany) and allowed to pass through via 1 mm mesh sized sieve shaker (Retsch, Germany). Lastly, the sample of powdered mango leaves as illustrated in Figure 1 (A) was stored in an air tight container at -20 °C chest freezer (REMI, Malaysia).

2.3 Moisture Content Analysis

The moisture content of mango leaves before grinding process was determined through oven drying method. Initially, the weight of the mango leaves was recorded using an analytical balance (Sartorius, Germany). Next, the leaves were well distributed on the aluminium tray and inserted into the oven (Memmert, Germany). Then, the sample was dried for an hour at 105°C to ensure complete removal of moisture. The dried leaves sample were measured for the moisture content analysis by using the formula as in e.q (1):

$$\text{Moisture content (\%)} = \frac{W_2 - W_3}{W_2 - W_1} \times 100 \qquad (1)$$

Where,

W_1= weight of aluminium tray, g
W_2= weight of aluminium tray + sample, g
W_3= weight of aluminium tray + oven dried sample, g

2.4 Extractives

For soxhlet, 2.5 grams of dried leaves samples was first loaded into thimble followed by filling up the round bottom flask with 150 mL of acetone. The soxhlet apparatus (DXY, China) was then assembled. Subsequently, the temperature was set up at 70°C for 4 hours. After extraction, the sample was air dried at room temperature for 10 minutes followed by conventional oven drying with a temperature of 105°C until a constant final weight was achieved. The sample free of extractive in demonstrated in Figure 1 (B). Lastly, the % (w/w) of extractive was calculated based on differences between initial (2.5 grams) and final weight.

2.5 Hemicellulose

1 gram of extract free dried biomass was transferred into 250 mL conical flask and added with 150 mL of 500 mol/m^3 sodium hydroxide (NaOH). The mixture was heated at 95 °C in water bath shaker (WiseBath, Korea) in static mode for 3.5 hours. After that, the mixture was neutralized by washing with distilled water. Subsequently, the mixture was separated into supernatant and pellet via centrifuge (Thermo Scientific, U.S.A) at 4000 rpm for 20 minutes. Finally, the residue (pellet) was dried at 105°C until a constant weight was attained. Figure 1 (C) shows the mango leaves powder with absence of hemicellulose content. The hemicellulose content % (w/w) was equalled to changes in weight of sample before and after sodium hydroxide treatment.

2.6 Lignin

0.3 grams of dried extract raw material was weighed and inserted into a test tube. Subsequently, 3 mL of 72 % of H_2SO was pipetted into the test tube. The sample was kept in room temperature for 2 hours. Throughout the incubation period, the test tube was vortexed by (Ika® Genius 3 Vortex, Sigma-Aldrich, U.S.A) every 30 minutes to complete the hydrolysis. Next, 87 mL of distilled water was poured into the test tube for an autoclave (Hirayama, Japan) session of 1 hour at 121°C. The slurry was cooled down at room temperature and then centrifuged at 4000 rpm for 30 minutes. Acid insoluble lignin was equivalent to the pellet dried at 105°C.The drying process was carried out until a constant weight of lignin was achieved. The remaining mango leaves powder with lignin content are shown in Figure 1 (D).

2.7 Ash

The sample hydrolysed to quantify the lignin content was further used for ash determination. The sample was incinerated at 575°C in a muffle furnace (Cole Parmer, U.S.A) with a heating rate of 20°C/min. Then, the sample was removed from furnace and placed in a desiccator until the sample cooled down. At last, the ash was determined by weighing the sample in the crucible. Figure 1 (E) demonstrates the left-over sample of mango leaves which represents the ash composition of the biomass.

2.8 Cellulose

The cellulose content was determined by elimination of amount of the hemicellulose, lignin, ash and extractives in term of percentage from 100% (total amount of biomass).

IOP Conf. Series: Materials Science and Engineering **318** (2018) 012001 doi:10.1088/1757-899X/318/1/012001

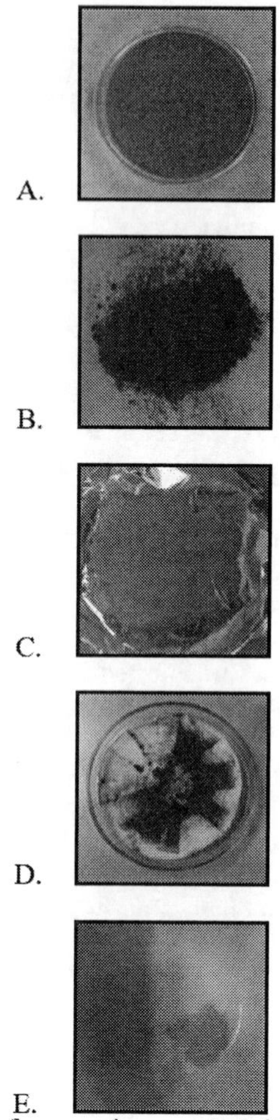

Figure 1: Biomass composition of mango leaves; A. Initial powder, B. Extractive free, C. Hemicellulose free, D. Acid insoluble lignin and E. Ash.

3. Result and Discussion

Figure 2 is a bar chart that illustrates moisture content and three main organic fractions (cellulose, hemicellulose and lignin) along with minor constituents (extractives and ash) of Harum Manis, Chokanan as well as Sunshine mango leaves. Each set of experiments was replicated thrice to obtain a more accurate and precise results. With this, the values reported is in the mean of 3 replicates ± standard deviation.

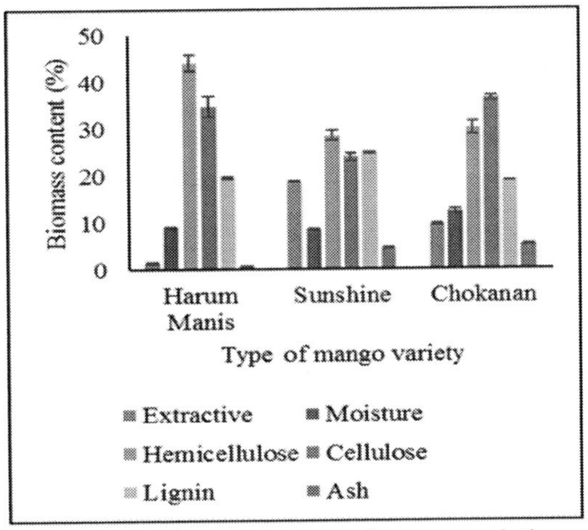

Figure 2: Biomass content of mango varieties

In term of bioethanol application, feedstock with minimal moisture content or hydrophilic nature is given priority as this is a promising feature for a substrate. With this, there is commonly an urge to determine hydrophilic traits of the leaves. Since, the moisture content of the biomass correlates to the hydrophilicity of the biomass, an analysis of the water content is crucial in the selection of an appropriate substrate for fuel biotransformation [2]. Hence, a moisture analysis was carried out whereby leaves of Chokanan mango have recorded maximum reading of 12.38±0.48% followed by Harum Manis (8.94±0.17%) and Sunshine (8.50±0.16%). Yet, Chokanan leaves with highest moisture adsorption are not preferred because the biomass will exhibit low density [15]. According to previous findings, biomass fuel with high density is more economical to be transported, stored and utilized compared to low density biomass. Thus, based on data presented above, both Harum Manis and Sunshine with relatively lower moisture content would satisfy the role as substrate to produce bioenergy in comparison to Chokanan.

Cellulose is known to be a linear sequence of homopolysaccharides made out of glucose units linked consecutively by β-(1,4) glyosidic bonds through condensation reaction [16]. According to Saifuddin and his co-researchers [5], the effectiveness of lignocellulosic crop to bioethanol majorly depends on the carbohydrate source which includes both cellulose and hemicellulose. These constituents of LCM are prominent extent of polymeric sugar that could be hydrolysed into monomers of pentose and hexose sugar to be promptly aided in biofuel production. Among three varieties of mango leaves investigated, Chokanan mango leaves resulted the greatest cellulose composition of 36.42±0.55%. However, the content of cellulose in Harum Manis mango leaves were only 1.72% lower compared to Chokanan. Hence both Chokanan and Harum Manis mango leaves would make a reliable candidate for bioethanol production.

Subsequent to cellulose, hemicellulose is the most inexhaustible (20-50%) sugar source in lignocellulosic biomass [17]. Hemicellulose is a configuration of branched heterogeneous polymer structured by hexoses (galactose, glucose and mannose), pentose's (eg: arabinose and xylose), as well as uronic acids (4.0- methyl galacturonic acids and galacturonic acid) [18]. Hemicellulose could be found in plant cell wall in the form of xylans, glucomannans, xyloglucans and mannans [19]. Through the experimental procedure for hemicellulose determination, the respective content for all three mango varieties were found to be in the range of 28-45%. Furthermore, this part of the study has also recommended that Harum Manis mango leave with the greatest hemicellulose content of 44.02±1.81% would aid the bioethanol production in a most significant manner. This is because apart of cellulose, biofuel orientated biomass is also highly dependent on its hemicellulose content for a worthy conversion process [4].

Lignin is comprised of coumaryl alcohol, sinapyl alcohol and coniferyl alcohol which are a part of phenyl propionic alcohol monomer units. These monomers are bonded with each other through aryl-aryl, alkyl-aryl and alkyl-alkyl linkage [20]. Besides that, lignin is an unstructured heteropolymer that contributes to the resistance, structural pillar and recalcitrance of lignocellulosic biomass [21]. Furthermore, owning to the fact that lignin's presence is the rationality to recalcitrance of lignocellulosic crops towards enzymatic attack for the release of reducing sugar. Thus, LCM with abundant of lignin content is not preferable [22]. In lignin analysis, Sunshine mango leaves have shown the highest composition of lignin (24.73±0.11%), this mango variety is assumed to be least favourable in production of bioethanol as compared with that of Harum Manis (19.45±0.23%) and Chokanan (18.83±0.09%).

Nitrate, chlorophyll, protein, waxes, steroids, fats, terpenes as well as phenolic compounds are known be the components of biomass extractive [17]. Extractives are non-structural constituents which are not linked chemically to the biomass. Furthermore, extractive actually is the most redundant composition of a biomass which interferes with the downstream analysis of a LCM sample [23]. The percentage of extractive in Sunshine, Chokanan and Harum Manis mango leaves were found to be 18.67±0.14%, 9.47±0.22% and 1.32±0.17%, respectively. Thus, through the finding obtained, Harum Manis could be the most competent LCM stock as these leaves have least percentage of extractive.

The final composition investigated in this research work was the ash content of the mango leaves. Ash could be defined as the non-flammable solid mineral element of a biomass which consist of Magnesium (MgO), Aluminium (AlO), Iron (FeO), Silica (SIO) and Calcium (CaO) [24]. In common, ash represents the most minimal percentage of a biomass constitution. This study has demonstrated that ash content was recorded in a range of 0.5-6%. Apart of that, the lowest amount of ash was recorded by Chokanan (5.28±0.13%). However, this criterion would not be supportive to account these leaves to be suitable for bioethanol production [2]. This is due to ash could not be converted into energy. Apart of it, dispersion of ash into the media could result the ash to exist as cations. This form of the ash would hinder the enzymatic saccharification of the LCM to fermentable sugars [24]. overall, it could be summarized that Harum manis with the smallest percentage of ash (0.5±0.21%) would be the most desired choice of substrate.

4. Conclusion

A reduction in lignin content of the mango leaves enhances a more proficient hydrolysis of cellulose and hemicellulose to its monomer that results a comprehensive yield of fermentable sugar. Consequently, the biotransformation of lignocellulosic crops to biofuel could become more productive. Hence, LCM with high cellulose and hemicellulose composition followed by low lignin content are currently taken into option to be more valuable for bioethanol production. Thus, through this study, Harum Manis mango leaves with the highest polysaccharide content of 78.73% (inclusive cellulose and hemicellulose) in comparison to Chokanan (66.42%) and (52.28%) has been suggested to be the most competitive biofuel crop. Besides that, Harum Manis have also recorded the least content of undesired substrate composition at 21.27% which comprises lignin, ash, moisture and extractive. Thus, by using Harum Manis leaves for bioethanol production, a higher yield of biofuel could be obtained. With this, a future research related production of bioethanol of Harum Manis leaves litters could developed to cut down air pollution caused by open burning. Finally, this study could be an upper hand to the nation because besides reducing the dependence of fossil fuel, bioethanol could also decrease the emissions carbon monoxide, particulate matter, toxic chemicals, and greenhouse gasses, creating a cleaner, safer motor fuel to improve the air quality.

5. Acknowledgement

The author would like to acknowledge the financial support granted by Universiti Malaysia Perlis (UniMAP) Short Term Grant (UniMAP/PPP1/STG/9001-00520) and Fundamental Research Grant Scheme (FRGS/1/2017/TK02/UNIMAP/02/7) from the Ministry of Higher Education Malaysia. Furthermore, sincere indebtedness and gratitude is addressed to Chemical Department of Faculty of Engineering Technology, Universiti Malaysia Perlis (UniMAP).

References

[1] E. D. Gutierrez, K. M. L. Amul, R. M. Carpio, and A. R. M. Toledo, "Effect of Selected Fermentation Parameters on Bioethanol Production from Ripe Carabao Mango (Mangifera indica) Peelings," *Asia Pacific J. Multidiscip. Res.*, **3**, 29–35 (2015).

[2] N. Akhtar, D. Goyal, and A. Goyal, "Physico-chemical characteristics of leaf litter biomass to delineate the chemistries involved in biofuel production," *J. Taiwan Inst. Chem. Eng.*, **62**, 239–246 (2016).

[3] S. Espinosa, L. Fernando, P. B. Y, D. Eapen, and P. J. Sebastian, "Evaluation of agro-industrial wastes to produce bioethanol : case study - mango (Mangifera indica L .)," *Energy Procedia*, **57**, 860–866 (2014).

[4] S. P. Das, R. Ravindran, D. Deka, M. Jawed, D. Das, and A. Goyal, "Bioethanol production from leafy biomass of mango (*Mangifera indica*) involving naturally isolated and recombinant enzymes," *Prep. Biochem. Biotechnol.*, **43**, pp. 717–734 (2013).

[5] M. Saifuddin, M. M. Khandaker, A. Hossain, S. Jahan, N. B. Mat, and N. Boyce, "Bioethanol Production from Mango Waste (*Mangifera indica* L . cv chokanan): Biomass as Renewable Energy," *Aust. J. Basic Appl. Sci.*, **8**, 229–237 (2014).

[6] A. . Shatalov, A. R. . Morais, L. C. Duarte, and F. Carcalheiro, "Selective single - stage xylan - to - xylose hydrolysis and its effect on enzymatic digestibility of energy crops giant reed and cardoon for bioethanol production," *Ind. Crops Prod.*, **95**, 104–112 (2017).

[7] R. Shaheen, M. Riaz, N. Jamil, H. Pervaiz, Z. Masood, F. Iqbal, and K. Nasir, "Determination of change in sugar content in healthy and diseased leaves of two mango varieties (Langra and Chaunsa) affected with quick decline disease, Department of Zoology , SardarBahadur Khan Women ' s Univeristy , Quetta , Pakistan," *Am. Eurasian J. Toxicol. Sci.*, **7**, 224–228 (2015).

[8] M. S. Nor Shazuani, "Variation in infestation and factors affecting oviposition of mango pulp weevil, *Sternochetus frigidus* (fabr.) (coleoptera:curculionidae) on mango *Mangifera indica* in northern peninsular malaysia," Universiti Sains Malaysia (2014).

[9] P. Ding and K. B. Darduri, "Morphology of Chok Anan mango flower grown in Malaysia," *African J. Agric. Res.*, **8**, 1877–1880 (2013).

[10] S. Kittiphoom, "Utilization of mango seed," *Int. Food Res. J.*, **19**, 1325–1335 (2012).

[11] M. T. Fernández-ponce, B. Razi, H. Nasri, L. Casas, C. Mantell, E. J. Martínez, and D. Ossa, "Pilot-plant scale extraction of phenolic compounds from mango leaves using different green techniques : Kinetic and scale up study," *Chem. Eng. J.*, **299**, 420–430 (2016).

[12] K. Vasnathi, K. Chairman, and A. J. A. Ranjit Singh, "Vermicomposting of leaf litter ensuing from the trees of Mango (*Mangifera indica*) and Guava (*Psidium guujuvu*) leaves," *Int. J. Adv. Res.*, **1**, 33–38 (2013).

[13] U. Chaiyo and S. Garivait, "Estimation of Black Carbon Emissions from Dry Dipterocarp Forest Fires in Thailand," *Atmosphere.*, **5**, 1002–1019 (2014).

[14] S. P. Das, A. Ghosh, A. Gupta, A. Goyal, and D. Das, "Lignocellulosic fermentation of wild grass employing recombinant hydrolytic enzymes and fermentative microbes with effective bioethanol recovery," *Biomed Res. Int.*, **2013**, 1–14 (2013).

[15] M. . Nimlos, E. Brooking, M. . Looker, and R. . Evans, "Biomass torrefaction studies with a molecular beam mass spectrometer," *Prepr Pap Am Chem Soc Div Fuel Chem*, **48**, 590–591 (2003).

[16] S. Sun, S. Sun, X. Cao, and R. Sun, "The role of pretreatment in improving the enzymatic hydrolysis of lignocellulosic materials," *Biore*, **199**, 49–58 (2016).

[17] D. Kumar, B. Singh, and J. Korstad, "Utilization of lignocellulosic biomass by oleaginous yeast and bacteria for production of biodiesel and renewable diesel," *Renew. Sustain. Energy Rev.*, **73**, 654–671 (2017).

[18] S. Angelini, D. Ingles, M. Gelosia, P. Cerruti, E. Pompili, G. Scarinzi, G. Cavalaglio, F. Cotana, and M. Malinconico, "One-pot lignin extraction and modi fi cation in g -valerolactone from steam explosion pre-treated lignocellulosic biomass," *J. Clean. Prod.*, **151**, 152–162 (2017).

[19] D. Barana, A. Salanti, M. Orlandi, D. S. Ali, and L. Zoia, "Biorefinery process for the simultaneous recovery of lignin , hemicelluloses , cellulose nanocrystals and silica from rice husk and Arundo donax," *Ind. Crop. Prod.*, **86**, 31–39 (2016).

[20] N. E.-A. N. El-Naggar, S. Deraz, and A. Khalil, "Bioethanol production from lignocellulosic feedstock based on enzymatic hydrolysis: Current status and recent developments," *Biotechnology*, **1**, 1–21 (2014).

[21] S. Jung, S. Kim, and I. Chung, "Biomass and Bioenergy Comparison of lignin , cellulose , and hemicellulose contents for biofuels utilization among 4 types of lignocellulosic crops," *Biomass and Bioenergy*, **83**, 322–327 (2015).

[22] A. Rangan, M. V Manchiganti, R. M. Thilaividankan, S. G. Kestur, and R. Menon, "Novel method for the preparation of lignin-rich nanoparticles from lignocellulosic fibers," *Ind. Crop. Prod.*, **103**, 152–160 (2017).

[23] A. O. Ayeni, O. A. Adeeyo, O. M. Oresegun, and E. Oladimeji, "Compositional analysis of lignocellulosic materials : Evaluation of an economically viable method suitable for woody and non-woody biomass American Journal of Engineering Research (AJER)," *Am. J. Eng. Res.*, **4**, 14–19 (2015).

[24] Y. Bin and C. Hongzhang, "Effect of the ash on enzymatic hydrolysis of steam exploded rice straw," *Bioresour. Technol.*, **101**, 9114–9119 (2010).

Zeolitic Imidazolate Framework-8 Membrane for H_2/CO_2 Separation: Experimental and Modeling

L S Lai[1], Y F Yeong[2], K K Lau[2], M S Azmi[2] and T L Chew[2]
[1]Department of Chemical Engineering, University of Illinois, Chicago, IL 60607.
[2]Chemical Engineering Department, Universiti Teknologi Petronas, Bandar Seri Iskandar, 32610 Perak, Malaysia.

E-mail: yinfong.yeong@utp.edu.my

Abstract. In this work, ZIF-8 membrane synthesized through solvent evaporation secondary seeded growth was tested for single gas permeation and binary gases separation of H_2 and CO_2. Subsequently, a modified mathematical modeling combining the effects of membrane and support layers was applied to represent the gas transport properties of ZIF-8 membrane. Results showed that, the membrane has exhibited H_2/CO_2 ideal selectivity of 5.83 and separation factor of 3.28 at 100 kPa and 303 K. Besides, the experimental results were fitted well with the simulated results by demonstrating means absolute error (MAE) values ranged from 1.13 % to 3.88 % for single gas permeation and 10.81 % to 21.22 % for binary gases separation. Based on the simulated data, most of the H_2 and CO_2 gas molecules have transported through the molecular pores of membrane layer, which was up to 70 %. Thus, the gas transport of the gases is mainly dominated by adsorption and diffusion across the membrane.

1. Introduction

H_2 has been well known as a carbon-free fuel [1], which can be used as an alternative energy source [2]. Separation of H_2 from CO_2 is important for steam reforming gas purification in order to obtain high purity of H_2 [3]. Meanwhile, membrane-based separation has become an alternative and attractive technology for gases separation due to its low energy consumption and compact design [4]. It involves relatively low cost if compared to conventional separation methods [5-7]. In addition, membrane based separation process is relatively simple with no moving parts and easy to be operated and controlled [8]. Therefore, rapid market growth for membrane system is reported, which is around 8 % to 10 % per year [9]. Besides, at present, large scale membrane systems have been installed for industrial application in United States and European countries [10], with as many as 100 membrane systems installed in the refineries for various gases separation processes [11]. Based on the various types of membranes for gas separation, polymeric [12, 13], inorganic [14, 15], mixed-matrix [16, 17] and organic-inorganic hybrid, such as metal organic framework (MOFs) [18-20], are most widely reported.

Metal organic framework (MOF) membrane has emerged as a potential application for gas separation attributed to their exceptionally high surface area, adsorption affinity, tailored porosity and frameworks [19, 21]. It is a relatively new type of nanostructured membranes, composing of metal clusters and organic linkers [22-23]. In comparison with MOFs membranes, ZIF-8 membranes exhibit relatively higher stability due to the strong bonding between metal cations and imidazolate anions [24]. In spite of the promising performance in gas separation, ZIF-8 membrane has shown advantages over the other types of ZIFs membranes with excellent chemical resistance, high thermal and hydrothermal stability [25]. Furthermore, in H_2/CO_2 gas separation, ZIF-8 membrane has shown higher performance over zeolite membranes, such as SAPO-34 [26], DDR [27] and FAU [38]. This is mainly because of the larger pore sizes of zeolite (larger than 3.6 Å), which can hardly separate small gaseous molecules of H_2 and CO_2 with the kinetic diameters of 2.9 Å and 3.3 Å, respectively [29].

Apart from that, the gas transport properties of the porous membrane can be affected by several factors, such as the gas diffusivity of the membrane layer, intercrystalline pores presented and mass transfer resistant of the support layer [30]. To date, the Maxwell-Stefan model has been widely used to study the intrinsic gas diffusivity through the membrane layers [30-31]. Meanwhile, the mass transport resistance caused by the support layer or intercrystalline pores in the membrane can be demonstrated through Binary-Friction model [32]. Overall, the combination of these factors has significantly affected the gas permeation and separation performance of the membrane. However, thus far, the gas transport properties of ZIF-8 membrane in gas separation of H_2/CO_2 are still scarcely reported. Therefore, in this study, ZIF-8 membrane was tested for H_2/CO_2 single gas permeation and binary gases separation. Subsequently, the gas transport properties of the membrane were studied by combining Maxwell-Stefan model and Binary Friction model.

2. Experimental

The ZIF-8 membrane synthesized in our previous work [33] using solvent evaporation secondary seeded growth was tested for H_2 and CO_2 using a constant pressure method at room temperature. The permeation flow rate was collected via bubble flow meter at the steady-state. For single gas permeation of H_2 and CO_2, the gas was fed at the total pressure difference ranged from 100 kPa to 700 kPa with 200 ml min^{-1} of flow rate. It is noted that, the blank support was tested using single gas permeation prior to the membrane growth for pore size characterization. The permeance of component i, P_i (mol m^{-2} s^{-1} Pa^{-1}), was calculated using (1) as follows [34]:

$$P_i = \frac{N_i}{\Delta p} \tag{1}$$

where N_i is the flux (mol m^{-2} s^{-1}) and Δp is the pressure difference (Pa). The membrane area was 6.36 $\times 10^{-5}$ m^2. Then, the ideal selectivity for gas component i and j was determined as the ratio of the respective gas permeance using (2) as follows:

$$\alpha_{i/j} = \frac{P_i}{P_j} \tag{2}$$

For binary gases separation, the collected gas from the permeate side was analyzed using gas chromatograph (GC, Perkin Elmer, Model 2103) in order to quantify the composition of the gases. The gas permeance for binary gas mixture was determined based on (1), with Δp referred to the partial pressure difference of the respective gas. Meanwhile, the selectivity for the binary mixed gas (α_{mixed}) was calculated based on the ratio of gas permeance also, as shown in (3):

$$\alpha_{mixed} = \frac{\left(\dfrac{N_T \times x_{i,p}}{p_f \times x_{i,f} - p_p \times x_{i,p}}\right)}{\left(\dfrac{N_T \times x_{j,p}}{p_f \times x_{j,f} - p_p \times x_{j,p}}\right)} \tag{3}$$

where x is the mole fraction, N_T is the total flux (mol m^{-2}s^{-1}), p_f is the pressure at the feed side (Pa) and p_p is the pressure at the permeate side (Pa).

The gas transport properties of the membrane were studied through a general model as shown in (4). The total gas permeance of the membrane (P_T) is calculated by the reciprocal resistance of the membrane layer (P_M) and the support layer (P_S). Meanwhile, for the ZIF-8 membranes synthesized in the present work, intercrystalline pores of the membrane (P_V) and membrane effective area (C) is also considered. The presence of pores or gaps on the membrane which was hardly to be avoided during the membrane synthesis has contributed towards the presence of intercrystalline pores on the membrane. Besides, the area of the membrane involved in the gas permeation might not be the same as the measured membrane area. Therefore, membrane effective area is required to be estimated based on the experimental results. The detailed solution methodologies for both single gas permeation and binary gases separation have been reported in our previous work [35].

$$P_T = \left[\left\{ \frac{1}{(P_M + P_V)C} \right\} + \left(\frac{1}{P_s} \right) \right]^{-1} \tag{4}$$

3. Results and discussion

Figure 1 exhibits the experimental and simulated H_2 and CO_2 single gas permeation data and ideal selectivities of the ZIF-8 membrane at 303 K. The simulation was performed based on the membrane properties stated in Table 1. It can be observed from Fig. 1 that the simulated data is fitted well with the experimental data with means absolute error (MAE) values ranged from 1.13 % to 3.88 %.

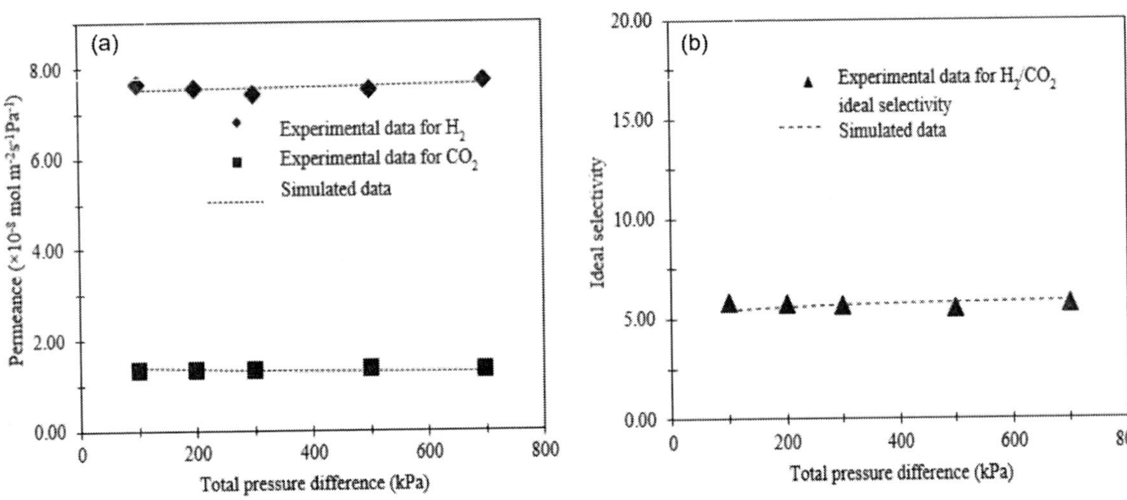

Figure 1. Experimental and simulated (a) single gas permeances and (b) ideal selectivity at 303 K for H_2 and CO_2 over ZIF-8 membrane.

Referring to figure 1 also, gas permeances of H_2 and CO_2 remain nearly unchanged with the increasing total pressure difference. Although the increase in pressure difference increases the fluxes, this increment is almost similar to the increase in pressure difference. Therefore, the gas permeances remain constant eventhough the flux increases with the total pressure difference. Besides, ideal selectivity of H_2/CO_2 remains almost constant, ranged from 5.72 to 5.82. This is attributed to the constant H_2 and CO_2 permeances when the total pressure differences increase.

Table 1. Structural properties for ZIF-8 membrane.

Properties	Value
Membrane layer	
Thickness (m)	5.0×10^{-5} [a]
Density (kg m^{-3}) [b]	924
q^{sat} for H_2 (mol kg^{-1}) [c] q^{sat} for CO_2 (mol kg^{-1}) [d]	30.000 10.070
K_{H2} (Pa^{-1}) [c] K_{CO2} (Pa^{-1}) [d]	5.000×10^{-8} 1.294×10^{-6}
D_{H2} (m^2 s^{-1}) [e] D_{CO2} (m^2 s^{-1}) [f]	9.5×10^{-9} 2.2×10^{-10}
Support layer	
Thickness (m)	2.0×10^{-3}
Pore radius (m)	5.5×10^{-8}
ε/τ ($\tau = 1$)	2.0×10^{-1}

[a]- Estimated from SEM image [33];
[b]- Obtained from the literature data [36];
[c]- Extrapolated based on the literature data [37];
[d] -Extrapolation based on literature data [38] using Langmuir adsorption model.
[e] -Estimated based on permeation data.
[f] -Retrieved from the literature data [38].

Meanwhile, Table 2 shows the simulated percentages of the gas molecules which transported through the membrane intercrystalline pores (P_V) across ZIF-8 membrane. It can be seen from Table 2 that the percentages of H_2 and CO_2 gases molecules permeate through the membrane intercrystalline pores are lower than 20 % and 35 %, respectively. These results show that most of the H_2 and CO_2 gases permeation is dominated by the surface dependency and diffusivity of the membrane. Nonethless, when the total pressure difference increases, percentage of gas permeated through the intercrystalline pores of the membrane also increases. This result has quantitatively implied that the effect of the intercrystalline pores of the membrane on its gases separation performance becomes more significant at higher total pressure difference.

Table 2. Simulated percentage of gases permeated through membrane intercrystalline pores (Pv) of ZIF-8 membrane at total pressure difference ranged from 100 kPa to 700 kPa at temperature of 303 K.

Total pressure difference (kPa)	**Percentage of P_V (%)***	
	H_2	**CO_2**
100	15.67	21.89
200	16.14	24.14
300	16.60	26.33
400	17.07	28.45
500	17.52	30.49
600	17.98	32.47
700	18.43	34.38

*-Calculated based on $[P_V/(P_V + P_M)]$ x 100%

The physical properties of ZIF-8 membrane shown in Table 1 were used to simulate the results for binary gases separation. Figure 2 shows the simulated and experimental data for binary gases separation of H_2/CO_2 under equimolar feed compositions of mixed gases at temperature of 303 K and total pressure difference ranged from 100 kPa to 700 kPa. The simulated data are matched with the experimental data with MAE values ranged from 10.81 % to 21.22 %. The values obtained are within the acceptable range (< 39%) as reported in the literature [31].

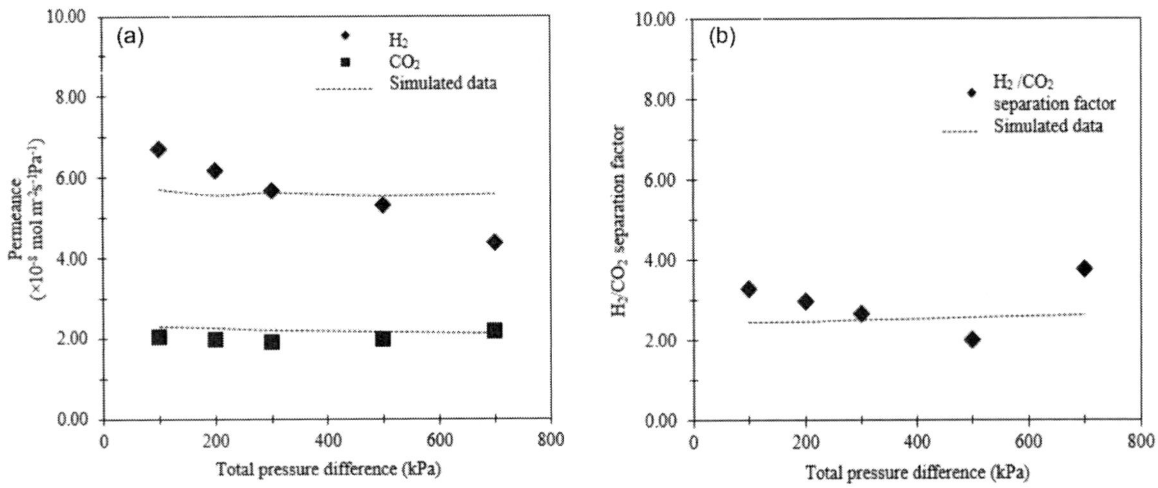

Figure 2. Experimental and simulated (a) binary gas permeances and (b) H_2/CO_2 separation factor for ZIF-8 membrane at equimolar feed composition of mixed gases of H_2/CO_2.

As observed from figure 2 also, H_2 permeances decrease with the increase in pressure difference. This is attributed to the presence of larger molecules, such as CO_2, in the mixed gases, which has created the drag force over H_2 gas molecules [39]. Therefore, the transport of H_2 molecules across the membrane is affected. Meanwhile, CO_2 permeance remains almost constant (varied from 2.05×10^{-8} mol m^{-2} s^{-1} Pa^{-1} to 2.17×10^{-8} mol m^{-2} s^{-1} Pa^{-1}) when the total pressure difference increases. As mentioned earlier, H_2 diffuses faster than CO_2 across the ZIF-8 membrane due to its lighter molar mass and lower viscosity as compared to CO_2 [40]. Therefore, highly mobile H_2 gas molecules will act as the sweeping gas to promote the permeation of CO_2 gas molecules transported through the ZIF-8 membrane [41]. Meanwhile, the kinetic diameter of CO_2 (0.33 nm) is comparable to the pores of ZIF-8 (0.34 nm). Hence, CO_2 molecules can be passed through ZIF-8 pore without much hindrance. Eventually, the CO_2 flux increases linearly with the partial pressure difference and resulted in the constant CO_2 permeance.

4. Conclusion

In conclusion, ZIF-8 membrane synthesized via solvent evaporation seeding method exhibited high quality in the membrane growth by showing H_2/CO_2 ideal selectivity of 5.83 at total pressure difference of 100 kPa and temperature of 303 K. In the binary gases separation, selectivity of 3.28 was obtained for mixed gases of H_2/CO_2 at equimolar feed composition, under total pressure difference of 100 kPa and temperature of 303 K. Besides, the simulated results are fitted well with the experimental results by achieving MAE values less than 21.22 %. Overall, the gas transport properties of ZIF-8 membrane in H_2/CO_2 gas separation have been successfully presented for single gas permeation and binary gases separation.

Acknowledgments
The financial and technical supports provided by CO_2 Management (MOR) research group, Universiti Teknologi PETRONAS are duly acknowledged.

References

[1] Adhikari S and Fernando S 2006 *I & EC Res.* **45** 875
[2] David W I F, Makepeace J W, Callear S K, Hunter H M A, Taylor J D and Wood T J 2014 *J. Am. Chem. Soc.* **136** 13082
[3] Rostrup-Nielsen J R and Rostrup-Nielsen T 2002 *Cattech*, **6**, 150
[4] Zhu W, Hrabanek P, Gora L, Kapteijn F and Moulijn J A 2006 *I & EC Res.* **45** 767
[5] Li S and Fan C Q 2010 *I & EC Res.* **49** 4399
[6] Freemantle M 2005 *Chem. Eng. News* **83** 49
[7] Granite E J and Brien T O 2005 *Fuel Process. Technol.* **86** 1423
[8] Venna S R, Doctor of Philosophy, Department of Chemical Engineering, University of Louisville, Louisville, Kentucky, 2010
[9] Strathmann H 2001 *AIChE J.* **47** 1077.
[10] Sridhar S, Bee S, Bhargava S K 2014 *Chem. Engr. Digest* 1
[11] Baker R W 2002 *I & EC Res.* **41** 1393
[12] Kim K -J, Park S -H, So W -W, Ahn D -J and Moon S-J 2003 *J. Membr. Sci.* **211** 41
[13] Khoshkam M, Sadeghi M, Chenar M P, NamaziFard M J and Baghersad S, 2011 *Proceedings of the Polymer Processing Society Asia/Australia Regional Meeting* Kish Island (Iran).
[14] Guo H, Zhu G, Li H, Zou X, Yin X and Yang W 2006 *Angewandte Chemie*, **118** 7211
[15] Chew T L, Ahmad A L and Bhatia S 2011 *Chem Eng. J.* **171** 1053
[16] Jusoh N, Yeong Y F, Cheong W L, Lau K K and Shariff A M 2016 *J. I & EC Res.* **44**, 164
[17] Ordonez M J C, Balkus K J, Ferraris Jr J P and Musselman I H 2010 *J. Membr. Sci.* **361**, 28
[18] Huang A, Chen Y, Wang N, Hu Z, Jiang J and Caro J 2012 *Chem. Commun.*, **48** 10981
[19] Bux H, Liang F, Li Y, Cravillon J, Wiebcke M and Caro J 2009 *J. Am. Chem. Soc.* **131**, 16000
[20] Venna S R and Carreon M A 2010 *J. Am. Chem. Soc.* **132** 76
[21] Huang H, Zhang W, Liu D, Liu B. Chen G and Zhong C 2011 *Chem. Eng. Sci.* **66** 6297
[22] Basu S, Cano-Odena A and Vankelecom I F J 2011 *Sep. Purif. Technol.* **81** 31.
[23] Hu Z, Zhang L, Jiang J 2012 *J. Chem. Phys.* **136** 244703
[24] Xu G, Yao J, Wang K, He L, Webley P A and Chen C S 2011 *J. Membrane Sci.* **385-386** 187
[25] Park K S, Ni Z, Côté A P, Choi J Y, Huang R, Uribe-Romo F J 2006 *Proceedings of the National Academy of Sciences of the United States of America*, **103** 10186
[26] Poshusta J C, Tuan V A, Falconer J L and Noble R D 1998 *I & EC Res.* **37** 3924
[27] Zheng Z, Hall A S, Guliants V V 2008 *J. Mater. Sci.* **43** 2499
[28] Weh K, Noack M, Sieber I and Caro J 2002 *Micropor. Mesopor. Mater.* **54** 27
[29] Li Y, Liang F, Bux H, Yang W and Caro J 2010 *J. Membr. Sci.* **354** 48
[30] Kangas J, Sandström L, Malinen I, Hedlund J and Tanskanen J 2013 *J. Membr. Sci.* **435** 186
[31] Grahn M and Hedlund J 2014 *J. Membr. Sci.* **471** 328
[32] Li H, Schygulla U, Hoffmann J, Niehoff P, Haas-Santo K and Dittmeyer R 2014 *Chem. Eng. Sci.* **108** 94
[33] Lai L S, Yeong Y F, Lau K K and Shariff A M 2017 *J. Chem. Technol. & Biotechnol.* **92** 420

[34] Li S, Martinek J G, Falconer J L and Noble R D 2005 *I & EC Res.* **44** 3220
[35] Lai L S, Yeong Y F, Lau K K and Shariff A M 2017 *Chem. Eng & Technol.* **40** 1031
[36] Chmelik C, van Baten J M and Krishna R 2012 *J. Membr. Sci.* **397-398** 87
[37] Battisti A. Taioli S and Garberoglio G 2011 *Micropor. Mesopor. Mater.* **143** 46
[38] Pusch A-K, Splith T, Moschkowitz L, Karmakar S, Biniwale R and Sant M 2012 *Adsorption,* **18** 359
[39] Huang A, Liu Q, Wang N, Zhu Y and Caro J 2014 *J. Am. Chem. Soc.* **136** 14686
[40] Krishna R and van Baten J M 2010 *J. Membr. Sci.* **360** 323
[41] Kong L, Zhang X, Liu Y, Li S, Liu H and Qiu J 2014 Mater. Chem. Phys. **148** 10

IOP Conf. Series: Materials Science and Engineering **318** (2018) 012003 doi:10.1088/1757-899X/318/1/012003

Corrosion Behaviour of Sn-based Lead-Free Solders in Acidic Solution

Nordarina J[1] , Mohd H Z[1*], Ahmad A M[2] and Muhammad F M N[1]

[1]Centre of Excellent Geopolymer & Green Technology (CEGeoGTech), School of Materials Engineering, Universiti Malaysia Perlis, 02600 Jejawi, Arau, Perlis, Malaysia
[2]School of Materials and Mineral Resources Engineering, Universiti Sains Malaysia, 14300 Nibong Tebal, Penang, Malaysia

E-mail: hafizzainol@unimap.edu.my

Abstract. The corrosion properties of Sn-9(5Al-Zn), Sn-Cu and SAC305 were studied via potentiodynamic polarization method in an acidic solution of 1 M hydrochloric acid (HCl). Sn-9(5Al-Zn) produced different polarization profile compared with Sn-Cu and SAC305. The morphological analysis showed that small, deep grooves shaped of corrosion product formed on top of Sn-9(5Al-Zn) solder while two distinctive structures of closely packed and loosely packed corrosion product formed on top of Sn-Cu and SAC305 solder alloys. Phase analysis revealed the formations of various corrosion products such as SnO and SnO_2 mainly dominant on surface of solder alloys after potentiodynamic polarization in 1 M hydrochloric acid (HCl).

1.Introduction

Rapid growth of science and technology has forced many researchers to ramp up scientific researches and investigations especially in electronic field. Not to be left behind, soldering technology are indeed need to catch up this rapid growth of the current situation of technology. From the usage of lead solder (Sn-Pb), the soldering technology has step into new paradigm towards greener soldering technology. This is mainly cause by law that has restricted the used of lead (Pb) due to high toxic content. It has been banned by the European Union back in February 2003 [1,2].

Joining the team to upgrade the solder performance, many efforts has been gathered all around the world to specify new recipe for new solder technology to change the usage of Pb in solder. Obviously addition and blending of new materials could specifically give solution to different problems. Example are addition of indium (In), copper (Cu), zinc (Zn), argentum (Ag) and gold (Au) [3,4]. However, addition of Zn and Cu has attract great interest towards research especially due to availability of both elements as well as promising future due to excellent properties of both Zn and Cu [5].

Sn-Ag-Cu (SAC) and Sn-Cu alloys are popular by researchers due to low melting temperatures and have excellent wettability as compared to Sn-Ag alone. Due to its popularity gained, SAC305 has been widely applied in microelectronic industry commercially [6]. Meanwhile, Sn-Zn captured the researchers as it also has low melting temperature that are close to Sn-Pb, good mechanical properties and low in cost [7,8]. In addition, addition of Al in Sn-Zn to produce Sn-9(5Al-Zn) solder has shown to passivation on the surface of solder [9]. However, there are limited research focuses on the corrosion behaviour especially in acidic solution. With most of recent studies limited to NaCl solution, it is worth to investigate other common corrosive solutions such as hydrochloric acid (HCl) to provide better understanding the reliability of lead-free solders.

Expanding demand for reduced size and compact electronic products has intrigued the degradation of solder materials. It embarks serious issues in term of product reliability and suggested that system failure could occur due to corrosion [10]. In this work, the corrosion performance of Sn-9(5Al-Zn), Sn-Cu and SAC305 in 1 M HCl will be studied.

2. Materials and Method

2.1 Sn-9(5Al-Zn) solders alloys preparation

The master alloy of 5Al-Zn was prepared by melting pure Al and Zn (Sigma-Aldrich). Mixture of 91% Sn and 9% 5Al-Zn were weighed and re-melted in a porcelain crucible using an induction furnace at 550 °C with the presence of nitrogen gas. The molten alloys were stirred to homogenize the chemical composition. The produced ternary Sn-9(5Al-Zn) solder was cast and air-cooled to room temperature. After cooling to room temperature, the alloy was pressed and mechanically punched to produce pellets with diameter of 5 mm and 3 mm thickness. Each solder pellet was attached to a copper (Cu) wire with 1 mm diameter to provide electrical connections. The alloys were cold-mounted with epoxy resin and were ground, polished and cleaned to obtain a mirror finish surface prior to electrochemical analysis.

2.2 Sn-Cu and SAC305 solders alloys preparation

Sn-Cu and SAC305 (Dagu Electronic) wire that are commercially available was re-melted in muffle furnace at 300°C. Then, the solder was cooled to room temperature and mechanically pressed by using hydraulic press to acquire constant thickness of 3mm. The solder plate was punched to pellet size with diameter of 5 mm each and was soldered to copper wire to provide electrical connection. Samples were later cold mounted and proceed to grind, polish and clean to mirror finish before corrosion analysis.

2.3 Electrochemical analysis

Potentiodynamic polarization were performed using three-electrode system consist of a platinum plate as counter electrode, a saturated calomel electrode (SCE) as reference electrode and the solder material as the working electrode. Potentiostat (AUTOLAB PGSTAT 30) was used to conduct the corrosion test of Sn-9(5Al-Zn), Sn-0.7Cu and SAC305 in 1 M HCl electrolyte. The scan potential range are set from -2.0 up to 2.0 V_{SCE} with the scan rate used at 1mVs.

2.4 Morphology and structural analyses

The microstructural characterization and corrosion products composition were analysed using field emission scanning electron microscope (NOVA NANOSEM 450). The phase characterization was conducted by X-Ray diffractometer (Shimadzu XRD 6000) at 2θ values ranging from 20 to 90 °. The XRD results were analysed by using X'pert HighScore Plus software to match the corresponding peaks with the standards from International Committee of Diffraction Data (ICDD) X-ray data file.

3. Results and Discussion

3.1 Electrochemical analyses

Polarization profiles for Sn-9(5Al-Zn) solder alloys were differed from Sn-Cu based solder. Sn-9(5Al-Zn) solder alloys showed distinctive active-passive behaviour with two stages of passivation occured. Meanwhle, Sn-Cu and SAC305 only consists of one passivation stage (Figure 1). Cathodic region of Sn-9(5Al-Zn), Sn-0.7Cu and SAC305 started at potential -2.0 V_{SCE}. From point A , the cathodic scan for Sn-9(5Al-Zn) ended at potential value -0.95 V_{SCE}. For Sn-Cu and SAC305, the potential ended at value of -0.64 and -0.67 V_{SCE} at point C. Region AB and AC mainly refers to reduction of dissolved oxygen O_2 into H_2O in aerated HCl electrolyte.

$$O_2 + 4H^+ + 4e^- \rightarrow 2H_2O \quad (1)$$

Starting from point B and C, current rapidly increased as potential increased. This is refers to active region. Here are the critical stages in polarization where the dissolution of metal occurs. For Sn-9(5Al-Zn), the current increased sharply from 2.5X 10^{-4} A. This value is larger as compared to 0.9 x 10^{-4} for Sn-0.7Cu and 1.7 x 10^{-4} for SAC305. In Sn-9(5Zn-Al) system, Zn is the most active material while for Sn-Cu and SAC305 solder alloys, Sn is the most active material [11]. This occurrence made Zn and Sn the first element to be removed during polarization. Since Zn is more reactive as compared to Sn, it is easier to dissolve from Sn-9(5Al-Zn) solder. Besides that, the galvanic corrosion effect is more severe due the presence of Zn/Sn and Zn/Al couplings in the system. However, in Sn-Cu based solder system, eventhough Sn is the most reactive material, it still has higher value of potential in electrochemical series, plus with abundant amount of Sn which helps to withstand degradation of the metal itself.

The first stage of passivation for Sn-9(5Al-Zn) took place at point D before re-activate and end up with passivation of second stage. This finding is similar with Nazeri et. al [11] where after primary dissolution occured, a short passivation will undergo before re-activation and re-passivation occured again. However, this phenomena cannot be found in Sn-Cu or SAC305 system. All three solders started to passivate at point E onwards. At the end of this passive region, the current has become almost independent with applied potential. Usually, the end product from This regions are reported as metal oxide such as SnO and ZnO [12].

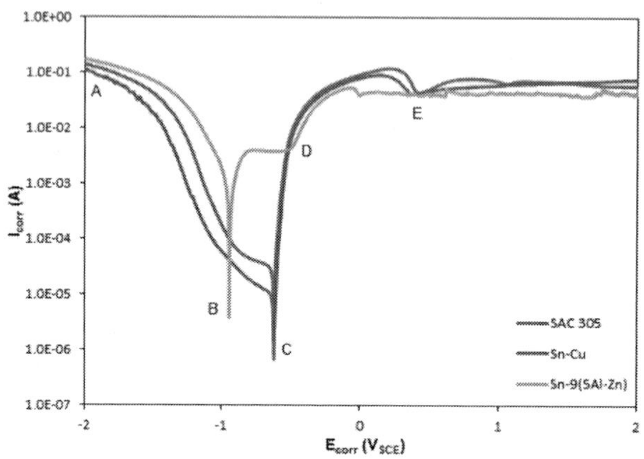

Figure 1. Potentiodynamic polarization of SAC305, Sn-Cu and Sn-(5Al-Zn) in 1 M HCl electrolyte.

Table. 1. E_{corr}, I_{corr} and I_{pass} of SAC305, Sn-Cu and Sn-9(5Al-Zn) in 1 M HCl electrolyte.

Solder	E_{corr}, (V_{SCE})	I_{corr} (A)	I_{pass} (A)
Sn-9(5Al-Zn)	-0.952	2.5 E-4	0.045
Sn-Cu	-0.643	0.9 E-5	0.041
SAC305	-0.673	1.7 E-5	0.059

3.2 Structural analyses

For ternary Sn-9(5Al-Zn) solder alloys, Sn are dominants with some Zn and Al found on the surface of the solder. No new compound formed due to limited solubility of all elements in each other. Sn-Cu consists of β-Sn and Cu_6Sn_5 while Ag_3Sn was detected for SAC305. The ICDD file no. of 00-004-

0673 and 00-001-0926 matched pure Sn, 00-004-0831 matched pure Zn, 01-089-2837 matched pure Al, 03-065-2303and 00-045-1488 matched Cu_6Sn_5 and 01-071-0530 matched Ag_3Sn. During the soldering process, Ag combined together with Sn to form Ag_3Sn while Cu_6Sn_5 is a result from precipitation of Cu in Sn matrix [13]. Some Sn peaks decreased due to formation of intermetallic compound that combined with Sn such as Cu_6Sn_5 and Ag_3Sn.

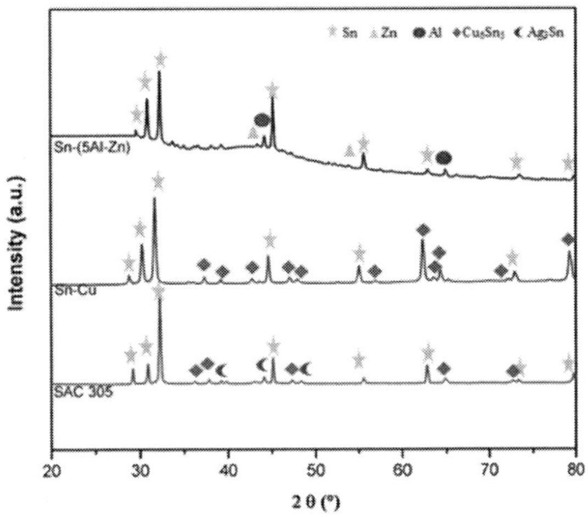

Figure 2. XRD profile for as-prepared of SAC305, Sn-Cu and Sn-9(5Al-Zn).

Figure 3. shows the XRD peaks of Sn-9(5Al-Zn), Sn-Cu and SAC305 after polarization in 1 M HCl. The corrosion product of all these three solder alloys only involved SnO, SnO_2 and ZnO. The ICDD file no. for SnO matched with 01-077-2296 and 00-024-1342 while SnO_2 matched with 00-033-1374 and 00-029-1484. As for ZnO, the ICDD file no. matched with 01-075-1533. As expected, no Al by-product were detected due to low amount of this element. Intermetallic compound formed also does not formed new corrosion products that enganging with oxides. It can be concludes that Sn oxide detected for all solders as Sn is the main element in all three solder compositions. ZnO formed as Zn is the first element to dissolve for Sn-9(5Al-Zn). Interestingly, Sn oxides are the dominant corrosion products and aligned with previous report of Mohamed et al. [14]. Formation of SnO_2 shows all the solder already achieved its stable formed [6].

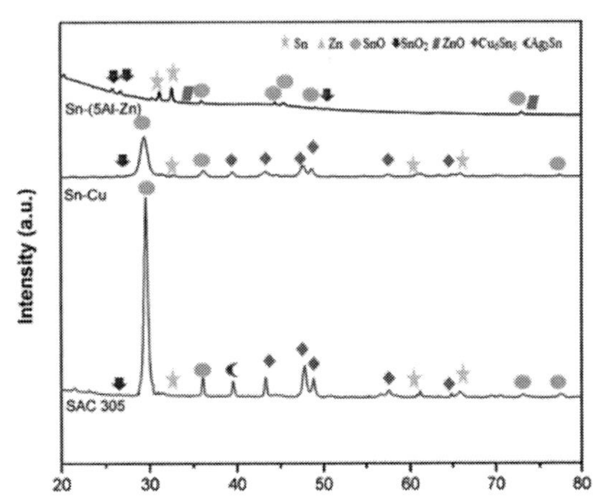

Figure 3. XRD profile after polarization of SAC305, Sn-Cu and Sn-9(5Al-Zn) in 1 M HCl electrolyte.

IOP Conf. Series: Materials Science and Engineering **318** (2018) 012003 doi:10.1088/1757-899X/318/1/012003

3.3 Morphology analyses

Morphological changes in Sn-9(5Al-Zn) solder before and after polarization showed in Figure 4. Needle-like Zn phase scattered in Sn matrix can be observed before corrosion. However, Al cannot be detected at the lining of Zn phase as it is very small in amount to independently be found on the surface of solder. It is clearly observed there are two distinctive phase between Zn and Sn matrix. After polarization, the surface of Sn-9(5Al-Zn) solder had become rough and small, deep grooves are all over the surface. Zn- rich phase are no longer detected. This groove possibly mades from dissolution process of Zn and Sn during polarization. The active materials react with oxide to form metal oxide and deposited on top of the surface in order to protect it from further reaction [15].

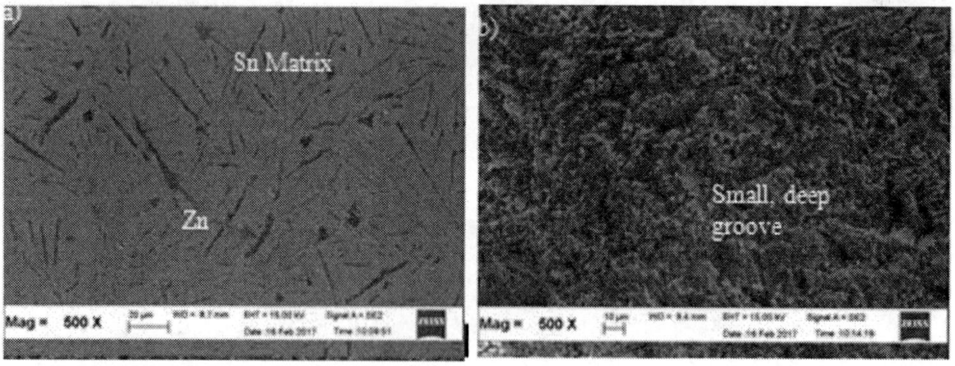

Figure 4. Microstructure of Sn-9(5Al-Zn) (a) as-prepared,(b) after polarization in 1 M HCl electrolyte.

Morphology of as-prepared Sn-Cu is almost similar to SAC305. The light gray area refers to Sn-matrix while the dark phase refer as Cu_6Sn_5 (Figure 5 and 6). However, the Ag_3Sn phase existed as darker gray area but not dominant on the surface of the SAC305 solder (Figure 6). The morphology of both solder changed after polarization in 1 M HCl electrolyte. The phase that can be found during as-prepared were diminished and replaced with the corrosion products throughout the surface. As predicted, the dissolution process of Sn in both type of solder has resulted in various shape of corrosion product while go throuh the passivation and pseudopassivation process. Due to random deposition of the by-products, uneven protective layer caused different shape o the corrosion product. This agreed the previous XRD results that protective layer might be build with diferent Sn oxides of SnO or SnO_2. Worse, continuous attack from Cl^- ions enhancing the formation of pitting. If this occurred, Sn chloride by products would be formed. However, in this research, only Sn oxides formed showed the degradation is not very severe. Different shape of corrosion product that formed on top of the surface allow gap to presence in between the corrosion product itself especially for Sn-Cu and SAC305. This is due to direct passivation after activation region. As Sn-9(5Al-Zn) has two stages of passivation, the first passivation layer already closed the gap in between of the corrosion products but reactivation made the second passivation layer to become uneven during deposited. This lead to small grooves formed on the surface of the solder. Compact shape corrosion product is much stable due to small gap and small chances to be attack further by Cl^- ions.

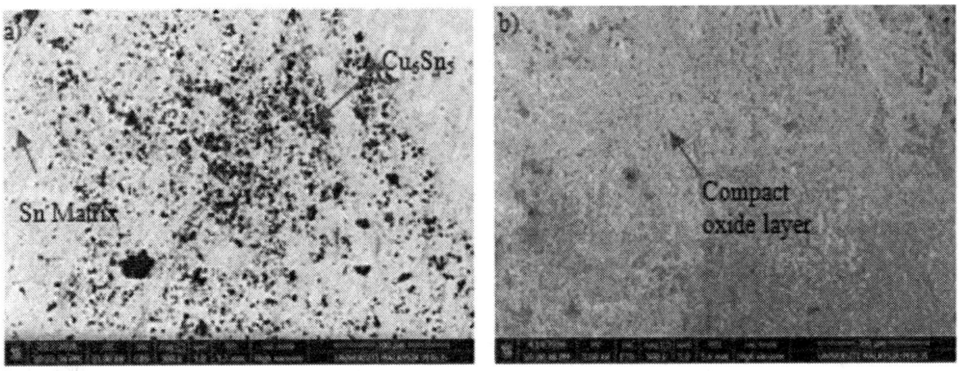

Figure 5. Microstructure of Sn-0.7Cu (a) as-prepared,(b) after polarization in 1 M HCl electrolyte.

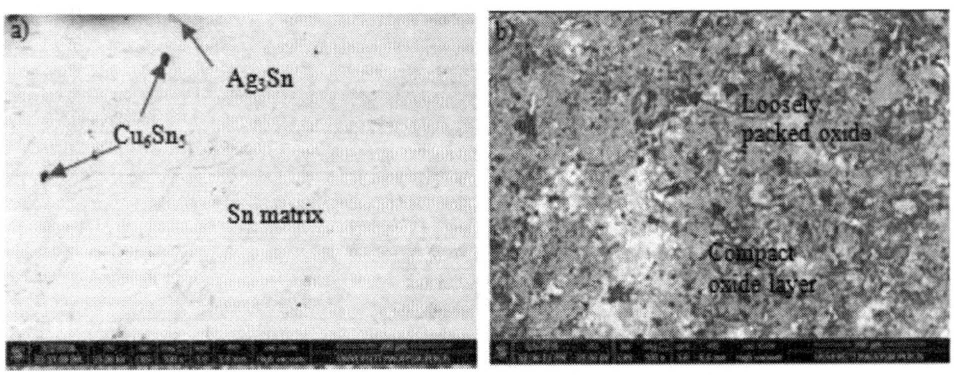

Figure 6. Microstructure of SAC305 (a) as-prepared,(b) after polarization in 1 M HCl electrolyte.

4. Conclusion

Corrosion properties of Sn-9(5Al-Zn) solder in 1 M HCl electrolyte are very differed in many aspects compared with Sn-Cu and SAC305. Sn-9(5Al-Zn) were prone to corrosion as compared to Sn-Cu based solder due to its larger I_{corr} and low value of E_{corr}. Removal of Zn is more rapid as Zn is easily degraded compared to Sn in the solder system. Structural and morphological results also proved that metal oxide of Sn are dominant on all three types of solder surface. Presence of mixed corrosion products after polarization confirmed that passivation has took place on the surface of solder. Sn-9(5Al-Zn) that has two passivation stages showed less corrosion resistance as compared to SAC305 and Sn-Cu that only consist one passivation region. Therefore, the reliability of Sn-based lead-free solder are significantly affected by the addition of element involved.

Acknowledgement
Authors would like to thank UniMAP Short Term Grant (900-0054 9) for financial support.

References

[1] M.Hindler, Z. Guo, A. Mikula 2012 *J.C. Thermod.*, **55**, 102-109
[2] The Euro. Parl. and counc. of the Euro. Union 2003 *J.Euro. Union L,* **37**, 19-23
[3] M.Abtew, G. Selvaduray 2000 J.*Mat. Sc. And Eng.* **27**, 95-141
[4] M. Liu, W. Yang, Y. Ma, C. Tang, H. Tang, Y. Zhan 2015 *J. Mat. Chem. Phys.* **168**, 27-34
[5] M. Islam, Y. Chan, M. Rizvi, W. Jillek 2005 *J. Alloy. Compd.* **400**, 136-144
[6] M. Wang, J. Wang, H. Feng, W. Ke 2012 *J. Corr. Sci.* **63**, 20-28

[7] K. Suganuma 2001 *Solid State Mater.* Sci. **5**, 55-64
[8] K. Suganuma, K.S. Kim 2007 *Springer* , 121-127
[9] U.S.Mohanty, K.L. Lin 2006 *J. Appl. Surface. Sci,* **252**, 5907-5916
[10] V. Chidambaram, J. Hald, R. Ambat, J.Hattel 2009 *JOM,* **61,** 59-65
[11] M.F.M.Nazeri, A.B. Ismail, A.A. Mohamad 2014 J. Alloy. Comp., **606**, 278-287
[12] N. Zaini, M.F.M. Nazeri 2016 *AIP Conference Proceedings*, **1756**
[13] L. Gao, S. Xue, L.Zhang, Z. Sheng, F. Ji, W. Dai, et al. 2010 *J. Microelectronic Eng.,* **87,** 2025-34
[14] M.N. Mohamed, N. A. Aziz, A.A. Mohamad, M.F.M. Nazeri 2015 *J. ElectroActive Mater.* **3**, 28-32
[15] M.F.M. Nazeri, A. A. Mohamad 2016 *J. Alloy. Compd.* **661**, 516-25

IOP Conf. Series: Materials Science and Engineering **318** (2018) 012004 doi:10.1088/1757-899X/318/1/012004

Corrosion Properties of SAC305 Solder in Different Solution of HCl and NaCl

Nurwahida M. Z.[1], Mukridz M. M.[1], Ahmad A. M.[2] and Muhammad F. M. N.[1]
[1]School of Materials Engineering, Universiti Malaysia Perlis, 02600 Jejawi, Kangar, Perlis, Malaysia.
[2]School of Materials and Mineral Resources Engineering, Universiti Sains Malaysia, 14300 Nobong Tebal, Penang, Malaysia.

E-mail: firdausnazeri@unimap.edu.my

Abstract. Potentiodynamic polarization was used to studied the corrosion properties of SAC305 solder in different solution of 1.0 M HCl and 3.5 wt.% NaCl using the same scanning rate of 1.0 mV/s. The polarization curves indicated that corrosion in NaCl was less severe than in HCl solution based on corrosion current and passivation behavior obtained. Morphology and phases obtained after corrosion using SEM and XRD were analyzed. Microstructure analysis shows the present of compact corrosion product with presence of larger flake for polarization in NaCl compared to HCl. Phases present in XRD analysis confirmed the present of SnO and SnO_2 corrosion product for sample from both solutions.

1. Introduction

Traditional Sn-Pb solder has been used as joining materials in many industries especially in electronic industries. However, international legislation on lead (Pb) due to high toxicity has leads to the industrial needs to find other new solder as its replacements [1-5]. The research and development were then conducted intensively to search for lead-free solder. Currently, applications of lead-free solders such as tin-silver (Sn-Ag) [6-8], tin-zinc (Sn-Zn) [4, 9], tin-copper (Sn-Cu) [10, 11] and tin-silver-copper Sn-Ag-Cu (SAC) [2, 6, 12-16] have been developed in electronic products. Among them, eutectic SAC305 is one of the most promising candidates to replace the traditional Sn-Pb solder alloys. Properties of SAC305 such as good thermal, mechanical, fatigue properties and low melting temperature make it as one of the most favorable lead-free solder candidates.

Reliability of solder is very crucial in determining performance and endurance of electronics devices during application [10, 17]. Among properties related to reliability includes coefficient of thermal expansion, intermetallic compound formation, fatigue strength, mechanical strength and corrosion [8, 11, 16]. For the solder to be considered as direct replacement for Sn-Pb, it must have oxidation or corrosion resistance in various conditions such as salt-water, moist or air contaminations environments [2, 3, 11, 18]. Since the uses of electrical devices are everywhere, the understanding of the effect of corrosive media to the corrosion of lead-free solder is needed.

Rosalbino et. al. [13] has reported the addition of copper (Cu) has improve the corrosion resistance of Sn-Ag solders alloy, where the passivity behavior of alloys was enhanced compared to traditional Sn-Pb solder alloy. Meanwhile, Wislei et.al [11] revealed that the distribution of fine and homogenous Ag_3Sn intermetallic particles in Sn-rich matrix plays major role in corrosion resistance, as galvanic coupling in between phases may dictate corrosion performance of SAC305. So far, most of corrosion performance studies of SAC305 are limited to NaCl [12] and KOH [19] solutions, where the corrosion of SAC305 solder in common acidic corrosive media such as hydrochloric acid (HCl) still remain limited. Thus, current study focused on the effect of 1 M HCl solution on the corrosion performance of SAC305. The corrosion characteristics will be compared with 3.5 wt % NaCl. Morphological, elemental and phases analyses were also discussed to support the findings.

2. Experimental

The solder wire of commercially supplied SAC305 (Dagu Electronic) was placed in clean crucible after rolled into coils. The crucible was then placed in a muffle furnace set with highest temperature of 300°C. The melted SAC305 solder then left to cool at room temperature. Hydraulic Press 20 Ton (10 Mpa) used to pressed solidified alloy to obtain constant thickness of 3 mm. Puncher was used to punch the solder into solder pellet with diameter of 5 mm and thickness of 3 mm. Copper wire with length of 3 cm was then solder with solder pellet to provide current flow. Wire-attached solder pellets were cold mounted with epoxy resin and then were ground and polished to mirror finish prior corrosion testing.

Acid solutions of HCl with concentration of 1.0 M and NaCl solution with concentration of 3.5 wt. % were prepared for the potentiodynamic polarization analysis. Standard three electrode system consists of platinum as counter electrode, Saturated Calomel Electroder (SCE) as reference electrode, and SAC305 solder as working electode were used to carry out potentiodynamic measurements. Potentiostat (Autolab PGSTAT 30) was used to undergo potentiodynamic polarization test. Scanning rate was set to 1.0 mV/s, with applied potential range of -2.0 to 2.0 V_{SCE} for both solutions. The electrochemical characterization was carried out three times to ensure reproducibility of the results.

Scanning electron microscope (NOVA NANOSEM 450) equipped by energy dispersive spectroscopy (X-Max 50, OXFORD) was used to determine the microstructure and corrosion products compositions of SAC305 solder after corrosion testing. Phase analysis was run by X-Ray Diffractometer (Shimadzu XRD 6000) at 2θ ranging from 20 to 90°. XPERT Highscore Plus software was then used to analyze the results obtained in accordance to International Committee of Diffraction Data (ICDD) X-Ray data files.

3. Results and Discussions

3.1 Potentiodynamic Polarization Analysis

Potentiodynamic polarization curve for SAC305 in 1.0 M HCl can be divided into two different regions namely cathodic and anodic regions (Figure 1). The scan starts at potential of -2.0 V in the cathodic region. In general, the cathodic region involved the process of oxygen and reduction [5]. As the potential increases, the current decrease rapidly. The point intersection between cathodic and anodic slopes represents the equilibrium corrosion potential (E_{corr}) and equilibrium corrosion current density (i_{corr}). Further increase in applied potential resulted in rapid increase of current density up to point "A" signifies anodic region where dissolution of Sn actively occurs.

As the potential continue to increases slightly, the current decreases reaching point B. The corresponding potential and current density are known as primary passive potential (E_{pp}) and critical current density (i_{crit}). At this point, the solder reaches the highest corrosion rate and the changed from active to passive state, where the oxide film of tin oxide starts to form at the solder surface due to reaction between Sn and O in the aerated solution that leads to the reduction of current density up to lowest passivation current value (i_p). Beyond i_p, a sharp increase in current density is observed normally caused by the breakdown of oxide film. On further scanning to region C, the opposition between the formation of oxide film and dissolution of Sn results in the current remains independent to applied potential, indicates the onset of pseudopassivation process.

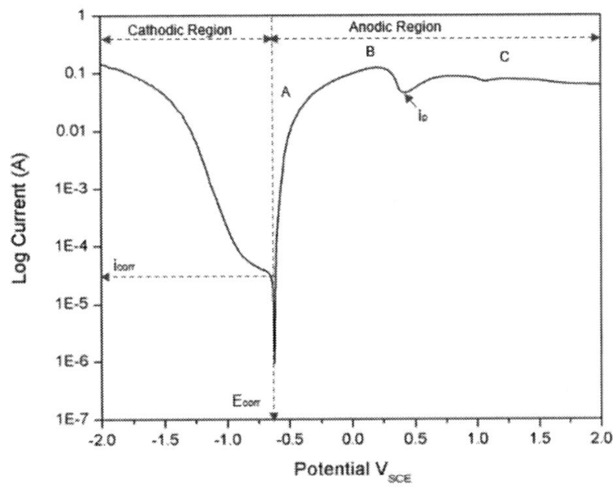

Figure 1. Potentiodynamic polarization curve of SAC305 solder in 1.0 M HCl

Comparing the polarization curves in between 1.0 M and 3.5 wt. % NaCl, almost similar behavior in the cathodic region was seen (Fig. 2). On the other hand, the differences start in the anodic region. For polarization in HCl solution, the value of E_{corr} obtain was -0.7133 mV which is higher than polarization in NaCl (-1.052 V_{SCE}). Meanwhile, the corrosion current density in HCl is 39.83 μA which higher than polarization in NaCl at value of 2.19 μA revealing that the kinetics of reaction in HCl is higher than in NaCl. The passivation in NaCl happen in three stages while in HCl only happen in two stages. This indicates that passivation in NaCl is better than in HCl. In addition, the value of i_p for polarization in NaCl at 0.0105 μA is lower than in HCl solution (0.0506 μA). This shows that less current corresponds to the thickening the oxide film at solder surface is required.

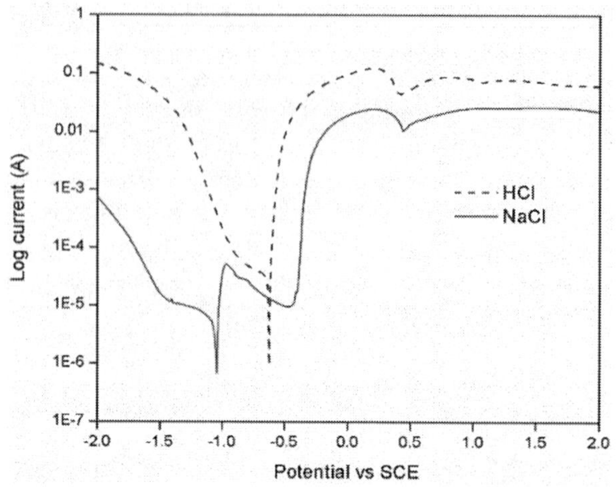

Figure 2. Potentiodynamic polarization curve of SAC305 solder in 1.0 M HCl and 3.5 wt.% NaCl solution at 1.0 mV/s scanning rate

Table 1. Corrosion parameters obtain from polarization graph

Solution	E_{corr}	i_{corr}	i_p
1.0 M HCl	-0.7133 V	39.83 μ A	0.0506 A
3.5 wt.% NaCl	-1.052 V	2.19 μ A	0.0105 A

3.2 Morphology analysis

Figure 3(a) shows the microstructure of as-prepared SAC305 solder. The results show the presence of typical three-phase appearance which were β-Sn matrix, Ag_3Sn (grey crystal) and Cu_6Sn_5 (black crystal). The results on the microstructure of SAC305 solder in HCl and NaCl solution shows in Figure 3(b) and 3(c). SAC305 polarized in HCl solution shows the formation of more loosely-packed oxide film with the present of pitting. Meanwhile, the microstructure of SAC305 solder polarized in NaCl solution shows more compact corrosion product (Figure 3(b)), of which is comparable with the results reported Rosalbino et.al [12] and Dezhi et.al. [14]. Presence of obvious pitting highlights that the corrosion attack in HCl is more severe that allows little recovery in the damaged passivation film.

Data obtained from the EDX analysis (Table 2) shows the addition presence of O at the surface which present by the formation of SnO and SnO_2 oxide film for solder in both solutions. Based on the data, the loss of Sn in HCl solution was higher compared in NaCl solution indicating that the dissolution of Sn is worse in HCl solution.

Table 2. Data on EDX analysis on SAC305 solder for before polarize and after polarize in HCl and NaCl solution

Sample SAC305 solder	Sn (wt.%)	Ag_3Sn (wt.%)	Cu_6Sn_5 (wt.%)	O (wt.%)
As-prepared	92.67	3.80	3.54	-
Polarization in 1.0 M HCl	27.21	49.17	14.30	9.32
Polarization in 3.5 wt.% NaCl	69.88	20.07	0.34	0.61

Fig. 3. Microstructure of (a) as prepared SAC305 solder, after potentiodynamic polarization in (b) 1.0 M HCl and (c) 3.5 wt.% NaCl

3.3 Phases analysis of SAC305 solder before and after polarization

Phases analysis was conducted on the sample of SAC305 solder for before and after polarization presented in Figure 4. For as-prepared sample, the XRD results shows the presence of three main phases which are β-Sn, Ag_3Sn and Cu_6Sn_5 attributed by the limited solubility of Ag and Cu in Sn as the main element, similar to the solubility of Cu in ε phase and Ag in η phase. This statement also in agreement with [12].

The XRD results for after polarize shows that SAC305 solder has the same corrosion product, which was SnO and SnO_2 in both solutions. As the most electrochemically active element in this solder system compared with Ag_3Sn and Cu_6Sn_5, Sn is first to dissolved. Therefore, the oxide film forms is Sn-based corrosion product. Furthermore, SEM and EDX analyses show confirmation of the presence of SnO and SnO_2 corrosion product at the surface which was consistent with XRD analysis. Cathodic region are generally correlated to the reduction of dissolved oxygen O_2 into H_2O [3, 5, 14]:

$$O_2 + 4H^+ + 4e^- \rightarrow 2H_2O \quad (1)$$

The anodic current begins at the turning point of the graph where in SAC305 solder, Sn was the most reactive material therefore Sn tends to dissolved [14]:

$$Sn \rightarrow Sn^{2+} + 2e^- \quad (2)$$

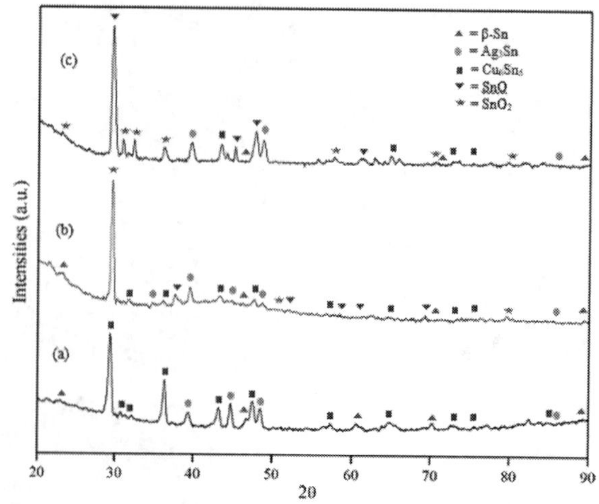

Figure 4. XRD of (a) as-prepared SAC305 solder and polarization of SAC305 in different solution (b) 1.0 M HCl and (c) 3.5 wt.% NaCl

As the solder polarize further, SnO forms from the dissolved Sn^{2+} ions to Sn $(OH)_2$. $m H_2O$, and Sn^{2+} ions are simultaneously oxidized gradually by oxygen to form SnO_2. This equation was also reported in other research [14, 20]:

$$Sn + 2H^+ \rightarrow Sn^{2+} + H_2 \quad (3)$$

$$2Sn^{2+} + O_2 + 6H_2O \rightarrow 2Sn(OH)_4 + 4H^+ \quad (4)$$

$$Sn(OH)_4 \rightarrow SnO_2 + H_2O \quad (5)$$

$$Sn^{2+} + 2H_2O \rightarrow Sn(OH_2) + 2H^+ \quad (6)$$

$$Sn(OH)_2 \rightarrow SnO + H_2O \quad (7)$$

Based on previous research of Rosalbino et.al [12] [13] and Dezhi et.al [14], the corrosion product at SAC solder surface were tin oxychloride or oxyhydrochloride. However, in present research, corrosion product of SnO and SnO_2 were the only corrosion product found at the solder surface. According to Mohanty [15] the other corrosion product forms was SnCl phase, but due to the machine or experimental error, SnCl phase is not detected.

4. Conclusions

As the conclusion, polarization in acidic solution (HCl) resulted in worse corrosion attack based on the higher value of i_{corr}. The passivation behavior of SAC305 in HCl was inferior compared in NaCl solution. The oxide film formed was SnO and SnO_2 which was the same for both HCl and NaCl solution although the solution is different.

5. Acknowledgement

Authors would like to thank UniMAP Short Term Grant (9001-00548) for financial support.

6. References

[1] E. Efzan and A. Marini, *Inter. J. of Eng.* vol. **1**, pp. 2305-8269, 2012.
[2] L. Gao, S. Xue, L. Zhang, Z. Sheng, F. Ji, W. Dai, *et al.* 2010 *Microe. Eng.* **87** 2025-34

[3] A. Wierzbicka-Miernik, J. Guspiel, and L. Zabdyr 2015 *Archives of Civil and Mecha. Eng.* **15** 206-13
[4] K. Pietrzak, M. Grobelny, K. Makowska, N. Sobczak, D. Rudnik, A. Wojciechowski, *et al.2012*, *J. of Mater. Eng. and Performance,* **21** 648-54
[5] Z. Ahmad 2006 *Principles of corrosion engineering and corrosion control* (Butterworth-Heinemann)
[6] Spotak Martin, Marian Drienovsky, T. Lydia, and P. Marian, 2015 Presented at the Metal 2015, Czech Republic, EU, 2015.
[7] A. Lis, M. S. Park, R. Arroyave, and C. Leinenbach, 2014 *J. of Alloys and Comp.* **617** 763-73
[8] W. R. Osorio, D. R. Leiva, L. C. Peixoto, L. R. Garcia, and A. Garcia 2013 *J. of Alloys & Comp.* **562** 194-204
[9] M. F. M. Nazeri and A. A. Mohamad 2014 *Measurement* **47** 820-6
[10] E. S. Freitas, W. R. Osório, J. E. Spinelli, and A. Garcia 2014 *Microelect.Relia.* **54** 1392-1400
[11] W. R. Osorio, L. C. Peixoto, L. R. Garcia, A. Garcia, and J. E. Spinelli 2012 " *Inter J. of Electrochem, Sci.* **7** 6436
[12] F. Rosalbino, E. Angelini, G. Zanicchi, R. Carlini, and R. Marazza 2009 *Electrochimica Acta* **54** 7231-35
[13] F. Rosalbino, E. Angelini, G. Zanicchi, and R. Marazza, 2008 *Mater. Chem. and Phy.* **109** 386-91
[14] D. Li, P. P. Conway, and C. Liu 2008 *Corr. Sci.* **50** 995-1004
[15] U. S. Mohanty and K.-L. Lin, 2008 *Corr. Sci.,* **50** 2437-43
[16] M. Yang, H. Ji, S. Wang, Y.-H. Ko, C.-W. Lee, J. Wu, *et al.*, 2016 *J. of Alloys and Comp.* **679** 18-25
[17] T. Laurila, V. Vuorinen, and M. Paulasto-Kröckel, 2010 *Mater. Sci. and Eng.: R: Reports,* **68** 1-38
[18] H. S. Mohran, A.-R. El-Sayed, and H. M. A. El-Lateef 2009 *J. of Sol. State Electrochem.,* **13** 1279-90
[19] M. C. Liew, I. Ahmad, L. M. Lee, M. F. M. Nazeri, H. Haliman, and A. A. Mohamad,2012 *Metallurgical and Mater. Transactions A,* **43** 3742-47
[20] L. Mei Lee, M. Firdaus Mohd Nazeri, H. Haliman, and A. Azmin Mohamad 2014 *Solder. & Surf. Mount Techno.,* **26** 79-86, 2014.

IOP Conf. Series: Materials Science and Engineering **318** (2018) 012005 doi:10.1088/1757-899X/318/1/012005

Finite Difference Formulation for Prediction of Water Pollution

Hanani Johari[1], Nursalasawati Rusli[2*], and Zainab Yahya[3]
Institute of Engineering Mathematics, Universiti Malaysia Perlis, Kampus Pauh Putra, 02600 Arau, Perlis, Malaysia

E-mail : nursalasawati@unimap.edu.my

Abstract. Water is an important component of the earth. Human being and living organisms are demand for the quality of water. Human activity is one of the causes of the water pollution. The pollution happened give bad effect to the physical and characteristic of water contents. It is not practical to monitor all aspects of water flow and transport distribution. So, in order to help people to access to the polluted area, a prediction of water pollution concentration must be modelled. This study proposed a one-dimensional advection diffusion equation for predicting the water pollution concentration transport. The numerical modelling will be produced in order to predict the transportation of water pollution concentration. In order to approximate the advection diffusion equation, the implicit Crank Nicolson is used. For the purpose of the simulation, the boundary condition and initial condition, the spatial steps and time steps as well as the approximations of the advection diffusion equation have been encoded. The results of one dimensional advection diffusion equation have successfully been used to predict the transportation of water pollution concentration by manipulating the velocity and diffusion parameters.

1. Introduction

Water is an important component of the earth. It covered two thirds of the earth's surface. Almost all earth's population, especially human being depend on freshwater to continue their growth. Nowadays, people are not only demand for the quantity of the water but also the quality of the water. Many aspect of social life depend on the economic availability and acceptable quantity of water. Therefore, water quality has become an issue of increasing by the years [1].

Water pollution can be defined in many ways. The pollution happened when there exist modification or changes in the physical, chemical and biological characteristics of water which gives bad effects on living things. Surface waters and groundwater are the two types of water resources which affected by pollution. Surface water such as rivers, lakes and oceans are the most useful sources to the human being and wildlife habitats [2]. Most of water pollutions are caused by human activities. Human activities can give bad effects on the quality of water environment. As stated by the World Health Organization (WHO) and United Nations Environment Programme (UNEP), some examples of the activities which can cause water pollution are chemical fertilizers and chemical release by smokestacks. Farmers fertilized the fields and the chemical from the fertilizer used are washed by rain

and moves to the surface water nearby. Pollution also happen when chemical released by the smokestacks enter the atmosphere and they fall back to the earth as rain and entering the rivers, ocean or lake [1].
Water pollution gives bad impacts especially to the human health because water is the important resources for human to use in daily life. Besides, water pollution also affects the animals live in the water when poisoned through pollutant. Furthermore, this pollution will result in a lot of cost to treat and recover the polluted water contents [2].

[3] stated that there are two techniques which can be applied to water transport problem, which are experimental and numerical modelling. Numerical modelling is considered to be more portable than experimental model. This is because, the parameters can be simply adjusted which make it easier to handle in industrial side. There are many numerical methods to solve the mathematical equations in order to deal with the water pollution problem. The mathematical models introduced can contribute to save the cost of human effort and material for a large number of chemical experiments to some degree. Moreover, it is inaccessible for on-site experiments in some cases due to environmental pollution issues [4].

Advection diffusion equation (ADE) is one of the most important partial differential equations that has been used in transport of pollutants. An appropriate numerical method can give an accurate information on the concentration of pollution at different times and at different location efficiently and quickly [5]. Therefore, many researchers have studied the numerical methods to solve the ADE in predicting the water pollution problem. One of the method to predict water pollution concentration is finite difference method (FDM). According to [6], FDM has been used to solve the flow equations in order to obtain the flow rate and flow direction of water pollution concentration. In this study, FDM will be used for solving 1D ADE because of its ability to predict the concentration of pollutant as stated by [7].

2. Mathematical Formulation

2.1. Advection Diffusion Equation

In this present paper, the mathematical formulation for advection diffusion equation will be presented.

$$\frac{\partial c}{\partial t}+u\frac{\partial c}{\partial x}=D\frac{\partial^2 c}{\partial x^2} \tag{1}$$

where, c is the concentration, x is the position on the water, t is the amount of time that passes, D is the diffusion rate in x direction, and u is the fluid velocity .

With an initial condition,

$$c(x,0)=f(x) \tag{2}$$

And boundary conditions,

$$c(0,t)=g_0(x) \tag{3}$$

$$\frac{\partial c(l,t)}{\partial x}=g_l \quad \text{for} \quad 0\leq x\leq l \tag{4}$$

3. The Finite Difference Method

3.1. The FTCS Techniques

The Equation (1) forms the Equation (5) which is a forward time approximation. Equation (6) and Equation (7) shows the discretization of the spatial derivative by using second ordered central difference approximation. Let the solution $c(x_i,t_j)$ be denoted C_i^j and its approximate value by c_i^j.

$$\frac{\partial c}{\partial t}=\frac{C_i^{j+1}-C_i^j}{\Delta t} \tag{5}$$

$$\frac{\partial c}{\partial x}=\frac{C_{i+1}^j-C_{i-1}^j}{2\Delta x} \tag{6}$$

$$\frac{\partial^2 c}{\partial x^2}=\frac{C_{i-1}^j-2C_i^j+C_{i+1}^j}{\Delta x^2} \tag{7}$$

Equation (5), (6) and (7) then are substituted into Equation (1) and form Equation (8). Equation (8) is a combination of the time and spatial derivative that have been discretized and substitute to form one explicit method.

$$\frac{c_i^{j+1}-c_i^j}{\Delta t}+u\frac{c_{i+1}^j-c_{i-1}^j}{2\Delta x}=D\frac{c_{i-1}^j-2c_i^j+c_{i+1}^j}{\Delta x^2} \tag{8}$$

After substitute and rearrange the equation according the time level $(j+1)$, lead to form equation (9),

$$c_i^{j+1}=c_i^j+D\frac{(\Delta t)}{(\Delta x^2)}(c_{i-1}^j-2c_i^j+c_{i+1}^j)-u\frac{(\Delta t)}{(2\Delta x)}(c_{i+1}^j-c_{i-1}^j)$$

$$c_i^{j+1}=(\alpha+\beta)c_{i-1}^j+(1-2\alpha)c_i^j+(\alpha-\beta)c_{i+1}^j \tag{9}$$

where,

$$\alpha=D\frac{(\Delta t)}{(\Delta x^2)}, \qquad \beta=u\frac{(\Delta t)}{(\Delta x)}$$

3.2. The Crank Nicolson Techniques

The implicit Crank Nicolson scheme approximations the Equation (1) by using forward finite differences approximation in time and the spatial derivatives by using second ordered central difference approximation are estimated by the average of their values at time step j and $j+1$.

$$\frac{\partial c}{\partial x}=\left(\frac{1}{2}\right)\left[\frac{C_{i+1}^{j+1}-C_{i-1}^{j+1}}{2\Delta x}+\frac{C_{i+1}^j-C_{i-1}^j}{2\Delta x}\right] \tag{10}$$

Then, discretization of $\frac{\partial^2 c}{\partial x^2}$ is obtained equation (11) from implicit second order centered difference in space.

$$\frac{\partial^2 c}{\partial x^2} = \left(\frac{1}{2}\right)\left[\frac{C_{i+1}^{j+1} - 2C_i^{j+1} + C_{i-1}^{j+1}}{(\Delta x^2)} + \frac{C_{i+1}^{j} - 2C_i^{j} + C_{i-1}^{j}}{(\Delta x^2)}\right] \tag{11}$$

Substituting equation (5, (10) and (11) into equation (1) and rearrange according to the time level $(j+1)$ and j, lead to equation (12)

$$\begin{aligned}\frac{C_i^{j+1} - C_i^j}{\Delta t} &= D\left[\frac{1}{2}\left(\frac{C_{i+1}^{j+1} - 2C_i^{j+1} + C_{i-1}^{j+1}}{(\Delta x^2)} + \frac{C_{i+1}^{j} - 2C_i^{j} + C_{i-1}^{j}}{(\Delta x^2)}\right)\right] \\ &\quad - u\left[\frac{1}{2}\left(\frac{C_{i+1}^{j+1} - C_{i-1}^{j+1}}{2\Delta x} + \frac{C_{i+1}^{j} - C_{i-1}^{j}}{2\Delta x}\right)\right]\end{aligned} \tag{12}$$

Rearrange equation (12) of the time level $(j+1)$ at the LHS and time level j at the RHS, then will obtained equation (13).

$$\begin{aligned}&(-2\alpha + \beta)C_{i+1}^{j+1} + (4 + 4\alpha)C_i^{j+1} + (-2\alpha - \beta)C_{i-1}^{j+1} \\ &= (2\alpha - \beta)C_{i+1}^{j} + (4 - 4\alpha)C_{i+1}^{j} + (2\alpha + \beta)C_{i-1}^{j}\end{aligned} \tag{13}$$

where,

$$\alpha = D\frac{(\Delta t)}{(\Delta x^2)}, \qquad \beta = u\frac{(\Delta t)}{(\Delta x)}$$

3.3. Error Estimation of the Numerical Scheme

In this section, the numerical solution of FTCS and CAN are validated with the analytical solution by [8]. The stability condition of the explicit and implicit difference scheme of 1D advection diffusion equation are: $0 \le u\frac{\Delta t}{\Delta x} \le 1$, $0 \le D\frac{\Delta t}{\Delta x^2} \le \frac{1}{2}$ and $\left(\frac{u\Delta t^2}{\Delta x}\right) \le 2D\frac{\Delta t}{\Delta x^2} \le 1$. The spatial grid, and time step, were provided by [8] are stated as follows,

$$\Delta x = 0.05 \text{ and } 0.025 \text{ m} \tag{14}$$

$$\Delta t = 0.01, 0.025 \text{ and } 0.005 \text{ s} \tag{15}$$

The selection of Δt and Δx has been tested based on the stability requirement. The Courant number, r must satisfy $r \leq 0.5$, if the Courant number greater than 0.5 the result is unstable. Thus, the values of Δx and Δt is set based on (14) and (15) which is more stable.

Then, the numerical solution will be compared with the analytical solution in equation (16) as follows,

$$C(x,t) = \left(\frac{1}{2}\right)\left[erfc\left(\frac{x-ut}{\sqrt{4Dt}}\right) + \exp\left(\frac{xu}{D}\right)erfc\left(\frac{x+ut}{\sqrt{4Dt}}\right)\right] \quad (16)$$

Based on these comparison, the error is determined as follow:

Absolute error = |Approximate value - Analytical value|

Figure 1 (a), (b) and (c) shows the error of the numerical solutions for this problem when compared with the analytical solution for $\Delta x = 0.05$ and $\Delta x = 0.025$. The values in Figure 3.18 are plotted at $t = 3s$ between $0 \leq x \leq 1$.

From Figure 1 (a), (b) and (c), we can see that the absolute errors for the FTCS and Crank Nicolson method are decreasing as the time step size decreases. However, the absolute errors for the FTCS is greater than the absolute errors for the Crank Nicolson method. Hence, the Crank Nicolson is the best method for solving ADE with as stated by the [8] and [9].

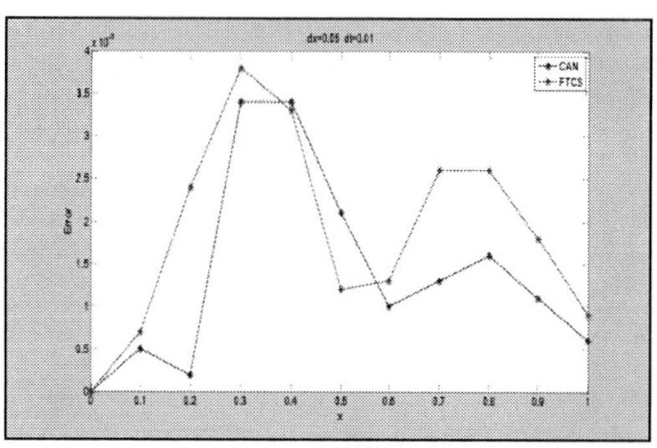

(a)(i)

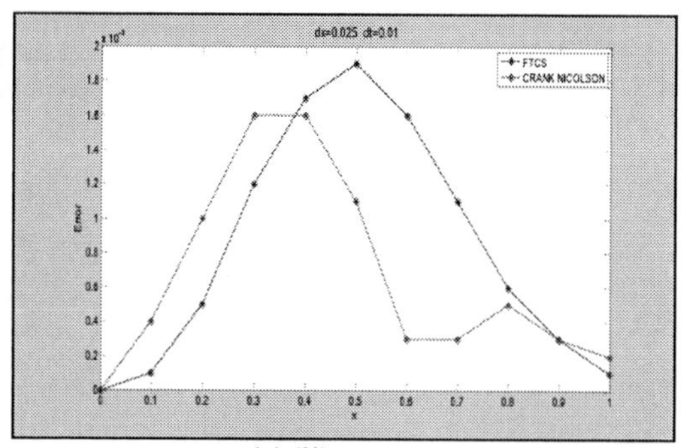

(a)(ii)

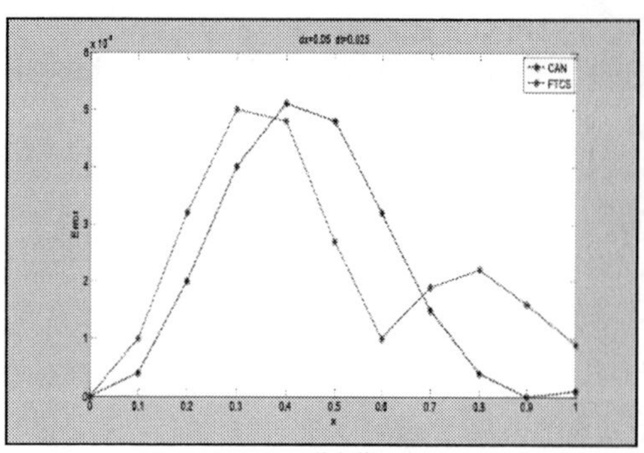

(b)(i)

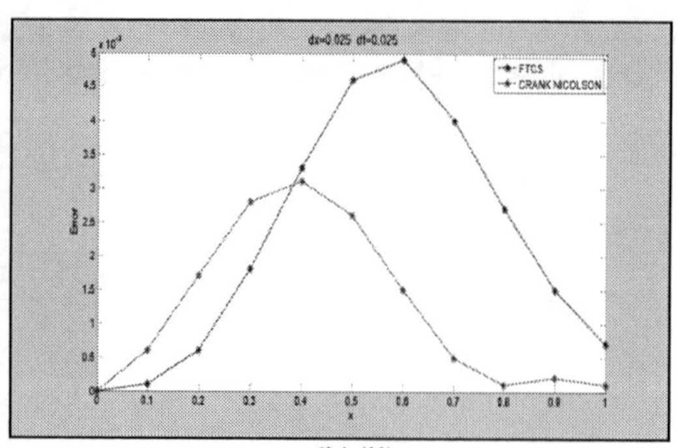

(b)(ii)

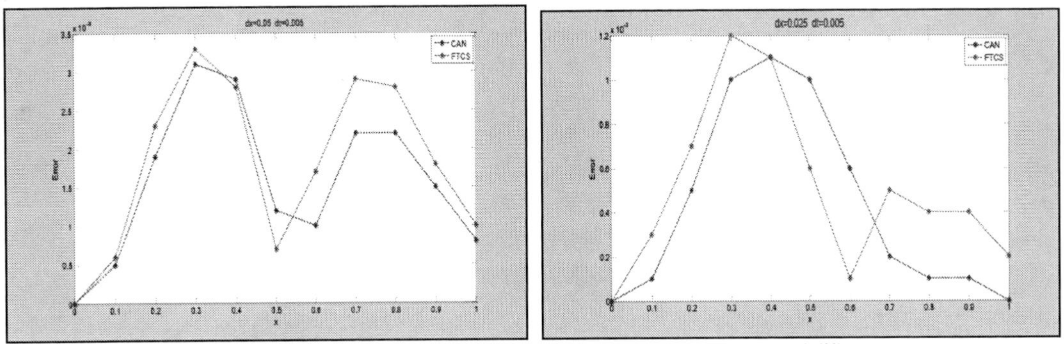

(c)(i) (c)(ii)

Figure 1. Comparison of the errors in the numerical solutions with analytical solution at $t = 3s$.
(a) $dt = 0.01$ [(i) $dx = 0.05$ (ii) $dx = 0.025$]
(b) $dt = 0.025$ [(i) $dx = 0.05$ (ii) $dx = 0.025$]
(c) $dt = 0.005$ [(i) $dx = 0.05$ (ii) $dx = 0.025$]

4. Results and Discussions

4.1. Numerical Simulation using Crank Nicolson Method

Based on the validation in section 3.3, it shows that the Crank Nicolson method is better than FTCS method. Therefore, the Crank Nicolson is used to predict the concentration of water pollution transport. By using the advection diffusion equation, which is varying the two parameters D and u that represent the diffusion rate and velocity rate respectively. In this study, it considered the specific 1D water pollution model problem as in Equation (17). Then it was discretized using the implicit method with an initial and boundary condition based on [10], which stated that,

$$\frac{\partial c}{\partial t} + u\frac{\partial c}{\partial x} = D\frac{\partial^2 c}{\partial x^2} \quad (17)$$

Initial condition: $c(x,0) = 0\ ,\quad 0 \le x \le 100$ (18)

Boundary condition:

$$c(0,t) = 1 \qquad 0 \le t \le 24\text{min} \quad (19)$$

$$c(100,t) = 0 \qquad 0 \le t \le 24\text{min} \quad (20)$$

By using the initial and boundary condition mention in Equation (17-20), the numerical simulation is presented in Figure 2. The Figure 2 shown the validation of the concentration distribution profile of water pollution between two difference methods. The calculated values are showed for times $0 \le t \le 24\text{min}$. The plotted graph show that the pollutant concentration distribution profile at different time for Crank Nicolson. It can be observed that the pollutant concentration is decrease with respect to distance. As mentioned before, the Crank Nicolson method gives better accuracy result compare to the FTCS method.

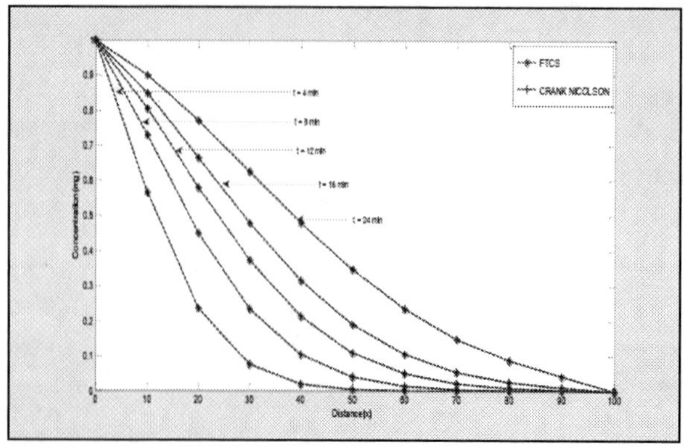

Figure 2. Validation of concentration distribution profile at different time between FTCS and Crank Nicolson method.

Based on the results in Figure 2, the concentration pollutant transport is changing within the time. When the time increased, the further the distance of concentration pollutant moves and the amount of water pollution concentration reduced. The transport of water pollution concentration involves velocity and diffusion rate. Therefore, these two parameters are manipulated to see the effect towards the speed of the pollutant transport and concentration of pollutant travel at a certain distance.

4.2 Result and Effect of Water Pollution Transport on Velocity and Diffusion Rate

The numerical simulation results are represented for estimating the water pollution concentration transport by varying the water flow velocity, u and the diffusion rate coefficient, D. The Crank Nicolson method is implemented to the advection diffusion equation that modelled the transport of the pollutant in the water. The aim of the result is to show that the advection term and diffusion term give the effect to the spreading of pollutant concentration in the water. Therefore, the varying of these two parameters, D and u need to satisfy the stability conditions as mentioned in section 3.2. From the stability conditions, the result is shown in Figure 3 by selecting the velocity rate and diffusion rate as follows, velocity rate, $u = 0.01ms^{-1}$ and $0.05ms^{-1}$ and diffusion rate, $D = 0.05m^2s^{-1}$ and $0.25ms^{-1}$.

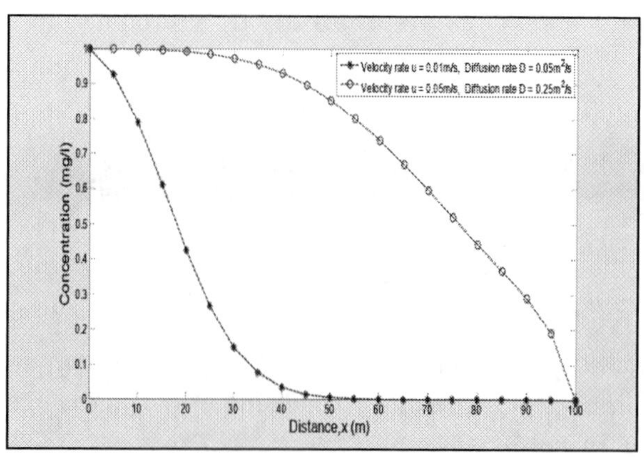

Figure 3. Concentration distribution for varying velocity rate and diffusion rate at time $t = 24\,\text{min}$.

Figure 3 shows the concentration distribution profile was calculated at $t = 24\text{min}$ by increasing the values of velocity rate from $u = 0.01\ \text{ms}^{-1}$ to $u = 0.05\ \text{ms}^{-1}$ and value of diffusion rate from $D = 0.05\ \text{m}^2\text{s}^{-1}$ to $D = 0.25\ \text{m}^2\text{s}^{-1}$. Figure 4 shows that, when the velocity and diffusion rate is high, the pollution concentration movement tend to spread faster. Besides that, the concentration of pollutant also decreases when it travels further along the distance of certain area.

It can be observed that at lower rate of velocity and diffusion, at concentration $0.9\ \text{mg/l}$, the distance of pollutant travelled is approximately 8 meters long and the concentration continue decreasing to $0.4\ \text{mg/l}$ when the distance travelled is approximately 22 meters. However, when the velocity and diffusion rate is higher, at concentration $0.9\ \text{mg/l}$ the pollutant spread further which is approximately 45 meters and concentration reduce to $0.4\ \text{mg/l}$, the distance of pollution travelled is approximately 85 meters.

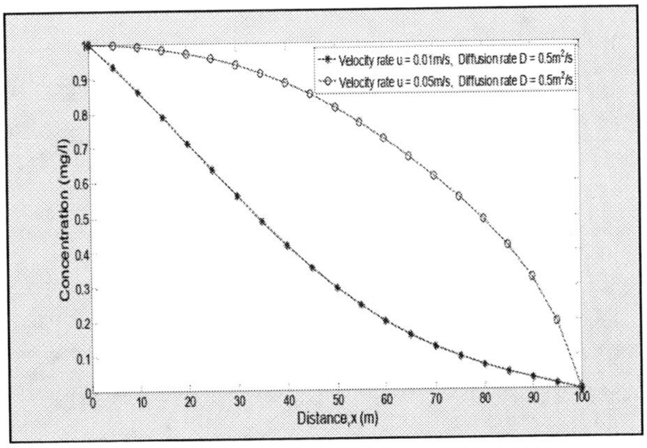

Figure 4. Concentration distribution for varying velocity rate at $t = 24\text{min}$.

Figure 4 shows the concentration distribution for each velocity rate at $t = 24\text{min}$. The velocity is varying at rate $u = 0.01ms^{-1}$ and $u = 0.05ms^{-1}$. Figure 5 shows the pollution concentration is spreading faster by increasing the velocity rate even though the diffusion rate is kept at constant rate, $D = 0.5m^2s^{-1}$. This is the same as the studies by [10], where the author fixed the value of diffusion rate and varying the value of velocity rate. The result gives faster rate of spreading of the pollutants on the water contents.

In that case, the water pollution concentration transport is predicted through the rate of velocity. At velocity $0.01\ \text{m/s}$, the concentration distribution is smooth and pollutant concentration slowly decreases. At $0.4\ \text{mg/l}$, the pollutant concentration travel approximately until 42 meters long. However, at high velocity $0.05\ \text{m}^2s^{-1}$ the pollutant concentration travel about 85 meters in distance and concentration of pollutant reduces rapidly. For this reason, the pollutant move in a longer more distance with higher rate of velocity than lower velocity, even though the diffusion rate is constant.

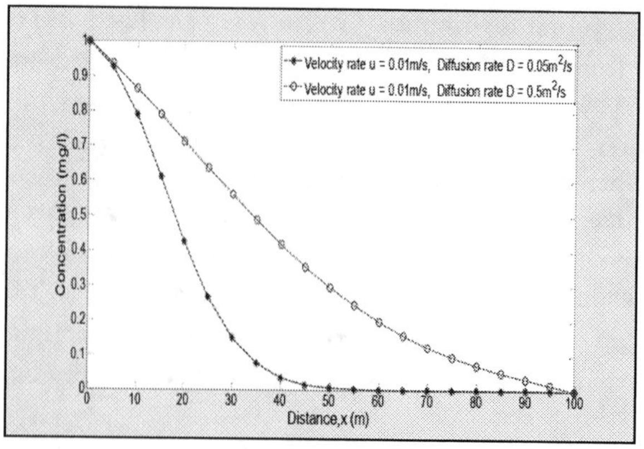

Figure 5. Concentration distribution for varying diffusion rate at $t = 24\text{min}$.

Figure 5 shows the concentration distribution for each velocity rate at $t = 24\text{min}$. The diffusion rate is varying at $D = 0.05m^2s^{-1}$ and $D = 0.5m^2s^{-1}$. Fig.4.4 shows when the diffusion rate is high, the pollution is spreading faster while the velocity rate is kept at constant rate, $u = 0.01ms^{-1}$. As discussed in [10], even the velocity is constant, the spreading of the pollutants is faster with higher rate of diffusion rate.

The diffusion rate also give the impact to the prediction of the water pollution concentration transport. At diffusion rate $0.05\ \text{m}^2 / \text{s}^{-1}$, for concentration 0.4 mg/l, the pollutant concentration travel approximately about 23 meters long. However, when the diffusion rate is high at $0.5\ \text{m}^2/\text{s}^{-1}$, the pollutant concentration travel approximately 43 meters in distance. For this reason, the concentration of pollutant spread further with the higher rate of diffusion than the lower rate, even though the velocity rate is constant. As a result, the concentration of pollutants is travel long distance then take slowly reduce its movements and existence. Based on the Figure 2, Figure 3 and Figure 4, the results show that when the two important parameters, velocity and diffusion rate in transport of pollutants are varying, the spreading of the pollutant will be affected as have been discussed in [10] and [11]. As a result, the water pollution concentration transport with higher rate of velocity or diffusion will spread faster than the lower rate.

5. Conclusions

In this paper, the prediction of the water pollution concentration transport has been studied through the numerical simulation. Based on the simulation results, the implicit Crank Nicolson method is successfully used to solve the 1D ADE. This method is great for solving the ADE after doing the error comparison with FTCS. It is successfully validated with the FTCS method on the water pollution concentration distribution. Therefore, this implicit Crank Nicolson method can be considered as one of the methods to solve one dimensional ADE specifically on water pollution problem. The water pollutants transport is affected by the advection and diffusion term when the velocity and diffusion rate are manipulated. In the real case, the water area such as river, lake and others inland have their own velocity and diffusion rate of pollutants moved along the streams. So, by these parameters the water pollution transport is predicted.

Acknowledgements

The authors would like to acknowledge the support from Research & Innovation, Universiti Malaysia Perlis for the financial funding through Short Term Grant (STG) 2016.

References

[1] Bartram, J. & Balance, R. (1997). *Water Quality Monitoring*.
[2] Jackson, C. (2003). Problems, Perceptions and Perfection - The Role of The Drinking Water Inspectorate in Water Quality Incidents and Emergencies. In *Internatinal Conference on Water Contamination Emergencies: Can We Cope?* (pp. 38–43).
[3] Gane, S. (2000). *Solution of The Advection Difference Finite Equation Using The Method of Schemes and Characteristics*.
[4] Wang, Q., Li, S., Jia, P., Qi, C., & Ding, F. (2013). A Review of Surface Water Quality Models, *2013*.
[5] Ahsan, M. (2012). Numerical Solution of The Advection Diffusion Equation Using Laplace Transform Finite Analytical Method. *River Basin Management*, *10*(June), 177–188.
[6] Magnus, U., Igboekwe, & Achi, N.J, (2011). Finite Difference Method of Modelling Groundwater Flow. *Journal of Water Resource and Protection*, *3*(3), 192–198.
[7] Landage, A. (2014). Groundwater Contaminant Transport 1-D FDM Approach. *Evolving Water Resources Systems*, *2014*(June), 1–9.
[8] Thongmoon, M., & Mckibbin, R. (2006). A Comparison of Some Numerical Methods for the Advection-Diffusion Equation. *Res. Lett. Inf. Math. Sci*, *10*(6), 49–62.
[9] Mohammadi, A., Manteghian, M., & Mohammadi, A. (2011). Numerical solution of one-dimensional advection-diffusion equation using simultaneously temporal and spatial weighted parameters. *Australian Journal of Basic and Applied Sciences*, *5*(6), 1536–1543.
[10] Rahaman, M. M., Andallah, L. S., & Begum, M. (2014). A Numerical Study For The Estimation Of The Water Pollution. *Research in Information Technology*, *2*(1), 131–143.
[11] Azad, T. M. A. K., & Andallah, L. S. (2015). An Explicit Finite Difference Method. *Science and Engineering*, *6*(1), 32–37.

IOP Conf. Series: Materials Science and Engineering **318** (2018) 012006 doi:10.1088/1757-899X/318/1/012006

Thermal Stability and Surface Wettability Studies of Polylactic Acid/Halloysite Nanotube Nanocomposite Scaffold for Tissue Engineering Studies

M Mohd Nizar[1], M S A Hamzah[2], S I Abd Razak[2,3,4] and N H Mat Nayan[1,4*1]
Faculty of Engineering Technology, Universiti Tun Hussein Onn Malaysia, 86400 Parit Raja, Johor Bahru, Malaysia
[2] Faculty of Biosciences and Medical Engineering, Universiti Teknologi Malaysia, 81310 Skudai, Johor, Malaysia
[3] IJN-UTM Cardiovascular Engineering Centre, Universiti Teknologi Malaysia, 81310 Skudai, Johor, Malaysia
[4] Medical Devices and Technology Group, Faculty of Biosciences and Medical Engineering, Universiti Teknologi Malaysia, 81310 Skudai, Johor, Malaysia

E-mail : nadirul@uthm.edu.my

Abstract. This paper reports the preliminary study about the incorporation of halloysite nanotubes (HNT) into polylactic acid (PLA) scaffold to improve the thermal resistance and surface wettability properties. The fabrication of the porous scaffold requires a simple yet effective technique with low-cost materials within freeze extraction method. The thermal stability of PLA/HNT scaffold compared to neat PLA scaffold achieved with increased content of HNT by 5 wt%. Moreover, the surface wettability of the scaffold also shows a positive impact with high content of HNT by 5 wt%. This new nanocomposite scaffold may have high potential as a suitable template for tissue regeneration.

1. Introduction

In medical technology, tissue engineering promising an alternative approach of malfunctioning or lost organs problems such as biodegradable porous scaffold, which patients cells itself can be used to treat them [1]. Tissue engineering emerged in the early 1990s aims to produce a successful scaffold that play a significant role by preserving tissue volume, providing temporary mechanical function, and delivering biofactors [2]. Since tissue and organ repair has been the ultimate goal of surgery from ancient time to the present, many attempts has been made by the tissue engineers to fabricate such ideal scaffold using various materials that can act as a template for the tissue regeneration. The success of scaffold for tissue regenerative mostly dependent on the properties of the materials which determine its morphology, biocompatibility, degradation rate, mechanical properties as well as its porosity.

Until recently, numerous synthetic polymers have found wider interest for tissue engineering as they have a range of degradation rates from very short (days) to long (several months) with good processing characteristics [3]. They also found to be more suitable for the replacement and repair of hard tissues by means of their inherent strength. Porous scaffold based on synthetic biodegradable polymer such as polylactic acid (PLA) appears to found a wide range of pharmaceutical applications for tissue regeneration of skin [4], cartilage [5], blood vessels [6], and cardiac valve [7]. Basically, PLA which can be derived from the renewable resources [8] such as corn starch is one of the interesting material for scaffold fabrication due to its synthetically controllable degradation rate [9], good mechanical properties [10], and biocompatibility [11]. However, PLA has low crystallization rate, yet contribute causes to some limitations of high-performance applications. This parameter is particularly essential to control PLA degradation rate, thermal resistance, as well as optical, mechanical and barrier properties [12]. Due to these limitations, fabrication of successful PLA scaffold is almost impossible in order to meet the requirements for tissue regeneration.

Alternatively, in order to maximize the benefits and versatility of PLA, the physical, rheological, optical or other properties of PLA compounds can be modified by adding nano-fillers into the PLA matrix [13]. Halloysite nanotubes (HNTs), a type of nanofiller have found its significant potential as a new type of additive for enhancing the mechanical, thermal and fire-retardant performances of polymers [14]–[19]. HNTs can be used as an alternative for cost reduction, improvement of mechanical parameters, e.g. increase of modulus/stiffness at room temperature or elevated temperature, especially in those PLA grades which allow a high degree of crystallinity. HNTs are a type of natural aluminosilicate nanotubes with a similar structure to kaolinite. However, halloysite nanotube differs from kaolinite because of its multilayer tubular structure that results from the wrapping of the 1:1 layers structures. The unit layers in HNTs are separated by a monolayer of water molecules as shown in figure 1.

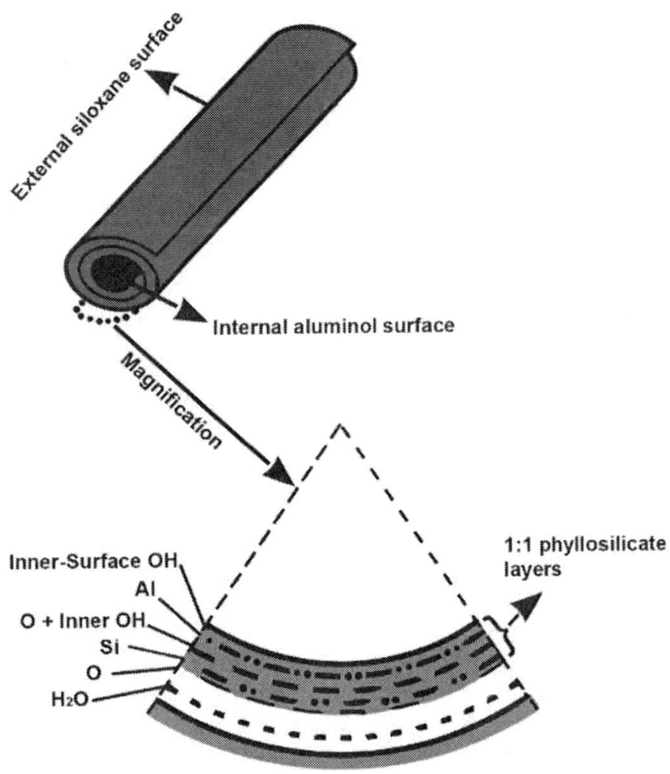

Figure 1. Structure of halloysite nanotube (HNT).

As reinforcing material, HNTs are considered as the ideal materials for preparing polymer composites. This is due to the fact that halloysites are rigid materials and that unique crystal structure of HNTs resembles that of carbon nanotubes (CNTs) in terms of aspect ratio. HNTs are more easily dispersed in polymer matrix by shearing due to their rod-like geometry and limited inter-tubular contact area, compared with other nanoparticles such as fumed silica, montmorillonite, and carbon nanotubes [20]. Moreover, due to several characteristic features such as nanoscale lumen, relatively high length-to-diameter ratio, and low hydroxyl group density on their surface, HNTs are considered as promising competitors and cheaper substitutes to both carbon nanotubes and organomodified layered silicates. Furthermore, incorporation of HNTs into various polymers including biopolymer can increase the strength and thermomechanical properties without loss if ductility [21].

Various fabrication techniques have been attempted to produce a 3D porous scaffold structure and acquire control over mechanical properties of the scaffold that may in turn contribute to define shape and function of the tissue [22]. This includes thermally induced phase separation [23], 3D printing [24], porogen leaching [25], freeze drying [26] and electrospinning [27]. Though freeze drying is the most popular applied technique, it is time and energy consuming. This technique also produced unwanted surface skin. However, there is another method to produce polymeric porous scaffold which is freeze extraction. Freeze extraction is a simple two-step process by which involves the freezing of polymer solvent allowing the phase separation process to occur, thus leading to formation of different phases of frozen solvent and concentrated polymer. In this method, the frozen solvent is removed which then resulted in formation of polymer scaffold [28].

Though there are few reports on inclusion of HNTs into PLA matrix, till today, there have been no study on the preparation of freeze extracted porous scaffolds of Therefore, this article aims in developing such PLA/HNTs nanocomposite scaffold by means of freeze extraction method. Inclusion of HNTs nanofiller in the PLA scaffold might open up opportunities in many biomedical applications such as tissue engineering. This preliminary study will investigate the thermal behavior and surface wettability of the nanocomposite scaffold.

2. Experimental

2.1. Scaffold fabrication

The porous PLA/HNTs scaffold was prepared by the solvent-freeze extraction method. PLA was dissolved in 1,4-dioxane for 24 hours to obtain a polymer solution. The required amount of HNTs powder (1, 3, 5, 7, 8 wt%) was added into the polymer solution and allowed to mixed by continuous stirring for 12 h. The solution was then poured into a glass round-shaped mold and cooled to -20°C until froze. The frozen PLA/HNTs polymer mixtures were then immersed in pre-cooled (-20°C) 80% ethanol for solvent extraction. During this process, the porous structure is obtained when the solvent was extracted out and replaced by the non-solvent (ethanol). The scaffolds were removed and rinsed with pre-cooled ethanol to remove excess solvent from the sample. Then, the samples were dried at room temperature. Following this procedure, neat PLA scaffolds and PLA/HNTs nanocomposite scaffolds with required HNTs filler loading were prepared. Figure 2 displays the freeze extraction method.

2.2. Thermogravimetric analysis

The aim thermogravimetric (TGA) analysis is to evaluate the effects of HNTs towards the thermal stability of PLA scaffold. Melting and crystallization behavior were studied using TGA analyser (TA Instruments Inc., New Castle, D.E. USA) under nitrogen gas flow over temperature range 30 to 600 °C at heating rate of 10 °C/min. The samples were prepared around 1 - 2 g each, and are heated. Differential peak temperature (DTp) was recorded as it represents the temperature derivative of the sample's weight changes.

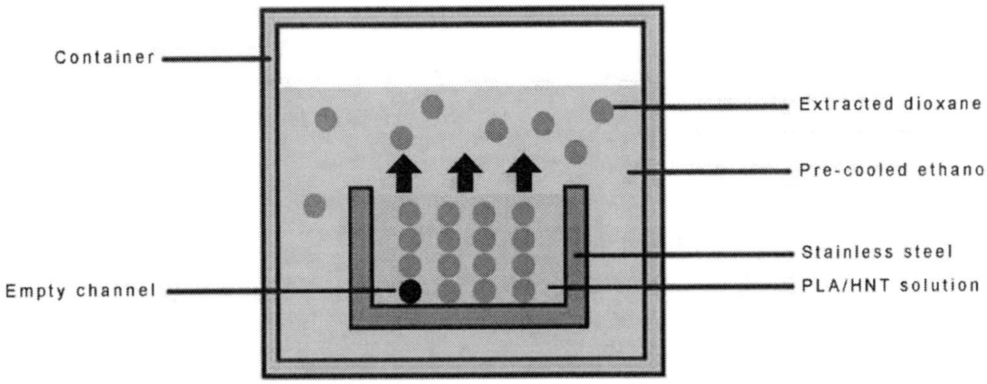

Figure 2. Freeze extraction method.

2.3. Surface wettability

The surface wettability was assessed using water contact angle applied sessile drop method. Contact angle measurement is basically used to measure wettability of solid surface when a drop of liquid is placed on it. In this procedure, the water contact angle was utilized with distilled water as the probe liquid. The contact angles were calculated from images captured by the camera of contact angle measuring system, VCA Optima, AST Products, Inc. Contact angle measurement is basically used to measure wettability of solid surface when a drop of liquid is placed on it. The result obtained may either spread out or remain in the droplet form. The contact angle depends on the hydrophilic moiety characteristic of the polymer either it is air interface or bulk of the solid oriented.

3. Results and discussion

3.1. Fabrication of scaffold

The PLA and PLA/HNTs scaffolds were successfully fabricated using freeze extraction method as can be captured in figure 3. The PLA/HNTs scaffold exhibit good porous structure based on the scanning electron microscope (SEM) as shown in figure 4. It is suggested that the successful inclusion of HNTs into the polymer scaffold during fabrication process may due to the formation of ionic and covalent bonds between both structures.

Figure 3. Fabricated of PLA/HNTs scaffold *via* freeze extraction method.

IOP Conf. Series: Materials Science and Engineering **318** (2018) 012006 doi:10.1088/1757-899X/318/1/012006

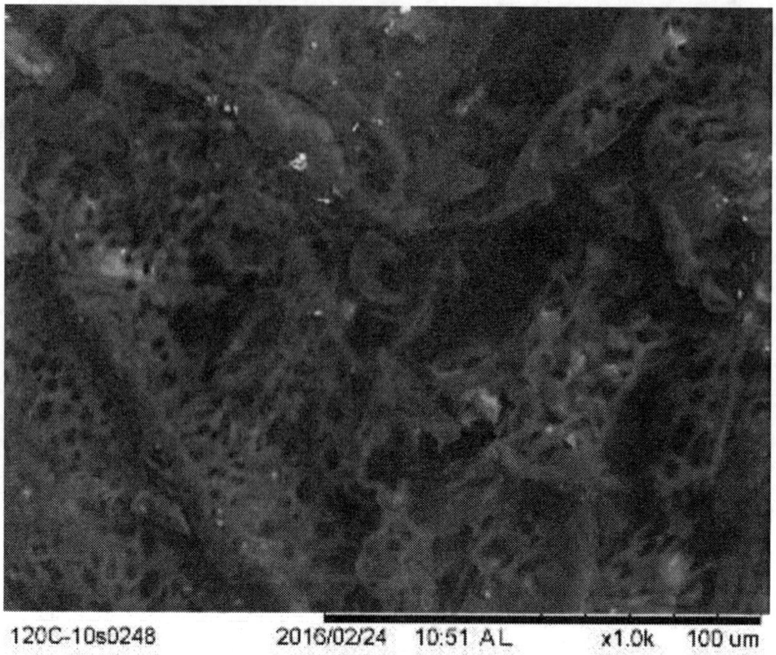

Figure 4. SEM of the cross-sectional of PLA/HNTs scaffold.

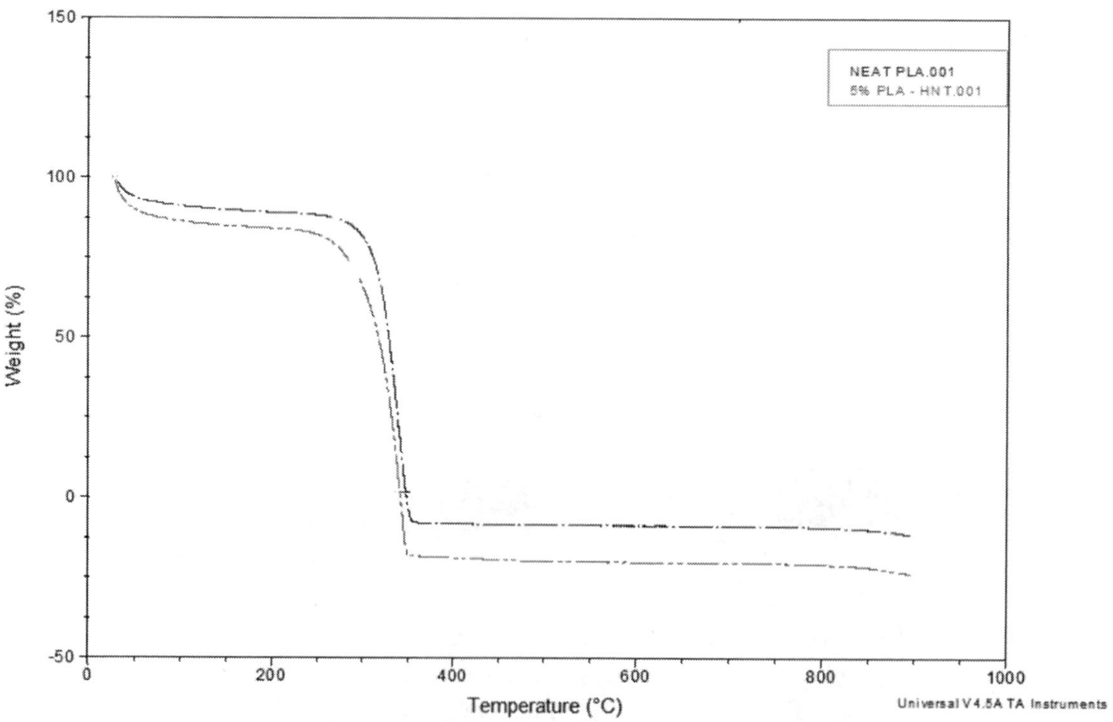

Figure 5. TGA curves of PLA/HNTs scaffolds.

3.2. Thermogravimetric analysis

The thermal stability of the PLA/HNTs scaffold is characterized by using the thermogravimetric (TGA) results as demonstrated in figure 5. Neat PLA degraded completely, while the PLA/HNTs composite degraded with forming residue. This residue can be assigned to the HNTs content of the sample, and it can be noticed that there was a gradual reduction in weight loss in PLA/HNTs composite between 350-450°C, which corresponds to the fact that HNTs starts degradation at approximately 400°C. The onset weight loss temperature for PLA was 304.5°C, which reduced to 295.07°C for PLA/5 HNTs (figure 5). As the onset weight loss temperature of the PLA/HNT composites decreased with the increase addition of HNTs, this indicated that the thermal stability of PLA/HNTs composites was reduced comparing with that of the neat PLA.

The reduction in thermal stability of PLA/5 HNTs compared to neat PLA may be due to presence of agglomerates in the scaffold. Compared to neat PLA, thermal stability of PLA/HNTs composites was achieved with high content of HNTs and uniformly dispersed HNTs in the polymer matrix. The uniformity dispersion is ascribed to the catalytic role of HNTs on the PLA pyrolysis with the existence of Si - OH and Al - OH acid sites on HNTs external surfaces [29], [30]. However, high content of HNTs in PLA/HNTs composite is not ideal for tissue engineering application. The reduction in thermal stability observed in current study may indicates that the space between PLA and HNTs allowed the volatile pyrolyzed products of PLA to accumulate, thereby accelerating degradation. Further analysis of other PLA/HNTs sample may need to be done to support this result.

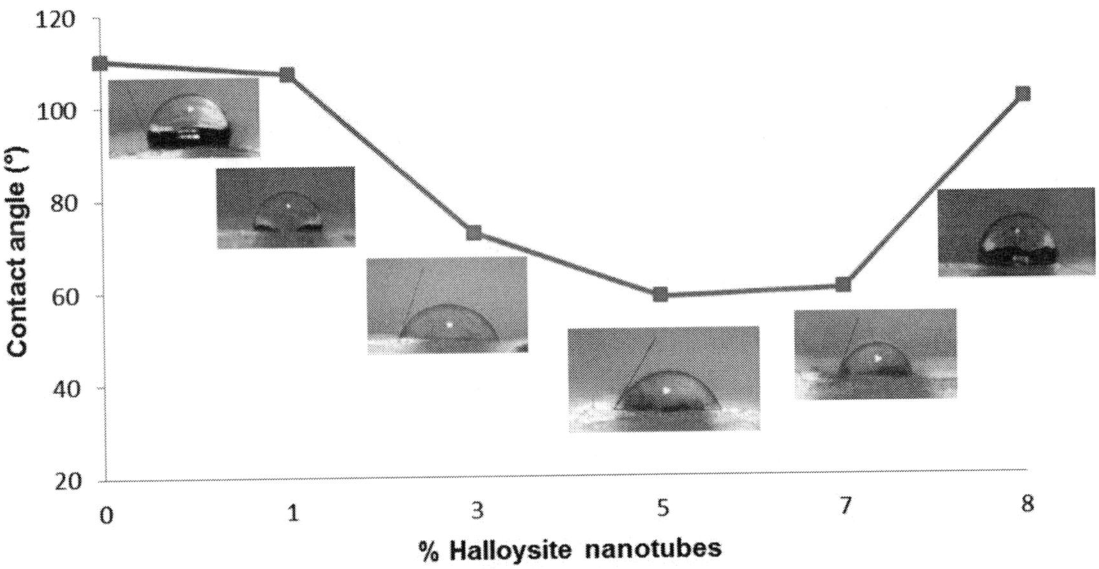

Figure 6. Water contact angle for each sample.

3.3. Surface wettability

The water contact angle testing was conducted to analyze the capability of the nanocomposite scaffold to absorb water. The ability of scaffold to absorb water is crucial to maintain the scaffold shape and play major roles during cell culture studies as it may affects the cell growth [31]. As shown in figure 6, the water contact angle of the nanocomposites has a declining trend by raising the HNT concentration. The contact angle on neat PLA was found to be 110.10° which, basically, assigned to have hydrophobic surface. As the HNTs was introduced to each scaffold formulation, the water contact angle of the scaffold surface reduces to 107.13°, 72.60° and 58.63°. Then, the contact angle increases back to 60.30° and 101.00° for PLA/1 HNT, PLA/3 HNT, PLA/5 HNT, PLA/7 HNT and PLA/8 HNT respectively. Therefore, based on the results, hydrophilicity properties of the scaffolds was improved. This is instigated by the capillary action of halloysite nanotubes and their good wettability properties [32]. This phenomenon can occur due to structure of HNTs that comprises hydrophilic groups (Al - OH and Si - OH) located on its surface beside alumina and silica groups at its crystal edges, thus ease the formation of hydrogen bonds between HNTs and water [33].

4. Conclusion

This study focused on the effects of HNTs on thermal behavior and surface wettability of fabricated PLA scaffold. PLA/HNTs nanocomposite scaffold was prepared by freeze extraction method. The introduction of HNTs into the PLA has increased the thermal stability of the scaffold which was achieved by the high inclusion of HNT and the well dispersion in the polymer matrix. Furthermore, the surface wettability of the PLA/HNTs scaffold also shows positive effect where the water contact angle on the surface increased with increasing of HNTs. Overall, this novel scaffold is easy to fabricate via freeze extraction method. This technique can be considered as the most cost-effective process since it requires only low-cost materials with minimum use of other utilities. In order to support the use of this nanocomposite scaffold in tissue engineering application, extension study such as the mechanical properties, porosity, and cytotoxicity test need to be conducted to ascertain its viability towards specific purpose.

Acknowledgement

The authors are indebted to FRGS (Fundamental Research Grant Scheme) for funding this research.

References

[1] Mironov V, Trusk T, Kasyanov V, Little S, Swaja R and Markwald R 2009 *Biofabrication* **1** 22001
[2] Hollister S J 2005 *Nat. Mater.* **4** 590
[3] Bonassar L J and Vacanti C A 1998 *J. Cell. Biochem. Suppl.* **30** 297-303
[4] Mohiti-Asli M, Saha S, Murphy S V, Gracz H, Pourdeyhimi B, Atala A and Loboa E G 2017 *J. Biomed. Mater. Res. Part B: Appl. Biomater.* **105** 327–339
[5] Hutmacher D W 2000 *Biomaterials* **21** 2529–43
[6] Li Z, Zhao X, Ye L, Coates P, Caton-Rose F and Martyn M 2015 *Chem. Eng. J.* **279** 767–776
[7] Iop L and Gerosa G 2015 *BioMed Res. Int.* **2015** 1–13
[8] Yang S, Madbouly S A, Schrader J A, Srinivasan G, Grewell D, McCabe K G, Kessler M R and Graves W R 2015 *Green Chem.* **17** 380–393
[9] Cui M, Liu L, Guo N, Su R and Ma F 2015 *Molecules* **20**, 595–607
[10] Shi H, Gan Q, Liu X, Ma Y, Hu J, Yuan Y and Liu C 2015 *RSC Adv.* **5** 79703–14
[11] Abdal-hay A, Hussein K H, Casettari L, Khalil K A and Hamdy A S 2016 *Mater. Sci. Eng. C* **60** 143–150
[12] Murariu M, Dechief A L, Ramy-Ratiarison R, Paint Y, Raquez J M and Dubois P 2015 *Nanocomposites* **1** 71–82
[13] Matusik J, Stodolak E and Bahranowski K 2011 *Appl. Clay Sci.* **51** 102-109
[14] Deng S, Zhang J, Ye L and Wu J 2008 *Polymer* **49** 5119–27
[15] Guo B, Lei Y, Chen F, Liu X, Du M and Jia D 2008 *Appl. Surf. Sci* **255** 2715–22
[16] Ismail H, Pasbakhsh P, Fauzi M N A and Abu Bakar A 2008 *Polym. Test.* **27** 841–850
[17] Liu M, Guo B, Zou Q, Du M and Jia D 2008 *Nanotechnology* **19** 205709
[18] Marney D C O, L. Russell L J, Wu D Y, Nguyen T, Cram D G, Rigopoulos N, Wright N and Greaves M 2008 *Polym. Degrad. Stab.* **93** 1971–78
[19] Ye Y, Chen H, Wu J and Ye L 2007 *Polymer* **48** 6426–33
[20] Du M, Guo B and Jia D 2010 *Polym. Int.* **59**, 574-582

[21] Schmitt H, Prashantha K, Soulestin J, Lacrampe M F and Krawczak P 2012 *Carbohydr. Polym.* **89**, 920-927
[22] Lee L, Zeng C, Cao X, Han X, Shen J and Xu G 2005 *Compos. Sci. Technol.* **65**, 2344–63
[23] Mannella G A, Conoscenti G, Carfì Pavia F, La Carrubba V and Brucato V 2015 *Mater. Lett.* **160** 31-33
[24] Rosenzweig D, Carelli E, Steffen T, Jarzem P and Haglund L 2015 *Int. J. Mol. Sci.* **16** 15118-135
[25] Choudhury M, Mohanty S, Nayak S 2015 *J. Biomater. Tissue Eng.* **5** 1-9
[26] Salerno A, Fernández-Gutiérrez M, San Román del Barrio J and Domingo C 2015 *J. Supercrit. Fluids* **97** 238-246
[27] Morelli S, Salerno S, Holopainen J, Ritala M and De Bartolo L 2015 *J. Biotechnol.* **204** 53-26
[28] Ho M H, Kuo P Y, Hsieh H J, Hsien T Y, Hou L T, Lai J Y and Wang D M 2004 *Biomaterials* **25** 129-138
[29] Kopinke F D, Remmler M, Mackenzie K, Möder M and Wachsen O 1996 *Polym. Degrad. Stab.* **53** 329–342
[30] Stoclet G, Sclavons M, Lecouvet B, Devaux J, Van Velthem P, Boborodea A, Bourbigot S and Salem-Idrissi N 2014 *RSC Adv.* **4** 57553–563
[31] Smetana K 1993 *Biomaterials* **14** 1046–50
[32] Abdullayev E, Price R, Shchukin D and Lvov Y 2009 *ACS Appl. Mater. Interfaces* **1** 1437–43
[33] Rawtani D and Agrawal Y K 2012 *Rev. Adv. Mater. Sci.* **30** 282–295

IOP Conf. Series: Materials Science and Engineering **318** (2018) 012007 doi:10.1088/1757-899X/318/1/012007

Substantially Stabilized Superacid Incorporated SBA-15 with Calcium Bridging for Selective Esterification of Glycerol

P Y Hoo[1], A Z Abdullah[2], S H Shuit[1], Y P Teoh[1], Q H Ng[1] and B Kunasundari[1]

[1] Department of Chemical Engineering Technology, Faculty of Engineering Technology, Universiti Malaysia Perlis (UniMAP), Kampus UniCITI Alam, Sungai Chuchuh, 02100 Padang Besar, Perlis, Malaysia

[2] School of Chemical Engineering, Universiti Sains Malaysia, Engineering Campus, Seri Ampangan, 14300 Nibong Tebal, Pulau Pinang, Malaysia.

Email: pengyong@unimap.edu.my

Abstract. The exploitation of the super acidity of heteropolyacids incorporated heterogeneous catalysts was only feasible if the heterogeneity of these catalysts was assured. To maintain the catalyst heterogeneity in polar medium, a novel two-step modification method was proposed to synthesize the highly active, yet stable heterogeneous catalyst, catered for selective esterification of monoglyceride. The surficial, structural and acidity properties of the modified catalysts were investigated via crucial characterization methods (N_2 BET, HRTEM, and FTIR). The collective evidences verified the predicted formation of calcium oxides (CaO) on the mesopores surfaces of the SBA-15 support after the first modification, and the successful subsequent 12-tungstophosphoric acid (HPW) functionalization. The superior stability of the synthesized catalysts (10wt%-HPW/CaSBA-15) was demonstrated (negligible change in both conversion – 75% and yield – 70%), without the need for catalyst regeneration. Such result was attributed to the strong interaction between HPW and SBA-15 via calcium bridging. Being alkaline in nature, calcium oxides in the inner pores readily reacted with highly acidic HPW introduced in the subsequent wetness incipient step, forming insoluble HPW acid sites on inner pore walls of SBA-15 via the calcium bridging. This modification deemed promising and other alkaline metals should be explored in the future.

1. Introduction

In the past three years, the drastic collapse of Crude Oil Brent price from as high as USD 115 per barrel in June 2014 to only USD 48 per barrel in July 2017 [1]. With no sign of recovery, such abrupt price drop adversely impacted the development and production of biodiesel [2, 3]. Thus, converting glycerol, the major waste from biodiesel production to other value added products (glycerides, acrolein, citric acid, etc.) became popular to reduce its production cost [4, 5]. Yet, selective production of monoglyceride from glycerol is challenging without a highly selective catalytic system as diglyceride and triglyceride would be formed at the same time. Also, the unfavourable environmental impact and costly separation and purification procedures due to the use of homogeneous acid has pushed the research direction to highly reusable heterogeneous catalysts in industrial monoglyceride synthesis [6, 7].

In the effort of searching the suitable catalyst for selective esterification of glycerol with fatty acid, a particular group of high performance catalysts, mesoporous SBA-15 supported with Keggin structured 12-tungstophosphoric acid (HPW, $H_3PW_{12}O_{40}$) was of interest in this study [8-10]. A benign combination of high Brønsted acidity from HPW and size selectiveness of SBA-15 was anticipated to be the key factors for such highly selective esterification process. However, due to its polarity, the leaching problem of HPW from its support significantly lowered the catalyst reusability. In fact, its reusability was so low that it could only be used once before the catalytic performance was significantly deteriorated, as shown in figure 1.

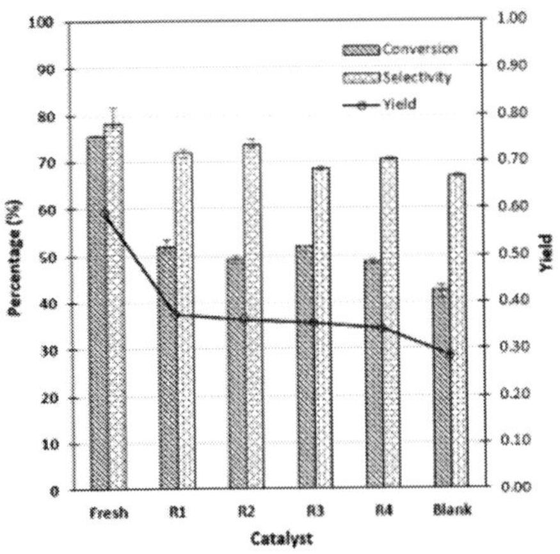

Figure 1. Conversion, Monolaurin Selectivity and Yield of the Fresh and Recycled HPW Incorporated SBA-15 in Esterification of Glycerol with Lauric Acid.

1.1. Reusability of HPW incorporated SBA-15

With reference to figure 1, along with many other available reports in the literature, the reusability, stability, and leaching study of HPW incorporated catalysts have become one of the most important and urgent matter to be addressed in order to continue exploiting the strong Brønsted acidity contributed by HPW [11, 12]. It was understood that only weak, physical interaction exists between the support and HPW acid sites [13]. The weak interaction resulted the vulnerability of these acid sites, causing these supported acid sites to be leached out into the reaction medium, ultimately contributed to the low reusability of the synthesized catalysts, or the need of catalysts regeneration [14]. Since then, researcher have proposed various solution to tackle such matter using different approach, including incorporation of insoluble HPW salts [15], "sandwiched" HPW method [10], or even on different supports [16-18].

1.2. A proposal to the matter of interest

Understanding the need for highly reusable yet performing heterogeneous catalyst in selective production of monoglyceride, this paper aims to improve the stability of the previously synthesized catalyst while retaining its already possessed superior catalytic performance. As far as the heteropolyacids are concerned, the super acidity is retained as long as the Keggin anion remains unchanged [19]. In this case, the conjugate base (the cation for the insoluble acid salt) must be chosen wisely in order to minimize the reduction of the acidity of the original acid which is in fact, responsible for the catalysis. As spontaneous reaction between alkaline and acidic component would produce the strong acid salt (hence the acid sites), it was hypothesized that by first modifying the mesopores surface within SBA-15 into mildly basic sites before functionalized it with HPW would yield a highly acidic yet stabilized heterogeneous catalyst with strong interaction between the support and acid. For such purpose, calcium, an element from the alkaline earth metals group (Group 2) was chosen due to its ability to form insoluble HPW salt ($CaHPW_{12}O_{40}$) and its low toxicity compared to the rest of the elements from the same group. Such criteria was of importance as the catalysts would be applied in food related process (monoglyceride as emulsifiers). A simple yet informative schematic diagram which illustrates the idea of this proposal can be found in figure 2.

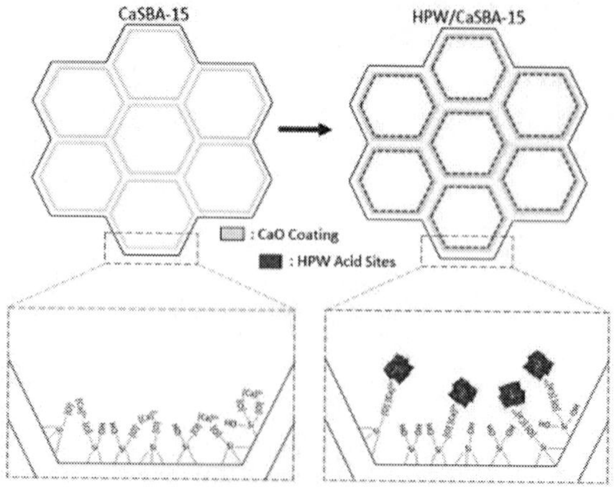

Figure 2. Proposed schematic diagram of the two-step calcium modified SBA-15 functionalized with HPW.

2. Material and methods

2.1. Synthesis of SBA-15

Initially, the typical SBA-15 was hydrothermally synthesized using Pluronic P123 as the organic template and tetraethyl orthosilicate (TEOS) as the silica source, as described in the established report by Zhao [20].

2.2. Calcium modification of SBA-15

SBA-15 was heated up to 378 K for 24 h to evacuate any possible moisture content. At room temperature, an appropriate amount of calcium nitrate tetrahydrate solution was added drop wise onto the SBA-15 as the calcium precursor via wetness incipient method with ethyl alcohol as solvent. To ensure the desired acid sites were formed, one to one molar ratio of calcium to HPW acid was followed. To synthesize CaSBA-15 for 10wt% HPW impregnation, 0.039 M of calcium nitrate solution was prepared. 1 ml solution per gram of SBA-15 was used based on the total pore volume of SBA-15 determined from BET analysis. The resulting sludge was slowly mixed (200 rpm) for 5 h. Then, the mixture was filtered, air-dried for 6 h followed by oven-dried at 353 K for 12 h. To convert the calcium precursors into basic calcium oxides, calcination of the powder was performed at a ramping rate of 2 K min^{-1} to 823 K and stay for 3 h. The resulting white solid was labelled as "CaSBA-15".

2.3. Functionalization of CaSBA-15 with HPW

The moisture of CaSBA-15 was evacuated using oven at 378 K for 24 h. Then, appropriate amount of HPW solution was introduced into CaSBA-15 via wetness incipient method using ethyl alcohol. The mixture was slowly stirred (200 rpm) for 5 h to allow acid-base reaction between HPW and surficial CaO on mesopores of the CaSBA-15. Then, the white powder was filtrated and dried at 298 K for 6 h followed by another 12 h in oven at 353 K. No calcination was followed to avoid disruption of characteristic SBA-15 p6mm hexagonal structure and the decomposition of HPW at high temperature [21]. The synthesized catalysts was labelled as "Xwt%-HPW/CaSBA-15", with X as the weightage of HPW impregnated on CaSBA-15. The HPW needed was calculated using Equation (1).

$$X\,(\%) = \frac{Weight\ of\ HPW}{Weight\ of\ HPW + Weight\ of\ CaSBA-15} \times 100\% \quad (1)$$

2.4. Material characterizations

The surficial, structural and chemical properties of the synthesized materials were investigated. By using Quanta-chrome Autosorb 1C automated gas sorption analyser, Nitrogen (N_2) adsorption and desorption on degassed synthesized materials (543 K, 6 h) enabled investigation on the type of materials, total surface areas, total pore volumes and the pore size distribution of the synthesized materials. Together with images analysis using TECNAI G2 20S-Twin high resolution transmission electron microscopy (HRTEM), the characteristic uniformed mesopores of pure, modified and functionalized SBA-15 could be verified. To verify key functional groups in pure, modified and functionalized SBA-15, the materials were subjected to Fourier Transformed Infrared (FTIR) analysis using KBr pellet method and IRPrestige-21 FTIR Spectrophotometer.

2.5. Catalyst performance and reusability tests

The developed catalysts were tested individually with selective monolaurin production from glycerol (99.8%) and lauric acid (99%). A temperature and pressure regulated batch reactor system made of magnetic stirring and heating mantle, a three-necked flask and a vacuum pump was used for such purpose. The catalytic performances of the HPW/CaSBA-15 catalysts were evaluated based on lauric acid conversion and monolaurin yield. The products sampled from specific reaction times were analysed using Agilent Technologies 7890A GC system equipped with CP-Sil 5CB (15 m x 0.32 mm x 0.1 mm) column. The lauric acid conversion and monolaurin selectivity were calculated as proposed by Pouilloux et al. [22]. The reusability test of the catalysts was accomplished by separating the used catalysts from the reaction mixture at the end of each runs. The whole reaction mixture was removed from reactor and centrifuged at high speed (4400 rpm, 5 mins). The liquid was removed from the vial and the remaining catalyst at the bottom of the vial was washed with acetone (99.8%) and deionized water to remove residual organic and polar compounds. The filtered, washed solid was oven-dried at 373 K for 24 h. The dried catalyst was then labelled R1 (R2 for second recycle) and reused in the same setup. The conversion and monolaurin yields were calculated.

3. Results and discussion

3.1. Surface analysis

From the Table 1, all catalysts exerted high specific total surface areas ranging from 564 to 766 m^2g^{-1}. Together with HRTEM structural analysis, the highly uniformed mesoporous SBA-15 was successfully synthesized and retained even after the two steps modification and functionalization. Further proof of such claim be seen in both nitrogen adsorption-desorption isotherms and the pore size distribution (PSD) of these materials in figure 3.

Table 1. Surface characteristics of SBA-15, CaSBA-15 and HPW/CaSBA-15.

Catalyst	Total Surface Area (m^2g^{-1})[a]	Mesopore Surface Area (m^2g^{-1})[b]	Total Pore Volume (cm^3g^{-1})[c]	Mesopore Volume (cm^3g^{-1})[d]	Average Pore Size (Å)[e]
SBA-15	766	578	0.99	0.90	70
CaSBA-15	629	468	0.67	0.60	63
10wt%-HPW/ CaSBA-15	604	459	0.87	0.79	63
20wt%-HPW/ CaSBA-15	564	408	0.69	0.61	58

[a,e]: From BET desorption method
[b]: From the difference between total surface area to micropore surface area (t-plot method)
[c]: From BJH desorption method
[d]: From the difference between total pore volume to micropore volume (BJH Desorption method)

Evidently, all of the modified catalysts exhibited high surface areas (564 – 629 m^2g^{-1}) with large amount of uniformed mesopores (58 – 63 Å) with narrow pore size distributions (50 – 100 Å), as shown in figure 3 (b). As shown in figure 3 (a), IUPAC Type IV isotherms and type H1 hysteresis loops were found for all materials, thus verifying their mesoporous nature [23]. Such observation verified the retention of highly unformed mesopores channels even after the two steps modifications. Quantitatively, as seen in Table 1, high mesopores contribution on both areas (76%) and volumes (91%) for HPW/CaSBA-15 has further strengthen such claim. With respect to SBA-15, the decreases in the total surface areas and average pore sizes were observed for all the modified catalysts. The decreases was found to be more significant with further modification performed on SBA-15. As such, the observation could be attributed to effect of the alternation on the surface characteristic of the supports [24]. For such phenomenon, it was suggested that inner pore coating of calcium or calcium oxides could have formed, causing wall thickening effect on the SBA-15. Further pore narrowing observed could be attributed to the functionalization using HPW, whereby we deduced the acid-base reaction could have occurred between the basic CaO and HPW on SBA-15 surfaces.

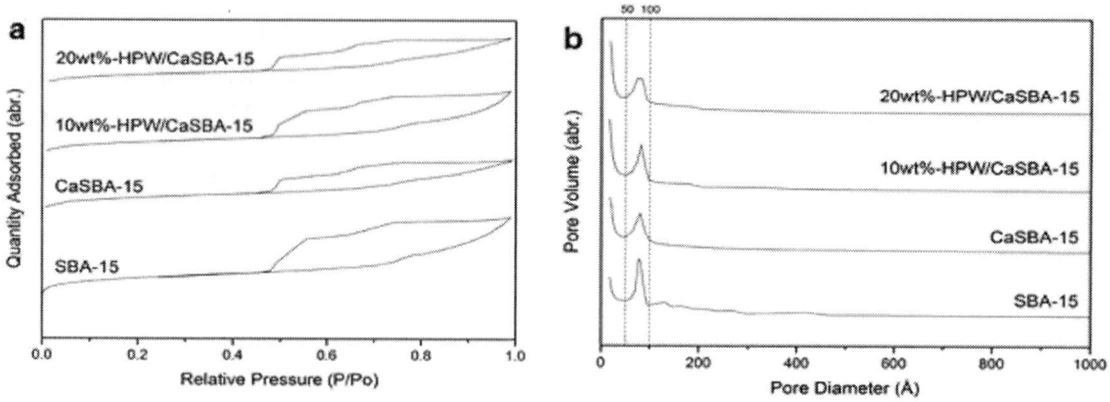

Figure 3. a) Nitrogen adsorption-desorption isotherms and b) BJH pore size distribution of SBA-15, CaSBA-15, 10wt%-HPW/CaSBA-15 and 20wt%-HPW/CaSBA-15

3.2. Structural analysis

As shown in figure 4, long, continuous mesopores were easily noticeable in all three HRTEM images of the modified materials. It was observed that the undisrupted long uniformed mesopores of SBA-15 was preserved even after 2 steps modifications [25]. At high magnification (195,000 X), the uniformity of the mesopores could be recognized. At the same time, the average mesopores sizes as predicted by BJH pore size analysis were justified. In fact, the trend of reduction of measured pore sizes were found to be in agreement with the trend showed in Table 1. These HRTEM images further strengthen the argument that the deposition or modification of SBA-15 with calcium oxide and later HPW acid actually occurred within these mesopores rather than on the externally available surface areas. No agglomeration of foreign particles could be detected, suggesting that well dispersion of both calcium oxide and HPW were achieved via the 2-step post impregnation method.

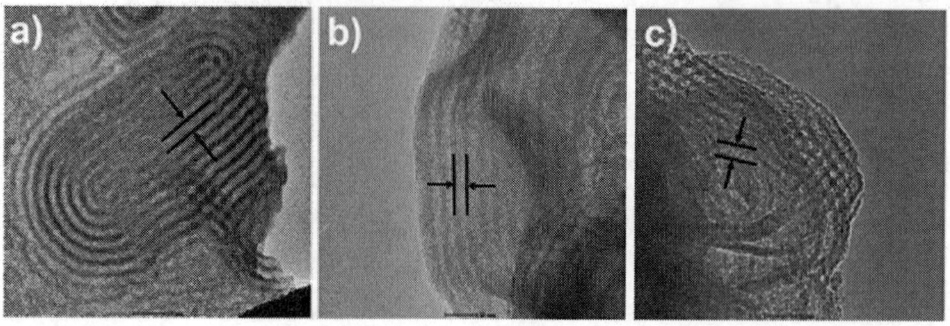

Figure 4. HRTEM images of a) CaSBA-15, b) 10wt%-HPW/CaSBA-15 and c) 20wt%-HPW/CaSBA-15 at magnification of 195,000X at 200kV

3.3. Chemical analysis

The FTIR scans (400 – 4000 cm^{-1}) for all materials were presented in figure 5 for the detection of characteristic bands of SBA-15, calcium oxides and HPW. In figure 5a, characteristic peaks attributed to Si-O-Si in the condensed silica network of SBA-15 could be identified. Specifically, these peaks include broad band of asymmetric stretching of Si-O-Si (1085 – 1220 cm^{-1}), symmetric Si-O-Si stretching (804 cm^{-1}), bending vibration of Si-O-Si (463 cm^{-1}), condensed Si-OH (3450 cm^{-1}) and non-condensed Si-OH (956 cm^{-1}) [26]. The characteristic bands of Keggin structured HPW, including asymmetric P-O (1087 cm^{-1}), terminal asymmetric oxygen W=O_d (960 cm^{-1}), corner shared asymmetric oxygen W-O_b-W (885 cm^{-1}) and edge shared oxygen W-O_c-W (801 cm^{-1}) could also be identified from figure 5b [27].

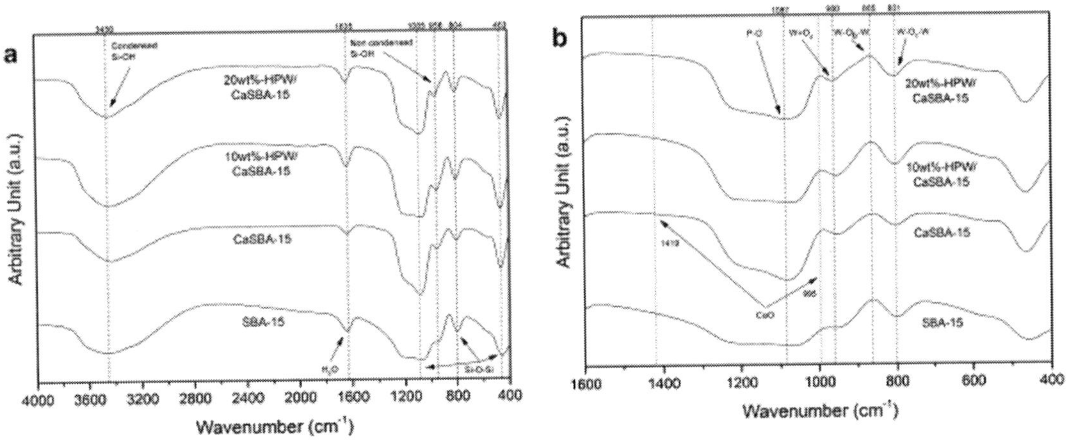

Figure 5. a) Full and b) detailed FTIR spectra of SBA-15, CaSBA-15, 10wt%-HPW/CaSBA-15 and 20wt%-HPW/CaSBA-15

Certainly, the detection of calcium oxide associated bands was of utmost interest for verification of the CaO formation as claimed. Due to crystal lattice vibration, CaO attributed bands could be detected sharply at 3640 cm^{-1}, 1000 cm^{-1}, 877 $cm^{-1}$and broadly at 1419 $cm^{-1}$and 450 cm^{-1} [28]. To our best effort, sharp band at 995 cm-1 and broad band at 1419 cm-1 were identified and labelled in figure 5b. Remarkably, obvious bands changes of these two bands were observed when comparing pure SBA-15 and CaSBA-15. In fact, when compared to SBA-15, CaSBA-15 showed band intensity increment from the broad band between 1600 and 1300 cm^{-1} and sharp band at 995 cm^{-1}. These two findings clearly indicated the possible formation of CaO. Curiously, the intensity of sharp CaO band at 995 cm^{-1} dropped when HPW was impregnated. It was proposed that the basic CaO would react with the acidic HPW to form insoluble salt. Should this happened, the amount of available CaO anchoring sites and hence the intensity of such band would be reduced. In agreement with the observation, it was very likely that the CaHPW salts were formed as intended as in the proposed scheme (figure 2).

3.4. Catalytic performance and reusability

Collective catalysts performances of each synthesized catalysts were illustrated in figure 6. From figure 6a, it was noted that the homogeneous HPW achieved very high lauric acid conversion (approximately 97%) in about 3 hours, while without any catalyst at all (blank), relatively low conversion were detected (59%) at the end of the experiment run. At the same time, both HPW impregnated catalysts showed lauric acid conversion in between the homogenously catalysed and blank experiment runs (73% for 10wt% and 81% for 20wt%-HPW/CaSBA-15). Certainly, these two catalysts showed improvement on the lauric acid conversion compared to that of the un-catalysed run with only small around of HPW acid impregnated on them (up to 20wt%). Next, the monolaurin yield profiles in figure 6b showed improvement in the monolaurin selectivity when these catalysts were used. Undoubtedly, highest monolaurin yield (68%) was achieved with the use of pure homogeneous

HPW. However, it was mainly due to the high lauric acid conversion (97%) and further esterification to dilaurin (24% selectivity) and trilaurin (9% selectivity).

In contrast, when 10wt%-HPW/CaSBA-15 was used, the selectivity of monolaurin, dilaurin and trilaurin were calculated to be 67%, 33% and 0%, respectively. Interestingly, no trilaurin could be detected with the use of heterogeneous catalyst. That is the similar for the case of 20wt%-HPW/CaSBA-15, too. Due to the mesopores, the formation of the much larger trilaurin could face steric hindrance and thus limiting its formation within the pore. As such, the size limiting role of SBA-15 came into effect in improving the selectivity towards monolaurin. The hindrance of trilaurin formation was achieved when HPW/CaSBA-15 catalysts were used. The selectivity improvement is undoubtedly beneficial to the research on selective esterification of glycerol with lauric acid.

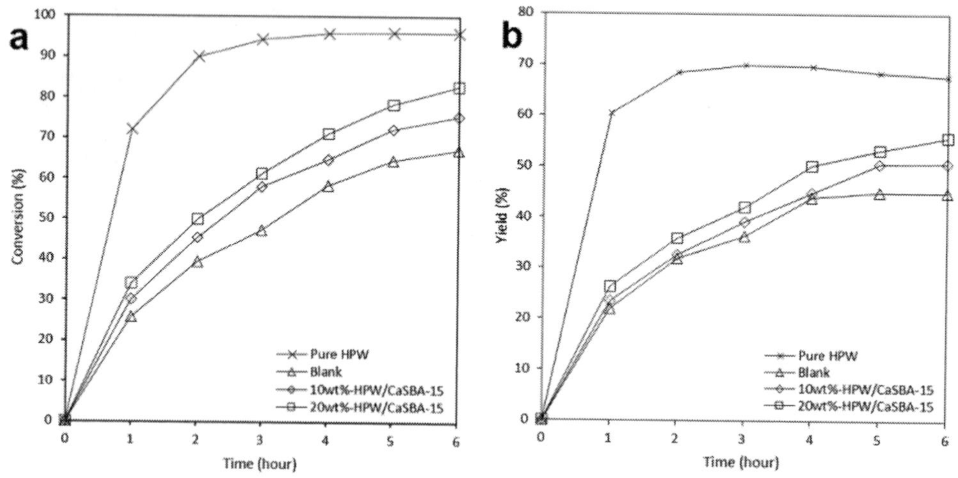

Figure 6. a) Lauric acid conversion profiles and b) monolaurin yield of Pure HPW, 10wt%-HPW/CaSBA-15, 20wt%-HPW/CaSBA-15 and blank (control)

While the strong catalysing power of HPW incorporated materials were commonly known, their reusability were not always assuring. As the most important investigation in this study, the reusability of HPW/CaSBA-15 catalysts was discussed next. To exemplify the stability and reusability of the material synthesized, 10wt%-HPW/CaSBA-15 was chosen for the reusability test in this section. The monolaurin selectivity and yield of fresh and recycled catalysts were plotted in figure 7.

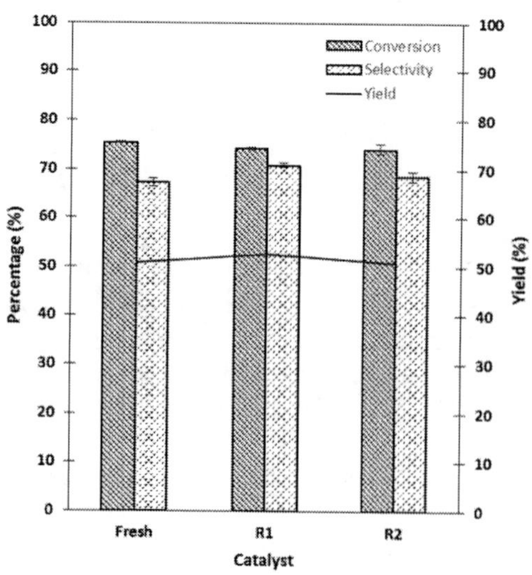

Figure 7. Lauric acid conversion, monolaurin selectivity and yield of the fresh and recycled 10wt%-HPW/CaSBA-15

When compared to figure 1, 10wt%-HPW/CaSBA-15 showed superior reusability up to at least 2 cycles while 40wt%-HPW/IM lost most of its catalytic performance right after the first recycle. 10wt%-HPW/CaSBA-15 demonstrated almost unchanged lauric acid conversion, monolaurin selectivity and also monolaurin yield up to the second recycle (R2). Such results were truly exciting as it provided direct evidence to the enhanced reusability and stability of the catalyst. It also further reinforced the argument of strong anchoring of HPW acid sites to the SBA-15 via calcium-bridge as proposed in figure 2. Without strong interaction, HPW acid would have been washed away from the support during the preparation phase before they can even be tested. With the strong interaction between SBA-15 and HPW via the calcium bridge, the reusability of the catalyst was greatly enhanced without compromising the catalytic performance.

4. Conclusions

By using the novel two-steps calcium modification and HPW functionalization, the highly active and selective yet reusable heterogeneous catalyst for esterification of glycerol was successfully synthesized. The reusability of the modified catalysts, HPW/CaSBA-15 was elucidated by the repeat usage of the recycled catalysts up to two runs and almost no change in terms of fatty acid conversion and monoglyceride yield was observed. When compared to the HPW incorporated SBA-15, the newly developed catalyst showed significant improvement in terms of reusability. Such reusability was attributed to the strong interaction between the support and HPW acid sites via the calcium-bridging, as illustrated in Fig. 2. Also, the formation of insoluble calcium HPW acid salt within the SBA-15 minimize the possible leaching into the polar reaction matrix, which have been a major issue in both research and industries.

Acknowledgments

The authors would like to acknowledge the financial assistance from a Research University Individual (RUI) grant (814181) from Universiti Sains Malaysia and Trans-Disciplinary Research Grant Scheme – TRGS (60702001) from Malaysian Ministry of Education. Faculty of Engineering Technology (FTK), Universiti Malaysia Perlis is also gratefully acknowledged for the financial, laboratory and technical supports.

References

[1] NASDAQ. Crude Oil Brent: Latest Price & Chart for Crude Oil Brent 2017; Available online: http://www.nasdaq.com/markets/crude-oil-brent.aspx?timeframe=3y.
[2] Reboredo F H, Lidon F, Pessoa F and Ramalho J C 2016 *Trends Biotechnol.* **34**(1) 3
[3] Wang Q and Li R 2016 *Renewable Sustainable Energy Rev.* **54** 925
[4] Johnson D T and Taconi K A 2017 *Environ. Prog.* **26**(4) 338
[5] Bagheri S, Julkapli N M and Yehye W A 2015 *Renewable Sustainable Energy Rev.* **41** 113
[6] Pérez-Pariente J, Diaz I, Mohino F and Sastre E 2003 *Appl. Catal., A* **254**(2) 173
[7] Carlos A 2009 *Catalytic processing of glycerol to fuels and chemicals*, in *Next Generation Biofuels and bio-based Chemicals*
[8] Gagea B C, Lorgouilloux Y, Altintas Y, Jacobs P A and Martens J A 2009 *J. Catal.* **265**(1) 99
[9] Dong B B, Zhang B B, Wu H Y, Chen X and Zhang K 2013 *Mater. Res. Bull.* **48**(7) 2491
[10] Hoo P Y, Abdullah A Z 2014 *Chem. Eng. J.* **250** 274
[11] Betiha M A, Hassan H M A, El-Sharkawy E A and Al-Sabagh A M 2016 *Appl. Catal. B* **182** 15
[12] Zhang D Y, Feng J L, Huo Q S, Melosh N, Fredrickson G H and Stuck G D 2016 *Fuel* **172** 293
[13] Baroi C and Dalai A K 2014 *Ind. Eng. Chem. Chem. Res.* **53**(49) 18611
[14] Gong S W, Lu J, Wang H H, Liu L J and Zhang Q 2014 *Appl. Energy* **134** 283
[15] Han X, Yan W, Chen K, Wu P H and Liu H B 2014 *Appl. Catal., A* **485** 149
[16] Chen Y, Cao Y, Zheng G, Dong B and Zheng X 2014 *Adv. Powder Technol.* **25**(4) 1351
[17] Caiado M, Machado A, Santos R, Matos I, Fonseca I Castanheiro J 2013 *Appl. Catal. A* **451** 36

[18] Brahmkhatri V and Patel A 2013 *J. Porous Mater.* **20**(1) 209
[19] Misono M *Catal. Today* **100**(1) 95
[20] Zhao D, Feng J, Huo Q, Melosh N, Fredrickson G, Chmelka B, Stucky 1998 *Sci.* **279**(5350) 548
[21] Rocchiccioli C, Fournier M, Franck R and Thouvenot R 1983 *Inorg. Chem.* **22**(2) 207
[22] Pouilloux Y, Abro S, Vanhove C and Barrault J 1999 *J Mol. Catal. A: Chem.* **149**(1) 243
[23] Thommes M, Kaneko K, Alexander V N and James P O 2015 *Pure Appl. Chem.* **87** 1051
[24] Webb J D, Seki T, Goldston J F Crudden C M 2015 *Microporous Mesoporous Mater.* **203** 123
[25] Álvarez M G, Chimentao R J, Llorca J, Medina F 2016 *Microporous Mesoporous Mater.* **219** 19
[26] Brahmkhatri V and Patel A 2011 *Appl. Catal., A* **403**(1) 161
[27] Yang L, Qi Y, Yuan X, Shen J and Kim J 2005 *J. Mol. Catal. A: Chem.* **229**(1) 199
[28] Balmain J, Hannoyer B and Lopez E 1999 *J. Biomed. Mater. Res.* **48**(5) 749

Immobilization of Cellulase from *Bacillus subtilis* UniMAP-KB01 on Multi-walled Carbon Nanotubes for Biofuel Production

Sandrasekaran Naresh[1], Siew Hoong Shuit[1,*], Balakrishnan Kunasundari[1], Peng Yong Hoo[1], Qi Hwa Ng[1] and Yi Peng Teoh[1]

[1]Department of Chemical Engineering Technology, Faculty of Engineering Technology, Universiti Malaysia Perlis (UniMAP), P.O Box 77, D/A Pejabat Pos Besar Kangar, Perlis, 01000, Malaysia.

E-mail: shshuit@unimap.edu.my

Abstract. *Bacillus subtilis* UniMAP-KB01, a cellulase producer was isolated from Malaysian mangrove soil. Through morphological identification it was observed that the *B. subtilis* appears to be in rod shaped and identified as a gram positive bacterium. Growth profile of isolated *B. subtilis* was established by measuring optical density (OD) at 600 nm for every 1 hour intervals. Polymath software was employed to plot the growth profile and the non-linear plot established gave the precision value of linear regression, R^2 of 0.9602, root mean square deviation (RMSD) of 0.0176 and variance of 0.0025. The hydrolysis capacity testing revealed the cellulolytic index of 2.83 ± 0.46 after stained with Gram's Iodine. The harvested crude enzyme after 24 hours incubation in carboxymethylcellulose (CMC) broth at 45°C and 100 RPM, was tested for enzyme activity. Through Filter Paper Assay (FPA), the cellulase activity was calculated to be 0.05 U/mL. The hydrolysis capacity testing and FPA shown an acceptable value for thermophilic bacterial enzyme activity. Thus, this isolated strain reasoned to be potential for producing thermostable cellulase which will be immobilized onto multi-walled carbon nanotubes and the cellulolytic activity will be characterized for biofuel production.

1. Introduction

Cellulase is an enzyme widely being utilized to breakdown cellulose. The abundance of the readily available lignocellulosic biomass such as agricultural waste drives the exploitation of the cellulase production in order to sustain the bioconversion of waste to the useful products [1]. Cellulase can be produced by microorganisms and plants. Nonetheless, microbial cellulase is much preferred as it easily available as extracellular enzyme in most of the system [2].

Currently, the commercial cellulase are being produced using the well-known *Trichoderma reesei* strain. *T. reesei* which has been referred as the best cellulase producer is a mesophilic fungal strain. Numerous studies have been carried out to enhance the cellulase production focusing on the optimization of fermentation conditions and recombinant technology to alter the gene of the mentioned fungal species [3]. Even so, bacterial enzyme was discovered to have some advantages compared to fungal enzyme as available in multi-enzyme complexes and the complexity is much higher than the latter [4]. This factors drive the interest of current research to take advantage of the bacterial enzymes.

The major bottleneck of using cellulase in bioconversion is the percentage enzyme recovery. Enzyme recovery is the process involving separation of enzymes from fermentation broth at the end of experiment for reusing it for subsequent batches. In order to do this, the enzyme immobilization technique has been introduced. Sodium alginate is the commonly used an entrapment technique to capture the enzymes thus enabling the recovery process [5]. Apart from this, there are a few other methods to immobilize the enzyme such as using the multi-walled carbon nanotube (MWCNT). Enzymes can be bounded on the MWCNTs using covalent and noncovalent approaches. The noncovalent approach has the advantage as less enzymes were denatured during the process compared to the covalent approach. Recently, the functionalized MWCNTs have gained wide interest among the researchers to serve as the base of immobilization for enzymes because of their intrinsic properties, such as high surface area and well-developed surface morphology and porosity. MWCNTs can be functionalized using organic, biological or polymeric molecules which enables the controlled immobilization and precision bounding of the enzymes on MWCNTs [6]. These properties made the recovery of enzyme immobilized onto the MWCNTs become possible.

B. subtilis is an aerobic, gram-positive, mesophilic bacterium commonly found in soil [7]. The *B. subtilis* UniMAP-KB01 was isolated from tropical mangrove soil at Taman Paya Bakau, Lumut, Perak, Malaysia. The strain is not fully tested for its capability for production of cellulase. Thus, this study will function as the preliminary data for better understanding of the isolated *B. subtilis* UniMAP-KB01 strain in production of cellulase which will be immobilized onto MWCNTs. The product from the enzymatic activity on cellulose (glucose) can be used as the main substrate of fermentation for future biofuel productions.

2. Materials and Methods

Bacillus subtilis UniMAP-KB01 was used throughout this study. The isolated bacterium was morphologically identified to verify it is free of contaminations of other microorganisms followed by construction of growth profile. Hydrolysis capacity testing and cellulase activity determination based on colorimetric assay were carried out using carboxymethylcellulose (CMC) as sole carbon source.

2.1. Morphological Identification

The morphology identification of bacteria was examined by using Gram staining method. Using a cooled sterile loop, a drop of sterile water was placed on a slide cleaned using 70% Ethanol. Then, the loop was sterilized again and cooled before picked a single colony bacteria from the agar plate. The colony then was gently stirred in the water drop to create a smear. Heat fixing was carried out by passing the slide a few times over the flame. A drop of crystal violet was placed onto the smear which covered the whole smear area. After 1 minute, using the distilled water the dye was washed before staining it with Gram's Iodine and left it for 1 minute. Then the dye gently rinsed before a few drops of ethanol were added to decolourize for 15 to 30 seconds. The smear was flooded with safranin and left for 45 seconds after rinsed of the ethanol. The safranin was washed with distilled water and the smear was dried prior observed under light microscope [8].

2.2. Growth Profiling

B. subtilis was cultured and maintained on the CMC agar plate containing of 1.36 g of potassium hydrogen phosphate (KH_2PO_4); 0.2 g of magnesium sulphate ($MgSO_4.7H_2O$); 2.0 g of sodium chloride (NaCl); 0.01 g of iron sulphate ($FeSO_4.7H_2O$); 1.0 g of ammonium sulphate ($(NH_4)_2SO_4$); 3.0 g of Carboxymethylcellulose (CMC); 1.0 g of yeast extract and 15.0 g of agar powder in 1 L of distilled water, pH 7.0 [9]. Inoculum was prepared by transferring two loopful of the bacterium into 30 mL of Nutrient Broth (NB) and incubated at 45°C, 100 RPM for 24 hours. Then, 3.0 mL of the 24 hours culture was pipetted transferred aseptically to 97.0 mL of NB in a 250 mL Erlenmeyer flask. Optical density at 600 nm was determined for every sample at 3 hours interval for 24 hours. The growth profile was established by plotting OD versus time for each sampled points [10].

2.3. Hydrolysis Capacity Testing

The CMC agar plate was divided into 4 quadrants before inoculating with a single colony of bacterium at the middle of each quadrants. The inoculated agar plate was incubated for 24 hours followed by staining with Gram's Iodine (2.0 g of potassium iodide and 1.0 g of iodine in 300.0 mL distilled water) for 5 minutes [11]. The cellulolytic index of *B. subtilis* was calculated by dividing clear zone diameter by colony diameter.

2.4. Cellulase Production

Cellulase production was carried out in a batch fermentation with 100 mL of working volume in 250 mL of Erlenmeyer flask. The inoculum was prepared as in section 2.2. Three milliliter of culture was transferred into flask containing 97.0 mL of CMC broth medium (section 2.2). The fermentation was carried out at 45°C, 100 RPM for 24 hours.

2.5. Determination of Cellulase Activity

Cellulase activity was measured using Filter Paper (FP) Assay. First, the fermentation broth was centrifuged at 4,000 RPM for 10 minutes. The supernatant was used as crude enzyme sample. An aliquot of 0.5 mL crude enzyme was incubated in a test tube containing filter paper (1 × 6 cm) and 1.0 mL of 0.05M Citrate buffer of pH 4.8 at 50°C for 60 minutes. Then, 3.0 mL of DNS reagent was added immediately to end the reaction and the test tubes were placed in the waterbath at 95°C for 5 minutes. The test tubes then were transferred to the ice bath and 1.5 mL of potassium sodium tartrate was added in each tubes. Then, 0.2 mL of the colour-developed mixtures were transferred to 2.5 mL of distilled water before measuring the Absorbance at 540 nm [5-6].

3. Results and Discussions

Figure 1 shows the *B. subtilis* UniMAP-KB01 cultured onto the CMC agar plate which was used throughout of the experiment.

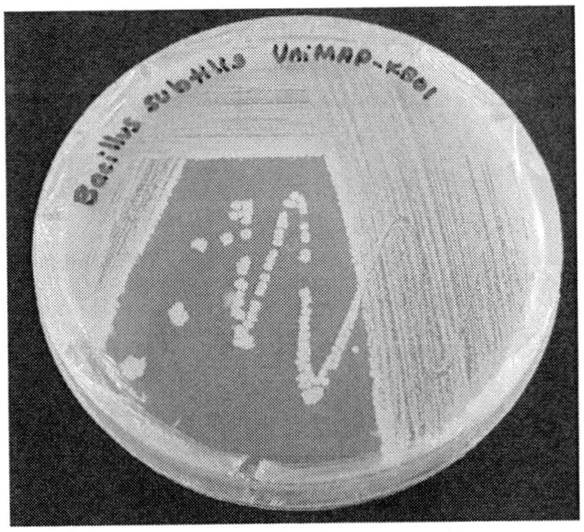

Figure 1. *Bacillus subtilis* UniMAP-KB01 cultured on CMC agar plate using streak plating technique.

3.1. Morphological Identification

Figure 2 shows that the stained culture is a gram positive bacterium as it appears to be in dark purple colour as it retained the crystal violet dye. The presence of the thick cell wall in the gram positive bacterium is the reason for retaining the purple coloured dye. Besides that, the culture appeared to be in a rod shaped and scattered individually. 16S rDNA sequencing analysis data from Macrogen Inc. shows that the strain is 99% similar to those *B. subtilis*. Thus, this strain was designated as *Bacillus subtilis* UniMAP-KB01. According to the literature, *B. subtilis* is a gram positive, aerobic bacterium and capable to produce spores in extreme conditions for survival. There are a few studies reported on the isolation of *B. subtilis* from gastrointestinal tract of pig [14], cow dung [15] and fish [16]. This cellulase producing microbe enables animals and fish to digest the cellulose present in the plant materials consumed. This supported the potential of *B. subtilis* to be considered as cellulase producer as it excrete mostly extracellular cellulase [17].

Figure 2. Gram staining of *Bacillus subtilis* UniMAP-KB01 viewed under light microscope.

3.2. Growth Profiling and Kinetic Studies

Figure 3 shows the growth profile of *B. subtilis* UniMAP-KB01. The lag phase was not observed in this growth curve as the optical density readings increased intensely from 0 until 12 hours of incubation time. This indicates the cells were in exponential phase. Then, the stationary phase was observed lasted for 9 hours before the optical density reading started to drop at 21 hours of incubation time. This showed the culture entered the death phase as the nutrients depleted and toxic accumulated. It was deduced that the starter culture for the rest of the experiment prepared at exponential phase.

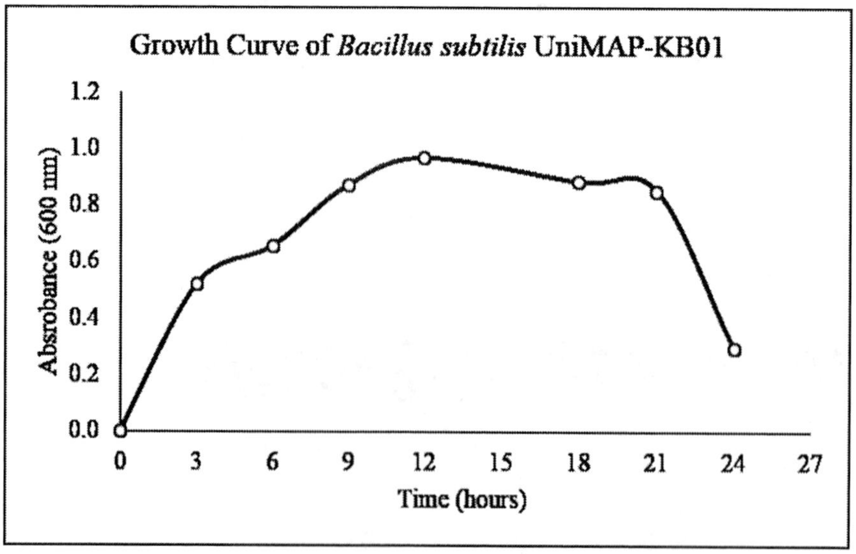

Figure 3. Optical density versus incubation time for 24 hours.

Figure 4 shows the non-linear growth kinetic graph of *B. subtilis* UniMAP-KB01 generated using POLYMATH software. X_t exp denotes the experimental values of microbial biomass concentration.

The X_t calc represent concentration of biomass at specific time calculated by the model equation as in Eq. 1, using $\mu = 0.0577$ and $X_0 = 0.5191$.

$$X_t = X_0 * e^{\mu * t} \tag{1}$$

X_t = microbial biomass concentration at specific time
X_0 = initial microbial biomass concentration
t = time
μ = specific growth rate

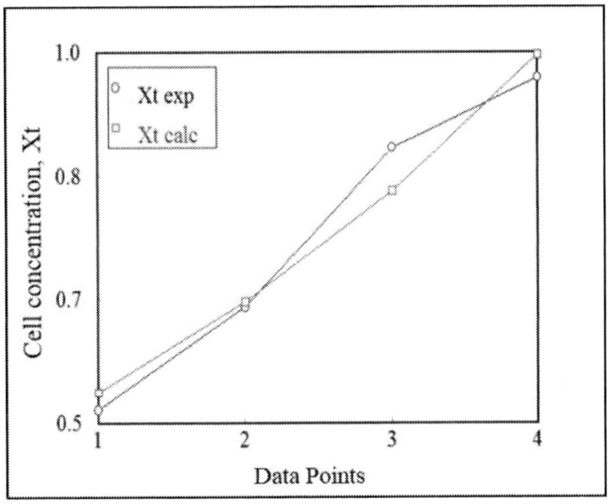

Figure 4. Non-linear growth kinetic graph of *B. subtilis* UniMAP-KB01 generated using POLYMATH software.

From the same plot, the linear regression, R^2 value was computed to be 0.9602, meanwhile, the adjusted R^2 was 0.9404 which were highly acceptable. Besides that, the Root Mean Square Deviation, RMSD was 0.0176 and variance was 0.0025 which verifies the data obtained is significant and the errors are insignificant. Table 1 shows the data precision of *B. subtilis* UniMAP-KB01 generated using POLYMATH software.

Table 1. The data precision of *B. subtilis* UniMAP-KB01 generated using POLYMATH software.

Data	Value/Precision
Linear Regression, R^2	0.9602
Adjusted Linear Regression, Adj R^2	0.9404
Root Mean Square Deviation, RMSD	0.0176
Variance	0.0025

3.3. Hydrolysis Capacity Testing

Hydrolysis capacity testing showed the capability of the microorganism to degrade the cellulose. Thus, indirectly the ratio of cellulase activity can be estimated as cellulolytic index. Figure 5 shows the clear zone formed when stained with Gram's Iodine. The measurements of the clear zone diameter (1.70 cm) and colony diameter (0.60 cm) was used to calculate the cellulolytic index of *B. subtilis* UniMAP-KB01 which was 2.83 ± 0.46. The purpose of this testing is to screen the ability of a bacterium to produce the cellulase enzyme thus serves as a preliminary data before performing enzyme activity assay.

Figure 5. Halozone formation by Gram's Iodine assay of *B. subtilis* UniMAP-KB01.

3.4. Cellulase Activity

The FP Assay was carried out with crude enzyme of B. subtilis UniMAP-KB01. The Absorbance obtained at 540 nm was 0.015 as y which was substituted into the straight line equation y = 0.1163x obtained from the glucose standard curve. The corresponding concentration of glucose produced was 0.129 mg/0.5mL. The concentration was less than 2.0 mg, thus using the Eq. 2, Filter Paper Unit (FPU) was calculated which was 0.05 U/mL.

$$FPU = \text{mg glucose released} \times 0.185 \quad (2)$$

Research by Andriani and Park in 2006, stated that after the optimization of pH, temperature, carbon source, CMC concentration, nitrogen source and tryptone concentrations, the enzyme activity of *B. subtilis* TD6 was calculated to be 4.1 U/mL. The difference of the current result is huge when compared to previous mentioned study. However, the prolonged incubation time (144 hours) and optimized condition that was used by Andriani and Park explains the differences [16]. Thus, it was taken into consideration to optimize the fermentation condition before using it for industrial applications.

4. Conclusion

Bacillus subtilis UniMAP-KB01 was screened for cellulase production and concluded as it has potential for scale up cellulase production. Supplementary studies on the optimization of media condition, fermentation parameters, enzyme kinetics and enzyme immobilization studies using MWCNTs will be carried out for bioconversion of lignocellulosic waste into glucose. The produce glucose which will then be used as a precursor for biofuel production such as bioethanol.

Acknowledgement

This study was financially supported by Universiti Malaysia Perlis Short Term Grant (STG) (A/C:9001-00519).

References

[1] R. C. Kuhad, D. Deswal, S. Sharma, A. Bhattacharya, K. K. Jain, A. Kaur, B. I. Pletschke, A. Singh, and M. Karp 2016 *Renew. Sustain. Energy Rev.* **55** 249–272
[2] R. K. Sukumaran, R. R. Singhania, and A. Pandey 2005 *J. Sci. Ind. Res. (India)* **64** 832–844
[3] M. Paloheimo, T. Haarmann, and S. Mäkinen 2016 *Gene Expression Systems in Fungi:*

Advancements and Applications (Switzerland: Springer International Publishing)
[4] S. S. Behera and R. C. Ray 2016 *Int. J. Biol. Macromol.* **86** 656–669
[5] C. Tang, C. D. Saquing, P. K. Sarin, R. M. Kelly, and S. A. Khan 2015 *J. Memb. Sci.* **472** 251–260.
[6] W. Feng and P. Ji 2011 *Biotechnol. Adv.* **29** 889–895
[7] Z. Baysal, F. Uyar, and Ç. , Aytekin 2003 *Process Biochem.* **38** 1665–1668
[8] D. Claus 1992 *World J. Microbiol. Biotechnol* **8** 451–452
[9] B. Kunasundari, Y. I. P. Teoh, and I. Roshita 2016 *Int. J. Adv. Sci. Eng. Technol.* **4** 155–157
[10] O. M. Ingraham, J. L. and F. C. Neidhardt 1983 *Growth of the bacterial cell.* (Sunderland, Massachusetts: Sinauer Associates Inc.).
[11] R. C. Kasana, R. Salwan, H. Dhar, S. Dutt, and A. Gulati 2008 *Curr. Microbiol.* **57** 503–507
[12] T. K. Ghose 1987 *Pure Appl. Chem.* **59** 2.
[13] G. L. Miller 1959 *Anal. Chem.* **31** 426–428
[14] L. Ma, W. Yang, F. Meng, S. Ji, H. Xin, and B. Cao 2015 *J. Taiwan Inst. Chem. Eng.* **56** 67–72
[15] S. Bai, M. Ravi, D. J. Mukesh, P. Balashanmugam, and M. D. Bala 2012 *Arch. Appl. Sci. Res.*, **4** 269–279
[16] D. Andriani and D. H. Park 2006 *Ann. Bogor.* **14** 31–37
[17] A. K. Ray, A. Bairagi, K. Sarkar Ghosh, and S. K. Sen 2007*Acta Ichthyol. Piscat.* **37** 47–53

IOP Conf. Series: Materials Science and Engineering **318** (2018) 012009 doi:10.1088/1757-899X/318/1/012009

Antimicrobial Activity of Plant Extracts from Aloe Vera, Citrus Hystrix, Sabah Snake Grass and Zingiber Officinale against Pyricularia Oryzae that causes Rice Blast Disease in Paddy Plants

M.N.A. Uda[1], N. Harzana Shaari [1], N. Shamiera.Said [1], Nur Hulwani Ibrahim[1], Maisara A. M. Akhir[1],Mohd Khairul Rabani Hashim[1], M. N. Salimi[1], M.A. Nuradibah[1],Uda Hashim[2], Subash C.B. Gopinath[1,2]

[1]School of Bioprocess Engineering,Universiti Malaysia Perlis, Kompleks Pusat Pengajian, Jejawi 3,02600 Arau, Perlis, Malaysia

[2]Institute of Nano Electronic Engineering,Universiti Malaysia Perlis (UniMAP), 01000 Kangar, Perlis, Malaysia

E-mail:nuraiman@unimap.edu.my

Abstract. Rice blast disease, caused by the fungus known as Pyricularia oryzae, has become an important and serious disease of rice worldwide. Around 50% of production may be lost in a field moderately affected by infection and each year the fungus destroys rice, which is enough to feed an estimated 60 million people. Therefore, use of herbal plants offer an alternative for the management of plant diseases. Herbal plant like Aloe vera, Citrus hystrix, Sabah snake grass and Zingiber officinale extracts can be used for controlling disease of rice blast. In this study, these four herbal plants were used for evaluating antimicrobial activity against rice plant fungus Pyricularia oryzae, which causes rice blast disease.

1.Introduction

Among issues raised by farmers are the epidemics of new and emerging plant diseases, threatening the food security. This issue will give huge impacts and attention on understanding for controlling the pathogens. Thus rates of introduction of exotic pathogens have increased in recent years, driven largely by the altered patterns and increasing rates of travel and trade. Under the economic transformation programmed (ETP), the Malaysian government wants to ensure that food security objectives are achieved. There is a need to struggle for invading this problem especially in the agriculture-based countries [1-2]. In this problem, rice was selected because it is an important crop worldwide, over half of the world population relies on it for food, including Malaysia where all the Malaysian are eating a rice as their staple food in daily life [3]. Rice cultivation is found most widely in Asia. The demand for rice is expected to rise more higher as the world's population increases. Rice is the most major and number one crop in Malaysia. It is the main source of energy for the people, as it contains higher value of carbohydrate, which provides more than 300 calories per 100 g of rice. Besides that, rice is rich with nutrients other than that it has lower value in fat and cholesterol, it actually contain higher value in starch. It is also one among many foods that are very easy to digest [4].However, there are many factors that make paddy rice production become slow and less productive nowadays. One of the main causative reasons is paddy disease [5]. Among the common diseases, rice blast disease is caused by P. oryzae [6] and Rice tungro disease (RTD) [7], which is the most destructive disease in rice-growing countries worldwide. Around 50% of production may be lost in a field moderately affected by infection and each year fungus destroys rice enough to feed an estimated 60 million people [8]. Rice blast disease is one of the major factors causing losses in quality and quantity of rice [9] where P. oryzae fungus that attacks the paddy plant can be found in all the regions of rice-growing environments. The fungus can attack all part of the paddy plant at different growth stages, either on leaf, grain and also the on neck of the paddy. Plant disease will reduce the production and quality of food crops and also causing significant losses to farmers and has big impact towards the economy as it threatening the food security. Concerns regarding food safety and the environment have led to the reduced use of agrochemicals and the development of sustainable agriculture. In this context, the focus of biological control studies reflects the desire of several sectors to develop sustainable methods for plant disease control.

Figure 1. (a) Sample of the infected paddy plant by P. oryzae from MARDI; (b) Symptoms on grain of paddy plant; (c) Infection on paddy leaf.

Herbal medicinal plants such as A. Vera, C. hystrix, Z. officinale and Sabah snake grass have been proved in the previous study that the plant has antifungal properties. A. vera from Aloe family has long association with herbal medicine and Aloe gel is the most widely recognized herbal in the world today, as it contains a wide range of enzyme which provides antiseptic, anti-inflammatory and also antifungal benefits [10]. Anthraquinones are major substances found in aloe, and are known for their laxative, anti-microbial and anti-cancer activities [11].C. hystrix, commonly known as kaffir lime, is a common tropical herb in the family Rutaceae found everywhere in Southeast Asia. C. hystrix is a thorny bush with aromatic leaves and dark green fruits with irregular bumpy surface. The valued parts of C. hystrix are the leaves and fruit peel which contain natural antioxidants and active compound that is important in fighting the diseases and cancer [12]. Z. officinale is belongs to the family of Zingiberaceae, a perennial herb with thick tuberous rhizomes. The gingerols is the most active compound found in ginger and showed that Z. officinale extracts have medicinal properties, and antimicrobial activity as reported [13] other than that as an antioxidant, anti-lipid, anti-diabetic, analgesic, and anti-tumor [14]. Sabah Snake Grass is suitable for plant growth in tropical weather such as Malaysia and Thailand. This plant also has antimicrobial activity similar to A. vera and suitable to use in this research. In Malaysia, the plant is popularly known as "Sabah snake grass" in English since it was found in Sabah of East Malaysia. The marginally curved stem supporting the leaves are look like the curve of an elephant's trunk and therefore it is also named as Daun Belalai Gajah (elephant's trunk) in Malay [15].

Figure 2. (a) A. vera leaf; (b) Z. officinale rhizomes; (c) C. hystrix leaf; (d) Sabah Snake Grass

2. Materials and Methods

2.1. Materials
Ethanol, Potato Dextrose Agar (PDA) media, Sodium Chloride, Tween-20 were purchased from Sigma-Aldrich. All analytical grade reagents were from Merck. All chemicals and reagents were stored and used as recommended by the supplier

2.2. Collection of infected paddy plant
The paddy plant which have been infected with pyricularia grisea are supplied by Malaysian Agricultural Research and Development Institute (MARDI) Seberang Perai . Then the fungus, Pyricularia oryzae was isolated for future analysis.

2.3. Preparation of A. Vera extraction
Firstly, the Aloe leaves are dried in the oven at 80OC for 48 hr. After that, the samples that have been dried are crushed into the small pieces using the blender until it became powdered. The solvent used is ethanol. For the preparation of ethanol, 20 g of the powder is soaked in 100 ml of the solvent and put on incubator shaker for 24 h. The mixed solution then is filtered and was kept in the refrigerator until further use.

2.4. Preparation of Z. officinale extraction
The rhizomes are washed, peeled and washed again in clean water. After washing, the rhizomes are cut into small pieces and were dried in the oven for 5 hours and 30 minutes. The gingers then are ground using an electric blender until it becomes powdered. 20 g of the ground material (Z. officinale) will be placed in a conical flask containing 100 ml of ethanol solvent and will be put on incubator shaker for 24 h. The mixed solution then is filtered and was kept in the refrigerator until further use.

2.5. Preparation of C. hystrix leaf extraction
The leaf of C. hystrix is washed with clean water for 3 times. After that, the leaves are cut into small pieces and dried it for a 30 seconds using microwave. After that, the leaves were machine using electric blender until it became powder. 20 g of the powder then are placed in conical flasks containing 100 ml of ethanol solvent and was put on incubator shaker for 24 h. Next, the mixed solution then is filtered and was kept in the refrigerator until further use.

2.6. Preparation of Sabah Snake Grass Extraction
For Sabah snake grass, the collected leaves were spread in a single layer on a tray and left at room temperature (25 ± 1°C) for 18 hours for withering process to obtain about 70% of moisture content. Withered leaves were grounded into relatively small particle size using a blender for 15 sec. The blended leaf will be soaked with ethanol and kept in incubator shaker to mix the solvent well for 48

hours. The extracts were filter using Whatman filter paper (No. 1) and then they were kept in refrigerated condition until use.

2.7. Preparation of Potato Dextrose Agar (PDA) media

PDA was prepared by dissolving 39 g of PDA with 1 L of distilled water in a scotch bottle. The bottle then is shaking for many times. After that, the solution inside the scotch bottle is autoclaved for 15 minutes. Then, the scotch bottle is taken out from the autoclave machine and placed it in the laminar flow and allows it to cooling down. Next, the PDA solution was poured into the empty petri dish and allowed it to solidify.

2.8. Preparation of microbial cultures

First step, the infected portion P. oryzae is cut into small pieces and sterilized. After that, rinsed it three times with sterile distilled water and was crushed using pestle and mortar, then it was transferred onto the surface of PDA agar. The second method is the infected grain paddy was taken, washed and dried and then transferred it on PDA agar. The fungus is incubated at 30°C for 7 to days. The mycelium growing out of the plant tissue is then subculture on new PDA media and incubated at 30oC for 7 to 10 days. The subculture process is repeated for three times to obtain the pure colony. The fungus then is identified based on the morphological and cultural characteristics based on the observation under the microscope.

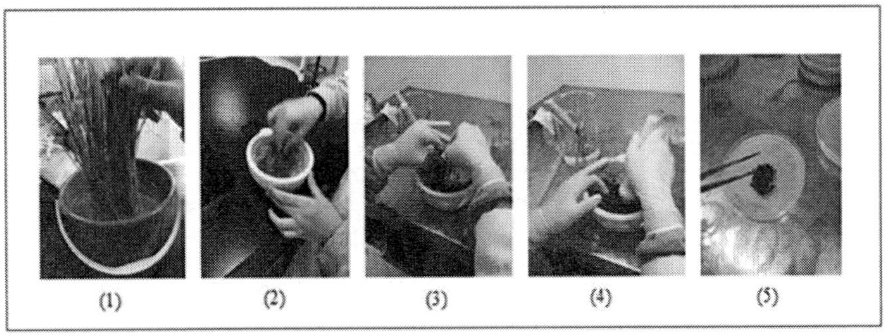

Figure 3. Process of preparation of microbial culture

2.9. Screening of antimicrobial activity

Screenings of antimicrobial activity are performed by standard disc diffusion method. The disc paper is used for evaluating the antimicrobial activity. The discs are prepared from Whatman No. 1 filter paper. The size of the disc is 6 mm. The discs are ensured to sterilized and autoclaved. After the sterilization the moisture discs are dried until further use. Then the three plants extracts discs and control discs are prepared.

2.10. Antimicrobial activity of A. Vera, C. hystrix, Z. officinale and Sabah Snake Grass

The antimicrobial activity is carried out by using disc diffusion technique. The fungal cultures are put on the PDA agar medium that has been prepared well and was grow for 7 days at the centre of the media. The disc is also put around the fungus and the three different plant extracts with three different concentrations (10, 20 and 30 mg/ml) was pipetted on it. The plates then are incubated at 30 °C for 7 days and the clear zone developments was closely monitor.

2.11. Determination of Minimal Inhibition Concentration (MIC)

The lowest concentration of the extractions where the effect is smaller is considered as the minimal inhibitory concentration.

2.12. Data Analysis
ANOVA analysis is used to test the significance different of the results between the three different of plants extracts and the differences in the concentrations.

3. Results and Discussions

3.1. Identification of P. oryzae
The measurement of mycelial growth and colony diameter were taken and growth pattern of P. oryzae on PDA was observed. The margin of colony of P. oryzae was like white, round, cottony and with fluffy mycelium surface shown on the PDA. Morphology of spores under microscope was also observed. Another main characteristic of P. oryzae is it pear-shaped of it conidia when observed under the microscope. The isolated fungus from paddy plant in this research, which was grown on PDA has a radial growth of approximately 40 mm to 50 mm after 8 days of growth as shown in the figure 4 below.

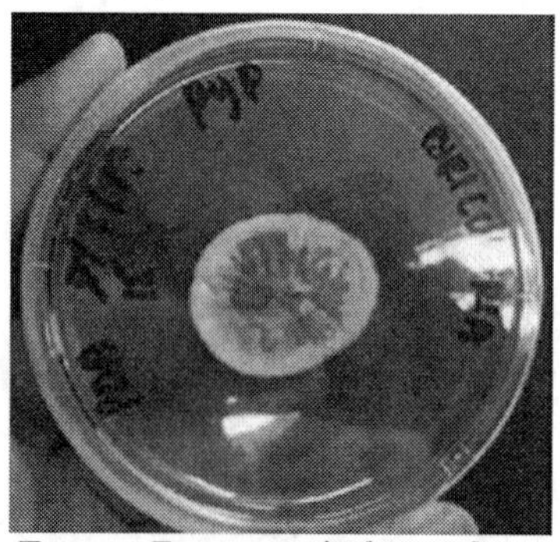

Figure 4. Fungus P. oryzae being cultured on PDA

The figure 5 below shows the result from the culture media, under 400X magnification (a) the microscopic morphology of P. oryzae (b) the colony morphology on PDA and (c) the conidia from the culture.

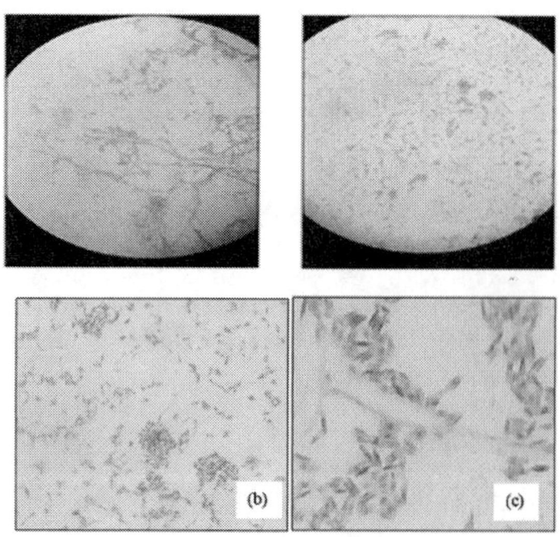

Figure 5. (a) The microscopic morphology of P. oryzae obtained, (b) P. oryzae colony under microscope (c) Conidia from the culture (pear-shaped)

After that, the step is continued with the process of subculture. Sub-culture was conducted at many times from each successive sub-culture before obtaining a completely pure culture. The figure 6 below showed the successive of sub-cultured fungus of P. oryzae.

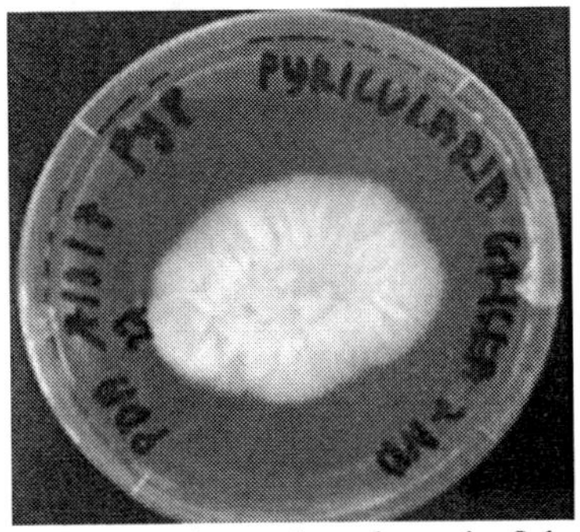

Figure 6. Successive sub-cultured of the fungus

3.2. In-vitro antimicrobial activity of A. Vera, C. hystrix, Z. officinale and Sabah Snake Grass extraction against P. oryzae fungus

The method used to test the antifungal activity towards the fungus P. oryzae was by using disc diffusion method. The zone of inhibition that was formed is recorded. Table 1 and figure 7 shows the response of the fungus towards the A. vera, C. hystrix, S. snake grass and Z. officinale plants extraction at three different concentrations which are 10 mg/ml, 20 mg/ml and 30 mg/ml in table and bar graph form. All the activities testing were repeated three times or triplicated. Figure 5 below showed the antifungal testing of the four plants extracts against the fungus.

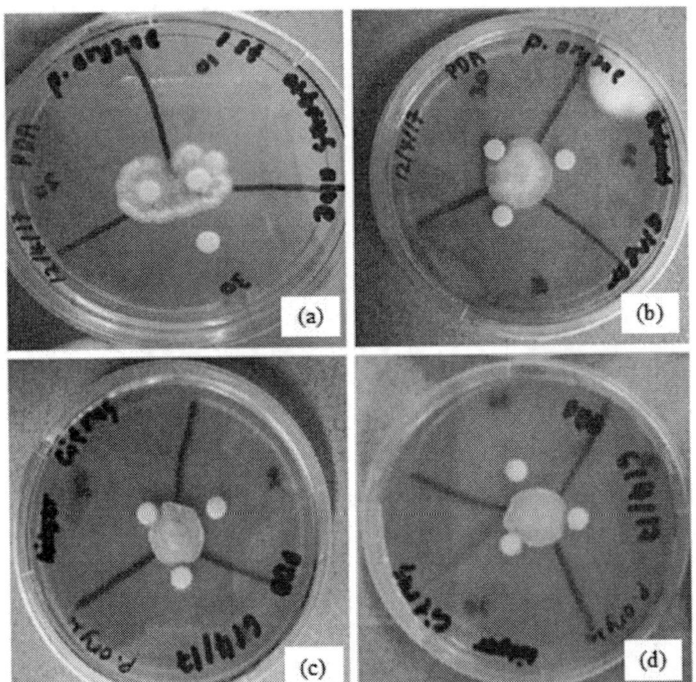

Figure 7. Antifungal testing (a) A. Vera (b) Z. officinale (c) C. hystrix (d) Sabah Snake Grass

Table 1. Zone of Inhibition on the fungus

Plant Extracts	Zone of Inhibitions (mm) Concentration (mg/ml)		
	10 mg/ml	20 mg/ml	30 mg/ml
A. vera	6.67	7.33	8.67
C. hystrix	6.33	6.67	7.33
Z.officinale	5.67	6.67	8.33
S. Snake Grass	10	10.8	12

The results obtained indicate that the concentrations of the tested plant extracts against P. oryzae had a positive effect in inhibiting the mycelia growth. From the results, Sabah snake grass extraction has the highest inhibition value at concentration 30 mg/ml, followed by A. Vera, Z. officinale and C. hystrix extraction. The higher the concentration, the bigger the size of the inhibition should be. The lowest inhibition zone recorded for four types of plant extracts was at the concentration of 10 mg/ml whereas the Sabah snake grass extract has diameter of 10 mm, A. vera extract has diameter of 6.67 mm and C. hystrix has diameter of 6.33 mm. As for Z. officinale, the lowest inhibition zone was recorded at concentration of 20 mg/ml which is 6.67 mm. The negative control used is distilled water. There is no inhibition zones for the negative control that showing in distilled water; this proves that the distilled water is not the antibacterial agent. For the positive control, which is DMSO, there is inhibition zone shown which concluded that DMSO can be used as an antifungal agent. From ANOVA analysis, there is no significant difference between the three concentrations and the inhibition zones of the four types of plant extractions as the value of F-calculated is smaller than F-critical value and p-value is bigger than 0.05. Different inhibition studies have been demonstrated in the past against different pathogens [16-19]. The current study is also shown to have inhibition against rice blast disease causing pathogen, similar to the past studies and complementing as a new addition.

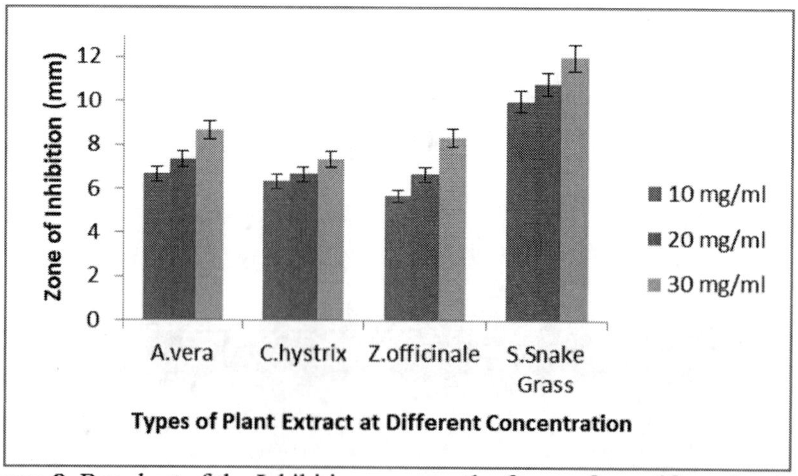

Figure 8. Bar chart of the Inhibition zone on the fungus for antifungal activity

4.Conclusion

In this study, the extraction of A.vera leaves, C. hystrix leaves, the rhizome of Z. officinale and Sabah snake grass proved that they have potential natural antimicrobial activity against the fungus of P. oryzae from the paddy plant that caused the rice blast diseases. The fungus of P. oryzae was successfully isolated from the paddy plant. Among all the parts from paddy plant, which are the neck of the paddy, leaf and the grain that were taken and isolated, the grain part is the part that has been successfully isolated and been identified as P. oryzae fungus. The main characteristics are its white colony colour, fluffy cottony look on PDA and pear-shaped conidia is observed under the microscope. In this experiment, three concentrations (10, 20 and 30 mg/ml) of each ethanol plants extract (A. vera, C. hystrix, Sabah snake grass and Z. officinale) were used to inhibit the growth of causative fungus of rice blast disease on paddy plant, P. oryzae and get the minimum inhibitory concentration (MIC). The inhibition zones formed are recorded. Basically, the stronger the inhibitory activity of the plant extracts, the larger inhibition zone is formed. The lower concentration of crude extract tested, the small inhibition zone is shown. From the results, Sabah snake grass extraction has the highest inhibition value at concentration 30 mg/ml with 12 mm of the inhibition zone formed, followed by A. vera with 8.67 mm, Z. officinale with 8.33 mm and C. hystrix extraction with the diameter of the inhibition zone of 7.33 mm.

Acknowledgement
The authors wish to thank Universiti Malaysia Perlis for providing financial and technical support to conduct this research. We acknowledge the support from "Geran Penyelidikan Jangka Pendek" (9001-00552).

References
[1] NKEA Agricultural, "ETP ANNUAL REPORT 2014," 2014.
[2] Uda, M. N. A., Adam, T., Hasfalina, C. M., Faridah, S., Zamri, I., Hashim, U., & Ariffin, S. A. B. (2014). "Reviewed Immunosensor Format Using Nanomaterial for Tungro Virus Detection," vol. 832, no. November 2013, pp. 410–414, 2014.
[3] S. S. Gnanamanickam, K. Krishnamurthy, A. Mahadevan, R. K. Upadhyay, K. G. Mukerji, and B. P. Chamola, "Biological control of major fungal diseases of rice and other food grains with bacterial antagonists," *Biocontrol potential its Exploit. Sustain. Agric. Vol. 1 Crop Dis. weeds, nematodes*, p. 267–278; 33 ref, 2000.
[4] Uda, M. N. A., Hasfalina, C. M., Samsuzanaa, A. A., Faridah, S., Zamri, I., Noraini, B. S.,Gopinath, S. C. B. "Immunosensor development formatting for tungro disease detection using nano-gold antibody particles application," vol. 20057, p. 20057, 2017.
[5] R. B. Zainon, "Paddy disease detection system using image processing," *J. about serious paddy Dis. worlwide*, vol. 1, no. 12, pp. 1–106, 2012.
[6] T. Namai, "Race differentiation of the rice blast fungus, Pyricularia oryzae, and environmentally friendly control of rice blast disease," *J. Gen. Plant Pathol.*, vol. 77, no. 6, pp. 350–353, 2011.
[7] Uda, M. N. A., Hasfalina, C. M., Samsuzana, A. A., Hashim, U., Ariffin, Shahrul A. B., I. Zamri, W., Nur Sabrina, B., B. Siti Noraini, S., Faridah, M., Mazidah, Gopinath, Subash C. B. "Immunosensor development for rice tungro bacilliform virus (RTBV) detection using antibody nano-gold conjugate," vol. 20058, p. 20058, 2017.
[8] S. Roy-barman and B. B. Chattoo, "Rice blast fungus sequenced," *Current*, vol. 89, no. 6, pp. 1–2, 2005.
[9] J. Hubert, R. B. Mabagala, and D. P. Mamiro, "Efficacy of Selected Plant Extracts against Pyricularia grisea ,Causal Agent of Rice Blast Disease," *Plant Extr. against Pyricularia grisea*, vol. 3, no. 2, pp. 602–611, 2015.
[10] A. Bele, A. A. Bele, and A. Khale, "Antimicrobial Activity of Aloe Vera Formulation," *Asian J. Pharm. Technol. Innov.*, vol. 4, no. 17, 2016.
[11] J. T. H. P.Young In, "Perspective of industrial application of Aloe vera," *New Perspect. Aloe*, vol. 6, no. 6, p. 197, 2006.
[12] V. Srisukh, C. Tribuddharat, V. Nukoolkarn, and N. Bunyapraphatsara, "Antibacterial activity of essential oils from Citrus hystrix (makrut lime) against respiratory tract pathogens," *Citrus hystrix*, vol. 38, no. 2, pp. 212–217, 2012.
[13] R. A. Valverde, L. Stanley, and P. Iii, "Preliminary Studies on Antibacterial and Antiviral Activities of Five Medicinal Plants," *J. Plant Pathol. Microbiol.*, vol. 4, no. 7, pp. 0–8, 2013.
[14] Hasan, "Chemical Composition and Antimicrobial Activity of the Crude Extracts Isolated from Zingiber Officinale by Different Solvents," *Pharm. Anal Acta*, vol. 3, no. 9, 2012.

[15] Yahaya et al, "Clinacanthus nutans (burm. F.) Lindau: An useful medicinal plant of south-east Asia," *Int. J. Pharmacogn. Phytochem. Res.*, vol. 7, no. 6, pp. 1244–1250, 2015.

[16] P. Velusamy, C-H. Su, G. Venkat Kumar, S. Adhikary, K. Pandian. Gopinath, S.C.B., Chen, Y., Anbu, P. "Biopolymers Regulate Silver Nanoparticle under Microwave Irradiation for Effective Antibacterial and Antibiofilm Activities". *PloS One* Vol. 11, page 0157612 (2016).

[17] S.C.B. Gopinath, K. Awazu, M. Fujimaki, and K.Shimizu. "Evaluation of Anti-A/Udorn/307/1972 antibody specificity to Influenza viruses using a waveguide mode sensor". *PLoS One* Vol. 8, page e81396 (2013).

[18] K. Hayashi, H. Onoue, K. Sasaki, J. Lee, P.K.R. Kumar, S.C.B. Gopinath, Y. Maitani, Y, T. Kai, and T. Hayashi. Topical application of polyethyleneimine as a candidate for novel prophylactic therapeutics against genital herpes caused by herpes simplex virus. *Arch. Virol.* vol. 159, pp. 425-435 (2014).

[19] S.C.B. Gopinath, K. Awazu, M. Fujimaki, K. Shimizu and T. Shima. Observations of immuno-gold conjugates on influenza viruses using waveguide-mode sensors. PLoS One vol. 8, p. e69

IOP Conf. Series: Materials Science and Engineering **318** (2018) 012010 doi:10.1088/1757-899X/318/1/012010

Effect of Packaging Materials on *Orthosiphon Stamineus* Dried-Leaf Quality During Storage

A.R. Norawanis *[1], *A.R. Shaari* [1] *and L.Y.Leng* [1]

[1]Faculty of Engineering Technology, Universiti Malaysia Perlis (UniMAP), Campus of UniCITI Alam Sungai Chuchuh, Padang Besar, 02100 Perlis, Malaysia.

E-mail: norawanis@unimap.edu.my

Abstract. The experiment was conducted to determine the effects on the total phenolic content, antioxidant capacity, moisture content and total different color (ΔE) when the *O. stamineus* dried whole-leaf were packed in different packaging materials (plastic bag, paper bag and glass container) and stored under room temperature (±25 °C) and relative humidity (±65 %RH) for 8 weeks. The total phenolic compounds and antioxidant activity were measured using the Folin-Ciocalteu method and 2,2-diphenyl-1-picrylhydrazyl (DPPH) free radical scavenging activity assay respectively, and analyzed using UV/VIS Spectrophotometer. The moisture content changes were examined using a moisture analyzer and the color changes were analyzed using colorimeter. The results showed that packing *O. stamineus* dried whole-leaf in different packaging materials significantly affected the herbal leaves quality. After 8 weeks of storage period, the total phenolic content and antioxidant capacity exhibited the increase values during storage. Meanwhile, the moisture content of the samples decreased by storage period for the samples packed in plastic bag and glass container. The moisture content of the samples packed in the paper bag fluctuated along the 8 weeks of storage period. The total different color (ΔE) of the *O. stamineus* dried whole-leaf increased by storage period. The highest changes of ΔE belonged to the samples packed in the glass container, followed by paper and plastic bags. The selection of the packaging materials can be considered as an important element to control the quality of raw herbal materials for further processing and the herbal finished products.

1. Introduction

The secondary metabolite compounds produced by herbal plants are widely used for medicinal purposes. The roles of the secondary metabolites are mainly for protection purposes in response to specific environmental stimuli such as attacked by pathogens or harmful microbes, insects and herbivores [1], [2]. The common examples of secondary metabolite groups are phenolic compounds, alkaloids, terpenoids and glycosides [3]. Some of these groups of compounds are important medicinal bioactive. Proper post-harvest handlings at the drying and storage stages are two important operations for maintaining the concentration of bioactive compounds [4], [5]. In actual practice most of the dried herbs are commonly stored for certain time before processing stages. The selection of proper storage conditions, including packaging, can be one of great importance. Packaging can directly influence food quality by protecting the product from both oxygen and light and also in maintaining it in optimum temperature and relative humidity [6]. *Orthosiphon stamineus* is known as one of the commercialized herbs in Malaysia. The plant is also known as misai kucing. It can grow to a height of 1.2 m and can be harvested in 2 to 3 months after transplanting. It is known that *O. stamineus* contains several chemically active components such as terpenoids like diterpenes and triterpenes, polyphenols like lipophilic flavonoidsand phenolic acids, and sterols [7]. It is also known that, leaves of this plant have been used traditionally for treating diuretic, and to treat rheumatism, abdominal pain, kidney and bladder inflammation, edema, gout and hypertension [8], [9]. To maintain the quality of herbal dried leave, the important factors that should be considered during storage are the types of packaging materials used, light intensity, relative humidity, oxygen and temperature [10]. Lack of the good post-harvest storage information on *O. stamineus* dried leaves can affect the commercial quality of the final products. Appropriate storage time and good choice of types of packaging container are important aspects for extending the storage life of *O. stamineus* dried leaves.

2. Materials and methods

2.1 Preparation of Raw Material

Orthosiphon stamineus plants were obtained from Sustainable Agrotechnology Institute, Universiti Malaysia Perlis (UniMAP) crop field. The cleaned plants were air dried at ambient temperature for 7 to 10 days. Only part of leaves was utilized in this experiment to control the consistency of further analysis.

2.2 Storage Treatment

To determine the effects of types of packaging materials on *O. stamineus* dried whole-leaf quality (moisture content, color, antioxidant capacity and total phenolic content), the different storage treatments were prepared by manipulating types of storage packaging materials and storage period. 10 g of *O. stamineus* dried whole-leaf samples were stored into three different types of storage packaging materials. The samples were prepared into three replication. The packaging materials used in this experiment were plastic bag, paper bag and glass container. The dried whole-leaf samples were stored for 8 weeks under room temperature (±25 °C) and relative humidity (±65 %RH). The samples quality changes were analyzed at 0, 4 and 8 weeks of storage period. Each sample was considered as independent samples.

2.3 Sample Extraction

1 g of *O. stamineus* dried whole-leaf was extracted by 100 ml of methanol for 3 hours at 40 °C and 200 rpm using a shaker water bath (Thermolab, Germany). The extracted solution was filtered using Whatman No.1 filter paper and then were sealed in the bottles and stored in a freezer (-20 °C) for chemical quality analysis.

2.4 Determination of Total Phenolic Content

200 μl of Follin-Ciocalteu reagent (FCR) and 200 μl of extract solution were mixed with 1.58 ml distilled water and shook rigorously for 4 minutes before adding 1 ml of 20 % sodium carbonate. The mixed solution was allowed to react for 2 hours in a dark place. The concentration of total phenolic content was quantified using UV/VIS spectrophotometer (Shimadzu. Japan) and the absorbance was read at λ=760nm. The caffeic acid was used as standard and the concentration of total phenolic content was expressed in caffeic acid equivalent (CAE).

2.5 Determination of Antioxidant Capacity

The antioxidant capacity of the extracts was determined using the modified DPPH method as described by Akowuah [11] with slight modification. About 2 ml of 2,2-diphenyl-1-picrylhydrazyl (DPPH) was mixed with 200 μl aliquots of samples. Methanol was used to mark up the mixture to 3 ml. The mixed solution was allowed to react in a room temperature for 1 hour. The control was also prepared. After 1 hour, the absorbance value was calculated using a UV/VIS spectrophotometer (Shimadzu, Japan) at λ=517nm. The antioxidant capacity of samples was estimated by utilizing the following equation:

$$\text{Antioxidant capacity} = \left[\frac{(A-B)}{A}\right] \times 100 \quad (1)$$

Where A and B are control absorbance and sample absorbance, respectively.

2.6 Determination of Moisture Content

The MS-70 moisture analyzer (A&D, Japan) was used to read the 1 g sample's moisture.

2.7 Determination of Color

An average of six readings was taken from individual sample of *O. stamineus* dried herbal leaves using colorimeter CR-400 (Konica Minolta, Japan). The collected data were available in the form of L*, a* and b* color space (CIELAB). The total color difference (ΔE) was calculated by using the following equation:

$$\Delta E = \sqrt{[(L^* - L_0)^2 + (a^* - a_0)^2 + (b^* - b_0)^2]} \quad (2)$$

Where L_0, a_0 and b_0 are the control values for the initial leaves color before storage.

2.8 Statistical Analysis

All measurements were carried out in triplicate and the results are statistically analyzed using JMP pro 11 package to determine the average value and standard error.

3. Results and discussion

The experiment was carried out to study the effect of packaging materials (plastic, paper and glass) on the *Orthosiphon stamineus* dried whole-leaf quality during storage. The samples were stored for 8 weeks and the changes of the *O. stamineus* dried whole-leaf quality (total phenolis content, antioxidant capacity, moisture content and total color different) were analyzed every 4 weeks. Figure 1 shows the changes of the total phenolic content of the *O. stamineus* dried whole-leaf during 8 weeks of storage period. The initial mean value of total phenolic content of the *O. stamineus* dried whole-leaf samples was 83.68 ± 2.93 mg/g. The total phenolic content of the samples packed in a plastic and paper bags increased insignificantly to 83.14 ± 5.59 mg/g and 84.72 ± 1.22 mg/g, respectively when stored at room temperature for 4 weeks. The increment of the total phenolic content of the samples packed in the plastic and paper bags could be seen for the next 4 weeks of the storage period. However, the contrary results was obtained when the samples were packed in the glass container. The total phenolic content of the samples kept in the glass container exhibited a decrement to 70.79 ± 4.30 mg/g from the initial value of 83.68 ± 2.93 mg/g after 4 weeks of the storage period. Nevertheless, the content of phenolic increased drastically to 108.30 ± 4.69 mg/g during 8 weeks of storage period. The quantity of the total phenolic content of the *O. stamineus* dried whole-leaf for all treatments (plastic, paper and glass) were insignificantly different at the end of experimental (8 weeks). According to the previous research, the usage of glass as a primary packaging material could sustain the quality of the food product compared to the others type of packaging material [12].

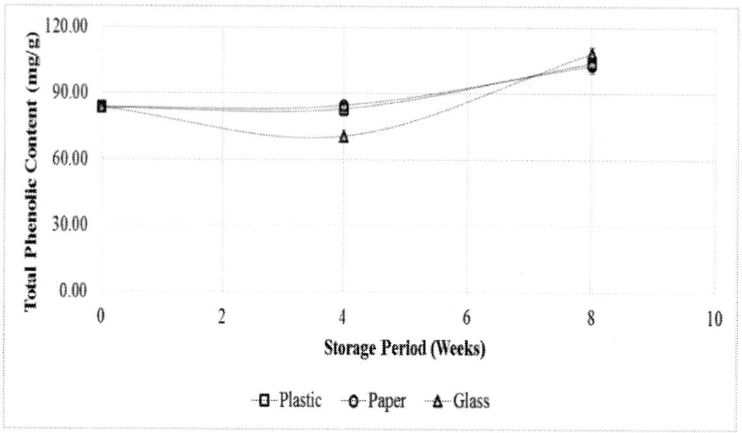

Figure 1. The changes of total phenolic content of *O. staminues* dried whole-leaf during storage

Figure 2 shows the changes of the antioxidant capacity of the *O. stamineus* dried whole-leaf during 8 weeks of storage period. The initial mean value of antioxidant capacity of the *O. stamineus* dried whole-leaf samples was 76.38 %. The samples packed in the plastic bag, paper bag and glass container exhibited a similar pattern throughout experimental storage period (8 weeks). The capacity of antioxidant increased steadily during 8 weeks of storage period. It was about 9 to 10 % increment for all samples. The other researcher also reported about the increment of the antioxidant capacity [13]. However, they found that the packaging materials not affected the quality during the storage.

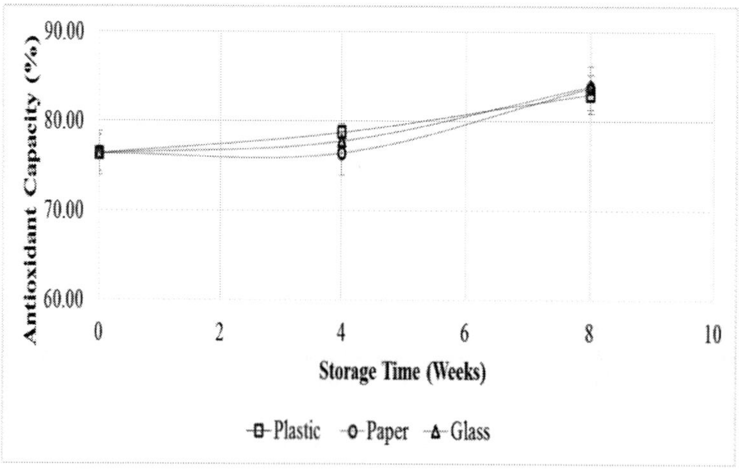

Figure 2. The changes of antioxidant capacity of *O. staminues* dried whole-leaf during storage

Figure 3 shows the changes of the moisture content of the *O. stamineus* dried whole-leaf during 8 weeks of storage period. The initial mean value of moisture content of the *O. stamineus* dried whole-leaf samples was 12.99 %. The samples packed in the plastic bag obviously exhibited the decrement of moisture content and the total mean of moisture content at 8 weeks of storage period was 11.40 %. It was followed by the samples packed in the glass container which the value mean of the moisture content at the end of storage period (8 weeks) was 11.69 %. The changes of the moisture content for the samples packed in the paper bag exhibited a dissimilar pattern compared to the samples packed in the plastic bag and glass container. The moisture content decreased drastically to 11.46 % at 4 weeks of storage compared to the initial mean value of 12.99 %. However, the moisture content of the samples packed in the paper bag significantly increased to 12.14 % at the end of storage period (8 weeks).

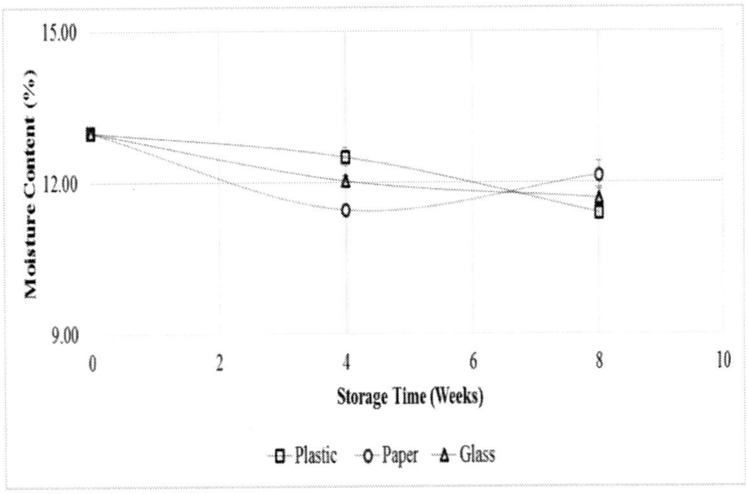

Figure 3. The changes of moisture content of *O. staminues* dried whole-leaf during storage

Figure 4 shows the changes of the total different color (ΔE) of the *O. stamineus* dried whole-leaf during 8 weeks of storage period. All the samples exhibited a similar trend of increment of the total different color (ΔE). The highest changes showed by the samples packed in a glass container (7.69 ± 2.01), followed by the samples packed in paper and plastic bags (6.98 ± 1.10 and 5.76 ± 1.37, respectively). The inceasing of the ΔE might be influenced by the adsorption and desorption process during storage. The dried whole-leaf turned out brown color. This results similar as reported by [5]. They reported that the changes of *Casia alata* color could be attributed by the conversion of chlorophyll to pheophytins at high relative humidity.

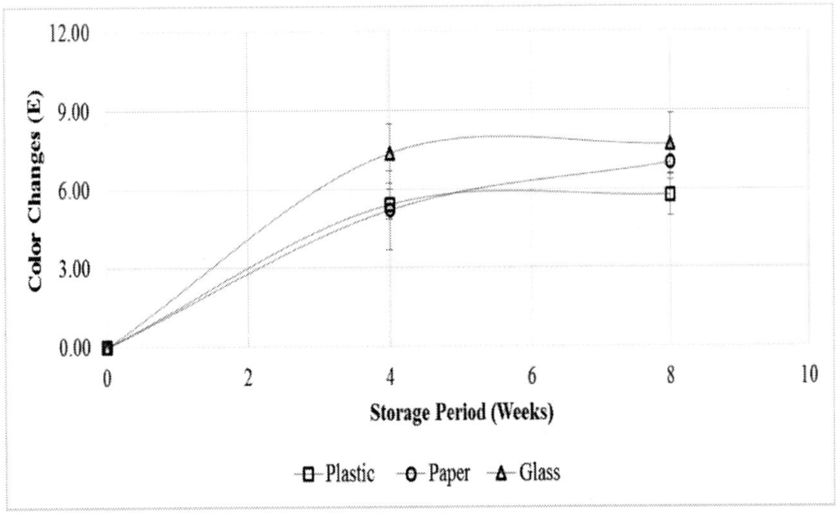

Figure 4. The changes of total different color (ΔE) of *O. staminues* dried whole-leaf during storage

4. Summary

As a conclusion, the data collected in this paper showed that the type of packaging material significantly affected the quality of *O. stamineus* dried whole-leaf herbal during 8 weeks of storage period. The total phenolic content, antioxidant capacity and total different color of the samples increased along the storage period. The moisture content of the samples decreased by storage period except for the samples packed in the paper bag. These finding might be used as an indicator for similar local dried herbal leaves to control the quality of raw herbal material in producing a great quality of herbal finished products.

5. Acknowledgement

This study was financially supported by the Universiti Malaysia Perlis (Short Term Grant, STG); the research facilities of the Institute of Sustainable Agrotechnology (INSAT) and Faculty of Engineering Technology, UniMAP.

6. References

[1] R. N. BENNETT and R. M. WALLSGROVE, "Secondary metabolites in plant defence mechanisms," *New Phytol.*, vol. 127, no. 4, pp. 617–633, 1994.
[2] M. Mazid, T. A. Khan, and F. Mohammad, "Role of secondary metabolites in defense mechanisms of plants," *Biology and Medicine*, vol. 3, no. 2 SPECIALISSUE. pp. 232–249, 2011.
[3] M. Wink, "Modes of Action of Herbal Medicines and Plant Secondary Metabolites," *Medicines*, vol. 2, no. 3, pp. 251–286, 2015.
[4] N. Abdul Razak, A. R. Shaari, I. H. Rukunudin, and M. S. Ahmad, "Quality Changes of Orthosiphon Stamineus Dried Herbal Leaves Under Extremely Low," vol. 8, no. November, pp. 99–102, 2014.
[5] N. Abdul Razak, A. R. Shaari, M. S. Nur Ain, and L. Y. Leng, "Effect of Moisture Sorption Behavior on Color Quality of Cassia Alata Leaves," vol. 1060, pp. 0–4, 2017.
[6] a a Kader, D. Zagory, and E. L. Kerbel, "Modified atmosphere packaging of fruits and vegetables.," *Crit. Rev. Food Sci. Nutr.*, vol. 28, no. 1, pp. 1–30, 1989.
[7] Y. Tezuka, P. Stampoulis, A. Banskota, S. Awale, K. Tran, I. Saiki, and S. Kadota, "Constituents of the Vietnamese medicinal plant Orthosiphon stamineus," *Chem. Pharm. Bull. (Tokyo).*, vol. 48, no. November, pp. 1711–1719, 2000.
[8] R. Hegnauer, "Biochemistry, distribution and taxonomic relevance of higher plant alkaloids.," *Phytochemistry*, vol. 27, no. 8. pp. 2423–2427, 1988.
[9] O. Z. Ameer, I. M. Salman, M. Z. Asmawi, Z. O. Ibraheem, and M. F. Yam, "Orthosiphon stamineus: traditional uses, phytochemistry, pharmacology, and toxicology.," *J. Med. Food*, vol. 15, no. 8, pp. 678–90, 2012.
[10] P. Prasad and A. Kochhar, "Active Packaging in Food Industry: A Review," *IOSR J. Environ. Sci. Ver. III*, vol. 8, no. 5, pp. 2319–2399, 2014.
[11] Akowuah, Ismail, Norhayati, and Sadikun, "The effects of different extraction solvents of varying polarities on polyphenols of Orthosiphon stamineus and evaluation of the free radical-scavenging activity," *Food Chem.*, vol. 93, pp. 311–317, 2005.
[12] A. Saint-Eve, C. Lévy, M. Le Moigne, V. Ducruet, and I. Souchon, "Quality changes in yogurt during storage in different packaging materials," *Food Chem.*, vol. 110, no. 2, pp. 285–293, 2008.
[13] F. J. García-Alonso, S. Bravo, J. Casas, D. Pérez-Conesa, K. Jacob, and M. J. Periago, "Changes in antioxidant compounds during the shelf life of commercial tomato juices in different packaging materials," *J. Agric. Food Chem.*, vol. 57, no. 15, pp. 6815–6822, 2009.

IOP Conf. Series: Materials Science and Engineering **318** (2018) 012011 doi:10.1088/1757-899X/318/1/012011

Effect of Filler Content and Chemical Modification on Mechanical Properties of Polylactic Acid/ Polymethyl Methacrylate/ Nypa Fruticans Husk Biocomposites

MR Mohamad Syahmie[1], T Pei Leng[1] and Zabidi Nurul Najwa[1]

[1]School of Materials Engineering, Universiti Malaysia Perlis, 02600, Jejawi, Perlis, Malaysia

E-mail: syahmie@unimap.edu.my

Abstract. The main purpose of incorporating Nypa fruticans husks (NFH) into Polylactic acid (PLA)/Polymethylmethacrylate (PMMA) is to decrease the costs and enhanced the properties of the biocomposites. 3-Aminopropyl Triethoxysilane (3-APE) was used as coupling agent. The effect of NFH content and 3-APE on the mechanical properties and morphology of the biocomposites were investigated. Results show that the effect of NFH content increased Young's modulus but decreased the tensile strength and elongation at break of PLA/PMMA/NFH biocomposites. However, silanized biocomposites using 3-APE) was found to enhanced the tensile strength and Young's modulus but decreased the elongation at break of the silanized biocomposites. Scanning electron microscopy (SEM) study of the tensile fracture surface of the biocomposites indicated that the used of 3-APE as couling agent improved the interfacial interaction netween NFH and PLA/PMMA blends.

1. Introduction

Plastic materials have found wide applications in every aspect of life and industries. However, most of the conventional plastics are nonbiodegradable, and their accumulation in the environment has been a threat to the planet. Environmental concerns and a shortage of petroleum resources have driven efforts aimed at the production of biodegradable materials [1].

The use of inorganic fillers is a usual practice in the plastics industry to improve the mechanical properties of thermoplastics, which are commonly known as polymer composites. These materials usually comprise of an effective polymeric matrix in which fibers or small filler particles are thoroughly dispersed in composite systems. The filler must be well dispersed in the matrix to avoid zones of weaker cohesion where flaws and other defects will be initiated upon stressing [2].

Nypa Fruticans (NF) is a monoecious palm with special characteristics. In Malaysia, NF palm can be found throughout the year and it can be considered an abundant source. [3]. Nypa, consisting of frond, shell, husk, and leaf, was chemically characterised for cellulose, hemicellulose, lignin, starch, protein, extractives, and inorganic constituents for its each part [4].

Polylactide (PLA) is biodegradable and biocompatible thermoplastic which can be fermented from renewable resources. It has proved to be a cost-effective alternative to traditional, commodity plastics for various end-use applications, due to its good mechanical properties, thermal plasticity, and facile fabrication [5]. Together with its biodegradable and compostable properties, PLA in principle holds great potential as an alternative to petroleum-based polymers. However, some of its applications are limited by its low elongation at break and poor toughness. PLA has a glass transition temperature ranging from 55 °C to 65 °C, and it is brittle at room temperature, fracturing via a crazing mechanism. To modify the mechanical brittleness of PLA, many strategies such as processing manipulation, copolymerization and blending have been employed [6].

Several approaches based on (nano)composites or direct chemical modifications of the polyester chains were previously explored [7–11], but melt-blending of PLA with synthetic and, at least, potentially/partly bio-based thermoplastic polymers represents a more robust and easy route for these technical applications. Among them, poly(methyl methacrylate) (PMMA) with its intrinsic high mechanical properties, high transparency, UV resistance and long-term stability is a serious candidate. Literature about PLA/PMMA blends is composed of several articles, each of them using different preparation techniques, conditions and raw materials, highlighting some tricky conditions to obtain PLA/PMMA miscibility. Miscibility in poly(lactic acid)/PMMA blends was first mentioned by Eguiburu et al. [12].

A coupling agent is a chemical that functions at the interface to create a chemical bridge between the filler and matrix. In order to enhance the behavior of these biocomposites, NFH have treated with 3-APE. The 3-APE has ethoxy groups that hydrolyzed in water or solvent producing silanol and next the silanol reacts with OH group of filler which form stable covalent bond onto the filler surface. Generally, the coupling agent improves the degree crosslinking in the interface region and offers a suitable bonding result, as well as creation of high filler surface area. The amine group from 3-APE can form hydrogen bonds to COO- sites on hydrolyzed PLA backbone [13].

In the present study the effect of NFH content and the silane coupling agent on mechanical properties and morphology of PLA/PMMA/NFH biocomposites were investigated.

2. Experimental

2.1. Materials

PLA supplied by TT Biotechnologies Sdn. Bhd, Penang and PMMA obtained from SLT Plastic Sdn. Bhd, Penang. The NFH was obtained from Kota Kayang, Perlis. The NFH was cleaned and dried at 80°C for 24 hours in an oven. Nypa fruticans was then crushed into small pieces and ground to a powder. The average particle size of NFH was below 75µm determined by using a sieve. 3-aminopropyl triethoxysilane (3-APE) was supplied by Fluka, PenangA subsubsection. The paragraph text follows on from the subsubsection heading but should not be in italic.

2.2. Filler Treatment

The silane coupling agent (3-APE), was dissolved in 3% (v/v) ethanol solution at room temperature. The NFH was slowly transferred into the 3-APE solution and stirred for 1 hour. The NF was filtered using filter paper and dried at 80 °C for 24 hours.

2.3. Preparation of Biocomposites

PLA/PMMA/NFH biocomposites were prepared using a twin-screw extruder. The biocomposites were extruded at 180 °C to 250 °C, feeder screw of 8 rpm and screw speed of 80 rpm. First, PLA and PMMA were inserted into the hopper until completely melted. The melted polymer were then extruded and pelleterized. The pellets were then dried in an oven at 80 °C for 24 hours to remove moisture.

The extrudate biocomposites were compressed using compression moulding machine, model GT 7014A. The compression was done at 180 °C with 4 min preheating, 2 min compression, and subsequent cooling for 4 min. A similar procedure was done for silanized PLA/PMMA/NFH biocomposites. Table 1 shows the formulation for untreated and treated PLA/PMMA/NFH biocomposites.

Table 1. Formulation of Untreated and Treated Biocomposites.

Material	**Untreated**	**Treated**
PLA (php)	70	70
PMMA (php)	30	30
NFH (php)	0, 10, 20, 30, 40	10, 20, 30, 40
3-APE (%)[a]	-	3

php = part per hundred of total polymer
[a] 3% from weight of NF

2.4. *Tensile Properties*
Tensile tests were carried out according to ASTM D638 using an Instron 5569. The crosshead speed for the biocomposites testing was 20 mm/min. An average of five samples for each formulation was tested. Tensile strength, elongation at break and Young's modulus were recorded and calculated by the instrument software.

2.5. *Morphological Studies*
The morphological study of the tensile fractured surfaces of the biocomposites was carried out using a scanning electron microscope (SEM), model JSM 6260 LE JOEL. The fractured ends of specimens were mounted on aluminium stubs and sputter coated with a thin layer of palladium to avoid electrostatic charging during examination.

3. Result and Discussion
The addition of NFH to the PLA/PMMA matrix resulted in changes of the tensile properties. The effect of NFH content on the tensile strength of untreated and treated PLA/PMMA/NFH biocomposites is illustrated in Figure 1. It can be seen that tensile strength of untreated PLA/PMMA/NFH biocomposites decreased with an increase of NFH content. Many studied reported that the tensile properties of the composites generally become poorer as the content of natural filler increased. The incorporation of NFH reduced the tensile strength due to poor adhesion between NFH and PLA/PMMA matrix. However, the treated biocomposites with 3-APE had higher tensile strength compared to untreated biocomposites. The increased in tensile strength of the silanazed biocomposites attributed by improved interfacial adhesion between NFH and PLA/PMMA matrix. The improved interfacial adhesion due to chemical bonding between the hydrophobic part of the silane on the NFH surface and PLA/PMMA matrix or van der Waals forces between them.

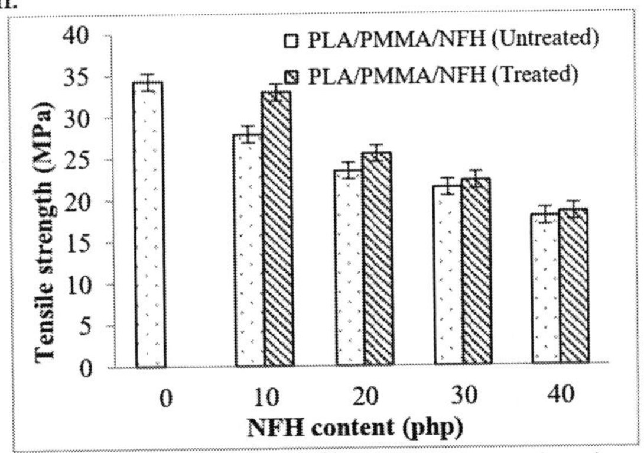

Figure 1. Effect of NFH content on tensile strength of untreated and treated PLA/PMMA/NFH biocomposites.

The effect of NFH content on elongation at break of untreated and treated PLA/PMMA/NFH biocomposites is present in Figure 2. The elongation at break of the untreated PLA/PMMA/NFH biocomposites decreased as the NFH content increased. The decreased of elongation at break may be attributed to a reduction in deformability of the rigid interface between fillers and matrix. The silanized biocomposites treated with 3-APE show lower elongation at break compared to untreated biocomposites. The silane treatment has increased the tensile strength of biocomposites with the enhancement in rigidity and decrement of the ductility of biocomposites, which consequently lowered the elongation at break of the biocomposites.

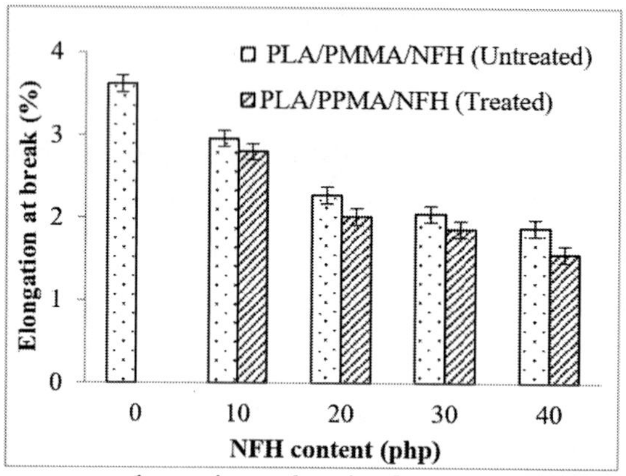

Figure 2. Effect of NFH content on elongation at break of untreated and treated PLA/PMMA/NFH biocomposites

Figure 3 illustrate the effect of NFH content on the Young's modulus of untreated and treated PLA/PMMA/NFH biocomposites. The result showed that the Young's modulus of untreated biocomposites increased with NFH content increases. Generally, filler that had higher stiffness than the matrix increased the Young's modulus of the biocomposites and reduced the elongation at break. The increasing of NFH content potentially increased the filler-matrix interaction, which led to higher efficiency of stress transfer from matrix to filler phase, thus increasing the Young's modulus. At similar NFH contents, treated biocomposites had higher Young's modulus than untreated biocomposites. It can be seen that the Young's modulus of treated biocomposites were increased slightly by silane treatment. The increases of Young's modulus of the silanized biocomposites ware most likely caused by improved interfacial adhesion between the NFH filler and PLA/PMMA matrix.

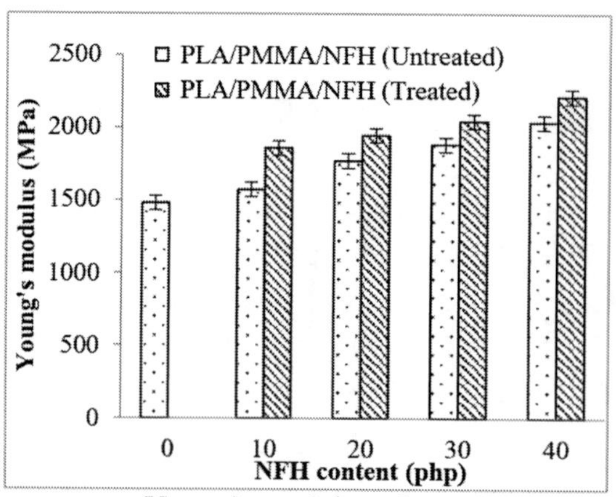

Figure 3. Effect of NFH content on Young's modulus of untreated and treated PLA/PMMA/NFH biocomposites

The SEM micrographs of the tensile fracture surfaces of untreated and treated PLA/PMMA/NFH biocomposites with silane coupling agent at 20 and 40 php NFH content were shown in Figure 4 and 5, respectively. The micrographs of the untreated biocomposites showed poor interfacial adhesion and dispersion of NFH content in the PLA/PMMA matrix and also exhibit NFH detachment and pull out from matrix. Meanwhile, the micrographs of treated PLA/PMMA/NFH biocomposites showed good interfacial bonding between the PLA/PMMA matrix and NFH. Moreover, the surface morphology of treated PLA/PMMA/NFH biocomposites exhibits less NFH detachment and pull out from PLA/PMMA matrix. It was apparent that the presence of 3-APE silane led to better interfacial interaction between NFH and PLA/PMMA matrix.

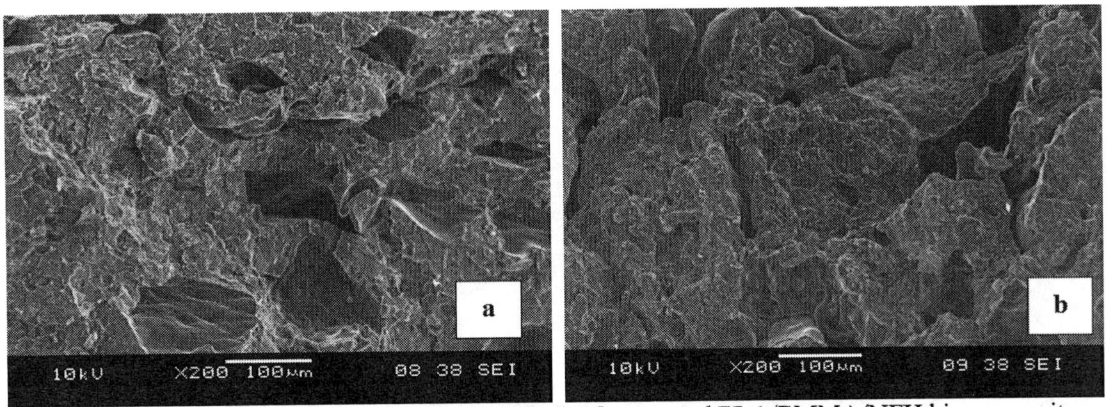

Figure 4: Scanning electron micrograph (SEM) of untreated PLA/PMMA/NFH biocomposites (a) (20 php of NFH) & (b) (40 php of NFH)

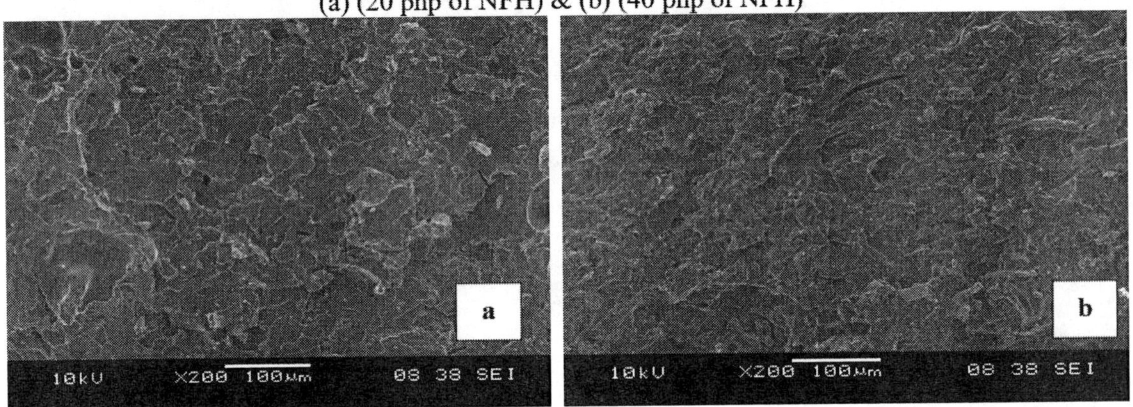

Figure 5: Scanning electron micrograph (SEM) of treated PLA/PMMA/NFH biocomposites (a) (20 php of NFH) & (b) (40 php of NFH)

4. Conclusion

The effect of NFH content had increased Young's modulus of the biocomposites, but decreased the tensile strength and elongation at break. The presence of silane coupling agent (3-APE) increased the tensile strength and Young's modulus but decreased the elongation at break of treated PLA/PMMA/NFH biocomposites. SEM micrographs of tensile fracture surface of silanized biocomposites indicated the improved of interfacial interaction between NFH and PLA/PMMA matrix.

References

[1] Liang H, Hao Y, Bian J, Zhang H, Dong L and Zhang H 2015 *Polym. Eng. & Sci.* **55**(2) 386
[2] Syahmie Rasidi M, Salmah H, Teh PL and Ismail H 2015 *Appl. Mech. Mater.* **754** 54

[3] Rasidi MS, Husseinsyah S and Leng TP 2015 *Polym. Eng. Sci.* **54**(3) 253
[4] Tamunaidu P and Saka S 2011 *Ind. Crops & Prod.* **34**(3) 1423
[5] Wang X, Zhuang Y and Dong L 2013 *J Appl. Polym. Sci.* **127**(1) 471
[6] Zhang C, Wang W, Huang Y, Pan Y, Jiang L, Dan Y, Luo Y and Peng Z 2013 *Mater. Design* **45** 198
[7] Dubois P and Marariu M 2008 *JEC Compos.* **45** 66
[8] Murariu M, Dechief AL, Bonnaud L, Gallos A, Fontaine G, Bourbigot S and Dubois P 2010 *Polym. Degrad. & Stab.* **95**(5) 889
[9] Murariu M, Doumbia A, Bonnaud L, Dechief AL, Paint Y, Ferreira M, Campagne C, Devaux E and Dubois P 2011 *Biomacromolecules* **12**(5) 1762
[10] Becker JM, Pounder RJ and Dove AP 2010 *Macromol. Rapid Commun.* **31**(22) 1923
[11] Michell RM, Müller AJ, Boschetti-de-Fierro A, Fierro D, Lison V, Raquez JM and Dubois P 2012 *Polym.* **53**(25) 5657

IOP Conf. Series: Materials Science and Engineering **318** (2018) 012012 doi:10.1088/1757-899X/318/1/012012

Phytosynthesis of Iron Nanoparticle from Averrhoa Bilimbi Linn.

Rosli, I. R.,[1], Zulhaimi, H. I. [1*], Ibrahim, S. K. M.[1], Gopinath[2], S. C. B., Kasim, K. F[1]., Akmal, H. M. [3], Nuradibah, M. A. [1] and Sam, T. S.[1]

[1]School of Bioprocess Engineering, Kompleks Pusat Pengajian Jejawi 3, Universiti Malaysia Perlis, 02600, Arau, Perlis, Malaysia.
[2]Institute of Nano Electronic Engineering, Universiti Malaysia Perlis, Lot 106, 108 & 110, Blok A, Taman Pertiwi Indah, Jalan Kangar-Alor Setar, Seriab, 01000 Kangar, Perlis, Malaysia
[3]Advanced Multidisciplinary MEMS-Based Integrated Electronics NCER Centre of Excellence (AMBIENCE), School of Microelectronic Engineering, Universiti Malaysia Perlis, SME Bank Industrial Estate, 02000 Kuala Perlis, Perlis, Malaysia

E-mail: hanna@unimap.edu.my

Abstract. This paper demonstrates iron nanoparticles (FeNP) was synthesized from natural sources of Averrhoa bilimbi Linn. The plant extracts act as natural reducing agent in producing FeNP. There is no addition of any surfactants during the nanoparticles formation. Gravimetric analysis is used to calculate the percentage yield of plant extracts. TPC and DPPH assay method were used to evaluate antioxidant activity in different A. bilimbi extracts and synthesized FeNP. Based on the analyses, it showed that fruit has the highest percentage yield and antioxidant activity followed by leaf, twig and bark. Analysis from TPC, fruit contains 27.26 mg GAE/g and 39.46 mg GAE/g for FeNP. DPPH assay showed fruit extract has the highest free radical antioxidant activity with 61.93% in A. bilimbi and 80.00% in FeNP. Phytosynthesis of FeNP were examine by using UV-Vis spectrophotometer. Based on the spectra, it showed that FeNP recorded peak absorbance at 465 nm, 450 nm, 460 nm and 440 nm for UAE-F, UAE-L, UAE-T and UAE-B, respectively. FTIR analysis shows the presence of strong alcoholic bond, aldehyde, stretch amine and alkene that was responsible in reduction process to form FeNP. The result of UV-Vis and FTIR showed that the existance of FeNP and involvement of functional group that were responsible on the formation of nanoparticles.

1. Introduction

Nanotechnology is one of the fastest developing research in this area of study caused by the improvement of their characteristics and properties [8]. Generally, nanotechnology able to measure ultrafine particles with their ranging size from 1-100 nanometers [14]. The researchers also have been progressively observed the shape and size of nanoparticles depends on the physiochemical properties by using advanced techniques [14].

Recently, many studies have proven that the plant extracts act as a potential precursor for the synthesis of the nanoparticles in non-hazardous ways [4]. The plants are used successfully in the synthesis of several greener nanoparticles such as cobalt, copper, silver, gold, palladium, platinum, zinc oxide and iron oxide [8]. During past decade, it has been demonstrated that many biological systems, plants able to transform inorganic metal ions into metal nanoparticles by the reductive capacities of the proteins and metabolites present in the extracts.

IOP Conf. Series: Materials Science and Engineering **318** (2018) 012012 doi:10.1088/1757-899X/318/1/012012

Synthesis using biological method is compatible with the green chemistry principles. Besides, biological method for synthesizing nanoparticles also makes use of environmental friendly, non-toxic and safe reagents. Nanoparticles synthesized using biological techniques or green technology havediverse natures, with greater stability and appropriate dimensions since they are synthesized using a one-step procedure [1].

2. Experimental works

2.1. Chemicals

Gallic acid, DPPH solution, Sodium carbonate, Folin-Ciocalteu reagent, Iron (II) Sulfate Heptahydrate, Iron(III) Chloride Hexahydrate.

2.2. Plant Materials

Fruits, leaves, twig and bark of *A. bilimbi* samples were obtained from Kg. Alor Ibus, Kedah. The *A. bilimbi* part of the plants which is fruits, leaves, twig and bark washed cleanly and dried in the oven. After that, samples were grinded into powder forms by using grinder and transfer into the beaker.

3. Methodology

3.1 Preparation of Plant Extract

The methods to prepare plant extract were followed Zulhaimi et. al. [17]. For the plant parts, bark was added along with fruits, leaves, twig and bark. Cold maceration (CM) and ultrasonic-assisted extraction (UAE) were utilized. Distilled water was used as solvent.

3.2 Percentage Yield of Plant Extract

In order to analyse the percentage plant yield extract, the extracts of evaporated dried extracts based on dry weight basis were calculated by using gravimetric analysis as the following equation:

$$\text{Yield (g/100 g of dry plant)} = (W_1 \times 100) / W_2 \quad (1)$$

Where W_1 is the weight of the extract after the solvent evaporation and W_2 is the weight of the initial powder used to extract.

3.3 Total Phenolic Content (TPC Assay)

1 mL extract solutions of different plant parts were added in the test tube following 1 mL of Folin-Ciocauteau reagent were added and mixed thoroughly. The mixtures were dismissed for 5 minutes at room temperature. Sequently, 4 mL of sodium carbonate (75 g/L) and 10 mL of distilled water were added and thoroughly mixed. The mixture were idle at room temperature for 2 hours. Standard curve was obtained using addition of 0.5 mL Gallic acid solution in extraction sample and the absorbance with UV-Visible Spectrophotometer at 765 nm was determined. total phenolic content was calculated using the following equation:

$$\text{TPC} = \frac{(\text{Absorbance of sample} - \text{Intercept}) \times \text{Dilution factor}}{\text{Slope} \times \text{Weight of extract}} \quad (2)$$

3.4 DPPH Assay

In this assay, purple radical DPPH was reduced by antioxidant samples to the corresponding pale yellow hydrazine. 0.5 mL of DPPH (25 g/L) solution was added to 1 mL of extraction sample solution (with different solvents). Mixtures were shake vigorously and kept at room temperature for 30 minutes in dark. Then, the absorbance determined using calibrated UV-Visible spectrometer at 517 nm. This scavenging activity was calculated as % inhibition using the formula:

$$\text{DPPH scavenging activity} = \left(\frac{\text{Absorbance control} - \text{Absorbance sample}}{\text{Absorbance control}}\right) \times 100 \quad (3)$$

IOP Conf. Series: Materials Science and Engineering **318** (2018) 012012 doi:10.1088/1757-899X/318/1/012012

3.5 Synthesis of FeNP

FeNP using *A. bilimbi* was synthesized using Iron (III) Chloride Hexahydrate ($FeCl_3.6H_2O$), Iron (II) Sulfate Heptahydrate ($FeSO_4.7H_2O$) and plant extract. $FeCl_3.6H_2O$ and $FeSO_4.7H_2O$ were prepared and mixed together in a beaker with a ratio 2 :1 by using magnetic stirrer. 5 mL of prepared plant extract was added to mix for 1 hour. The formation of extract colour to dark green or black colour observed indicating the formation of FeNP [6]. The samples were tested using UV-Vis Spectrophotometer and FTIR.

3.6 UV-Vis Spectrophotometer

The blank and sample solutions were put in a Quartz cuvette and the baseline corrected using plant extracts as the blank. The spectra of FeNP obtained in the wavelength range of 400 – 600 nm.

3.7 Fourier Transform Infrared (FTIR)

Fourier Transform Infrared (FTIR) spectrometry is a complex method of spectroscopy with the ability to identify materials and determine the quality of a sample. FTIR spectroscopy was performed to determine the functional groups of *A. bilimbi* that appeared in FeNP. The functional groups that present in the synthesized nanoparticles were found at a region of 650 cm^{-1} to 4000 cm^{-1}.

4. Results and discussion

4.1 Percentage Yield of A. bilimbi Extracts

Figure 1 showed the percentage yield of CM and UAE using water as solvent for each parts of *A. bilimbi*. The percentage yields of CM for fruit (CM-F), leaf (CM-L), twig (CM-T), and bark (CM-B) by using water as solvent were 49.77%, 29.88%, 24.93% and 19.95%, respectively.

The percentage yields for UAE for fruit (UAE-F), leaf (UAE-L), twig (UAE-W), and bark (UAE-B) by using water solvent were 49.87%, 22.25%, 20.40% and 19.92%, respectively. From the graph, UAE-F showed highest percentage of plant extract compared to the others plant parts. The result also give the same to CM-F.

Percentage yield of extract of CM-F and UAE-F was higher than other parts, showed that fruits powder has high solubility than the other parts [5]. Moreover, the preliminary phytochemical studies by revealed the presence of carbohydrates, proteins, amino acids, flavonoids, tannins, bitter principles, essential oil, valepotriates, coumarin, and terpenes in *A. bilimbi* fruits [2].

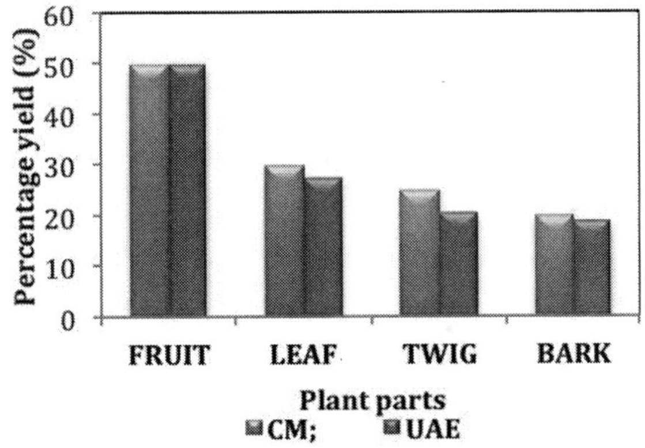

Figure 1. The percentage yields of CM and UAE of *A. bilimbi*.

4.2 Total Phenolic Content assay (TPC assay)

The Figure 2 showed TPC of FeNP on different plant parts of *A. bilimbi*. UAE method was preceeded due to simpler process yet produced high percentage yield of extract. Based on figure in 4.2, TPC for FeNP from UAE process for all plant parts has higher value in comparison to the plant extract, with 32.22 mg GAE/g for UAE-FeNP-F, 27.04 mg GAE/g for UAE-FeNP-L, 7.12 mg GAE/g for UAE-FeNP-T and 4.27 mg GAE/g for UAE-FeNP-B. It was noticed that TPC value of plant extract is lower with 19.26 mg GAE/g for UAE-F, 12.26 mg GAE/g for UAE-F, 12.26 mg GAE/g for UAE-L, 9.11 mg GAE/g for UAE-T and 2.17 mg GAE/g for UAE-B.

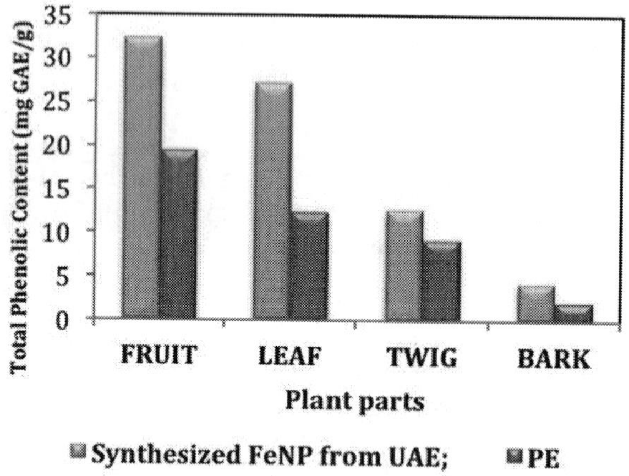

Figure 2 The total phenolic content of FeNP and *A. bilimbi* extracts.

The phenolic compound such as phenolic acids, flavonoids and tannins are considered to be major influenced to the ability as antioxidant plants. The higher content of total phenolic content in *A. bilimbi* extract facilitates the reduction of iron ions to nano scale sized of iron particles [14]. The nano size of FeNP resulted to higher total phenolic compounds value [14].

4.3 DPPH assay

Figure 3 shows the result of DPPH scavenging activity of *A. bilimbi* on different plant parts. The DPPH activity result of UAE-F, UAE-L, UAE-T and UAE-B were expressed in percentage which is 61.93%, 43.54%, 25.12% and 11.00%, respectively. Meanwhile, the result of DPPH activity of FeNP using plant extracts also compared in Figure 4.3. It was found that DPPH value is drastically high, with 80.00% for UAE-FeNP-F, 53.75% for UAE-FeNP-L, 31.5% for UAE-FeNP-T and 17.78% for UAE-FeNP-B.

DPPH (1,1-diphenyl-2-picrylhydrazyl) was one of the free radical scavenging activity that use to measure antioxidant activity. This is the simplest and most widely reported method for screening antioxidant activity in foods and plants [9]. Besides, DPPH scavenging activity of antioxidants is due to their hydrogen donating ability. Therefore, there is correlation reduction between the numbers of DPPH molecules with the number of hydroxyl groups [9]. Furthermore, the radical scavenging activities of FeNP also appear to be regenerative making it more effective and allowing scavenging of multiple free radical by antioxidant nanoparticles [9]. Considering the high surface area to volume ratio, it appears that FeNP, showed higher tendency to interact to reduce DPPH free electron [9].

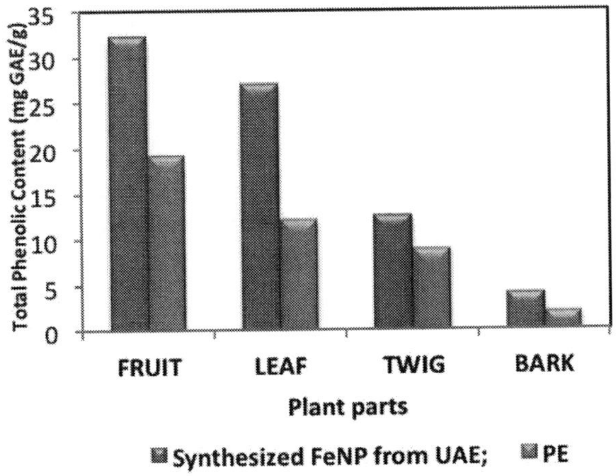

Figure 3. The DPPH scavenging activity of FeNP and *A. bilimbi* extracts.

4.4 Phytosynthesis of FeNP

The reduction of iron ions and formation of FeNP occurred instantly after mixing of ferric chloride and ferric sulphate with plant extracts. This can be examined from the colour changing upon addition of each plant parts. It was observed that the solution turned the colour from yellow to brownish for fruit. Meanwhile, other plant parts for leaf, twig and bark were all turned into black, which indicates the formation of FeNP.

4.5 UV-Vis Spectrophotometer

UV-Vis Spectrophotometer was carried out to confirm the formation of FeNP. Light sources will emit over ultraviolet and visible light spectra [7]. The samples solutions of transition metals tend to have distinct colour characteristics with certain corresponding peaks. Previous studies suggested FeNP that formed from plant extract showed a band peaking in between the range of 400 nm to 600 nm. The spectra clearly depicted maximum absorption peak which indicated the formation of FeNP. The Figure 4 showed the peaks for UAE-FeNP-F, UAE- FeNP-L, UAE-FeNP-T and UAE- FeNP B.

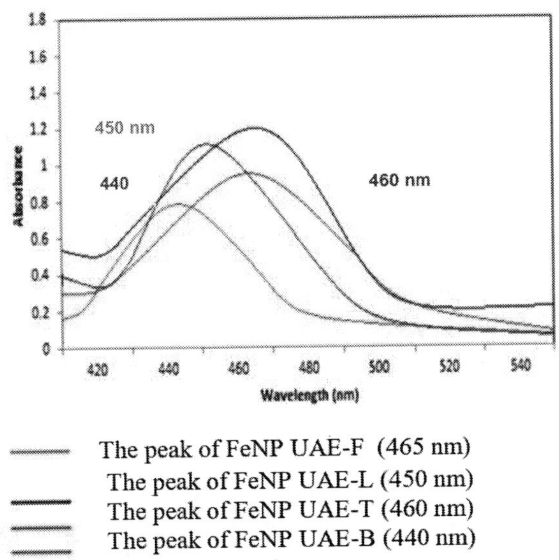

Figure 4. The peak of FeNP in UV - Vis Spectrophotometer

IOP Conf. Series: Materials Science and Engineering **318** (2018) 012012 doi:10.1088/1757-899X/318/1/012012

From the result in Figure 4, it was observed that the most broad absorption peak was found at the range of 440 and 465 nm. This could be due to the excitation of nanoparticles from ground state to the excite state [12]. For FeNP-UAE-F, broad peak absorbance was recorded is at 465 nm. For FeNP-UAE-L, it was observe that the peak is at 450 nm, which evidenced that phytochemical content extract act as reducing agent to metal precursor to the formation of nanoparticle.

Meanwhile, the most broad absorbance peak at 460 nm and 440 nm were recorded for twig and bark. All of the most broad absorbance peaks were shown at different wavelength for different type of reducing agent. This may occur due to agglomeration and settling of nanoparticles in a cuvette which shown different broad peak in each result.

4.6 Fourier Transform Infrared (FTIR)

The Figure 5 interpreted the FTIR spectra of synthesized FeNP from (a) UAE-F and (b) UAE-L. The absorption peaks in FTIR spectra of nanoparticles for fruit were located mainly at 3384 cm^{-1} and 1640 cm^{-1}. Meanwhile, Figure 4.6 depicted the FTIR spectra of FeNP using (a) UAE-T and (b) UAE-B. The Figure 4.9 (a) showed peak at 3368 cm^{-1} and 1644 cm^{-1}. The Figure 4.6 (b) showed peak at 3383 cm^{-1} and 1639 cm^{-1}. The peak at 3384, 3368 cm^{-1} are due to strong alcoholic bond like aramanthine and phenolic compound presence in the aqueous phase. The peak can also be found in Sathya *et* al. that studied on iron oxide nanoparticles [18]. 1641, 1644 and 1639 cm^{-1} indicate the presence of (C=O) bond that were accountable for the formation of FeNP.

(a)

(b)

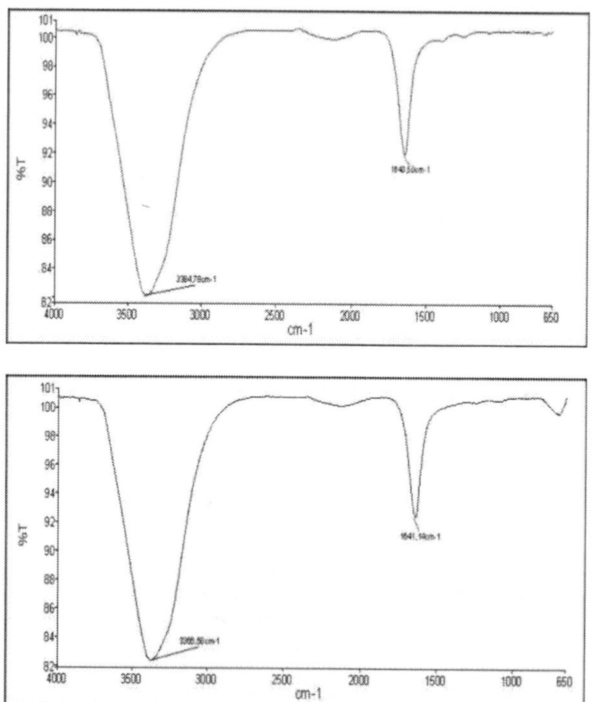

Figure 5. The FTIR spectra of FeNP synthesis using (a) UAE-F and (b) UAE-L.

The representative peaks in FTIR spectra of FeNP for fruit, leaves, twig and bark have similar characteristic stretching vibration as shown in Figure 5 and Figure 6. The band at same range of absorption where the functional group was corresponded indicated the involvement of amine (N-H) at band 3366, 3384, 3368 and 3383 cm^{-1}. There may have possibility that amine group played an important role in the reduction of metal ion. Besides, band 1640, 1641 and 1644 cm^{-1} indicate the presence of alkene group (C=C) stretching frequency that may responsible in reduction process [7]. It was noted that the spectra shared same type of stretch vibration with variable type of intensity [7].

(a)

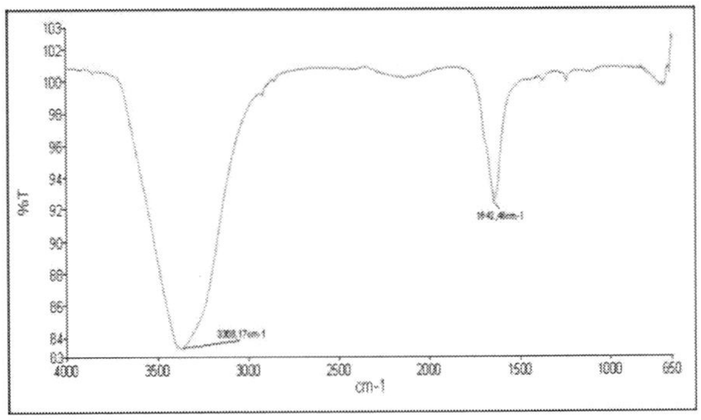

(b)

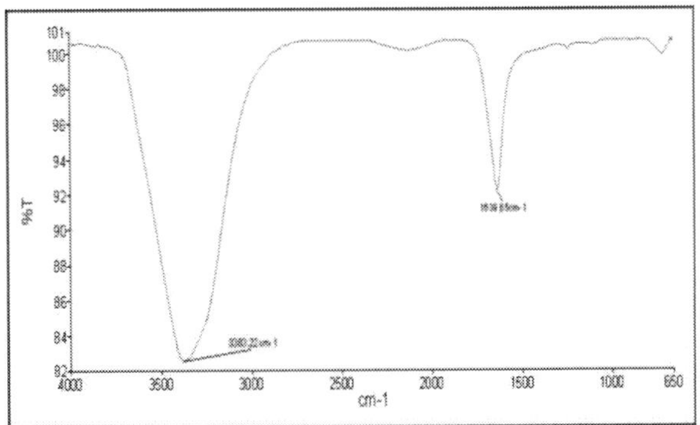

Figure 6. The FTIR spectra of FeNP synthesis using (a) UAE-T and (b) UAE-B.

5. Conclusion

The green approach of preparation FeNP are able to be produced by using different parts of *A. bilimbi*. The *A. bilimbi* is very suitable as natural reducing agent to produce FeNP . CM and UAE technique were carried out to extract A. bilimbi plant parts which are fruit, leaves, twig and bark. Water was used as the solvent. The antioxidant activity were examined by using total phenolic content (TPC) assay and DPPH assay showed fruit has the highest antioxidant activity. Based on the two techniques, it showed UAE gave highest result for antioxidant activity. UAE were then chosen to synthesize FeNP. FeNP that formed from ferric chloride and plant extract were analysed by using UV-Vis Spectrophotometer and FTIR spectra analysis. The result showed that the existance of FeNP and possibility of functional group that were responsible in the reduction process of FeNP.

Acknowledgement

Authors would like to thank Universiti Malaysia Perlis, School of Bioprocess Engineering, Universiti Malaysia Perlis and Institute of Nano Electronic Engineering, Universiti Malaysia Perlis for the support of laboratory works. This study was funded by Universiti Malaysia Perlis under STG 9001-00554.

References

[1] Balamurughan, M. G., Mohanraj, S., Kodhaiyolli, S & Pugalenthi, V. (2014). Ocimun sanctum Leaf Extract Mediated Green Synthesis of Iron Oxide Nanoparticles. *Journal of Spectroscopic and Microscopic Studies, 4*, 201-204.

[2] Dangat, B., Shinde, A., Jagtap, D., Desai, V., Shinde, P. & Gurav, R. (2013). Mineral Analysis of Averrhoa Bilimbi L. a Potential Fruit. *International Journal of Pharmacy and Pharmaceutical Science Research, 3*,136-139.

[3] Guo, L., Huang, Q., Li, X. & Yang, S. (2001). Iron nanoparticles: Synthesis and Application in Surface Enhanced Raman Scattering and Electrocatalysis. *Journal of Physical Chemistry Chemical Physics, 3*, 1661-1665.

[4] Imtiyaz, H., Singh, N., Singh, A., Himani, S. (2016). Green Synthesis of Nanoparticles and its Potential Application. *Journal of Biotechnology, 13,* 2638-2650.

[5] Kumar, K., Gousia, S. & Anupama. (2013). Review on Phytochemical Constituents and Biological assays of Averrhoa bilimbi Linn. *Journal of Pharmacy and Pharmaceutical Science Research, 3*, 136-139.

[6] Luo, S., Huang, Q., Li, X., Weng, Z., Chen, M., Megharaj, R. & Naidu, R. (2014). Green Synthesis of Iron Nanoparticles by Various Tea Extracts: Comparative Study of The Reactivity Spectrochim. *Journal of A Molecule Biomolecule Spectro, 130*, 295-301.

[7] Mahdavi, M., Namvar, F., Ahmad, M., B., & Mohamad, R. (2013). Green Biosynthesis and Characterization of Magnetic Iron Oxide (Fe3O4) Nanoparticles using Seaweed Sargassum multicum Aqueous Extract. *Journal of Molecules, 18*, 5954-5964.

[8] Makarov, V., Makarova, S., S., Love, A., J., Sinitsyna, O., V., Dudnik, A., O. & Yaminsky. (2014). Biosynthesis of Stable Iron Oxide Nanoparticles in Aqueous Extracts of Hodeum vulgae and Rumex acetosa Plants. *Journal of Langmuir, 30*, 5982-5988.

[9] Marinova, G. & Batchvarov, V. (2011). Evaluation of the Methods for Determination of the Free Radical Scavenging Activity By DPPH. *Journal of Agriculture Science, 17*, 11-24.

[10] Maynard, A., D. (2007). Overview of Nanotechnology Toxicological Issues and Environmental Safety. *Journal of Springer, 3*, 1-14.

[11] Mittal, A., K., Cristi, Y. & Banerjee, U., C. (2013). Synthesis of Metallic Nanoparticles using Plant Extracts. *Journal of Biotechnology Advertisement*, 31, 346-356.

[12] Njagi, E., C., Huang, H. & Stafford, L.(2011). Biosynthesis of Iron and Silver Nanoparticles at Room Temperature using Aqueous Sorghum bran Extracts. *Journal of Langmuir, 27*, 264-271.

[13] Nuxoll, E., E., T., Shimotori, W., A., Arnold, E., L. & Cussler. (2003). Iron Nanoparticles in Reactive Environmental Barriers. Presentation at the AICHEM Annual Meeting, November 20, 2003.

[14] Shah, S., Dasgupta, S., Chakraborty, M., Vadakkekkara, R. & Hajoori, M. (2014). Green Synthesis of Iron Nanoparticles using Plant Extracts. *Journal of International Biological Pharmacy Research, 5,* 549-552.

[15] Siddiqi, K., S. & Husen, A. (2016). Fabrication oF Metal and Metal Oxide Nanoparticles by Algae and Their Toxic Effects. *Journal of Nanotechnology Research, 11*, 363.

[16]. Sreeja Narayanan, Binulal N. Sathy, Ullas Mony, Manzoor Koyakutty, Shantikumar V. Nair and Deepthy Menon (2012). Biocompatible magnetite/Gold Nanohybrid Crast Agents via Green Chemistry for MRI and CT Bioimaging. *Applied Materials and Interfaces*, 4, 251-260.

[17]. Zulhaimi, H. I., Rosli, I. R., Kasim, K.F., Akmal, H. M., Nuradibah, M. A., Sam, S. T. (2017). A Comparative Study of Averrhoa bilimbi Linn Extraction Method. AIP Conference Proceedings 1885, 020178 (2017).

[18]. K. Sathya, R. Saravanathamizhan and G. Baskar (2017). Ultrasound Assisted Phytosynthesis of Iron Oxide Nanoparticle.Ultrasound-Sonochemistry, 39, 446-451.

IOP Conf. Series: Materials Science and Engineering **318** (2018) 012013 doi:10.1088/1757-899X/318/1/012013

Supersaturation Control using Analytical Crystal Size Distribution Estimator for Temperature Dependent in Nucleation and Crystal Growth Phenomena

Zakirah Mohd Zahari, Siti Zubaidah Adnan, Ramesh Kanthasamy, Suriyati Saleh, and Noor Asma Fazli Abdul Samad

Faculty of Chemical & Natural Resources Engineering, Universiti Malaysia Pahang, Lebuhraya Tun Razak, 26300 Kuantan, Pahang, Malaysia

E-mail: asmafazli@ump.edu.my

Abstract. The specification of the crystal product is usually given in terms of crystal size distribution (CSD). To this end, optimal cooling strategy is necessary to achieve the CSD. The direct design control involving analytical CSD estimator is one of the approaches that can be used to generate the set-point. However, the effects of temperature on the crystal growth rate are neglected in the estimator. Thus, the temperature dependence on the crystal growth rate needs to be considered in order to provide an accurate set-point. The objective of this work is to extend the analytical CSD estimator where Arrhenius expression is employed to cover the effects of temperature on the growth rate. The application of this work is demonstrated through a potassium sulphate crystallisation process. Based on specified target CSD, the extended estimator is capable of generating the required set-point where a proposed controller successfully maintained the operation at the set-point to achieve the target CSD. Comparison with other cooling strategies shows a reduction up to 18.2% of the total number of undesirable crystals generated from secondary nucleation using linear cooling strategy is achieved.

1. Introduction

In crystallisation processes, the need to achieve the desired crystal products, particularly crystal size distribution (CSD) and to consistently improve the product quality requires an understanding of the fundamentals of crystallisation processes [1-2]. The fundamental driving force of the crystallisation process is supersaturation, where the most common method to achieve supersaturation is by cooling operation. During the cooling process, it is also essential to maintain operation within the metastable zone which is bounded by the saturation and metastable concentrations. This is due to the fact that if the crystallisation operation exceeds the metastable concentration then, the high supersaturation is expected which results in excessive nucleation. The excessive nucleation is undesirable in the crystallisation because the new crystals produced usually are considered fine particles at relatively low size. The production of these new crystals also affects the downstream process, such as fouling problem [2]. Therefore, it is necessary to generate necessary set-point for crystallisation operation.

Two approaches could be considered for generating set-point, namely the model-based design approach and direct design control approach [3]. In the model-based design approach, the set-point profile is obtained using dynamic optimisation, where it requires a detailed first principle model with accurate model parameters, such as nucleation and growth kinetics. Thus, these approaches depend on the model accuracy which may not be possible to obtain. In direct design control, the optimal set-point is obtained from analytical CSD estimator and subsequently a feedback control system can be used to maintain the operation at its set-point. This approach does not require a detailed first principle model and close to optimal set-point can be obtained relatively fast [2].

The analytical CSD estimator originally developed by Nagy and Aamir [3] is able to generate the set-point that consists of both the supersaturation and the total batch time. It is computationally efficient and can be applied for size independent or size dependent crystal growth rates [3-4]. However the crystal growth kinetic neglects the dependence of temperature and thus the parameters are assumed constant. In the pharmaceutical crystallisation, the crystal growth kinetic is dependent on the temperature of crystallisation operation and thus the kinetic must be extended. Therefore the objective of this work is to extend the current analytical CSD estimator to cover the temperature dependence crystal growth rate. The application of this extended analytical CSD estimator is highlighted using potassium sulphate crystallisation where the set-point needed to achieve the desired target CSD is generated using the extended estimator. The performance of closed-loop simulation is also evaluated in terms of its ability to maintain the operation at the generated set-point and to achieve the target CSD. In addition, the performance comparison with other cooling strategies is also demonstrated.

2. Analytical CSD Estimator

Analytical CSD estimator has been developed based on the assumptions of constant supersaturation throughout the entire batch operation and based on the crystal growth dominant phenomena [2-4]. It consists of two main equations, firstly being the final CSD, followed by final characteristic length. Due to the size independence crystal growth, the shape of CSD is not changed but only the size of CSD will be increased. This estimator is applied to generate a supersaturation set-point that yields a target CSD on the condition the initial seed distribution and growth rate kinetics of the crystallisation are specified. However, the crystal growth kinetic neglect the dependence of temperature and thus the parameters are assumed constant. In this study, the Arrhenius expression is included in the crystal growth kinetic for the temperature dependent cases. The analytical CSD estimator model equations for temperature independent and temperature dependent crystal growths are shown in table 1.

Table 1. Analytical CSD estimators for the case of temperature independence/dependence crystal growths.

Characteristic	Final CSD	Final Characteristic Length
Crystal growth temperature independence $G_x = k_{gx}S^{gx}$	$f_{n,i} = f_{n0,i}$	$L_{x,i} = L_{x0,i} + k_{gx}S^{gx}t_c$
Crystal growth temperature dependence $G_x = k_{gx0}\exp(-E_g/RT)S^{gx}$	$f_{n,i} = f_{n0,i}$	$L_{x,i} = L_{x0,i} + k_{gx0}\exp(-E_g/RT)S^{gx}t_c$

The set-point candidates can be generated from the analytical CSD estimators in table 1 and the optimal set-point is then obtained by solving the model-based optimisation problem in Eqs. (1) to (2). The objective function is to minimise the sum of squares of relative errors between the predicted CSD obtained from the analytical CSD estimator and the desired target CSD.

Minimise
$$F_{obj} = \sum_{i=1}^{N}\left(\frac{f_{n,i} - f_{n,i,target}}{f_{n,i,target}}\right)^2 \quad (1)$$

Subject to:
$$S_{sp,\min} \le S_{sp} \le S_{sp,\max};\ t_{c,\min} \le t_c \le t_{c,\max};\ c_f \le c_{f,batch} \quad (2)$$

Where N is the number of discretisation points, $f_{n,i}$ is the predicted CSD that is obtained from the analytical CSD estimator and $f_{n,i\ target}$ is the desired target CSD, $c_{f,batch}$ is the expected solute concentration at the end of the batch and $c_{f,batch}$ represents the maximum acceptable solute concentration at the end of the batch to achieve the required yield [2].

3. Model Applications: Potassium Sulphate Crystallisation Case Study

3.1 Modelling of a Batch Crystallisation Process

The mathematic model of potassium sulphate crystallisation process is generated from the generic multi-dimensional model-based framework [5]. The population balance equation (PBE) is obtained based on these assumptions; the volume is constant, no effects of agglomeration and breakage are considered. The PBE obtained is solved using the method of classes [5] as shown in Eqs. (3)–(5).

For $i = 1$;
$$\frac{dN_1}{dt} + \frac{G_x}{2\Delta Cl} N_1 = B_{nuc} \tag{3}$$

For $1 < i < n$;
$$\frac{dN_i}{dt} + \frac{G_x}{2\Delta Cl} N_i - \frac{G_x}{2\Delta Cl} N_{i-1} = 0 \tag{4}$$

For $i = n$;
$$\frac{dN_n}{dt} + \frac{G_x}{2\Delta Cl} N_n - \frac{G_x}{2\Delta Cl} N_{n-1} = 0 \tag{5}$$

The overall mass balance for potassium sulphate concentration in the solution is shown below:

$$\frac{dc}{dt} = -\frac{\rho_c k_v V}{m_w}\left(\sum_{i=1}^{i=N} S_{xi}^3 \frac{dN_i}{dt}\right) \tag{6}$$

The energy balance used for this operation is shown in Eq. (7).

$$\rho V c_p \frac{dT}{dt} = -\Delta H_c \rho_c k_v V\left(\sum_{i=1}^{i=N} S_{xi}^3 \frac{dN_i}{dt}\right) - U_1 A_1 (T - T_w) \tag{7}$$

The saturation and metastable concentrations for potassium sulphate are generated based on the following polynomial expressions as the function of the solution temperature [6].

$$c_{sat} = 6.29\times10^{-2} + 2.46\times10^{-3}T - 7.14\times10^{-6}T^2 \tag{8}$$

$$c_{met} = 7.76\times10^{-2} + 2.46\times10^{-3}T - 8.10\times10^{-6}T^2 \tag{9}$$

The temperature dependence nucleation and crystal growth rates are shown in Eqs. (10)–(11):

$$B_{nuc} = k_b \exp(-E_b / RT) S^b \tag{10}$$

$$G_x = k_g \exp(-E_g / RT) S^g \tag{11}$$

Where the relative supersaturation (S) is applied as $S = (c - c_{sat})/c_{sat}$. The initial seed distribution of the seeded batch crystalliser is assumed to be a parabolic distribution as shown in Eqs. (12)–(13).

For 250 μm ≤ L ≤ 300 μm:
$$f(L,0) = 0.032(300 - L)(L - 250) \tag{12}$$

For L < 250 μm, L > 300 μm:
$$f(L,0) = 0 \tag{13}$$

The mathematical model for the potassium sulphate crystallisation process is developed in MATLAB 2014a software and is solved using the "ode15s" solver. Table 2 shows the parameter values used to solve the mathematical model. The open-loop simulation is validated with literature data [6-7] using the same operating conditions and parameters. Although the mathematical model used in this study is similar as published in the literature [6-7], in this work the PBE is solved using the method of classes.

Table 2. Parameter values for the seeded batch cooling crystalliser [6-7]

Parameters	Values	Units
Nucleation order, b	1.45	-
Kinetic coefficient for nucleation, k_b	285.01	$\mu m^3/s$
Growth order, g	1.5	-
Kinetic coefficient for crystal growth, k_g	1.44 x 10^8	$\mu m/s$
Arrhenius equation for nucleation, E_b / R	7517	K
Arrhenius equation for crystal growth, E_g / R	4859	K

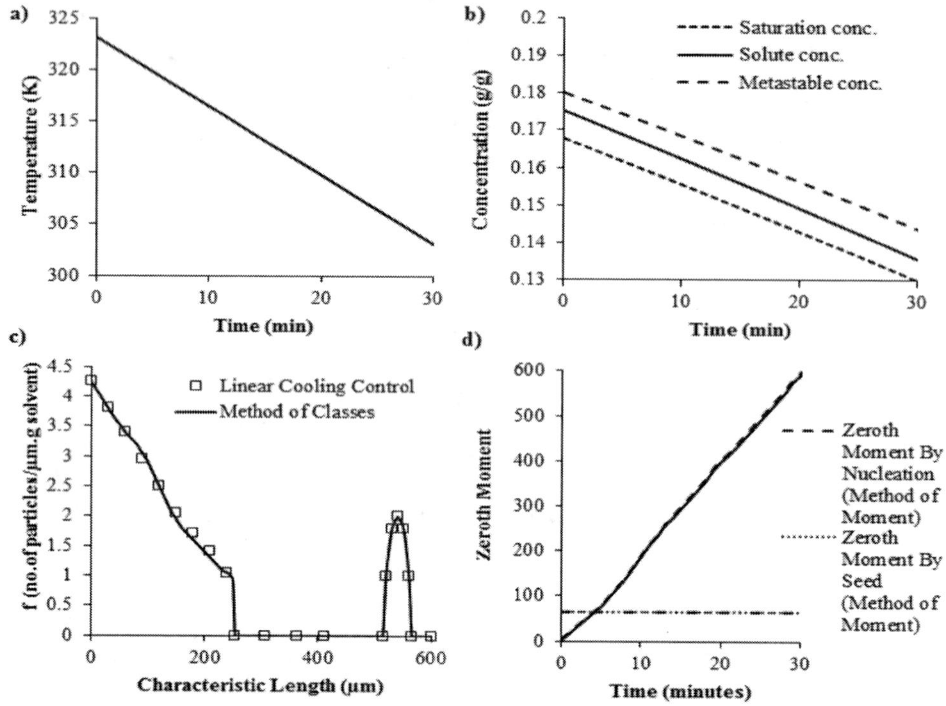

Figure 1. Open-loop simulation for a) Temperature profiles, b) Concentration profiles, c) CSD comparison, d) Zeroth moment comparison

Based on the simulation results, figure 1a) shows the temperature is cooled down linearly from 323.15 K to 303.15 K. The potassium sulphate concentration in figure 1b) also follows the same trends where the concentration initially at 0.1742 g/g and is dropped steadily until 0.1372 g/g at the end of operation. Figure 1c) shows the final CSD obtained from method of classes is in good agreement with literature data, indicating the method of classes is indeed reliable. As shown in figure 1c), there are two peaks existing where the peak on the right is the crystals grown from initial seed and the other peak indicates the crystals formed by secondary nucleation. Although the seed of CSD has been grown from mean characteristic length of 275 µm to 512 µm, but the number of crystals produced from secondary nucleation is relatively high. The number of crystals produced from secondary nucleation is able to be determined by method of classes considering there is a clear gap between those peaks as shown in figure 1d). The total number of crystal seeds remains constant at 70 until the end of operation. This is due to no agglomeration and breakage is considered and thus the total number of crystal seeds is not changing. Meanwhile the total number of crystals formed from secondary nucleation is approximately 600 indicating the crystallisation process is also dominated by nucleation phenomena. Nevertheless, the model has been tested by comparing with the results of Shi et al. [6] and a good agreement has been achieved between both methods in terms of the final CSD.

3.2 Direct Design Control

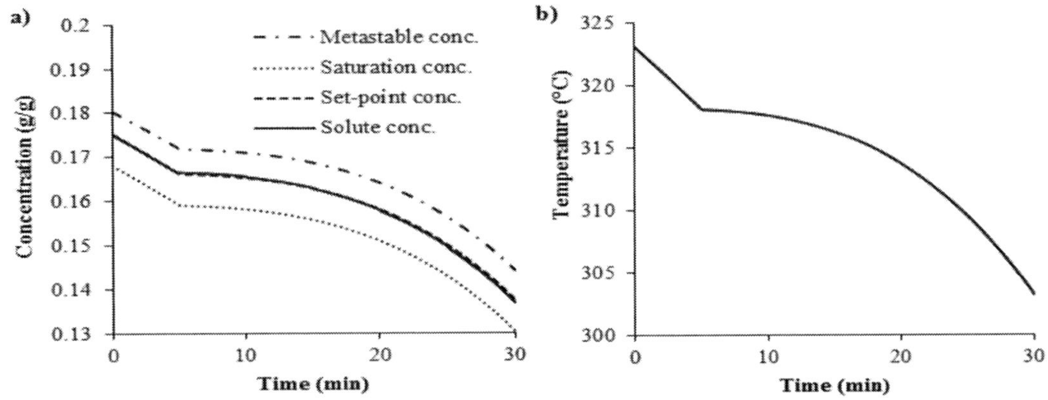

Figure 2. Closed-loop simulation for a) Concentration profiles, b) Temperature profiles

Based on the open-loop simulation, it is clearly shown that the linear cooling profile causes a production of a large number of crystals at relatively low characteristic length, which is undesirable in crystallisation process. Therefore, the direct design control using analytical CSD estimator is employed in order to reduce the total number of crystals produced from secondary nucleation while maintaining the desired final CSD. In this study, the final CSD obtained in figure 1c) is chosen as the target CSD. The model-based optimisation problem in Eqs. (1) to (2) is solved using a sequential quadratic programming (SQP) in order to obtain the optimal set-point. The optimal set-point consisting of the supersaturation set-point of 0.042 g/g that is to be maintained and the total crystallisation time of 30 minutes are obtained. A closed-loop simulation is then performed, where a proportional-integral (PI) controller has been employed in order to maintain the potassium sulphate concentration at the generated optimal set-point. Based on the closed-loop simulation results, it can be concluded that the potassium sulphate concentration initially started at 0.1742 g/g was successfully maintained at the set-point using the PI controller in figure 2a). Approximately, 0.1367 g/g of potassium sulphate concentration remains by the end of the operation. Figure 2b) shows the temperature initially decreased linearly from 323.15 K to 318.15 K and then further decreased steadily to 303.15 K at 30 minutes.

The evolution of CSD is shown in figure 3. Figure 3a) shows the seed of CSD at the start of the crystallisation operation where the aims to achieve the target CSD. At time of 20 minutes, the seed of CSD has been grown from initially mean characteristic length of 275 μm to 415 μm in figure 3b). Due to the sharp increase of temperature profiles in figure 2b), the potassium sulphate concentration is also decreased where the solute has been transferred into the seeds of crystal which explains the growth of crystal seeds. Meanwhile the nucleation effect is also dominant which explains the increment of the secondary peak from time frame of 0 to 30 minutes. Figure 3c) shows the CSD obtained from simulation is very close to the desired target at the final time of 30 minutes. This indicates the generated set-point is reliable for achieving the target CSD. The final CSD obtained is also compared with different cooling strategies; the linear cooling strategy and dynamic optimisation strategy. Figure 2d) shows the final CSD obtained is approximately similar except the secondary peak produced by the secondary nucleation. Among these three different cooling strategies, the optimal cooling strategy from direct design control shows the secondary peak has been reduced up to 18.2% compare to the peak produced from linear cooling strategy but approximately similar with dynamic optimisation. In this case, the lower secondary peak is preferred due to the fact that less number of new crystals has been produced. However, the CSD obtained from dynamic optimisation only gives the optimal cooling profile without the implementation of the feedback control and thus the CSD might differ. Meanwhile the direct design control considers both implementation covering the optimal cooling profile and closed-loop implementation involving PI control indicating its practicality.

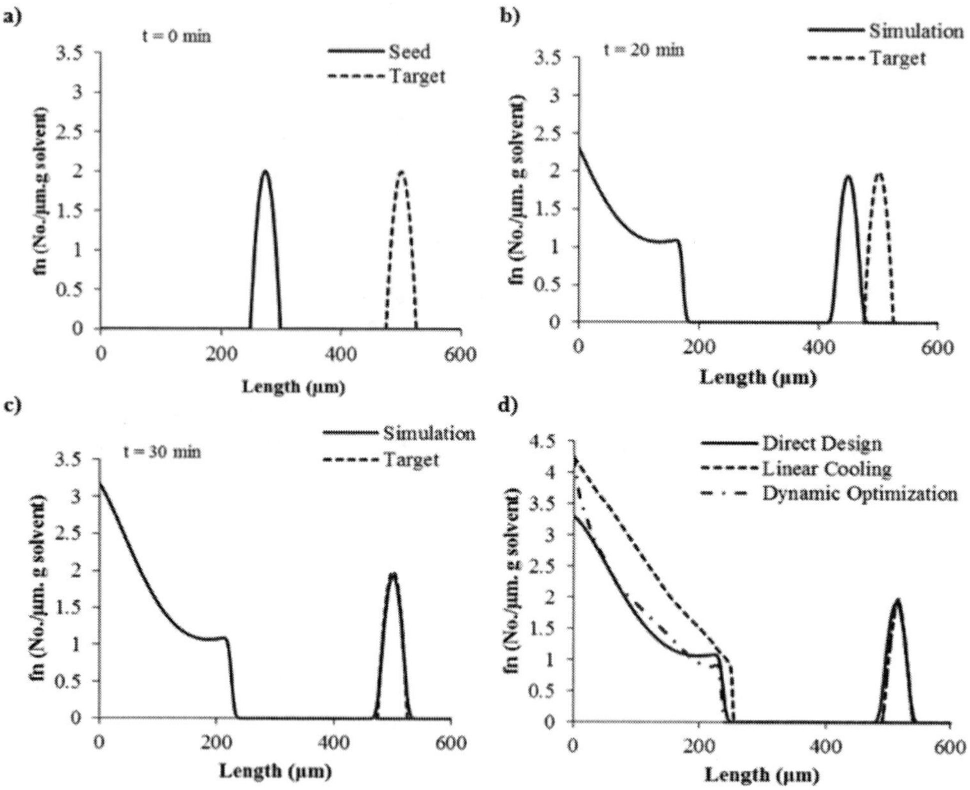

Figure 3. Evolution of CSD for a) At the start of the process, b) At time of 20 minutes, c) At the final time, d) Comparison using different cooling strategies

4. Conclusions

The application of extended analytical CSD estimator for achieving the desired target CSD is highlighted through potassium sulphate crystallisation. The extended estimator is covering the effects of temperature on the crystal growth rate. The open-loop simulation has been performed for potassium sulphate crystallisation and a good agreement has been obtained with literature data. The final CSD obtained in open-loop simulation is then used as the target CSD and the optimal set-point consisting of supersaturation set-point of 0.042 and 30 minutes crystallisation time has been generated from extended estimator. In the closed-loop simulation, the PI controller is successfully maintaining the operation at its set-point which results in achieving the desired target CSD. Comparison with other cooling strategies also indicated the optimal cooling strategy using direct design control manages to reduce the total number of crystals from secondary nucleation compared to linear cooling strategy.

References

[1] Vetter T, Burcham CL and Doherty MF 2014 *Chem. Eng. Sci.* **106** 167–180
[2] Samad NAFA, Sin G, Gernaey KV and Gani R 2013 *Comput. Chem. Eng.* **54** 8–23
[3] Nagy ZK and Aamir E 2012 *Chem. Eng. Sci.* **84** 656–670
[4] Zahari ZM, Saleh S and Samad NAFA 2017 *Chem. Eng. Trans.* **56** 205–210
[5] Samad NAFA, Singh R, Sin G, Gernaey KV and Gani R 2011 *Comput. Chem. Eng.* **35 (5)** 828–843
[6] Shi D, El-Farra NH, Li M, Mhaskar P and Christofides PD 2006 *Chem. Eng. Sci.* **61 (1)** 268–281
[7] Paengjuntuek W, Kittisupakorn P and Arpornwichanop A 2008 *J. Chin. Inst. Chem. Eng.* **39** 249–256

IOP Conf. Series: Materials Science and Engineering **318** (2018) 012014 doi:10.1088/1757-899X/318/1/012014

Anhydrous Weight Loss Prediction of Meranti Sawdust during Torrefaction using Rousset Model

Nur Hazirah Huda Mohd Harun, Noor Asma Fazli Abdul Samad, and Suriyati Saleh
Faculty of Chemical & Natural Resources Engineering, Universiti Malaysia Pahang, Lebuhraya Tun Razak, 26300 Kuantan, Pahang, Malaysia

E-mail: suriyati@ump.edu.my

Abstract. In torrefaction, the mass loss distribution is evaluated in terms of anhydrous weight loss (AWL). Since temperature gives significant effects on AWL and the behaviour of biomass is highly associated with the AWL, therefore a suitable model for estimating the reaction kinetics is necessary for describing the thermal degradation and predicting the AWL in order to improve its process. In this study, the kinetic parameters of Meranti sawdust are estimated by applying three-parallel reaction models namely the Rousset Model for torrefaction of Meranti sawdust at temperatures of 240°C, 270°C and 300°C. All kinetic parameters are estimated according to the degradation of biomass constituents which are lignin, cellulose and hemicellulose by following the Arrhenius Law. The result shows that AWL estimation using the kinetic parameters predicted from the Rousset model is in good agreement with the experimental result as the R^2 value obtained is 0.99. It shows that the Rousset Model successfully described the degradation of lignin, cellulose and hemicellulose as well as the formation of char, volatile, tar and intermediate compound. Therefore it can be concluded that the Rousset Model is applicable to represent the torrefaction behaviour.

1. Introduction

Malaysia is blessed with huge amounts of biomass which comes from logging, agriculture and forestry sectors. Other than palm oil wastes, saw mill industries also produce a large amount of waste in the form of sawdust. Meranti sawdust is one of the most available sawdust because a huge amount of Meranti is used for producing furniture. One of the attractive options to utilise the available biomass is by converting it to the alternative biofuel or pellets. One of the promising techniques that can be used to convert biomass into valuable biofuel products is by using torrefaction.

Theoretically, torrefaction is a thermal treatment process of biomass in the inert environment at atmospheric pressure. It is a mild pyrolysis process where biomass is heated up to 200°C–300°C in order to improve the properties of the biomass. This process upgrades the fuel properties of biomass by reducing the moisture content, increasing heating value, improving grindability, increasing carbon content as well as changing the characteristic of biomass to become more brittle [1]. In recent years, many studies on the torrefaction process have been conducted, but little attention has been paid to the kinetic parameters study of biomass torrefaction. Most researchers are focusing more on the effects of torrefaction on the physical and chemical property changes of the biomass [1-3]. However, the fundamental knowledge of thermal behaviour and decomposition kinetics of biomass are important to understand the process of torrefaction for further improvement. Here the kinetic study of torrefaction process is crucial in optimising the yield of the desired products. Usually, thermal degradation in torrefaction is best described by anhydrous weight loss (AWL) of the particular biomass which is highly dependent on the composition of the three main constituents of biomass, namely lignin, cellulose and hemicellulose. Typically, biomass is a build-up of 40–50 wt% cellulose, 25–35 wt% hemicellulose and 15–35 wt% lignin [3]. Among the three constituents, hemicellulose decomposes at the lowest temperature within 220°C–315°C. Meanwhile cellulose decomposes at a temperature range of 315°C–400°C, whereas lignin decomposes slowly within the range of 160°C–900°C [2].

Therefore, three-parallel reactions using the Rousset model can be employed to predict the AWL of Meranti sawdust torrefaction. The Rousset model is used to describe the decomposition of lignin, cellulose and hemicellulose [4]. However, the current Rousset model has been developed by considering the isothermal phase of torrefaction only. Usually, considerable mass loss is produced during heating process before it reaches the desired torrefaction temperature. In order to obtain an accurate result that can be used to represent the real process, both heating and isothermal phases of torrefaction are taken into account when deriving the kinetic parameters. Based on both phases, the kinetic parameters for the Rousset model were estimated and fitted against the residual mass obtained from thermogravimetric analysis (TGA). The applicability of these models to predict the residual mass and the mass yield for each constituent were also discussed.

2. Methodology

2.1. Thermogravimetric Analysis

The sample used for the torrefaction process was Meranti sawdust, a type of logging residue which was collected from a saw mill at Gambang, Pahang, Malaysia. After the collection, the samples were oven-dried at 105°C for 4 hours and were grinded and sieved. Only particles in the range of 500 μm to 1 mm were collected and used in this work. Torrefaction of Meranti sawdust was carried out using a TGA/DSC 1 Mettler Toledo analyser with nitrogen flow of 30 ml/min at three different temperatures: 240°C, 270°C and 300°C at a heating rate of 10°C/min. Prior to the analysis, 2 mg of samples were weighed and pre-heated from room temperature to 105°C at 10°C/min for 4 minutes in order to remove all moisture trapped in the samples. Later on, the samples were heated to the desired temperature. Once the desired temperature is reached, the temperature was maintained for 90 minutes. AWL is then calculated by using the mass loss data obtained from the TGA experiment.

2.2. Mathematical Model for AWL

The mathematical model for AWL is useful to describe the mass loss behaviour of Meranti sawdust during torrefaction. In this work, the Rousset Model was chosen to determine the decomposition kinetic parameters of Meranti torrefaction in order to evaluate the decomposition rate of Meranti sawdust during the process. The Rousset Model is a three–parallel steps kinetic model which demonstrates the wood degradation as three independently decomposition kinetic mechanisms [4]. Degradation of wood is related to decomposition of its constituents, which are lignin, cellulose and hemicellulose. Table 1 shows the mathematical modelling using the Rousset model. There are two sets of equations for heating phase and isothermal phase. As shown in table 1, there are 3 main reactions for each constituent. The first reaction involves the decomposition of lignin into char and volatile. Cellulose is transformed into char, volatile and tar components in the second reaction. Meanwhile hemicellulose is decomposed through two reactions in series where firstly intermediate component and volatile are produced from hemicellulose, and subsequently this intermediate component is decomposed into char and volatile. All equations are assuming to follow first order reaction where k is the rate constant for each reaction (1/s) and β = heating rate (°C/s) in the heating phase only.

Table 1. Mathematical modelling for AWL

Reactions	Heating Phase	Isothermal Phase
$Lignin[L] \rightarrow Char1[C1] + Volatile1[V1]$ $Cellulose[CL] \rightarrow Tar[T]$ $Cellulose[CL] \rightarrow Char2[C2] + Volatile2[V2]$ $Hemicellulose[H] \rightarrow Intermediate[B]$ $Hemicellulose[H] \rightarrow Volatile3[V3]$ $Intermediate[B] \rightarrow Char[C]$ $Intermediate[B] \rightarrow Volatile4[V4]$	$\frac{d[L]}{dt} = \frac{1}{\beta} \cdot \{-k_L[L]\}$ $\frac{d[C1]}{dt} = \frac{1}{\beta} \cdot \left\{\frac{1}{2} \cdot k_L[L]\right\}$ $\frac{d[CL]}{dt} = \frac{1}{\beta} \cdot \{-(k_T + k_{CC})[CL]\}$ $\frac{d[T]}{dt} = \frac{1}{\beta} \cdot \{k_T[CT]\}$ $\frac{d[C2]}{dt} = \frac{1}{\beta} \cdot \left\{\frac{1}{2} \cdot k_{CC}[CL]\right\}$ $\frac{d[H]}{dt} = \frac{1}{\beta} \cdot \{-(k_{V3} + k_{HB})[H]\}$ $\frac{d[B]}{dt} = \frac{1}{\beta} \cdot \{k_{HB}H - (k_{C3} + k_{V4})[B]\}$ $\frac{d[C3]}{dt} = \frac{1}{\beta} \cdot \{k_{C3}[B]\}$	$\frac{d[L]}{dt} = -k_L[L]$ $\frac{d[C1]}{dt} = \frac{1}{2} \cdot k_L[L]$ $\frac{d[CL]}{dt} = -(k_T + k_{CC})[CL]$ $\frac{d[T]}{dt} = k_T[CT]$ $\frac{d[C2]}{dt} = \frac{1}{2} \cdot k_{CC}[CL]$ $\frac{d[H]}{dt} = -(k_{V3} + k_{HB})[H]$ $\frac{d[B]}{dt} = k_{HB}H - (k_{C3} + k_{V4})[B]$ $\frac{d[C3]}{dt} = k_{C3}[B]$

3. Results and discussion

Figure 1a) shows the mass loss profiles obtained from TGA for Meranti torrefaction at 240°C. Thermogravimetric (TG) and differential thermogravimetric (DTG) reflect the degradation process that occurs during torrefaction of Meranti sawdust. As described by Bergman et al. [5], torrefaction can be divided into three main stages; non-reactive drying that occurs from 50°C to 150°C, reactive drying where hydrogen and carbon bonds start to break which occurs at 150°C to 200°C and destructive drying (from 200°C to 300°C) in which the chemical composition of biomass are altered and destructed. As shown in figure 1a), the removal of moisture (non-reactive and reactive drying) accounted about 8.04% and the destructive drying accounted around 9.12% of the Meranti decomposition. Based on the established thermal degradation (TDT) temperature of biomass chemical constituents and two TG curves in figure 1a), the Meranti torrefaction at 240°C can be best described by dehydration (up to 105°C) followed by hemicellulose decomposition (220°C–315°C).

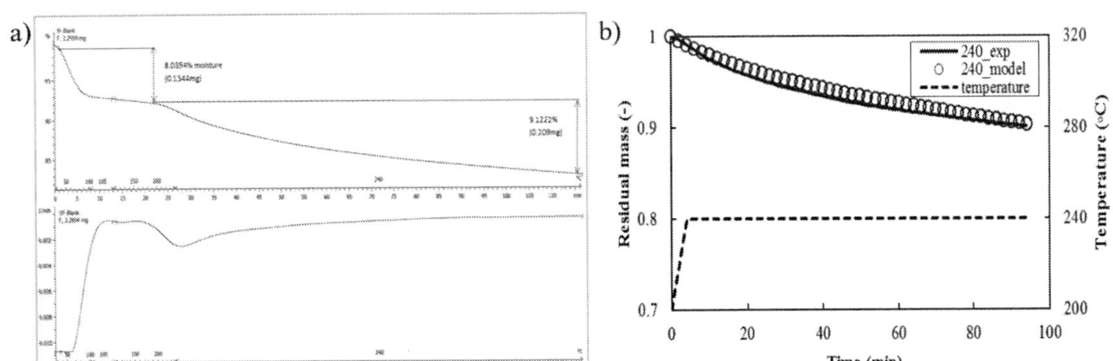

Figure 1. a) Mass loss profiles from TGA and DTG, b) Residual mass of sawdust at 240 °C

Figure 1b) shows the torrefaction process of Meranti sawdust obtained from TGA. It can be divided into two main stages which are heating and isothermal stages. The first stage is the heating stage where sample is heated from drying temperature (200°C) up to the torrefaction temperature (240°C). During this phase, Meranti sawdust sample is heated up at a rate of 10°C/min for 4 minutes until it reached the desired torrefaction temperature. The second stage is the isothermal stage. At this stage, the sample is continuously heated at torrefaction temperature for 90 minutes. Both stages need to be considered in deriving kinetic parameter because there will be weight loss at both stages which significantly contribute to the entire degradation process, especially at higher heating rate and temperature [6-7]. Therefore, two sets of equation are needed to derive the AWL of Meranti torrefaction using the Rousset model as shown in table 1. The kinetic parameters in the mathematical model in table 1 can be estimated by assuming the initial conditions based on the mass loss fraction of the biomass. The initial conditions used for all the three main constituents [8] for Meranti sawdust are used as follows: Lignin = 0.34, Cellulose = 0.42, Hemicellulose = 0.33.

In addition, the initial conditions for other components in the Rousset model are assumed to be zero. The AWL model consists of seven reactions which contribute to seven kinetic rates. As the reactions are assumed to follow the Arrhenius Law ($k = A\ exp\ (-Ea/RT)$), both pre-exponential factor (A) and activation energy (Ea) need to be considered in the parameter estimation. Thus, it gives a total amount of 14 parameters that need to be estimated for the whole model. The pre-exponential factors (A) and activation energies (Ea) for every temperature were calculated by using Arrhenius Law and the iteration was repeated until constant values of both A and Ea were reached. The iteration was done using MATLAB (R2014a) software and '*lsqcurvefit*' routine was selected for nonlinear optimisation due to its ability to minimise the variance between experimental and simulated data in order to find the optimal parameters that best fit the data. The kinetic parameters obtained by fitting the experimental data at temperatures of 240°C, 270°C and 300°C are shown in table 2.

Table 2. Kinetic parameters for Rousset model

Parameters	A (s^{-1})	Ea ($Jmol^{-1}$)
k_L	3.21×10^7	116500
k_T	1.57×10^9	179000
k_{CC}	5.83×10^{15}	240500
k_{V3}	2.31×10^5	91400
k_{HB}	2.14×10^3	66.1
k_{CI}	1.97×10^4	56300
k_{V4}	55.70	52600

Figure 2a) shows the residual mass of Meranti sawdust torrefaction for both experimental and modelled data for 240, 270 and 300°C respectively. Here, the residual mass is obtained based on the total main constituents degradation (lignin, cellulose and hemicellulose). As shown in Figure 2a), the modelled data at all torrefaction temperatures fitted well with the experimental data. This proves the estimated kinetic parameters used in the three-parallel reactions are accurately describing the residual mass from torrefaction process. Residual mass is the amount of sample retained after the torrefaction. A small amount of residual mass indicates a huge amount of mass is loss in the form of volatiles due to the heating effect. At a high temperature (300°C), the residual mass obtained at the end of operation is about 40%, while at a low temperature (240°C) the residual mass obtained is about 70%. This indicates around 60% of mass is loss at a temperature of 300°C but only 30% mass is loss at a temperature of 240°C as char, volatiles, tar and intermediate compounds. The high amount of mass loss at high temperature is due to the destruction of biomass structure caused by extensive heating of torrefaction [9]. Figures 2b) – d) demonstrates the Meranti sawdust decomposition based on the degradation of lignin, cellulose and hemicellulose. Based on the model, lignin was degraded into char and volatiles by following a one step reaction model. As shown in Figure 2b), the composition of lignin at 240°C is decreased to 0.332115 from its initial state which is 0.34. Loss of 0.0007885 of lignin contributed to the equal formation of char and volatile. Meanwhile, cellulose was decomposed into tar, char and volatiles by two separate reactions, following the Broido-Shafizadeh model [10]. In figure 2c), cellulose is slightly decomposed during the torrefaction at 240°C. Initially, its composition is 0.42 and to the end of the process, 0.419988 of it is still retained. So, only 0.000025 of cellulose is decomposed to tar, char and volatile. From the amount, 0.0000125 is tar while the rest is divided in half between char and volatile.

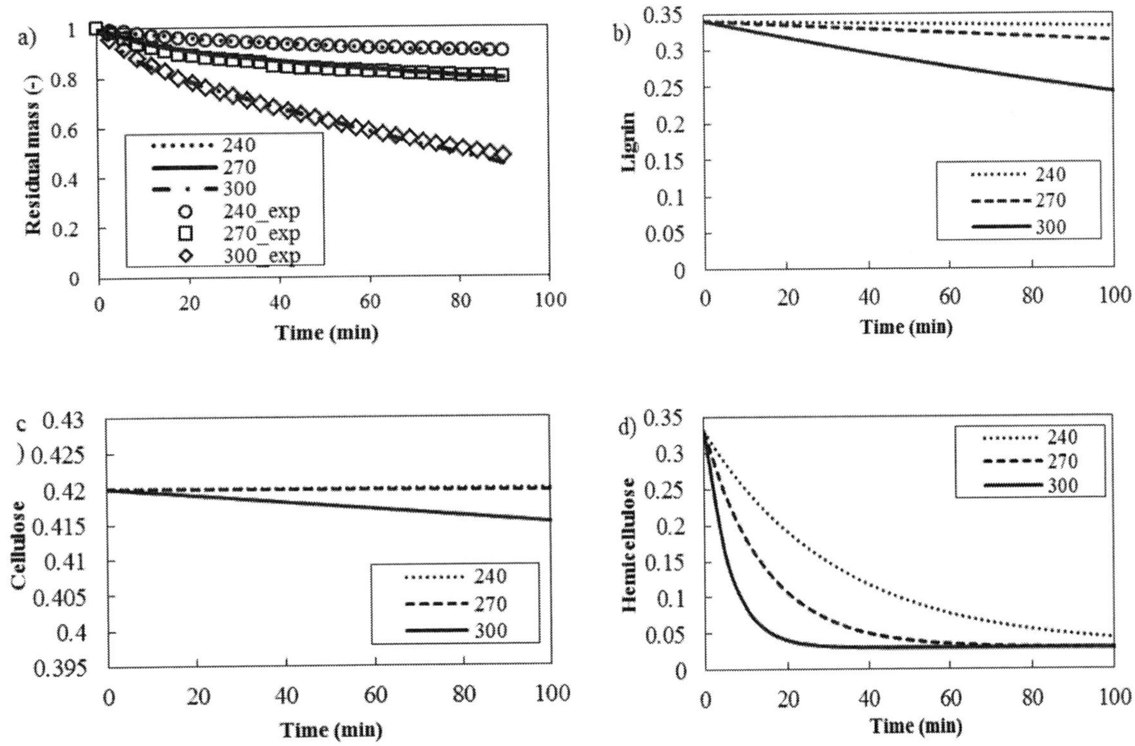

Figure 2. Residual mass and constituents of Meranti sawdust at three different temperatures

Hemicellulose, corresponding to the Di Blasi-Lanzetta model was decomposed into char and volatiles with formation of intermediate compound. In contrast to lignin and cellulose, hemicellulose is totally decomposed at the end of the torrefaction. It was decomposed to five respective components in two steps. The first step is the decomposition of hemicellulose to intermediate and volatile. As shown in figure 2d), from initial composition of 0.33 at 240°C, hemicellulose was continuously decomposed into intermediate compound and volatiles until it diminished at the end of the reaction. Figure 3a) shows the composition of intermediates increased to 0.0326 at the beginning but decreased towards the end of operation. This is due to the fact that decomposition of intermediates led to formation of char and volatiles at the end of the reaction. It is also important to note that char, volatile and tar were also formed throughout the torrefaction process. Char was formed from the decomposition of lignin, cellulose and intermediate compounds while volatiles were released from the decomposition of lignin, cellulose, hemicellulose and intermediate compounds. In addition, tar was solely formed from the first reaction of cellulose decomposition. Among the volatile matter produced, the amount of char and volatile are the highest as shown in figures 3b) and c). Torrefaction temperature of 300°C recorded the highest amount of char and volatile produced which are approximately 0.8 and 0.7 respectively. At temperature of 270°C about 0.4 of char and 0.3 of volatile has been obtained while at the temperature of 240°C, only 0.3 char and 0.15 volatile is produced with small amount of tar and intermediate compounds. Furthermore, the yield of tar is not significant in the Meranti sawdust torrefaction due to the fact that around 0.02 tar is produced at all temperature. The slow decomposition rate of cellulose is found to be a contributing factor for insignificant tar production. Overall, the degradation of main components and formation of constituents show similar trends over the temperatures. However, the rate of decomposition and formation at a temperature of 300°C is higher compared to temperatures of 240°C and 270°C. This trend is supported by the fact that high decomposition rate for Meranti sawdust approximately began at a temperature of 250°C [9].

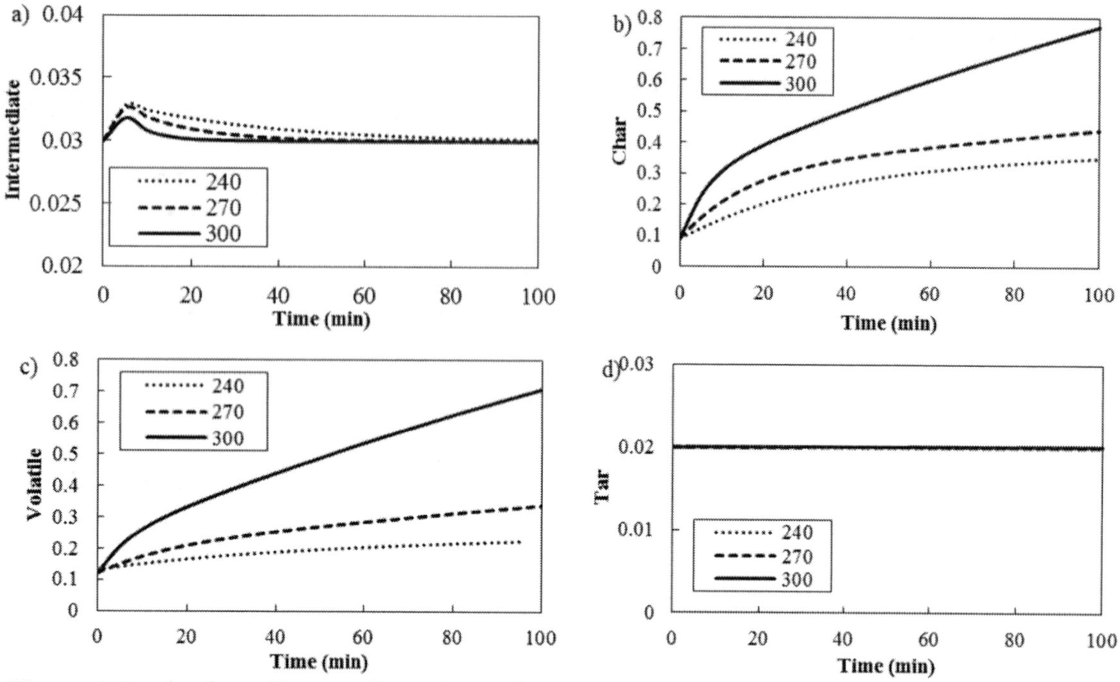

Figure 3. Production of intermediate, char, volatile and tar components from decomposition of lignin, cellulose and hemicellulose

4. Conclusions

In this study, three-parallel reactions, namely the Rousset Model was used to evaluate the kinetic of Meranti sawdust torrefaction. In contrast to other available kinetic models, this model evaluates the torrefaction of biomass with respect to the individual composition of biomass which are lignin, cellulose and hemicellulose. Based on the reactions model, the kinetic parameters for Meranti sawdust are estimated by considering both heating and isothermal period during torrefaction process. Then, the result is validated against the experimental data. The result shows that the mathematical model using the estimated kinetic parameters is accurately describing the AWL of the Meranti sawdust torrefaction between the temperatures 240°C and 300°C as it shows good agreement between experimental and modelled data, which subsequently can be used to predict the behaviour of the individual constituents. In addition, it was also found that hemicellulose is the dominant constituent in the Meranti sawdust torrefaction due to its significant decomposition rate compared to the lignin and cellulose constituents.

References

[1] Saleh S, Hansen BB, Jensen PA and Dam-Johansen K 2013 *Energy & Fuels* **27 (12)**, 7541–7548
[2] Chen WH, Peng J and Bi XT 2015 *Renew. Sustainable Energy Rev.* **44** 847–866
[3] Chew JJ and Doshi V 2011 *Renew. Sustainable Energy Rev.* **15** 4212–4222
[4] Rousset P, Turner I, Donnet A and Perre P 2006 *Ann. For. Sci.* **63** 213–229
[5] Bergman PCA, Boersma AR, Zwart RWH and Kiel JHA 2005 ECN Report ECN-C--05-013
[6] Harun NHHM, Samad NAFA and Saleh S 2017 *Energy Procedia* **105** 744–749
[7] Harun NHHM, Wahid FRAA, Saleh S and Samad NAFA 2017 *Chem Eng Trans.* **56** 1195–1200
[8] Ahmad ZS, Munaim MSA and Said FM 2016 The National Conference for Postgraduate Research, Universiti Malaysia Pahang
[9] Mazlan MAF, Uemura Y, Osman NB and Yusop S 2015 *J. Phys.: Conf. Ser.* **622** 1–7
[10] Bradbury AGW, Sakai Y and Shafizadeh F 1979 *J. Appl. Polym. Sci.* **23 (11)** 3271–3280

Malachite Green Adsorption by Spent Coffee Grounds

Siti Syamimie Atirah Mat[1], Sharifah Zati Hanani Syed Zuber[1], Siti Kartini Enche Ab Rahim[1], Khairunissa Syairah Ahmad Sohaimi[1], Noor Amirah Abdul Halim[1], Nor Fauziah Zainudin[1], Nor Aida Yusoff[1], Nor Munirah Rohaizad[1], Noor Hidayah Ishak[1], Adilah Anuar[1], Mohd Sharizan Md Sarip[1]
[1]Faculty of Engineering Technology, Universiti Malaysia Perlis, Sungai Chuchuh, 02100 Padang Besar, Perlis, Malaysia.

E-mail: sharifahzati@unimap.edu.my

Abstract. In this work, the ability of spent coffee grounds (SCG) as a low-cost adsorbent to remove malachite green (MG) from aqueous solutions was studied. Batch adsorption tests were carried out to observe the effect of various experimental parameters such as contact time, initial concentration of malachite green and adsorbent dosage on the removal of dye. The results obtained show that the percentage of dye removal will decreased with the increased of initial concentration of dye in the range of 50 mg/L to 250 mg/L. Besides, percentage removal of dye was also found to be increased as the contact time increased until it reached equilibrium condition. The results also showed that the adsorbent dosage in range of 0.2 g to 1.0 g is proportional to the percentage removal of malachite green dye. Study on the kinetic adsorption and isotherm adsorption has also been investigated. The adsorption isotherm data were described by Langmuir isotherm with high-correlation coefficients while the experimental data showed the pseudo-second-order kinetics model was the best model for the adsorption of MG by SCG with the coefficients of correlation $R^2 > 0.9978$.

1. Introduction

Dyeing stuff has been popular in many of production industries nowadays. Therefore, the uses of synthetic dyes in worldwide advanced technology such as in the production of textile [1], leather [2], paper [3], printing [4] and food manufacturing industries [5] are very well-known. As a consequences from this, the natural streams and water body turned into a place where large amount of colored wastewater been discarded. Dye waste water can cause water pollution and at the same time it will disrupt the aquatic life as dye are stable to light and oxidation, as well as resistant to aerobic digestion [6]. Malachite green which is one of common dye used in industries that already been discarded into the water bodies such as rivers can affect the aquatic life and cause detrimental effects in liver, kidney, intestines, and gonadotropic cells [7]. Besides that, malachite green is carcinogenic, has mutagenic characteristics and reduced fertility in humans [8]. Hence, there is a need to treat the dye containing water before being disposed to the environment.

There are many types of techniques or processes that can be used to remove dye from aqueous solution such as advance oxidation [9], combining chemical and biochemical process [10], aerobic and anaerobic digestion [11] and adsorption [12]. However, adsorption is the most commonly used process since the cost required is low and it is also an efficient method for the treatment of dye-containing waste water [12]. The most commonly used adsorbent in industry for removal dyes is activated carbon

[7], but the inherently high cost affected it widespread use. Therefore, various low cost waste materials from industry and agriculture such as paper waste sludges [3], broad bean peels [13] and rice hull [14] have been widely investigated as low-cost adsorbent.

Spent coffee grounds (SCG) are one of main residues of the coffee industry. Some of the SCG has been composted and used as animal feed [15] meanwhile another half amount of SCG are burned as wastes [16] which will contribute to the production of carbon dioxide to the atmosphere and then lead to global warming. Hence, another alternative way to reduce the abundant amount of SCG as wastes from coffee industry is by turning it into an adsorbent in dye removal. Moreover, SCG is not economically expensive and easily available adsorbent for the removal of cationic dyes in water treatments [17]. Thus, the purpose of present work is to evaluate the adsorption potential of SCG for the removal of malachite green from the aqueous solution.

2. Experimental Method

2.1 Preparation of Adsorbent

The spent coffee grounds (SCG) were collected from the coffee shop located in Pasir Puteh, Kelantan. The collected spent coffee grounds (SCG) were washed with distilled water to remove any adhering dirt. Then, the SCG had been dried at 60°C for 24 hours in oven [11]. After drying, grinder was used to grind SCG to obtain particle size <250 μm. Next, the product had been stored in air-tight container for further use. No other physical and chemical treatments were applied prior to adsorption experiments.

2.2 Preparation of Stock Solution

One gram of malachite green (MG) purchased from Sigma Aldrich (M) Sdn Bhd, Malaysia was dissolved in one liter of distilled water to obtain the stock solution of 1000 mg/L. Serial dilution were made by diluting the stock solution to the desired concentration.

2.3 Characterization of Adsorbent

The Fourier Transform Infrared Radiation (FT-IR) spectrophotometer model Perkin Elmer Spectrum 65, Perkin Elmer Inc., USA was used to determine the functional group present in SCG, where the spectra of SCG were recorded in the range of 4000-650 cm^{-1}. Besides that, Scanning Electron Microscopy (SEM) was carried out to examine the surface morphology of the adsorbent.

2.4 Batch Adsorption Experiment

The batch experiment was conducted by agitating 0.1 mg of the adsorbent and 100 ml of dye solution (50 – 250 mg/ L) without changing pH at the temperature of 30 °C and rotation speed of 150 rpm in a water-bath shaking incubator for 90 minutes to ensure equilibrium was reached. The concentration of the MG after the equilibrium adsorption was measured by using UV-vis spectrophotometer (Shimadzu, Model UV-1800) at 616 nm. The amount of adsorption at equilibrium, q_e (mg/g) and the dye percentage removal was calculated by using equation 1 and 2 below, respectively:

$$q_e = \frac{(C_0 - C_e)\,V}{m} \tag{1}$$

$$Removal\ percentage = \left(\frac{C_0 - C_e}{C_e}\right) \times 100 \tag{2}$$

where $C_0$and C_e(mg/ L) are the liquid-phase concentration of dye at initial and equilibrium, respectively. V is the volume of solution (L) and m is the mass of adsorbent used (g).

Experiments to investigate the effect of contact time, initial dye concentration and adsorbent dosage on adsorption rate were carried out using the procedure described above except that the contact time, initial MG concentration and SCG dosage were varied from 10 to 120 min, 50 to 250 mg/L and

0.2 to 1.0 g, respectively. All the experiments were performed in duplicate and the average values were reported.

3. Results and Discussions

3.1 Characterization of adsorbent

The FT-IR spectrum of SCG is given in Figure 1 which shows a broad band centered at 3357.95 cm^{-1} which might be attributed to bonded hydroxyl groups (O-H) on the surface of the spent coffee ground [18]. Meanwhile, the peak which in range of 2850 cm^{-1} to 3000 cm^{-1} can be due to the presence of C-H stretching of alkanes group in SCG [19]. The peak at 1746 and 1662 cm^{-1} can be assigned to C=O stretching of carboxyl linkage derived from xanthine derivatives of caffeine [20]. Besides, the strong and sharp peak at 1026 cm^{-1} can be attributed to C-OH stretching of polysaccharide in SCG [19]. From the FT-IR spectrum, it is confirmed that SCG has possible surface functional group which are able to interact with MG dye in adsorption process.

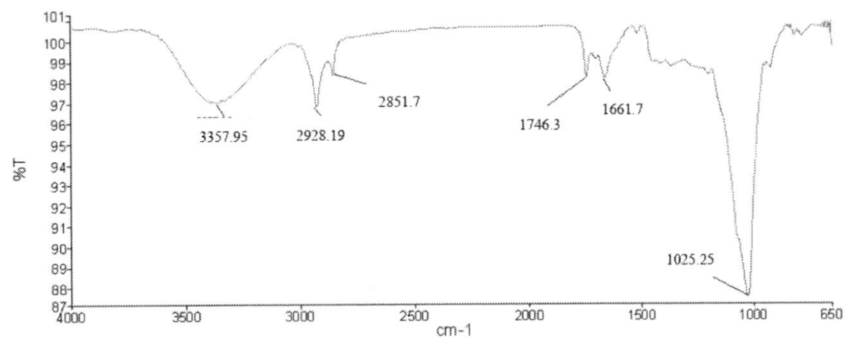

Figure 1. FT-IR Spectrum of SCG

The SEM micrographs of SCG at the magnification of 500x is as shown in Figure 2. The SEM image reveals the surface of SCG is not smooth, but scraggy with cavities of different sizes. These cavities can be characterized as channels onto the surface of untreated coffee materials which are useful for dye adsorption [21].

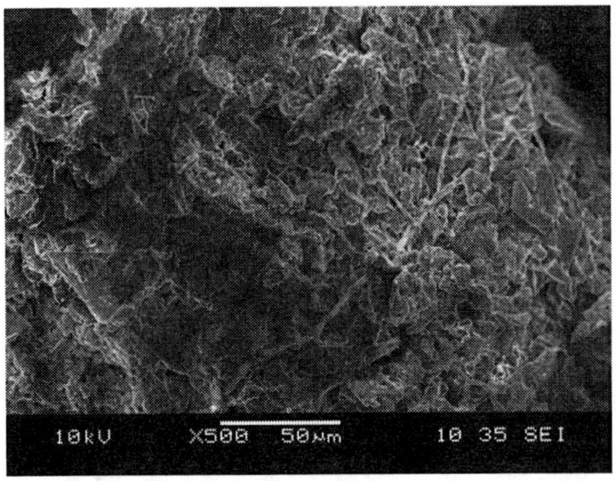

Figure 2. SEM morphology of SCG

3.2 Batch Adsorption Experiment

3.2.1 Effect of Contact Time

As shown in Figure 3, the adsorption increases more rapidly during the initial time. This pattern occurred because during the initial stage, a large number of active sites were available resulting in the quick adsorption of malachite green on the surface of SCG [22]. After that, the adsorption process becomes slower due to gradual occupancy of those active sites on the surface of adsorbent. As the time increase, the tendency of occupying the remaining vacant sites was more difficult caused by the increases of the repulsive forces bulk solution and molecules of dye [22]. The percentage removal of dye also increases as the time increases until it reached the equilibrium. The equilibrium can be assumed to have achieved after 90 minutes, because the final dye concentration does not seem to increase much at that limit. This may be caused by the active site of the adsorbent already reached the limit or saturated. Thus, further adsorption of adsorbate onto the active sites of the SCG surface will be slow and stop when all the surfaces already occupied [10]. Based on the result shown, the percentage removal of dye proportional to contact time but will be constant when the equilibrium achieved.

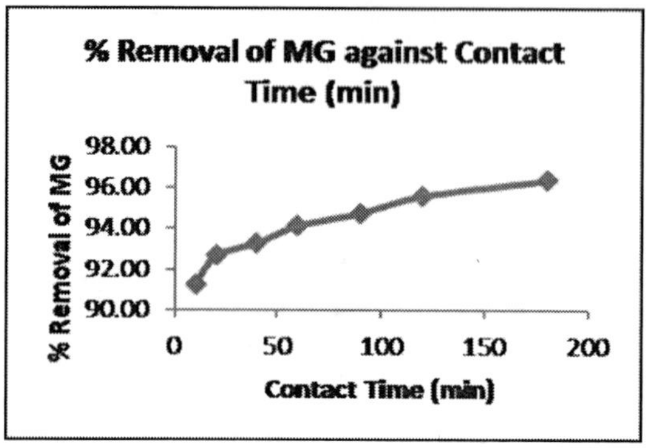

Figure. 3. Effect of contact time on adsorption of dye solution by SCG.

3.2.2 Effect of Initial Dye Concentration

Results obtained from the experiment show that the percentage removal of adsorbate is decreases when the initial concentration of the malachite green dye is increases. This is caused by the surface adsorption sites are limited due to the fixed amount of adsorbent which is the SCG. The increases amount of the adsorbate cannot be supported by the fixed dosage of adsorbent [23]. Based on the Figure 4, the percentage removal are decreases from 99.63% to 98.69% with an increase in initial malachite green concentration from 50 mg/L to 250 mg/L.

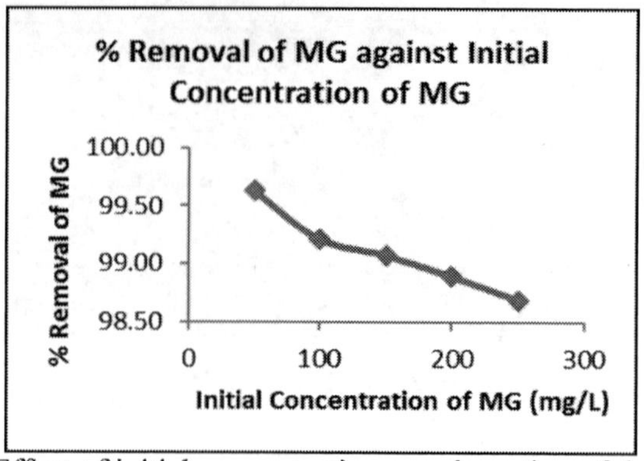

Figure 4. Effect of initial concentration on adsorption of dye by SCG.

3.2.3 Effect of Adsorbent Dosage

The result of this experiment shows that the adsorbent dosage is proportional to the percentage removal of malachite green dye. The increases of surface area resulted from the increases of the adsorbent mass. Thus, the number of active sites will also increases [12]. Figure 5 shows the percentage removal of dye is increases from 77.35% to 94.47% as the dosage of adsorbent increases.

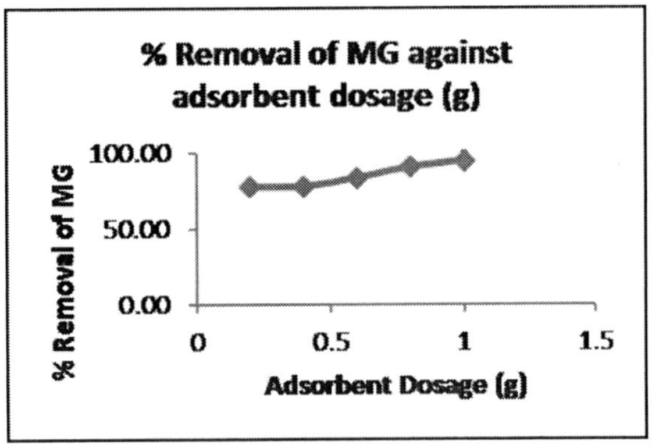

Figure. 5. Effect of adsorbent dosage on adsorption of dye solution by SCG.

3.3 Adsorption Kinetics

Adsorption kinetics related to the process of adsorption such as chemical reaction and mass transfer. Thus, pseudo-first-order model and pseudo-second-order model have been applied on the dye adsorption kinetic studies.

Pseudo-first-order kinetic model is given by:

$$\ln(q_e - q_t) = \ln q_e - K_1 t \qquad (3)$$

where, qe is the amounts of dye adsorbed (mg/g) at equilibrium while qt is the amounts of dye adsorbed (mg/g) at time, t. K1 is the rate constant of pseudo-first order adsorption.
Pseudo-second-order kinetic model is given by:

$$\frac{t}{q_t} = \frac{1}{K_2 \times q_e^2} + \frac{t}{q_e} \qquad (4)$$

where, qe is the amounts of dye adsorbed (mg/g) at equilibrium while qt is the amounts of dye adsorbed (mg/g) at time, t. K_2 is the rate constant of pseudo-second order adsorption.
From the results obtained in Figure 6, the data plotted is obviously shows that the coefficient of determination (R^2) for the pseudo-first-order model is less than the pseudo-second-order model. The values of R^2 for pseudo-second-order model (R^2=0.999) are higher than those of pseudo-first-order model (R^2=0.909). Besides that, the calculated amount of dye adsorbed, q_e,cal of pseudo-second-order model is more closer to the experimental amount of dye adsorbed, q_e,exp compare to the pseudo-first-order model. In view of these data, it is obvious that the experimental results are well represented by the pseudo-second-order model.

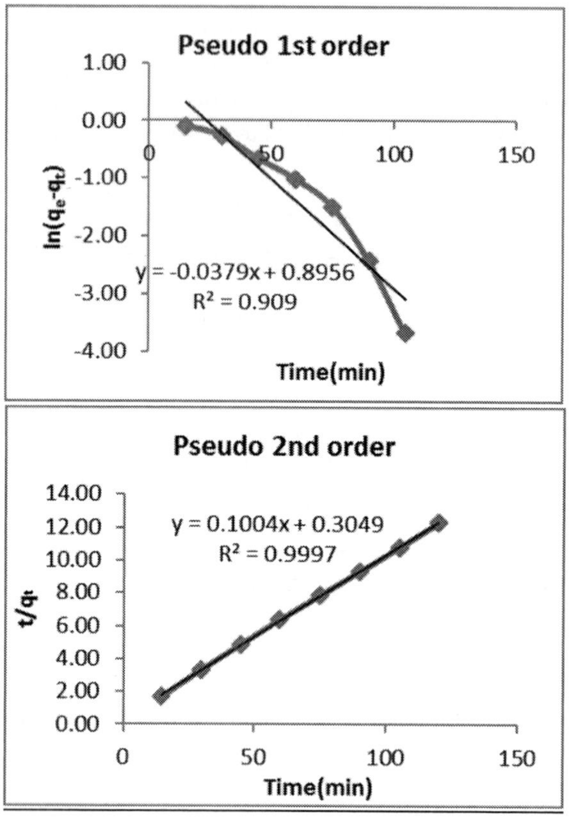

Figure 6. Pseudo-first and pseudo-second-order curve

3.4 Adsorption Isotherm

Adsorption isotherm is significant to explain on how the interaction between adsorbent and adsorbate happens during the adsorption process [1]. There are two models used to explain the findings of this experiments which are Langmuir and Freundlich isotherm model.

Langmuir isotherm curve were plotted based on the below equation;

$$\frac{1}{q_e} = \frac{1}{bq_oC_e} + \frac{1}{q_o} \tag{5}$$

where, C_e is the equilibrium concentration of dye solution (mg/L), q_e is the equilibrium capacity of the dye on adsorbent (mg/g), q_o is the monolayer adsorption capacity of adsorbent (mg /g), and b is the Langmuir adsorption constant (L/mg)

On the other hand, Freundlich isotherm curve were plotted based on the below equation;

$$\ln q_e = \ln K_f + \frac{1}{n}\ln C_e \tag{6}$$

where, C_e is the equilibrium concentration of dye solution (mg/L), q_e is the equilibrium capacity of the dye on adsorbent (mg/g), K_f is the Freundlich's adsorption capacity (mg /g), and n is the adsorption intensity constant.

In this observation, the determination coefficients show that Langmuir isotherm model has the best fitting with the experimental data R^2=0.997 meanwhile for Freundlich isotherm model the value of R^2 is 0.964. The Langmuir equation assumes that there is no interaction between the adsorbate molecules and that the sorption is localized in a monolayer [24]. It is then assumes that once a dye molecule occupies a site, no further adsorption can take place at that site. The results for malachite green dyes show that Langmuir isotherm also fitted more on the data obtained comparing to the Freundlich isotherm.

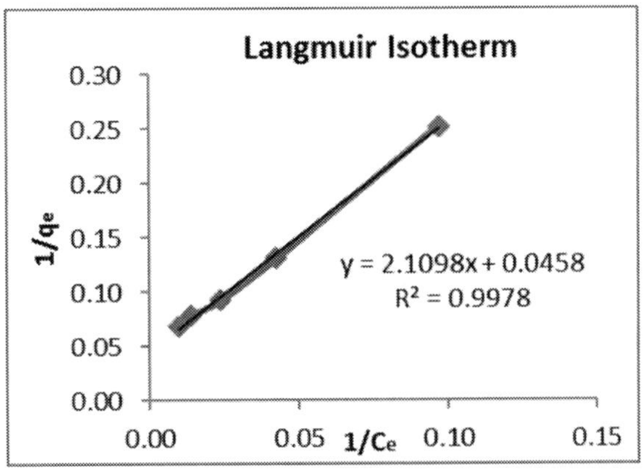

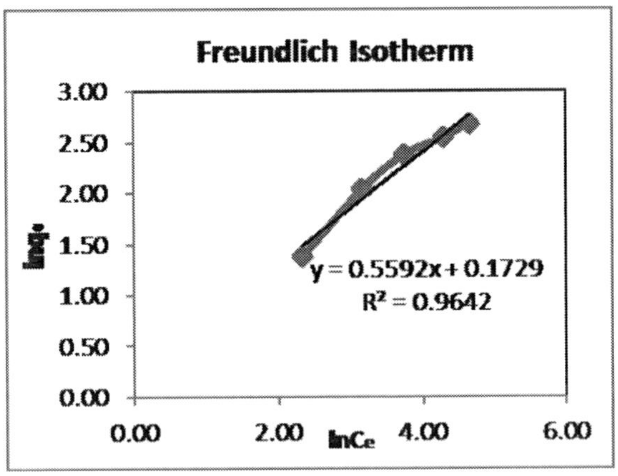

Figure. 7. Langmuir and Freundlich isotherm curve

4. Summary

This study shows that the spent coffee grounds (SCG) can be used as low-cost adsorbent for the removal of malachite green dye from aqueous solutions. From the data obtained, it showed that the adsorption of malachite green on SCG was well-fitted to the pseudo-second-order model while the adsorption equilibrium was best described by the Langmuir isotherm model.

References

[1] Yagub M T, Sen T K, Afroze S. and Ang H M 2014 *Adv. Colloid Interface Sci.* **209**, 172–184
[2] Sivashankar R, Sathya A B, Vasantharaj K, and Sivasubramanian V 2014 *Environ. Nanotechnol. Monit. Manage.* **1-2**, 36–49
[3] Méndez A, Barriga S, Saa A and Gascó G 2010 *J. Therm. Anal. Calorim.* ***99*(3)**, 993–998
[4] Dawood S, Sen T K and Phan C 2014 *Water, Air, Soil Pollut.* ***225*(1)**, 1–16
[5] Bensalah N, Alfaro M A Q and Martínez-Huitle C A 2009 *Chem. Eng. J.* ***149*(1-3)**, 348–352
[6] Gupta V K, Ali I., Suhas and Mohan D 2003 *J. Colloid Interface Sci.* ***265*(2)**, 257-264
[7] Joseph F, Agrawal Y K and Rawtani D 2013 *Front. Life Sci.* ***7*(3-4)**, 99–111
[8] Baek M H, O, Ijagbemi S J and Kim D S 2010 *J. Hazard. Mater.* ***176*(1-3)**, 820–828
[9] Deilami M and Fallah N 2015 *Biological Forum – An International Journal* ***7*(1)**, 1490–1498

[10] Hassan H S, Elkady M F, El-Shazly A. H and Bamufleh H S 2014 *J. Nanomater.* **1-14**
[11] Ahmed M, Idris A and Adam A 2007 *J. Eng. Sci. Technol.* ***2*(1)**, 55–69
[12] Lafi R, ben Fradj A, Hafiane A and Hameed B H 2014 *Korean J. Chem. Eng.* ***31*(12)**, 2198–2206
[13] Hameed B H and El-Khaiary M 2008 *J. Hazard Mater.* ***154*(1-3)**, 639-648
[14] Ong S T, Lee C K and Zainal Z 2007 *Bioresour. Technol.* ***98*(15)**, 2792-2799
[15] Kyzas G Z 2012 *Mater.* ***5*(10)**, 1826–1840
[16] Safarik I, Horska K, Svobodova B and Safarikova M 2011 *Eur. Food Res. Technol.* ***234*(2)**, 345–350
[17] Mussatto S I, Machado E M S, Martins S and Teixeira J A 2011 *Food and Bioprocess Technol.* ***4*(5)**, 661–672
[18] Imessaoudene D, Hanini S, Bouzidi A, Araremi A and Desalin 2016 *Water Treat.* ***57*(13)**, 6116-6123
[19] Nitayaphat W, Jintakosol T, Engkaseth K and Wanrakakit Y 2015 *Chiang Mai J. Sci.* ***42*(2)**, 407–416
[20] Kumar B and Kumar U 2014 *Korean J. Chem. Eng.* ***32*(8)**, 1655–1666
[21] Kyzas G Z, Lazaridis N K and Mitropoulos A C 2012 *Chem. Eng. J.* **189-190**, 148–159
[22] Song Y, Ding S, Chen S, Xu H, Mei Y and Ren J 2015 *Korean J. Chem. Eng.* ***32*(12)**, 2443–2448
[23] Kobiraj R, Gupta N, Kushwaha A K and Chattopadhyaya M C 2012 *Indian J. Chem. Technol.* ***19*(1)**, 26–31
[24] Low L W, Teng T T, Ahmad A, Morad N and Wong Y S 2010 *Water, Air, Soil Pollut.* ***218*(1-4)**, 293-306

IOP Conf. Series: Materials Science and Engineering **318** (2018) 012016 doi:10.1088/1757-899X/318/1/012016

Spider Silk Processing for Spidroin Recovery from *Crossopriza Lyoni* Web

J A Mohtar[1], W L Ooi[2], and F Yusuf[3]
[1]Department of Chemical Engineering Technology, Faculty of Engineering Technology, Universiti Malaysia Perlis, Uniciti Alam Campus, 02100 Sungai Chuchuh, Padang Besar, Perlis, Malaysia
[2]Department of Chemical Engineering Technology, Faculty of Engineering Technology, Universiti Malaysia Perlis, Uniciti Alam Campus, 02100 Sungai Chuchuh, Padang Besar, Perlis, Malaysia
[3]Department of Biotechnology Engineering, Kulliyah of Engineering, International Islamic University Malaysia, Gombak, Malaysia

E-mail: johanariff@unimap.edu.my

Abstract. Spider silk is a potential biomaterial that can be used in various applications for its outstanding physicomechanical properties attributed by the spidroin composition. Efforts for commercializing spider silks have been mainly focused on the characterization of spidroins from the Entelegyne spiders for exceptional fibre construction. Hence, studies on silk proteins from the Haplogyne species remain neglected. The aim of this study is to isolate spidroin from *Crossopriza lyoni* web. Silk processing involved the pretreatment of fibres for the shell layer removal from the surface. A screening study was conducted to analyze the effect of temperature, incubation time and agitation speed on spidroin extraction using Ajisawa's reagent by OFAT analysis followed by statistical optimization of the extraction process via RSM for maximal protein recovery. All parameters exerted significant effect on spidroin recovery ($p<0.05$) in which the maximum protein concentration (451.78 ± 0.110 µg/ml) was obtained at optimal condition of 70°C, 350 rpm and 1.25 hours. The discovery of spidroin from this study provides a basic platform for engineering spider silk to meet the demand for a variety of silk-based products in the near future.

1. Introduction

Nature readily contains plethora of beneficial materials in order to fulfill the very needs of humans. With the endowment of intelligence, men have been able to make clothes out of natural fibers from the silkworm, *Bombyx mori*. The legacy has kept flourishing for thousands of years when the silk thread finally found its applications not only in textiles but also in medicals, military, automobile and building construction [1]. However, silkworm-derived fibers often lack the required strength for such purposes. Thus, researches have steered the direction to the use of spider silk due to its superior properties that are twice stronger than that of silks from the silkworms [2] in order to meet the increasing demand for diversified silk-based products [3].

Spider silk is a proteinaceous biopolymer that is composed of two distinct protein layers; the core and the shell layer [4]. The core contains the silk proteins known as spidroins with molecular weight between 200 kDa to 350 kDa while the shell layer is composed of lipid, skin and glycoprotein. Spidroin is the only protein of interest and highly demanded for a wide range of applications due to its contribution for tensile strength and elasticity to silk fibers. Moreover, spidroins can be re-constructed into desired products by mechanical means. Spiders are known to spin up to seven types of silk fibres with the constitution of two types of spidroins namely Spidroin-I (for silk strength enhancement) and Spidroin-II (for elasticity and flexibility) [5]. Specialized abdominal glands are the major site of spidroin production in the form of high concentrated fluid (dope) prior to spinning into functional fibres.

Spidroins can be harvested via silks from the spider web [6], silks from forcible silking [7] and from the silk glands [8]. Silk fibres can be dissolved to recover spidroins in mixed solutions or harsh aqueous solutions such as Ajisawa's solution, lithium bromide and lithium thiocynate [9].

All spiders produce various types of silk fibres that can be classified into two major groups; the Entelegyne and the Haplogyne species, respectively. Entelgynes such as the Nephilids are entirely more advanced in spinning orb-webs using strong threads [10] whereas Haplogynes constitute mostly the basal lineage spiders that build sheet webs with relatively fragile silks [11]. In ancient times, spider silks have been extensively used in South Pacific areas as fishing lures and knitting etiquette frock [12] as they exhibit a variety of outstanding physicomechanical properties such as high tensile strength, elasticity and toughness.

At present, companies such as Kraig Biocraft Lab and Araknitek have successfully commercialized exceptionally synthetic fibres using recombinant spidroins from the Nephilids. However, such effort is likely unknown in the study of spidroin from the Haplogyne species; causing possible innovations to remain unexploited. Previous studies have shown that spidroins from Haplogynes have a unique molecular architecture of the repetitive amino acid motifs; A_n, S_n, $(GX)_n$, and $(AQ)_n$ (S, serine; Q, glutamine) which differ from the silk proteins found in Entelegynes [11]. This provides the potential wealth of protein designs.

Thus, the aim of this research was to isolate spidroins from *Crossopriza lyoni* silk web. *C. lyoni* is a haplogyne that is regarded as sedentary web-bulding spider. It is a cosmopolitan synanthropic cellar spider distributed widely in Malaysia [13]. Individual spiders can be easily found in large numbers on huge layer of webs in close proximity to humans and they are harmless for handling.

2. Materials and methods

2.1. Sampling of spider webs

Crossopriza lyoni webs were collected in the presence of the adult spiders using a net in large amount from abandoned houses as well as in construction sites in Perlis. The collected webs were stored in a plastic container regardless the age and stage-specific.

2.2. Pretreatment of spider silk fibres

Spider webs were initially washed under the running tap water and dipped overnight in distilled water to remove debris. They were dried in hot air oven at 60°C for 1 hour. The whole mass was then boiled in aqueous solution of 0.02 M sodium carbonate for 20 minutes at 100 ° C with stirring to remove the tightly bound shell layer, rinsed with distilled water and dried in hot air oven. The dried degummed fibres were readily to be used in the extraction process of spidroins. The extract was subjected to Bradford assay to quantify the protein amount. The degumming step was originally adopted from [14] to remove the sericin layer in silkworm fibres.

2.3. Scanning electron microscopy analysis

Scanning electron microscopy (SEM0 analysis was conducted at UKM-MTDC Smart Technology Centre, UKM, Bangi. SEM SU1510 (Hitachi) was used to perform microstructural study of pretreated silk fibers in order to validate the degumming process of shell layer removal from the fibres. The morphological structures of the fibres were observed by using backscattered electron imaging without coating. Initially, the sample was mounted on the stub and screwed to the base stand firmly. Upon loading the sample into the chamber, it was slid onto the holder receptacle on the stage and evacuation process ensued. Accelerating voltage was set at 15 kV. With the adjusted electron optical system and the mode of detector opted for backscattered electron, the images were captured and recorded at 15,000 magnification with a scale of 3 μm.

2.4. *Preliminary extraction of spidroins*

10 mg of degummed silk fibers was dissolved in 20 ml Ajisawa's reagent (calcium chloride: ethanol: water; 1/2/8 molar ratio) at optimal conditions of 250 rpm, 2.5 hours and 70°C [14-15]. The protein dope was dialyzed in a visking tube with molecular weight cutoff of 12-14 kDa against deionized water for three days with six consecutive buffer exchanges. Following dialysis, the dope was centrifuged at 4°C at 4000 rpm for 20 minutes. The dope was concentrated in vivaspin tube with molecular cutoff membrane of 30 kDa at 4°C at 1500 rpm for 20 minutes to a volume of 2 ml. The extract was subjected to Bradford assay to determine the protein concentration [16]. The protein solution was stored at -20°C until further use.

2.5. *Screening of significant factors*

The effects of three parameters on the extraction of spidroins were analyzed in this experiment namely the incubation time, temperature and agitation speed. Previously, these parameters have shown significant effect on the extraction of silk proteins from *Bombyx mori* as reported by [14]. The experiment was carried out by one-factor-at-a time (OFAT) analysis to determine their significance on the spidroin extraction. The range values for each parameter were initially set at five levels as shown in table 1. However, a slight modification to the value range for agitation speed was performed at values that fixed to the digital hot plate available in the laboratory. The values of each factor that resulted in maximum silk protein concentration were selected for the subsequent OFAT analysis. Each extract was subjected to Bradford assay to determine the protein concentration.

Table 1. The value range of the selected parameters.

Parameters	Range	Reference
Incubation time (hr)	0.5, 1.5, 2.5, 3.5, 4.5	[14]
Temperature (°C)	50, 60, 70, 80, 90	[14]
Agitation speed (rpm)	60, 125, 350, 700	-

2.6. *Statistical optimization of spidroin extraction*

From the OFAT analysis, the magnitude of each parameter and the range were determined based on the presence of curvature [17]. The significant parameters were subjected to optimization study. They were optimized using response surface methodology (RSM) via Box-Behnken Design in order to obtain an optimal extraction condition that yields in maximal recovery of spidroins from the web fibres.

The parameters include the incubation time, temperature and agitation speed. In order to predict the coefficients of the model, a total of 17 runs with 5 replicated center points (runs 3, 6, 7, 10 and 11) were provided by the design. The runs were executed in order to contradistinguish the experimental values and actual values obtained from model. Table 2 shows the experimental range and levels of each parameter and table 3 demonstrates the Box-Behnken Design matrix of the three parameters in non-coded values with the protein concentration as the observable response. Equation (1) involves in the Box-Behnken Design is as follows:

$$Y = \beta_0 + \sum_{i=1}^{k} \beta_i X_i + \sum_{i=1}^{k} \beta_{ii} X_i^2 + \sum_{j=1}^{k} \beta_{ij} X_i X_{ij} + \varepsilon \quad (1)$$

where the input variables were X_1, X_2,....., X_k, β_i (i = 1,2,...., k) and β_{ij} (i = 1,2,...., k; j = 1,2,...., k) were unknown parameters and the random error was ε. Each extract was subjected to Bradford assay to determine the protein concentration.

Table 2. Experimental levels of independent parameters.

Parameter	Units	Level	
		(-)1	(+)1
Incubation time	(hr)	0.5	2.0
Temperature	(°C)	70	90
Agitation speed	(rpm)	60	350

2.6.1. Validation of extraction model. Following optimization process, validation of the model was performed in order to ensure an accurate model could be predicted by solving the regression equation [16]. Verification study was carried out under the conditions of 70°C and 350 rpm for 1.25 hours. The three replicates for protein yield were supplied with the extraction conditions to affirm the prediction of model. The percentage error between theoretical data and experimental data was calculated and a percentage error less than 1% indicates a high reliability of RSM model. Thus, the response model is appropriate to reflect the anticipated optimization.

2.7. Statistical analysis

All experiments were conducted in triplicates and the data was analyzed using one-way ANOVA with post-hoc Duncan Multiple Range Test. ANOVA was performed using SigmaStat (version 3.1) by Systat Software Inc. For optimization study, the experiment was designed by using Box-Behnken Design (BBD) via Design Expert 7.1.5 software.

Table 3. Box Behnken Design matrix of the three parameters in coded values with the protein concentration in μg/ml.

Run	A[a] (hr)	B[b] (°C)	C[c] (rpm)	Protein Concentration (μg/ml)
1	2.00	70	205	
2	0.50	90	205	
3	1.25	80	205	
4	0.50	80	350	
5	0.50	70	205	
6	1.25	80	205	
7	1.25	80	205	
8	1.25	70	60	
9	0.50	80	60	
10	1.25	80	205	
11	1.25	80	205	
12	2.00	80	350	
13	1.25	90	60	
14	1.25	70	350	
15	2.00	80	60	
16	1.25	90	350	
17	2.00	90	205	

Note: [a]-Incubation time, [b]-Temperature, [c]-Agitation speed

3. Results and discussion

The silk pretreatment was aimed to remove the tightly bound shell layer from the fibres. In order to ensure the efficiency of the shell layer removal, the supernatant extract from the pretreated fibres was subjected to Bradford assay to estimate for its protein content. It was mixed with Bradford reagent and a concentration of 53.83 µg/ml of protein was determined. This indicates a small amount of proteins that were unbound from the threads; presumably the shell layer proteins. Sponner *et al.* reported that spider silk fibre naturally contains two prominent protein layers; the shell layer and the core spidroin [4]. The former contains a number of glycoproteins and skin type proteins that act as a protective layer to the core spidroin.

In addition, SEM analysis was conducted to compare the morphological differences of the silk fibre before and after the degumming process. The microstructure of the degummed fibre was observed on a SEM SU1510 (Hitachi) and photographed at a 15 kV as shown in figure 1. It can be observed that the silk fibre before the degumming process appeared to be covered with a layer of grooving lines (shell layer) whereas for the degummed fibre, it appeared clean and smooth. This qualitative result corroborates the findings that the degumming pretreatment with 0.02 M sodium carbonate was able to remove the shell layer from the silk fibres.

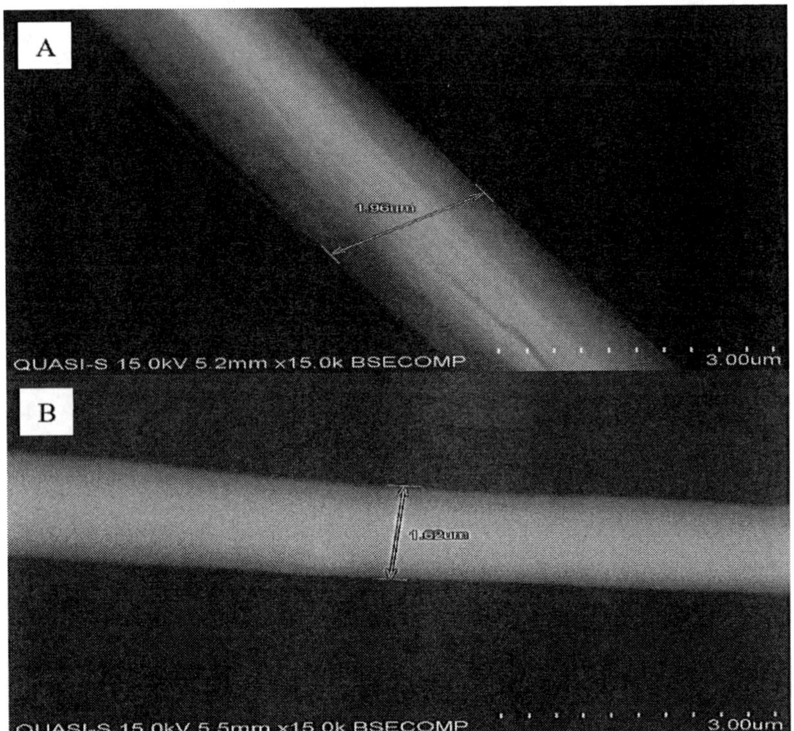

Figure 1. SEM analysis of the silk fibre surface before (A) and after pre-treatment (B).

The pretreated fibers were subjected to solvent-salt based extraction using Ajisawa's reagent encompassing calcium chloride, ethanol and water (1:2:8 molar ratio) according to the optimum condition set by [14-15] at 250 rpm, 70°C for 2.5 hours. It can be seen from the Bradford assay data that only 20 µg/ml of protein concentration was obtained. The dope is believed to only compose of the shell layer void-spidroins as it was removed during the degumming process. From the preliminary study, it is inferred that the spidroin extraction method was successfully established and this method was used throughout the whole experiment.

Meanwhile, the undissolved fibres were centrifuged and dried in the oven and the weight was recorded. Approximately 26% of undissolved fibres were obtained and suspected to belong to various types of spider silk fibres other than the dragline silk thread of the web. This is in agreement with the fact that spider web is spun with more than one type of silk thread. In addition, many previous studies that used Ajisawa's solution for spidroin recovery was performed on the dragline silk fibre. Therefore, the incorporation of Ajisawa's solution in the present study may favour the extraction of spidroins from the dragline silk thread. However, further investigation is required to support the hypothesis.

Due to the small amount of silk proteins recovered, an optimization study of the extraction condition was carried out to overcome the problem in order to yield maximal protein concentration. Initially, a screening study of the effect of significant parameters on spidroin extraction was conducted. Temperature, incubation time and agitation speed were selected and tested in the study. OFAT analysis was conducted in order to screen for the significant factors affecting the spidroin recovery. From figure 2, all of the three parameters exerted statistically significant effect on spidroin extraction from the web fibre ($p < 0.05$) by one-way ANOVA. From the post hoc analysis of comparisons using the Duncan Multiple Range test, it is indicated that the mean score of protein concentration at 80°C (27.95 ± 0.147 µg/ml), 125 rpm (383.61 ± 0.4430 µg/ml) and 1.5 hours (117.22 ± 0.11 µg/ml) were significantly higher compared to others ($p < 0.05$) in each experiment, respectively. Thus, they were opted as the best level of treatments for the optimization study as they led to the highest maximal spidroin recovery.

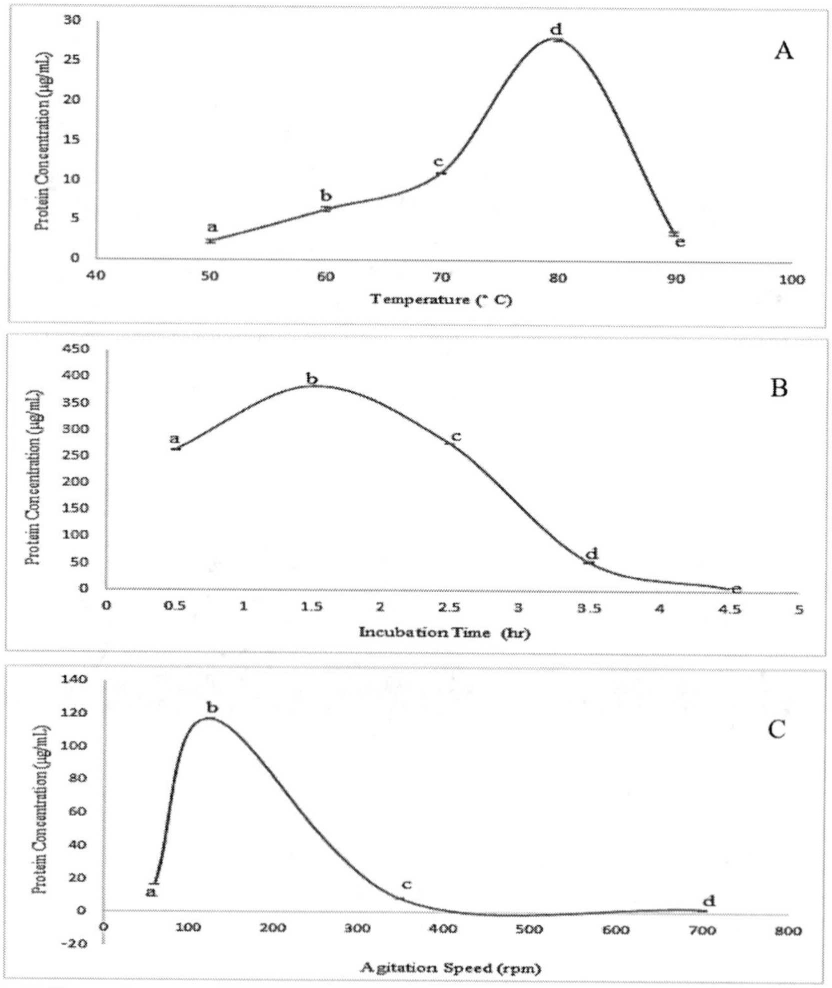

Figure 2. A. Effect of temperature [F(4,10) = 2295.304, p = <0.001], B. Effect of incubation time [F(4,10) = 2688235.49, p = <0.001] and C. Effect of agitation speed [F(3,8) = 122566.659, p = <0.001] on spidroin extraction from *C. lyoni* web.

Interestingly, the present finding is in contrast with [14] which 70°C was the best temperature that led to the maximum amount of protein recovered. Nevertheless, it can be comprehended that the higher the exposing temperature, the more the spidroins can be extracted from the fibre. This is because at high temperature, the protein molecules could move faster as they possessed high kinetic energy to overcome the strong intermolecular force imposed on them.

Meanwhile, 1.5 hours was observed to exert significant effect on the extraction of spidroins. This implies that incubation time is one of the important factors that contribute to silk protein extraction. In contrast to [14], 2.5 hours was the best time which produced the maximum concentration of spidroin. This phenomenon may possibly due to the high hydrolysis rate of interlinking bonds between fibroins themselves and more fibroins were able to be dissolved in the extraction solution at 90 minutes compared to others. Moreover, 125 rpm was observed to exert significant effect on the extraction of spidroins. At present, few screening studies have been performed on agitation speed, but most of the experiments divulged that the higher the agitation speed, the more proteins could be recovered [15]. This is because of the high rate of solubility of silk spidroin at higher agitation speed.

The optimization of extraction conditions within the selected range was conducted using Box-Behnken Design (BBD). In statistical modelling, regression analysis is a process performed by RSM to estimate the relationships among parameters shown in table 2. The dependence of spidroin concentration on the extraction condition was shown by the second order regression equation whereas the multiple regression analysis of the experimental data generated the parameters of the equation. These parameters are presented in the second order polynomial equation as indicated in Equation (2).

$$Y = 1447.27 - 49.31 * A + 7.70 * B + 79.93 * C - 0.11 * A * B - 4.44 * A * C + 0.14 * B * C + 0.44 * A^2 + 2.83 \times 10^{-3} * B^2 + 78.72 * C^2 \quad (2)$$

where Y (response) represents the spidroin concentration and A, B, C refer to temperature, agitation speed and incubation time, respectively.

From table 4, both actual and predicted values of spidroin concentration are presented. It is obviously shown that the maximum protein concentration is obtained by run number 14 with 451.78 ± 0.110 μg/ml at optimal conditions of 70°C, 1.25 hours and 350 rpm. Meanwhile, run number 16 contributes to the lowest protein concentration (10.11 ± 0.0567 μg/ml) under the conditions of 90°C, 1.25 hours and 350 rpm.

Table 4. Box-Behnken Design matrix of the three parameters with the protein concentration in μg/ml.

Run	A^a (hr)	B^b (°C)	C^c (rpm)	Protein Concentration (μg/ml)	
				Actual	Predicted
1	2.00	70	205	220.17	217.23
2	0.50	90	205	166.67	169.61
3	1.25	80	205	20.22	71.48
4	0.50	80	350	250.61	249.75
5	0.50	70	205	219.56	226.34
6	1.25	80	205	84.11	71.48
7	1.25	80	205	105.83	71.48
8	1.25	70	60	26.22	28.30
9	0.50	80	60	85.56	176.70
10	1.25	80	205	52.22	71.48
11	1.25	80	205	95.00	71.48
12	2.00	80	350	195.17	204.03
13	1.25	90	60	213.44	219.36
14	1.25	70	350	451.78	445.86
15	2.00	80	60	70.00	70.86

16	1.25	90	350	10.11	8.02
17	2.00	90	205	33.94	27.16

Note: [a]-Incubation time, [b]-Temperature, [c]-Agitation speed

Based on the analysis of variance of the model by Fisher's F-test, it is noted that the least square regression line (LSRL) is a good model for the data fitting with Model F-value of 30.07 with the probability value of less than 0.0001. Generally, *p*-value is the probability to verify the impact of coefficients in which a value less or equal to 0.05 denotes statistical significant at 95 % confidence level. The response surface model is deemed adequate due to the insignificance ($p = 0.9592$) of the lack of fit with F values of 0.094 relative to pure error for all variables. The adequacy of the model is also corroborated by its determination coefficient, R^2 value of 0.9748.

Generally, 3D surface profiler has been used widely in the analysis of the model graph due to the different effects and interactions that can be easily visualized based on the curvature patterns. From the curvatures in figure 3, it can be inferred that the highest spidroin concentration could be obtained if the temperature, incubation time and as well as agitation speed are kept at higher treatment levels. This is possibly because the proteins are less stable at high temperature and can be dissolved more into the extraction solution; the longer the incubation time, the higher the hydrolysis rate of the interlinking bonds between spidroins themselves. Meanwhile, the higher the agitation speed, the more bonds that linked silk proteins together can be broken by strong forces.

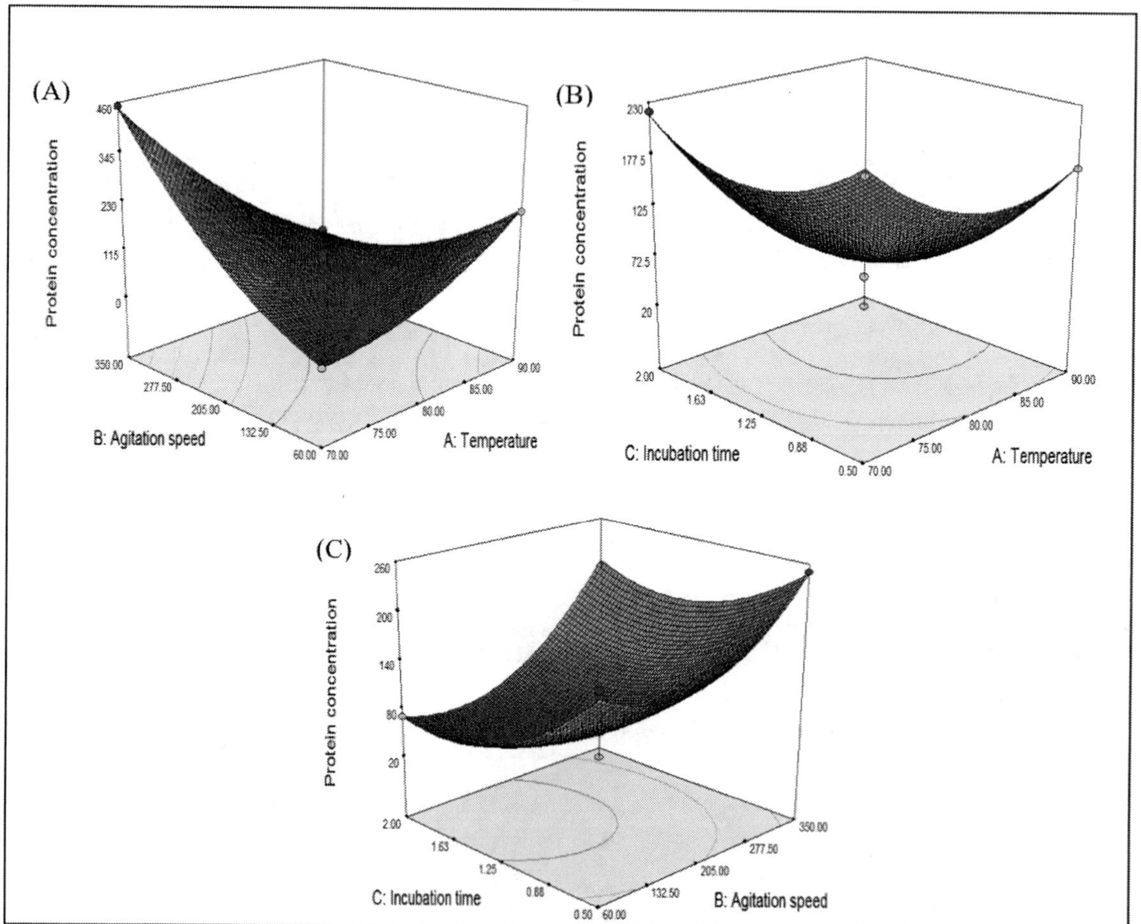

Figure 3. 3D surface of spidroin concentration as a function of (A) temperature and agitation speed, (B) temperature and incubation time and (C) agitation speed and incubation time.

Validation experiment was conducted in order to ensure an accurate model could be predicted as well as the predicted value would similar to the experimental value. The optimal values of the selected variables were obtained by solving the regression equation (Equation 2). The optimal condition for protein yield estimated by the model equation was given at 70°C, 350 rpm and 1.25 hours. The theoretical protein yield predicted under the above condition was 445.862 μg/ml. Verification of the model was performed by comparing the percentage error between the theoretical and the experimental spidroin concentrations. The average protein concentration obtained was 447.81 μg/ml with the percentage error of 0.44% (< 1%). Thus, this indicates a figure well within the estimated value of the model equation. RSM model is highly reliable since the predicted value is in the agreement with the experimental value. This also indicates the RSM model is valid and hence, the response model is appropriate to reflect the anticipated optimization. Furthermore, the results also suggested that the model in Equation (2) is satisfactory and accurate.

4. Conclusion

In this study, spider silk processing from *Crossopriza lyoni* web using Ajisawa's reagent under optimized condition of 70°C, 1.25 hours and 350 rpm has successfully recovered a maximal spidroin amount of 451.78 ± 0.110 μg/ml. Spider silk is a potential biopolymer that is at least five times as strong as steel and twice as elastic as nylon and Kevlar synthetic fibre. Thus, the extracted silk protein from *Crossopriza lyoni* from the present study will address future investigations on the purification and characterization of the protein. With the endowment of high biodiversity of Haplogyne spiders in Malaysia, this study may likely become the first attempt at national level towards the development of novel biomaterials from spider silk in order to meet the ever increasing demand for silk-based products in the near future.

Acknowledgement

This work was supported by Universiti Malaysia Perlis (UniMAP), Malaysia through the Short Term Grant (STG) (Grant No.: 9001-00499) and we gratefully acknowledge the support.

References

[1] Tokareva O, Michalczechen-Lacerda V A, Rech E L and Kaplan, D L 2013 *Microb. Biotechnol.* **6** 651-63.
[2] Hillyard P 1994 *The book of the spider* 1st ed (New York: Random House).
[3] Hsia Y, Gnesa E, Tang S, Jeffery F and Vierra C 2011 *Appl. Phys. A.* **105** 301-09.
[4] Sponner A, Schlott B, Vollrath F, Unger E, Grosse F and Weisshart K 2005 *Biochem.* **44** 4727-36.
[5] Gatesy J, Hayashi C, Motriuk D, Woods J and Lewis R 2001 *Science.* **291** 2603-05.
[6] dos Santos-Pinto J R A, Arcuri H A, Lubec G and Palma M S 2016 *Biochem. et Biophys Ac.* **1864** 1444-54.
[7] dos Santos-Pinto J R A, Lamprecht G, Chen W Q, Heo S, Hardy J G, Priewalder H, Scheibel T R, Palma M S and Lubec G 2014 *J. Proteomics.* **105** 174-85.
[8] Lane A K, Hayashi C Y, Whitworh G B and Ayoub N A 2013 *BMC Genomics.* **14** 1-12.
[9] Kundu B, Kurland N E, Yadavalli V K and Kundu S C 2014 *Int. J. Biol. Macromolec.* **70** 70-77.
[10] Sensenig A, Agnarsson I and Blackledge T A 2010 *J. Evol. Biol.* **23** 1839-56.
[11] Boutry C and Blackledge T A 2010 *J. Exp. Biol.* **213** 3505-14.
[12] Romer L, Scheibel T and Landes 2008 *Bioscience.* **2** 154-61.
[13] Nasir D and Singla R 2015 *Common Malaysian spiders* (Serdang: Universiti Putra Malaysia press).
[14] Sah M and Pramanik K 2010 *Int. J. Environ. Sci. Dev.* **1** 404-08.
[15] Mirghani M, Kabbashi N, Elfaki F and Zulkifli M 2010 *Biotechnol.* **201** 1-5.
[16] Sah M N, Kumar A and Pramanik K 2010 *Int. J. Bioinform. Res.* **2** 33-41.
[17] Bezerra M, Santelli R, Oliveira E, Villar L and Escaleira L 2008 *Talanta.* **76** 965-77.

The Potential of Biodiesel Production derived from Fish Waste

Amira Farzana Samat [1], Nor Amirah Safiah Muhamad [1], Nur Aziera Abd Rasib [1], Siti Aminah Mohd Hassan [1], Khairunissa Syairah Ahmad Sohaimi [1], and Nur Izzati Iberahim [1]

[1] Faculty of Engineering Technology, Universiti Malaysia Perlis, UNICITI Alam Campus, 02100 Sungai Chucuh, Padang Besar, Perlis, Malaysia.

E-mail: amirafarzana@unimap.edu.my

Abstract. Petroleum based diesel is one of the largest greenhouse emitters in the worlds based on its contribution to more likely of all carbon, methane and other greenhouse emissions. Besides, the depletion of fossil fuel that indirectly increased its price has force the global oil industry not to be so dependent on the fossil fuel but instead start focusing on alternative sources. Biodiesel is recognized as a clean alternative fuel or as a fuel additive to reduce pollutant from combustion equipment. In this study, the discarded parts of mixed marine fish species were used as the raw material to produce biodiesel. Marine fish oil was extracted from the discarded part of fish and if refined through a series of pretreatment process. The refined marine fish oil undergoes esterification process to reduce the amount of free fatty acid. The oil was then transesterified with methanol and sodium hydroxide as an alkaline catalyst that will speed up the conversion of oil to methyl ester. The three process parameters considered for this study were reaction time, reaction temperature and methanol to oil molar ratio. Biodiesel obtained was then analyzed using gas chromatography (GC). Statistical analyses were performed using SPSS software. The data obtained was analyzed by using one way analysis of variance (ANOVA) repeated measure. The results obtained showed that the conversion of FAME yield is the highest at reaction time 180 minutes, reaction temperature 60°C and methanol to oil molar ratio at 15:1 with FAME yield 80.16%, 80.03% and 80.39%. Thus, it can be concluded that the conversion of biodiesel increased as the reaction time, temperature and

1. Introduction

The emission of greenhouse gases worldwide contributed from the utilization of fossil fuel at the present rate will deteriorate local air quality severely and the global warming will increase beyond a repairable extent. Besides these negative environmental impacts, the depletion of fossil fuel that indirectly increased its price has force the global oil industry not to be so dependent on the fossil fuel but instead start focusing on looking at other alternative source especially from renewable energy that is biodiesel ([2]).

1.1. Biodiesel as alternative resources

Biodiesel is better for the environment because it is made from renewable resources, overall lower emission of greenhouse gases and most other toxin (excluding nitrogen oxides) ([1]). It is completely nontoxic and biodegrades rapidly. Besides, it can be produced domestically using natural resources and its uses will also decreases our dependency on imported fuel and contribute to our own economy ([5]). Currently there are many claims and problems arise due to the usage of edible oil (corn, coconut, peanut, rapeseed oil) that exceed 95% of overall production of biodiesel ([5]). The usage of edible oils in biodiesel production causes food resources to be converted into automotive tools. Thus, the biodiesel production from edible oils in large scale may cause imbalance in the food supply and demand market worldwide ([2]).

1.2. Fish waste as raw materials for biodiesel

In order to overcome this problem, there are suggestions that biodiesel must be produce by using other alternative oil resources such as non-edible oils ([2]) or from waste. Extracted oil from fish waste is an example of raw material for biodiesel production. Fish has high protein content, low saturated fat and also contains fatty acids known for its great benefit to human health. However, research show that the remaining materials that is transformed into waste exceeds 50% from the total of fish capture ([6]). Before, fish wastes are always considered as loss. Nowadays, there are several ways to utilize fish waste to produce renewable energy which are production of biodiesel, manure composting, and production of biogas and burning of fish wastes ([7]). By using fish processing industry waste as a raw material for renewable energy source, it can be turned into a business operation that gives profit to the processing team.

2. Materials and Methods

2.1. Raw materials preparation

The fish wastes were used as the source raw material for the production of fish oil biodiesel. The fish waste included viscera, fish head and other discarded part of fish were collected from the fish market. To prepare the fish oil, the fish waste/gut was treated with a series of process. The fish waste/gut (around 10 kg) were first cut into smaller sizes in order to enhance a speedy oven drying. Then, they were dried by using oven drying at temperature 70°C for 3 days to reduce the moisture content in the fish waste. The fish wastes were then undergo further size reduction and were pounded in a mortar into finer form.

2.2. Fish oil extraction

Then, by using soxhlet extractor the fish oil was extracted with n-hexane. The solid sample (fish waste that already in finer form) was placed in a cotton cloth. Before the sample was placed in the inner tube of soxhlet, the weight of the sample (50 g) was taken and then it was fitted to a round bottom flask of appropriate size that contains the 150 ml n-hexane. Heat was applied for 3 hours at 60°C. After 3 hours, the mixture of the solvent and the extracted fish oil was cooled at room temperature for few minutes before it was filtered to remove the remaining impurities in the mixture ([7]).

Then, the obtained mixture was separated by using rotary evaporator in order to get crude fish oil. The temperature of the water bath was set at 35°C. For each run, 200 ml of the obtained mixture of extracted fish oil and n-hexane was placed inside the rotary bottle for 15 minutes. After 15 minutes, the crude fish oil was obtained inside the rotary bottle. The fish oil extracted was then placed in a freezer ready to be used for transesterification process.

2.3. Esterification reaction of fish oil

Esterification reaction to remove free fatty acid in the fish oil was carried out as fish oil was mixed with methanol and 1.0% w/w sulphuric acid (H2SO4) that act as acidic catalyst. 100 g fish oil was weight and poured in a 2 neck round bottom flask. Methanol to oil ratio used during this process was 20:1. Thus, 100 g of fish oil with 80.91 ml of methanol was measured and was pour into a beaker together with 6 g of sulphuric acid. A beaker fill with water was placed on top of the hot plate. The 2 neck round bottom flask that was filled with 100 g fish oil was put inside the beaker. The fish oil was heated until it reached 60°C. The mixture of sulphuric acid and methanol was added into the 2 neck round bottom flask. This esterification process was run for 3 hours. After 3 hours, the oil obtained was washed with distilled water. The distilled water was heat until the temperature reached 50°C and the oil was added and mixed by using magnetic stirrer until bubbles formed. The mixture was poured into separatory funnel and allowed to separate into two different layers. The oil was collected and sodium anhydrous was added to remove the remaining water. The oil was then filtered to remove sodium sulphate anhydrous. The oil obtained was ready to be used in transesterification process ([8]).

After the reaction was completed, the mixture obtained was cool down for few minutes in room temperature and then was pour into separatory funnel to separate the mixture. The mixture was allowed to separate for one night. After few hours, two different layers can be observed. The upper layer was biodiesel and the bottom layer was glycerol. The bottom layer was removed and the upper layer was collected and ready for biodiesel purification process.

IOP Conf. Series: Materials Science and Engineering **318** (2018) 012017 doi:10.1088/1757-899X/318/1/012017

The oil obtained was washed with distilled water to remove the remaining methanol, unreacted catalyst and free glycerol. The distilled water was heated until the temperature reached 50°C and the oil was added and mixed by using magnetic stirrer until bubbles form. The mixture was pour into separatory funnel and allowed to separate into two different layers. The oil was collected and sodium sulphate anhydrous was added to remove the remaining water. The oil was then filtered using filter paper to remove sodium sulphate anhydrous ([9]).

2.4. Gas chromatographic (GC) method to analyze fatty acids methyl ester
A gas chromatographic (GC) method was used to analyze fatty acids methyl ester content. The biodiesel was mixed with heptane in a ratio of 50:50. 2 ml of biodiesel and 2 ml of heptane was mix in a beaker. The mixture was then transferred into a vial by using syringe and 0.45µm whatman column filter. The 30 parameters used in biodiesel GC analysis were as follow: detector used was flame ionization detector (FID), type of column was Bp × 70 (30cm length × 0.25µm thick ×0.32mm diameter), injector temperature was 245°C, detector temperature was also 245°C, column temperature was 120-245°C (4°C min-1), carrier gas was nitrogen and its flow rate was 20 cm/min. Each run of biodiesel sample take about 35 minutes to complete ([10]).

Statistical analyses were performed using SPSS software. The data obtain was analyzed by using one way analysis of variance (ANOVA) repeated measure. Significant values were assessed with Tukey's test and p-values less than 0.05 were considered significant.

3. Results and Discussion

Discarded part of fish waste was used as the source of raw material in this research producing crude fish oil that was converted into fatty acid methyl ester. The main problem with processing these low quality animal fats is that they contain high amount of free fatty acids that could not be converted to biodiesel using an alkaline catalyst due to the formation of fatty acids salts (soap). Soap can prevent separation of biodiesel from glycerin fraction thus lowering the biodiesel conversion yield. To overcome this problem, two-steps esterification reaction was used. First, the extracted crude fish oil that is high in free fatty acid was treated with acid-catalyzed esterification in order to reduce the free fatty acids content. Free fatty acids content in the fish oil must be less than 2% in order for transesterification reaction to take place

Transesterification reaction covert fatty acid into fatty acid methyl ester in the presence of methanol and sodium hydroxide (NaOH) as alkaline catalyst. The transesterification reaction takes place with methanol to oil molar ratio, reaction temperature and reaction time as the process parameters. After transesterification reaction take place, the after product, biodiesel and glycerin was separated. The desired product, biodiesel was then undergo gas chromatography (GC) analysis to determine the compound contain in fish oil biodiesel.

In this research, effect of reaction time on fatty acid methyl esters (FAMEs) yield was investigated. The reaction times used in this study during transesterification reaction for conversion of oil into FAME were 90, 120, 150 and 180 minutes. Figure 1 showed the result obtained during transesterification reaction with methanol to oil molar ratio and temperature used were constant at 15:1 and 60°C respectively.

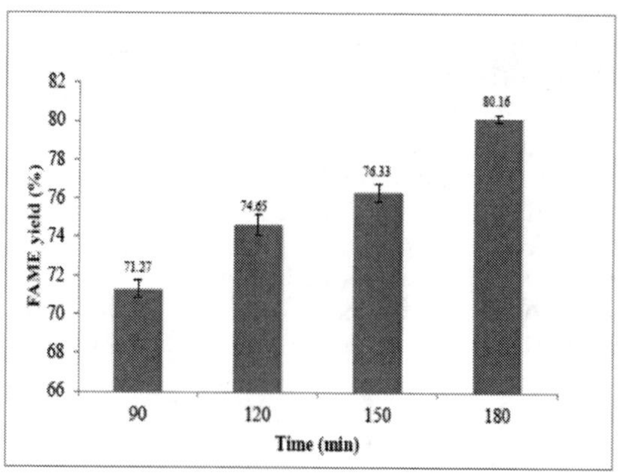

Figure 1. FAME yield versus reaction time.

It was observed that the ester yield slightly increased with the increased in reaction time ([4]). FAME yield is at the lowest at reaction time 90 minutes (71.27%) and increased as the reaction time increased. This was because, after 90 minutes, methanol has not yet completed the conversion of oil to ester. So, as the reaction was stopped after 90 minutes, not all of the oil has been fully converted into esters. FAME yield continued to increase at reaction time 120 and 150 minutes with 74.65% and 76.33% respectively. The conversion of FAME yield is at the highest at reaction time 180 minutes with 80.16%. Thus, it was observed that the conversion of oil into ester increased as the reaction time increased ([3]).

The reaction temperatures used were 30, 40, 50 and 60°C. The reaction temperature used is below the boiling point of methanol which is 68°C. During the transesterification reaction, methanol to oil molar ratio and reaction time is kept constant at 15:1 and 180 minutes. Figure 2 showed result obtained during transesterification reaction with reaction temperature as the parameter.

Based on ANOVA analysis, there was a significant difference at the $p < .000$ level in FAME yield for four reaction time $F(3, 8) = 77.797$, $p < .000$. Despite reaching statistical significance, the actual difference in mean yield between groups was quite large (0.97). The Tukey HSD test indicated that the mean yield for reaction time at 90 minutes ($M = 71.3$, $SD = 0.77$) was significantly different from reaction time at 120 minutes ($M = 74.6$, $SD = 0.91$), reaction time at 150 minutes ($M = 76.3$, $SD = 0.77$) and reaction time at 180 minutes ($M = 80.2$, $SD = 0.34$). There was no significant difference in mean yield between reaction time at 120 minutes and reaction time at 150 minutes.

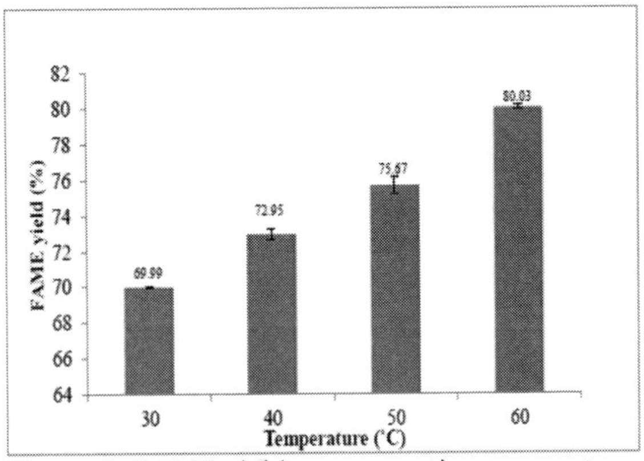

Figure 2. FAME yield versus reaction temperature.

From Figure 2, it was observed that the ester yield increased as the reaction temperature increased. The minimum yield of ester was obtained at the temperatures 30°C at 69.99%. The yield of methyl ester continued to increase at temperature 40 and 50°C with 72.95% and 75.67% respectively. The highest methyl ester yield obtained was at temperature 60°C with 80.03%. As the temperature increased, the kinetic energy also increased. The increased in kinetic energy will cause the molecules of the oil and methanol as well as sodium hydroxide to moves more vigorously and results in more contact with one another and increase the conversion rate of oil to methyl ester. Thus, FAME yield increased as the reaction temperature increased ([3]).

Based on ANOVA analysis, there was a significant difference at the $p < .000$ level in FAME yield for four reaction temperature $F(3, 8) = 213.127$, $p < .000$. Despite reaching statistical significance, the actual difference in mean yield between groups was quite large (0.98). The Tukey HSD test indicated that the mean yield for reaction temperature at 30°C ($M = 69.9$, $SD = 0.11$) was significantly different from reaction temperature at 40°C ($M = 72.9$, $SD = 0.57$), reaction temperature at 50°C ($M = 75.7$, $SD = 0.80$) and reaction temperature at 60°C ($M = 80.0$, $SD = 0.22$).

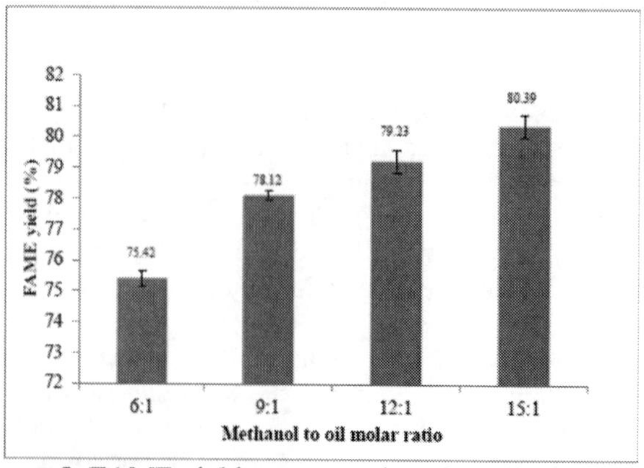

Figure 3. FAME yield versus methanol to oil molar ratio.

It was observed that FAME yield increases with increase in molar ratio. From the figure 3, it can be observed that the molar ratio of methanol to oil at 6:1 has the lowest FAME yield percentage with 75.42%. The molar ratio of methanol to oil at 9:1 and 12:1 has the slightest increase in FAME yield with 78.12% and 79.23% respectively. Molar ratio of methanol to oil at 15:1 recorded the highest conversion of FAME with 80.39%. This is because lower molar ratios mean that methanol used during transesterification reaction is also low. The lack of methanol prevents smooth conversion of oil into methyl ester thus lowering the percentage yield ([3]).

Based on ANOVA analysis, There was a significant difference at the $p < .000$ level in FAME yield for four methanol to oil molar ratio $F (3, 8) = 51.239$, $p < .000$. Despite reaching statistical significance, the actual difference in mean yield between groups was quite large (0.95). The Tukey HSD test indicated that the mean yield for methanol to oil molar ratio at 6:1 (M =75.4, SD =0.43) was significantly different from methanol to oil molar ratio at 9:1 (M =78.1, SD = 0.27), methanol to oil molar ratio at 12:1 (M =79.2, SD = 0.64) and methanol to oil molar ratio at 15:1 (M =80.4, SD = 0.63). There were no statistically significant different in mean yield between methanol to oil molar ratio at 9:1 and methanol to oil molar ratio at 12:1 and also between methanol to oil molar ratio at 12:1 and methanol to oil molar ratio at 15:1.

Fatty acid methyl ester (FAME) analysis by gas chromatography (GC) shows the presence of palmitic acid methyl ester (C-16:0) at retention time 10.38, stearic acid methyl ester (C-18:0) at retention time 12.10 and 30.95, oleic acid methyl ester (C-18:1) at retention time 17.48 and eicosanoic acid methyl ester (C-20:0) at retention time 30.95 in the fish oil biodiesel. These fatty acids influence the amount of cetane number of diesel fuel. Figure 4 showed the chromatogram obtained from gas chromatography (GC) analysis and Table 1 showed the list of compound presence in the fish oil biodiesel.

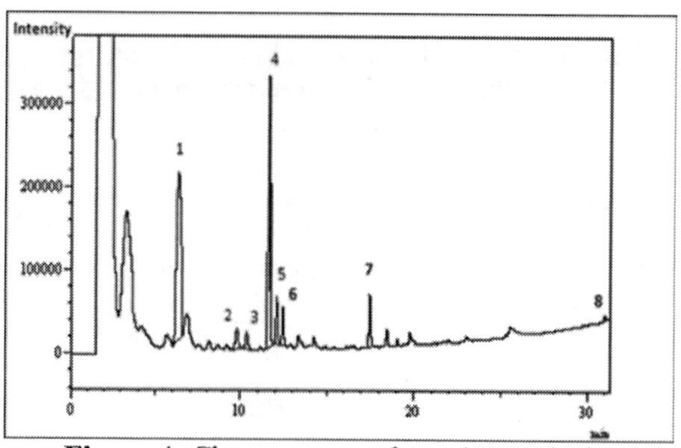

Figure 4: Chromatogram from GC analysis.

Table 1: List of compound in fish oil biodiesel.

Peak	Retention time	Carbon number	Compound
3	10.38	C-16:0	Palmitic acid methyl ester
5	12.10	C-18:0	Stearic acid methyl ester
6	12.44	C-18:1	Oleic acid methyl ester
7	17.48	C-20:0	Eicosanoic acid methyl ester
8	30.95	C-18:1	Oleic acid methyl ester

4. Conclusion

The present study indicates that instead of dumping, utilization of discarded part of fish waste as a low cost feedstock is a better option. Animal fats characterized by high free fatty acids (FFAs) have been deemed suitable as a feedstock in biodiesel production. The two-step transesterification process was used in order to convert oil with high FFAs into methyl esters. The first step in this process which is acid catalyzed esterification reduced the FFA content of the oil until the FFA content is less than 2%. Then, alkaline catalyst transesterification process is used to convert the products of esterification process into methyl esters and glycerol. The effects of methanol to oil molar ratio, reaction temperature and reaction time were analyzed in this researched. From this research, it was proven that methanol to oil molar ratio, reaction temperature and reaction time greatly affects the conversion yield of biodiesel. Fatty acid methyl ester (FAME) yield increased as the methanol to oil molar ratio, reaction temperature and reaction time increased. FAME yield has the highest percentage when methanol to oil molar ratio, reaction temperature and reaction time were the highest which are 15:1 for methanol to oil molar ratio, 60°C for reaction temperature and 180 minutes for reaction time. Overall, this two-steps transesterification method reduced the production cost and is a suitable method for conversion of oil that is high in free fatty acids into biodiesel.

5. Acknowledgement

Authors would like to thank School of Bioprocess and Faculty of Engineering Technology, University Malaysia Perlis (UniMAP) for supporting this work through the Research Acculturation Grant Scheme (RAGS) and lab accommodation purpose.

References

[1] Edlund M, Visser H, and Heitland P, 2002 *J. Anal. Atom. Spectrom.*, **17(3)**, *232-235*
[2] Gui M M, Lee K T, and Bhatia S, 2008 *Energy*, **33(11)**, *1646-1653*
[3] Mrad N, Varuvel E G, Tazerout M, and Aloui F, 2012 *Energy*, **44(1)**, *955-963*
[4] Ramadhas A S, Jayaraj S, and Muraleedharan C, 2005 *Fuel*, **84(4)**, *335-340*
[5] Ushakov S, Valland H,and Æsøy V,2012 *Energ. Convers. Manage*, **65**, *228-238*
[6] Varuvel E G, Mrad N, Tazerout M, and Aloui F,2012 *Appl. Energ.*, **94(0)**, *224-231*
[7] Yahyaee R, Ghobadian B, andNajafi G, 2011 Renew. *Sust. Energ. Rev.*, **17(0)**, *312-319*
[8] Encinar J M, Sánchez N, Martínez G, and García L, 2011 *Biores. Technol.*, **102(23)**, *10907-10914*
[9] Chai M, Tu Q, Lu M, and Yang Y J, 2014 *Fuel Process. Technol.*, **125**, *106-113*
[10] Meher L C, Sagar D V, and Naik S N, Renew. 2006 *Sust. Energ. Rev.*, **10(3)**, *248-268*

IOP Conf. Series: Materials Science and Engineering **318** (2018) 012018 doi:10.1088/1757-899X/318/1/012018

Extraction and Production of Omega-3 from UniMAP Puyu (Jade Perch) and Mackarel

I Nur Izzati[1], H Zainab[1], M Nornadhiratulhusna[1], Y Chee Hann[1], A. S Khairunissa Syairah[1], S Amira Farzana[1]

[1]Department of Chemical Engineering Technology, Faculty of Engineering Technology, Uniciti Alam Campus, Sg. Chuchuh, Universiti Malaysia Perlis

Email: izzatiiberahim@unimap.edu.my, zainab@unimap.edu.my, ycheehann@gmail.com, syairah@unimap.edu.my, amirafarzana@unimap.edu.my

Abstract. Extraction techniques to extract fish oil from various types of fish are numerous but not widely accepted because of the use of chemicals that may be harmful to health. In this study, fish oil is extracted using a technique of Microwave-Assisted Extraction, which uses only water. The optimum conditions required for the production of fish oil for extraction is carried out by examining three parameters such as microwave power (300-700W), extraction time (10-30 min) and amount of water used (70-190ml). Optimum conditions were determined after using design of experiments (DOE). The optimum condition obtained was 300 W for microwave power, 10 minutes extraction time and 190 milliliter amounts of water used. Fourier transform infrared spectroscopy (FTIR) was used to analyze the functional groups of fish oil. Two types of fish such as Jade Perch or UniMAP Puyu and Indian Mackerel were used. A standard omega-3 oil sample (Blackmores) purchased from pharmacy was also determined to confirm the presence of omega-3 oil in these fishes. Similar compounds were present in Jade Perch and Indian Mackerel as compared to the standard. Therefore, there were presence of omega-3 fish oil in the two types of fish. From this study, omega-3 in UniMAP Puyu fish was higher compared to Indian Mackerel fish. However, based on the FTIR analysis, besides the presence of omega-3, the two fishes also contain other functional groups such as alkanes, alkenes, aldehyde, ketones and many others. The yield of fish oil for the Jade Perch was low compared to Indian Mackerel which was 9% while Indian Mackerel was 10 %.

1. Introduction

Omega-3 is consequential to avail avert cardiovascular disease, hypertension, inflammatory and autoimmune disorders, dejection and certain disrupted neurological functions [1]. It additionally reduced the symptoms of attention deficit hyperactivity disorder (ADHD), amend lung function, phrenic adeptness and also reduce against Alzhaimer's disease and dementia. Omega-3 consists of α-linolenic acid (ALA), eicosapentaenoic acid (EPA) and docosahexaenoic acid (DHA) where the ALA from the plant and DHA and EPA from the fish.

Many research studies have found that omega-3 is good for health and can obviate many diseases. In addition, it is suitable for all walkas of life to be consumed. Today, omega-3 is utilized as a supplements in the form of tablets and gel-capsules but the gel-capsules has many advantages such as omega-3 oil are quite astronomically immense and the capsules cannot be divided and can be arduous to swallow. Omega-3 in the form of tablets is used for the supplements to the people to keep them in good conditions. In order to make omega-3 capsule as supplements, the oil is extracted out from the fish such as salmon and sardines or from plant sources such as chia seed, to get the omega-3.

Nowadays, studies have found that the content of omega-3 is higher in fish oil compared to the plant. Additionally, it was also found that UniMAP Puyu or Jade Perch, a type of fish under family of Terapontidea which have higher omega-3 as compared to the other fish [2]. Therefore, studies were conducted by utilizing UniMAP Puyu and Mackerel to compare the contents of omega-3 in the two fishes as UniMAP Puyu is known to have the high omega-3 contents (3.87% of fish flesh) known in any fish. The mackerel fish utilized in this research is Indian Mackerel (commonly called Kembung) because they are abundant in Malaysia. The incipient engenderment of omega-3 were conducted utilizing UniMAP Puyu because the content of omega-3 is higher than mackerel.

The content of the omega-3 is quite high in fish oil compared to the fish itself. Extraction method was utilized in order to obtain the fish oil. There are several extraction methods that can be used such as microwave assisted ectraction and hexane extraction. Microwave assisted extraction (MAE) was utilized in this research due to the high effiency and yield of the final compounds, which increases the integrated value of the final products and their price in the international market. Microwave assisted extraction is non-conventional extraction method in which it works with the same basic principle but improved by heating the moisture inside the cells. In such a situation, the water compound evaporated resulting high pressure on the cell membrane. High pressure on the cell membrane enchance the porosity of the biological matrix and thus increase the extraction efficiency on fish oil releasing to the extracting solvent. Beside that, MAE enables faster extraction time, lower energy and solvent consumption as well as lower environmental pollution because water is used rather than other harmful solvents such as hexane, methanol in this extraction method. Therefore, it consideras a green technology to be explored in this research.

2. Materials and Methods

2.1. Materials

UniMAP puyu (Jade Perch) and Mackerel were used as the sample in this experiment.

2.2. Preparation of Sample

The samples were gutted, and washed with sufficient amounts of water. The fishes were stored at -20°C, and then freeze dried at a drying temperature of -47°C, under a vacuum of 0.133 bar. The dried samples were then kept in desiccators for analysis.

2.3. Extraction by using MAE

MAE used was the greatest working frequency which was 2450 MHz. 2 g of test sample was weighed and placed in round bottom flask with addition of water (70, 130 and 190 ml) was added to every vessel. The incubating periods used were 10 minutes, 20 minutes and 30 minutes at different power of the microwave of 300 W, 500 W and 700 W. The mixture was cooled and the solvent was separated from the solid material by using a centrifuge and rotary evaporator [3].

2.4. Optimization via RSM

An experimental design consisted of 3 coded factors including volume of water, extraction time and microwave power was used to investigate and validate the extraction parameters for screening purposes. Software was employed for the design of experiments, model building and data interpretation. The three-dimensional response surface plots were generated by keeping one response variable at its optima and plotting the other two independent factors with the response. In all the optimization experiments using MAE, 2 g of the sample was used and the 3 coded factors generated by the software included: volume of water of 70 ml, 130 ml and 190 ml; microwave power 300 W, 500 W and 700 W. A summary of the extraction parameters with 20 runs in a quadratic design mode was employed. After the microwave treatment, the extract was analyzed by Fourier Transform

Infrared Spectroscopy (FTIR). The predicted solution by the software was confirmed by actual experimental runs in duplicate.

2.5. Functional Group Analysis by using FTIR

About 100-200 mg of sample was mixed with powdered KBr and placed in the mortar. An unknown sample was added about 2 spatula tips (not more than 1-2 mg). The samples were mixed well and ground together till uniform. As the KBr absorbed water from the atmosphere, it was done as fast as possible. KBr pallet was used to determine the functional group using FTIR technique. 30-50 mg of the mixture was added into the cavity. The second bolt was screwed and tighten with the ratchets. The bolts was removed and loosed, leaving the pallet place. To make a new pallet, the press was rinsed with water and followed by acetone and the process was repeated. A spectrum was recorded by placing pallet holder with pellet in place into the transmission cell holder in sample chamber of the FTIR instrument.

3. Result and Discussion

3.1. Functional Group Analysis by FTIR

Fourier Transform Infrared Spectroscopy (FTIR) was used to determine the functional group of the UniMAP Puyu fish and Indian Mackerel fish. To determine the presence of omega-3 in these fishes, the functional groups were compared with omega 3 bought at the pharmacy which is use as standard omega 3. Figure 1 (a) shows the spectrum functional group analysis of UniMAP Puyu using the FTIR techniques, Figure 1 (b) shows the spectrum functional group analysis of Indian Mackerel using the FTIR techniques, and Figure 1 (c) shows the spectrum functional group analysis of standard fish oil using the FTIR techniques.

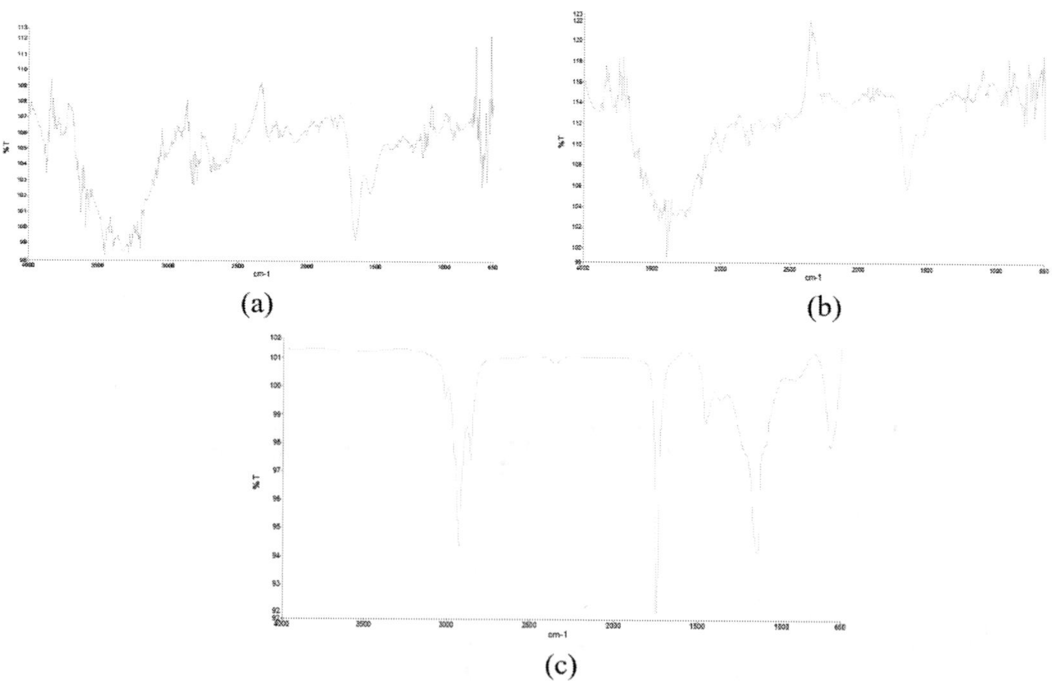

(a) (b) (c)

Figure 1. Spectra functional group analysis (a) UniMAP Puyu, (b) Indian Mackerel and (c) Standard fish oil.

3.2. Screening of Fish Oil Production from UniMAP Puyu and Indian Mackerel

Design of Experiment (DoE) software was used to determine the optimum condition for the fish oil production for UniMAP Puyu and Indian Mackerel. A 2-level factorial design, with 3 factors which are A for microwave power from 300 to 700 W, B for extraction time from 10 to 30 min and C for amount of water from 70 to 190 ml were employed. The run was set for 2 replicates and 3 center point per block. These 3 factors were used to determine the relationship on a response, which is the amount of fish oil extracted, Yb. The 3 factors which was replicated twice and 3 center points corresponded to 19 experimental runs were suggested by the DoE software. Table 1 shows the factors and the range that were set in DOE software for UniMAP Puyu and Indian Mackerel.

Table 1. Parameters and the range set in DOE software for UniMAP Puyu and Indian Mackerel

Parameter	Range	
	Low Actual	High Actual
Microwave power (W)	300	700
Extraction time (min)	10	30
Amount of water (ml)	70	190

3.3. DOE Software for UniMAP Puyu

The screening study on UniMAP Puyu was analyzed using ANOVA as shown in Table 2.

Table 2. Analysis of Variance (ANOVA) for fish oil productions

Source	Sum of Square	df	Mean Square	F Value	p-Value Prob>F	Remarks
Model	20.98	4	5.17	47.29	<0.0001	Significant
A-power	9.77	1	9.77	89.29	<0.0001	Significant
B-time	1.89	1	1.89	17.29	0.0011	Significant
C-Amount water	8.27	1	8.27	75.57	<0.0001	Significant
AC	0.77	1	0.77	7.00	0.0202	Significant
Curvature	2.07	1	2.07	18.97	0.0008	Significant
Residual	1.42	13	0.11			
Lack of Fit	0.30	2	0.099	0.88	0.4840	Not significant
Pure Error	1.13	10	0.11			
Cor Total	24.18	18				
R-Squared	0.9357					

The factors used in the design functioned to analyze and determine the interactions between microwave power (A), extraction time (B) and amount of water added (C). The interaction of the factors that were significantly affected the response has been identified which were AC. Figure 2 shows the interaction of microwave power (A) and amount of water added (C) towards the yield of fish oil.

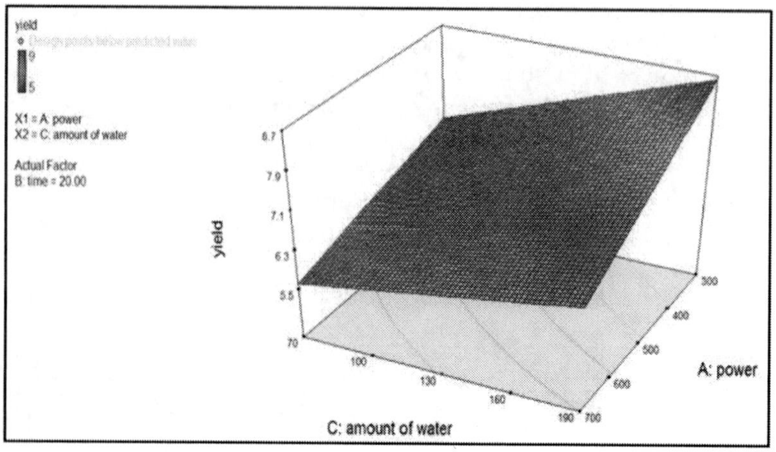

Figure 2. The interaction of microwave power (A) and amount of water added (C) and the yield of fish oil.

3.4. DOE Software for Mackerel

The Analysis of Variance (ANOVA) from the experimental results was determined. Table 3 shows the Analysis of Variance (ANOVA) for fish oil productions.

Table 3. Analysis of Variance (ANOVA) for fish oil productions

Source	Sum of Square	df	Mean Square	F Value	p-Value Prob>F	Remarks
Model	52.19	6	8.70	49.92	<0.0001	Significant
A-power	18.06	1	18.06	103.66	<0.0001	Significant
B-time	20.25	1	20.25	116.22	<0.0001	Significant
C-Amount water	7.56	1	7.56	43.40	<0.0001	Significant
AB	1.00	1	1.00	5.74	0.0355	Significant
AC	3.06	1	3.06	17.58	0.0015	Significant
ABC	2.25	1	2.25	12.91	0.0042	Significant
Curvature	0.13	1	0.13	0.76	0.4015	Not Significant
Residual	1.92	11	0.17			
Lack of Fit	0.000	1	0.000	0.000	1.000	Not significant
Pure Error	1.92	10	0.19			
Cor Total	54.24	18				
R-Squared	0.9646					

The factors used in this design were analysis to determine the interactions between microwave power (A) and extraction time (B) and microwave power (A) and amount of water added (C). The interaction of the factors that were significantly affected the response were analysed, which were AB. Figure 3 shows the interaction of microwave power (A) and extraction time (B) towards the yield of fish oil.

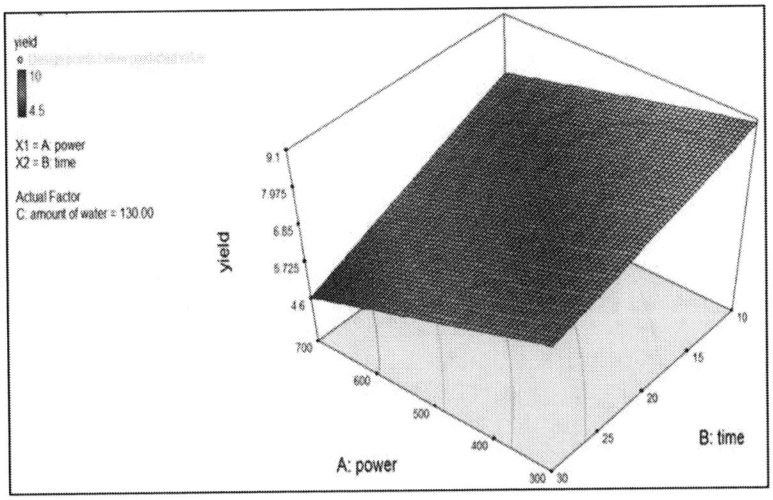

Figure 3. The interaction of microwave power (A) and extraction time (B) and the yield of fish oil.

In this study, the contents of omega-3 of fish oil from UniMAP Puyu and Indian Mackerel was compared and showed that UniMAP Puyu contains higher amount of omega-3 compared to Indian Mackerel. FTIR analysis proved that UniMAP Puyu contains more carboxylic acid than Indian Mackerel. Omega-3 consists of eicosapentaenoic (EPA) acid and docosahexaenoic (DHA) acid in the fish oil which is types of carboxylic acid. Standard sample of omega-3 (Blackmore) purchased from pharmacy was compared to the fish oil obtained from UniMAP Puyu and Indian Mackerel. Similar compounds were obtained for the standard and the fish oil extracted from the two types of fishes. Therefore, UniMAP Puyu and Indian Mackerel contained omega-3. Consumption of fishes such as UniMAP Puyu and also the cheaper variety Indian Mackerel could reduce or prevent heart disease, alleviate cardiovascular disease. Previous research done prove that omega-3 improve neural health, resolution of chronic inflammatory and prevent certain disrupted neurological functions due to their antioxidant activity where the antioxidants act as scavenging molecules [4]. So, it can aid prevention and treatment for Alzheimer's disease, autoimmune diseases and Attention Deficit Hyperactivity Disorder (ADHD). Omega-3 also explained the free radicals that are composed in the body during the cellular processes. Otherwise, elevated levels of free radicals can damage major cell components, such as DNA, proteins, and cell membranes. Damage caused by these free radicals, categorically to DNA, plays a role in the development of cancer. Since UniMAP Puyu contain high amount of omega-3 oil it could be commercialized in capsule form for easy consumption.

The omega-3 of fish oil was extracted using microwave-assisted extraction (MAE) for both fishes (UniMAP Puyu and Indian Mackerel). The method used was suitable for the extraction of the fish oil and safe for the consumer to intake orally the product produced. Microwave-assisted extraction method only used water for the extraction. This method is cost effective as only water was used instead of chemicals but it is but also environmental friendly. It could be considered as a green extraction technique.

From the DOE for the UniMAP Puyu, only the interaction of microwave power and amount of water added was significant. The yield generated was increased when the value for microwave power was decreased and high amount of water added. Highest yield was generated to maximum amount, which was approximately 9 % when the microwave power was at 300W and the amount of water added was 190 ml. However, for Indian Mackerel, the interaction of microwave power and extraction time and interaction of microwave power and amount of water added were significant. When the value for microwave power was reduced and the extraction time was lower, yield was increased. The yield was highest at approximately 10 % when the microwave power was set at 300 W and the extraction time 10 min uses. For the second interaction, the yield generated was increased as the value for microwave power was decreased and high amount of water added. Approximately 9 % yield produced when the maximum amount used was at microwave power of 300 W and the amount of water added was 190 ml.

In conclusion, MAE method considers as a green technology for the extraction of omega-3. The parameter of microwave power, extraction time and amount of water added for yield of fish oil was evaluated by using Design of Experiment (DOE). The optimum condition for the yield of fish oil produced was 9.75 % when the microwave power was at 300 W, extraction time was 10 min and the amount of water added was 190 ml.

Acknowledgement
The author gratefully acknowledges the financial support from Minister of Higher Education (MoHE) Malaysia and UniMAP under the support of Fundamental Research Grant Scheme (FRGS) and Short Term Grant (STG).

References

[1] M. Li, P. J. Pham, T. Wang, C. U. Pittman, & Li, T. (2009). Separation and Purification Technology, 66(1), 1-8.

[2] S. Osman, (21 December 15). Puyu komersial Puyu Kukum atau Jade Perch beri pendapatan lumayan kepada penternak. Utusan Online. Retrieved from http://www.utusan.com.my/bisnes/usahawan/puyu-komersial-1.17140.

[3] H. Zainab, N. Nurfatirah, A. Norfaezah, & H. Othman, (2016). Green bio-oil extraction for oil crops. IOP Conference Series: Materials Science and Engineering, 133, 012053.

[4] O. A. Byelashov, A. J. Sinclair, G. Kaur, (2015). Dietary sources, current intakes and nutritional role of omega-3 docopentaenoic acid. Lipid Technology, 27(4), 79-82.

IOP Conf. Series: Materials Science and Engineering **318** (2018) 012019 doi:10.1088/1757-899X/318/1/012019

Kinetic Parameters of Non-Isothermal Thermogravimetric Non-Catalytic and Catalytic Pyrolysis of Empty Fruit Bunch with Alumina by Kissinger and Ozawa Methods

Alina Rahayu Mohamed[1], Nurfahani Li[1], Khairunissa Syairah Ahmad Sohaimi[1], Nur Izzati Iberahim[1], Nor Munirah Rohaizad[1] and Rosniza Hamzah[1]
[1]Department of Chemical Engineering Technology, Faculty of Engineering Technology, Universiti Malaysia Perlis (UniMAP), P.O Box 77, D/A Pejabat Pos Besar Kangar, Perlis, 01000, Malaysia

E-mail: alina@unimap.edu.my

Abstract. The non-isothermal thermogravimetric non-catalytic and catalytic empty fruit bunch (EFB) pyrolysis with alumina were performed at different heating rates of 10, 15, 20, 25, 30 and 40 K/min under nitrogen atmosphere at a flow rate of 100 ml/min under dynamic conditions from 301 K to 1273 K.The activation energy were calculated based on Kissinger and Ozawa methods. Both reactions followed first order reactions. By Kissinger method, the activation energy and Ln A values for non-catalytic and catalytic EFB pyrolysis with alumina were 188.69 kJ mol^{-1} and 201.67 kJ/mol respectively. By Ozawa method, the activation energy values for non-catalytic and catalytic EFB pyrolysis with alumina were 189.13 kJ/mol and 201.44 kJ/mol respectively. The presence of catalyst increased the activation energy values for EFB pyrolysis as calculated by Kissinger and Ozawa methods.

1. Introduction

The In 2014, Malaysia had produced 19.67 million tonnes of crude palm oil. The crude palm oil processing industry created huge amount of biomass waste such as empty fruit bunch (EFB) and palm shell which required good waste management system. Due to the rapid reduction of fossil fuel, therefore developed countries pursued biomass as optional method of power generation [1]. Biomass was viewed as a potential source of renewable energy [2]. Therefore, empty fruit bunch (EFB) was targeted to replace non-renewable energy since the amount of EFB was abundant and easy to be collected. One of the promising methods for EFB utilization is via pyrolysis. Pyrolysis is a thermochemical processing of biomass under inert environment at moderate temperature of 400-700 °C to produce bio-oil, bio-char and bio-gas [3]. Thermogravimetric pyrolysis is a route to investigate the mass loss of biomass and produce information regarding the profiling of thermochemical breakdown of biomass [4;5]. The values of Ea and Ln A can be calculated from thermogravimetric data using several methods. These values are significant since the empirical information about the reaction can be elucidated. The information can be utilized in the design of reactor suitable with such reactions. Dynamic thermogravimetric pyrolysis of several biomass such as miscanthus [6], poplar wood [7], wheat straw lignin [8], rice straw [9], chestnut shells, cherry stones and grape seeds [10] and a mixture of plants tems such as cotton, wheat, rich and corn [11] had been reported. The kinetics studies were investigated by several methods such as Coats-Redfern, Kissinger and Ozawa methods. Thermogravimetric catalytic pyrolysis of biomass is an interesting process that can be manipulated in order to modify the biomass pyrolysis process at fundamental level in the presence of catalyst added [12; 13].

The pyrolysis of EFB in reactors is a common process whereby the process parameters were optimized towards maximizing bio-oil [14]. Fast pyrolysis of empty fruit bunches [15]. Bio oils from pyrolysis of oil palm empty fruit bunches [16]. Optimum conditions for maximizing pyrolysis liquids of oil palm empty fruit bunch. A more recent interests were towards addition of various types of catalyst during EFB pyrolysis in order to improve the bio-oil properties towards reduction of oxygen content and maximization of hydrocarbon compounds. Most of the catalytic pyrolysis studies on EFB pyrolysis reported using different types of reactors such as fixed-bed [17;18], tube furnace [19] and semi-batch reactor [20].The catalytic pyrolysis of empty fruit bunch is a complex reaction involving lignocellulosic decomposition assisted by catalysts. The dynamic non-isothermal catalytic pyrolysis of EFB in thermogravimetric analyser and in the presence of alumina catalyst with its relevant kinetics parameters has not been reported yet. Therefore, this paper is aimed to report the kinetics of thermogravimetric catalytic pyrolysis of EFB with alumina by Kissinger and Ozawa methods.

2. Materials and methods

2.1. EFB pretreament

EFB was obtained from North Star Palm Oil Mills, located in Kuala Ketil, Kedah, Malaysia. The biomass was rinsed with tap water to remove impurities and chopped manually. It was dried in the oven for 24 h at 343 K. The dried sample was shredded using shredding machine and sieved using the Retsch sieve shaker. The sieved EFB with particle size of 250–500 μm was selected for the pyrolysis process.

2.2. Catalyst preparation

Approximately 10 g of alumina (Al_2O_3) catalyst support, grade 135 (Sigma- Aldrich, Germany) was calcined in a furnace at 923 K for 5 h at a ramp rate of 3 K/min under static atmosphere and the catalyst was placed in a desiccators prior to application.

2.3. EFB pyrolysis at linear heating rates

The non-catalytic and catalytic pyrolysis of EFB were conducted using Mettler Toledo thermal analyser TGA/DSC 1 equipped with Star[e] software. In non-catalytic pyrolysis condition, approximately 10.00 mg of sample were placed in a 70 μl alumina crucible.

2.4. Catalytic EFB pyrolysis at linear heating rates

In catalytic condition, approximately 1.00 mg of alumina catalyst was weighed and placed inside the crucible containing 10.00 mg EFB to produce 9.09 wt% catalyst addition to biomass. The nitrogen gas was used as the carrier gas at a flow rate of 100 ml/min under dynamic conditions from 301 K towards final pyrolysis temperature of 1273 K at different rates of 10, 15, 20, 25, 30 and 40 K/min. Each heating rate produced data Excel spreadsheet and the data were calculated using Excel software with Ozawa and Kissinger methods.

2.5. Kinetic analysis using Kissinger and Ozawa method

Eq. (1) is the Kissinger equation whereas Eq. (2) is the Ozawa equation. Both Eq. (1) and (2) provided the formula for calculating the apparent activation energy (E) and the constant of ideal gas law, R. g(α) was the integral form of the kinetics mechanism function in Eq. (2) [21].

$$\text{In}\left(\frac{\beta}{T^2 p}\right) = -\frac{E}{RT} + \text{In}\left(\frac{AR}{E}\right) \quad (1)$$

$$\text{In}\,(\beta) = \text{In}\,\frac{AE}{Rg(a)} - 5.331 - 1.052\,\frac{E}{RT} \quad (2)$$

3. Results and Discussions

3.1. TG and DTG Analysis

The TG curves for both non-catalytic and catalytic EFB pyrolysis with alumina, obtained at different heating rates of 10, 15, 20, 25, 30 40 K/min, are plotted separately in Figures 1(a) and 1(b). The curves showed the thermal degradation of EFB that shifted towards the right as the heating rate was increased. The shift in the TGA curves at higher heating rates was due to different heat transfer and kinetic rates delaying the sample decomposition [22].

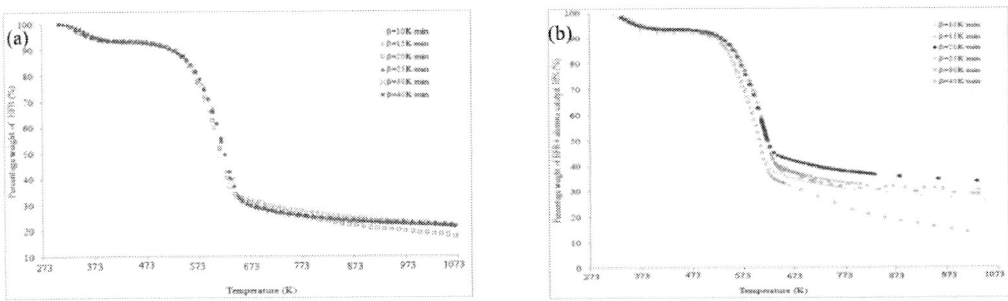

Figure 1. TG curves for (a) EFB pyrolysis and (b) catalytic EFB pyrolysis with Al.

The DTG curves for both EFB pyrolysis and catalytic EFB pyrolysis are shown in Figures 2(a) and 2(b). DTG indicates the corresponding rate of weight loss. The DTG peaks differed in their position and height, because the heating rate acted as a significant factor on lignocellulosic thermal decomposition [21].

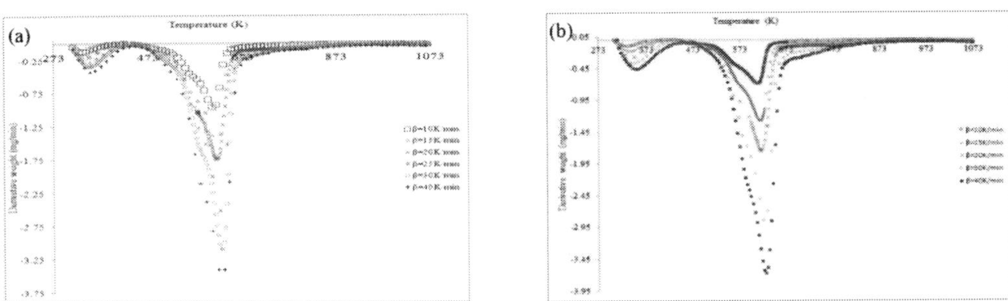

Figure 2. DTG curves for (a) EFB pyrolysis and (b) catalytic EFB pyrolysis with Al.

3.2. Kinetics analysis based on Kissinger and Ozawa method

Kissinger and Ozawa methods were respectively differential and integral thermal analysis methods which were used to determine activation energy of degradation at different heating rates [8].

3.2.1 Kissinger method. The graph of ln (β/T^2) against ($1/T$) for non-catalytic EFB pyrolysis and catalytic EFB pyrolysis with alumina are plotted in Figure 3 and Figure 4 respectively.

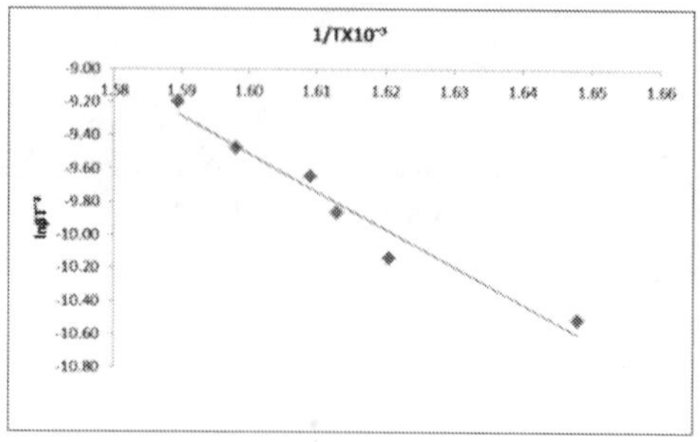

Figure 3. Linearization curves of Kissinger method for non-catalytic EFB pyrolysis.

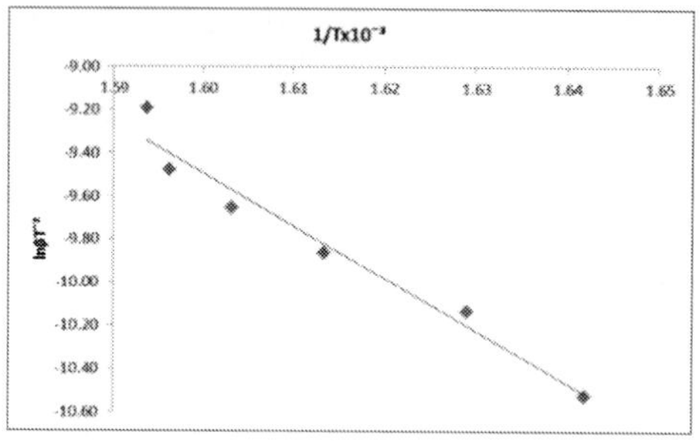

Figure 4. Linearization curves of Kissinger method for catalytic EFB pyrolysis with alumina.

3.2.2 Ozawa method.

The graph of ln β against 1/T for non-catalytic EFB pyrolysis and catalytic EFB pyrolysis with alumina are plotted in Figure 5 and Figure 6 respectively.

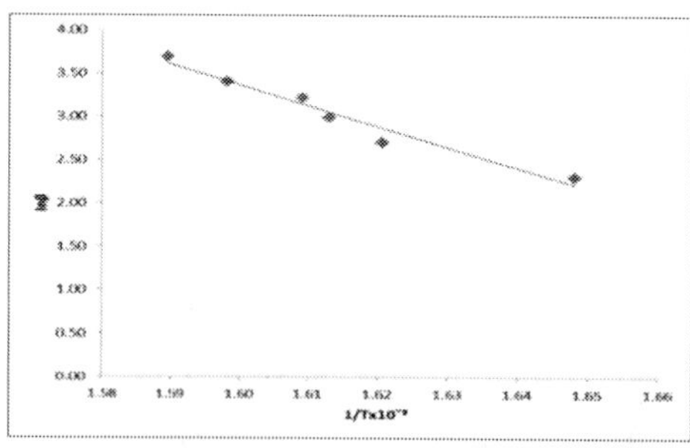

Figure 5. Linearization curves of Ozawa method for non-catalytic EFB pyrolysis.

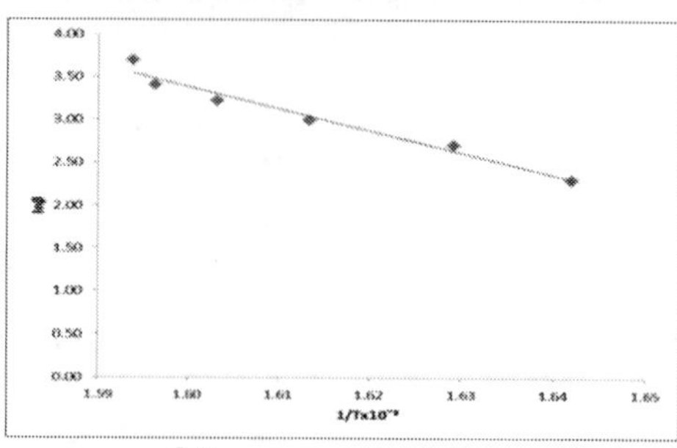

Figure 6. Linearization curve of Ozawa method for catalytic EFB pyrolysis with Al.

From Table 1, by Kissinger method, non-catalytic and catalytic EFB pyrolysis with alumina have good R^2 values which were 0.9561 and 0.9637 respectively. And by Ozawa method, both non-catalytic EFB pyrolysis and catalytic EFB pyrolysis with alumina have good R^2 values which are 0.9602 and 0.9670 respectively. The high values of coefficient of determination, R^2 indicated that there were strong correlation between the dependable and non-dependable parameters.

The kinetic parameters between non-catalytic and catalytic EFB pyrolysis with alumina between both Kissinger and Ozawa methods are also shown in Table 1. By both Kissinger and Ozawa methods, the activation energy values for non-catalytic EFB pyrolysis are 188.69 and 189.13 kJ/mol respectively which are much lower compared to activation energy values in catalytic EFB pyrolysis with alumina which are 201.67 and 201.44 kJ/mol respectively. The presence of alumina at about 9.09 wt% has increased activation energy values for EFB pyrolysis by 12.98 and 12.31 kJ/mol as calculated by Kissinger and Ozawa methods respectively. Therefore, the catalytic EFB pyrolysis has followed different pathway for reaction to proceed of which its mechanism is difficult to define. This could be due to the phase of alumina catalyst used which is suspected in the α-phase. The x-ray diffraction (XRD) analysis is in progress and will reveal the crystalline phase of alumina catalyst. It is of our interest to investigate the values of apparent activation energy by the isoconversional model-free methods such as Flynn-Wall-Ozawa method in our subsequent analysis.

Table 1. Kinetics Parameters Values for Both Kissinger and Ozawa Method.

Method	Reaction	Equation	R^2	Ea (kJ/mol)	Ln A
Kissinger	Non-catalytic EFB pyrolysis	$y = -22.70 \times 10^3 x + 26.81$	0.9561	188.69	36.84
	Catalytic EFB pyrolysis with alumina	$y = -24.26x + 29.32$	0.9637	201.67	39.42
Ozawa	Non-catalytic EFB pyrolysis	$y = -23.93 \times 10^3 x + 41.66$	0.9602	189.13	36.96
	Catalytic EFB pyrolysis with alumina	$y = -25.49x + 44.17$	0.9670	201.44	39.41

4. Conclusion

Non-isothermal thermogravimetric non-catalytic and catalytic EFB pyrolysis with alumina were conducted at different heating rates.. The values of activation energy and Ln A from non-catalytic and catalytic EFB pyrolysis with alumina catalyst were calculated by using Kissinger and Ozawa methods. By using Kissinger method, the Ea values calculated for non-catalytic is lower than catalytic EFB pyrolysis with alumina which are 188.69 kJ mol^{-1} and 201.67 kJ/mol^{-1} respectively. The results show the same pattern with Ozawa methods. The activation energy values calculated for non-catalytic is lower than catalytic EFB pyrolysis with Al which are 189.13 kJ mol^{-1} and 201.44 kJ mol^{-1} respectively.

Acknowledgement

The authors would like to express their gratitude to Malaysian Ministry of Higher Education for for the financial assistance in terms of Fundamental Research Grant Scheme (FRGS) with the number of 9003-00294.

References

[1] Olisa Y P and Kotingo K W 2014 *Eur. J. Eng. Tech.* **2** 1-7
[2] Shuit S H, Tan K T and Lee K T 2009 *Energy* **34** 1225-35
[3] Sukiran M A, Kheang L S, Bakar N A and May C Y 2011 *Am. J. Appl. Sci.* **8** 984-988
[4] Dupont C, Chen L, Cances J, Commandre J M, Cuoci A, Pierucci S 2009 *J. Anal. Appl. Pyrol.* **85** 260-267
[5] Wang Q, Zhao W, Liu H, Jia C and Xu H 2012 *Energy. Procedia.* **17**, 869 – 875
[6] Ana M C and Bridgwater A V 2015 *Fuel Process. Technol.* **138** 184–193

IOP Conf. Series: Materials Science and Engineering **318** (2018) 012019 doi:10.1088/1757-899X/318/1/012019

[7] Katarzyna S, Pietro B and Francesco F 2012 *Appl. Energy* **97** 491–497
[8] Qing Y and Shubin W 2009 *Cellul. Chem. Technol.* **43** 133-139
[9] Nithitorn K, Witchaya P and Suthum P 2015 *Energy Procedia.* **79** 663 – 670
[10] Gamzenur O and Ayse P E 2017 *Waste. Manage.* **64** 315–326
[11] Xun W, Mian H, Wanyong H, Zhihua, C, Shiming L, Hiquan H and Bo X 2016 *Bioresour. Technol.* **219** 510–520
[12] Yi Y, Tan L, Shiping J, Yixin L and Haiping Y 2011 *Bioresour. Technol.* **102** 11027–33
[13] Naqvi S R, Uemura Y, Osman N and Yusup S 2015 *Res. Chem. Intermed.* **41** 9743-55
[14] Abdullah N, Gerhauser H and Sulaiman F 2010 *Fuel* **89 (8)**, 2166-69
[15] Sukiran A, Chin C M, Bakar N K A 2009 *Am. J. Appl. Sci.* **6(5)** 869-875
[16] Sulaiman F and Abdullah N 2011 *Energy* **36** 2352-59
[17] Lim X Y and Andresen J M 2011 *Fuel Process. Technol.* **92** 1796-1804
[18] Auta M, Ern L M and Hameed B H 2014 *J. Anal. Appl. Pyrol.* **107** 67-72
[19] Yi H C, Khanh V D, Suzana Y, Mook T L, Asna M Z and Yoshimitsu U 2014 *J. Energy. Inst.* **87** 227–234
[20] Misson M, Haron R, Kamaroddin M F A and Amin N A S 2009 *Bioresour. Technol.* **100** 2867-73
[21] Yang Q, Wu S, Lou and Lv G 2010 *J. Anal. Appl. Pyrol.* **87** 65-69
[22] Seo D K, Park S S, Hwang J andYu T U 2010 *J. Anal. Appl. Pyrol.* **89** 66-73
[23] Mohamed A R and Hamzah Z 2015 *React. Kinet. Mech. Catal.* **114** 529-545

Critical Parametric Study on Final Size of Magnetite Nanoparticles

A H M Yusoff[1,*], M N Salimi[1], and M F Jamlos[2]
[1]School of Bioprocess Engineering, Universiti Malaysia Perlis, 02600 Arau, Malaysia
[2]School of Computer and Communication Engineering, Universiti Malaysia Perlis, 02600 Arau, Malaysia

E-mail: huzaifah.usm@gmail.com

Abstract. The great performance of magnetite nanoparticle in varsity field are mainly depended on their size since size determine the saturation magnetisation and also the phase purity. Magnetite nanoparticles were prepared using a simple co-precipitation method in order to study the influence of synthesis condition on the final size. Variable parameters include stirring rate, reaction temperature and pH of the solution can finely tuned the size of the resulting nanoparticles. Generally, any increase in these parameters had a gently reduction on particle size. But, the size was promoted to increase back at certain point due to the specific reason. Nucleation and growth processes are involved to clarify the impact of synthesis condition on the particle sizes. The result obtained give the correct conditions for pure magnetite synthesis at nanoscale size of dimensions less than 100 nm.

1. Introduction

Magnetite nanoparticles are by and large widely looked as a potential nano candidate materials for various application notably for biomedicine. Remarkable properties of magnetite that arise at nanoscaled level provide a promising achievement in biological field. The excellent magnetism properties, good biocompatibility and the suitability to conjoint with other materials to form various nanoformulation strategies (such as liposome, micelle and dendrimer) attains a successful transportation of therapeutic agent by lessened the reticuloendothelial system (RES) clearance [1,2]. The effectiveness of nanoparticles in biomedicine field are mainly governed by their physicochemical properties. Enormous quantity of literature demonstrated that a massive portion of magnetite properties rely upon their nanoscale size. Nanoscale magnetite particles with a range of size 1-100 nm becomes the prerequisite for excellent performance in magnetofection and drug and gene delivery application [3,4,5].

Nanoparticle sizes is possible to finely control by means of both synthesis technique and varying the synthesis condition. Among the countless techniques employed for magnetite production, co-precipitation technique is the promising choice regarding of fast preparation, economical and the simplicity of the technique. However, the difficulties to tailoring the particles' sizes is the important challenge. Thus, critical process parameter such as reaction temperature, pH of mixture and stirring rate should be considered for the production of magnetite nanoparticles with a particular final size.

2. Experimental

2.1 Materials

Iron (III) chloride ($FeCl_3$, 98%pure, anhydrous), iron (II) chloride ($FeCl_2.4H_2O$, 99%), sodium hydroxide (NaOH) were obtained from Acros Organics and were used without further purification. Deionized (DI) water was used throughout the experiment.

2.2 Synthesis of magnetite nanoparticles

Generally, the chemical reaction involve for co-precipitation of magnetite from ferrous and ferric ions precursor can be represented as:

$$Fe^{2+} + 2Fe^{3+} + 8OH^- \rightarrow Fe_3O_4 + 4H_2O \quad (1)$$

Synthesis of magnetite were carried out using stoichiometric amount of $FeCl_3$ and $FeCl_2.4H_2O$ at a mole ratio of 2:1. The ratio was maintain throughout the study to avoid phase transformation. Iron salts was dissolved in treated DI water for 20 minutes under continuous incoming nitrogen gas that gently bubbled at the bottom of the mixture. Influence of size under different stirring rate was studied from 300 rpm to 1100 rpm. The reaction mixture was heated up to desired temperature (30 °C, 50 °C, 70 °C and 90 °C) prior the addition of NaOH. The necessary alkaline pH of the solution was manipulated by NaOH. The mixture immediately turn into black suspension indicated that magnetite was formed. The mixture was left to react completely for 30 minutes. Final pH was recorded for each synthesis process. The precipitated magnetite was collected by centrifugation at 3000 rpm for 15 minutes and re-dispersed with DI water. The re-dispersed magnetite was again centrifuged for at least 3 times to remove unreacted ions. The as-synthesized magnetite was dried in vacuum oven for overnight at 60 °C for characterization.

2.3 Characterization

The samples crystal structure were determine by Panalytical X'pert Pro X-ray diffractometer using Cu Kα radiation (λ = 0.15406 nm) at room temperature. The samples powder were analysed over 2θ range from 20° to 80° at a scan rate of 1°/min.

The trend of change of crystallite size is determined by a well-known Scherrer formula. Values of full width at half maximum (FWHM) of diffraction peaks at $2\theta = 35.4°$ was chosen for calculation of the crystallite size using Scherrer's formula:

$$D_S = \frac{K\lambda}{\beta_D \cos\theta} \quad (2)$$

where D_s is the crystallite size (nm), K is the shape parameter which is 0.94 for spherical shape, λ is X-ray wavelength (λ = 0.15406 nm for Cu Kα radiation), β_D is the full width at half maximum for the diffraction peak under consideration (rad) and θ is the corresponding Bragg angle (°). For each case, the calculated crystallite sizes were tabulated in Table 1.

3. Results and Discussion

3.1 Phase identification analysis

Figure 1 depicts the XRD pattern of prepared samples at different processing conditions. All the detectable diffraction peaks were compared with the standard reference data of magnetite (JCPDS card No. 89-0691). Generally, the diffraction data obtained were in good agreement with the certified standard, indexed to the inverse spinel structure with a face-centred cubic unit cells of magnetite. The formation of magnetite normally corresponded by the diffraction peaks occurring at $2\theta = 35.4°$ (311) which appeared in all samples. The narrow and sharp peaks at this diffraction peaks suggest that the magnetite nanoparticles had a good crystallize structure. No other extraneous peaks is detected except for S4 sample which consist of (211) peak of maghemite. S4 is the smallest size found in the present study where it was noticed to easily transform into maghemite phase due to the oxidation of unstable Fe^{2+} in the octahedral sublattice into Fe^{3+} [6]. Maghemite and magnetite vary slightly in their structure where they share the same inverse spinel structure, but the formation of vacancies in the octahedral sites of maghemite make them different. The maghemite phase transformation experimentally reveal three low intensity additional peaks of (110), (210) and (211) [7]. In present case, partial oxidation particularly at the surface of the particles is the reason for the only (211) peaks is observed in S4 sample.

Exclude S4, the XRD patterns of the samples prepared at different stirring rate confirmed the existence of magnetite only. The observed peaks were narrower and more intense at higher agitation rate and slightly shifted to higher Bragg angle. Same issue also happened at higher temperature and molarity. The broadening of the peaks can be attributed to the ultra-small particles and imperfection in the polycrystallinity of Fe_3O_4 NPs whereas Bragg angle change was assigned to the lattice constant expansion [8].

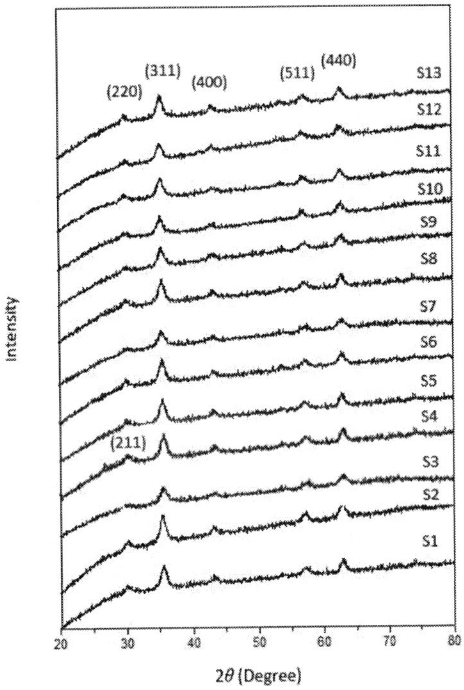

Figure 1. X-ray diffraction pattern of synthesized nanoparticles at different processing parameters.

3.2 Effect of stirring rate on magnetite particle size

In this study, the crystallite size decreased linearly from 13.25 nm to 9.64 nm and then increased to 11.78 nm for the stirring rate from 300 to 1100 rpm. Mostly, bigger and agglomerated particles are formed at a low stirring velocities. However, with a further increase in the stirring rate speed, significant effect on the particle size was observed. We observed that at higher mechanical stirring allowed the mixture for greater seed formation and consequently a smaller final particle size was obtained. At low stirring rate, the uniformity and regularity of the reaction solution are deteriorated. The nucleating species are not dispersed equally throughout the mixture and prone to aggregate [9]. The mobility and homogeneity are improved at elevated degree of agitation by hastening the transport of the growth species to the nanoparticle's surface [9]. The high energy transferred to the suspension medium caused the solution to disperse well into smaller droplets [10]. Higher shear forces resulted from the mechanical stirring greater than 900 rpm, created a lot of bubbles and splashing of the solution which cause an anomalous diffusion.

The monodispersity and growth of nanoparticles can be further explained via the growth mechanism. Essentially, size distribution of nanoparticles is controlled by their nuclei growth process where diffusion-controlled growth reaction is desired instead of adsorption-controlled growth. Therefore, increasing the stirring rates enhanced the diffusion transport of solute to the growing particle [11]. Through LaMer mechanism [12], there is a minimum degree of supersaturation (S_c) of the monomer species for nucleation to occur. By increasing the stirring rate will provide a rapid increase of the monomer concentration higher than S_c to all location of the solution and abundance of nuclei generation are instantly formed. The blowout nucleation generation rapidly depleted the free monomers bringing the concentration back down below S_c. After this stage, the nucleation is almost stopped leaving the surviving particles proceed to growth. Thus, by separating the nucleation and growth stage, each nucleus is produced at just about the same time and all particles have almost indistinguishable size [13]. Consequently, monodispersed nanoparticles are yield. Generally speaking, the degree of nucleating agent mobility are enhance by raise up the stirring rate to extend their distance travel. However, further increase of stirring rate has resulted the particles which have the size below the critical nucleus to collided and suspended to make a new nuclei and increase the crystallite size [14].

Table 1. Crystallite size of magnetite nanoparticle at variable parameter.

Sample	Synthesis condition	Size (nm)
S1	300 rpm	13.25
S2	500 rpm	13.25
S3	700 rpm	10.60
S4	900 rpm	9.64
S5	1100 rpm	11.78
S6	30 °C	21.18
S7	50 °C	11.78
S8	70 °C	10.60
S9	90 °C	30.26
S10	pH 11.33	17.66
S11	pH 11.94	13.15

S12	pH 12.15	10.10
S13	pH 12.25	17.67

3.3 Effect of solution temperature on magnetite particle size

The effect of the solution's temperature is another parameter to take into consideration. The use of temperature from 30 to 70 °C led to a particle size decrease down from 21.28 to 10.60 nm. Low temperature strongly influence the rate of particle growth through the effect on supersaturation [15,16]. The threshold precipitation temperature for instantaneous spinel magnetite formation is at 20 °C, whereas goethite compound is the main phase found at below 20 °C instead [17]. The only magnetite phase form in this study at selected temperature range showed that the result are in agreement as previously reported [17]. We suggested that at low temperatures, insufficient energy own by the nucleating species resulted a slow diffusion rate. Thereby, allowed the nucleating species to grow larger through a mass number of aggregated ions. Gradual reduction of the size obtained when the temperature was keep increasing can be attributed to the expedited collision rate of nuclei that derived from low solution viscosity [15]. Besides that, there is more energy within the particles that can reduce the extent of aggregation.

The changes in temperature up to 90 °C alter the mean sizes of the magnetite from 10.60 nm to 30.26 nm. The temperature above 70 °C might be the point where the deviation from a normal growth mechanism to Ostwald ripening-type phenomenon. The size of the nanoparticles are getting bigger at temperature of 90 °C that resulted from rapid growth of nanoparticles and afterwards anomalous aggregation are formed. The mechanism of growth is caused by the increment in critical radius requirement and surface energy of particles [14]. The dissolution of unstable particles smaller than the critical radius and the redeposition of particles onto stable nuclei leads to an increased in particle size. Earlier studies also reveal that the final size was dramatically increase for the temperature of 90 °C [18] and these findings are consistent with our results. The significant size changes at elevated temperature can be hindered by introduce oleic acid as the surfactant agents during the initial crystallization phase of magnetite. Oleic acid impede the particle growth by provide the esterification process between its carboxylic acid group and the hydroxyl group on the magnetite surface [19].

3.4 Effect of pH of solution on magnetite particle size

As indicated in Table 1, the size of magnetite nanoparticles can be further tuned by varying the pH of solution. The crystallites sizes of the nanoparticles reduced from 17.66 to 10.10 nm when the pH of solution was keep increasing from 11.33 to 12.15. Darminto and his co-workers suggested that the predominant growth mechanism at low pH is the well-known Ostwald ripening process [20]. The physics driving for dissolution-crystallization process during crystal growth is the interfacial tension that results in greater final diameter [21]. On increasing the pH of the solution, excess ions are available on the growth phase which leads to lower the interfacial tension. The reaction rates are faster at this state and more initial nuclei are able to precipitate to form slightly smaller particles.

The pH of solution also controlled the protonation-deprotonation equilibria of surface introduction of hydroxyl ions [22]. Its effect was obviously seen at pH beyond 12.15 with a significant size increment. The continuous raise of pH during synthesis period promotes the diabetic barrier of electron transfer which favour the Fe^{3+} ions to have higher affinity for hydroxide ions relative to Fe^{2+} [23,24]. In terms of chemical reaction euphemistically the reaction followed different reaction mechanism instead of Eq. 1. Lian *et al* [25] proposed a new reaction pathway for magnetite production at high pH:

$$Fe^{3+} + 3OH^{-} \rightarrow Fe(OH)_3 \quad (3)$$

$$Fe(OH)_3 \rightarrow FeOOH + H_2O \quad (4)$$

$$Fe^{2+} + 2OH^{-} \rightarrow Fe(OH)_2 \quad (5)$$

$$2\,FeOOH + Fe(OH)_2 \rightarrow Fe_3O_4 + 2H_2O \quad (6)$$

The stabilization of Fe^{3+} ions resulted for $Fe(OH)_3$ to first precipitated (Eq. (3)). $Fe(OH)_3$ also can be translated into FeOOH (Eq. (4)) or well known as goethite. Then, $Fe(OH)_2$ reach their solubility product due to the excessive hydroxyl ions in the system (Eq (5)). Lastly, $Fe(OH)_2$ that afterwards diffuse into the surface of FeOOH nucleates to complete the reaction (Eq. (6)). Thus, the explanation for the larger particles is the proposed reaction speed up the goethite formation as compared to Eq. 1 at the same reaction time. Larger goethite nucleate was yield before introduction of $Fe(OH)_2$. Some literature reported [24,26,27] that goethite was present at high pH which also can be explained using the same reaction pathways. Inadequate reaction time for enough $Fe(OH)_2$ to react with FeOOH to completely form into magnetite leaves a noticeable amount of unreacted FeOOH.

4. Conclusion

The experimental results suggest that the process parameters greatly control the final size of the magnetite nanoparticles ranging from 9.64 nm to 30.26 nm. Respect to this study, it can be concluded that the mean crystallite size decreases with increasing stirring rates and at some point the crystallite size was reported to raise up back. Same behaviour also observed for changes of reaction temperature and pH of solution where each findings has its particular reason. Roughly, the outcome size in the present study is suitable for biomedical application in the fact that all sizes are below 100 nm. However, phase changes can evolve such as maghemite and goethite which is due to the susceptible to oxidation for certain small size and high pH effect, respectively. In conclusion, these finding are useful for the synthesis of magnetite nanoparticles at desired size, while preserve the pure phase of magnetite for adoption of various application.

Acknowledgements

We would like to express sincere appreciation to the Ministry of Higher Education (MOHE) for the scholarship and also the RAGS grant given that funded this research.

References

[1] Hillaireau H and Couvreur P 2009 *Cellular and molecular life sciences : CMLS* **66** 2873
[2] Thomsen L B, Lichota J, Eskehave T N, Linemann T, Mortensen J H, du Jardin K G and Moos T 2012 *Curr Pharm Biotechnol* **13** 2349
[3] Faraji A H and Wipf P 2009 *Biorg. Med. Chem.* **17** 2950
[4] Marcu A, Pop S, Dumitrache F, Mocanu M, Niculite C M, Gherghiceanu M, Lungu C P, Fleaca C, Ianchis R, Barbut A, Grigoriu C and Morjan I 2013 *Applied Surface Science* **281** 60
[5] Jeannot V, Mazzaferro S, Lavaud J, Vanwonterghem L, Henry M, Arboleas M, Vollaire J, Josserand V, Coll J L, Lecommandoux S, Schatz C and Hurbin A 2015 *Nanomedicine*
[6] Sidhu P S, Gilkes R J and Posner A M 1977 *J. Inorg. Nucl. Chem.* **39** 1953
[7] Serna C J and Morales M P 2004 *Maghemite (γ-Fe2O3): A Versatile Magnetic Colloidal Material* (Madrid: Springer)
[8] Cullity B D and Stock S R 2001 *Elements of X-ray diffraction*, (New York: Prentice-Hall)

[9] Prabu S B, Karunamoorthy L, Kathiresan S and Mohan B 2006 *Journal of Materials Processing Technology* **171** 268
[10] Sun J, Zhou S, Hou P, Yang Y, Weng J, Li X and Li M 2007 *Journal of biomedical materials research. Part A* **80** 333
[11] Cao G 2004 *Nanostructures & Nanomaterials: Synthesis, Properties & Applications* (London: World Scientific)
[12] LaMer V K and Dinegar R H 1950 *Journal of the American Chemical Society* **72** 4847
[13] Wen T, Brush L N and Krishnan K M 2014 *Journal of Colloid and Interface Science* **419** 79
[14] Kim T-H, Lim D-Y, Yu B-S, Lee J-H and Goto M 2000 *Industrial & Engineering Chemistry Research* **39** 4702
[15] Salimi M N, Bridson R H, Grover L M and Leeke G A 2012 *Powder Technol.* **218** 109
[16] Myerson A S and Ginde R 2002 *Crystals, crystal growth, and nucleation* (Woburn: Butterworth-Heinemann)
[17] Belin T, Guigue-Millot N, Caillot T, Aymes D and Niepce J C 2002 *J. Solid State Chem.* **163** 459
[18] Petcharoen K and Sirivat A 2012 *Materials Science and Engineering: B* **177** 421
[19] Liu X, Kaminski M D, Guan Y, Chen H, Liu H and Rosengart A J 2006 *J. Magn. Magn. Mater.* **306** 248
[20] Darminto, Cholishoh M N, Perdana F A, Baqiya M A, Mashuri, Cahyono Y and Triwikantoro 2011 *AIP Conference Proceedings* **1415** 234
[21] Cao G and Wang Y 2011 *Nanostructures and Nanomaterials: Synthesis, Properties, and Applications* (London: World Scientific)
[22] Vayssières L, Chanéac C, Tronc E and Jolivet J P 1998 *J. Colloid Interface Sci.* **205** 205
[23] Rustad J R, Rosso K M and Felmy A R 2004 *The Journal of Chemical Physics* **120** 7607
[24] Gnanaprakash G, Mahadevan S, Jayakumar T, Kalyanasundaram P, Philip J and Raj B 2007 *Mater. Chem. Phys.* **103** 168
[25] Lian S, Wang E, Kang Z, Bai Y, Gao L, Jiang M, Hu C and Xu L 2004 *Solid State Commun.* **129** 485
[26] Andrade Â L, Souza D M, Pereira M C, Fabris J D and Domingues R Z 2010 *Quim. Nova* **33** 524
[27] Mahadevan S, Gnanaprakash G, Philip J, Rao B P C and Jayakumar T 2007 *Physica E: Low-dimensional Systems and Nanostructures* **39** 20

IOP Conf. Series: Materials Science and Engineering **318** (2018) 012021 doi:10.1088/1757-899X/318/1/012021

Incremental Dynamic Analysis of Koyna Dam under Repeated Ground Motions

Nik Zainab Nik Azizan[1,2], Taksiah A. Majid[2], Fadzli Mohamed Nazri[2], Damodar Maity[3], Junaidah Abdullah[2]
[1]School of Environmental Engineering, Universiti Malaysia Perlis, Kompleks Pusat Pengajian Jejawi 3, Universiti Malaysia Perlis, 02600 Arau Perlis, Malaysia
[2]School of Civil Engineering, Engineering Campus, Universiti Sains Malaysia, 14300 Nibong Tebal, Penang, Malaysia
[3]Department of Civil Engineering, India Institute of Technology, Kharagpur, West Bengal 721302, India

E-mail: nikzainab@unimap.edu.my

Abstract. This paper discovers the incremental dynamic analysis (IDA) of concrete gravity dam under single and repeated earthquake loadings to identify the limit state of the dam. Seven ground motions with horizontal and vertical direction as seismic input considered in the nonlinear dynamic analysis based on the real repeated earthquake in the worldwide. All the ground motions convert to respond spectrum and scaled according to the developed elastic respond spectrum in order to match the characteristic of the ground motion to the soil type. The scaled was depends on the fundamental period, T_1 of the dam. The Koyna dam has been selected as a case study for the purpose of the analysis by assuming that no sliding and rigid foundation, has been estimated. IDA curves for Koyna dam developed for single and repeated ground motions and the performance level of the dam identifies. The IDA curve of repeated ground motion shown stiffer rather than single ground motion. The ultimate state displacement for a single event is 45.59mm and decreased to 39.33mm under repeated events which are decreased about 14%. This showed that the performance level of the dam based on seismic loadings depend on ground motion pattern.

1. Introduction

Every dam location has its unique geological characteristics. The high cost is required for carrying out a geological survey to understand the geological characteristics of an area before deciding on a suitable location to construct a dam. A dam is usually built in a valley area that can be at active fault areas, such as the Himalayas, Southwest China, Iran, Turkey, and Chile [1]. Dams are damaged by earthquakes exceeding 6 M (Richter magnitude) occurred in Hsinfengking, China in 1962; Kariba, Zambia–Zimbabwe Border in 1963; Kremasta, Greece in 1966; and Koyna, India in 1967 [2]. In 1999, serious damage to the Shih-Kang dam caused the dam body to crack and separation of dam body from foundation bedrock to occur due to Chi-chi earthquake with 7.6 M [3] as shown in Figure 1. Severe damage to the structure of the dam would cause flooding to occur in the D/S direction and electricity and water supply will be interrupted for the area involved. Flooding is also able to cause the loss of life and a lot of the other structures and infrastructure damages. The person involved in the design, construction and operation of a dam should be sensitive to earthquake issues.

Seismic sequences characterised by the repetition of medium-strong earthquake ground motions occurring in a short time after other shocks have been observed in many parts of the world. The characteristic of repeated earthquake studied by Allen (1968), Sieh (1978) and Schwartz and Coppersmith, (1984), identifies the repeated earthquake events based on fault, location and magnitudes for the San Andreas fault and Wasatch faults. The structural damage has occurred in the middle of Italy caused by repeated earthquake events in 1997 investigated by Decanini et al. (2000), the primary characteristic and the impact of seismic sequences. Elnashai et al. (1998) expected the repeated earthquake event able considerably higher in ductility demand instead of a single earthquake event. This has been ratified by Faisal (2012), identify the ductility demand increased by 1.4 and 1.3 times when considering double and triple earthquake events. The dispersion of response demand of SDOF structure increased at the strong aftershock ground motions with respect to mainshock ground motions alone [10].

Figure 1. Shih-Kang dam damage due to earthquake

From the repeated earthquakes considered in the previous studies is a time gap between two events is considered as zero acceleration, which means to stop the motion of any structure due to the damping before the start of the next event. The time gap 100 s has been used a few researchers [11–13] compared to 40 s introduced by Amadio et al. (2003). This is caused by non-movement on the structure before the next event. The examples of repeated earthquake shown are Figure 2 with a different time gap.

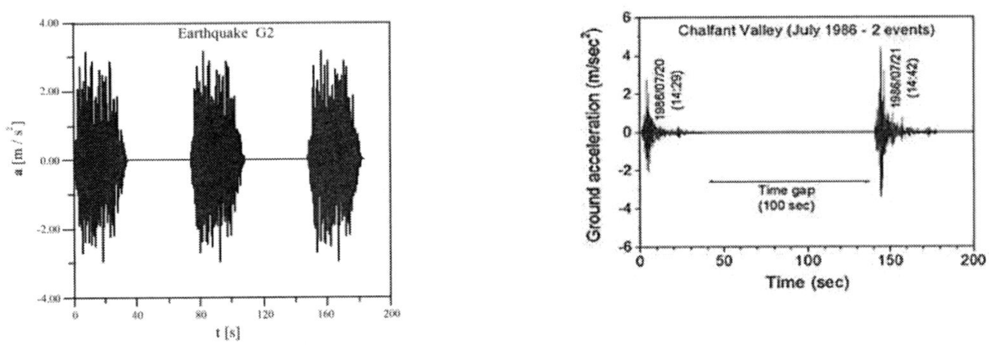

a) Amadio et al. (2003) b) Efraimiadou et al. (2013)

Figure 2. Examples of repeated earthquake

The dam structure has already been damaged during the first individual seismic event, and the damage that is not yet repaired may increase the damage to the dam at the end of the seismic sequences [15]. Seismic sequences will lead to greater accumulated damage to the structure; and therefore, the seismic design should be given significant attention [15–18].

The performance of structure under seismic loads can be approximated more details by using IDA which is a parametric analysis method. The ground motion record equals to or more than one, scale to multiple levels of intensity based on several intensities of response spectral applied to the structural model and hence producing the curves of response parameterized vs intensity level [19]. This concept has been implemented by U. S. Federal Emergency Management Agency (FEMA) guidelines and establish as the method to determine global collapse capacity [20]. IDA curve of the structure response is normally plotted as damage measure (DM) versus intensity measure (IM). The DM can be identified as peak roof drift ratio θ_{roof} or θ_{max} while the IM as peak ground acceleration (PGA), peak ground velocity (PVA) or the 5% damped the first mode spectral acceleration ($S_a(T_1, 5\%)$).

Concrete dams are one of the largest structures constructed by humanity, which their seismic response can be affected by the presence of various contingency. Since limited data is available for the actual concrete dams on the field and laboratory tests, the limit state of concrete dams in light of uncertainties calculation is based on theoretical and numerical methods [21]. Alembagheri and Ghaemian (2012) introduced a new concept to verify the limit state of the damage in concrete dams by using IDA and SPO analysis that not only consider initial loads but also the hydrodynamic loads. The methods can be applied to obtain indexes for seismic performance evaluation and damage assessment.

The effects of the repeated earthquake to the structures have been studied since 1990's. The damage of the structure due to the first earthquake event will be accumulated after the second earthquake happened. Most of the researchers focus on the structure of buildings and bridges to define the performance of structure due to the repeated earthquake. Since limited data is available for seismic sequences, the previous study creates the artificially repeated earthquake as seismic loading was not adequate. Therefore, the repeated earthquake should be significant attention to the concrete dam as analysed the structure of buildings and bridges subject to real seismic sequences, for example, Chi-Chi earthquake in 1999 cause the serious damage to Shih-Kang dam. The seismic sequences data have been recorded within short duration with the same station, direction and fault distance.

2. Numerical modeling

Koyna dam is located on Koyna River in the west of the Indian Peninsula. The dam has experienced with the crack damage caused by Koyna earthquake in December 1967 with magnitude 6.5 [23,24]. However, the dam not in danger from a major failure and only required of repairs and permanent strengthening. The structure of the dam has an elongated geometry with constant cross section. The structure of the dam can be determined by plane strain condition because the loading position does not differ along the trend. The depth of the reservoir for Koyna dam is 91.8m and the height of the dam is 103m. The full dimension of Koyna dam as illustrated in Figure 3. The rigid foundation has been considered for this model and the material properties for this model are tabulated in Table 1. The present study uses the concrete damage plasticity (CDP) model to evaluate the nonlinear seismic performance of the concrete gravity dam [25,26]. The concrete damage plasticity model is to simulate the nonlinear constitutive behaviour of concrete by presenting scalar damage variable to describe the irreversible damage during loading process [27–30].

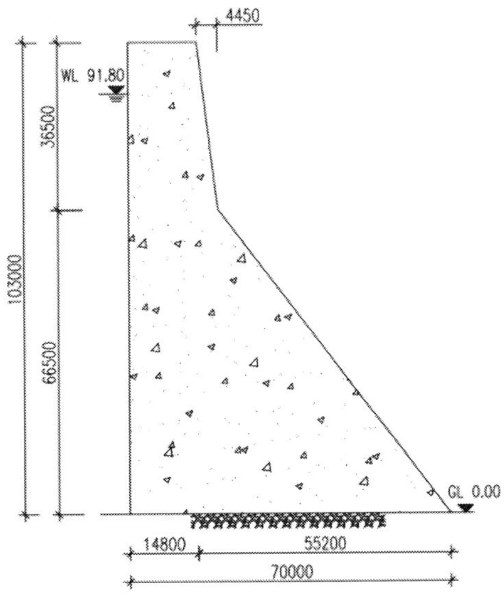

Figure 3. Dimension of Koyna Dam

Table 1 Concrete properties

Material parameter	Value
Modulus of elasticity (E)	31513 MPa
Poisson's ratio (ν)	0.2
Density (ρ)	2643 kg/m^3
Dilation angle (ψ)	36.31°
Compressive initial yield stress (σ_{c0})	13.0 MPa
Compressive ultimate stress (σ_{cu})	24.1 MPa
Tensile failure stress (σ_{t0})	2.9 MPa
Damping for the first mode vibration	3%

3. Incremental dynamic analysis (IDA)

IDA is the performance-based earthquake engineering approach for seismic performance evaluation. IDA involves applying a series of nonlinear time history analysis to a structure for numerous ground motion records by scaling every record to several levels of intensity to determine the full range of the structure's performance from elastic to yielding, nonlinear inelastic and ultimately leading to global instability [19].

The real ground motion is scaled with the design respond spectrum according to Eurocode 8 (2004). The concrete dam is subjected to a set of ground motion which are scaled to multiple levels of intensity until the structure failed [22,32]. The IDA curve represented the maximum crest displacement to monitors the dynamic response of the structure with different concrete strength caused by degradation process. In this study, IDA carried out has covered: (i) fourteen ground motion which is seven real repeated and seven real single ground motions, and (ii) twenty types peak ground accelerations. Total numbers of 196 nonlinear time history analysis have been carried out in the IDA.

The seismic input in this study is divided into two categories which are a single ground motion; and the real repeated ground motion. The criteria for selecting the ground motion are as follows: (i) the distance from the epicenter is considered less than 15km, (ii) the events more than one in the same direction and fault distance (repeated earthquake), (iii) the magnitude is equal to or greater than 5.5, and (iv) the PGA is equal to or greater than 0.15g. Seven earthquake records based on these criteria are downloaded from the strong-motion database of the Pacific Engineering Research Center [33] as shown in **Table 2**. The single ground motion is picked the highest PGA for repeated events.

Table 2 Summary parameters of the selected ground motion

No	Seismic	Date	M	PGA-H (g)	PGA-V (g)
1	Mammoth Lakes	25 May 1980	5.7	0.485	0.312
		27 May 1980	5.9	0.315	0.190
2	Chalfant Valley	20 July 1986	5.8	0.270	0.201
		21 July 1986	6.2	0.447	0.324
3	Northwest - China	5 April 1997	5.9	0.274	0.192
		6 April 1997	5.9	0.125	0.081
4	Northridge	17 Jan 1994	6.7	0.474	0.265
		17 Jan 1994	6.0	0.235	0.031
5	Coalinga	22 July 1983	5.8	0.124	0.083
		25 July 1983	5.8	0.143	0.153
6	Chi-chi	20 Sept 1999	7.6	1.009	0.320
		20 Sept 1999	6.2	0.139	0.050
7	Loma Prieta	18 Oct 1989	6.9	0.179	0.095
		18 Oct 1989	6.9	0.484	0.089

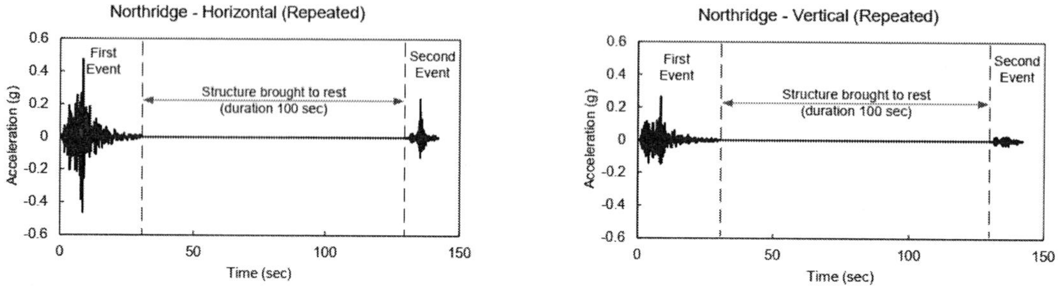

Figure 4. Repeated earthquake with 100 seconds gap

The repeated earthquakes in this study are based on the real repeated events in worldwide with the same seismic station and the duration of the events between 2 to 3 days are as tabulated in Table 2. The gap between two consecutive ground motions is 100 seconds with zero acceleration of amplitude Figure 4. This gap has been suggested by Hatzigeorgiou and Liolios (2010) and is absolutely enough to achieve for the structure to be brought to rest after vibration due to its inherent damping. The single earthquake input for this study was chosen based on the maximum horizontal PGA of each seismic event. For example, the PGA for Mammoth Lake event on 25 May 1980 was 0.485g and 0.315g on 27 May 1980. The single event for this case was 0.485g.

The response spectrum of the 14 ground motion input data which are single and repeated events as tabulated in Table 2, has to be scaled according to the site elastic response spectrum to correspond with the characteristics of the ground motion records to the site condition. The scaling method utilised the spectrum acceleration at fundamental period of the structure, $S_e(T_1)$ as the intensity measurements of the ground motion as suggested by Shome et al. (1998). The time history record of each ground motions is converted to the acceleration response spectrum (g) by using the SeimoSignal software. Then, the acceleration response spectrum of each ground motion is scaled to the same pseudo-spectrum acceleration at the fundamental period of the structure, T_1 in the elastic response spectrum Figure 5.

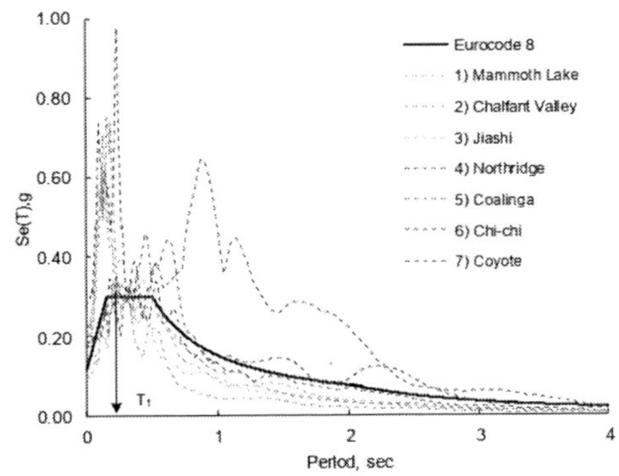

Figure 5. Scaling of ground motions to the same spectrum acceleration at T_1

Plotting an IDA curve required scaled up or scaled down of the ground motion intensity measure (IM) and structural state variable which is damage measure (DM) or engineering demand parameter. An IM is the intensity of a ground motion record by increasing the scalar factor and using peak ground acceleration (PGA) for this study. A DM is the state of the structural model to a specified seismic loading that can be determined from the output of the nonlinear dynamic analysis [19]. The output considered in this study is the maximum crest displacement of the dam to identify the limit states of the concrete dam. The limit state was based on the structural behaviour from the early elastic to the failure.

4. Results and discussion

4.1 Validation

The dam has been analysed using two-dimensional plane strain formulations. To validate the accuracy of the numerical model, the results available for the Koyna dam were considered [35]. The assumed material properties as tabulated in Table 1. The comparisons of the results indicate the validity of the present numerical model based on the natural period as tabulating in Table 3 which is the percentage difference is less than 3 percent.

Table 3 Natural frequencies (rads/sec) of Koyna dam

Mode	Present	Nayak and Maity (2013)	% of deviation
1	18.87	19.27	2.08
2	50.09	51.50	2.74
3	68.17	67.56	-0.90
4	98.70	99.73	1.03

4.2 Comparison single and repeated ground motions

The original model which is an unscaled ground motion for single and repeated earthquakes are shown in Figure 6. The cracking trend performs at the neck of the dam from the D/S face and the cracking appeared at the heel of the dam from the U/S face. The displacement under repeated ground motion increase about 3% to 51% compared with the single event. The highest displacement increase is Chalfant Valley event where the single event is 40.87 mm and increases to 82.62 mm for the repeated events. However, there is some ground motion event shows the displacement decreased under repeated ground motions which are Northridge and Coyote events. This happened cause of the PGA value for the second ground motion event lower than the first event. Damage accumulation on structures under repeated ground motion as mention by Zhang et al.

4.3 Incremental dynamic analysis (IDA)

A total number of 198 analysis was conducted to develop the IDA curve for Koyna dam. The cracking patterns were recorded for each analysis with different ground motions. Figure 7 and Figure 8 shows the cracking patterns for Koyna dam under single and repeated ground motions event with different intensity measure (IM).

The cracking began at the dam body from the neck of the dam on the downstream (D/S) face. Later on, a partial crack occurred at the base of the dam at the upstream (U/S) face and as a general trend. When the PGA of the motion increased, the cracking was created at the neck of the dam from the U/S face and then moved forward from the both face until cracks were joined, that can be concluded as collapse. In the meantime, the cracking at the heel area increased around 15% to 40% from the width of the dam. The first crack occurred on the neck at displacement around 20.72mm to 27mm with PGA around 0.100g to 0.175g which is considered as a yielding state. The ultimate state of the dam with 36.27mm to 47.31mm displacement and the PGA value between 0.225g to 0.275g. Meanwhile, the yielding and ultimate state under repeated ground motions, the maximum crest displacement are significantly different with a single ground motion which is the values between 20.72mm to 29.43mm and 36.27mm to 40.14mm with the same PGA value under single ground motions.

The IDA curves have been plotted in Figure 9 for a single ground motion and the repeated in Figure 10. The slope of the graphs start from 0 to 0.05g are shown the maximum crest displacement are same values for 7 ground motions. Later on, the curves start change according to the ground motion pattern. The yielding and ultimate region have been marked in the graphs Figure 9 and Figure 10 for single and repeated ground motions. It is shown that the yielding state under single ground motion is the lower displacement values rather than repeated. However, the ultimate state results under repeated ground motions outstrip a single earthquake loading. Figure 11 shows the IDA curves for Chalfant Valley to comparing the repeated and single ground motions. The maximum displacement under repeated ground motion increased about 7% for the yielding state while 14% decreased for the ultimate state. That is mean, the ultimate state of repeated ground motion reached earlier than single ground motion.

Earthquake Events	Mammoth Lake	Chalfant Valley	Northridge	Chi-chi	Coyote
Single	 42.04 mm	 40.87 mm	 29.85 mm	 165.71 mm	 43.64 mm
Repeated	 42.04 mm	 82.62 mm	 28.81 mm	 172.27	 42.13 mm

Figure 6. Cracking patterns and maximum crest displacement results with original ground motions

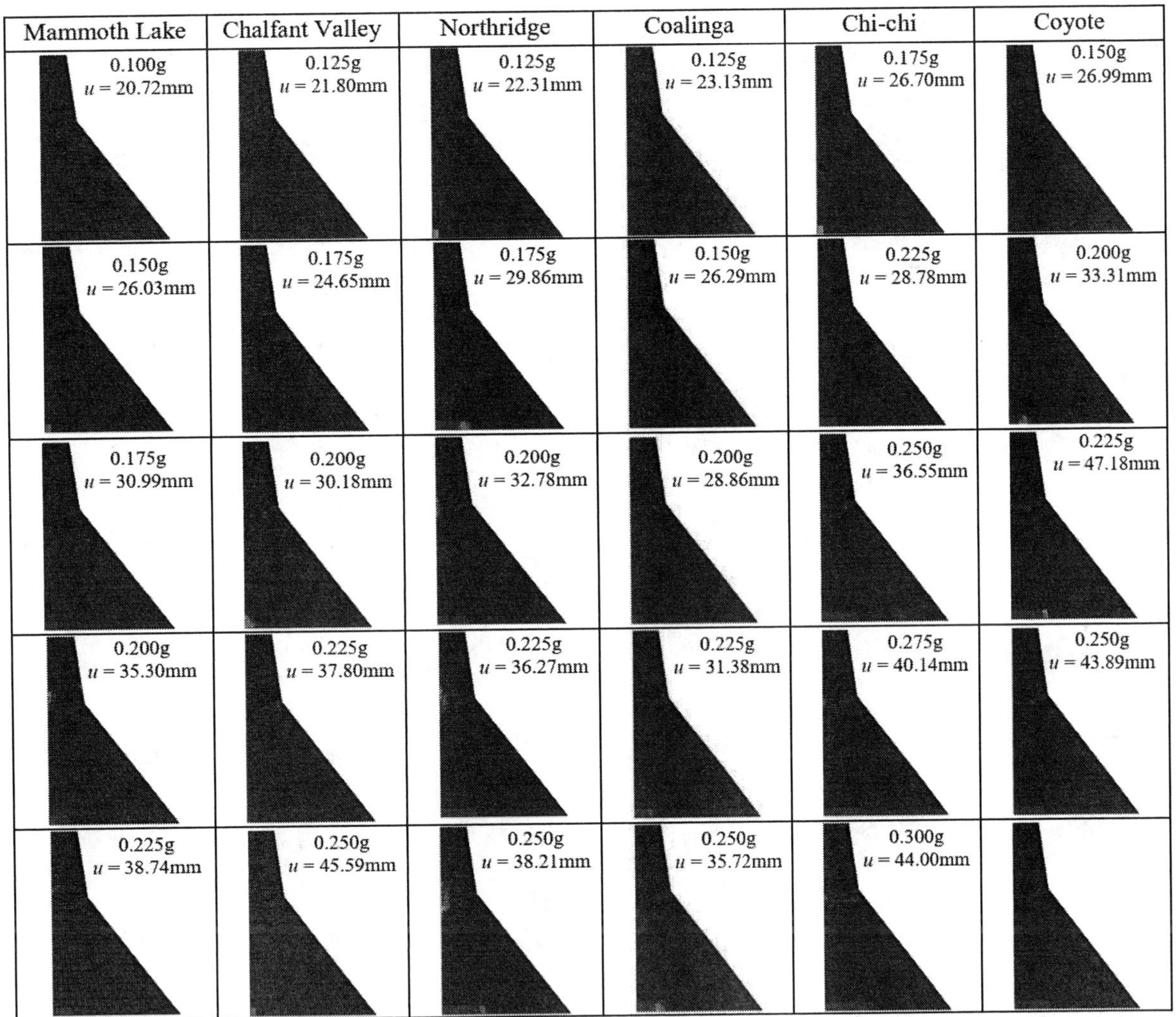

Figure 7. Cracking patterns and maximum crest displacement from IDA for Koyna dam under single ground motion

Figure 8. Cracking patterns and maximum crest displacement from IDA for Koyna dam under repeated ground motion

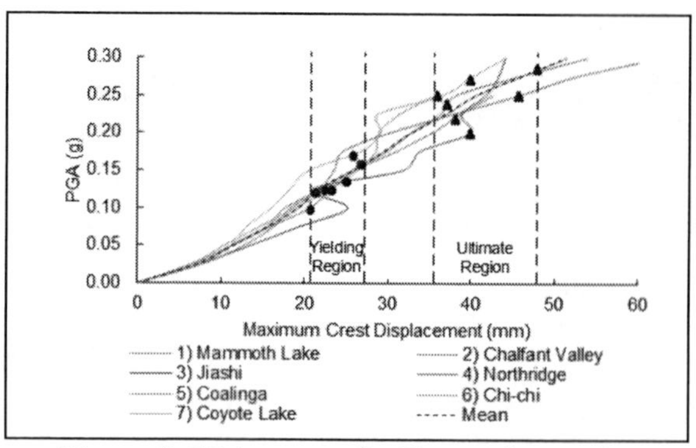

Figure 9. IDA curves for Koyna dam under single ground motion

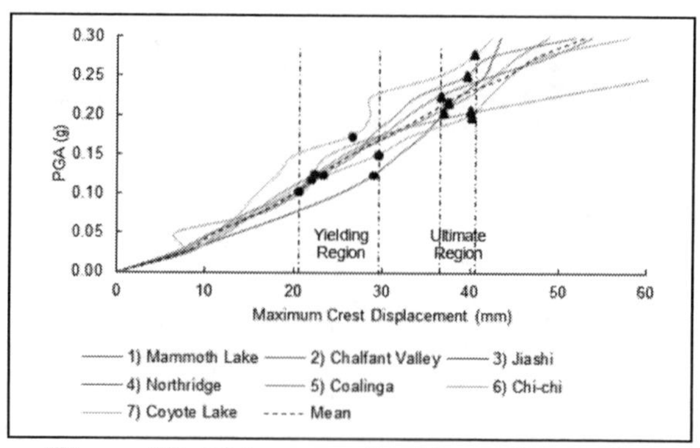

Figure 10. IDA curves for Koyna dam under repeated ground motions

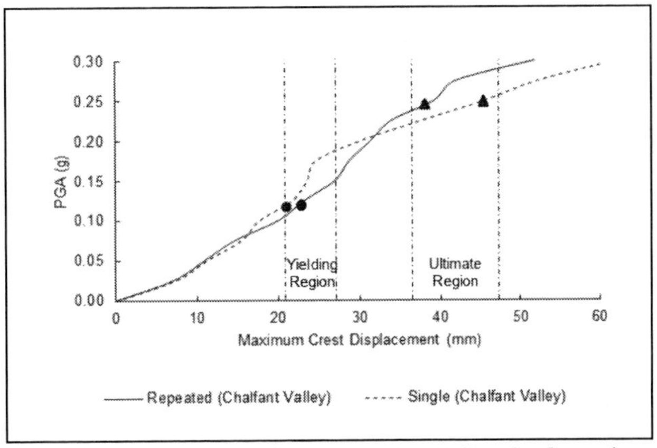

Figure 11. Comparison IDA curve for single and repeated ground motions

5. Conclusions

Incremental dynamic analysis (IDA) is the performance-based earthquake engineering approach for seismic performance evaluation. IDA involves applying a series of nonlinear time history analysis to a structure for numerous ground motion records by scaling every record to several levels of intensity until an ultimate condition is reached. It was observed that the dam started to perform at the neck area on the downstream (D/S) face. Later on, a partial crack occurred at the base of the dam at the upstream (U/S) face and as a general trend. When the PGA of the motion increased, the cracking was created at the neck of the dam from the U/S face and then moved forward from the both face until cracks were joined, that can be concluded as collapse. In the meantime, the cracking at the heel area increased around 15% to 40% from the width of the dam. The maximum displacement under repeated ground motion increased about 7% for the yielding state while 14% decreased for the ultimate state. That mean, the ultimate state of repeated ground motion reached earlier than a single ground motions event in the same PGA. The results show that the repeated ground motion should not be neglected in analysed of a concrete gravity dam. Further study will have to be investigated by comparing the IDA curves and static pushover analysis to identify the minimum values reached for yielding and ultimate state for concrete gravity dam.

Acknowledgment

The authors wish to thank the Universiti Malaysia Perlis (Unimap) and Minister of High Education (MOHE) for supporting the research. We would like to thank the anonymous reviewers for suggestions and comments.

References

1. International Rivers. Earthquakes Triggered by Dams. *International Rivers* 2008. https://www.internationalrivers.org/earthquakes-triggered-by-dams [accessed October 14, 2015].
2. Gupta HK. A review of recent studies of triggered earthquakes by artificial water reservoirs with special emphasis on earthquakes in Koyna, India. *Earth-Science Reviews* 2002; **58**(3–4): 279–310. DOI: 10.1016/S0012-8252(02)00063-6.
3. Chen-Shan K, Wei-Pin N, Yun-Jen C. Damage and Rehabilitation Work of Shih-Kang Dam. *Seismic Fault-Induced Failures* 2001(January): 33–48.
4. Allen CR. The tectonic environments of seismically active and inactive areas along the San Andreas fault system. *Proceedings of Conference of Geologic Problems of San Andreas Fault System*, vol. 5, 1968.

5. Sieh KE. Prehistoric large earthquakes produced by slip on the San Andreas Fault at Pallett Creek, California. *Journal of Geophysical Research* 1978; **83**(B8): 3907. DOI: 10.1029/JB083iB08p03907.
6. Schwartz DP, Coppersmith KJ. Fault behavior and characteristic earthquakes - examples form the Wasatch and San-Andreas fault zones. *Journal of Geophysical Research* 1984; **89**(NB7): 5681–5698. DOI: 10.1029/JB089iB07p05681.
7. Decanini L, Gavarini C, Mollaioli F. Some remarks on the Umbria-Marche earthquakes of 1997. *European Earthquake Engineering* 2000; **3**: 18–48.
8. Elnashai AS, Bommer JJ, Martinez-Pereira A. Engineering implications of strong-motion records from recent earthquakes. In: Bisch P, Labbe P, Pecker A, editors. *Eleventh European Conference on Earthquake Engineering*, Paris, France: A.A Balkema; 1998.
9. Faisal A. Influence of Repeated Earthquakes on the Ductility Demand of Inelastic RC Buildings. Universiti Sains Malaysia, 2012.
10. Zhai C, Wen W, Li S, Chen Z, Chang Z, Xie LL. The damage investigation of inelastic SDOF structure under the mainshock-aftershock sequence-type ground motions. *Soil Dynamics and Earthquake Engineering* 2014; **59**: 30–41. DOI: 10.1016/j.soildyn.2014.01.003.
11. Hatzigeorgiou G. Behavior factors for nonlinear structures subjected to multiple near-fault earthquakes. *Computers and Structures* 2010; **88**(5–6): 309–321. DOI: 10.1016/j.compstruc.2009.11.006.
12. Hatzigeorgiou G. Ductility demand spectra for multiple near- and far-fault earthquakes. *Soil Dynamics and Earthquake Engineering* 2010; **30**(4): 170–183. DOI: 10.1016/j.soildyn.2009.10.003.
13. Efraimiadou S, D. Hatzigeorgiou G, E. Beskos D. Structural pounding between adjacent buildings subjected to strong ground motions . Part II : The effect of multiple earthquakes. *International Association for Earthquake Engineering* 2013(February): 1529–1545. DOI: 10.1002/eqe.
14. Amadio C, Fragiacomo M, Rajgelj S. The effects of repeated earthquake ground motions on the non-linear response of SDOF systems. *Earthquake Engineering and Structural Dynamics* 2003; **32**(2): 291–308. DOI: 10.1002/eqe.225.
15. Zhang S, Wang G, Sa W. Damage evaluation of concrete gravity dams under mainshock–aftershock seismic sequences. *Soil Dynamics and Earthquake Engineering* 2013; **50**: 16–27. DOI: 10.1016/j.soildyn.2013.02.021.
16. Abdelnaby AE, Elnashai AS. Performance of Degrading Reinforced Concrete Frame Systems Under the Tohoku and Christchurch Earthquake Sequences. *Journal of Earthquake Engineering* 2014; **18**(7): 1009–1036. DOI: 10.1080/13632469.2014.923796.
17. Hatzivassiliou M, Hatzigeorgiou GD. Seismic sequence effects on three-dimensional reinforced concrete buildings. *Soil Dynamics and Earthquake Engineering* 2015; **72**: 77–88. DOI: 10.1016/j.soildyn.2015.02.005.
18. Faisal A, Majid TA, Hatzigeorgiou GD. Investigation of story ductility demands of inelastic concrete frames subjected to repeated earthquakes. *Soil Dynamics and Earthquake Engineering* 2013; **44**: 42–53. DOI: 10.1016/j.soildyn.2012.08.012.
19. Vamvatsikos D, Cornell CA. Incremental dynamic analysis. *Earthquake Engineering and Structural Dynamics* 2002; **31**(3): 491–514. DOI: 10.1002/eqe.141.
20. FEMA-356. *Prestandard and Commentary for the Seismic Rehabilitation of Building*. Federal Emergency Management Agency. Washington, D.C; 2000.
21. Alembagheri M, Seyedkazemi M. Seismic performance sensitivity and uncertainty analysis of gravity dams. *Earthquake Engineering & Structural Dynamics* 2015; **44**(3 July 2014): 41–58. DOI: 10.1002/eqe.
22. Alembagheri M, Ghaemian M. Seismic assessment of concrete gravity dams using capacity estimation and damage indexes. *Earthquake Engineering & Structural Dynamics* 2012; **42**(1): 123–144. DOI: 10.1002/eqe.
23. Chopra AK, Chakrabarti P. The Koyna Earthquake and the Damage to Koyna Dam. *Bulletin of the*

Seismological Society of America 1973; **63**(2): 381–397.
24. Anderson C, Mohorovic C, Mogck L, Cohen B, Scott G. *Concrete Dams Case Histories of Failures and Nonfailures with Back Calculations*. 1998.
25. Mridha S, Maity D. Experimental investigation on nonlinear dynamic response of concrete gravity dam-reservoir system. *Engineering Structures* 2014; **80**: 289–297.
26. Oudni N, Bouafia Y. Response of concrete gravity dam by damage model under seismic excitation. *Engineering Failure Analysis* 2015; **58**: 417–428. DOI: 10.1016/j.engfailanal.2015.08.020.
27. ABAQUS. *Analysis User's Manual Volume 3: Materials (Abaqus 6.12)*. Version 6. SIMULIA, Dassault Systèmes Simulia Corp., Providence, RI, USA.; 2012.
28. Lee J, Fenves GL. Plastic-damage model for cyclic loading of concrete structures. *Journal of Engineering Mechanics* 1998; **124**(8): 892–900.
29. Sümer Y, Aktaş M. Defining parameters for concrete damage plasticity model. *Challenge Journal of Structural Mechanics* 2015; **1**(3): 149–155.
30. Zappitelli MP, Villa EI, Fernández-Sáez J, Rocco C. Cracking Development Prediction in Concrete. *Mecánica Computacional* 2014; **XXXIII**: 23–26.
31. Eurocode8. *Design of structures for earthquake resistance . Part 1: General rules, seismic actions and rules for buildings*. vol. 3. Brussels; 2004.
32. Feyza Soysal B, Arici Y. Incremental dynamic analysis of a gravity dam. *Tenth U.S National Conference on Earthquake Engineering*, Achorage, Alaska: Fortiers of Earthquake Engineering; 2014. DOI: 10.4231/D39Z90C6T.
33. Ancheta TD, Darragh RB, Stewart JP, Seyhan E, Silva WJ, Chiou BSJ, *et al.* Pacific Earthquake Engineering Research Center - PEER, Ground Motion Database. *Pacific Earthquake Engineering Research Center* 2015. http://ngawest2.berkeley.edu/spectras/new?sourceDb_flag=1 [accessed July 25, 2015].
34. Hatzigeorgiou GD, Liolios AA. Nonlinear behaviour of RC frames under repeated strong ground motions. *Soil Dynamics and Earthquake Engineering* 2010; **30**(10): 1010–1025.
35. Nayak P, Maity D. Seismic damage analysis of aged concrete gravity dams. *International Journal for Computational Methods in Engineering Science and Mechanics* 2013; **14**(5): 424–439.

IOP Conf. Series: Materials Science and Engineering **318** (2018) 012022 doi:10.1088/1757-899X/318/1/012022

Spatial and Temporal Flood Risk Assessment for Decision Making Approach

Nazirah Azizat[1], Wan-Mohd-Sabki Wan Omar[1,*]
[1]School of Environmental Engineering, Universiti Malaysia Perlis, 02600 Arau, Perlis, Malaysia

E-mail: wansabki@unimap.edu.my

Abstract. Heavy rainfall, adversely impacting inundation areas, depends on the magnitude of the flood. Significantly, location of settlements, infrastructure and facilities in floodplains result in many regions facing flooding risks. A problem faced by the decision maker in an assessment of flood vulnerability and evaluation of adaptation measures is recurrent flooding in the same areas. Identification of recurrent flooding areas and frequency of floods should be priorities for flood risk management. However, spatial and temporal variability become major factors of uncertainty in flood risk management. Therefore, dynamic and spatial characteristics of these changes in flood impact assessment are important in making decisions about the future of infrastructure development and community life. System dynamics (SD) simulation and hydrodynamic modelling are presented as tools for modelling the dynamic characteristics of flood risk and spatial variability. This paper discusses the integration between spatial and temporal information that is required by the decision maker for the identification of multi-criteria decision problems involving multiple stakeholders.

1. Introduction

Flooding is widely natural hazard that noticeably affects social, economic and environmental issues. Flood risk management is of utmost importance to identify the design, planning and methods of flood protection to ensure the sustainability of community development. However, the traditional flood protection method only focuses on structural flood mitigation, rather than non-structural measures. In the perspective of flood risk management, a flood protection method should incorporate stakeholders to develop and implement the adaptation alternative to minimize the effects of flood impacts, mainly in floodplain areas.

However, due to the uncertainty of flood prediction, many vulnerability and adaptation assessments should be considered in the research of floods. There are many types of uncertainty in flood management, from hydrologic, hydraulic, geotechnical, and structural uncertainties, to economic, environmental, ecological, social and political uncertainties[1]. According to Baecher (2009), flood risk management related to disasters involves hazard, vulnerability and consequence assessments [2]. However, vulnerability and adaptation assessments cause difficulty in making decisions, due to the uncertainty of flood predictions. Dynamic and spatial characteristics of flood should be considered in decision making. Spatial and temporal dynamics of parameters related to floods were analyzed in prediction of where and when a flood might occur [3].

For this reason, a continual process in the vulnerability of flood risk management is significant to identify the weaknesses of the current situation, and suitable methods for the selection, exploration and monitoring of the best adaptation strategies and measures that can be applied in the decision-making process. Nevertheless, time is not taken into account in the vulnerability assessment for flood prediction. Thus, temporal dimension should be considered in vulnerability assessment studies, in order to capture changes dynamically. In this research, value of critical rainfall is used for the temporal prediction of a flood.

The framework was prepared for the comprehension of the flood-related decision-making process. The aim of the study is to predict flood vulnerability in terms of spatial and temporal, as well as to provide information and guidance to the decision maker to take the right action towards adaptation to flood problems. Thus, integration between flood vulnerability assessment and decision making is needed to develop adaptation strategies in decreasing the adverse effects of flooding.

2. Flood Mitigation

Due to the recurrent nature and adverse impacts of flooding, flood mitigation is necessary to be carried out, involving the management and control of floodways. The management of a flood disaster is a multidiscipline approach, involving hydrology, management of water resources, economics, statistics, population, public policy and planning [4]. The regulation and flood controlling works should be implemented in a floodplain area by improving flow of floodways, and constructing a continuous system of dykes, in order to minimize loss of properties, damage, and human lives.

The efficiency of a flood hazard assessment in producing flood mitigation requires flood conditions such as historical data, depth of inundation, flow velocities, and extension area where effected by the flood. Some of the flood characteristics can be obtained from a flood model. Hydrodynamic modelling is a practical method in flood assessment resulting from a flood prediction event, that can be developed using historical events, condition of previous events, and modification conditions for future events. Hydrodynamic models are used to estimate design floods in a study area, and generate a flood inundation model for an early warning system. This modelling assists the understanding of flood hazards, and effects of mitigation including structural (curative) and non-structural (preventive) measures to alleviate flood problems.

3. Modelling Approach

In Malaysia, research on flooding is mainly based on the development of hydraulic models, with an emphasis on a structural approach, using engineering measures including building of seawall, revetments levees, and river embankment. The combination of various flood frequencies and various mean annual floods by region have been used in developing a hydrological model to estimate a design flood for Peninsular Malaysia[5]. The prediction of flood frequency using the artificial neural network (ANN) and statistical theory is the frequent method to use instead of stochastic models [6]. However, this model can only be applied if the recurrence intervals of the highest discharge do not exceed the lengths of recorded highest discharges data without considering the maximum flood hydrograph.

Various hydrological models have been applied to flood design. Source code is one of the important criteria of flood design, which has the capability to support the full parameters and dimensional overland routing approach. Nowadays, these models are available in different types of dimensions, from 1D dimension models, to complex 3D models which take flow direction into inundation of flood.

However, the changes of the environmental process in four dimensions of time and space must be considered in the study. Therefore, a combination simulation model and GIS can be used as a plausible solution in providing the potential to analyze the environmental process in spatial and temporal aspects. In addition, environmental problems are different in all places. Hence, generic solutions developed by using the interaction of the dynamic spatial model can be applied to each society and location, which provides relevant information for the decision-making process. The interaction of spatial, temporal and decision making should be measured and addressed in the process of environmental modelling.

4. Methodology

Figure 1 represents the input, research activity and expected output for each step of this research, which is a systematic assessment of flooding. The integration of approaches combines many disciplines which will be used to assess the vulnerability and adaptation to flood. The development of the hybrid spatio-temporal model using a combination of the Geographic Information System (GIS) and the System Dynamics (SD) modelling approaches, meanwhile a multi-criteria decision model will be used in assessing adaptation alternatives.

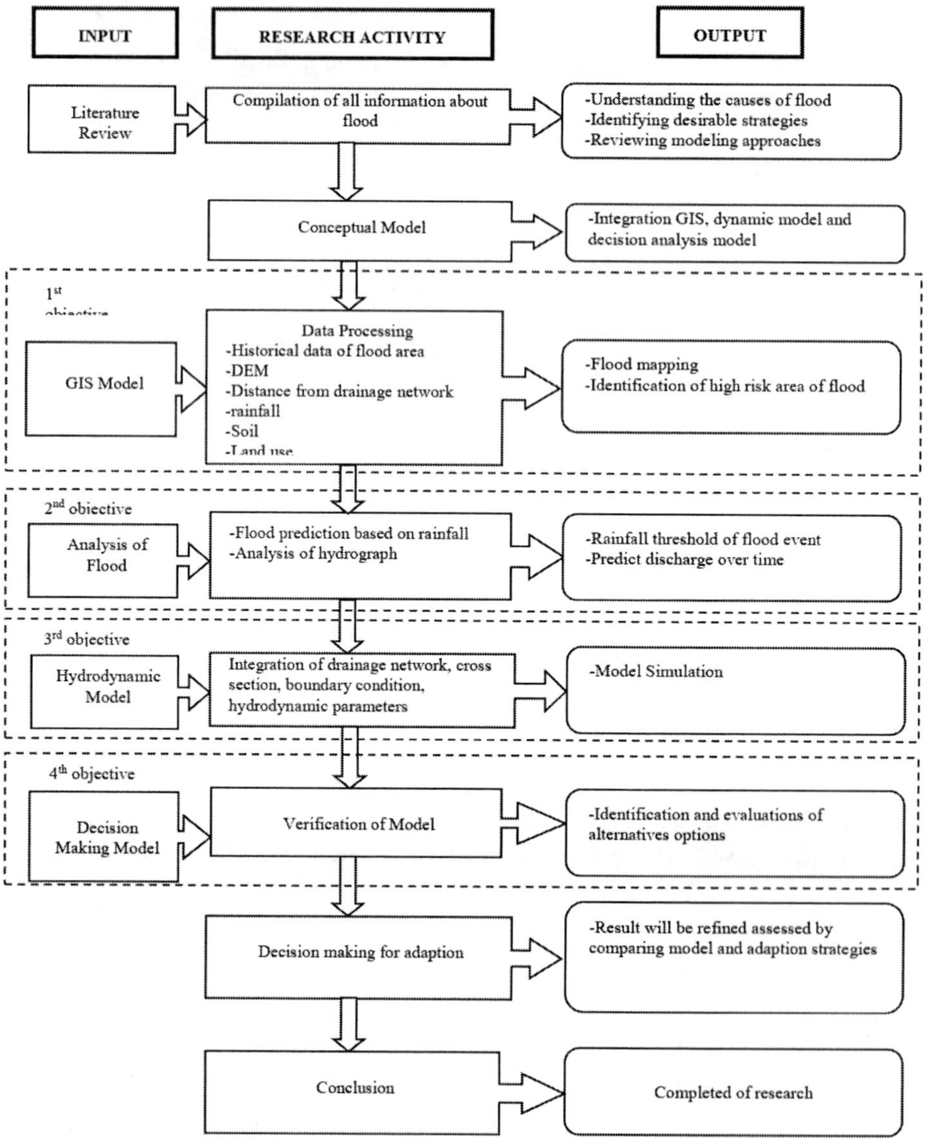

Figure 1. Flowchart of the research

4.1. GIS Model

Spatial analysis is an important component of the integration of the spatial and temporal model to organize all initial data required for the flood assessment process. Additionally, it was developed in support of data visualization and data conversion by the Geographic Information System (GIS).

Flood hazard mapping will be produced using a frequency ratio model to evaluate the influence of classes of each conditioning factor on flood occurrence. Frequency ratio models are used based on the assumption that the recurrent flood will occur under a similar condition as the previous flood, using relationships between flood distribution and each related parameter. In calculation of the Flood Hazard Index, a summation of values of each parameter's frequency ratio is used to produce a flood prone area, as seen in Equation 1:

$$Fr_i = \frac{N_{(Si)}/N_{pix(Ni)}}{\sum N_{pix(Si)}/\sum N_{pix(Ni)}} \quad (1)$$

Where; $N_{(Si)}$ is the number of pixels of flood area in a certain parameter class; $N_{pix(Ni)}$ is the number of pixels in a certain parameter class; $\Sigma N_{pix(Si)}$ is the total number pixels of a flood area in a certain parameter class; and $\Sigma N_{pix(Ni)}$ is the total number of pixels in a certain parameter class.

The higher pixel values of the flood hazard index are the higher hazard to flood occurrence, and the lower values are the lower hazard to flood occurrence.

4.2. Flood Analysis

Regarding the results from the spatial analysis, several flood areas will be used to perform a rainfall analysis. For the temporal analysis, rainfall and stream data from the flood report will be calculated to make several time series showing the circumstances of the flood event described by the respective flood reports.

The rainfall threshold of a flood based on the cumulated volume of antecedent rainfall during a storm event resulting in a critical water stage (or discharge) at a specific river section are estimated by hydrological modelling [7,8]. Analysis is undertaken of rainfall threshold related to rainfall rate, duration and inundation area for the river system critical state. Figure 2 illustrates the steps of the rainfall threshold evaluation procedure.

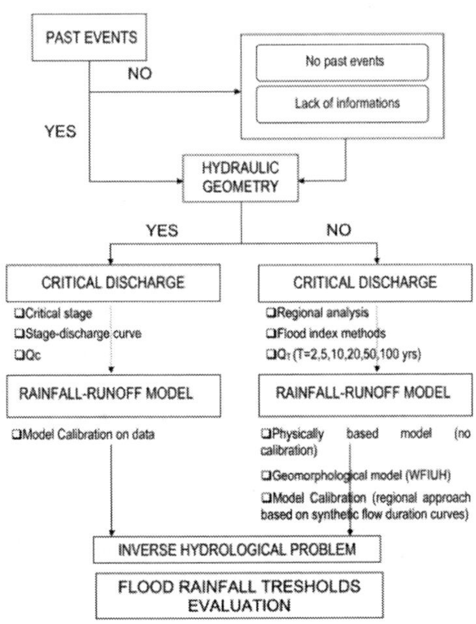

Figure 2. Flood Rainfall Threshold Evaluation [9]

IOP Conf. Series: Materials Science and Engineering **318** (2018) 012022 doi:10.1088/1757-899X/318/1/012022

4.3. Hydrodynamic Modelling

Hydrodynamic modelling is a common model in flood assessment, due to temporal data for numerical modelling using Saint-Venant equations becoming more accurate, and the detailed data collection of river investigation has been improved in areas such as high resolution topographic, rainfall, water level, river velocity and water discharge. Several studies of the commercial tools of the hydrodynamic model using HEC_RAS for 1D river modelling [10], MIKE 11[11] and SOBEK [12]. For 2D overland flow, modelling had been developed via TUFLOW , and RMA2 [12]. The integration of two or more hydrodynamic models to improve the modelling capability had been studied by Smith and Rowland [13].

Flood inundation simulation in dynamic system software depends on depth values at the cross section of the river. The hydrological model approach uses the Saint-Venant equation for open channel flow. The Saint-Venant equations are two coupled partial derivative equations.

$$\frac{\partial A(x,t)}{\partial t} + \frac{\partial Q(x,t)}{\partial x} = 0 \qquad (2)$$

$$\frac{\partial A(x,t)}{\partial t} + \frac{\partial}{\partial x}\left[\frac{Q^2(x,t)}{A(x,t)}\right] + gA(x,t)(\frac{\partial Y(x,t)}{\partial x} + S_f(x,t) - S_b(x)) = 0 \qquad (3)$$

$$S_f = \frac{Q^2 n^2}{A^2 R^{\frac{4}{3}}} \qquad (4)$$

Where, Q is the total flow down to reach; A is the cross sectional area of flow in channel (Ac) and floodplain (Af); P is wetted perimeter; R is the hydraulic radius (A/P); n is the Manning roughness value; and S_f is the friction slope.

4.4. Decision Model

The decision making process involved making choices regarding to appropriate mitigation, resulting in an effective outcome in managing the wide impact of the flood. An Analytical Hierarchy Process (AHP) model technique will be developed for the evaluation of user preferences for many criteria to adapt to the adverse impact of flooding. The Analytic Hierarchy Process (AHP) relates the decision with a pairwise comparison of alternative adaptation, based on judgements of the decision maker to derive priority scales. As Saaty [14] contends, decision-making is an organized technique in order to generate priorities which breaks down the decision into several steps:

1. Identification of problem and knowledge.
2. Goal of decision as the top of a decision hierarchy, then objectives from a various perspective, via the intermediate to the lowest level
3. Set of pairwise comparison matrixes is developed by valuation of each element at upper level and the level immediately below.
4. Comparison between evaluation of the priorities in the level immediately below for every element used to obtained priorities. A summation weighing up the values of each element in the level below will produce a general priority. The weighing process was continued and added to, until reaching final alternative priorities in the bottom most level.

5. Conclusion

In conclusion, an effective assessment of flood risks integrates spatial and temporal factors in the decision modelling approach. The dynamic and spatial flood model involved the analysis of floods in terms of area and time. An availability of comprehensive flood assessment studies can assist a decision maker in restructuring the method design adaptation strategies at a local level. It is essential to develop a flexible and well-structured method which enables the collection of information to design the most effective adaptation in reducing the impacts of flooding. An integration of Geographic Information System (GIS), system dynamic (SD) and multiple criteria decision aid modelling approach provides a powerful methodology to assist the decision maker in selecting the appropriate adaptation strategies for the prevention of flooding in the future.

References

[1] Ahmad S S and Simonovic S P 2015 System dynamics and hydrodynamic modelling approaches for spatial and temporal analysis of flood risk *Int. J. River Basin Manag.* **5124** 1–19
[2] Baecher G B 2009 Quantifying Flood Risk *ASFPM Found. Symp. 1 Defin. Meas. Flood Risks Floodplain Resour.* 1–24
[3] Hu Y, Huang J, Du Y, Han P and Huang W 2015 Monitoring spatial and temporal dynamics of flood regimes and their relation to wetland landscape patterns in Dongting Lake from MODIS time-series imagery *Remote Sens.* **7** 7494–520
[4] Mohit M A and Sellu G M 2013 Mitigation of Climate Change Effects through Non-structural Flood Disaster Management in Pekan Town, Malaysia *Procedia - Soc. Behav. Sci.* **85** 564–73
[5] Malaysia D of I and D (DID) 1987 Hydrological Procedure No. 4 (Magnitude and Frequency of Floods in Peninsular Malaysia (Revised and Updated) *Publ. Unit, Minist. Agric.*
[6] D/iya S G, Gasim M B, Toriman M E and Abdullahi M G 2014 Floods in Malaysia: Historical Reviews, Causes, Effects and Mtigations Approach *Int. J. Interdiscip. Res. Innov.* **2** 59–65
[7] Montesarchio V, Napolitano F, Rianna M, Ridolfi E, Russo F and Sebastianelli S 2015 Comparison of methodologies for flood rainfall thresholds estimation 909–34
[8] Martina M L V, Todini E and Libralon A 2009 Rainfall Thresholds for Flood Warning Systems : A Bayesian Decision Approach 203–27
[9] Montesarchio V, Lombardo F and Napolitano F 2009 Rainfall thresholds and flood warning : an operative case study 135–44
[10] Thakur P K and Sumangala A 2000 Flood Inundation Mapping and 1-D Hydrodynamic Modeling Using Remote Sensing and Gis Technique
[11] Gaitan S, van de Giesen N C and ten Veldhuis J A E 2016 Can urban pluvial flooding be predicted by open spatial data and weather data? *Environ. Model. Softw.* **85** 156–71
[12] Shafapour Tehrany M, Shabani F, Neamah Jebur M, Hong H, Chen W and Xie X 2017 GIS-based spatial prediction of flood prone areas using standalone frequency ratio, logistic regression, weight of evidence and their ensemble techniques *Geomatics, Nat. Hazards Risk* **5705** 1–24
[13] Smith J and Rowland J 2005 Temporal analysis of floodwater volumes in New Orleans after Hurricane Katrina *Sci. Storms USGS Response to Hurricanes 2005* 57–60
[14] Saaty T L 2008 Decision making with the analytic hierarchy process *Int. J. Serv. Sci.* **1** 83

IOP Conf. Series: Materials Science and Engineering **318** (2018) 012023 doi:10.1088/1757-899X/318/1/012023

Review of Repair Materials for Fire-Damaged Reinforced Concrete Structures

MZA Mohd Zahid[1], BH Abu Bakar[2*], FM Nazri[2], MM Ahmad[1], K Muhamad[1]
[1]Department of Civil Engineering Technology, Faculty of Engineering Technology, Universiti Malaysia Perlis, Uniciti Alam Campus, 02100, Padang Besar, Perlis, Malaysia
[2]School of Civil Engineering, Universiti Sains Malaysia, Engineering Campus, 14300, Nibong Tebal, Pulau Pinang, Malaysia

E-mail: mohdzulham@unimap.edu.my, cebad@usm.my*, cefmn@usm.my, munsif@unimap.edu.my, khairunnisa@unimap.edu.my

Abstract: Reinforced concrete (RC) structures perform well during fire and may be repaired after the fire incident because their low heat conductivity prevents the loss or degradation of mechanical strength of the concrete core and internal reinforcing steel. When an RC structure is heated to more than 500 °C, mechanical properties such as compressive strength, stiffness, and tensile strength start to degrade and deformations occur. Although the fire-exposed RC structure shows no visible damage, its residual strength decreases compared with that in the pre-fire state. Upon thorough assessment, the fire-damaged RC structure can be repaired or strengthened, instead of subjecting to partial or total demolition followed by reconstruction. The structure can be repaired using several materials, such as carbon fiber-reinforced polymer, glass fiber-reinforced polymer, normal strength concrete, fiber-reinforced concrete, ferrocement, epoxy resin mortar, and high-performance concrete. Selecting an appropriate repair material that must be compatible with the substrate or base material is a vital step to ensure successful repair. This paper reviews existing repair materials and factors affecting their performance. Of the materials considered, ultra-high-performance fiber-reinforced concrete (UHPFRC) exhibits huge potential for repairing fire-damaged RC structures but lack of information available. Hence, further studies must be performed to assess the potential of UHPFRC in rehabilitating fire-damaged RC structures.

1. Introduction

Reinforced concrete (RC) exhibits fire resistance as a result of its low heat conductivity that prevents the loss or degradation of the mechanical strength of concrete core and internal reinforcing steel; RC exhibits lower heat conductivity of about 1.6 W/mK than that of steel (45 W/mK), according to EN1994-1-2 [1] and EN1993-1-2 [2], respectively. According to EN1992-1-2 [3], the conductivity of concrete decreases as the surrounding temperature increases. Under prolonged exposure to high temperatures, the mechanical strength of concrete and the properties of steel reinforcement bar inside it would reduce. Chan et al. [4] reported that the compressive strength of concrete may reduce up to 50% when the concrete is exposed to 600°C. At 800°C, the residual compressive strength of the concrete is only 20% of the original value. Furthermore, normal-strength concrete experiences a sharper loss in tensile splitting strength than compressive strength at 600°C. Exposure to high temperatures may also change the pore structure of concrete through pore structure coarsening, which leads to increased permeability but reduced durability of the concrete. After exposure to 600°C, the cumulative pore volume in normal-strength concrete increases twice. At temperatures higher than 600°C, extreme C-S-H gel dehydration and pore structure coarsening contribute to the strength loss of the concrete [4].Fires in concrete structures rarely result in serious global structural damage, and most of the damaged structures can be successfully reinstated. Therefore, repair of fire-damaged concrete structures is a more viable and economical option than demolition and rebuilding [5,6].

Various materials and methods can be used to rehabilitate fire-damaged RC structures. As such, appropriate repair materials and techniques must be selected. Repair materials must have good bond strength with the substrate and restore the original bearing capacity and fire resistance of the RC structures, considering that a second fire cannot be ruled out [7]. Moreover, repair techniques must involve least interventions to reduce the cost due to delayed use of the facilities. This paper discusses different repair materials such as fiber-reinforced polymer (FRP),glass fiber-reinforced polymer (GFRP),carbon-fiber reinforced polymer (CFRP), normal strength concrete (NSC), fiber-reinforced concrete (FRC), shotcrete, and ferrocement. This paper also elaborates available repair techniques for applying these materials and their corresponding performance.

2. Repair Materials and Methods

Selecting an appropriate repair material is a critical step in repair process. Repair materials must be compatible with the base material, project needs, available technical resource, financial constraint, and other specific requirements of the project [8]. In general, repair materials should be similar to the original construction material. In a previous study, NSC was used to repair fire-damaged concrete members. Lin et al. [9] investigated the behavior of RC columns repaired with cast-in-place NC after being damaged by fire. However, the performance of NC as repair material for fire-damaged concrete structures has been rarely investigated.

Shotcrete is a cementitious material used by engineers to repair fire-damaged RC members [8].In contrast to NC, the use of shotcrete omits the need for formworks and the time for their construction and the need to pump and consolidate the concrete in relatively blind formworks. Shotcrete exhibits short setting time and high early-age mechanical properties; this material is a popular choice for various applications, such as tunnel support, rapid repair, slope support, gas and oil wells, and other underground structures [10]. Prugar et al. [11] used shotcrete to repair fire-damaged RC joist of high school buildings. Gosain et al. [8] used shotcrete to repair fire-damaged RC slab and joist of a stadium. Epoxy resin mortar is another commercial cementitious material that does not require formworks and is easy to apply to damaged RC members. Yaqub and Bailey [6] used epoxy resin mortar to repair RC columns severely damaged by fire. However, data on the performance of shotcrete and epoxy resin in repairing fire-damaged structure remain limited.

Yaqub et al. [12] and Haddad et al. [13] used another cementitious material, namely, ferrocement, which consists of wire meshes and mortar, to repair fire-damaged RC structures. Applying ferrocement is labor intensive, but the workmanship required to construct it is quite low and the raw materials are locally available. Repairing structures using ferrocement is cheap and ideal for residential buildings in developing countries, such as Malaysia. Although ferrocement is fairly an old technology and has been used as a construction material for more than 50 years [14], its application in repairing fire-damaged RC members is limited and has been rarely reported. Ferrocement is a highly-versatile RC material that can be used to effectively wrap structural members with various cross sections; as such, this material exhibits considerable potential for structure repair. The strength of ferrocement depends on two factors, namely, sand to cement ratio and quantity of reinforcing material (wire mesh). However, the use of ferrocement requires installation and tying of rods and mesh together, and these processes are tedious and time consuming. Therefore, ferrocement is not recommended to be used as repair material for commercial buildings and bridges because of time constraint and economic implications. A large amount of time required to repair these facilities will increase the financial burden by the owner and users.

CFRP and GFRP, in either plate or sheet types, are the most popular materials used to repair defective RC structures [5, 6, 15 - 19]. The advantages of these materials include ease of application, high strength to weight ratio, excellent mechanical strength, and resistance to corrosion and chemical attack. These materials also exhibit low density, which is important because they add less weight to the repaired structure and can be conveniently applied without using heavy machineries. The tensile strength of CFRP and GFRP is extremely high, which is about three to eight times higher than steel reinforcements in RC (Table 1). FRP exhibits lower density than concrete and steel (Table 1), and the thickness of FRP plate and sheet is around 1.5 and 0.17 mm, respectively, which is very low; hence, FRP can be conveniently applied without requiring heavy machineries to handle and install it to the structure. FRP is also suitable to structures of various geometry types or cross section. Yaqub and Bailey [5] showed the more effective confinement of FRP sheet for fire-damaged circular cross section RC column than for square column.

In current concrete technology development, fibers are introduced into concrete mixtures to overcome the limitations of low tensile strength and brittle failure of plain concrete. Many types of fiber can be used, and steel fiber is the most popular choice. Steel fiber-reinforced concrete (SFRC) is a suitable candidate for repairing fire-damaged RC members. Haddad et al. [20] explored the effectiveness of using SFRC to repair fire-damaged RC slab.

Recent advancement in SFRC has resulted in the development of UHPFRC, which is a cementitious material with higher compressive strength (> 150 MPa), tensile strength (>10 MPa), and ductility than NC. UHPFRC can be fabricated by adding fine aggregates, steel fibers, and water-reducing agent into the concrete mixture with a low water-cement ratio (<0.2). The actual maximum value of the compressive strength of cast in place UHPFRC is 115 - 120 MPa [21]. Compressive strength is not the primary feature of UHPFRC because tensile properties are often more important, especially for rehabilitation work or cast-on-site application. UHPFRC exhibits not only high tensile strength (>10 MPa) but also strain-hardening properties. According to Naaman and Reinhardt [22], strain hardening occurs when tensile strength σ_{pc} exceeds cracking stress, σ_{cc}. This property may prevent the formation of shrinkage cracks due to different expansion properties between the old and new concrete-based materials. Hence, UHPFRC is a suitable candidate material for repairing structures. Moreover, UHPFRC exhibits higher density and lower porosity, resulting in improved durability, than ordinary concrete or high-performance concrete (HPC). For instance, Charron et al. [23] revealed that the equivalent water permeability of UHPFRC with residual tensile deformation until 0.13% is similar to that of uncracked NC. As a result, no carbonation, chloride ion ingress, or sulfate reaction occurred because the material is waterproof. Therefore, the use of UHPFRC will protect steel bars against corrosion [24].

Table 1. Comparison of concrete, CFRP and GFRP properties [5-6, 15-19]

Material	Concrete	Steel	CFRP	GFRP
Tensile Elastic Modulus (GPa)	≈25	≈180	165– 238	76
Tensile strength (MPa)	3-6	500	2,700-4,100	1,518-2,300
Density (kg/m^3)	2,240 – 2,400	7,850	≈1,500	≈1,800

Selecting a suitable method for repair depends on the type of affected RC member. For flexural members, such as slabs and beams, the repair material is either applied at the soffit or tensile side only or to the soffit and side surfaces. Scholars have also attempted to apply the repair material at the compressive side (top surface), but the effect is insignificant [25]. For compressive members, such as columns, jacketing method is employed, regardless of the type of repair material.

3. Factors Affecting the Performance of Repair Material

Several factors affect the performance of repair materials. First, the damage severity of the post-heated RC members affects the performance of repair materials. Yaqub and Bailley [5] reported that RC columns heated up to 500 °C experienced two different types of damage, namely, crack and crack with significant spalling. The load carrying capacity of column that experienced significant spalling repaired with epoxy resin and wrapped with GFRP sheet, increases only up to 110% of the original capacity. Meanwhile, for the fire-damaged RC columns without significant spalling and repaired with GFRP only, the load carrying capacity increased up to 129% of the original capacity.

Another critical factor that influences the performance of repair materials is the bond strength between the repair material and the old material [26, 27]. In general, concrete-based repair materials exhibit good bond strength with the old concrete material; nevertheless, the bond strength between the old concrete substrate and other materials, which are bound with agents, such as epoxy, should be considered. Toutanji and Gomez [19] reported that concrete beams repaired with FRP subjected to wet/dry conditions showed worse performance caused by epoxy degradation than those kept at room temperature. Hence, FRP is unsuitable to be used in regions with hot climate, such as Malaysia. Farhat et al. [28] observed no bond degradation between UHPFRC used as repair material and concrete beam under thermal cycling test. Yaqub et al. [12] reported that ferrocement exhibits good bond with fire-damaged concrete and observed no de-bonding between the ferrocement jacket and the damaged substrate.

The bond strength between the cementitious material and the old concrete substrate depends on the surface treatment method used and the surface condition of the substrate. Surface treatment is used to clean the surface from loose materials and dust, which may jeopardize the interface bond strength between the repair material and the substrate. Tayeh et al. [29] found that sand blast method is the optimal surface treatment method. Baharudin et al [30] suggested that the substrate surface should be wetted with water prior to applying repair materials. In the works of Tayeh et al. [29] and Baharudin et al. [30], UHPFRC with steam curing was used as repair material; however, this method is impractical and costly and should be replaced by cast-in-place UHPFRC for repair works [31]. The bond strength between cementitious repair material and fire damaged concrete substrate also depend of the maximum temperature experienced by the substrate. Recent study found that, the bond strength decrease as the heat temperature experienced by the substrate increases [32]. Therefore, the bond strength of cast-in-place UHPFRC and fire-damaged concrete must be further investigated.

Cross-section shape also plays a vital role in increasing the strength of fire-damaged RC members, primarily, the column. Yaqub and Bailey [5] found that fire-damaged circular cross-section wrapped with FRP exhibits increased axial strength and ductility than the square cross-section. This finding could be attributed to the fact that square cross sections contain ineffectively confined concrete regions and intensification of stresses at their corners, whereas circular cross sections contain fully effective confined concrete regions with FRP jackets.

The performance of repair materials, such as FRP, depends on the type of structural members to be repaired (i.e., either flexural or compression member). Haddad et al. [20] used FRP to repair fire-damaged RC slab and found that FRP improves the stiffness but fails to improve the ductility of the slab. Other studies that used FRP to repair fire-damaged RC columns indicated that FRP can improve the ductility but fails to restore the stiffness of the column [5, 6, 12, 15, 33]. Therefore, the behavior of FRP in strengthening flexural and compression governing members must be further investigated.

Proper curing method may positively affect the performance of repair materials, primarily cementitious materials. Curing of fire-damaged concrete substrates may also improve the mechanical properties, such as compressive strength and stiffness, of repair materials. Haddad et al. [20] showed that water curing significantly recovered the compressive strength and modulus elasticity of fire-damaged concrete cylinder at about 30% and 60%, respectively, of the undamaged samples. A similar study reported the significant effect of curing on the stiffness of the investigated RC slab; that is, the stiffness of fire-damaged RC slabs increased by 56%. Therefore, fire-damaged substrates should be water cured after removing all loose substances prior to applying repair materials.

4 Findings of Previous Works

In general, repair materials are considered effective if they can improve the original capacity (i.e.: strength, stiffness and ductility) of fire-damaged RC members or structure systems. Studies reported that FRP may increase the strength, ductility, and energy dissipation of fire-damaged concrete up to or more than the original capacity [5, 6, 12, 15, 33]. However, FRP fails to improve the elastic stiffness of the fire-damaged concrete members due to the huge difference in the strength-to-weight ratio or stiffness between FRP and the concrete material. Scholars have attempted to overcome this problem by increasing the stiffness or FRP confinement via increasing the thickness of the material; this method slightly increased the column stiffness but adversely affected the ductility of the repaired column [15]. Haddad et al. [13] used two CFRP layers to repair fire-damaged RC beams and found that the stiffness can be restored up to 160% of the original undamaged sample; however, the repaired samples exhibit brittle failure. Furthermore, Yaqub et al. [12] found that ferrocement is more effective than FRP in improving the stiffness of fire-damaged concrete members but it cannot restore the original stiffness of the member. This finding may be partly due to the increased cross-sectional area after ferrocement jacketing. Haddad et al. [20] reported that SFRC may restore the original stiffness of fire-damaged RC slab but cannot restore the load carrying capacity. Therefore, ferrocement or SFRC should be combined with FRP for repairing fire-damaged concrete members to restore their original strength, ductility, and stiffness. Hence, the potential of the composite materials of FRP and SFRC or ferrocement for repairing fire-damaged RC members must be assessed in future studies.

Li et al. [9] found that fire-damaged RC columns repaired with fresh cast-in-place NSC could attain their original or even higher strength and stiffness than those of the undamaged columns but cannot improve the ductility. The ductility of structural members is important to evaluate the performance of a structural system subjected under extreme loading [12]. Under this condition, the ductile structural members will show sign, such as crack and deformation, before failure. The building occupants or facilities' users should be warned to give time for them to leave the facilities. In comparison with normal concrete, shotcrete is preferred as repair material for fire-damaged concrete members because it omits the requirement of formwork and compaction work. Few studies have successfully used shotcrete for repair [8, 11]. Although the mechanical properties of shotcrete and normal concrete are identical [34], the high velocity pressure during placing of shotcrete makes it relatively denser than normal concrete and is thus more vulnerable to explosive spalling defect under fire. Therefore, the behavior of shotcrete must be studied under fire because the repaired structure should exhibit not only sufficient bearing capacity under ordinary conditions but also an adequate fire resistance because a second fire cannot be ruled out.

Leonardi et al. [7] and Abdullah et al. [35] proposed the use of a relatively new developed concrete known as UHPFRC to repair fire-damaged RC members. In the numerical study conducted by Leonardi et al. [7], UHPFRC jacketing technique was used to repair and improve the strength of fire-damaged RC beams and columns, which were exposed to standard fire for 120 and 180 minutes. The performance of UHPFRC jacket under fire was also investigated, and the material may resist second fire up to 180 minutes. Despite these promising results, further studies on the thermal behavior of normal concrete and UHPFRC must be conducted. Kahanji et al. [36] performed an experimental study and reported that UHPFRC beams experience severe explosive spalling after being exposed to 60 minutes of standard fire. Leonardi et al. [7] also reported that the axial strength of the RC column decreased by18% after 30 minutes of fire exposure under the standard fire curve of ISO 834 [37], which is equivalent to about 800 °C. Al-Salloum et al. [38] found that the axial strength of RC column reduced by about 60% after exposure to 800 °C; this finding is supported by other works [39, 40, 41].

The production cost of UHPFRC is three times higher than that of normal concrete due to high cement content, very fine aggregates, and additional admixtures, such as silica fume, steel fiber, and high-range water reducer [42]. The high production cost will directly affect the total cost of repair works. Scholars have used UHPFRC of different thicknesses in jacketing method (Table 2). Generally, the recovery percentage of the load bearing capacity of structural members increases with increasing thickness of UHPFRC. In the computational study by Leonardi et al. [7], fire-damaged RC members with 40 mm-thick UHPFRC recover 300% of its load bearing capacity, which is extremely high and sufficient. To optimize the cost of UHPFRC in repair works, researchers must determine the optimum UHPFRC thickness for recovering the original load bearing capacity of the repaired structural members.

The tensile properties of UHPFRC (i.e., first cracking stress, σ_{cc} and strain, ε_{cc}, elastic modulus, E_{cc}, strain hardening modulus, E_{pc}, composite strength, σ_{pc}, and energy dissipation capacity, g) significantly improved as the fiber content increased from 1.0% to 3.0% (by volume); thus far, the sensitivity of the tensile properties of UHPFRC to fiber type (i.e., smooth, twist, and hook) remains debatable among researchers [43, 44]. Haddad [13] reported that the load bearing capacity of RC beam decreased by 28% after heating at 600 °C for 3 hours. Yaqub and Bailey [5] reported that the load bearing capacity of RC columns decreased by about 45% after uniform heating to 500 °C. Table 3 shows that the load carrying capacity of existing beams increases as the tensile strength or fiber content of UHPFRC increases. In the numerical analysis conducted by Lampropoulus et al. [25], UHPFRC exhibits huge potential to recover the original capacity of fire-damaged RC members. Lampropoulus et al. [25] conducted a uniaxial tensile test on a dog bone specimen containing UHPFRC with 3% (by volume) steel fiber content; the tensile strength obtained is about 12 MPa, which was then used in numerical analysis to predict the performance of RC beams strengthened with UHPFRC. According to Wille et al. [44], the tensile behavior of UHPFRC in dog bone size specimen may substantially differ if large specimens are used. Therefore, an experimental study must be conducted to determine the actual performance of UHPFRC with various fiber contents in improving the capacity of existing or damaged RC members.

Table 2. Thickness of UHPFRC layer used in strengthening works

Study	Investigated structural member	UHPFRC Thickness (mm)	Recovering (%)
Lampropoulus et al, [25]	Beam(existing)	50	300
Leonardi et al, [7]	Beam and column (fire damaged)	40	300
Martinola et al, [45]	Beam (under designed: low compressive strength and reinforcement ratio)	40	200
Farhat et al, [28]	Beam (under designed in flexural)	16	145

Yaqub and Bailey [5] and the present study indicated that cross section influences the performance of repair materials. As such, the performance of UHPFRC in rehabilitating RC members with different geometries or cross section must be determined. To the best of authors' knowledge, only Leonardi et al. [7] investigated the potential of using UHPFRC to repair fire-damaged RC columns and considered square columns in the numerical analysis. Therefore, further experimental works must be conducted to assess the actual performance of UHPFRC in repairing RC members of different geometries and cross sections.

Table 3. Tensile strength of UHPFRC and its performance in improving load bearing capacity of existing RC beams [25]

Tensile Strength (MPa)	Increment (%)
8	119
12	200
16	244

5. Conclusion

Selecting appropriate materials for repairing fire-damaged RC members is a crucial task. Repair materials must be compatible with damaged concrete substrate and must restore the original capacity of damaged structural members or systems. This paper reviews materials that can be used to repair fire-damaged concrete. These materials include CFRP, GFRP, FRC, ferrocement, epoxy resin mortar, and UHPFRC. Based on the information presented in this paper, the following conclusions are drawn:

1. Engineers prefer to use shotcrete as repair material because it omits the need for formworks, the time to construct formworks, and the need to pump and consolidate the concrete in relatively blind formworks. However, limited data are available with regard to the performance of shotcrete in repairing fire-damaged RC members.
2. Previous studies investigated the potential of using CFRP and GFRP, which can restore the load carrying capacity and ductility of fire-damaged RC members but fail to improve their stiffness.
3. FRC and ferrocement may improve the stiffness of fire-damaged RC members but cannot restore the load carrying capacity and ductility.
4. Epoxy resin exhibits good bond strength with fire-damaged concrete but cannot improve the capacity of fire-damaged RC members.

UHPFRC has a set of special traits, such as exceptional durability, outstanding mechanical properties, and ease of application, making it a potential candidate for successful repair of fire-damaged RC structures. Leonardi et al. [7] performed a numerical study to explore the potential of UHPFRC. However, few aspects, such as fire behavior, of NSC and UHPFRC must be further clarified through experimental works to determine their actual performance. Additional parameters, such as fiber content, UHPFRC thickness, and geometry of fire-damaged RC members, should also be evaluated to comprehensively explore the potential of UHPFRC as a repair material for fire-damaged RC members.

Acknowledgement
The authors would like to thank Universiti Malaysia Perlis (UniMAP) Ministry of Higher Education of Malaysia for providing financial assistant in publishing this article through Short Term Grant 9001-00506 and Fundamental Research Grant Scheme (FRGS) 9003-00510.

References

1. EN1994-1-2. European Committee for Standardization. (2005)
2. EN1993-1-2. European Committee for Standardization. (2005)
3. EN1992-1-2. European Committee for Standardization. (2004)
4. Y. N. Chan, G.F. Peng, M. Anson, Cem. Concr. Comp. **21**. 23 (1999)
5. M. Yaqub and C.G. Bailey, Constr. Build. Mater. **25**. 359 (2011a)
6. M. Yaqub and C.G. Bailey, Composite Structures. **93**. 1103 (2011b)
7. A. Leonardi, A. Meda, Z. Rinaldi, Strain. **47**. 28-35 (2011)
8. N. K. Gosain, R.F. Drexler, D. Chouduri, Struct. Eng. Mag. Struct. Forensic. 18 (2008).
9. C. Lin, S. Chen, C. Yang, ACI Struct Journal. **92**. 406 (1995)
10. K. Xiang, G.H. Wang. Procedia Engineering. **52**. 446 (2013)
11. Prugar J.F., Osowski S.M., and Brighton J., Strengthening of Fire Damaged Concrete Joists at High School. Proceeding of 4th Congress on Forensic Engineering (American Society of Civil Engineering, USA, 2006) pp. 502-515
12. M. Yaqub, C.G. Bailey, P. Nedwell, Q.U.Z. Khan, I. Javed. Composites: Part B. **44**. 200 (2013)
13. R.H. Haddad and M. J. Shannag, Mater and Structr. **41**. 287 (2008)
14. A.E. Naaman. Ferrocement and Laminated Cementatious Composites (Techno Press 3000, Ann Arbor, 2000), pp. 1-5.
15. A. Hannan, M. Jawarneh. Mater Structr. **50**. 95 (2017)
16. R.Z. Al-Rousin, R.H. Haddad, A. O. Swesi. Composites: Part B. **70**. 20 (2015)
17. Xiang K., Wang G., Procedia Engineering. **52**. 446-452(2013)
18. L.A. Brisby, J.F. Chen, S.Q. Li, T.J. Stratford, N. Cueva, K. Crossling. Engr. Structr. **33**. 3381 (2011)
19. H. A. Toutanji and W. Gómez. Cement and Concrete Composites. **19**. 351 (1997)
20. R.H. Haddad, N. Al-Mekhlafy, A.M. Ashteyat, Constr. Build. Mater. **25**. 1213 (2011).

21. F. de Larrard, T. Sedran. Cement and Concrete Research. **24**. 997-1009(1994)
22. A.E. Naaman, H.W. Reinhardt. Setting the stage: toward performance based classification of FRC
composites. In: Proceedings of 4th RILEM symposium on high performance fiber reinforced cement, composites (HPFRCC4), (2003) pp. 1-4.
23. J.P. Charon, E. Denarié, E. Brühwiller, Mater. Structr. **40**.269 (2007)
24. H. Martin-Sanz, E. Chatzi, E. Denarie, "The use of ultra high performance fiber reinforced cement-based composites in rehabilitation project: A review" in 9th International Conference on Fracture Mechanics of Concrete and Concrete Structures. Framcos-9 conference proceeding, edited by V. Souma et al (University of California, Berkeley, USA, 2016) pp. 1-12.
25. A.P. Lampropoulus, S. A. Paschalis, O. T. Tsioulou, S. E. Dritsos. Engineering Structures. **106**. 370-384 (2016)
26. N.J.S. Gorst and L.A. Clark, Cem. Concr. Compos. **25**, 1089 (2003)
27. B. Mu, C. Meyer, and S Shimanovich, Cem. Concr. Res. **32**, 783(2002).
28. F. A. Farhat, D. Nicolides, A. Kanellopoulos, B.L. Karihaloo. Engineering Fracture Mechanics. **74**. 151 (2007)
29. B.A Tayeh, B.H. Abu Bakar, M.A. Megat Johari, and Y.L. Voo. Procedia Eng. **54**, 554 (2013).
30. N. K. Baharuddin, F. M. Nazri, R Putra Jaya, B. H. Abu Bakar. World Journal of Engineering. **13(5).** 461(2016)
31. E. Denarie, "Recommendation for tailoring of UHPFRC recipes for rehabilitation," ARCHES Report, Lausanne, 2009.
32. M.Z.A. Mohd Zahid and K. Muhammad, "The assessment of bond strength between heat damaged concrete and high strength fiber reinforced concrete" in AIP Conf. Proc. 1885, 020117-020117-6; doi: 10.1063/1.5002311
33. A. Hanan, R. Haddad, S. Afram, M. Abdel-Halim, Materials and Structures. **46**. 1843-1860 (2013)
34. D. Wang, C. Shi, Z. Wu, J. Xiao, Z. Huang, Z. Fang. Construction and Building Materials. 96. 368-377 (2015)
35. H. Abdullah, M.Z.A. Mohd Zahid, B.H. Abu Bakar, F. M. Nazri, and A. Ayob, Appl. Mech. & Mater. **802**, 283 (2015)
36. C. Kahanji, F. Ali, A. Nadjai, Journal of Structural Fire Engineering. **7(4).** 328-348(2016)
37. ISO 834-1. Fire resistance test – element of building construction. Switzerland. (1999)
38. Y. A. Al Salloum, T. H. Al Musallam, H. M. Elsanadedy, R. A. Iqbal. Constr. Building Materials. **115**. 345-361(2016)
39. V. Kodur. ISRN Civil Engineering. 1-15(2014)
40. O. Arioz. Fire Safety Journal. **42**. 516-522(2007)
41. U. Shneider. Fire Safety Journal. **13**. 55-68(1988)
42. M. N. Fardis. Innovative materials and techniques in concrete construction: ACES Workshop. Springer
Dordrecht Heidelberg London New York: University of Patras; 2012
43. J. Liu, F. Han, G. Cui, Q. Zhang, J. Lv, L. Zhang, Z. Yang. Construction and Building Materials. **121**. 310-318 (2016)
44. K. Wille, S. El-Tawil, A. E. Naaman. Cement & Concrete Composites. **48.** 53-66 (2014)
45. G. Martinola, A. Meda, G. A. Plizzari, Z. Rinaldi. Cement & Concrete Composites. **32**. 731-739(2010).

Flood Inundation Modelling in the Kuantan River Basin using 1D-2D Flood Modeller coupled with ASTER-GDEM

Z F Ng[1*], J I Gisen[1,2] and A Akbari[3]

[1] Faculty of Civil Engineering and Earth Resources, Universiti Malaysia Pahang, Malaysia.
[2] Centre for Earth Resources Research & Management, Universiti Malaysia Pahang, Malaysia.
[3] Department of Civil Engineering, Khavaran Institute of Higher Education, Mashhad, Iran.

E-mail: ngzonefhong@outlook.com

Abstract. Topography dataset is an important input in performing flood inundation modelling. However, it is always difficult to obtain high resolution topography that provide accurate elevation information. Fortunately, there are some open source topography datasets available with reasonable resolution such as SRTM and ASTER-GDEM. In Malaysia particularly in Kuantan, the modelling research on the floodplain area is still lacking. This research aims to: a) to investigate the suitability of ASTER-GDEM to be applied in the 1D-2D flood inundation modelling for the Kuantan River Basin; b) to generate flood inundation map for Kuantan river basin. The topography dataset used in this study is ASTER-GDEM to generate physical characteristics of watershed in the basin. It is used to perform rainfall runoff modelling for hydrological studies and to delineate flood inundation area in the Flood Modeller. The results obtained have shown that a 30m resolution ASTER-GDEM is applicable as an input for the 1D-2D flood modelling. The simulated water level in 2013 has NSE of 0.644 and RSME of 1.259. As a conclusion, ASTER-GDEM can be used as one alternative topography datasets for flood inundation modelling. However, the flood level obtained from the hydraulic modelling shows low accuracy at flat urban areas.

1. Introduction

Flood leads to property damages, loss of life and economic crisis [1]. The main reasons causing flood to happen are such as heavy and high intensity rainfall, ineffective drainage system, and high volume of runoff due to increasing impervious ground surface [2]. In Kuantan, monsoon season largely affects the weather. During the monsoon season (the northern east monsoon), major floods occurred. Furthermore, the land use changes for urbanisation has worsen the flood condition. In 2013, there are about 14,044 people separated and major damages in terms of electricity supply, road's structure, buildings, and belongings. Due to these disasters, the Malaysian government should allocate substantial amount of money to repair and manage the damages [1]. Therefore, good flood risk management and planning must be implemented to reduce floods [3]. Since it is not possible to avoid rapid urbanisation, efficient flood management practices are advisable [2]. The objectives of this study are: a) to investigate the suitability of ASTER-GDEM to be applied in the 1D-2D flood inundation modelling for the Kuantan River Basin; b) to generate flood inundation map for Kuantan river basin.

The Geospatial Information System (GIS) and Remote Sensing (RS) datasets have been introduced few decades ago and widely used to facilitate flood hazard modelling [4]. There are several GIS applications available such as ArcGIS and QGIS while RS datasets such as ASTER-GDEM and SRTM to be integrated in the hydrology and hydraulic studies. In hydrology studies, topography dataset is used for watershed delineation where all the sub basins are to be identified and the stream line can be defined. For hydraulic modelling, topography dataset acts as the base elevation map in the flood inundation modelling. After the hydraulic analysis, the flood level and flood extent can be identified. For better result, high resolution topography for example LiDAR, IFSAR, IKONOS and Spot DEM are recommended. However, the acquisition of these datasets is costly.

In this study, ASTER-GDEM with 30m resolution was chosen to perform flood inundation modelling in the Flood Modeller. The selected topography dataset and Flood Modeller software are available and downloadable from the NASA Land Data Products and Services, U.S. Geological Survey (USGS) and the Flood Modeller main website respectively without any charges.

1.1. Study area

The study area selected for the research is the Kuantan River Basin (KRB). KRB is one of the main river basin in the Pahang state. It has a watershed boundary of 1630 km^2. Sg. Kuantan is the main river in KRB begins from Mukim Ulu Kuantan and ends at the Perkampungan Tanjung Lumpur outlet point towards the South China Sea. The KRB contains several tributaries which flow along the rural, agricultural, urban, and industrial areas of Kuantan district into the Sg. Kuantan [1]. This study area has been chosen due to the frequent flood occurrence along the Sg. Kuantan in the basin. Figure 1 shows the boundary of the study area displaying the streamflow station, rainfall stations, and the main river.

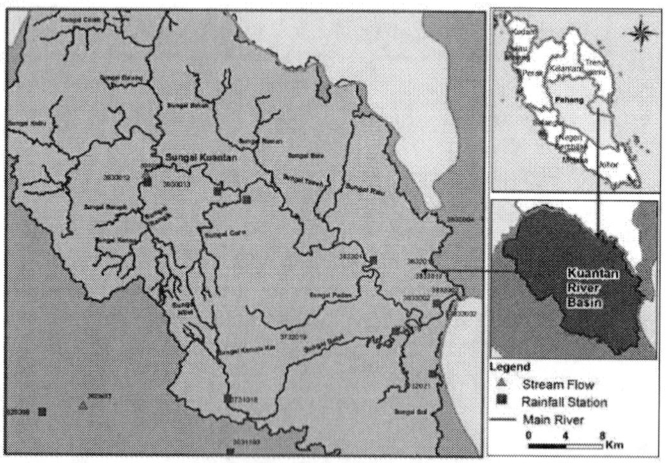

Figure 1. Boundary of KRB displaying the streamflow station, rainfall stations, and the main river.

1.2. Topography datasets

Topography dataset which stores elevation data serve as the main source for representing river bathymetry profile and floodplain topography. It is very common to extract river cross sections from topography dataset when there is lack of survey information. The selection of suitable Digital Elevation Model (DEM) resolution for the representation of river cross section profile is crucial for flood inundation mapping [5]. Table 1 shows some guidance on the selection of cell size [6].

Table 1. Recommended DEM cell sizes and their range of applications.

Cell Size (m)	Watershed Area (km^2)	Typical
30	5	Urban watersheds
90	40	Rural watersheds
460	1000	River basins
930	4000	Nations
5600	150,000	Continents
9300	400,000	Global

ASTER is an advanced multispectral imager that has been launched on board NASA's Terra spacecraft in December 1999 [7]. The ASTER-GDEM is a joint product developed and made available to the public by the Ministry of Economy, Trade, and Industry (METI) of Japan and the NASA. These global data are free source which are available at the NASA Land Data Products and Services, U.S. Geological Survey (USGS) website [6]. ASTER-GDEM has several merits such as low cost, high spatial resolution, good correlation over vegetated areas but it may prone to masking by clouds [7]. According to previous studies, making an accurate flood map is not easy and cheap. Hence, ASTER-GDEM can become the alternative mean to produce flood map.

1.3. Flood inundation modelling

Flood inundation modelling is a complex process where topography dataset, hydrological model, and hydraulic model are combined to perform final analysis. Through these processes, the potential flood water level and flow pattern can be predicted from time to time even if the river system is hindered by man-made structures or vegetation. Moreover, the flood extent area can be detected to ease the local authorities to trace and perform maintenance based on the affected zone. In time, many researchers, water authorities, and environmental state agencies have turn their focus on developing flood inundation map instead of traditional ground survey solution [3].

There are four key data items for flood inundation modelling such as topography dataset to construct the model grid; discharge flow or stage data to provide model inflow and outflow boundary conditions; an estimation of the effective friction parameter for each model cells; and a source of validation data [8]. Generally, flood modelling consists of two components which are the hydrological simulation where hydrographs of the flood event were generated, and the hydraulic simulation where the influence of flood wave across the river channel and the mapping of flood extent regions [9]. For flood routing models, there are one-dimensional (1D), two-dimensional (2D) or coupled 1D-2D (also referred as quasi-2d) numerical solutions available for steady and unsteady flow conditions [10]. 2D hydrodynamic models discuss about the variation of flow in both, the longitudinal and the transverse directions of river channel towards the floodplain [5].

1.4. Flood Modeller

There are many hydraulic modelling have been done using different software such as Flood Modeller, MIKE-Flood, and HEC-RAS to solve the full momentum St. Venant equations [8]. Flood Modeller has been developed to solve one-dimensional (1D) channel flow model (Saint-Venant equations) and two-dimensional (2D) shallow-water equation solver [5]. In simple term, flood routing models solve part or the full one-dimensional 1D Saint-Venant continuity and momentum depth-averaged equations in the longitudinal direction 1D model. The 1D Saint-Venant continuity Equation (1) and momentum equations Equation (2) are shown as below:

$$\frac{\partial Q}{\partial x}+\frac{\partial A}{\partial t}=q \tag{1}$$

$$\frac{\partial Q}{\partial t}+\frac{\partial}{\partial x}\left(\frac{\beta Q^2}{A}\right)+gA\frac{\partial H}{\partial x}-gAS_f=0 \tag{2}$$

$$S_f=\frac{Q|Q|}{K^2} \tag{3}$$

$$K^2=\frac{A^2R^{\frac{4}{3}}}{n^2} \tag{4}$$

$$R=\frac{A}{P} \tag{5}$$

where Q [m^3/s] is the discharge; A [m] is the wetted area; q [m^3/s] is the lateral inflow; g [m^2/s] is the gravity acceleration, Sf [-] is the friction slope; K [-] is the channel conveyance calculated according Manning's equation; R [m] is the hydraulic radius; P [m] is the length of the wetted perimeter; and n [-] is the Manning's roughness coefficient.

The equations are described in mathematical terms as a pair of one-dimensional non-linear hyperbolic partial differential equations. In general, the solution of any system of differential equations depends on the existence, uniqueness, and stability conditions of the solution itself. Furthermore, the numerous assumptions for the simplification of the Navier-Stokes equations to the 1D Saint-Venant equation, the mathematical integration difficulties as well as the computational cost are still high, especially for representing the inundation of a flood wave across a longer river and bigger area. However, it could be enough to further simplify the momentum equation to the steady dynamic wave, diffusive wave, or kinematic wave [10]. Flood Modeller also incorporates an integrated Muskingum-Cunge flow routing solver and a comprehensive range of hydraulic structures, including gates, abstractions and pumps operated by logical rules. Besides that, it also provides three integrated 2D numerical solvers, tried and tested on projects throughout the world for various hydraulic and environmental studies as listed in Table 2.

Table 2. Three 2D solvers in Flood Modeller.

2D Solvers	Description
Alternating Direction Implicit (ADI) solver	First developed in the 1980s, the quick, accurate and robust Alternating Direction Implicit (ADI) solver can be applied to riverine, overland, estuarine, and coastal modelling problems where the flow is not rapidly changing.
Total Variation Diminishing (TVD) solver	Providing accurate and stable results for complex hydraulics, the Total Variation Diminishing (TVD) solver provides accurate representation of two-dimensional 'shocks' (rapid changes in water surface profile).
FAST solver	The FAST solver enables rapid assessments of flooding, providing results up to 1,000 times quicker than traditional 2D solvers. It produces results in seconds or minutes, as opposed to hours or days

Flood Modeller 2D solver, and many other models of its type, represent shallow water hydraulics. These can be described by the shallow water equations as below:

$$\frac{\partial h}{\partial t}+\frac{\partial hu}{\partial x}+\frac{\partial hv}{\partial y}=0 \tag{6}$$

$$\frac{\partial u}{\partial t}+u\frac{\partial u}{\partial x}+v\frac{\partial u}{\partial y}+g\frac{\partial (h+z)}{\partial x}+\frac{gn^2u\sqrt{u^2+v^2}}{h^{4/3}}-\frac{v}{h}\left(2\frac{\partial^2 hu}{\partial x^2}+\frac{\partial^2 hu}{\partial y^2}+\frac{\partial^2 hv}{\partial x\partial y}\right)=0 \tag{7}$$

$$\frac{\partial v}{\partial t}+u\frac{\partial v}{\partial x}+v\frac{\partial v}{\partial y}+g\frac{\partial (h+z)}{\partial y}+\frac{gn^2v\sqrt{u^2+v^2}}{h^{4/3}}-\frac{v}{h}\left(2\frac{\partial^2 hv}{\partial y^2}+\frac{\partial^2 hv}{\partial x^2}+\frac{\partial^2 hu}{\partial x\partial y}\right)=0 \tag{8}$$

where u [m/s] and v [m/s] are the two components of the horizontal velocity; h [m] is the flow depth; z [m] is the bed elevation; x [m] and y [m] are the horizontal distances in the x [m] and y [m] directions; t [s] is the time; g [m^2/s] is the acceleration due to gravity; n [-] is the Manning's coefficient of roughness; and v [m/s] is the kinematic eddy viscosity used to parameterize horizontal turbulent momentum transport, with a value between 0 and 1.

These equations describe the motion of water in terms of a depth-averaged 2D velocity and the water depth, in response to the forces of gravity and friction. These equations typically represent situations where flow is approximately horizontal, is uniform with depth, and where vertical accelerations are small. The velocity components represent the water velocity averaged either over the depth of the water column, ignoring variations in flow direction and magnitude with depth or over time, ignoring short-term turbulent velocity variations. One important property of shallow water flows is the different behaviors of subcritical and supercritical flows. Supercritical flows tend to develop jumps (sudden changes in velocity and water level), which are difficult to represent in the model without causing instability. The difficulty of modelling these jumps is the reason 2D solver has the TVD solver as well as the more commonly used ADI solver.

2. Methods

In this study, the required primary data is ASTER-GDEM in ASCII file format as the floodplain topography and cross section survey data as the 1D schematic model. Next, the simulated discharge which were calculated by rainfall runoff modelling were used as the inflow and outflow boundaries condition in the Flood Modeller. The calibrated hydrological parameters for the flood event in the year 2010 was validated with the flood event in the year 2013.

For the hydraulic modelling, the river cross-sections derived from the ASTER-GDEM was not as accurate as the survey data. Thus, the survey data were used for cross section generation in Flood Modeller. The cross sections in the format of comma-separated values (CSV) were imported into Flood Modeller. Then, the calculated discharge for the year 2010 were added as input boundaries to simulate the 1D steady and unsteady flow. Steady flow was performed to make sure that the river profile generated was appropriate before the unsteady flow modelling was carried out. Next, the simulated water level data was calibrated with the observed water level (Kuantan bypass) which is located near the downstream part of Sg. Kuantan.

For 2D modelling, an active area was created which represents the flood prone area along the main river. Next, the 1D-2D link line was generated in Flood Modeller which connects the 1D open channel flow to the 2D floodplain topography. Finally, 2D modelling with 2D ADI solver was conducted and flood inundation map for the year 2010 was obtained.

3. Results and discussion

For the 1D unsteady flow modelling, the calibration was done using the data from 2010. The calibration parameters were obtained by fitting the simulated water level against the observed as shown in Figure 2. Then, the same parameters were applied for the validation process. The validation result for the year 2013 is displayed in Figure 3.

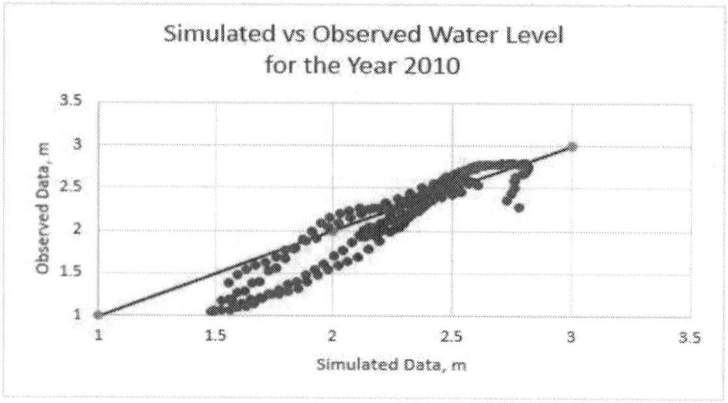

Figure 2. Level of fitness between the calibrated water level to the observed for the year 2010.

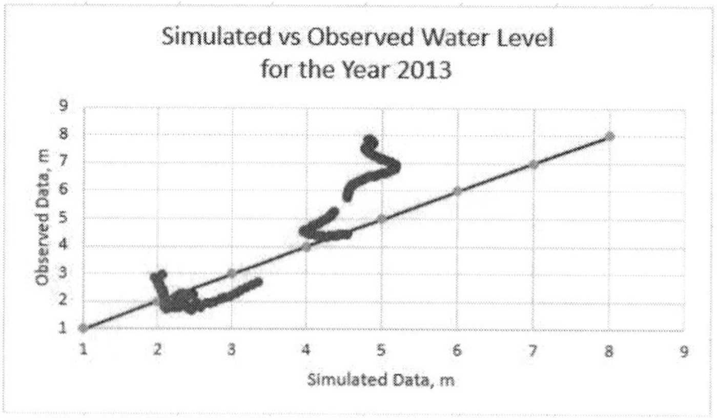

Figure 3. Level of fitness between the validated water level to the observed for the year 2013.

From the error analysis, the simulated water levels for the year 2010 has a Nash–Sutcliffe efficiency (E) of 0.802 and Root-mean-square error (RSME) of 0.227. Both results indicate high accuracy of efficiency, Meanwhile, the validated results for year 2013 applying the same parameters value obtained from the calibration process show NSE of 0.644 and RSME of 1.259. Compare to the calibration, both errors obtained for the validation result are higher, indicating that the calibration parameters may need to be improved.

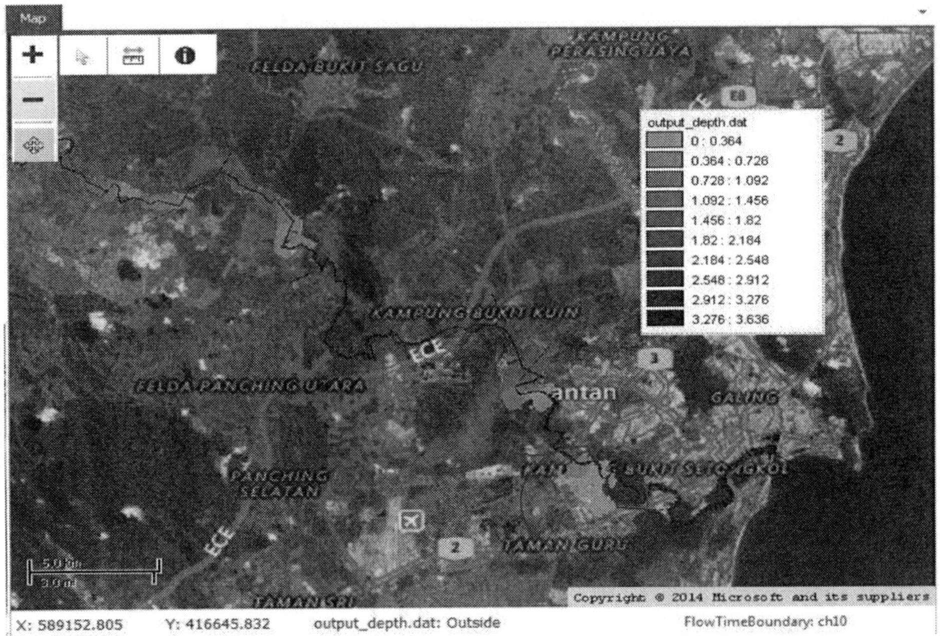

Figure 4. Result from the 2D flood inundation map obtained from the calibration process for the year of 2013.

Figure 4 indicates the flood inundation map for the year 2013. The maximum flood level range was 3.276–3.636m while the minimum flood level range was 0–0.364m. The maximum flood occurred at downstream at the low land area near the river mouth while the minimum flood level occurred at the upstream. Since the observed flood studies were only available at the circled region in Figure 4, the discussion focused only at this location. The others flood zones will be discussed in future after all the information are completely obtained.

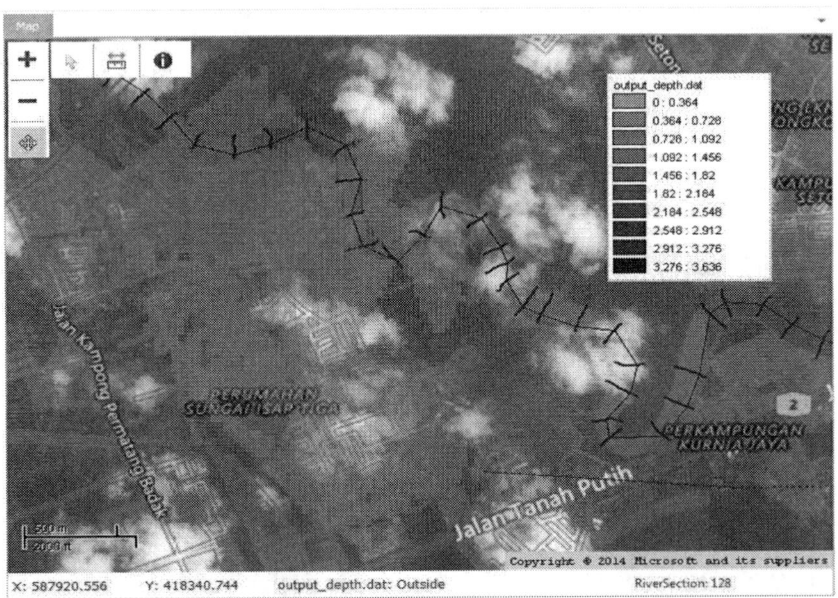

Figure 5. Simulated flood map of the focus area based on the variable flood studies.

Figure 5 displays the flood extend area (highlighted in blue) or the floodplain area simulated by the Flood Modeller for the calibration result for year 2013. From the figure, the flood level range was 0 – 1m. Based on the hatched areas, the affected region was at Perumahan Sungai Isap Tiga. Besides that, the road branches within the villages were affected. In addition, there was water spilled over the swampy area as circled in Figure 5. The simulated flooding areas obtained were underestimated compare to the flood report in the year of 2013 from Department of Irrigation and Drainage (DID).

The cause of flood occurred along the Kuantan river was due to continuous development of the Kuantan city which leads to the increasing imperviousness of land areas. More impervious areas have reduced the infiltration of water into the ground and consequently increase the runoff volume during rainy season. Therefore, it is important to consider the land use effect in the urban area when designing flood mitigation measures in the Kuantan River Basin.

4. Conclusion

In all, the 30m resolution of ASTER-GDEM is suitable to be integrated with the Flood Modeller to create flood inundation map for the KRB. In this study, the 1D channel model and 2D floodplain model were presented well by using Flood Modeller. The study has shown that the performance of the model to simulate the flood level is adequate with NSE of 0.644 and RSME of 1.259.

However, the generated flood map does not provide sufficient extension because the open source Flood Modeller applied in this study has limitation in term of the number of nodes and the 2D cells extend. Hence, it is recommended to use full version of Flood Modeller if possible. Moreover, if higher resolution of topography dataset is available such as LiDAR, more accurate result can be obtained. Nevertheless, the adequate result obtained in this study has shown that the 30m ASTER-GDEM coupled with the open source Flood Modeller can be used as preliminary flood inundation study.

By having the flood inundation map, the flood level and flood areas can be predicted and this information are useful for the water authorities in decision making such as evacuation during flood event, and selecting flood mitigation options.

Acknowledgements
The authors would like to thank University Malaysia Pahang and Ministry of Higher Education, for supporting this research through the grant numbers of RDU150127 and FRGS/1/2015/WAB08/UMP/02/1. Also, the authors would like to show their appreciation to Department of Irrigation and Drainage and the Centre for Earth Resources Research & Management for the technical support.

References

[1] Zaidi S M, Akbari A and Ishak W M F 2014 A critical review of floods history in kuantan river basin: challenges and potential solutions *IJCEG* **5**

[2] Sahoo S N and Sreeja P 2015 Development of flood inundation maps and quantification of flood risk in an urban catchment of brahmaputra river *ASCE-ASME j. risk uncertain. eng. syst. Part A. Civ. eng.* **3**

[3] Manfreda S, Leo M D and Sole A 2011 Detection of flood-prone areas using digital elevation models *JHE* **16** 781–90

[4] Hamblen C S, Soong D T and Cai X M 2006 Digital floodplain mapping and an analysis of errors involved *World Environmental and Water Resource Congress* (Omaha, Nebraska, United States)

[5] Tarekegn T H, Haile A T, Rientjes T, Reggiani P and Alkema D 2010 Assessment of an ASTER-generated DEM for 2D hydrodynamic flood modelling *Int J Appl Earth Obs Geoinf* **12** 457–65

[6] Akbari A, Ramli N I B and Ngien S K 2016 Application of public domain satellite-based DEMs in natural hazard modelling *Int J Appl Earth Obs Geoinf* **7** 140–44

[7] Pryde J K, Osorio J, Wolfe M L, Heatwole C, Benham B and Cardenas A 2007 *Comparison of watershed boundaries derived from SRTM and ASTER digital elevation datasets and from a digitised topographic map* ASABE (Minneapolis Convention Center, Minneapolis, Minnesota)

[8] Wright N G, Asce M, Villanueva I, Bates P D, Mason D C, Wilson M D, Pender G and Neelz S 2008 Case study of the use of remotely sensed data for modelling flood inundation on the river severn *J. Hydraul. Eng.* **134** 533–40

[9] Santillan J R, Ramos R V, David G and Recamadas S M 2013 *Development, calibration and validation of a flood model for marikina river basin, philippines and it's applications for flood forecasting, reconstruction, and hazard mapping* (University of the Philippines, Diliman, Quezon City 1101, MetroManila, Philippines)

[10] Dimitriadis P, Tegos A, Oikonomou A, Pagana V, Koukouvinos A, Mamassis N, Koutsoyiannis D and Efstratiadis A 2016 Comparative evaluation of 1D and quasi-2D hydraulic models based on benchmark and real-world applications for uncertainty assessment in flood mapping *J Hydrol (Amst)* **534** 478–92

IOP Conf. Series: Materials Science and Engineering **318** (2018) 012025 doi:10.1088/1757-899X/318/1/012025

Basic and Morphological Properties of Bukit Goh Bauxite

Muzamir Hasan[1,2*], Ahmad Amirul Faez Ahmad Nor Azmi[1,2], Weng Long Tam[2], Biao Yu Phang[2] and M. Azizul Moqsud[3]

[1]Centre for Earth Resources Research & Management, Universiti Malaysia Pahang, 26300 Gambang, Kuantan, Malaysia

[2]Faculty of Civil Engineering & Earth Resources Research & Management, Universiti Malaysia Pahang, 26300 Gambang, Kuantan, Malaysia

[3]Department of Civil and Environmental Engineering, Yamaguchi University, Ube Shi, Tokiwadai 1, Japan.

E-mail : muzamir@unimap.edu.my

Abstract. Investigation conducted by International Maritime Organization (IMO) concluded that the loss of the Bulk Jupiter that carrying bauxite from Kuantan has uncovered evidence to suggest liquefaction led to loss of stability. This research analysed Bukit Goh bauxite and comparison was made with International Maritime Solid Bulk Cargoes (IMSBC Code) standard. To analyse these characteristics of the bauxite, four samples were selected at Bukit Goh, Kuantan ; two of the samples from the Bukit Goh mine and two samples from the stock piles were tested to identify the bauxite basic and morphological properties by referring to GEOSPEC 3 : Model Specification for Soil Testing ; particle size distribution, moisture content and specific gravity and its morphological properties. Laboratory tests involved including Hydrometer test, Small Pycnometer test, Dry Sieve test and Field Emission Scanning Electron Microscop (FESEM) test. The results show that the average moisture content of raw Bukit Goh bauxite is 20.64% which exceeded the recomended value of maximum 10%. Average fine material for raw bauxite is 37.75% which should not be greater than 30% per IMSBC standard. By that, the bauxite from Bukit Goh mine do not achieved the minimum requirements and standards of the IMSBC standard and need to undergo beneficiation process for better quality and safety.

1. Introduction

Bauxite is the principal ore of alumina (Al_2O_3), which is used to produce aluminium. The main compund of bauxite are hydrated aluminium oxides, hydrated aluminosilicates, iron oxides, hydrated iron oxides, titanium oxide and silica [1]. There are approximately 165 million tonnes of bauxite are mined every year around the world. The bauxite is transported to refineries by conveyor ship, or rail [2]. Bauxite is formed by the effects of sedimentation and compression miles of rock that takes a long period perhaps millions of years and mix with elements of dehydration by hot temperatures. Bauxite usually found in the soil compared to oil and gas as well as some iron ore and tin, which requires certain very deep excavation. In Kuantan, bauxite reserve has been found and discovered and due to high demanding from China the bauxite mining has started. Hence, Malaysia has increased the production of bauxite. Investigation conducted by International Maritime Organization (IMO) concluded that the loss of the Bulk Jupiter that carrying bauxite has uncovered evidence to suggest liquefaction led to loss of stability [3][4]. Liquefaction is a phenomenon where the soil experience a process by which water-saturated sediment temporarily loses strength and acts as a fluid [5]. Hence, in other to lower the risks of the liquefaction phenomenon, there are some standard and regulation that need to be follow; which is the International Maritime Solid Bulk Cargoes (IMSBC) code in order to determine the safety of bulk cargoes carrying bauxite and lower the risk of the cargoes from the liquefaction phenomenon to occur. Fig. 1 shows the classification of Bauxite in Group C which it is neither liable to liquefy (Group A) nor to possess chemical hazards (Group B) [6]. If bauxite samples are complied with the specifications as stated in IMSBC Code, the samples are allowed be shipped and exported.

BAUXITE

DESCRIPTION

A brownish, yellow claylike and earthy mineral. Moisture content: 0% to 10%. Insoluble in water.

CHARACTERISTICS

ANGLE OF REPOSE	BULK DENSITY (kg/m^3)	STOWAGE FACTOR (m^3/t)
Not applicable	1190 to 1389	0.72 to 0.84
SIZE	**CLASS**	**GROUP**
70% to 90% lumps: 2.5 mm to 500 mm 10% to 30% powder	Not applicable	C

Figure 1. Bauxite's Group C Classification.

There are a very close relationship between the moisture content and particle size distribution as the presence of finest particle will affect the water holding properties. Despite on identifying the standard quality of Kuantan bauxite, its basic and morphological properties of Kuantan bauxite also identified to determine the particles present and its effect. If the particles are found to be harmful, prevention and safety measure should be applied and discussed.

2. Methodology

2.1. Preparation of samples

The basic properties of bauxite have been studied to identify Bukit Goh bauxites fulfill the requirement standard in IMSBS Code. There are some standard and regulations that need to be follow by using IMSBC Code to determine the standard quality of bauxite. Thus, this study is carried out to determine does the bauxite production is categorized as Group C in the IMSBC Standard or not for exporting. All samples are undergoing six laboratory testing and analysis; 2 samples from Bukit Goh mine (M2L2B1 and M2L2B2) and 2 samples from stock pile (PTSTL1B1 and PTSTL2B1). The laboratory tests involved are Hydrometer test, Wet Sieving Analysis, Dry Sieving Analysis, Small Pycnometer test to determine specific gravity and moisture content and SEM test. Hydrometer test, Wet Sieving Analysis and Dry Sieving Analysis is done to determine the size distribution of bauxite samples (Table 2). The specific gravity and moisture content of bauxite can be determine Small Pycnometer Test and Moisture Content Determination. All the testing are based on Geospec 3: Model Specification For Soil Testing. The analysis data from the result will be compared with IMSBC Code.

IOP Conf. Series: Materials Science and Engineering **318** (2018) 012025 doi:10.1088/1757-899X/318/1/012025

Table 1. Laboratory test and standards.

Laboratory Tests	Standards
Moisture Content Test	Geospec 3: Part 2; 5 Clause 3.2
Specific Gravity Test	Geospec 3: Part 2; 7 Clause 3.4
Particle Size Distribution	Geospec 3: Part 2; 8 Clause 3.5
X-Ray Fluorescence (XRF)	Quantexpress (Full Analysis) by XRF S8 Tiger

3. Experimental results and analysis

3.1. Properties of main materials

A summary of the properties of bauxite were shown in Table 2.

Table 2. Summary of properties of bauxite.

Basic Propertie s	**Particle Size Distributio n**	**Bulk Density**	**Moitur e content**
M2L2B1	<70%	2887	23.14
M2L2B2	<70%	2807	21.90
OPSTL1B 1	<70%	2823	21.76
OPSTL2B 1	<70%	2828	23.18
IMSBC Code	70% to 90% lumps ; 2.5mm to 500mm	1190kg/ m^3 to 1389kg/ m^3	0-10

Based on the table above, it shows that the result from the particle size distribution, bulk density and, moisture content is compared and after comparing the bauxite in Bukit Goh do not fulfilled the requirement needed in the standard IMSB Code. This is because based on the analysis and result from the laboratory test all the particle size distribution is below 70% while the moisture content from all the bauxite samples is high which 20% above are. Hence, the results compared with the IMSBC Code are exceeding the specified value states in the code.

3.2. Particle size distribution

There are few test had been done to determine particles size distribution, 6.3 mm to pan sieve size is used to separate the particle according to the size. In Table 4 shows the percentage of passing for fine particle determination for the samples by using Wet Sieving Method. In Table 5 shows the percentage passing 2.5 mm for sample M2L2B1 is 36%, M2L2B2 is 44%, PTSTL1B1 is 31% and PTSTL2B1 is 40%. Based on the result analysis prove that the particle size distribution Bukit Goh bauxite's not in range the requirement size in IMSBC Code. The result proved that Bukit Goh bauxite's in average consist more than 30% fine particle and less than 70% coarse particle. This situation will resulting the moisture content of the sample increase due to higher percentage of fine particle.

Table 3. Percentage passing.

Sieve size (mm)	M2L2B1 (%)	M2L2B2 (%)	PTSTL1B1 (%)	PTSTL2B1 (%)
6.30	73.96	88.18	60.76	73.08
5.00	59.49	72.61	53.78	58.61
3.35	44.10	55.61	39.38	43.49
1.18	22.67	25.35	16.26	20.70
0.60	17.84	20.59	12.13	14.98
0.30	13.88	16.96	8.87	10.89
0.15	7.84	10.73	5.21	6.73
0.0063	0.04	0.64	0.17	0.03
Pan	0.00	0.00	0.00	0.00

Table 4. Particle size distribution.

Samples	Particle size distribution (%)
M2L2B1	36
M2L2B2	44
PTSTL1B1	31
PTSTL2B1	40

3.3. *Specific gravity*

The result from Small Pycnometer Test was collected and tabulated in Table 6 where the specific gravity for PTSTL1B1 and PTSTL2B1 are just the same. Meanwhile, there is slightly difference in specific gravity for M2L2B1 and M2L2B2. This is because the sample M2L2B1 and M2L2B2 are from different locations of the same mining area while samples PTSTL1B1 and PTSTL2B1 are from the stockpiles in Bukit Goh area. Based on studied carried out by [7], specific gravity of bauxite in India is higher compared to Bukit Goh bauxite. The soil'specific gravity largely depends on the density of the mineral making up the individual soil particle.

Table 5. Results of specific gravity.

Samples	Specific gravity of container 1	Specific gravity of container 2	Average specific gravity
M2L2B1	2.965	2.808	2.887
M2L2B2	2.871	2.742	2.807
PTSTL1B1	2.892	2.753	2.823
PTSTL2B1	2.897	2.758	2.828

3.4. *Moisture content*

Moisture content of the sample is determined by air-dry method and oven-dry method. The oven-dry method can be further divided into two which is oven-dry hot method and oven-dry cool method. The results are shown in Table 7, Table 8 and Table 9. From the table below show percentage moisture content for oven-dry is higher compared to air-dry. According to IMSBC Code, allowable average moisture content is 0%-10% to make sure the bauxite is save to be exported. The result shows moisture content of Bukit Goh bauxite is higher compared to

IOP Conf. Series: Materials Science and Engineering **318** (2018) 012025 doi:10.1088/1757-899X/318/1/012025

recommended percentage in the IMSBC Code. Due to higher moisture content, it clearly show that Bukit Goh bauxite have large amount of fine particle compared to coarse particle.

High level of moisture content lead the liquefaction of mineral ores to be occurs resulting the cargo loss of stability during the voyage. In cargoes loaded with too high a moisture content, liquefaction may occur without warning at any time during the voyage. Some cargoes have liquefied and caused catastrophic shifting of cargo almost immediately upon departure from the load port whilst others have liquefied after several weeks of apparently uneventful sailing.

Table 6. Average moisture content (air-dry method).

Samples	Moisture content container 1 (%)	Moisture content container 2 (%)	Average moisture content (%)
M2L2B1	24.71	24.39	24.55
M2L2B2	19.03	19.39	19.21
PTSTL1B1	19.74	19.31	19.53
PTSTL2B1	18.93	19.62	19.28

Table 7. Average moisture content (oven-dry hot method).

Samples	Moisture content container 1 (%)	Moisture content container 2 (%)	Average moisture content (%)
M2L2B1	23.49	22.43	22.96
M2L2B2	22.59	25.25	23.93
PTSTL1B1	23.92	23.10	23.51
PTSTL2B1	26.07	25.83	25.95

Table 8. Average moisture content (oven-dry cool method).

Samples	Moisture content container 1 (%)	Moisture content container 2 (%)	Average moisture content (%)
M2L2B1	22.37	21.46	21.91
M2L2B2	21.29	23.84	22.57
PTSTL1B1	22.57	21.89	22.23
PTSTL2B1	24.42	24.18	24.30

3.5. Morphological properties of bauxite

The morphological properties of Bukit Goh bauxite had been studied by using SEM Test as in Figure 2 and Figure 3 where magnification for each figure 5000x and 10 000x respectively. Closer inspection of the particles shows a layer of material coating most of the particle surfaces. The different sizes of particles can be observed with clear image of lump particles and powdery like-structure of fine. Clear image of particles started to be seen at 5000x magnification and under 10 000x magnification, fine particles attached to the bauxite sample are clearly can be seen.

This proved the main cause of high percentage of moisture content as well as the big number of bulk density of the bauxite sample because of many fine particle attached at the sample. Bauxite samples collected from Bukit Goh mine are disturbed sample, hence the tendency for this sample to liquefy is higher than undisturbed soil because of shear force of anti-liquefaction of undisturbed soil is 1.5 to 2 times greater than disturbed soil.

Figure 2. 5000x magnification of bauxite sample.

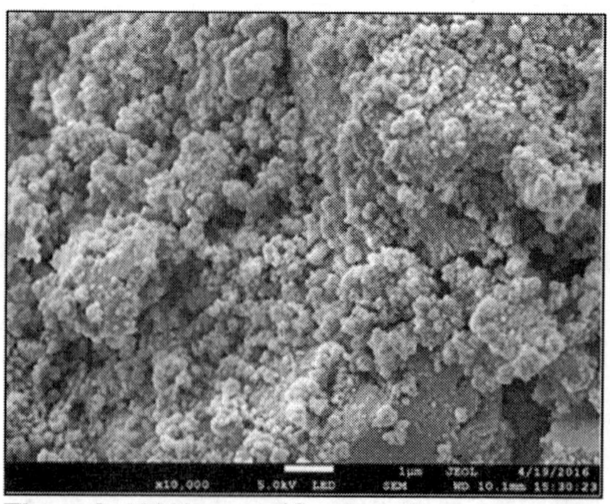

Figure 3. 10000x magnification of bauxite sample.

4. Conclusion

Results from the basic properties study is conclude that the average particle size distribution for percentage passing 2.5 mm to 500 mm for the samples is 37.75%. The average Bulk Density for the samples is 2836.25 kg/m^3 and the average percentage of moisture content of the samples is 20.64%. The results obtained was compared with IMSBC Code where each of this basic properties are exceeding the specified value stated in the code. All this basic properties are lead for liquefaction to occur during bauxite cargoes transportation, hence it is important to identify the properties before it is being exported. As results comparing to IMSBC Code, Bukit Goh bauxite's are cannot be categorized as Group C because the basic properties obtained are not fulfilling the requirement in the standard IMSBC Code and thus therefore it is not suitable to be exported. It will be high risk to transporting bauxite due to the waves of the ocean.

Study on morphological properties of Bukit Goh bauxite shows percent of fine particles is higher than the maximum limit set by the IMSBC Code. Clearly seen that the fine particle attached to the bauxite ore resulting in higher percentage of moisture content, low percentages of coarse particle as well as high value of bulk density. Due to the presence of high fine particle, it will absorb more water compared to granular particles.

Acknowledgements

The authors would like to acknowledge the Universiti Malaysia Pahang (UMP) for financing this research through the Research University Grant Scheme, Project Number RDU1603122. The cooperation given by all parties involved in this research is greatly acknowledged.

References

[1] Shakhashiri, BZ 2012 Chemical of The Week: Aluminium. *University of Wisconsin: SciFun.org*

[2] Donoghue, AM 2014 Bauxite Mining and Alumina Refining: Process Description and Occupational Health Risk. *Journal of Occupational and Environmental Medicine,* **56** 5

[3] International Maritime Organization 2015 September 18 IMO warns on bauxite luquefaction dangers. Retrieved from http://www.imo.org/en/MediaCentre/PressBriefings/Pages/38-bauxite-CCC.aspx

[4] World Maritime News 2015 September 21 IMO : Bauxite liquefaction sank Bulk Jupiter. Retrieved from http://worldmaritimenews.com/archives/172059/imo-bauxite-liquefaction-sank-bulk-jupiter/

[5] Seed, HB 1996 Liquefaction of Saturated Sands During Cyclic Loadin. *Journal of the Soil Mechanics and Foundations Division*, 105-134

[6] IMSBC Code 2013 *International Maritime Solid Bulk Cargoes Code and Supplement. International Maritime Organization*

[7] Deewal, K., Dharavath, K., Kulshreshta, M 2014 Evaluation of Characteristic Properties of Red Mud for Possible use as a Geotechnical Material in Civil Engineering. *International Journal of Advances in Engineering & Technology*

IOP Conf. Series: Materials Science and Engineering **318** (2018) 012026 doi:10.1088/1757-899X/318/1/012026

Strengthening Performance of PALF-Epoxy Composite Plate on Reinforced Concrete Beams

Siew C Chin[1,2*], Foo S Tong[1], Shu I Doh[1], Jolius Gimbun[2], Huey R Ong[3] and Januar P Serigar[4]

[1] Faculty of Civil Engineering and Earth Resources, Universiti Malaysia Pahang, 26300 Gambang, Pahang, Malaysia
[2] Centre of Excellence in Advanced Research in Fluid Flow (CARIFF), Universiti Malaysia Pahang, 26300 Gambang Pahang, Malaysia
[3] Faculty of Chemical Engineering and Natural Resources, Universiti Malaysia Pahang, 26300 Gambang, Pahang, Malaysia.
[4] Faculty of Mechanical Engineering, Universiti Malaysia Pahang, 26600 Pekan, Pahang, Malaysia

E-mail: scchin@ump.edu.my

Abstract. This paper presents the effective strengthening potential of pineapple leaves fiber (PALF)-epoxy composite plate on reinforced concrete (RC) beam. At first the PALF is treated with alkali (NaOH) and its morphology is observed via scanning electron microscope (SEM). The composite plates made of PALF and epoxy with fiber loading ranging from 0.1 to 0.4 v/v was tested for its flexural behaviour. The composite was then used for external RC beam strengthening. The structural properties of RC beams were evaluated and all the beams were tested under four-point bending. It was found that the flexural strength increased as the fiber volume ratio increases. The maximum flexural strength (301.94 MPa) was obtained at the fiber volume ratio of 40%. The beam strengthened with PALF-epoxy composite plate has a 7% higher beam capacity compared to the control beam. Cracks formed at the edge of the plate of PALF-strengthened beams resulted in diagonal cracking. Result from this work shows that the PALF-epoxy composite plate has the potential to be used as external strengthening material for RC beam.

1. Introduction

Reinforced concrete (RC) beam often suffers from degradation, deterioration, acid attack and seismic event. Rehabilitation of RC beams using an external strengthening method by installing the FRP plastic composite to the beam is one of the common retrofits method [1]. Besides the synthetic fiber, the natural fiber is getting a great attention in research since it is cost-effective and environment friendly. There are many natural fibers that can be used as external strengthening material such as sisal, jute, coir, kenaf, bamboo, oil palm fiber, sugar cane and others.

Abundant of pineapple leaf fibers are being produced and abandoned every year, with only very small portions are used as feedstock and for energy production [2]. Due to its high specific strength and stiffness behaviour, as well as abundantly available in Malaysia, pineapple leaf fiber (PALF) has a eat potential as an external strengthening material. PALF has large amount of α-cellulose (81.27%), low quantities of hemicelluloses (12.31%) and lignin content (3.46%) [3]. Due to its high compositions of α-cellulose and low microfibrillar angle, it can be reinforced in composite matrix [4]. PALF has been combined with thermoset, thermoplastic, biodegradable plastics such as polyester, polypropylene, epoxy, polycarbonate, and with natural rubber in the past.

Various investigations have been conducted to study the physical and mechanical properties of PALF fiber; however, limited study on the application of PALF-epoxy composite plate for external strengthening of RC beams [5]–[8]. Hence, this paper aims to elucidate the physical and mechanical properties of PALF-epoxy composite plates for use as external strengthening of RC beam.

2. Methodology

2.1. Preparation of PALF

The fibers were washed with running water at room temperature in order to remove the impurities in the fibers. After washing, the fibers were dried in oven at 70 °C for 24 hours.

2.2. Chemical treatment

The fibers were then alkali-treated with three levels of NaOH concentration (2%, 5% and 8%), respectively at 30 °C for an hour as shown in figure 1. The ratio of the fibers and the solution was 1:20 (w/v). After treatment, the fibers were washed and rinsed with tap water several times. The fibers were then dried in an oven at 70 °C for 24 hours.

2.3. Fabrication of composite samples with different fiber volume ratios

The composite samples with fiber volume ratios of 0, 0.1, 0.2, 0.3 and 0.4 were fabricated (25 mm wide x 6 mm thick x 200 mm long), respectively by hand-laying technique as illustrated in figure 2. The specimens were tested under ASTM D790. The crosshead speed was set at 5 min^{-1}.

Figure 1. Fibers treated in NaOH solution.

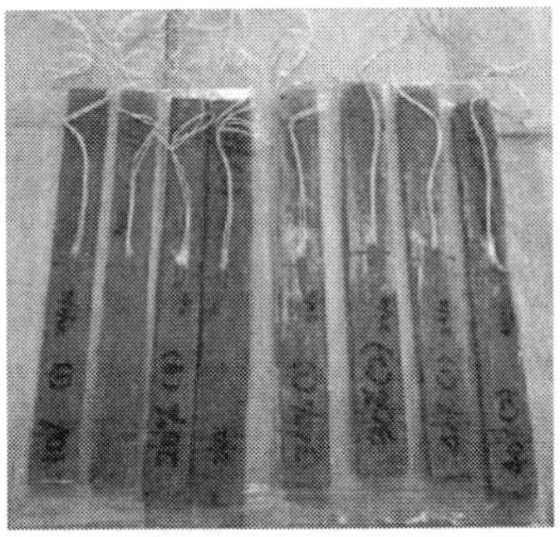

Figure 2. Composite samples with different fiber volume ratios.

2.4. Fabrication of PALF composite plate

The fibers were treated with 4% NaOH, washed and dried in oven. The fibers were placed into a mould with a size 100 mm width, thickness of 8 mm and 600 mm in length by hand-layup method as shown in figure 3.

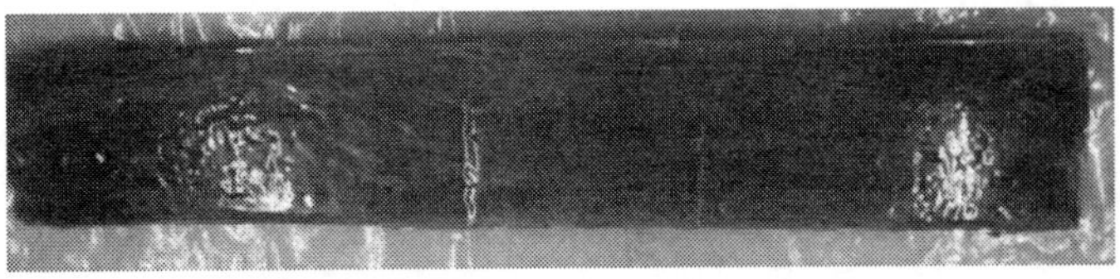

Figure 3. PALF-epoxy composite plate.

2.5. Beam preparation
A total of three beams were cast with the concrete grade 25 MPa in a dimension of 100 mm width x 130 mm height and 1600 mm in length. Two steel bars of diameter 10 mm were used as tension and compression reinforcement, respectively. Steel bars of diameter 6 mm with spacing 300 mm centre to centre were used as the shear links. The fibers were washed with running water at room temperature in order to remove the impurities in the fibers. After washing, the fibers were dried in oven at 70 °C for 24 hours.

2.6. Fiber morphology analysis
The treated fibers were observed under SEM to study the changes in fiber surface morphology. The test was conducted using TM3030, HITACHI. The acceleration voltage used was 5 kV.

2.7. Flexural strength test
The composite samples at various fiber volume ratios described earlier in section 2.1.3 was tested for its flexural strength. The specimens were tested under ASTM D790. The crosshead speed was set at 5 min^{-1}.

2.8. Four-point bending test
The fabricated composite plate was bonded at the mid-span of the bottom soffit of the beam with Sikadur 30 epoxy adhesive. All the beams were tested to failure under four-point loading using Universal Testing Machine (UTM) at 100 kN as illustrated in figure 4.

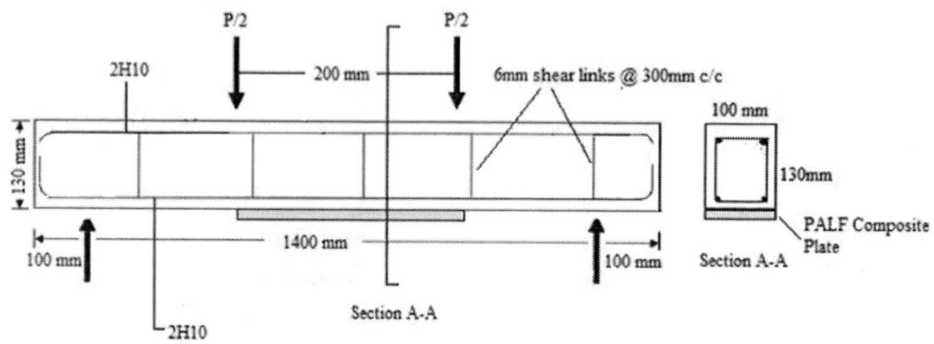

Figure 4. Schematic diagram of four-point bending test.

3. Results and discussion
3.1. Fiber morphology
The untreated fibers contained impurities such as dust as shown in the red circle in figure 5(a). NaOH treatment removed the impurities from the surface of the fibers and also partially removed the hemicellulose, lignin and other soluble materials [9]. It was observed that the presence of voids inside the fiber increases with increasing concentration of NaOH treatment.

3.2. Flexural strength of PALF-epoxy composite plate
Figure 6 shows the relationship between the flexural strength versus the fiber volume ratio. The flexural strength increases as the fiber volume ratio increased from 0.1 to 0.4. However, the fiber volume ratio of 0.1 did not exhibit significance difference compared to the neat epoxy sample.

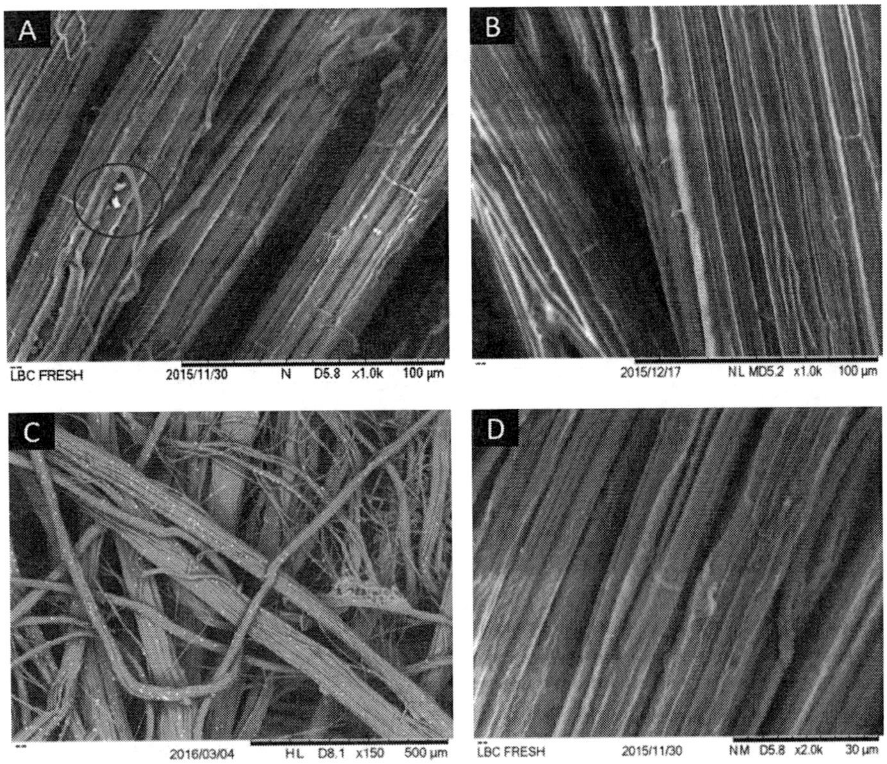

Figure 5. SEM result: (a) untreated PALF; (b) 2% NaOH treated PALF; (c) 5% NaOH treated PALF; (d) 8% NaOH treated PALF.

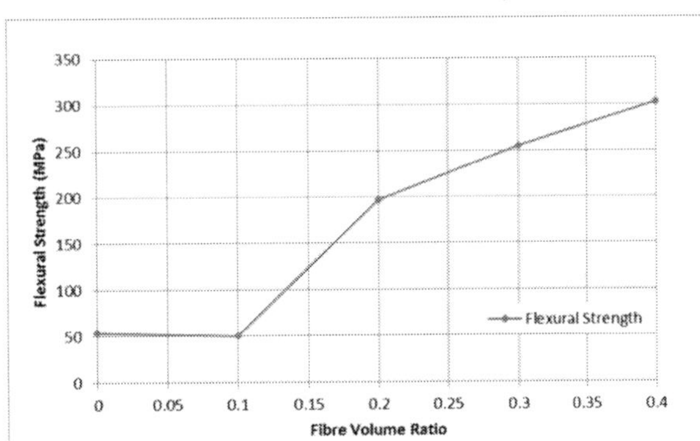

Figure 6. Load versus strain of composite plates.

The flexural strength is calculated according to equation (1).

$$\text{Flexural strength} = 3PL/2bd^2 \quad (1)$$

where L is the support span; b, the specimen width; d, the thickness; and P, the maximum load.

Table 1 summarizes the maximum peak load and flexural strength of plates reinforced with different PALF to epoxy fiber volume ratio. It was found that the flexural strength increased with the increase of PALF content. In the case of pure epoxy, the flexural strength is 54.06 MPa. The flexural strength of composite with a fiber volume ratio of 0.1 is 50.11 MPa, which decreased 7.31% as compared to the neat epoxy, due to higher elasticity of the pure epoxy plate. The flexural strength of the composite with 0.2 fiber volume ratio increased 292% as compared to the composite with 0.1 fiber volume ratio. The flexural strength of the composite with 0.3 fiber volume ratio increased 30% as compared to composite with 0.2 fiber volume ratio. Meanwhile, the flexural strength of the composite with 0.4 fiber volume ratio increased 19% as compared to the composite with 0.3 fiber volume ratio. In addition, the increase in fibre content from 20% to 40% increased the flexural strength in the range of 260% to 460% as compared to the neat epoxy. The maximum flexural strength was obtained at volume ratio 40%.

Table 1. The maximum peak load and flexural strength of plates reinforced with different PALF/Epoxy fibre volume ratio.

PALF/Epoxy Fibre Volume Ratio	Maximum Peak Load (N)	Flexural strength (MPa)
0	202.71	54.06
0.1	187.90	50.11
0.2	737.23	196.60
0.3	954.50	254.54
0.4	1132.28	301.94

3.3. Strengthening of RC beam under four-point bending

The ultimate load of the control beam and beams strengthened with PALF-epoxy composite plate is listed in table 2. The ultimate load of the control beam was attained at 15.04 kN, whereas the ultimate load of the beam specimens strengthened with PALF-epoxy composite plates, B1 and B2 were 16.14 kN and 14.68 kN, respectively, which demonstrated an increment in the beam strength by 7.3% in PALF B1 as compared to the control beam.

Table 2. Ultimate load of beam specimens.

Beam	Ultimate Load (kN)	Strengthening Percentage (%)	Strengthening Ratio
CB	15.04	-	
PALF B1	16.14	+7.3	1.07
PALF B2	14.68	-2.4	0.98

The load-deflection curve in figure 7 shows that the curve lines of the beam with PALF-epoxy composite plate exhibited a steeper slope compared to the control beam at the early stage of loading in the elastic phase. This indicates that the PALF-epoxy composite plates have a better stiffness as compared to the control beam, and hence the composite plates can resist the deformation of RC beam.

The deflection at ultimate load of the control beam was obtained at 21.30 mm, whereas the deflection at ultimate load of beam with PALF-epoxy composite plate B1 and B2 was 20.35 mm and 11.82 mm, respectively, as shown in table 3. PALF B1 strengthened beam shows about 4.5% of the mid-span deflection while PALF B2 strengthened beam shows about 45% reduction in deflection compared to the control beam. Generally, the result demonstrates that the beams strengthened with PALF-composite plates exhibited a lower deflection compared to the control beam.

Table 3. Deflection at ultimate load.

Beam	Deflection at Ultimate Load (mm)	Percentage of Deflection Reduction (%)
CB	21.30	-
PALF B1	20.35	4.5
PALF B2	11.82	45

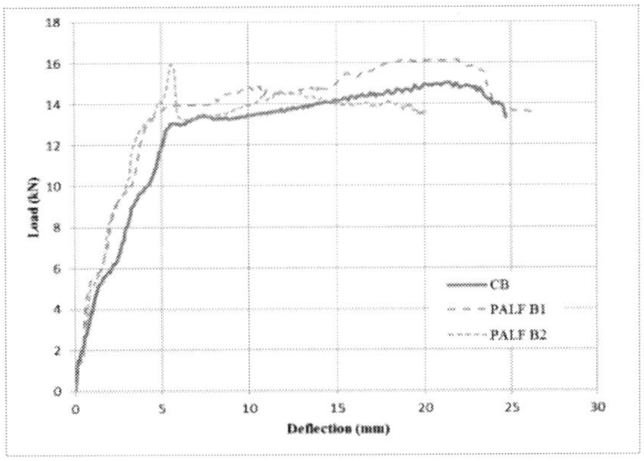

Figure 7. Load versus deflection of beam specimens.

Figure 8 shows the crack pattern of the control beam. When a load was applied to the control beam, cracks started to appear starting from the soffit of the beam and then widened when the load increased as the bottom surface of the beam was weak in tension. The first crack of the control beam appeared at the mid-span of the beam. The first crack occurred at the load of 5.72 kN. Cracks started to appear continuously as the load increased. The appearance of cracks and propagation of cracks continued to the point load with the increased of load. It was observed that a large amount of crack propagation around 13 kN of load. The propagation of cracks continued until the failure of the beam. The cracks happened at the middle of the tension zone. All the cracks were vertical flexural cracks.

Figure 8. Crack patterns of control beam.

Meanwhile, the crack pattern of the strengthened beam with PALF-epoxy composite plate, PALF B1 is shown in figure 9. It was observed that the first cracks appeared at the load of 9 kN which can be seen at the end of the plate. The cracks at both ends continued to extend and penetrate and the crack width became wider as the load increased. As the load reaching 13.4 kN the cracks had penetrated towards the neutral axis of the beam in diagonally. However, there were very little flexural cracks formed at the mid-span of the beam at the same time. As the load continually increases, the cracks widened and two main diagonal cracks can be seen clearly at both ends of the plate. The crack width increased before beam failure. This signifies that the PALF-epoxy composite plate had resisted all the tensile stress in the mid-span and directed the cracks in the flexural zone at the end of the plate forming diagonal crack failure. Hence, the cracks in the mid-span of the beam were lesser than the control beam.

Figure 10 shows the crack pattern of beam, PALF B2. The first crack appeared at the load of 10 kN and followed by the increasing of diagonal cracks as the load increased. The cracks propagated through the neutral axis of the beam in a diagonal direction at the load of 13 kN. More cracks were developed and appeared at both ends of the plate. Very little vertical cracks were seen at the flexural zone. Wider cracks formed as load applied continuously. The beam fails with a detached of PALF-epoxy composite plate from the concrete surface as depicted in figure 10. The plate has the ability to resist and transfer the load to the region without strengthening. This caused fewer flexural cracks found at the mid-span of the beam. Large crack width identified upon beam failure at the left end of the plate was possibly due to plate end interfacial debonding.

Figure 9. Crack patterns of beam PALF B1.

Figure 10. Crack patterns of beam PALF B2.

4. Conclusions

It can be concluded that the alkali treatment of fibers managed to remove impurities and binding materials, which then increased the interaction of fiber and matrix. The PALF-epoxy composites with the volume ratio 0.4 showed the best flexural properties compared to other volume ratios. Strengthening of PALF-epoxy composite plate has increased the beam strength and reduced the beam deflection up to 45% compared to the control beam. The main contribution of this research work is to study the effectiveness of using natural fiber; pineapple leaves fiber composite plate as external strengthening material in the strengthening of RC structures. From the positive aspects, it was found that this cost-effective and environmental friendly composite plate managed to increase the beam stiffness and ultimate load as well as reduce the beam deflection and cracking compared to the control beam. In terms of deficiencies, beam with PALF B2 attained an early ultimate load of 15.98 kN at a deflection of 5.59 mm as the beam experience yielding and eventually strain hardening takes places before plate end interfacial debonding. It is recommended that this work can be improved with proper surface preparation to ensure sufficient surface bonding as well as to increase the length of the PALF-epoxy composite plate.

Acknowledgement

We acknowledge the research funding from Universiti Malaysia Pahang (PGRS 160326 and RDU1703119) as well as RAGS/1/2015/TK01/UMP/02/1 (RDU151409) from Ministry of Higher Education Malaysia. We thank Shu Qian, Ting for helping to prepare the sample and the technicians who involved in assisting in the experimental testing.

References

[1] Hemaanitha R and Kothandaraman S 2014 *Int. J. Polym. Sci.* **5** 3

[2] Asim M, Abdan K, Jawaid M, Nasir M, Dashtizadeh Z, Ishak M R and Hoque M E 2015 *Int. J. Polym. Sci.* **2015**.

[3] Rahman M A 2011 *J. Text. Apparel, Technol. Manag.* **7** 2

[4] Panyasart K, Chaiyut N, Amornsakchai T and Santawitee O 2014 *Energy Procedia* **56** 406

[5] Tong F S, Chin S C, Doh S I and Gimbun J 2017 *Indian J. Sci. Technol.* **10**, 2

[6] Tan K F, Chin S C, Doh S I, Gimbun J and Tamizi M 2017 Potential Use of Bamboo Reinforced Concrete Beams Towards Sustainable Construction *Civil Engineering and Urban Planning 2017* ed A Mebarki (Singapore: World Scientific) pp 456-467

[7] Bhutta M A R, Nur Hafizah A K, Jamaludin M Y, Warid M H, Ismail M and Azman M 2013 August. Strengthening reinforced concrete beams using kenaf fiber reinforced polymer composite laminates. In *Proceeding: Third International Conference on Sustainable Construction Materials and Technologies* pp 18-21

[8] Alam M A, Nouri K, Jumaat M Z and Muda Z C 2015 Flexural strengthening of reinforced concrete beam using jute rope composite plate. In *The 3rd National Graduate Conference, Putrajaya* pp 8-9

[9] Mwaikambo L Y and Ansell M P 2002 *J. Appl. Polym. Sci.* 84 2222

IOP Conf. Series: Materials Science and Engineering **318** (2018) 012027 doi:10.1088/1757-899X/318/1/012027

Water Supply Treatment Sustainability of Semambu Water Supply Treatment Process - Water Footprint Approach

Edriyana A Aziz[1,2], Marlinda Abdul Malek[3], Syazwan N. Moni[1], Iqmal H. Hadi[1], and Nabil F. Zulkifli[1]

[1] Faculty of Civil Engineering & Earth Resources, Universiti Malaysia Pahang, Lebuhraya Tun Razak, 26300, Kuantan, Pahang, Malaysia.

[2] Center for Earth Resources Research & Management (CERRM), Universiti Malaysia Pahang, Lebuhraya Tun Razak, 26300, Kuantan, Pahang, Malaysia.

[3] The Institute of Energy, Policy and Research (IEPRe), Universiti Tenaga Nasional, Malaysia.

Email: edriyana@ump.edu.my

Abstract. In this study, the assessment by using Water Footprint (WF) approach was conducted to assess water consumption within the water supply treatment process (WSTP) services of Semambu Water Treatment Plant (WTP). Identification of the type of WF at each stage of WSTP was carried out and later the WF accounting for the period 2010 – 2016 was calculated. Several factors that might influence the accounting such as population, and land use. The increasing value of total WF per year was due to the increasing water demand from population and land use activities. However, the pattern of rainfall intensity from the monsoonal changes was not majorly affected the total amount of WF per year. As a conclusion, if the value of WF per year keeps increasing due to unregulated development in addition to the occurrences of climate changing, the intake river water will be insufficient and may lead to water scarcity. The findings in this study suggest actions to reduce the WF will likely have a great impact on freshwater resources availability and sustainability.

1. Introduction

In this study, water consumption of WSTP needs to be determined. WSTP is a process of water treatment through few steps begins with the water intake, aeration, mixing, flocculation, sedimentation, filtration and backwash. WF approach was used as a tool to assess the amount of overall water consumption for the WTP.

For years, Life Cycle Assessment (LCA) has been used to evaluate a wastewater treatment plant and all the impacts are evaluated from the construction process to operation and until the dismantling process [1]. In this assessment, the most importance phase that also an adapted method from LCA of all 4 phases [2] is the impact assessment phase. However, LCA is a broad impact assessment tool and all environmental impacts connected with a product or service have to be assessed [3] from the early stage which is before the construction begin until the demolishing and transportation of waste materials stage [4]. Basically, LCA is used to estimate the broad environmental impact such as global warming, eutrophication, acidification, human toxicity, pollution and many more impacts based on standard methodology named ISO 14040 [2]. However, the importance of water uses and impacts are being assessed in LCA but LCA does not quantify and map indirect water use that involved along the supply chain in water resources management [5]. In term of sustainability of water resources, it is not fare if not assessing on overall water resources.

Hence, the aim of this study is to introduce a WF approach in order to determine the sustainability of the services category through the WF accounting for WSTP. Therefore, the sustainability of WSTP will be able to be identified.

2. Methodology

In this study, Semambu WTP is located at Kuantan (3° 52' N) has been chosen. The site selection was based on its capacity to supply treated water to sub-district of Sg Karang and Beserah with the average population of 84930 people. Meanwhile, the area for both sub-districts are 30 300.00 Hectare (Ha).

2.1. Data Collection

In this study, data such as rainfall intensity, river flowrate, temperature and water intake volume were collected from Department of Drainage and Irrigation (DID), Malaysia Meteorological Department and Pengurusan Air Pahang Berhad (PAIP) respectively. Meanwhile the land use development of Kuantan district was collected from the Majlis Perbandaran Kuantan (MPK).

2.2. Water Footprint Accounting

WF accounting can be divided into three types of water; blue water, green water and grey water. The unit of the green water footprint process is water volume per unit of time. Time refers either to the day, week or year. Meanwhile, backwash water from the filtration process was considered as WFgrey in this study.

2.3. River Water Availability

The WF of water supply treatment process (WSTP) is defined as the summation of all the water consumed in whole stages of process and can be expressed as the following equation [6]:

$$WF_{blue} = BWE + BWI + LRF \tag{1}$$

BWE= Blue water evaporation
BWI = Blue water Incorporation
LRF = Lost Return flow

The unit of the blue process water footprint is water volume per unit of time. In this study, the LRF was not considered in the calculation.

The WFgreen in other hand refers to the precipitation that does not runoff or recharge to the groundwater but is stored in the soil or temporarily stays on top of the soil. However in this study, the WF green is considered as a water that being incorporated in the open tank area of the WTP. TheWFgreen can be represented with the following equation:

$$WF_{green} = GWE + GWI \tag{2}$$

GWE= Green water evaporation
GWI = Green water Incorporation

Water intake resource globally is mostly from the river. Semambu WTP water intake is from Sg. Kuantan. The river water availability is important to ensure the continuity of water supply to consumer. Water availability is referring to the existing volume of the river. Climate change is a factor that contributes to the changes of river water volume. In this study, the volume of river was obtained from the volumetric flowrate (Q) provided by DID.

2.4. *Effect of Land Use*

Over the past five years, the population of Kuantan has increased rapidly. Semambu WTP distributes to 2 sub-districts which Sungai Karang and Berserah. Table 1 and 2 show the land use of Sungai Karang and Berserah.

Table 1. Land use of Sungai Karang sub-district

	Existing Landuse (2004)		Proposal Area	Landuse (2015)	
	Area (ha)	%	Area (ha)	Area (ha)	%
Residential	1272	25	644	1,916	38
Commercial	103	2	90	193	3
Industrial	63	1	561	624	12
Institutional	118	2	13	131	2
Open Space	95	1	80	175	3
Infrastructure	158	3	43	201	4
Transport	419	8	81	500	10
Agricultural	1209	24	-603	605	12
Forestry	599	12	-72	527	10
Water Body	57	1	-	57	-
Beach Area	43	1	-	-	-
Empty Land	794	16	-	-	
	4935	99	-	4,935	98

Table 2. Land use of Beserah sub-district District

	Existing Landuse(2004)		Proposal area Area (ha)	Landuse (2015)	
	Area (ha)	%		Area (ha)	%
Residential	1,412	5	1,169	2,582	10
Commercial	362	1	554	916	3
Industrial	1,984	7	1,929	3,914	15
Institutional	370	1.46	30	400	1
Open space	204	0.8	29	233	1
Infrastructure	289	1	120	409	1
Transport	623	2.45	2,477	3,110	1
Agricultural	3,216	12	-1,028	2,188	9
Forestry	13,434	53	-2,401	11,032	43
Water Body	577	2.28	-	577	2
Beach Area	158	0.62	-158	-	-
Empty Land	2,731	10	-2,731	-	-
	25,365	100	-2,731	25,365	99

3. Result and Discussion

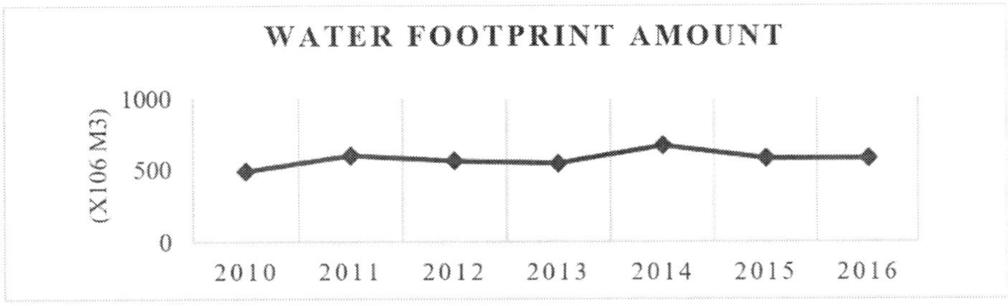

Figure 1 : Water Footprint per Year Amount

Figure 2 shows the WF amount from year 2010 – 2016 at Semambu WTP. In 2014, the WF amount shows the highest with the value of 668 913 249.9 m^3/year due to high precipitation occurrences as shown in Fig. 3. During 2014, Kuantan district received a major flood event.

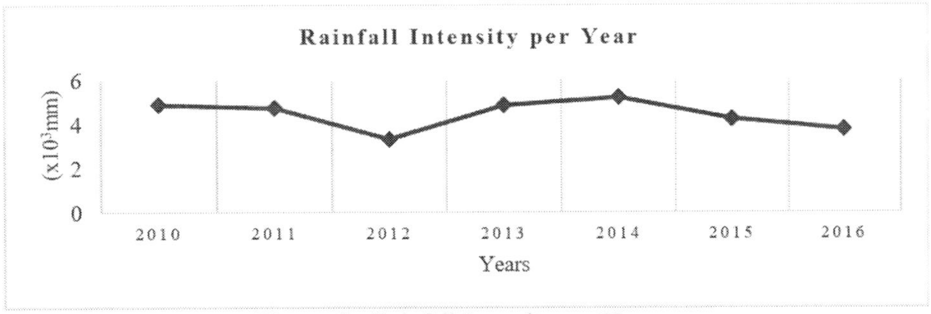

Figure 2 : Rainfall Intensity per Year

Table 3. Rainfall Intensity

Years	Rainfall Intensity (mm)
2010	4895.962
2011	4744.416
2012	3331.5
2013	4868.2
2014	5224
2015	4246
2016	3771.41

3.1. Water Availability

Table 4. Total Water Availability

Year	Total water availability per year (m^3)
2010	68 387 169.54
2011	141 745 382.8
2012	94 936559.23
2013	101 101 078.9
2014	233 734 173.7
2015	156 029 090.4
2016	154 894 943.4

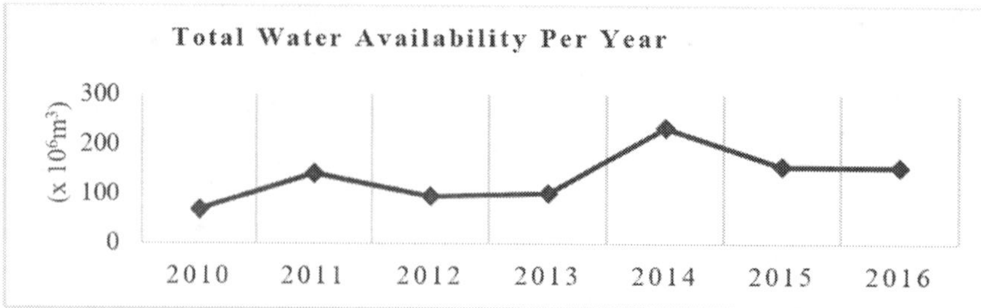

Figure 3 : Total Water Availability per Year

In 2014 it is clearly shows that the water availability of Sungai Kuantan is the highest as during that year Kuantan received a massive flood due to the high rainfall intensity.

This graph shows the water availability in 2011 is increasing but decrease back when it reaches 2012 and 2013. 2013 is the year where there is dry season (El Nino) happen and 2016 where wet season happen.

3.2. Land Use Development

From Table 1 and 2, it is clearly shows that there was a multiplication of growth in residential and industrial activities. It is believed that the increased of land used activities affected the water demand for the Panching WTP. Figure 5 shows the total water demand in year 2015 and 2016.

Table 5. Percentage water footprint on water demand

Year	Water Demand $x10^6$ (m^3/yr)	Water Footprint $x10^6$ (m^3/yr)	Total Water Intake $x10^6$ (m^3/yr)	Percentage Water Footprint Based on Water Demand (%)
2010	9.0	491.57	108.1	98.17
2011	9.3	600.24	118.1	98.5
2012	9.6	563.21	124.8	98.30
2013	10.0	545.84	114.9	98.17
2014	10.4	668.91	103.0	98.45
2015	10.7	579.43	98.1	98.15
2016	10.9	579.85	105.0	98.12

In this study, the water footprint of Semambu WTP was highly dependent to water intake. Table 3 shows that when the water intake increased the water footprint was also increased. Meanwhile, every year, water demand keep increasing due to increasing population and land use development. The WF amount increased from 2010 to 2012. However, the percentage decreased from 2012 to 2015. This result shows that the overall water consumption during the water supply treatment process was beginning to reduce although the water intake and water demand were increasing. Therefore, the results show that reducing amount of WF based on water demand is good but WF was still greater than water intake value. This shows that WSTP in Semambu WTP is not sustainable since the percentage of water intake to water demand is very high.

The following Table 5 shows water availability amount to WF at Sg. Kuantan intake. In 2010 - 2016, percentage of WF to water availability was 86.09%, 76.39%, 83.13%, 81.48%, 65.06%, 73.08% and 73.29% respectively. Those values were all greater than the Sg. Kuantan water availability. This shows unsustainable WSTP occurrences at Semambu WTP.

Table 6. Water Availability and WF

Year	Water Availability x10^6 (m^3/yr)	Water Footprint x10^6 (m^3/yr)	Percentage Water Footprint Based on Water Availability (%)
2010	68.4	491.57	86.09
2011	141.7	600.24	76.39
2012	95.0	563.21	83.13
2013	101.1	545.84	81.48
2014	233.7	668.91	65.06
2015	156.0	579.43	73.08
2016	154.9	579.85	73.29

4. Conclusion
As a conclusion, objectives are all achieved. One conclusion can be made from the study. From the trend of water footprint and water demand as keep increasing year by year thus action must be taken. We can conclude from objective 2 that water availability is not available at the river and new intake location should be considered. There are also new WTP constructed at new river intake at Kuantan district since the development of Kuantan is rapidly growth yearly. WF approach can be used to assess water consumption of WSTP as this approach indicates the overall amount of water according to WFblue, WFgreen and WFgrey. All types of water are important and need to be considered in order to ensure the sustainability of water supply management of Semambu WTP.

5. Acknowledgement
This study will not be able to succeed without committed authorities who supplied all the information needed. Thus, an appreciation should be given to Pengurusan Air Pahang Berhad (PAIP), Majlis Perbandaran Kuantan (MPK) and Department of Statistic, Malaysia. Last but not least, a special thanks to Fundamental Research Grant Scheme (FRGS) for funding my research. This study was conducted under FRGS Grant, RDU160155.

References

[1] L. Corominas *et al.*, "Life cycle assessment applied to wastewater treatment: State of the art," *Water Res.*, vol. 47, no. 15, pp. 5480–5492, 2013.
[2] The International Standards Organisation, "Environmental management — Life cycle assessment — Principles and framework," 2006.
[3] W. Klöpffer, "Life cycle assessment: From the beginning to the current state.," *Environ. Sci. Pollut. Res. Int.*, vol. 4, no. 4, pp. 223–228, 1997.
[4] L. F. Cabeza, L. Rincon, V. Vilarino, G. Perez, and A. Castell, "Life cycle assessment (LCA) and life cycle energy analysis (LCEA) of buildings and the building sector: A review," *Renew. Sustain. Energy Rev.*, vol. 29, pp. 394–416, 2014.
[5] A. Boulay, A. Y. Hoekstra, and S. Vionnet, "Complementarities of Water-Focused Life Cycle Assessment and Water Footprint Assessment Anne-Marie," *Environ. Sci. Technol.*, vol. 47, pp. 11926–11927, 2013.
[6] S. N. Moni, E. A. Aziz, and M. A. Malek, "Introduction of Water Footprint Assessment Approach to Enhance Water Supply Management in Malaysia," *AIP Conf. Proc.*, vol. 70001, 2017.

IOP Conf. Series: Materials Science and Engineering **318** (2018) 012028 doi:10.1088/1757-899X/318/1/012028

Water Supply Treatment Sustainability of Panching Water Supply Treatment Process - Water Footprint Approach

Edriyana A Aziz[1,2], Marlinda Abdul Malek[3], Syazwan N. Moni[1], Nabil F. Zulkifli[1], and Iqmal H. Hadi[1]
[1] Faculty of Civil Engineering & Earth Resources, Universiti Malaysia Pahang, Lebuhraya Tun Razak, 26300, Kuantan, Pahang, Malaysia.
[2] Center for Earth Resources Research & Management (CERRM), Universiti Malaysia Pahang, Lebuhraya Tun Razak, 26300, Kuantan, Pahang, Malaysia.
[3] The Institute of Energy, Policy and Research (IEPRe), Universiti Tenaga Nasional, Malaysia.

Email: edriyana@ump.edu.my

Abstract. In many parts of the world, freshwater is scarce and overexploited. The purpose of this study is to determine the water footprint of Water Supply Treatment Process (WSTP) at Panching Water Treatment Plant (WTP) as well as to identify the sustainability of the Sg. Kuantan as an intake resource due to the effect of land use development. The total water footprint (WF) will be calculated by using WF accounting manual. The results obtained shows that the water intake resource is still available but it is believed that it will not be able to cope with the increasing WF. The increment of water demand percentage by 1.8% from 2015 to 2016 has increased 11 times higher of the water footprint percentage, 19.9%. This result shows that the water consumption during the water supply treatment process is two times higher than the demand thus it shows the inefficient of the water management

1. Introduction

Human activities consume and pollute a lot of water [1]. Globally, most of the water use occurs in agricultural sector, but there are also extensive water consumed in the industrial and domestic sectors [2]. Irrigation, bathing, washing, cleaning, cooling and so on are specific activities of water consumption and pollution [1]. Increasing total water consumption and pollution are generally because of increasing population [3].

Nowadays, in Malaysia, population growth and increasing urbanization, industrialization and agriculture have led to high demand for water, therefore, contributes to water pollution, and also would create water scarcity [4]. Thus, in moving towards a developed nation status, the country's water system should be jointly headed in that direction. Besides, urban areas are likely to face water stress driven by various factors such as water pollution, population growth, poor area planning, economic growth that leads to increasing demand, climate change and inefficient infrastructure (Childers, Pickett, Grove, Ogden, & Whitmer, 2014).

Over the past decades, WF concept which has been developed by Water Footprint Network (WFN) significantly is developing plenty of publications and researches which are used in water resource management and other journals [3], [5]–[11].

Thus, using WF as a tool is important to highlight the amount of water uses and the connection to global water resources in accordance to sustainability distresses. Therefore, to date, water supply treatment process has never been assessed its overall water consumption to produce a treated water from the water intake. In particular to the WF of other product that may be taking place in the WTP covered area, this study will be a baseline or first data set of a production chain. This is because, product's WF is the sum of WF of the various processes involved to produce product either goods or services. Hence, WF of water supply treatment process will be assessed significantly to improve water resources management in accounting overall water consumption according to all types of water. Currently, in water management, volume of blue water intake is recorded before this resource undergoing the series of in-line processes to produce a treated water. However, the recorded value is only blue water intake and not considering the overall water consumption along the process. Therefore, WF will act as a tool to account overall water consumption which includes blue, green and grey WF concurring to WF definition in the WF Manual [12].

2. Methodology

Kuantan is the state capital of Pahang that have increasing population. Thus, this study was conducted at the 2nd major WTP in Kuantan district; Panching WTP which is located at 3.834534, 103.201831, where its water intake is from Sg. Kuantan as shown in Figure 1. The Panching WTP has a capacity of 160 million litres per day, covering around residential area as well as 5600 hectares of industrial area.

2.1. Data Collection

In this study, data such as rainfall intensity, river flowrate, temperature and water intake volume were collected from Department of Drainage and Irrigation (DID), Malaysia Meteorological Department and Pengurusan Air Pahang Berhad (PAIP) respectively. Meanwhile the land use development of Kuantan district was collected from the Majlis Perbandaran Kuantan (MPK).

2.2. Water Footprint Accounting

The aim of water footprint accounting in this study is to quantify water footprint in each single process of water supply treatment at Panching Water Treatment Plant (WTP). In general there are three types of water footprint which is WFgreen, WFblue and WFgrey. Before the water footprint calculation, type of water footprint involves at each processes was identified.

The WF blue indicates 'consumptive water use' or in the other words, fresh surface water or groundwater. The WF blue can be represented as the following equation:

$$WF_{blue} = BWE + BWI + LRF \quad (1)$$

BWE= Blue water evaporation
BWI = Blue water Incorporation
LRF = Lost Return flow

The unit of the blue process water footprint is water volume per unit of time. In this study, the LRF was not considered in the calculation.

The WFgreen in other hand refers to the precipitation that does not runoff or recharge to the groundwater but is stored in the soil or temporarily stays on top of the soil. However in this study, the WF green is considered as a water that being incorporated in the open tank area of the WTP. The WFgreen can be represented with the following equation:

$$WF_{green} = GWE + GWI \tag{2}$$

GWE= Green water evaporation
GWI = Green water Incorporation

The unit of the green water footprint process is water volume per unit of time. Time refers either to the day, week or year.

The degree of freshwater pollution that can be associated with the process involves in the WTP is called grey water footprint (WFgrey). It refers to the volume of water that is required to assimilate pollutants to such an extent that the quality of ambient water remains above water quality standard. The WFgrey can be calculated by dividing the pollutant load (L, in mass/time) by difference between the ambient water quality standard for that pollutant.

2.3. *River Water Availability*

The river water availability is crucial in order to ensure the sustainability of water supply to consumer. Water availability is referring to the existing volume of the river. Climate change is a factor that contributes to the changes of river water volume. In this study, the volume of river was obtained from the volumetric flowrate (Q) provided by DID.

2.4. *Effect of Land Use*

Land use decisions can potentially affect both water quantity and water quality. Nevertheless the decisions are not always well-coordinated. The strong connection between water availability and land use decisions, including the need to develop land use regulations that can be used to reduce future water consumption. Major change in land use development such as consequence of population growth, increasing and changing food demands, and expanding biofuel cultivation need to regulate to control the water supply sustainability.

Over the past two years, the population of Kuantan has increased rapidly. Treated water from Panching WTP is distributed to 2 sub-districts which are Penor and Kuala Kuantan. Table 1 and 2 show the land use of Penor and Kuala Kuantan.

Table 1. Land use of Penor sub-district

Type of land use	Hectre 2004- 2010 (Ha)	Land use 2004-2010 (%)	Hectre 2004-2010(Ha)	Land use 2010-2015 (%)
Residential	1243.63	5.57	2853.18	18.37
Forest Tree	12437.45	55.77	11188.84	50.17
Agricultural	4688.56	21.02	3517.16	15.77
Tourism	1132.97	5.08	1240.70	5.56
Others	2797.39	12.56	3500.12	10.13
Total	22300.00	100.00	22300.00	100.00

Table 2. Land use of Kuala Kuantan sub-district

Type of land use	Hectre 2004- 2010 (Ha)	Land use 2004-2010 (%)	Hectre 2004-2010(Ha)	Land use 2010-2015 (%)
Residential	3358.03	8.94	8567.16	22.83
Forest Tree	7668.61	20.43	5238.29	13.95
Agricultural	15590.23	41.54	12925.05	34.44
Commercial	132.67	0.35	647.08	2.07
Industrial	1498.8	3.99	1677.66	4.47
Others	9275.66	24.75	8468.76	22.24
Total	37524.00	100.00	37524.00	100.00

In this study, the effect of land use development to the sustainability of Panching Water Supply Treatment Process (WSTP) was obtained by the comparison between water availability and water footprint.

3. Result and Discussion

3.1. Water Footprint Accounting

Table 3 shows the types of water footprint involved in each process at Panching WTP.

Table 3. Water footprint types for each process at Panching WTP

Process	WF_{blue}	WF_{green}	WF_{grey}
Screening	✓	✓	
Aeration	✓	✓	
Mixing Chamber	✓	✓	
Flocculation	✓	✓	
Sedimentation	✓	✓	
Filtration	✓	✓	✓

At this WTP, only filtration process having all three types of WF due to pollutants discharge (backwash).

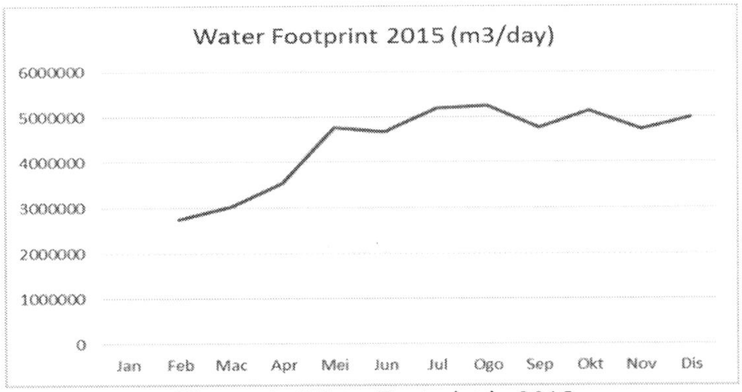

Figure 1 : Water Footprint in 2015

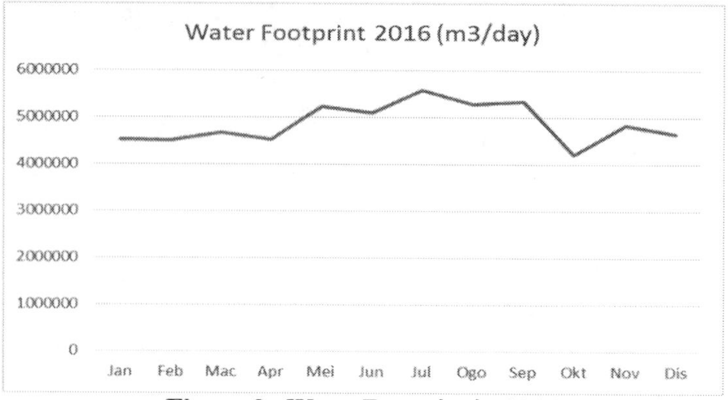

Figure 2. Water Footprint in 2016

In February 2015, the lowest water footprint amount was calculated with the value of 2,760,322 m^3/day as the plant was at the early stage of its operation and no backwash was indicated as WFgrey. Thus, WFblue and WFgreen were involved at that time. From the both graphs as shown in Figure 1 and Figure 2, the trend was gradually increased until the middle of the year and remain constant towards the end of year 2015 however there was a slightly decreased in year 2016. The decrement towards the end of year 2016 might due to the decreasing precipitation, thus affecting the water intake level.

3.2. Water Availability

Figure 4 shows the water availability of the Sg Kuantan in year 2015 and 2016 respectively. It shows that the water availability decreased from 2015 to 2016 with amount of reduction 1,134,147 m^3/day. This might due to the decreasing amount of precipitation, thus the volume of Sg Kuantan was affected.

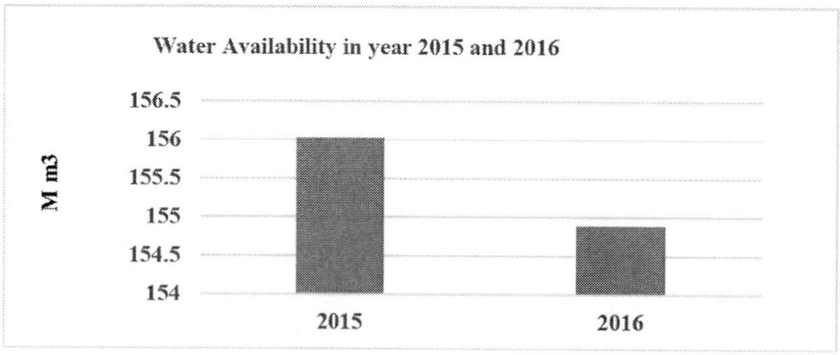

Figure 3 : Water Availability of Sg Kuantan 2015 and 2016

3.3. Land Use Development

From Table 1 and 2, it is clearly shows that there was a multiplication of growth in residential and industrial activities. It is believed that the increased of land used activities affected the water demand for the Panching WTP. Figure 5 shows the total water demand in year 2015 and 2016.

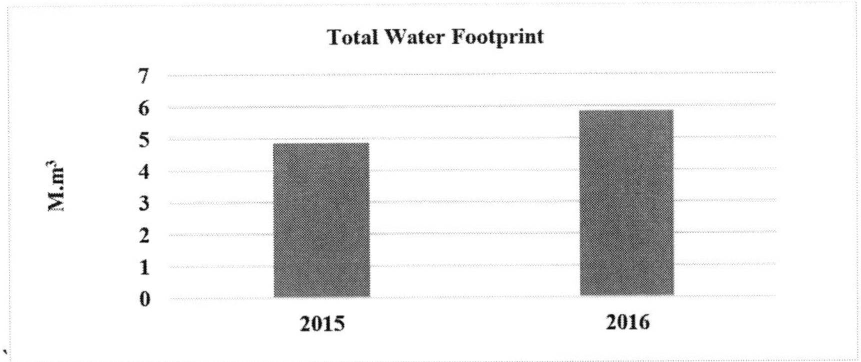

Figure 4 : Total water demand in year 2015 and 2016

Table 4. Percentage of water footprint based on water demand

	Total Water footprint x10^6 (m^3/yr)	Total Water demand x10^6 (m^3/yr)	Total Water Intake x10^6(m^3/yr)	Percentage of water footprint based on water demand (%)
2015	48. 7	44.3	44.2	9.93%
2016	58.4	45.1	52.0	29.5%

In this study the WF of Panching WTP was highly dependent to water demand. Table 4 shows that when the water demand increased the water footprint was also increased. The increment of water demand percentage by 1.8% from 2015 to 2016 has increased 11 times higher of the water footprint percentage to 19.9%. This result shows that the water consumption during the water supply treatment process is two times higher than the demand thus it shows the inefficient of the treatment process involved in the Panching WTP; the water intake value was not in accordance to the water demand.

4. Conclusion

As a conclusion, although the volume of Sg. Kuantan is available as intake resource for Panching WTP, but the volume is decreasing from year 2015 to 2016. Unfortunately, the WF is increasing from year 2015 to 2016. If this scenario keeps continue to occur, the water resource will not be able to sustain. Land use development must be well-regulated in order to ensure the sustainability of water supply in Kuantan district. In addition to that, the consumer should aware to their daily water consumption as it affecting the water demand. Furthermore, the management of water supply treatment process must be managed efficiently.

5. Acknowledgement

This study will not be able to succeed without committed authorities who supplied all the information needed. Thus, an appreciation should be given to Pengurusan Air Pahang Berhad (PAIP), Majlis Perbandaran Kuantan (MPK) and Department of Statistic, Malaysia. Last but not least, a special thanks to Fundamental Research Grant Scheme (FRGS) for funding my research. This study was conducted under FRGS Grant, RDU160155.

References

[1] M. M. Mekonnen, M. Pahlow, M. M. Aldaya, E. Zarate, and A. Y. Hoekstra, "Sustainability, Efficiency and Equitability of Water Consumption and Pollution in Latin America and the Caribbean," pp. 2086–2112, 2015.

[2] J. S. Wallace, "Increasing agricultural water use efficiency to meet future food production," *Agric. Ecosyst. Environ.*, vol. 82, no. 1–3, pp. 105–119, 2000.

[3] A. E. Ercin and A. Y. Hoekstra, "Water footprint scenarios for 2050 : A global analysis," *Environ. Int.*, vol. 64, pp. 71–82, 2014.

[4] R. Z. R. Z. Abidin, "Water Resources Management In Malaysia – The Way Forward," *Board Eng. Malaysia*, p. 56, 2003.

[5] D. Chico, M. M. Aldaya, and A. Garrido, "A water footprint assessment of a pair of jeans : the in fl uence of agricultural policies on the sustainability of consumer products," *J. Clean. Prod.*, vol. 57, pp. 238–248, 2013.

[6] M. M. Aldaya, P. Martínez-Santos, and M. R. Llamas, "Incorporating the water footprint and virtual water into policy: Reflections from the Mancha Occidental region, Spain," *Water Resour. Manag.*, vol. 24, no. 5, pp. 941–958, 2010.

[7] J. S. Li and G. Q. Chen, "Water footprint assessment for service sector: A case study of gaming industry in water scarce Macao," *Ecol. Indic.*, vol. 47, pp. 164–170, 2014.

[8] F. Abnisa *et al.*, "How to get published," *Biomass and Bioenergy*, vol. 5, no. 1, pp. 1–10, 2014.

[9] L. Ruini, M. Marino, S. Pignatelli, F. Laio, and L. Ridolfi, "Water footprint of a large-sized food company: The case of Barilla pasta production," *Water Resour. Ind.*, vol. 1–2, pp. 7–24, 2013.

[10] P. Chooyok, N. Pumijumnog, and A. Ussawarujikulchai, "The Water Footprint Assessment of Ethanol Production from Molasses in Kanchanaburi and Supanburi Province of Thailand .," *APCBEE Procedia*, vol. 5, pp. 283–287, 2013.

[11] R. P. Marano and R. A. Filippi, "Water Footprint in paddy rice systems . Its determination in the provinces of Santa Fe and Entre Ríos , Argentina," *Ecol. Indic.*, vol. 56, pp. 229–236, 2015.

[12] A. Y. Hoekstra, A. K. Chapagain, M. M. Aldaya, and M. M. Mekonnen, *The Water Footprint Assessment Manual*. 2011.

IOP Conf. Series: Materials Science and Engineering **318** (2018) 012029 doi:10.1088/1757-899X/318/1/012029

Hydrological Modelling using HEC-HMS for Flood Risk Assessment of Segamat Town, Malaysia

N S Romali[1], Z Yusop[2], and A Z Ismail[3]

[1]Faculty of Civil Engineering and Earth Resources, Universiti Malaysia Pahang, Lebuhraya Tun Razak, 26300 Gambang, Pahang, Malaysia
[2]Centre of Environmental Sustainability and Water Security, Universiti Teknologi Malaysia, 81300 Skudai, Johor, Malaysia
[3]Faculty of Civil Engineering, Universiti Teknologi Malaysia, 81300 Skudai, Johor, Malaysia

E-mail: zulyusop@utm.my

Abstract. This paper presents an assessment of the applicability of using Hydrologic Modelling System developed by the Hydrologic Engineering Center (HEC-HMS) for hydrological modelling of Segamat River. The objective of the model application is to assist in the assessment of flood risk by providing the peak flows of 2011 Segamat flood for the generation of flood mapping of Segamat town. The capability of the model was evaluated by comparing the historical observed data with the simulation results of the selected flood events. The model calibration and validation efficiency was verified using Nash-Sutcliffe model efficiency coefficient. The results demonstrate the interest to implement the hydrological model for assessing flood risk where the simulated peak flow result is in agreement with historical observed data. The model efficiency of the calibrated and validated exercises is 0.90 and 0.76 respectively, which is acceptable.

1. Introduction

Flood caused severe impact to human and properties, hence is known as a destructive natural disaster in the world [1, 2]. Flooding pose a serious threat to people and cause major damage to properties, infrastructures, agricultural production, give serious impact to socio-economic activities and may caused the loss of human life and decrease the quality of human health [3, 4, 5]. The consequences of natural disaster may lead to a great monetary loss to the economy of a region. Furthermore, on economic perspectives, natural disasters are also known as economic disaster [6]. In Europe, river flooding had caused about 100 billion euros of damage over 20 years' period from 1986 to 2006 [7]. Whereas according to the Emergency Disasters Database (EM-DAT) [8], flood in Malaysia had caused an approximate loss of RM 41.9 million within the years of 1995 to 2005.

Flood risk management is a non-structural flood mitigation measures that could be significantly reduce the impact of flooding [9]. This risk-based approach is contradicting to the conventional flood control approach which focusing on structural flood mitigation measures. The assessment of flood risk is the initial step in flood risk management [10]. Flood risk assessment consist of two main elements i.e. flood hazard and flood vulnerability [7, 11, 12, 13]. Flood hazard assess the probability of a flood event to take place i.e. the extent or magnitude of flood while flood vulnerability assesses the potential flooding impacts to community and assets (flood damage) [14, 15].

Flood hazard normally represented by flood mapping. Generation of flood mapping needs the information of peak flood as it is the necessary input to hydraulics model [16]. Besides that, the estimation of flood damage for flood risk vulnerability assessment requires the combination of four main elements i.e. flood characteristics, exposure, value of elements at risk, and the susceptibility of the elements at risk to particular hydrologic conditions [17, 18]. The characteristics of flood such as flood depth and the information of affected area may be obtained from flood modelling.

Numerous flood modelling studies had been done by researchers worldwide (for example [16, 19, 20, 21]). Ullah et al. [19] applied the combination of remote sensing, geographical information system (GIS), HEC-RAS (1D) and HEC-Geo RAS to model a flood inundation forecasting of Kalpani River. River modelling using HEC-RAS requires two types of data, i.e., geometric data and hydrological data. Similarly, Khattak et al. [16] also utilised the application of HEC-RAS and ArcGIS to develop floodplain maps for the the part of Kabul that lies in Pakistan. Peak flows from selected flood event and various return periods were the input into the HEC-RAS model to find the corresponding flood level expected along the study area. The information of the hydrologic condition needed in the hydraulic modelling can be generated from hydrologic modelling. For example, Oleyiblo and Li [22] had applied HEC-HMS to obtain simulated peak discharge for the purpose of flood forecasting in Misai and Wan'an cathments in China. The application of HEC-HMS also can be observed in the study by Bhuiyan et al. [23]. In the study, the HEC-HMS was applied to simulate the hydrological processes of Sturgeon Creek watershed system in Manitoba, Canada.

The intent of this study is to assist in the assessment of flood risk for the urban area of Segamat town, Johor. An attempt has been made to provide peak flood discharge of Segamat River, as the input to the hydraulic model for the generation of flood mapping. For the purpose, a hydrological modelling using HEC-HMS is applied. This paper presents the applicability of the hydrological results to be used for the further flood risk assessment works.

1.1. Description of study area

Segamat River is one of the tributaries of the Muar River that flows through the town of Segamat [24]. The location of Segamat River is shown in figure 1. The river is approximately 23 km in length, with average width of 40 m and is 14 m above sea level [25].

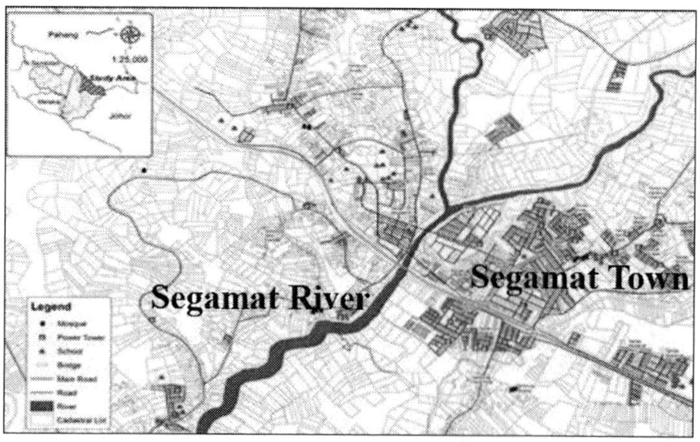

Figure 1. Location of study area

Segamat is a district in Johor which located in the north part of the state. Segamat is facing flooding problems in some areas in the last few decades, that caused displacement of people and severe properties damages. Historical records showed a series of flood occurred since 1950s, where the recent worse flood took place in 2006 and 2011. During the 2011 flood, about 31,000 people were evacuated and cause several deaths [26].

2. Methodology

The purpose of conducting hydrological modeling is to estimate flow hydrograph from tributary catchments. The estimated flow hydrograph serves as an input to hydraulic modeling. The methodology used in the hydrological modeling in this study is shown in figure 2.

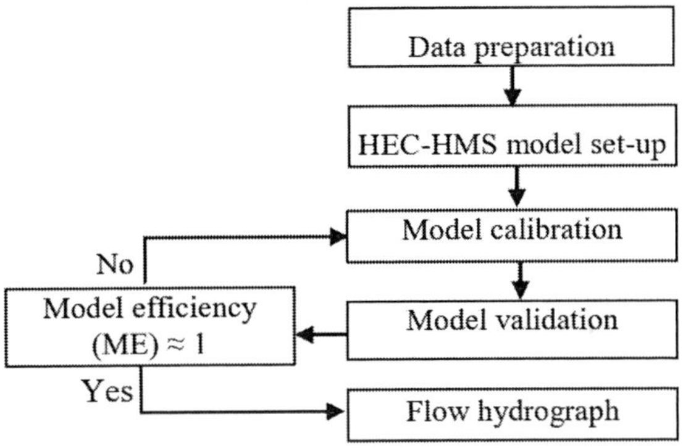

Figure 2. Study flow chart.

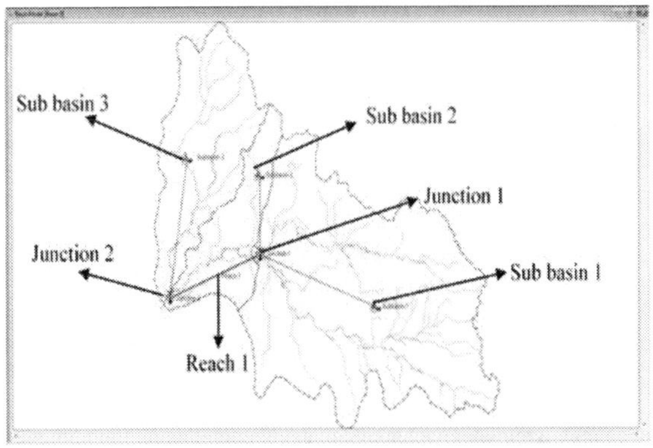

Figure 3. Conceptual of hydrological model at Segamat River

The first phase of hydrological modeling was a HEC-HMS model set-up. HEC-HMS used for this analysis is basically a hydrologic model which provides various options for calculating hydrologic losses, transforming excess rainfall and estimating base flow. The details of HEC-HMS model set-up are described in the next section. While in the second phase, the resulted hydrological model was calibrated and validated. The calibration model efficiency was verified using Nash-Sutcliffe model efficiency coefficient.

2.1. HEC-HMS model set-up

The hydrological analysis used the model set up of HEC-HMS. The catchment characteristics data, area of catchment, river length and slope are needed in the hydrological analysis. The main parameters used were hydrological parameters while the losses methods used were Soils Moisture Accounting (SMA). The parameters required for the model analysis are initial loss, constant rate, impervious, time of concentration, storage coefficient, and Muskingum's K and X value.

The output from this hydrological analysis is the hydrograph of Segamat River. The conceptual of hydrological model and the list of the rainfall station used for the hydrological modeling is shown in figure 3 and table 1 respectively. While the element parameter and method applied in the HEC-HMS set up is shown in table 2.

2.2. Model calibration and validation

A model calibration process involved adjusting or tuning model parameters until the model simulation result closely match the observed behavior. Meanwhile, model validation is basically a similar process to model calibration but using another set of hydrological data [27]. The purpose of model calibration is to establish the model parameter values to be used for model validation and subsequently for simulating hydrograph using defined data input.

The calibration and validation model is verified using model efficiency (ME) in order to define the absolute square differences between the predicted and observed value. The ME is calculated based on Nash-Sutcliffe model efficiency coefficient shown in equation (1) according to Yusop et al. [28]:

$$ME = 1 - \frac{\sum_{i=1}^{n}(Q_{Oi} - Q_{Si})^2}{\sum_{i=1}^{n}(Q_{Oi} - Q_{a})^2} \quad (1)$$

Where *ME* is the model efficiency (efficiency index), Q_{Oi} is observed flow at time *i*, Q_{Si} is simulated flow at time *i*, Q_a is an average observed flow, and *n* is the number of data points.

Table 1: The list of rainfall station for hydrological modelling

Sub basin	Station	Station name	River
1	2529001	Kemelah	Segamat River
2	2428011	Ldg Segamat	Segamat River
3	2528002	Stesen Telemetrik	Kapeh River

Table 2. Element parameter and method setup in HEC-HMS model

Model Parameter	Method
Canopy method	Simple canopy
Surface method	Simple surface
Loss method	Soil Moisture Accounting
Transform method	Clark Transform
Base flow method	None
Routing method	Muskingum

3. Results and discussion

3.1. Model calibration and validation

The results of simulated peak flow and model efficiency (ME) for model calibration and validation are presented in table 3. The simulated peak flow is compared with the observed peak flow where the difference is 9% and 3% for calibration and validation results respectively. The model efficiency is 0.90 and 0.76 for model calibration and validation respectively, which is considered satisfactory.

Table 3: Observed and simulated peak flow for model calibration and validation

Run	Peak flow (m^3/s)		ME
	Observed	Simulated	
Calibration	23.30	25.40	0.90
Validation	79.90	77.50	0.76

Table 4. Tested model parameters for 2011 Segamat flood event (Sub basin 1)

Parameter	Unit	Value
Initial storage (simple canopy)	mm	8
Initial storage (Simple surface)	mm	8
Soil losses	%	18
Imperviousness	%	10
Time of concentration	hour	30
Storage coefficient	hour	55

The HEC-HMS model was calibrated using rainfall and flow data of Segamat River gauging station from 26th May to 7th June 2010. Figure 4 shows the observed and calibrated hydrograph where the simulated peak flow for the calibration event is 25.40 m^3/s, which is close to the observed peak flow of 23.30 m^3/s. Table 4 shows the model parameter values adjusted for each of the components of sub basin 1. The Muskingum *K* and *X* applied at Reach 1 is 2.15 hr and 0.199 respectively. The parameters must be calibrated simultaneously through adjustment of their values until the model simulation replicate observed data [22].

For the model validation process, a set of storm event data was used to validate the HEC-HMS model based on the optimized model parameters obtained during calibration. Rainfall and flow data of Segamat River gauging station dated 4th August to 9th September 2010 was used for the validation analysis. The simulated peak flow for the validation event is 77.50 m^3/s, which is close to the observed peak flow of 79.90 m^3/s as illustrated in hydrograph in figure 5.

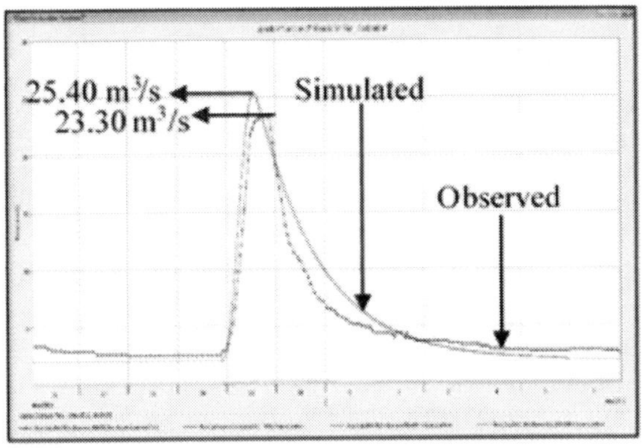

Figure 4. Observed and calibrated hydrograph for 26th May – 7th June 2010 data

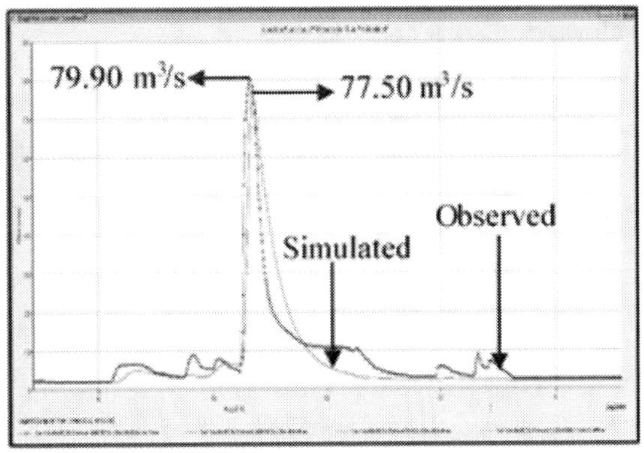

Figure 5. Observed and validated hydrograph for 4th Aug. to 9th Sept. 2010 data

3.2. 2011 Segamat Flood Event

The HEC-HMS model was set up for a 20 January 2011 (01:00) to 14 February (01:00) flood event. The event hydrological modelling was performed using hourly time steps. The event simulation result is presented in figure 6. From the flow data obtained from Drainage and Irrigation (DID) Malaysia, the peak flow (Q_P) is observed as 5.1 m³/s which is not presenting the major flood event occurred on 2011. However, the simulated Q_p from the analysis gives a flow rate of 1238.2 m³/s which is a reasonable flow value for a major flood event in Segamat. As can be seen in figure 7, other major floods occurred in Segamat recorded a high peak flow more than 1000 m³/s. For example, all the four major flood events in year 1969, 1979, 1983 and 2007 (labelled as 1, 2, 3 and 4 respectively in figure 7), occurred with peak flow more than 1000 m³/s.

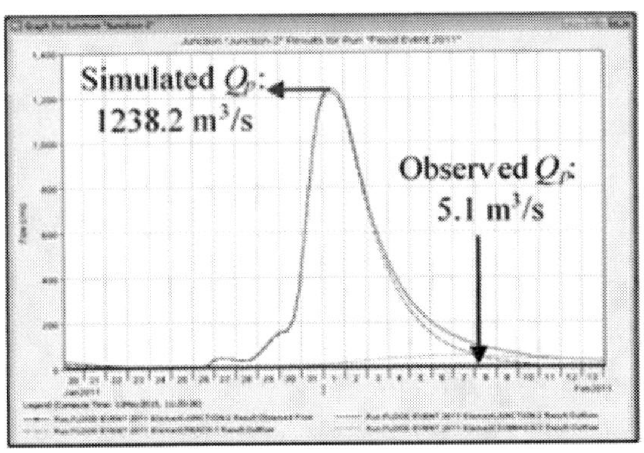

Figure 6. Hydrograph for 2011 Segamat flood event

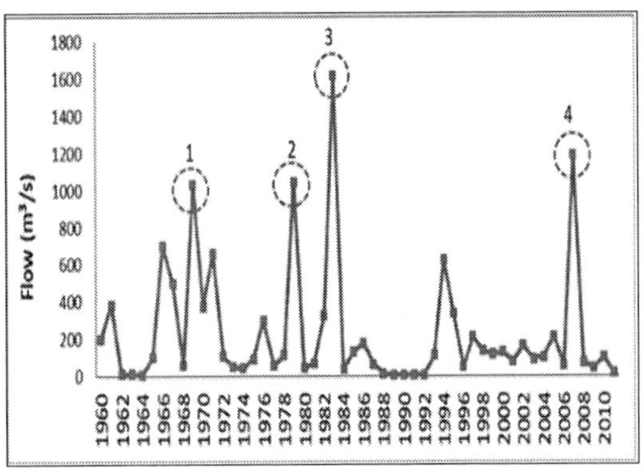

Figure 7. The historical annual peak flow at Segamat River gauging station from 1960 to 2011

4. Conclusion

Hydrological modeling using HEC-HMS was performed to obtain peak flow hydrograph of Segamat River. The information is useful as the input for the flood mapping of Segamat town. The peak flow of 1238.2 m³/s hydrograph was obtained for 2011 Segamat flood event using a calibrated and validated HEC-HMS model. The simulated peak flow result is in agreement with historical observed data and the model efficiency of 0.90 and 0.76 is acceptable for the calibrated and validated HEC-HMS model. Hence, the hydrologic model is recommended for future works of flood modeling and flood risk assessment of the study area.

Acknowledgement

The authors would like to acknowledge Universiti Teknologi Malaysia and Universiti Malaysia Pahang for the financial support through research grant RDU1703115.

References

[1] Messner F, Penning-Rowsell E, Green C, Meyer V, Tunstall S and van de Veen A 2007 *FLOODsite Project Deliverable D9.1.* (Wallingford, UK: Floodsite Consortium, 2007)
[2] Eleuterio J, *PhD thesis* (University of Strasboug, France, 2012)
[3] Abdul Rahman H 2009 *Malaysian Journal of Environmental Management* **10(2)** p 17-32
[4] Chang L, Lin C and Su M 2008 *Water SA* **34(2)**, p 209–216
[5] Jonkman S N 2005 *Nat. Hazards* **34(2)** p 151–175
[6] Padli J and Habibullah M S 2009 *Asian Social Science* **Vol.5** No. 4
[7] De Moel H, Van Alphen J and Aerts J C J H 2009 *Nat. Hazards Earth Syst. Sci.* **9(2)** p 289–301
[8] EMDAT 2000 *The OFDA/CRED International Disaster Database* Unversite Catholicque de Louvan, Brussels, Belgium
[9] Merz B, Kreibich H, Schwarze R and Thieken A 2010 *Nat. Hazards Earth Syst. Sci.* **10(8)** p 1697–1724
[10] Gain A K, Mojtahed V, Biscaro C, Balbi S and Giupponi C 2015 *Nat. Hazards* **79(3)** p 1499–1530
[11] Apel H, Thieken A H, Merz B and Blöschl G 2004 *Nat. Hazards Earth Syst. Sci.* **4(2)** p 295–308
[12] Kreibich H, Seifert I, Merz B and Thieken A 2010 *Hydrological Sciences Journal* **55(8)** p 1302–1314
[13] Ward P J, De Moel H and Aerts J C J H 2011 *Nat. Hazards Earth Syst. Sci.* p 3181–3195
[14] Apel H, Merz B and Thieken A H 2008 *Int. J. River Basin Management* **6(2)** p 149–162
[15] Hudson P, Botzen W J W, Kreibich H, Bubeck P and Aerts J C J H 2014 *Nat. Hazards Earth Syst. Sci.* **14** p 1731–1747
[16] Khattak M S, Anwar F, Saeed T U, Sharif M, Sheraz K and Ahmed A 2016 *Arab J Sci Eng* **41** p 1375-1390
[17] Meyer V and Messner F 2015 *FLOODsite Project Report*
[18] De Moel H and Aerts J C J H 2011 *Nat. Hazards* **58(1)** p 407–425
[19] Ullah S, Farooq M, Sarwar T, Tareen M J and Abdul Wahid M 2016 *Arab J Geosci* **9** p 439
[20] Sidhu K and Durga Rao K H V 2016 *Geocarto International* DOI: 10.1080/10106049.2016.1178818
[21] Grimaldi S, Petroselli A, Arcangeletti E and Nardi F 2013 *Journal of Hydrology* **487** p 39-47
[22] Oleyiblo J O and Li Z 2010 *Water Science and Engineering* **3(1)** p 14-22
[23] Bhuiyan H A K M, McNairn H, Powers J and Merzouki A 2017 *Hydrology 2017*, **4, 9**
[24] Teo F Y, Liew Y S, Falconer R A and Lin B L 2013 *Proceedings of 2013 IAHR World Congress*
[25] Gasim M B, Surif S, Mokhtar M, Toriman M E, Abd Rahim S and Bee C H 2010 **39(3)** p 353-361
[26] Segamat District. In Wikipedia. Retrieved May 11, 2017, from https://en.wikipedia.org/wiki/Segamat_District#Segamat_District_Facts
[27] Ismail A Z 2015 Master thesis (Universiti Teknologi Malaysia, 2015)
[28] Yusop Z, Chan C H and Katimon A 2007 *Water Science and Technology* **Vol 56 No 8** p 41-48

IOP Conf. Series: Materials Science and Engineering **318** (2018) 012030 doi:10.1088/1757-899X/318/1/012030

Compressive and Flexural Tests on Adobe Samples Reinforced with Wire Mesh

G A Jokhio[1,*], Y M Y Al-Tawil[2], S M Syed Mohsin[1], Y Gul[3], and N I Ramli[1]
[1]Senior Lecturer, Faculty of Civil Engineering and Earth Resources, University Malaysia Pahang, Malaysia
[2]Student, Faculty of Civil Engineering and Earth Resources, University Malaysia Pahang, Malaysia
[3]PhD Scholar, Faculty of Built Environment, University Technology Malaysia, Malaysia

Email: jokhio@ump.edu.my

Abstract. Adobe is an economical, naturally available, and environment friendly construction material that offers excellent thermal and sound insulations as well as indoor air quality. It is important to understand and enhance the mechanical properties of this material, where a high degree of variation is reported in the literature owing to lack of research and standardization in this field. The present paper focuses first on the understanding of mechanical behaviour of adobe subjected to compressive stresses as well as flexure and then on enhancing the same with the help of steel wire mesh as reinforcement. A total of 22 samples were tested out of which, 12 cube samples were tested for compressive strength, whereas 10 beams samples were tested for modulus of rupture. Half of the samples in each category were control samples i.e. without wire mesh reinforcement, whereas the remaining half were reinforced with a single layer of wire mesh per sample. It has been found that the compressive strength of adobe increases by about 43% after adding a single layer of wire mesh reinforcement. The flexural response of adobe has also shown improvement with the addition of wire mesh reinforcement.

1. Introduction

The word 'adobe' originates from Arabic but has been extensively used in Spanish to mean building material made from earth or mud, possibly mixed with some organic material. The predominant use of adobe is found in the Arabic, Persian, and Spanish regions of the world. It is understood to be a readily available material without requiring extensive skills for its use and is commonly associated with low-cost construction [1]. Generally, it is believed that the involvement of engineers and architects, and detailed designs, is not required when it comes to building with adobe.

The historical use of adobe as a building material has been documented in several research articles. It has been reported, for example, that the natural soil, earth etc. have been used as building material for over 11,000 years [2-4]. From the ancient city of Jericho to the Mesopotamian Ziggurats and Athens, and from Great Wall of China in the east to the Andean cities in the West, we can observe the use of earth as a construction material. Moreover, the use of earth as a building material can also be found in the civilization of the Indus, Egypt, and Greece. Vaulted structures can be found in the Central Asia dating from 4th century BC [5]. In Central Asia, the use of adobe masonry has been observed in buildings of importance such as monumental or religious nature [6], especially domed structures as well regular traditional houses [5-11]. Mud bricks have been used in the construction of shelters for several millennia [12], and approximately 30% of the human population lives in earthen structures to the present day [13]. The city of Shibam in south Yemen and the walls of Marrakech in Morocco are also mainly constructed with adobe. A very rich cultural heritage of earth building can be found in the present world, notably; Africa, Iran, Afghanistan, Yemen, Iraq, and Syria. Moreover, the use of earth buildings can also be found in Europe including Spain, Germany, England, France, Portugal, Italy, Denmark, and Sweden [14].

There has been an increased interest in this construction material and method by scientific and engineering community over the past 3 decades [15] as it can be witnessed that the published research in this field has increased about ten folds in the past decade and a half compared to the previous decade [16]. This is partly due to the fact that earth building provides a sustainable alternative to other construction materials and techniques, which are relatively more polluting. However, the ubiquitous acceptance of earth as a primary building material is hindered by certain issues such as vulnerability of this kind of construction to extreme actions such as earthquakes [17-18]. Another important challenge facing earth building is that the participation of skilled technicians, engineers and architects is generally deemed unnecessary. This results in non-engineered construction invoking insurance providers to set very restrictive conditions for subscription to the insurance coverage for earthen dwellings [14]. Last but not the least, there is a notion of class associated with this kind of construction as it is considered to be only for the very poor or the very rich and the 'Middle Class' rarely uses it [19].

The need of energy-efficient sustainable housing development cannot be overestimated. It is needed that materials and building technologies evolve to be good to the environment, energy efficient, affordable and fit in the contemporary context [20-22]. Therefore, sustainable and economic construction by utilizing earth as the primary building material, such as adobe, needs to be promoted [23] as much as possible and resources need to be allocated for further research in this field.

2. Mechanical Properties of Adobe

Mechanical properties of adobe are difficult to ascertain due to their wide ranges owing to several factors involved in the preparation of the material. Adobe not being a factory material and not having strict guidelines for its preparation, and the absence of skilled personnel during its preparation results in this wide range of its mechanical properties. Several researchers have investigated and reported in the literature the mechanical properties of adobe.

Some researchers have focused on the determination of mechanical properties of adobe from historical buildings and archaeological sites such as the in-situ characterization of the adobe masonry of a two thousand years old building in Turkmenistan [6]. In another study, Bronze Age earthen construction materials from East Crete were analysed scientifically for their constituent properties [24]. The authors pointed out the importance of including the scientific analysis of construction materials used in deteriorated prehistoric earthen structures into the practice of archaeological site excavations. The mechanical properties of adobe walls in a Roman Republican Domus at Suasa were investigated by reproducing and mechanically testing wall samples [25].

For the experimental computation of mechanical properties, several standards for earth construction can be found in the literature [26-30]. Different standards, however, provide different procedures and specimen dimensioning criteria making it difficult to comply with them [26, 31-32]. Another study [33] focused on the post-peak strain behaviour of traditional earthen construction material. The study was based on compression and shear test results of adobe masonry and reported the lack of significant influence on the shear strength by the application of two different pre-loads during the drying phase. The same research group in an earlier study [34] compared earth block masonry, rammed earth and cob. They reported that building technique practice is one of the crucial parameters affecting performance of earth block masonry. Leaving the earth blocks dry or otherwise wetting them prior to use strongly affected results in shear tests. In other studies, some new interpretations of the '3 points bending test' have also been proposed for compressed earth blocks that give the compressive strength directly [32,35]. These models, however, suffer from unavailability of validation data and cannot be readily implemented.

The compressive response of adobe in laboratory tests depends upon the specimen form and size. Strength values derived from cubes and cylinders after application of shape correction factors were reported to range from 0.6 to 1.75 MPa. Prisms, on the other hand, tend to overestimate the compressive strength due to platen restrain effects [36]. The large variance in laboratory test results has been attributed to the inherent inhomogeneity and natural randomness of earthen materials as well as a lack of internationally accepted standardized testing procedures. In a study of the influence of the testing procedures in the mechanical characterization of adobe bricks [37], it was found, as expected, that the compressive and tensile splitting strength of cylindrical specimens, 0.58 and 0.16 MPa respectively, are close to those values obtained in an earlier study [38]. Several other authors have also performed studies aimed at calculating the modulus of elasticity of adobe [13, 25, 39-40].

Clay is the most important component of mud bricks, since it provides the dry strength of the blocks. Excessive clay content, however, increases drying shrinkage, and thus micro-cracking of the mortar and blocks [17]. Traditionally, in order to activate the bonding properties of the clay, mud is soaked 24 hours before use, which has been found to be beneficial [41]. Various stabilized soil applications including the use of blended binders also have been found beneficial [42]. Strength, durability and shrinkage characteristics of cement stabilized soil blocks were studied separately [43].

Coarse sand or straw is generally added to mud for making adobe bricks in order to control drying shrinkage [17]. Moreover, similar to concrete and other such materials, adobe is stronger in compression whereas its tensile strength is low resulting in efforts directed at its improvement. The mechanical properties of adobe are affected by the fibre contents [44]. Mud-brick makers of Turkey and the Middle East, for example, have long been using fibrous ingredients such as straw for this purpose [12]. In a study [45], the researchers investigated the effects on the compressive strength of adobe when different types of fibres are added. They concluded that as opposed to average strength of 2 MPa achieved by traditional mud bricks, those reinforced with plastic fibres, straw, and polystyrene along with a mix of clay, pumice, cement, lime, gypsum and water produced strengths up to 6.5, 5.4, and 4.3 MPa, respectively. The authors claimed that these fibre reinforced mud bricks fulfil the compressive strength requirements of the ASTM and Turkish Standards.

Some tests have used dynamic analysis by applying cyclic displacements to straw reinforced adobe. The straw fibres produce elongated softening branches of the stress-strain curve, whereas the increased aspect ratio makes the specimens less ductile [46]. The compressive strengths and moduli of elasticity of cubic and prismatic specimens were reported to be 1.57 and 148.08, and 1.7 and 130.22 MPa, respectively. In an independent study [47], it was concluded that addition of hibiscus cannabinus fibres (Kenaf) contributed to a homogenous microstructure with reduced pore sizes having positive effect on the mechanical properties of adobe. According to another study, the addition of straw acts as shear reinforcement and increases energy absorption [48]. Fly ash as an additional material also exhibits similar effects. The straw mix gave the highest compressive strength of 3.99 MPa for the straw mix ratio of 33.3% [49]. Maximum flexural strength, however, occurred at 25% straw mix ratio and was measured to be 0.82 N/mm2. Sheep's wool was added as a natural fibre to clay in another study [50] to find out that it increases the compression strength with the highest value reported as 4.44 MPa for a specimen with 19.5% alginate, 0.5% lignum, 0.25% wool and 0.25% water. The same specimen exhibited a flexural strength of 1.45 MPa in a 3-point bending test. The compressive strength of lateritic adobe was reported in another study [51].

The experimental analysis and modelling of the mechanical behaviour of earthen bricks were investigated in a research [52]. The bricks and blocks under consideration were prepared by manual compaction and consisted of clay, coarse sand and straw. The compressive strength of bricks was reported to range from 5.15 to 8.29 MPa with the range of modulus of elasticity being from 59 to 94 MPa, whereas the blocks exhibited lower strength, ranging from 2.14 to 2.88 MPa and interestingly higher modulus of elasticity, ranging from 98 to 211 MPa.

Knowledge of the stress-strain behaviour laws of adobe is important, because these curves express essential information about the properties and mechanical behaviour of adobe [53]. The quasi-brittle behaviour of adobe and other concrete-like materials can be well modelled by using a material response curve such as that by Popovics [54] for implementation into standard finite element codes. One such example is a constitutive model developed for describing adobe's stress-strain behaviour under compression [36]. It is based on third order polynomials derived from data obtained from cylindrical specimens and includes relations for the pre-peak and post-peak ranges. The authors reported the coefficients of variation with average and highest values of; 15.38% and 27.5% for cylindrical specimens, 19% and 42.5% for prismatic specimens, and 24.23% and 76.8% for cubical specimens. Due to this large variation, despite having the highest compressive strengths of 1.41, 3.31 and 1.75 MPa, the average of the same were 0.99, 1.54 and 1.15 MPa for cylindrical, prismatic, and cubic specimens respectively. The proposed stress-strain relationship, therefore, was obtained through fitting a curve using optimization routines and consisted of two distinct cubic relations for ranges of strain between 0 to 1.07 and 1.07 to 4. The proposed normalized stress-strain relationship is shown in Fig 1. In another study [55], strength and stress-strain characteristics of traditional adobe block and masonry were documented by conducting uniaxial compressive tests on adobe blocks and masonry prisms with different constituents. The average unconfined compressive strength of block and mortar was reported to range from 1.39 to 1.7 MPa. Stress-strain characteristics for adobe masonry were also presented such as average initial tangent modulus ranging from 32.61 to 36.51 MPa and average tangent modulus at 50% of peak stress ranging from 81.51 to 114.18 MPa. The coefficient of variation for all these values, however, was considerably high.

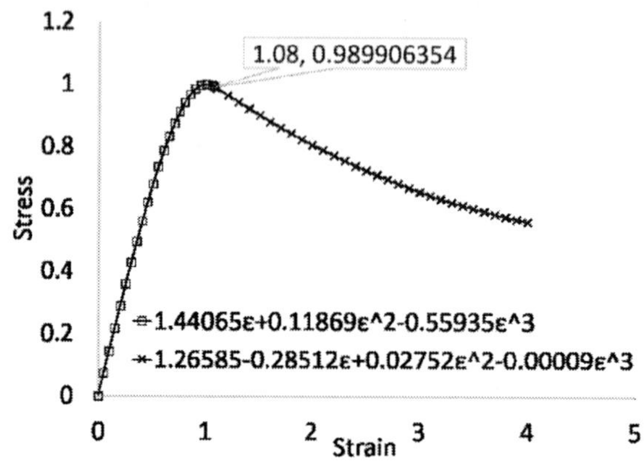

Figure 1. Stress strain curve for adobe [36]

3. Steel Wire Mesh as Reinforcement

Steel wire mesh has traditionally been used as a reinforcement along with cement sand mortar in the preparation of ferrocement, for e.g. [56-57]. Wire mesh generally reduces cracking in ferrocement, thus allowing for using much smaller cross-sections and very thin structural elements compared to conventional reinforced concrete structures. When it comes to adobe, wire mesh has generally been used as external reinforcement in order to improve the seismic response of adobe as well as provide an improved hold for the plastering and other finishing applied to adobe walls [1]. However, the use of steel wire mesh as reinforcement in adobe cubic or prismatic samples has not been previously reported in the literature. Therefore, the present study is unique in this sense. The findings of the present research are of preliminary in nature and provide a strong foundation for future researches to build upon it.

There are several types of steel wire mesh that can be used as reinforcement in adobe or any other cementitious composite. The main variations of wire mesh include chain link mesh, chicken wire mesh (named so because of its extensive use in making chicken coops), and expanded metal mesh. These three types of wire mesh have been illustrated in Fig 2. Chain link mesh has a range of opening sizes from 5 to 25 mm with wire diameters ranging from 0.5 to 2 mm in size. The chicken mesh is generally larger with its opening and wire diameters ranging from 13 to 50 mm and from 0.6 to 1.2 mm, respectively. The size ranges of the expanded metal mesh are similar to those of the chicken wire mesh. In the present study, chain link mesh, which is also known as the square mesh has been used.

Several researchers have investigated the effectiveness of one type of steel wire mesh over others, for example [58], however, it is not the focus of this study. Therefore, no attempt was made to optimize the selection of wire mesh as a reinforcement for adobe samples. This aspect can be studied later in another study. In the present study, chain link or square wire mesh was used only on the basis that it is easily available and relatively simpler to work with. A wire mesh reinforcement layer formed for a cubic specimen is shown as an example in Fig 3.

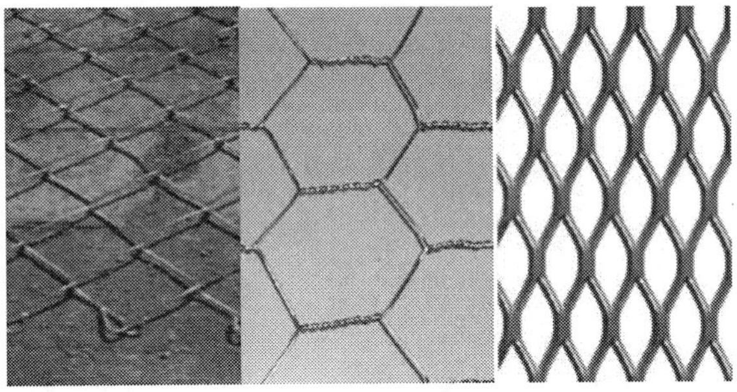

Figure 2. Three main variations of steel wire mesh from left: chain link mesh, chicken mesh and expanded metal mesh.

Figure 3. Wire mesh reinforcement layer prepared for a cubic specimen.

4. Experimental Setup and Methodology

The experimental setup for the present study consisted of two types of tests i.e. compressive strength test and flexural strength test. The details of specimens, materials, preparation procedures and tests conducted are provided as the following.

4.1 Details of Specimens

A total of 22 samples were prepared and tested in both categories. There were 12 cubes for testing of compressive strength, 6 out of which were control specimens whereas the remaining 6 were reinforced with wire mesh. For the purposes of identification, the cube samples without any reinforcement were named CP-1 to CP-6, whereas those reinforced with wire mesh were named CW-1 to CW-6. Similarly, 5 out of the 10 prismatic specimens for the flexural strength tests were control whereas the remaining 5 were reinforced with a single layer of wire mesh. The control prismatic specimens were named PP-1 to PP-5 and those prismatic elements that were reinforced with wire mesh were named PW-1 to PW-6. The details of all the samples are presented in Table 1. The cubes and prismatic specimens ready for testing are shown in Fig 4.

Figure 4. Cube and prismatic samples ready for testing

Table 1. Details of specimens.

S. No.	Sample Name	Sample Type	Test Type
1	CP-1	Cubes without any reinforcement (Size: 150 mm × 150 mm × 150 mm)	Compressive strength test
2	CP-2		
3	CP-3		
4	CP-4		
5	CP-5		
6	CP-6		
7	CW-1	Cubes reinforced with wire mesh (Size: 150 mm × 150 mm × 150 mm)	
8	CW-2		
9	CW-3		
10	CW-4		
11	CW-5		
12	CW-6		
13	PP-1	Prismatic specimens without any reinforcement (Size 100 mm × 100 mm × 500 mm)	Flexural strength test
14	PP-2		
15	PP-3		
16	PP-4		
17	PP-5		
18	PW-1	Prismatic specimens reinforced with wire mesh (Size 100 mm × 100 mm × 500 mm)	
19	PW-2		
20	PW-3		
21	PW-4		
22	PW-5		

4.2 Adobe Mix Preparation
Since the present research focused on the effect of wire mesh reinforcement on the mechanical properties of adobe, the mixture used for preparing the adobe material was the same for all the samples. The ingredients of the mixture consisted of sand, silt, clay, and Kenaf (hibiscus cannabinus) fibre. The proportion of different constituents used in the preparation of the mixture is given in Table 2. In one batch of mixture, 0.012 m^3 volume was prepared, which required a total of 4 gallons of water. The mix was prepared in a concrete mixer in order to make sure that the ingredients, especially the Kenaf fibres are distributed well throughout the material matrix. Dry ingredients were mixed first and then water was added gradually. This process ensured the preparation of a homogenous adobe mixture.

Table 2. Proportions of constituents in adobe mixture used

S. No.	Name of Constituent	Average Size of Particles	Proportion
1.	Sand	2 mm	50%
2.	Silt	0.06 mm	20%
3.	Clay	0.002 mm	25%
4.	Kenaf fibre	-	5%

4.3 Sample Preparation
Once the adobe material mix was ready, the specimens were prepared by pouring the mixture into moulds. The pouring process was carried out in steps with every next addition of material done only after the previous material was well compacted with the help of applying sufficient pressure.

4.4 Compressive Strength Tests
The compressive strength tests were carried out on the Universal Testing Machine. For this purpose, the British standard for compressive strength tests for concrete cubes (BS EN 12390:3-2002) was adapted for adobe. The sample preparation, especially the compacting process was modified accordingly, whereas the remaining aspects of the standard were adhered to. The compressive strength test setup is shown in Fig 2. After the placement of the sample in the machine, the load was applied and the maximum value of the load observed on the digital display was recorded as the breaking load for that particular sample. Compressive strength test setup along with a specimen being tested is shown in Fig 5.

4.5 Flexural Strength Tests
The test setup for flexural strength test consisted of a system that applied a single point load at the mid-span of the prismatic specimen. The two end supports were placed about 50 mm inside the edge of the beam resulting in an effective span of 400 mm. The test setup was mounted on the Universal Testing Machine. Deflection sensors were used to measure the beam deflection in order to plot the load-deflection curve. The load and deflection values were recorded for each specimen from the start of the test until failure at regular intervals. Fig 6 shows the flexural testing setup along with a prismatic specimen being tested.

5. Results and Discussion

5.1 Compressive Strength Test Results
Adobe has been reported in literature not as a very strong material. There is a high variation reported in the mechanical properties of adobe. The present research focused on the effects of wire mesh reinforcement on the mechanical properties of adobe. While the compressive strength of all the samples considerably varied, a significant improvement was noted when the samples were reinforced with wire mesh. On average, the total load taken by the samples without any reinforcement was about 4.97 kN resulting in a compressive strength of 0.22 MPa. This strength is on the lower side compared

to the values reported in literature. This is due to the fact that the present study did not focus on the optimization of adobe strength. The inclusion of the primary parameter of wire mesh reinforcement for this study, however, increased the average total load to 7.12 kN resulting in compressive strength of 0.31 MPa. Therefore, the addition of a single layer of wire mesh reinforcement increases the compressive strength of adobe samples by approximately 43%. The summary of the results of the compressive strength test is presented in Table 3.

Figure 5. A cube sample being tested for compressive strength

Figure 6. A prismatic sample being tested for flexural strength

Out of interest, it was checked what happens if the outer samples in both categories i.e. samples showing the highest and the lowest compressive strength were ignored. The average after ignoring these samples for un-reinforced cubes was 4.9 kN and that for the samples reinforced with wire mesh was 6.85 kN, which means an increase of over 45%. Although this may be a more accurate approach, the main finding of an increase of 43% has been reported conservatively after including all the six samples in both categories.

An interesting observation was made in that the cubes reinforced with wire mesh retained their shape even after failure and did not total dismantle. This behaviour is different from the unreinforced cubic samples, which after failure are totally destroyed. A cubic sample after failure is shown in Fig 7. It is hypothesized that the walls made with adobe bricks reinforced with wire mesh will tend to retain their shape even after failure, resulting in an improved response to earthquake and other similar disasters.

Figure 7. A cubic sample after failure

5.2 Flexural Strength Test Results

The flexural strength of adobe has traditionally been observed as almost non-existent. For the un-reinforced samples, it was observed in this study that the flexural strength of adobe is negligible. The comparison of the normalized stress-strain curve of un-reinforced prismatic specimens was made with the stress strain curve for adobe suggested by Illampas et al. [36]. The experimental results were taken as the average load of all the 5 samples against each deflection value. This comparison has been shown in Fig 8. It can be seen that while the average of the experimental results does display the tendency of the proposed model, the variation throughout the load-deflection history is relatively large, thus reducing the reliability of this association.

Table 3. Summary of compressive strength test results

S. No.	Sample Name	Maximum Load (kN)	Compressive Strength (MPa)
1	CP-1	5.3	0.2356
2	CP-2	4.9	0.2178
3	CP-3	4.7	0.2089
4	CP-4	4.6	0.2044
5	CP-5	5.6	0.2489
6	CP-6	4.7	0.2089
Average for Unreinforced Samples		4.9667	0.2207
Standard Deviation for Unreinforced Samples		0.3983	0.0177
7	CW-1	5.8	0.2578
8	CW-2	6.8	0.3022
9	CW-3	6.9	0.3067
10	CW-4	9.5	0.4222
11	CW-5	6.8	0.3022
12	CW-6	6.9	0.3067
Average for Samples Reinforced with Wire Mesh		7.1167	0.3163
Average for samples Reinforced with Wire Mesh		1.2416	0.0552
% Increase for Samples Reinforced with Wire Mesh		43.2883%	

Figure 8. Stress-strain response for unreinforced adobe beams

It can be observed in Fig 6 that the initial pre-peak response is varying highly, whereas, the post-peak softening is closer to the proposed model. The overall value of coefficient of determination between the two data sets is only 27.35%, whereas the coefficient of determination for the softening branch only is 71%. This can in part be attributed to the fact that the overall flexural strength of adobe is very low resulting in some discrepancies in the initial measurement of the same.

The addition of a single layer of wire mesh reinforcement, however, significantly improves the flexural response of the prismatic adobe specimens. The comparison can be seen in Fig 9, where the average of actual load-deflection response of the unreinforced specimens is compared with that of the specimens reinforced with wire mesh. The addition of wire mesh reinforcement has not only increased the flexural strength of adobe samples to about 3 times the original values, the response of the reinforced specimens also appears more reliable and can be represented by an idealized tri-linear curve. The slope and the relevant coefficients of determination for the three parts of the idealized curve are given in Table 4.

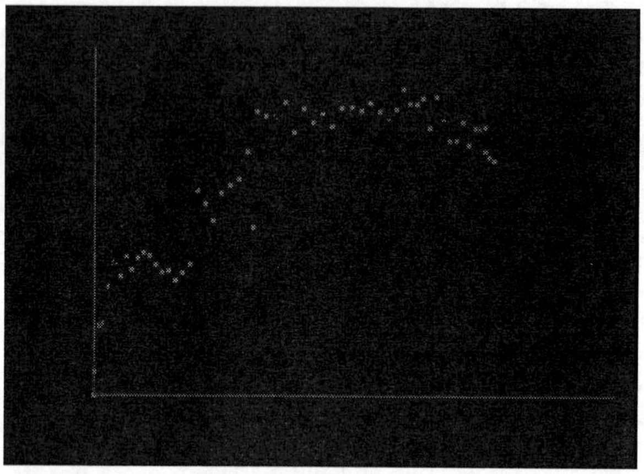

Figure 9. Comparison of load-deflection response of unreinforced samples and samples reinforced with wire mesh along with idealized tri-linear load-deflection curve

Table 4. Summary of compressive strength test results

Stage of Load-Deflection Curve	Slope (Modulus of Elasticity in MPa)	Coefficient of Determination (%)
Stage 1	121.98	93.4
Stage 2	16.13	81.2
Stage 3	0	26.5
Overall		90

Although the coefficient of determination for the stage 3 or the proposed tri-linear model is numerically low, it is due to the nature of this calculation that in the case of zero or a very small slope, the coefficient of determination is generally low. It does not directly indicate the lack of significance of the relationship.

6. Conclusions and Recommendations

From the results obtained through experiments in the present research, the following conclusions and recommendations are made:

- A total of 22 adobe samples consisting of 12 cubes and 10 prismatic samples were tested for compressive and flexural strength, respectively.
- It has been found that the addition of a single layer of wire mesh reinforcement increases the compressive strength of adobe samples by about 43%.
- It is hypothesized that the adobe walls made with bricks reinforced by wire mesh will tend to retain their shape post-failure.
- The load-deflection response of adobe in case of flexural loading is generally highly unreliable, however, with the addition of wire mesh reinforcement, the flexural strength of adobe can be increased about 3 times.

- It has been indicated through the research findings that the load-deflection curve for adobe reinforced with wire mesh and subjected to flexure can be approximated as a tri-linear curve.
- It is recommended that the same study may be repeated for an adobe mixture designed for optimized compressive strength.
- The same study may also be repeated by increasing the wire mesh layers to 2 and 3 in order to study the resulting effects.

The authors acknowledge the research environment and infrastructure provided by University Malaysia Pahang. The research funding for the present study was also provided by University Malaysia Pahang through research grant number RDU160370, for which the authors are highly grateful.

References

[1] M. Blondet and G. V. Garcia M., *Adobe Construction, Tech. Rep.*, (2003).

[2] Q. Angulo-Ibanez, A. Mas-Tomas, V. Galvan-LLopis and J. L. Santolaria-Montesinos, Const & Bldg Mat, **30**, 389-399 (2012).

[3] L. Keefe, *Earth Buildings - Methods, materials, repair and conservation,* (Taylor & Francis, 2005).

[4] T. Morton, *Earth masony - design and construction guidelines,* (HIS BRE Press, 2008).

[5] E. Fodde, Int J of Arch Heritage, **3**, 2, 145-68, (2009).

[6] E. Adorni, E. Coisson and D. Ferretti, Const & Bldg Mat, **40**, 1-9, (2013).

[7] A. Chasagnoux, *L'architecture voute iranienne-modelisation et simulation par elements finis*. PhD thesis (in French), (Universite de Nantes, 1996).

[8] R. Ghirshman, S. Gilbert and J. Emmons, *Persian Art, The Parthian and Sassanian Dynasties, 249{BC}-{AD}651.*, (Golden Press, 1962).

[9] M. Hejazi, *Historical buildings of Iran: their architecture and structure.*, (Queen Mary and Westfield College University of London, 1997).

[10] A. U. Pope, *Persian Architecture,* (George Braziller, 1965).

[11] H. E. Wulff, *The traditional crafts of Persia,* (MIT Press, 1966).

[12] H. Binici, O. Aksogan, D. Bakbak, H. Kaplan and B. Isik, Const & Bldg Mat, **23**, 2, 1035-1041, (2009).

[13] F. Fratini, E. Pecchioni, L. Rovero and U. Tonietti, App Clay Sc, **53**, 3, 509-516, (2010).

[14] M. C. J. Delgado and I. C. Guerrero, Const & Bldg Mat, **20**, 9, 679-690, (2006).

[15] R. Illampas, I. Ioannou and D. C. Charmpis, Int J of Arch Heritage, **7**, 2, 165-188, (2013).

[16] F. Pacheco-Torgal and S. Jalali, Const & Bldg Mat, **29**, 512-519, (2012).

[17] M. Blondet and G. V. Garcia M., *Earthquake Resistant Earthen Buildings?*, (2004).

[18] Y. Gul, G. A. Jokhio and E. Kakar, *6th Int Civ Engg Cong*, (2013).

[19] L. Baker, *Mud,* **1**, (2015).

[20] K. M. Roufechaei, A. H. Abu Bakar and A. A. Tabassi, J of Cleaner Prod, **65**, 380-388, (2014).

[21] S. H. Sameh, J of Cleaner Prod, **65**, 362-373, (2014).

[22] G. Seyfang, En Pol, **38**, 12, 7624-7633, 2010.

[23] H. Niroumand, M. F. Zain and M. Jamil, *Proc - Soc and Behav Sc,* **89**, 0, 231-236, (2013).

[24] E. Nodarou, C. Frederick and A. Hein, J of Arch Sc, **35**, 11, 2997-3015, (2008).

[25] E. Quagliarini and S. Lenci, J of Cul Heritage, **11**, 3, 309-314, (2010).

[26] R. Walker and H. Morris, *Proc of the Aust Str Engg Conf, Auckland,* vol. 1, pp. 477-84, 1998.

[27] P. Walker, *The Australian earth building handbook,* (Standards Australia, 2002).

[28] NZS, *Materials and workmanship for earth buildings,* (Standards New Zealand, 1998).
[29] NMAC, *New Mexico earthen building materials code,* (New Mex Admin Code, 2009).
[30] P. Walker, Masonry Int, 1-6, (1996).
[31] B. Briccoli, L. Rovero and U. Tonietti, *Proc of the 10th Int Conf on the Study of Conserv of Earthen Arch Heritage, Mali,* 253-8, (2008).
[32] J. C. Morel, A. Pkla and P. Walker, Const & Bldg Mat, **21**, 2, 303-309, (2007).
[33] L. Miccoli, A. Garofano, P. Fontana and U. Muller, Engg Str, **104**, 80-94, (2015).
[34] L. Miccoli, U. Muller and P. Fontana, Const & Bldg Mat, **61**, 327-339, (2014).
[35] J. Morel and A. Pkla, Const & Bldg Mat, **16**, 5, 303-310, (2002).
[36] R. Illampas, I. Ioannou and D. C. Charmpis, Const & Bldg Mat, **53**, 83-90, (2014).
[37] D. Silveira, H. Varum and A. Costa, Const & Bldg Mat, **40**, 719-728, (2013).
[38] D. Silveira, H. Varum, A. Costa, T. Martins, H. Pereira and J. Almeida, Const & Bldg Mat, **28**, 1, 36-44, (2012).
[39] A. Eslami, H. Ronagh, S. Mahini and R. Morshed, Const & Bldg Mat, **35**, 251-260, (2012).
[40] P. Gavrilovic, V. Sendova, W. Ginell and L. Tolles, *11th Eur Conf on EQ Engg*, (2002).
[41] J. Vargas, J. Bariola and M. Blondet, *Resistencia Sismica de la Mampotera de Adobe*, (2000).
[42] A. S. Muntohar, Const & Bldg Mat, **25**, 11, 4215-4220, (2011).
[43] P. Walker, *Cem & Conc Comp,* **17**, 4, 301-310, (1995).
[44] U. Yetgin, O. Cavdar and A. Cavdar, Const & Bldg Mat, **22**, 3, 222-227, (2008).
[45] H. Binici, O. Aksogan and T. Shah, Const & Bldg Mat, **19**, 4, 313-318, (2005).
[46] S. Lenci, Q. Piattoni, F. Clementi and T. Sadowski, Int J of Frac, **172**, 193-200, (2011).
[47] Y. Millogo, J. C. Morel, J. E. Aubert and K. Ghavami, Const & Bldg Mat, **52**, 71-78, (2014).
[48] L. Turnali and A. Saritas, Const & Bldg Mat, **25**, 4, 1747-1752, (2011).
[49] P. Vega, A. Juan, M. Ignacio Guerra, J. M. Moran, P. J. Aguado and B. Llamas, Const & Bldg Mat, **25**, 7, 3020-3023, (2011).
[50] C. Galan-Marin, C. Rivera-Gomez and J. Petric, Const & Bldg Mat, **24**, 8, 1462-1468, (2010).
[51] E. A. Okunade, J of Engg & App Sc, **2**, 9, 1455-59, (2007).
[52] Q. Piattoni, E. Quagliarini and S. Lenci, Const & Bldg Mat, **25**, 4, 2067-2075, (2011).
[53] J. Gere and B. Goodno, *Mechanics of materials, Brief Ed.*, (Cengage Learning, 2011).
[54] S. Popovics, Cem & Conc Res, 583-99, (1973).
[55] F. Wu, G. Li, H. N. Li and J. Q. Jia, Mat & Str, **46**, 9, 1449-1457, (2013).
[56] S. H. Lodi, S. F. A. Rafeeqi, G. A. Jokhio and A. J. Sangi, J of Ferrocement, **36**, 2, 792-798, (2006).
[57] S. H. Lodi, S. F. A. Rafeeqi and G. A. Jokhio, NED Uni J of Res, **7**, 2, 137-151, (2010).
[58] S. P. Shah, M. Sarigaphuti and M. E. Karagular, ACI Sp Pub, **142**, 1-18, (1994).

IOP Conf. Series: Materials Science and Engineering **318** (2018) 012031 doi:10.1088/1757-899X/318/1/012031

The ZnO-FET Biosensor for Cardiac Troponin I

M F M Fathil[1], **M K Md Arshad**[1,2], **M N M Nuzaihan**[1], **Subash C B Gopinath**[1,3], **A R Ruslinda**[1] and **U Hashim**[1,2]

[1] Institute of Nano Electronic Engineering, Universiti Malaysia Perlis, 01000 Kangar, Perlis, Malaysia

[2] School of Microelectronic Engineering, Universiti Malaysia Perlis, 02600 Arau, Perlis, Malaysia

[3] School of Bioprocess Engineering, Universiti Malaysia Perlis, 02600 Arau, Perlis, Malaysia

E-mail : mohamadfaris@unimap.edu.my

Abstract. This paper investigates the influence of substrate-gate coupling on the ZnO-FET biosensor's sensitivity for detection of cardiac troponin I (cTnI), a 'gold standard' biomarker for acute myocardial infarction (AMI). The FET-based device with introduction of substrate-gate coupling on *p*-type silicon-on-insulator (SOI) substrate is fabricated using conventional lithography processes. An *n*-type zinc oxide (ZnO) thin film deposited via electron-beam evaporator is used as transducer for bridging the source and drain regions. Surface modifications via functionalization with 3-aminopropyltriethoxysilane (APTES) and glutaraldehyde (GA) as chemical linkers, followed by immobilization of cTnI monoclonal antibody (MAb-cTnI) as bio-receptor on the ZnO thin film allow different concentration of cTnI detection with high selectivity. The device's sensitivity increases up to 9 %·(g/ml)$^{-1}$ with the increase of the substrate-gate voltage (V_{SG}) up to -10 V at very low limit of detection (LOD) down to 1.6 fg/ml.

1. Introduction

Cardiovascular disease (CVD) is the leading cause of death worldwide with 17.3 million casualties in 2008 [1]. Of those, a large proportion are caused by heart attacks, also known as acute myocardial infarction (AMI). AMI can be defined as a process where heart muscle cells died due to insufficient blood flow when small arteries are blocked causing sudden complete blockage of coronary arteries [2,3]. The permanently damaged heart cells released several cardiac biomarkers (i.e. myoglobin, creatine-kinase, cardiac troponin, and etc.) into the bloodstream [4]. Among these biomarkers, cardiac troponin I (cTnI) is considered as the 'gold standard' due to early concentration to reach peak (2-4 h) [5] and prolonged release in the blood (more than 10 days) [6]. A normal patients exhibits 1 pg/ml of cTnI in their blood, but can escalates to 100 ng/ml during AMI [7] as in Figure 1.

Field-effect transistor (FET)-based biosensors (Figure 2) are presently at the centre of interest for biomolecules detection [8]. The channel's surface, which is directly exposed to the surrounding can sense different surface potential upon biological interaction and modulates the current from drain to source region [9]. Specific immobilized bio-receptors (i.e. antibody, deoxyribonucleic acid (DNA), or enzyme) on the channel through covalent binding with suitable chemical linkers allows biomolecules detection with high selectivity [8]. The influence in electrical behaviour of the channel depends whether positive or negative charge biomolecules and the properties of the semiconductor material used as channel, which caused increment or reduction in current flow from source to drain [8].

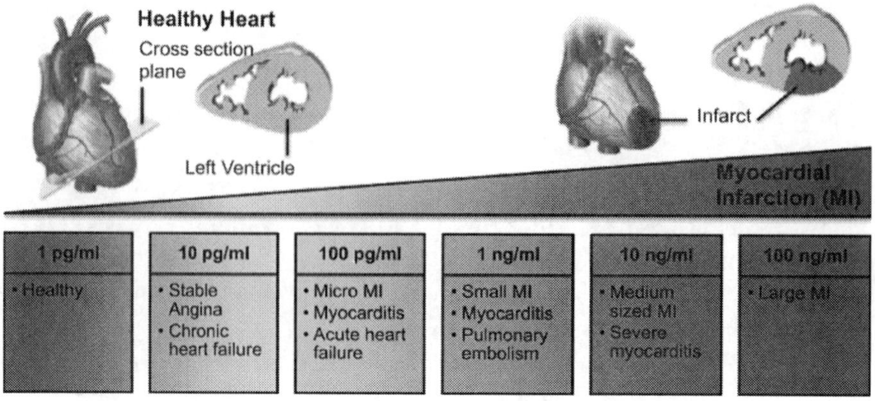

Figure 1. Severity of AMI at different concentration of cTnI [10].

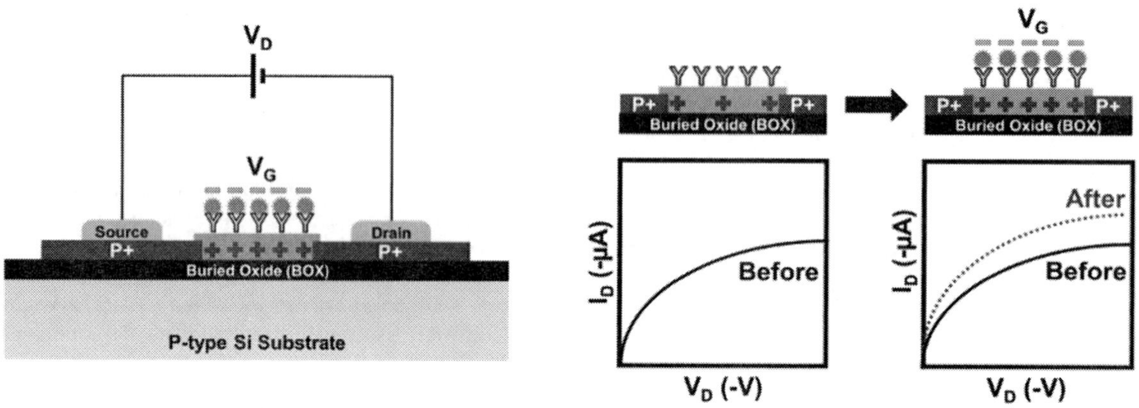

Figure 2. The architecture of the FET-based biosensor with the electrical characteristics during operation [9].

In the recent years, several reports on integration of additional gate in FET-based biosensor had been demonstrated for enhancing performance in biomolecules detection with higher sensitivity [11–14]. Thus, this work presents the impact of the substrate-gate coupling on the sensitivity of zinc oxide (ZnO)-FET biosensor for cTnI target biomarker detection. The device was fabricated on a *p*-type silicon-on-insulator (SOI) substrate through conventional photolithography processes. The surface of *n*-type ZnO thin film as channel was immobilized with cTnI monoclonal antibody (MAb-cTnI), which allow specific capturing of cTnI target biomarker at various concentration. By sweeping the substrate-gate voltage (V_{SG}) with negatively value, the drain current (I_D) values before and after detection of several concentration of cTnI biomarker are measured and analysed to compare the sensitivity of the device at selected V_{SG}. Limit of detection (LOD) of the biosensor is then determined based on the V_{SG} which demonstrates the best sensitivity.

2. Research methodology

2.1. Fabrication of biosensor

The biosensor was fabricated on a 2 × 2 cm silicon-on-insulator (SOI) substrate with thickness of 70 nm, 145 nm, and 1 mm for top-silicon (Si), buried oxide (BOX), and Si substrate layers, respectively

IOP Conf. Series: Materials Science and Engineering **318** (2018) 012031 doi:10.1088/1757-899X/318/1/012031

as shown in **Error! Reference source not found.**(a). The source and drain regions of the FET biosensor were formed through conventional lithography processes (**Error! Reference source not found.**(b)). It was started with positive photoresist (PR) spin-coating on the surface of the SOI substrate via a spin coater at 3000 rpm for 40 s. The coated substrate was then soft-baked on a hot plate at 110°C for 40 s to improve the PR adhesion towards the surface of the SOI substrate. The process was continued by aligning and transferring the source and drain patterns on the chrome mask to the coated substrate using ultraviolet (UV) light for 10 s via a mask aligner system. Then, the substrate was dipped into a PR developer solution for 35 s and rinsed with deionized (DI) water in order to develop the source and drain patterns. The substrate was hard-baked on the hot plate for 2 min at 110°C to evaporate the remaining moisture and increased the PR adhesion after development process. By using inductive couple plasma reactive ion etching (ICP-RIE), the top-Si layer was dry etched for 27 s to form the source and drain regions according to the previously patterned photoresist. The process was followed by stripping the PR from the substrate by using acetone, and subsequent rinsing with DI water.

The substrate-gate region of the biosensor was exposed from the top side of the substrate (**Error! Reference source not found.**(c)). Thus, another photolithography processes were repeated by using a second chrome mask for exposing the substrate-gate region. This time, the etching process was replaced by using wet etching inside buffered oxide etch (BOE) solution for 150 s to selectively remove the undesirable BOX layer, followed by rinsing with DI water.

The ZnO thin film with thickness of 200 nm, which acts as a transducer was deposited on the wafer by using electron-beam evaporator, annealed inside a muffle furnace at 500°C for 2 h, and was slowly cooled down to room temperature for 24 h (**Error! Reference source not found.**(d)). Additional photolithography processes using third chrome mask and wet etch by using hydrochloric acid (HCl) in DI water with ratio of 1:900 were conducted to form the ZnO thin film only at the specific gap between source and drain regions (**Error! Reference source not found.**(e)). The PR was remained on the ZnO thin film in order to protect the film from the next process.

Finally, the nickel/gold (Ni/Au) thin film as electrode pads were deposited on the substrate via a thermal evaporator. The photolithography processes of fourth mask were followed. The Ni/Au thin film was wet-etched by using "aqua regia" solution and rinsed with DI water. The remaining PR was stripped from the surface of the substrate by using acetone and rinsed with DI water (**Error! Reference source not found.**(f)).

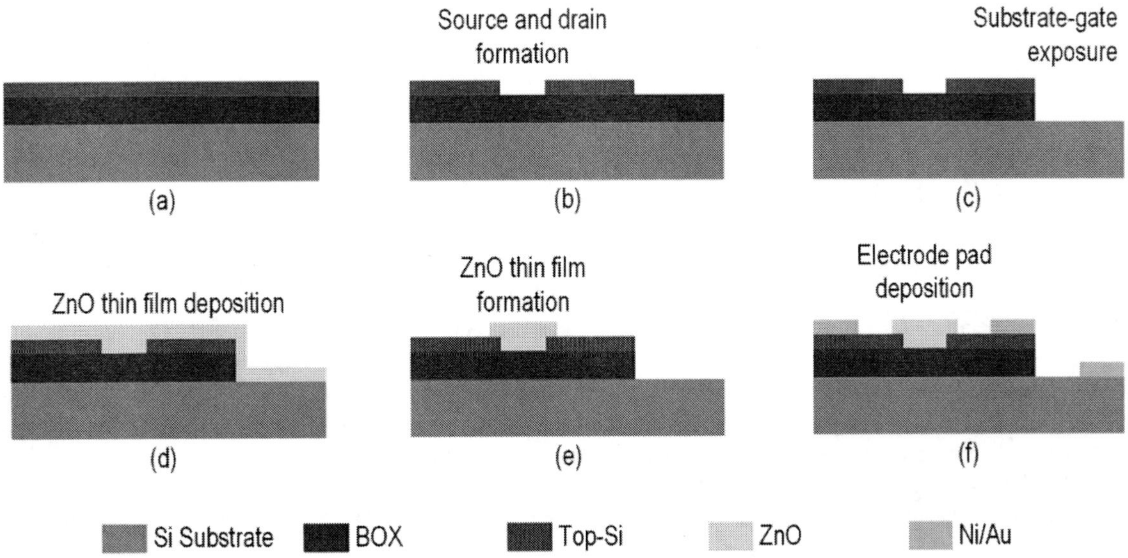

Figure 3. The fabrication process flow of ZnO-FET biosensor via conventional lithography processes. (a) Initial SOI substrate. (b) Source and drain formation via ICP-RIE of Top-Si. (c) Substrate-gate exposure by wet-etching of BOX layer. (d) Deposition of ZnO thin film. (e) Formation of ZnO thin film as channel by wet-etching. (f) Ni/Au deposition and formation as electrode pads.

2.2. ZnO thin film surface modification
The highly selective detection of the cTnI target biomarker requires the surface of ZnO thin film to be modified by using chemistry linkers (i.e. 3-aminopropyltriethoxy-silane (APTES) and glutaraldehyde (GA)) and bio-receptor (cTnI monoclonal antibody (MAb-cTnI)) as in **Error! Reference source not found.**.

Hydroxyl functional groups available on the surface of ZnO were allowed to covalently bind with silane groups from 2% APTES in DI water via immersion for 120 min in room temperature. The surface was then rinsed with DI water to remove remaining unbound APTES. Next, 1 µl of 2.5% glutaraldehyde (GA) in DI water was dropped on the previously produced amine-terminated surface and left in room temperature for 1 h, allowing the free end of APTES contained amine functional groups to covalently bind with aldehyde group in GA. Again, the surface was rinsed with DI water to remove remaining unbound GA.

Finally, 1 µl of highly selective 10 µg/ml MAb-cTnI in 0.01 M phosphate buffered saline (PBS) as bio-receptor was dropped on the functionalized ZnO thin film and left at room temperature for 1 h, thus allowing the amine functional group from the antibody to covalently bind with aldehyde-terminated surface on the ZnO thin film. To remove remaining unbound MAb-cTnI, the surface of the thin film was rinsed with 0.01 M PBS. The biosensor was stored in PBS at 4°C before use in experiment.

2.3. Detection of cTnI
The biosensor was dropped with 1 µl of cTnI target biomarker at several concentrations ranging from 10 pg/ml to 1 µg/ml. Each drop required a waiting time of 10 min prior to washing the device with 0.01 M PBS to remove any unbound cTnI biomarker.

2.4. Electrical characterization
The current versus voltage (I-V) characteristics of the device before and after detection of different cTnI target biomarker concentrations were measured by using Keithley 4200-SCS semiconductor parametric analyser. The drain voltage (V_D) was bias with -1 V, the substrate-gate voltage (V_{SG}) was sweep from 0 to -10 V at step voltage of 0.1 V, while the source was grounded as in Figure 5. The I_{DS} were measured and analysed to understand the effect of (V_{SG}) on the sensitivity of the device's detection.

3. Results and discussions
3.1. Electrical characteristics
The electrical detection of several cTnI target biomarker concentrations by using ZnO-FET biosensor was demonstrated as in Figure 6. By sweeping the V_{SG} from 0 to -10 V, the output I_D had significantly increased.

This is related to the device architecture, which utilized *p-n*-p junctions of drain, channel, and source, respectively from FET devices. The current can only flow from p-type drain to p-type source region when hole conduction layer forms in the n-type ZnO thin film [15,16]. The *p-n*-p junctions allow the detection of biomolecule charges at the n-channel region and increase the current modulation with the aid of gate bias (i.e. substrate-gate) and, hence amplify the output signal.

With the introduction of negatively V_{SG} to the substrate-gate region [11–14], electron carriers accumulate at the BOX/Si substrate interface, attract hole carriers in the ZnO thin film, and follow by formation of hole conduction layer across the channel as in Figure 7(b). This event allows more current flow from drain to source region with the increase of V_{SG} [17,18].

Upon detection of different concentration of cTnI biomarker concentrations, significant reduction in I_D can be observed from -5.04 to -1.35 μA at V_{SG} = -10 V and V_D = -1 V with the increase target concentrations from 10 pg/ml to 1 μg/ml, compared to the initial I_D before detection (I_{D0}) at 7.55 μA. The cTnI target biomarker has an isoelectric point (P_I) of 9.87, which exhibits positive charge when diluted in PBS solution at pH 7.4 [19]. As the cTnI was captured by MAb-cTnI (the bio-receptor, which was previously immobilized on the ZnO thin film), it caused repulsion of hole carriers and diminished the hole conduction channel in the thin film as in Figure 7(c). As a result, the increase concentration of cTnI target biomarker causes reduction of I_D.

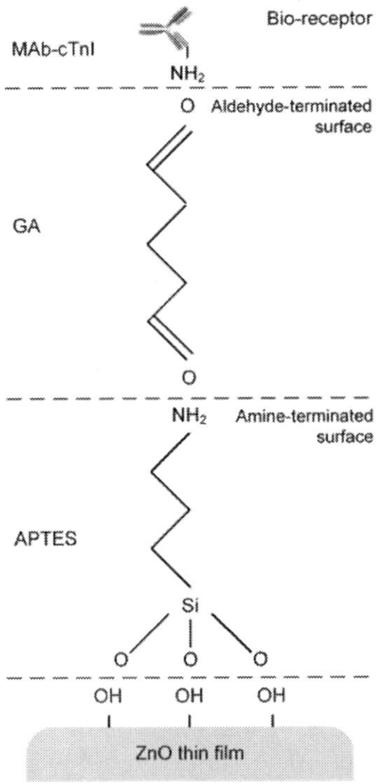

Figure 4. Surface modification of the ZnO thin film with chemical linkers (APTES and GA) and bio-receptor (MAb-cTnI).

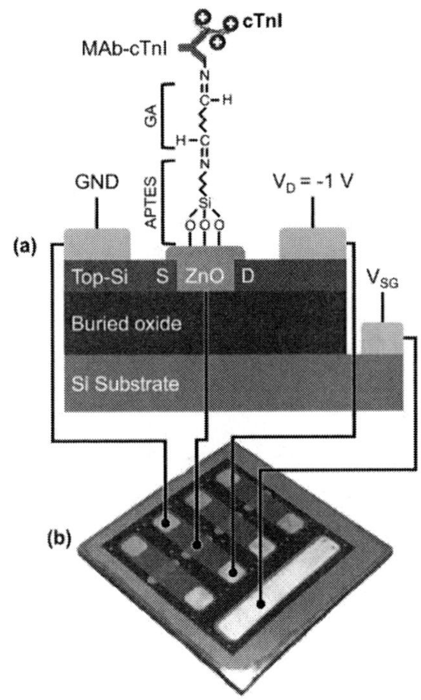

Figure 5. Device architecture of ZnO-FET biosensor. (a) Cross-section view of the device with surface modification. (b) The fabricated biosensor in array.

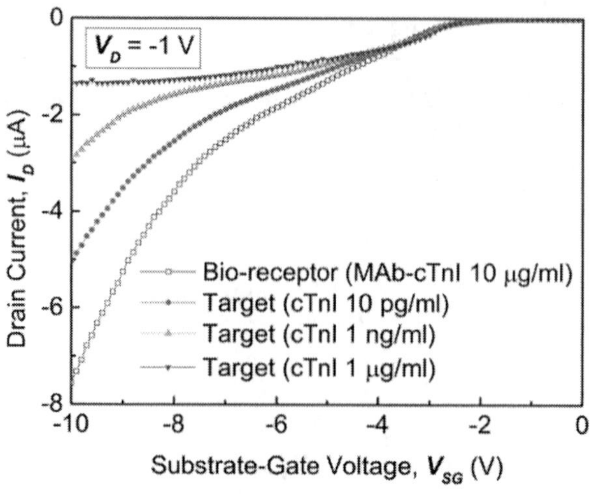

Figure 6. I_D versus V_{SG} of different cTnI target biomarker concentrations ranging from 10 p to 1 μg/ml at V_D = -1 V.

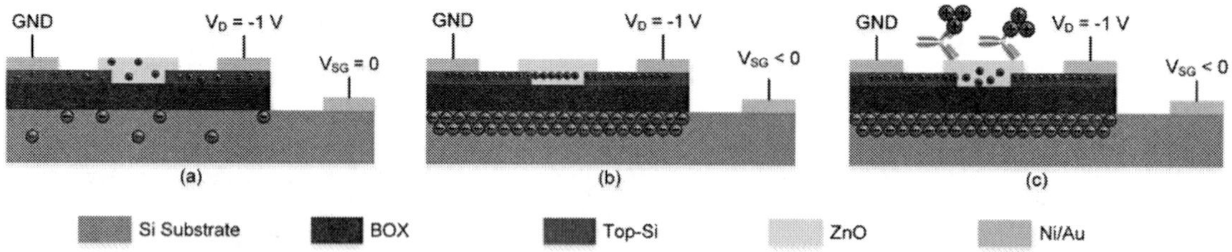

Figure 7. The impact of V_{SG} on the formation of hole conduction layer at the channel region. (a) Without V_{SG}, (b) With $V_{SG} < 0$ V, and (c) during detection of positively charged cTnI target biomarker with $V_{SG} < 0$ V.

3.2. Sensitivity

With more V_{SG} supplied to the device from -6 V to -10 V as in Figure 6, the reduction in I_D also become more apparent. Several I_D values at V_{SG} = -6, -8, and -10 V from Figure 6 were extracted and inserted into Equation (1) to calculate the relative change in I_D as plotted in Figure 8.

$$\text{Relative change in } I_D = \frac{\Delta I_D}{I_{D0}} \times 100 \quad (1)$$

ΔI_D represent the difference between I_D and I_{D0}. From the relative change in I_D, the sensitivity of the device at different V_{SG} can be calculated based on the slope of the calibration curve as in Equation (2).

$$\text{Sensitivity} = \Delta(\Delta I_D/I_{D0})/\Delta(\text{cTnI Concentration}) \quad (2)$$

The results are plotted as in Figure 9 for easier interpretation. The sensitivity of the device had clearly increased double from ~4 to 9.57 %·(g/ml)$^{-1}$ with the increase of V_{SG} from -6 to -10 V. This exceptional result suggest that the V_{SG} biasing has significant impact toward the sensitivity of the ZnO-FET biosensor for the detection of cTnI biomarker.

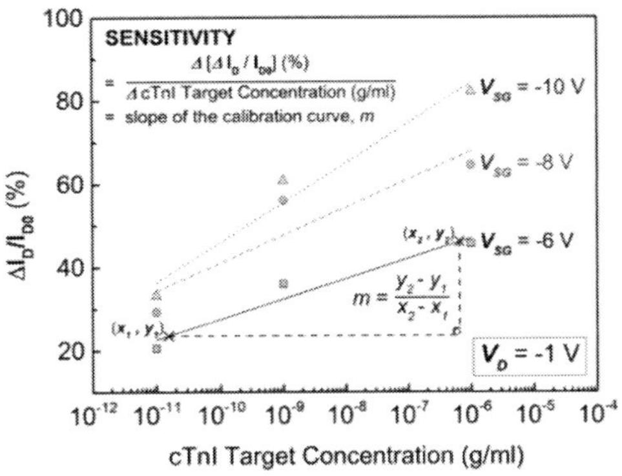

Figure 8. The sensitivity determination of the ZnO-FET biosensor at V_{SG} = -6, -8, and -10 V. The relative change in I_D versus cTnI target biomarker concentrations.

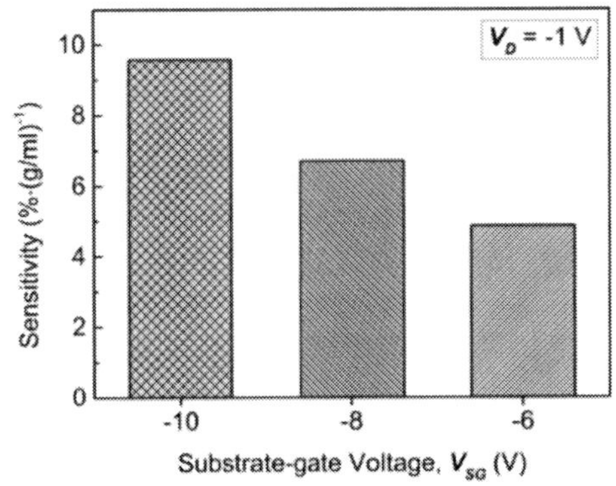

Figure 9. Sensitivity versus V_{SG} of the cTnI target biomarker.

3.3. Limit of detection

The LOD of the device is determined at V_{SG} = -10 V as in Figure 10**Error! Reference source not found.** due to high sensitivity detection as previously demonstrated based on the result from Figure 9. A device with higher sensitivity should give more accurate determination of LOD. It was calculated by replacing the $y_{(LOD)}$ as in Equation 3 into the straight-line equation of the calibration curve as in Equation 4 in order to get the $x_{(LOD)}$, which is the LOD of the device.

$$y_{(LOD)} = 3\sigma \tag{3}$$

$$y_{(LOD)} = m \cdot \log_{10} x_{(LOD)} + C \tag{4}$$

σ, m, and c represent the standard deviation of the y-intercept, the slope, and the y-intercept of the calibration curve, respectively. Thus, LOD of the device is 1.6 fg/ml, which can be considered as among the lowest LOD attained for the detection of cTnI compared to the other previous reports.

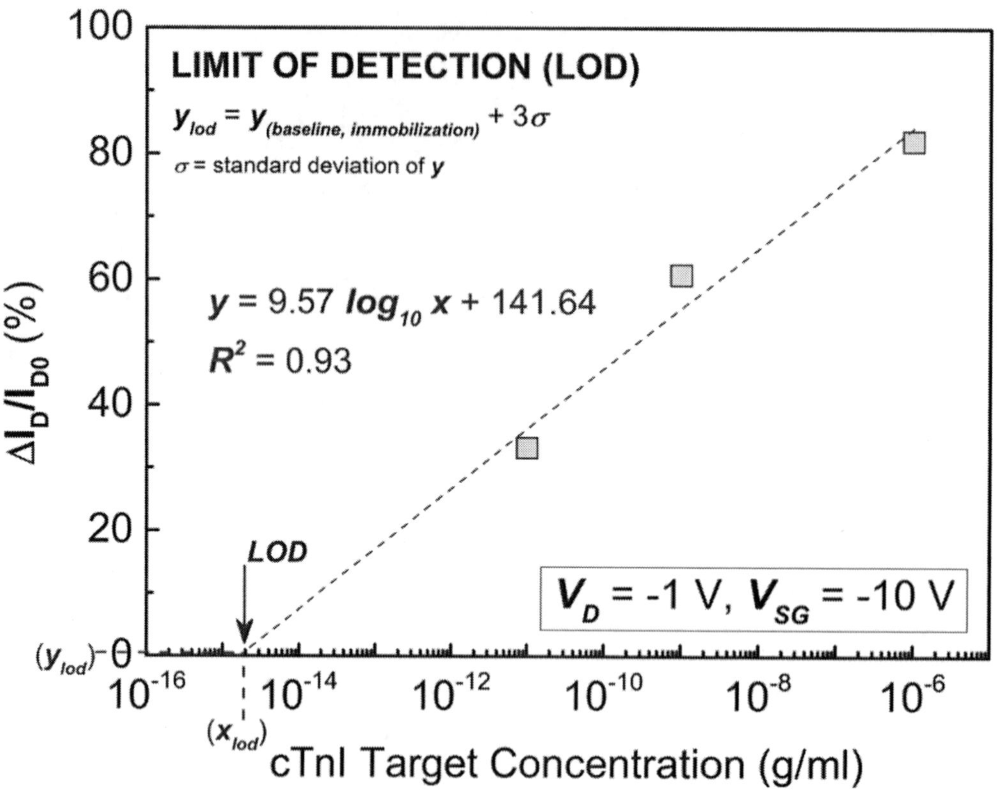

Figure 10. Limit of detection (LOD) of the ZnO-FET biosensor at V_{SG} = -10 V and V_D = -1 V.

4. Conclusion

The substrate-gate coupling has significant impact on the sensitivity of the ZnO-FET biosensor. With the increase of V_{SG} supplied to the substrate-gate region, more current flowed from drain to source region due to formation of better hole conduction layer in the ZnO thin film. As different cTnI biomarker concentrations were captured on the modified surface of ZnO thin film, reduction in I_D were obtained as a result of diminished hole conduction channel due to repulsion with positively charged cTnI captured on the ZnO thin film. The changes are more apparent with the introduction of higher V_{SG}. The device sensitivity had increased double with V_{SG} of -10 V compared to lower V_{SG} of -6 V, suggesting improvement in sensitivity with the introduction of substrate-gate biasing. With better sensitivity, the LOD of the device had managed to reach down to 1.6 fg/ml. This device shows promising potential in detection of cTnI biomarker for determination of AMI.

References

[1] WHO 2014 Cardiovascular diseases *World Heal. Organ.*
[2] Herrick J B 1912 Clinical features of sudden obstruction of the coronary arteries *J. Am. Med. Assoc.* **LIX** 2015
[3] Thygesen K, Alpert J S, Jaffe A S, Simoons M L, Chaitman B R, White H D, Thygesen K, Alpert J S, White H D, Jaffe A S, et al 2012 Third Universal Definition of Myocardial Infarction *J. Am. Coll. Cardiol.* **60** 1581–98
[4] Rosalki S B, Roberts R, Katus H A, Giannitsis E and Ladenson J H 2004 Cardiac Biomarkers for Detection of Myocardial Infarction: Perspectives from Past to Present *Clin. Chem.* **50** 2205–13
[5] Burcu Bahadır E and Kemal Sezgintürk M 2015 Applications of electrochemical immunosensors for early clinical diagnostics *Talanta* **132** 162–74

[6] Wright R S, Anderson J L, Adams C D, Bridges C R, Casey D E, Ettinger S M, Fesmire F M, Ganiats T G, Jneid H, Lincoff A M, et al 2011 2011 ACCF/AHA Focused Update of the Guidelines for the Management of Patients With Unstable Angina/ Non-ST-Elevation Myocardial Infarction (Updating the 2007 Guideline): A Report of the American College of Cardiology Foundation/American Heart Association *Circulation* **123** 2022–60

[7] Agewall S, Giannitsis E, Jernberg T and Katus H 2011 Troponin elevation in coronary vs. non-coronary disease *Eur. Heart J.* **32** 404–11

[8] Sarkar D, Liu W, Xie X, Anselmo A C, Mitragotri S and Banerjee K 2014 MoS 2 Field-Effect Transistor for Next-Generation Label-Free Biosensors *ACS Nano* **8** 3992–4003

[9] Liu S and Guo X 2012 Carbon nanomaterials field-effect-transistor-based biosensors *NPG Asia Mater.* **4** e23

[10] Fathil M F M, Md Arshad M K, Gopinath S C B, Hashim U, Adzhri R, Ayub R M, Ruslinda A R, Nuzaihan M.N. M, Azman A H, Zaki M, et al 2015 Diagnostics on acute myocardial infarction: Cardiac troponin biomarkers *Biosens. Bioelectron.* **70** 209–20

[11] Dai P, Gao A, Lu N, Li T and Wang Y 2013 A back-gate controlled silicon nanowire sensor with sensitivity improvement for DNA and pH detection *Jpn. J. Appl. Phys.* **52** 121301

[12] Buitrago E, Badia M F-B, Georgiev Y M, Yu R, Lotty O, Holmes J D, Nightingale A M, Guerin H M and Ionescu A M 2014 Electrical characterization of high performance, liquid gated vertically stacked SiNW-based 3D FET biosensors *Sensors Actuators B Chem.* **199** 291–300

[13] Fernandes P G, Chapman R A, Seitz O, Stiegler H J, Wen H-C, Chabal Y J and Vogel E M 2012 Effect of Back-Gate Biasing on Floating Electrolytes in Silicon-on-Insulator-Based Nanoribbon Sensors *IEEE Electron Device Lett.* **33** 447–9

[14] Park J-K, Jang H, Park J and Cho W 2014 SOI dual-gate ISFET with variable oxide capacitance and channel thickness *Solid. State. Electron.* **97** 2–7

[15] Bergveld P 2003 Thirty years of ISFETOLOGY *Sensors Actuators B Chem.* **88** 1–20

[16] Sze S M and Ng K K 2006 *Physics of Semiconductor Devices* (Hoboken, New Jersey: John Wiley & Sons)

[17] Fathil M F M, Arshad M K M, Ruslinda A R, Adzhri R, Hashim U, Nuzaihan M.N. M and Ibau C 2016 The effect of substrate-gate bias on the zinc oxide field-effect transistor for biosensing application *2016 IEEE International Conference on Semiconductor Electronics (ICSE)* (IEEE) pp 125–8

[18] Fathil M F M, Md Arshad M K, Ruslinda A R, Gopinath S C B, Nuzaihan M.N. M, Adzhri R, Hashim U and Lam H Y 2017 Substrate-gate coupling in ZnO-FET biosensor for cardiac troponin I detection *Sensors Actuators B Chem.* **242** 1142–54

[19] Eriksson S 2003 Negative Interference in Cardiac Troponin I Immunoassays from a Frequently Occurring Serum and Plasma Component *Clin. Chem.* **49** 1095–104

Acknowledgments
The authors are grateful to the Department of Higher Education, Ministry of Higher Education, (KPT) for funding this research through the Fundamental Research Grant Scheme (FRGS; with the Grant number 9003-00536). The authors also would like to acknowledge all the team members in Institute of Nano Electronic Engineering (INEE), Universiti Malaysia Perlis (UniMAP) for their guidance and help.

IOP Conf. Series: Materials Science and Engineering **318** (2018) 012032 doi:10.1088/1757-899X/318/1/012032

A Low Power 2.4 GHz CMOS Mixer Using Forward Body Bias Technique for Wireless Sensor Network

C.J. Yin[1], S.A.Z Murad[1], A. Harun[1], M.M. Ramli[1], T.Z.A Zulkifli[2] and J. Karim[3]
[1]School of Microelectronic Engineering, Campus Pauh Putra, Universiti Malaysia Perlis, 02600 Arau, Perlis, Malaysia
[2] Electrical and Electronic Engineering Department, Universiti Technologi PETRONAS, 32610 Bandar Seri Iskandar, Perak, Malaysia
[3]Fakulti Kejuruteraan Elektrik, Universiti Teknologi MARA (UiTM), 40450 Shah Alam, Selangor, Malaysia

E-mail: sohiful@unimap.edu.my

Abstract. Wireless sensor network (WSN) is a highly-demanded application since the evolution of wireless generation which is often used in recent communication technology. A radio frequency (RF) transceiver in WSN should have a low power consumption to support long operating times of mobile devices. A down-conversion mixer is responsible for frequency translation in a receiver. By operating a down-conversion mixer at a low supply voltage, the power consumed by WSN receiver can be greatly reduced. This paper presents a development of low power CMOS mixer using forward body bias technique for wireless sensor network. The proposed mixer is implemented using CMOS 0.13 μm Silterra technology. The forward body bias technique is adopted to obtain low power consumption. The simulation results indicate that a low power consumption of 0.91 mW is achieved at 1.6 V supply voltage. Moreover, the conversion gain (CG) of 21.83 dB, the noise figure (NF) of 16.51 dB and the input-referred third-order intercept point (IIP3) of 8.0 dB at 2.4 GHz are obtained. The proposed mixer is suitable for wireless sensor network.

1. Introduction

Wireless sensor network (WSN) have grown enormously in the technology markets due to the rapid evolution of portable light weight devices. It can be made to real time monitoring to gather all of the required information of an object at a certain area. WSN are motivated based on its reliability, accuracy, flexibility, expense and power consumption. Battery usage does play a major concern in the WSN devices [1]. Therefore, power consumption reduction is needed to be made to enhance battery endurance ability as it is important in WSN technology for allowing long operating lifetime [2]. A proper architecture for the main building block is needed in order to achieve low power consumption for the overall WSN.

A mixer is one of the circuit blocks in RF transceiver. The mixer is employs to convert high frequency to low frequency spectrum or low frequency to high frequency spectrum that generated by the multiplication of two signals [2]. Nowadays, a low power mixer design is needed to accomplish the desired of the future market for WSN technology. However, when the circuit design is moving towards power consumption, the other parameters will generally impact the overall performance of the system. Therefore, other important parameters such as conversion gain, linearity, port-to-port isolation and noise figure need to be considered in order to design high performance mixers [3].

2. Design procedure

Figure 1 shows the steps of designing a CMOS 2.4 GHz downconversion mixer using forward body bias technique for wireless sensor network in 0.13 μm Silterra technology. The schematic design and simulation will be done using Cadence software. The first stage in the design process is to estimate the values for each components element which are the design specification and component parameters of the mixer. Basically, the design procedure of the mixer comprise of optimizing several simulations until a desired mixer performance is achieved. The elements that are reflecting the mixer performance are conversion gain, linearity, power consumption and noise figure [4]. The circuit will be redesigned and/or properly readjusted the parameters if it does not meet the specification. By adjusting the circuit design as to optimize a particular performance parameter may serve to unintentionally degrade the other performance parameters. It is therefore important to monitor all of the performance parameters throughout the design process [5]. It is then followed by layout design implementation.

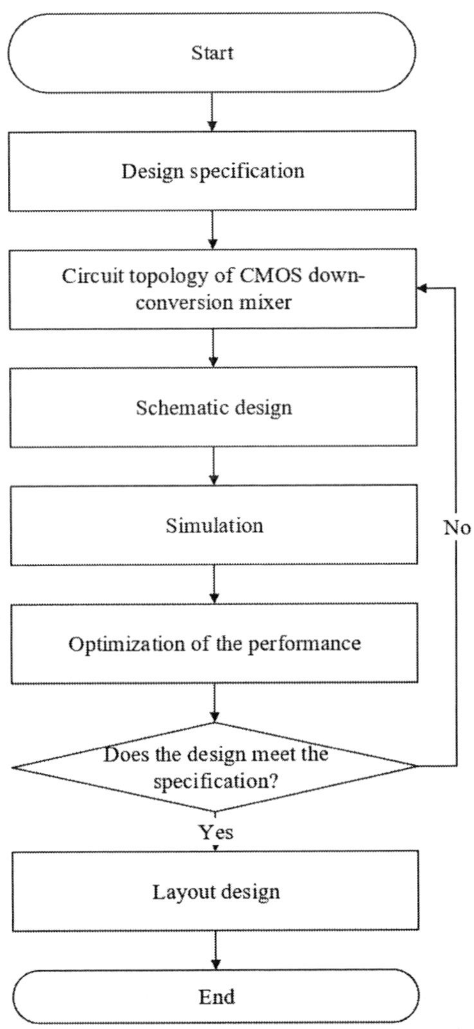

Figure 1. Design and simulation flow chart.

IOP Conf. Series: Materials Science and Engineering **318** (2018) 012032 doi:10.1088/1757-899X/318/1/012032

3. Forward body bias technique

The transistor can operate under lower voltage using a forward body bias technique. This is an effective ways to consume less power. The voltage accros the gate and the source of a MOS transistor would drop by having a low supply voltage. Therefore, low threshold are needed to operate MOSFETs at low supply voltage. A low threshold voltage transistors utilizing current CMOS technologies can be obtained by circuit design techniques. A reverse or forward bias between the body (substrate) and source of a MOSFET can be applied when it is used as a four-terminal device [6].

Under the forward body bias technique, the MOSFET threshold voltage is reduced. Hence, the body-source junction of the NMOS transistor in the mixer will have a negligible leakage current. Body bias is the voltage source that is connected to the body terminal of transistors rather than to power or ground [7]. Figure 2 shows the forward body biase connection for NMOS and PMOS transistor. For this technique, the body is connected to a positive voltage source which allows for the reduction of the threshold voltage of the NMOS transistors. In this design, the biasing voltage will be set to a value that allows the transistors to operate in saturation region. The transistors with forward body bias would enter strong inversion region, while those with zero body bias is in weak inversion region. It is therefore resulting in low supply voltage and subsequently low power consumption [8].

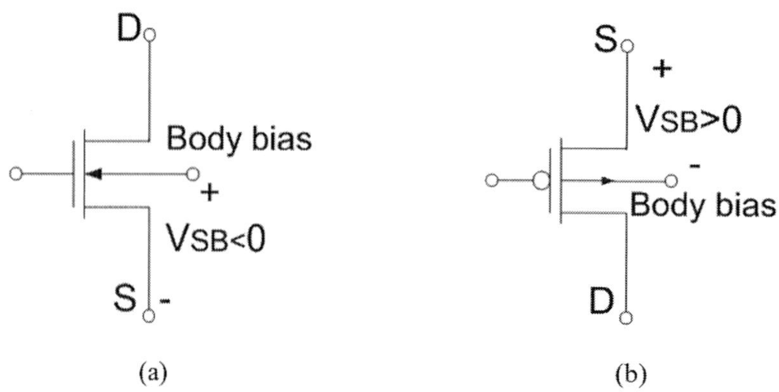

Figure 2. A forward body biased (a) NMOS transistor, (b) PMOS transistor.

4. Circuit implementation

Figure 3 shows the proposed schematic of 2.4 GHz CMOS down-conversion mixer. The proposed circuit consists of three stages which are transconductance stage, switching stage and output stage.

Transistor MI and M2 represent the transconductance (RF) stage. The transconductance stage obtains an RF signal and the output current is matching to a voltage of the RF signal. Transistors M3 to M6 represent the switching (LO) stage. The purpose of having switching stage as a frequency translation of RF signal into an intermediate frequency (IF) signal by converting the current which produce from transconductance stage of the circuit in response to local oscillator (LO) signal. The switching NMOS transistors' body is connected to a bias voltage (VBULK) in the circuit for forward body bias technique. The purposed technique is used to decrease the threshold voltage of the transistors and thus reduce the supply voltage and subsequently power consumption

Inductors (L1, L2, L3 and L4) and capacitors (C1, C2, C3 and C4) act as input matching of 50 Ω in order to match the impedance between input port and transconductance stage to force the load impedance to be the complex conjugate of the source impedance. Therefore, the maximum power can be transferred to the load. Furthermore, low power consumption and high conversion gain can be optimized at the load stage through resistor R1 and R2.

Cascade LC network, which is also known as resonant circuit, consists of inductors (L5-L8) and capacitor (C5-C12) are connected at the IF output stage to function as low pass filter (LPF). The undesired high frequency at the output can be removed through LPF at the output stage to allow signal to pass through with low frequency to produce low level power consumption. Voltage headroom problems occur through higher current at the switching transistor. The large current reduces load size and thus reduces the voltage conversion gain of the mixer [7]. Therefore, current-bleeding PMOS transistors (M7 and M8) are introduced in LO stage to reduce the bias current of the LO switches. The current-bleeding also can increase the current flow in RF transconductance stage to improve the conversion gain [8]. PMOS transistors are used instead of NMOS due to the fact that PMOS is less prone to noise. Inductor L13 and L14 are placed at the source of RF input transistor to function as source degeneration inductor. The inductive source degeneration technique is utilized to improve linearity performance and provide the input impedance matching [9]. Transistor M0 is used as a current source in a double-balanced differential topology to control mixer tail current so as to adjust conversion gain of transconductance stage.

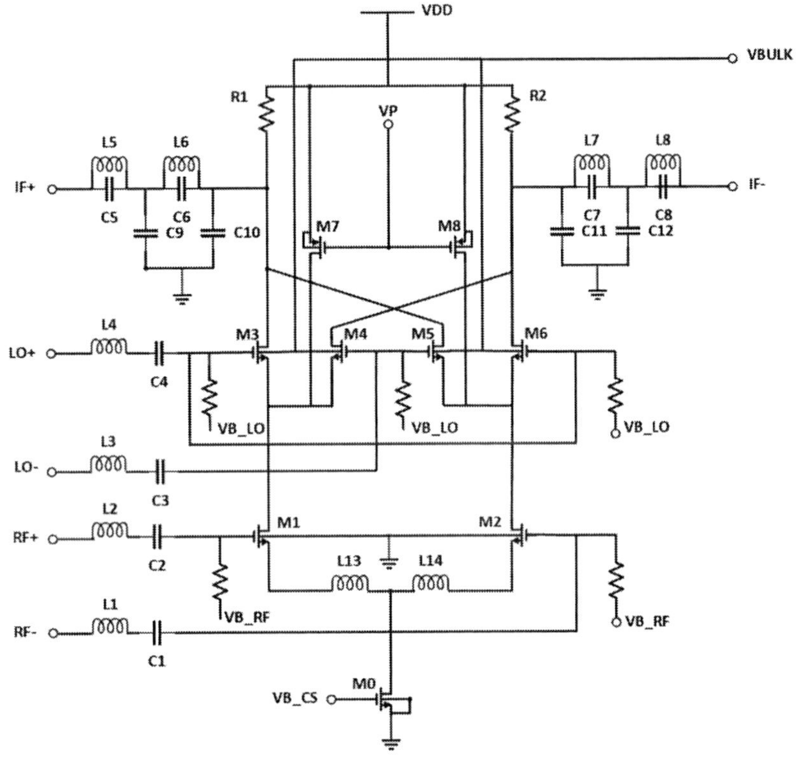

Figure 3. Schematic of the proposed 2.4 GHz CMOS mixer using forward body bias technique.

Table 1 shows the list of biasing voltage, transistors' size and all components values of the proposed mixer. The supply voltage 1.6 V is chosen to obtain good conversion gain. The biasing is selected below 1 V to ensure low power consumption. The forward body bias technique for M3, M4, M5 and M6 is connected with V_BULK 0.6 V. The value of R1 and R2 is optimized at 2 kΩ for good conversion gain. The value of C1, C2, C3 and C4 are good enough as an input matching and DC block.

Table 1. List of biasing, transistors size and component.

Components	Parameters
VDD	1.6 V
VBULK	0.6 V
VB_RF, VB_LO	0.8 V
VP, VB_CS	0.8 V
R1, R2	2 kΩ
C1, C2, C3, C4	10 pF
L1, L2	20 nH
L3,L4	1 nH
M0	2.5 μm / 0.13 μn
M1, M2	6.0 μm / 0.13 μn
M3,M4,M5,M6	3.5 μm / 0.13 μn
M7,M8	2.5 μm / 0.13 μn

The transistors size of transconductance stage is chosen large than switching stage transistors size as a tradeoff between conversion gain and linearity. The size of transistors M7 and M8 are optimized for providing enough bias current to switching stage.

5. Simulation result

A 2.4 GHz RF input signal mixer is down-converted to a lower frequency IF output signal through a LO signal of 2.3 GHz using CMOS 0.13 μm process. All the parasitic effects from resistors, capacitors and inductors are taken into consideration during simulation.

The conversion gain (CG) corresponding to the LO input power is obtained by using a small signal analysis in Cadence SpectreRF. Figure 4 shows the CG of the proposed mixer. It is observed that the mixer provides the conversion gain when the LO input power is varied between −20 dBm to 10 dBm. The higher conversion gain of 21.83 dB is obtained at LO input power of 1 dBm. The linearity of the switching transistor pairs can become worse for large LO power. Therefore, a LO input power of 0 dBm is chosen.

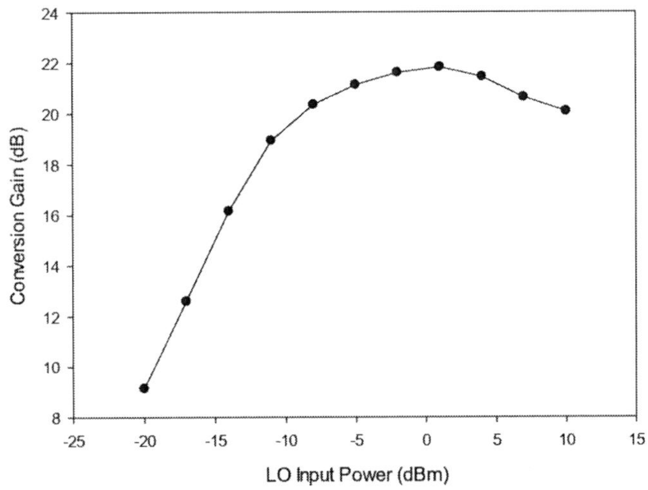

Figure 4. Conversion gain of the proposed mixer

Figure 5 shows the simulated noise figure of the proposed mixer from 100 Hz to 100 MHz. It can be seen that the noise figure of 16.51 dB is obtained at 10 MHz frequency.

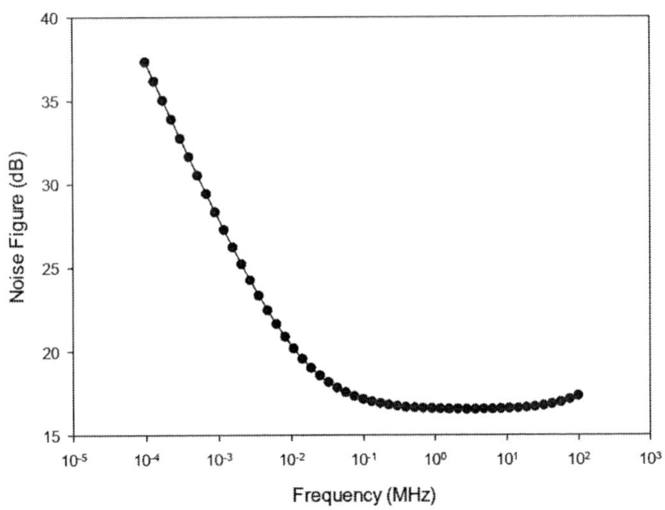

Figure 5. Noise figure of the proposed mixer.

The performances of linearity are measured by P1dB compression point and the third-order intercept point (IIP3). P1dB and IIP3 are simulated by using small signal analysis with the RF input power is swept from −70 dB to 10 dB. Figure 6 depicts the extrapolated P1 dB compression point of −10 dBm is achieved. The two tone test of IIP3 performance is employed. The second tone that is a small tone close to the RF frequency is applied. A 2.40 GHz and 2.41 GHz input frequency is choose at 10 MHz separation with -20 dBm RF input power and 0 dBm LO power is applied, respectively. Figure 7 represents the simulated result of IIP3. The IIP3 of 8 dBm is achieved.

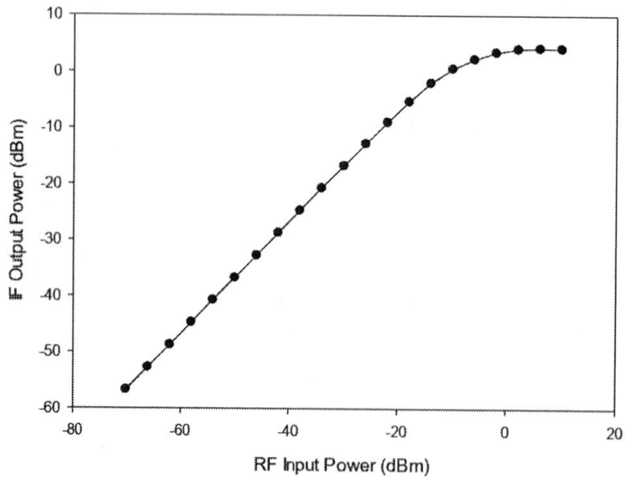

Figure 6. 1-dB compression point.

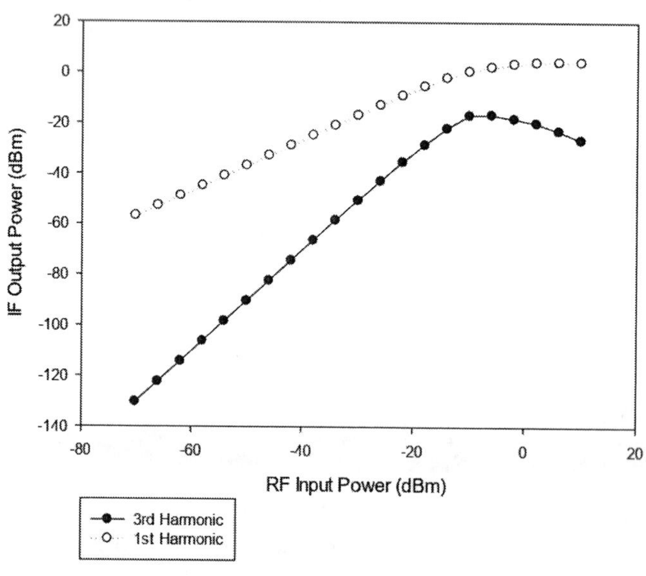

Figure 7. Third-order intercept point (IIP3).

Figure 8 depicts the proposed mixer layout which performed in Cadence Virtuoso Layout. The complete layout is designed symmetrically for differential operation to avoid mismatches in the differential circuit. The parasitic capacitance from the input and output interconnects can be

minimized using top 6 metal. A low parasitic effect can be obtained for longer distance from a substrate. The proposed mixer layout occupied a total die area of 1.96 mm x 1.71 mm including the pads. The fabrication process could take place in the near future.
In the case of RF circuits, the performance of the implemented design is to a great extent dependent on the way the circuit is laid out. This is mainly due to the fact that the parasitic coupling between interconnects and devices and/or the substrate is relatively high. In addition, RF circuits are more sensitive to parasitic resistance, inductance and capacitances associated with the interconnects. These undesired couplings and parasitics would affect the performance of the circuit [12].

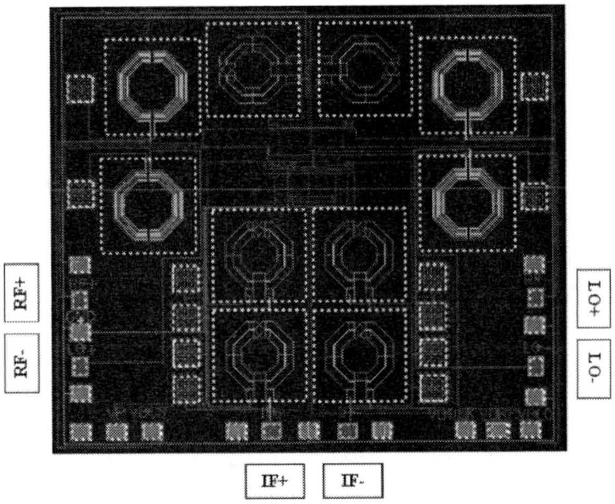

Figure 8. Layout of the proposed mixer.

Table 2 shows the proposed mixer performances are compared with the previously published works. As can be seen, the forward body bias technique could achieve low power consumption with good CG and linearity as compare with the previous work. However, the proposed mixer's noise figure is slightly higher. Therefore, the proposed mixer can obtain low power, high CG and good linearity with other performances are comparable.

Table 2. Performance comparison.

Reference	[13]	[14]	[15]	[16]	[17]	This work
Technology (μm)	0.13	0.18	0.13	0.13	0.18	0.13
Supply voltage (V)	1.2	1.8	1.3	1.8	1.0	1.6
RF (GHz)	2.4	2.4	2.4	2.4	2.4	2.4
LO power (dBm)	0	5	4	N/A	-13	5
Conversion gain (dB)	15.2	25	12.9	8.8	13.8	21.8
IIP3 (dBm	-6.7	-24.7	0	0	-7.2	8
Noise figure (dB)	16.3	19.1	7.40	5.2	13.3	16.5
Power consumption (mW)	6.0	8.7	2.8	1.6	0.5	0.9
Chip area (mm^2)	0.12	0.05	N/A	1.13	N/A	3.35

6. Conclusion

A 2.4 GHz low power CMOS mixer using forward body bias technique for wireless sensor network is presented. The proposed mixer is implemented using CMOS 0.13 μm Silterra technology. A low power consumption is obtained using forward body bias technique. The simulation results show that a low power consumption of 0.91 mW is achieved at 1.6 V supply voltage with the conversion gain (CG) of 21.83 dB, the noise figure (NF) of 16.51 dB and the input-referred third-order intercept point (IIP3) of 8.0 dB at 2.4 GHz are obtained. The proposed mixer is suitable for wireless sensor network.

Acknowledgments

This work is supported and contributed from FRGS grant (9003-00387) that enabled the production of this article.

References

[1] T. H. Lin, W. J. Kaiser and G. J. Pottie 2004 Integrated Low-Power Communication System Design for Wireless Sensor Networks *IEEE Comm. Magazine* 142-150

[2] Murad, S. A. Z., et al. 2013 High gain 2.4 GHz CMOS low noise amplifier for wireless sensor network applications. *Conf. IEEE Inter. RF and Microwave (RFM)*

[3] Murad, S.A.Z, Shahimin, M.M., Pokharel, R.K., Kanaya, H. and Yoshida, K. 2012 Linearity improvement of 5.2-GHz CMOS up-conversion mixer for wireless applications. *Micro. and Opt. Tech. Lett.* **54** 923-925

[4] Jouri. M, Golmakani Abbas, Yahyabadi. M and Khosrowjerdi. H. 2010 Design and simulation of a down-conversion CMOS mixer for UWB Applications *Proc. Electrical Engineering/Electronics Computer Telecommunications and Information Technology (ECTI-CON)*

[5] Y. H. Dai and Christina F. Jou 2013 A 5.15-5.825 GHz high-gain and low-noise cmos down-conversion mixer with low-power for WLAN 802.11a *Proc. IEEE MTT-S International Microwave Workshop Series on RF and Wireless Technologies for Biomedical and Healthcare Applications (IMWS-BIO)*

[6] Anishaziela Azizan and Sohiful Anuar Zainol Murad 2016 A 0.3 mW 2.4 GHz Low Power Low Noise Amplifier using Forward Body Bias Technique for Wireless Sensor Network *J. Teknologi (Sciences & Engineering)* 31-37

[7] Murad, S. A. Z., et al. 2017 Ultra-Low Power CMOS RF Mixer for Wireless Sensor Networks Application: A Review *MATEC Web of Conf. 97, EDP Sciences*

[8] Azizan, A., et al. 2015 Design of a 2.4 GHz CMOS LNA using two-stage forward body bias technique for WSN application *Conf. IEEE Research and Development (SCOReD)*

[9] F. Jiang 2013 *Broadband Low-Noise CMOS Mixers for Wireless Communications* (Ontario, Canada)

[10] Ramiah, Harikrishnan, Jeevan Kanesan, and Tun Zainal Azni Zulkifli 2013 A CMOS up-conversion mixer in 0.18 μm technology for IEEE 802.11 a WLAN application. *IETE J. of Res.* **59**, 454-460

[11] T.-Y. Y. Myoung-Gyun Kim 2014 Analysis and Design of Feedforward Linearity-Improved Mixer Using Inductive Source Degeneration *IEEE Trans. On Micro The. and Tech.* **62**, 323-331

[12] Y. H. Chang, C. Y. Huang and Y. C. Chiang 2012 A 24 GHz down-conversion mixer with low noise and high gain. *Conf. Microwave Integrated Circuits (EuMIC)*

[13] Chong, W. K., et al. 2014 Design of inductorless, low power, high conversion gain CMOS subharmonic mixer for 2.4 GHz application. *Conf. IEEE Inter. Microwave and RF*

[14] Patil, Santosh B., and Rajendra D. Kanphade 2015 A 2.4 GHz double balanced differential input single output low power transmitting mixer in TSMC 180nm CMOS RF process. *Int. Conf. on IEEE Electronics and Communication Systems*

[15] Mahmoud, Fatma M., and M. A. Abdelghany 2016 Low flicker-noise RF CMOS gilbert-cell mixer for 2.4 GHz wireless communication systems. *Int. Conf. on IEEE Electrical, Electronics, and Optimization Techniques*

[16] Murad, S. A. Z., et al. 2016 Low noise figure 2.4 GHz down conversion CMOS mixer for wireless sensor network application. *IEEE Student Conf. on Research and Development (SCOReD)*

[17] Harisankar, P. S., Mayank Chakraverty, and Vaibhav Ruparelia 2016 Analysis of process and temperature variations on a 2.4 GHz RF CMOS down conversion subthreshold mixer *Int. Conf.on IEEE Energy Efficient Technologies for Sustainability (ICEETS)*

IOP Conf. Series: Materials Science and Engineering **318** (2018) 012033 doi:10.1088/1757-899X/318/1/012033

Frequency Selective Surface for Structural Health Monitoring

Farzana Hazira Wan Mustaffa, Saidatul Norlyana Azemi, Mohd Faizal Jamlos, Azremi Abdullah Al-Hadi, Ping Jack Soh

Advanced Communication Engineering Centre (ACE) CoE, School of Computer and Communication Engineering, Universiti Malaysia Perlis (UniMAP), Pauh Putra, Arau 02600, Malaysia.

E-mail: snorlyana@unimap.edu.my

Abstract. Structural health monitoring (SHM) technologies have attained attention to monitor civil structures. SHM sensor systems have been used in various civil structures such as bridges, buildings, tunnels and so on. However the previous sensor for SHM is wired and encounter with problem to cover large areas. Therefore, wireless sensor was introduced for SHM to reduce network connecting problem. Wireless sensors for Structural Health monitoring are new technology and have many advantages to overcome the drawback of conventional and wired sensor. This project proposed passive wireless SHM sensor using frequency selective surface (FSS) as an alternative to conventional sensors. The electromagnetic wave characteristic of FSS will change by geometrical changes of FSS due to mechanical strain or structural failure. The changes feature is used as a sensing function without any connecting wires. Two type of design which are circular ring and square loop along with the transmission and reflection characteristics of SHM using FSS were discussed in this project. A simulation process has shown that incident angle characteristics can be use as a data for SHM application.

1. Introduction

The importance of monitoring the health of civil structures has attained attention over the past two decades. Extreme catastrophes such as earthquakes can cause massive damage to the civil structures which resulting in life threatening conditions evolving in the structures either immediately after the impact or long after the actual calamity has happened [1]. To overcome this problem, structural health monitoring (SHM) systems have been introduced since the systems have great potential to enhance the regular operation and maintenance of structures like tunnels, buildings, bridges and others. The purposes of the SHM are damage analysis, damage localization and estimating the remaining lifetime of the structure [2]. The damage was determined by changes in the modal parameters like frequencies which were gained from measured time histories. From this, any incidence can be quickly detected and evaluated. In addition, SHM is an arising technology which combines advanced sensor technology with intelligent algorithms to interrogate the health condition of structures in real time or whenever necessary [3].

Currently, the common sensor that has been found in SHM system is fiber optic sensors. This type of sensor can sense temperature and normal strain. Due to small in size, light weight, durability and other characteristics, fiber optic sensors have been widely used in SHM system. Some other potential sensor technologies are using piezoelectric materials. This type of sensor can be directly integrated into a structure. However, considering the complexity and high cost to support equipment, both of these sensors are not the suitable sensors to cover large area and harsh environment. As an improvement to the presently available sensor, this project purpose Frequency Selective Surface as a passive sensor for SHM.

In recent times, frequency selective surfaces (FSSs) have shown potential as structural health monitoring (SHM) sensors. The structure of an FSS is made from an array of metallic elements that provides a filtering response to pass or reject electromagnetic waves at a particular band of frequency. Filtering response was caused by capacitive and inductive coupling between the elements which depends on the FSS geometry such as element shape, dimensions and spacing. It is also depend on local environment of FSS will change by geometrical variation of FSS due to structural failure or mechanical strain. Henceforth, the dependency of FSS can be employ for SHM sensing.

In this paper, passive wireless sensor for strain and structural failure are made using a frequency selective surface (FSS). Frequency Selective Surfaces (FSSs) are made up of planar periodic structures with slot component placed on a substrate. The surface implemented with identical patch or aperture conducting elements that repeating periodically in either a one- or two-dimensional array [4]. It is capably acted as band pass or band stop filter respectively to electromagnetic waves which allowing waves at only the desired incidence angles to propagate.

The electromagnetic characteristic of FSS was studied based on geometric shape changes. The proposed design will able to monitor any structural movement based on the incident angle. An important requirement for both dependent and independent incident angle was taken in two polarizations either Transverse Electric (TE) or Transverse Magnetic (TM).

2. Frequency Selective Surface as Sensor

Frequency Selective Surface (FSS) are planar periodic structures of identical metal-dielectric in either one or two dimensional array structures which perform spatial filtering of microwave energy. The array of element can determine a spatial electromagnetic filter and exhibits capacitive and inductive frequency characteristics. FSS are normally composed of a conducting sheet periodically with apertures or an array of metallic patches that are designed to reflect and transmit signals at the surface from certain frequencies.

Aperture FSSs, dealing as high pass filters because it transmit at higher frequencies and reflect back the signals at lower frequencies radiation which produce capacitive response [5]. Regardless of patch type FSSs, it transmitting at lowers frequencies and reflects back with higher frequencies radiation and act as low pass filters illustrated in Figure 1. Therefore it has shown the acquirement of low pass filter. Geometry of the structure in single period known as unit cell used to determine the resonant frequency response.

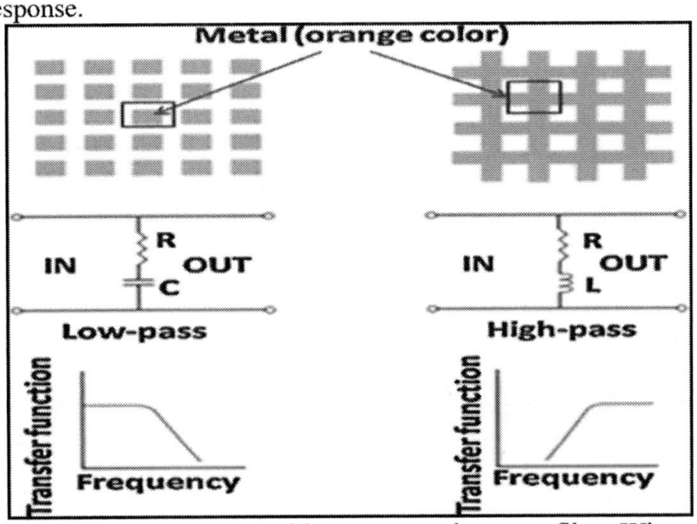

Figure 1. The patch-array produces a capacitive response, low pass filter. Whereas b) the slot-array in inductive, high pass filter. [6]

The electromagnetic performance of FSS might also rely upon the polarization and the angles of the electromagnetic waves arrive at the structure [6]. According to broad variety of possible elements used to realize FSS arrays such as square, hexagon, cross dipole and others that having different shape of element will make FSS tend to have different frequency respond characteristics either low-pass, high-pass, band-pass and band-stop filters as shown in Figure 1.

Presently, passive wireless sensor using frequency selective surfaces (FSS) has been proposed. Since various structural health monitoring (SHM) technologies are using wired system, the wireless SHM sensor using frequency selective surfaces (FSS) is believed to surpass available sensor. FSS have filtering properties to certain frequencies of incident electromagnetic radiation and when the electromagnetic characteristic of FSS is changed by geometric shape change of FSS by mechanical strain or structural failure, the information of those changes can be assured by monitoring the filtering response changes.

Other than that, the advantage of FSS over other wireless sensors is that it is totally passive because it only consists of conductive elements which can give advantages to energy storage [7]. This will make FSS having simple fabrication. The FSS sensor also capable to operate in harsh environment unlike the wired sensor that easily damaged since they are have complicated system.

3. Sensor Design and Fabrication

The design for FSS element shape are square loop and circular ring with the operating frequency 2.45 GHz for angle of incidence of the electromagnetic. Both of these two shape have been chosen since they have uncomplicated modelling and low cost manufacturing processed in large quantities. Moreover, square shape element has outstanding performance compared to others design in the matter of cross-polarization level, angular response stability and broad operating bandwidth [8]. At the same time, circular ring shape element was chosen to compare their characteristic performance with square loop shape in order to choose the best performance of Frequency Selective Surfaces.

Several parametric studies must be consider for both design in order to achieve the independent and dependent for both polarization TE incidents and TM incidents such as differences unit cell size, differences in outer length of conducting element and differences of inner length of conducting element.

The design was using the computer simulation technology (CST) microwave studio suite. This is because of the capability of CST and it is suitable to design the proposed design for this project. When all the simulation process already meets the FSS requirements, the project proceed with fabrication process of the proposed FSS. FR-4 has been chosen as the substrate for both design of FSS and the fabrication of FSS undergoes printed circuit board process, PCB.

A prototype of square loop and circular ring FSS was fabricated by using FR-4 material with the height of conducting element 0.035mm and 6 x 6 of unit cell is shown in Figure 2 and Figure 3.

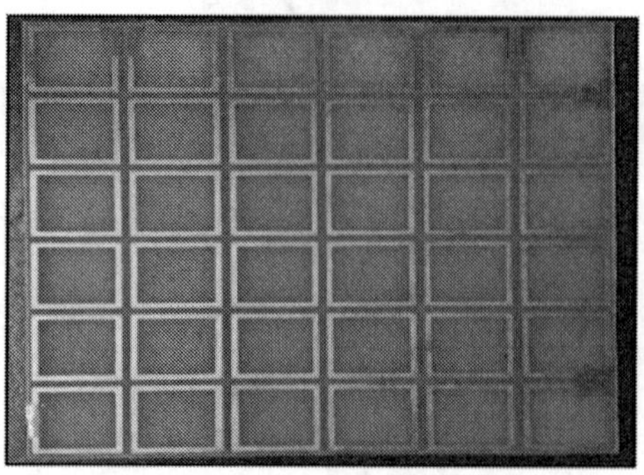

Figure 2. Fabricated Square Loop FSS

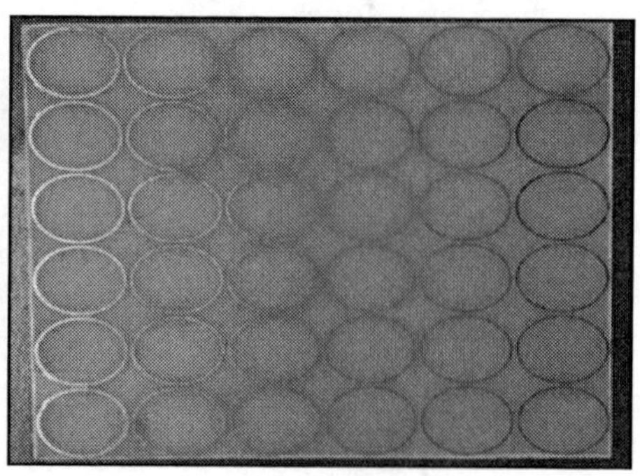

Figure 3. Fabricated Circular Ring FSS

4. Simulation Result

4.1. Circular Ring

Based on the parametric analysis, final Circular Ring dimension are chosen. Figure 4 is the final design of Circular Ring with its parameter and Figure 5 below shows the simulation results of S_{21} and S_{11}.

As shown in Figure 5, the Circular Ring FSS designed in this section is a band stop filter and the operating frequency at 2.45 GHz. The frequency responses at -10 dB are between 1.84 GHz and 3.12 GHz, therefore the bandwidth is 52.24% approximately. One of the important characteristics to study for FSS to act as a sensor for SHM is the changes in incident angle. In this project design, there are four values of angles taken from 0 degree up to 60 degree for both TE and TM mode that have been simulated.

Figure 6 (a) and (b) show the simulation result as the incident angle change (from 0 to 60 degree) the frequency response change accordingly. As for TE incident, the operating frequency shifted to lower frequency. Meanwhile for TM incident, the operating frequency shifted to higher frequency and the bandwidth is not much change as the incidence angle changed. The deviation of the frequency response for TE and TM will be used as a data for SHM. The changes in angle are described as the changes of building structure. For example, when the FSS shows reading at 2.62 GHz, it means that the structure is already tilted around 40 degree at TM direction.

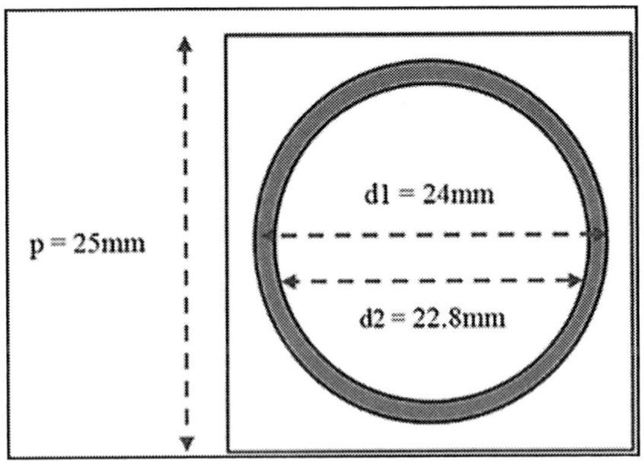

Figure 4. Final design of Circular Ring FSS.

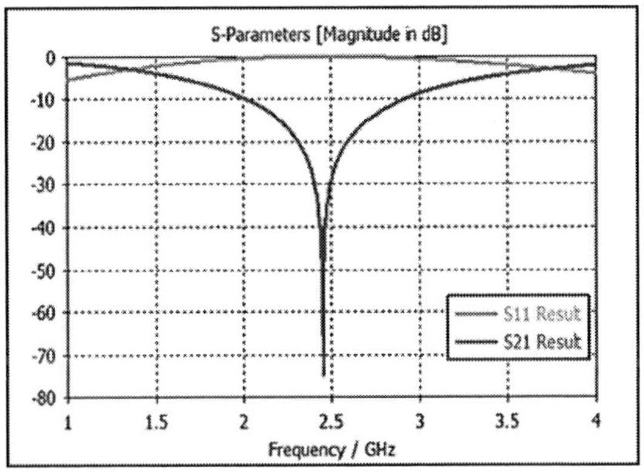

Figure 5. S_{11} and S_{21} Circular Ring simulation result.

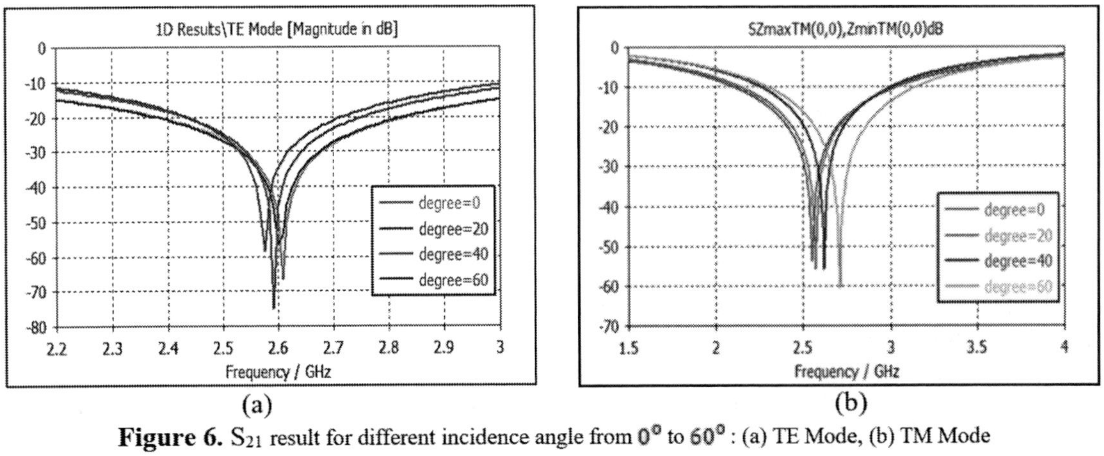

Figure 6. S_{21} result for different incidence angle from 0° to 60° : (a) TE Mode, (b) TM Mode

IOP Conf. Series: Materials Science and Engineering **318** (2018) 012033 doi:10.1088/1757-899X/318/1/012033

4.2. Square Loop

Final Square Loop parameter and dimension are chosen and shown in Figure 7. Figure 8 shows the result S_{21} and S_{11} simulation result. Figure 8 shows S_{21} and S_{11} simulation result of the Square Loop FSS. It is designed to act as band stop filter with the operating frequency 2.45 GHz. For Square Loop design, frequency responses at -10 dB are between 1.84 GHz and 3.124 GHz, hence the bandwidth is 52.24% approximately.

From Figure 9, the simulation result have shown that the operating frequency shifted to lower frequency for TE incident. The bandwidth is decreased as well as the incident angle change. Meanwhile for TM incident, the operating frequency shifted to higher frequency and the bandwidth narrower as the incidence angle changed. The deviation of the frequency response for TE and TM will be used as a data for SHM. The changes in angle are described as the changes of building structure. For example, when the FSS shows reading at 2.6 GHz, it means that the structure is already tilted around 20 degree at TM direction.

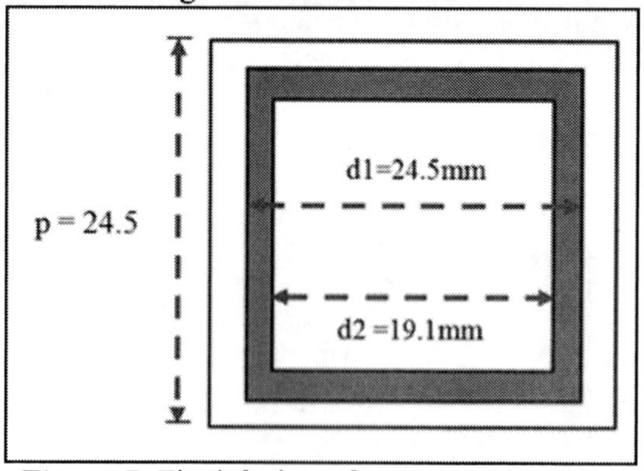

Figure 7. Final design of Square Loop FSS.

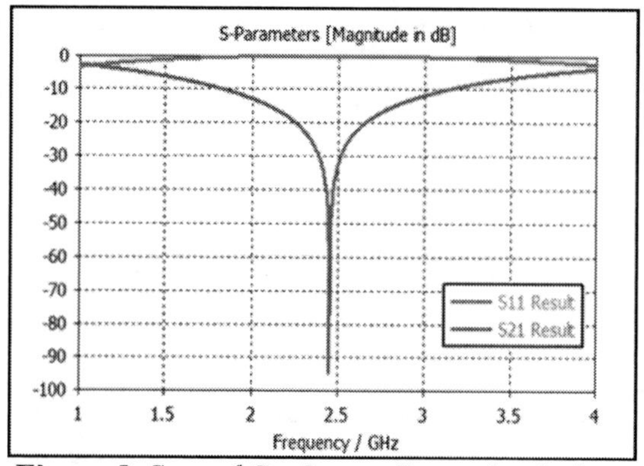

Figure 8. S_{11} and S_{21} Square Loop simulation result.

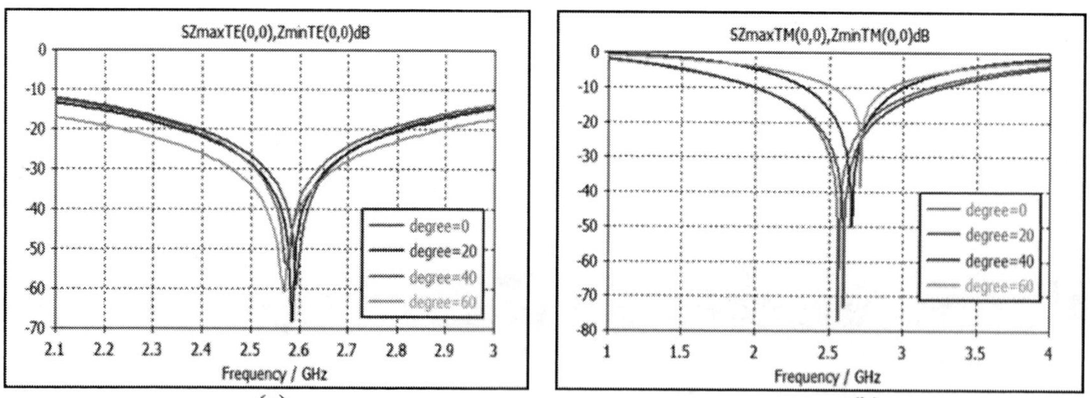

(a) (b)

Figure 9. S_{21} result for different incidence angle from 0° to 60° : (a) TE Mode, (b) TM Mode

5. Measurement Result

Keysight Technologies E5071C Economical Network Analyzer have been used for measurement of transmission coefficient (S_21) of the proposed FSS. The two horn antenna is connected to the both port on the ENA and shows the result of transmission coefficient directly. A FSS prototype will be placed between two horn antennas with appropriate distance. The example measurement setup were done in the free space for both FSS is shown in Figure 10.

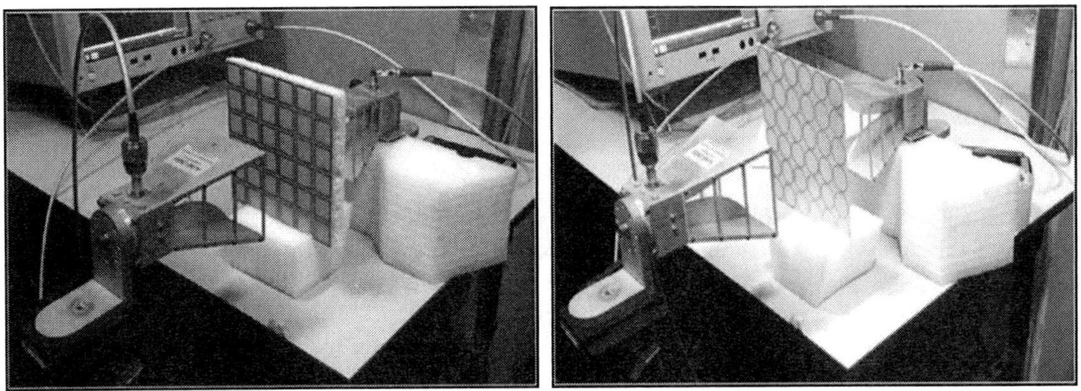

(a) (b)

Figure 10. Free space measurement setup: (a) Square Loop FSS, (b) Circular Ring FSS

5.1. Circular Ring

Figure 11 shows the comparison between simulation and measurement result of Circular Ring. The measurement was done to evaluate the simulation design that has been done in CST. The measurement design was measured with 25mm x 6 unit cell particularly at 0 degree angle. From the measurement graph the result of band pass frequency for Circular Ring was operating at 2.42 GHz with approximately -10dB.

5.2. Square Loop

The measurement design in Figure 12 was measured with 24.5mm x 6 unit cell at 0 degree angle. From the measurement graph in Figure 10, the result of band pass frequency for Square Loop was operating at 2.82 GHz with approximately -10dB.

Although the frequency bandwidth is different from simulation and measurement, nevertheless the operating frequency is approximately the same. The significant different is due to in simulation the FSS is define as infinitely array, hence produce a very good depth of insertion loss. However, the FSS is fabricated with 6 x 6 unit cell hence produce a less value of insertion loss compare to the simulation result.

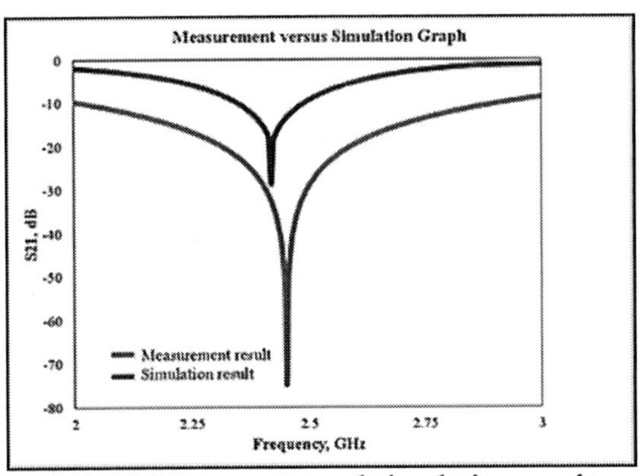

Figure 11. Measurement and simulation results of Circular Ring FSS

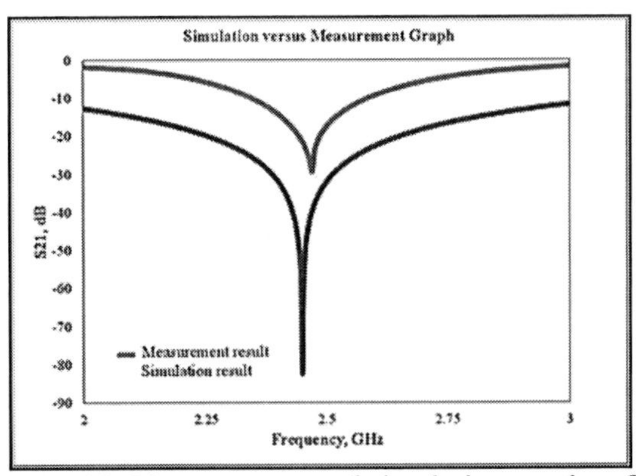

Figure 12. Measurement and simulation results of Square Loop FSS

IOP Conf. Series: Materials Science and Engineering **318** (2018) 012033 doi:10.1088/1757-899X/318/1/012033

6. Parametric Analysis

Both design was analyze with the different unit cell size, outer and inner length of conducting element by maintaining the height of conductor element at 0.035mm.

6.1 Circular Ring

Frequency responses for the Circular Ring design in different unit cell size in Figure 13 shows that by increasing the unit cell size, the operating frequency shifted to higher frequency. In term of bandwidth responses, there is not much different result and almost same bandwidth % as the unit cell size is changes.

Figure 14 shows differences in outer length of conducting element by increasing the length of outer conducting element, the operating frequency shifted to lower frequency. In a meanwhile, the bandwidth is getting wider from 32% to 40%.

As the length increase in Figure 15, the frequency response shifted to lower frequency and when the length decrease, frequency shifted to higher frequency. At the same time, the bandwidth also is getting narrower from 42% to 38%.

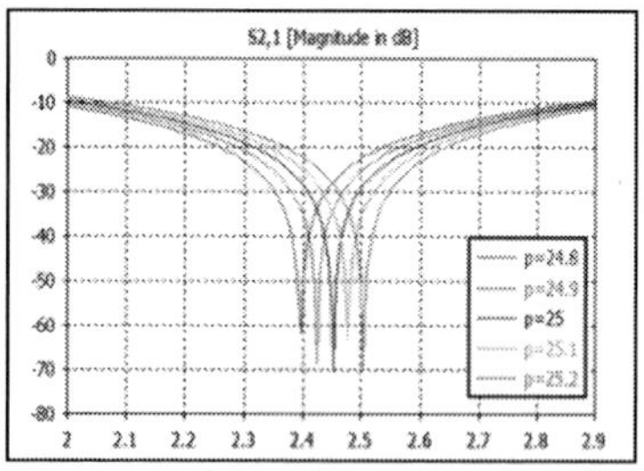

Figure 13. S_{21} result for different unit cell size

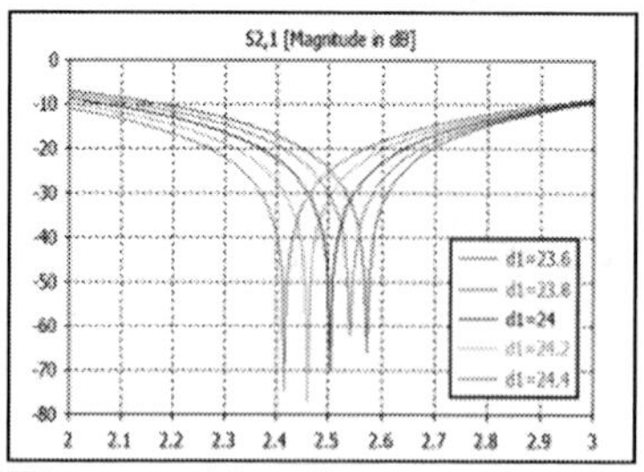

Figure 14. S_{21} result for different outer length of conducting element

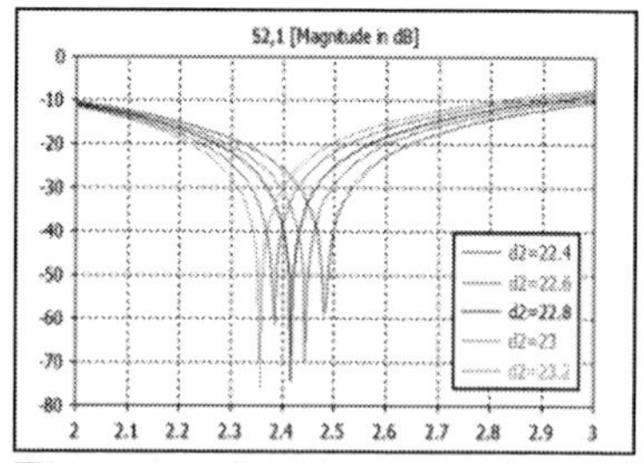

Figure 15. S_{21} result for different inner length of conducting element

6.2 Square Loop

Figure 16 shows that by increasing the unit cell size, the operating frequency shifted to higher frequency. In a meanwhile, the bandwidth is getting narrower from 55% to 51%. Simulation result for Figure 17, frequency responses shows as the length increase, the frequency response shifted to lower frequency and when the length decrease, frequency shifted to higher frequency. In term of bandwidth responses, there is not much different result and almost same bandwidth % as the inner length conducting material is changes.

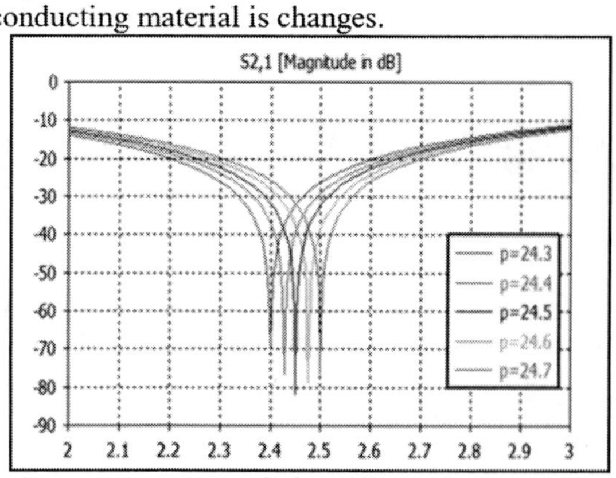

Figure 16. S_{21} result for different unit cell size

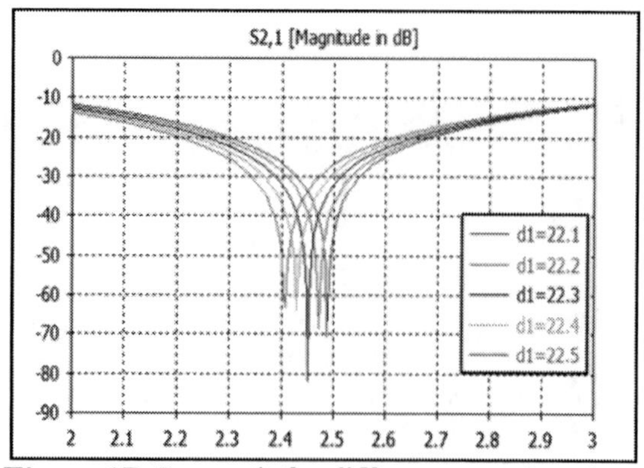

Figure 17. S_{21} result for different outer length of conducting element

7. Conclusion

Two different types of FSS have been chosen to design and compare their performance. From the simulation process it can be identify that Square Loop FSS is the most suitable design to be apply as a sensor for SHM system application. The simulation result shows that Square Loop have better incident angle compare to Circular Ring design. The measurement result from fabrication process also shows that Square Loop FSS have better reflection coefficient. Thus, Square Loop is the suitable design for FSS to use as a sensor, where it has moderate sensitivity toward incident angle and produce a wide bandwidth as well to cover large frequency range. It is also been analyze that differences in each parameter measured will give effect towards frequency responses and as well as incident angle. FSS have been design to provide data in structural changes for monitoring system. The measured changes such as strain and moving of structure can be wirelessly transmitted to the base station for data analysis. By using passive and completely wireless FSS, the cost of monitoring will be saved and less power consumption needed.

Acknowledgement

This project is been funded by Fundamental Research Grant Scheme (FRGS) 9003-00545 with a title of A fundamental study on the applicability using Flexible 3D Frequency Selective Surface for Structural Health Monitoring System.

References

[1] V. Kottapalli, A. Kiremidjian, J. Lynch, E. Carryer, T. Kenny, K. Law and Y. Lei, "Two-tiered wireless sensor network architecture for structural health monitoring", in SPIE's 10th Annual International Symposium on Smart Structures and Materials, San Diego (2003)

[2] A. Araujo, J. Garcia-Palacios, J. Blesa, F. Tirado, E. Romero, A. Samartin and O. Nieto-Taladriz, "Wireless Measurement System for Structural Health Monitoring With High Time-Synchronization Accuracy", IEEE Transactions on Instrumentation and Measurement, 61, no. 3, pp. 801-810, (2012)

[3] R. Di Sante, "Fibre Optic Sensors for Structural Health Monitoring of Aircraft Composite Structures: Recent Advances and Applications", Sensors, 15, no. 8, pp. 18666-18713 (2015)

[4] Fallahi, A., Mishrikey, M., Hafner, C., & Vahldieck, R. Efficient procedures for the optimization of frequency selective surfaces. Antennas and propagation, IEEE Transactions on, 56(5), 1340-1349 (2008)

[5] I.Ullah, "Measuring and filtering microwave radiations using frequency selective surface through energy saving glass", (2012)

[6] P. Kala Monica and M. Susila, "Study of Frequency Selective Surface for Wireless Communication", Middle-East Journal of Scientific Research, 24, no. 5, pp. 1673-1678 (2016)

[7] A.Deivasigamani, A. Daliri, C. Wang and S. John, "A review of passive wireless sensors for structural health monitoring", Modern Applied Science, 7, no. 2, pp. 57-76, (2013)

[8] Munk, B.A., "Frequency selective surfaces: theory and design". Wiley-Interscience. (2005)

IOP Conf. Series: Materials Science and Engineering **318** (2018) 012034 doi:10.1088/1757-899X/318/1/012034

Early Wheel Train Damage Detection Using Wireless Sensor Network Antenna

[3]A F M Fazilah, [1,3]S N Azemi, [1,3]A. A. H. Azremi, [1,3]P.J. Soh and [1,2]L. M. Kamarudin[2]

[1]Advanced Communication Engineering Centre (ACE) CoE,
[2]Centre of Excellence for Advanced Sensor and Technology,
[3]School of Computer and Communication Engineering, University of Malaysia Perlis (UniMAP), Pauh Putra, Arau, 02600

E-mail: ainurfasihah17@gmail.com.my, lyana05@gmail.com

Abstract. Antenna for a wireless sensor network for early wheel trains damage detection has successfully developed and fabricated with the aim to minimize the risk and increase the safety guaranty for train. Current antenna design is suffered in gain and big in size. For the sensor, current existing sensor only detect when the wheel malfunction. Thus, a compact microstrip patch antenna with operating frequency at 2.45GHz is design with high gain of 4.95dB will attach to the wireless sensor device. Simulation result shows that the antenna is working at frequency 2.45GHz and the return loss at -34.46dB are in a good agreement. The result also shows the good radiation pattern and almost ideal VSWR which is 1.04. The Arduino Nano, LM35DZ and ESP8266-07 Wi-Fi module is applied to the core system with capability to sense the temperature and send the data wirelessly to the cloud. An android application has been created to monitor the temperature reading based on the real time basis. The mainly focuses for the future improvement is by minimize the size of the antenna in order to make in more compact. In addition, upgrade an android application that can collect the raw data from cloud and make an alarm system to alert the loco pilot.

1. Introduction

In the transportation sector especially in railway systems, it is a must to take into account when revolves to the predictive diagnostics and security. The most common reasons of railways accidents are derailment of a freight car caused by wheel train damage [1]. Due to the numbers of coach of the train and immense in level of noise, the train crew will be unnoticeable about the derailment. Derailment will cause severely damage kilometers of track and hit the train at opposite site of track [2]. To improve the railway security, a lot of creations have been done in designing and implementing a detector device of the wheel trains. It can then possible to maintain before a major breakdown occurred.

The traditional approached is a device that purely in mechanical technique that located in the two ends of train coaches and touched with the train brake line. But, these types of detector are come out with drawbacks which are very costly and complexity in their installation [2]. Thus, to improve the traditional method, an onboard wireless sensor networks (WSNs) was introduced. This method is capable in monitoring the functioning conditions of the train's wheel bearing and provides the indicator of realtime warnings in case of faults. This system is consisted of sensors which notified about diagnostic information in wireless mode, the process will operates the same from the whole carriage and all the information will collected by the central router. The sensors that placed in pair are attached inside the wheel's bearing housing of the train [3].

This project is focus on one part of wireless sensor network which is antenna where it will be implemented in communication part. Antenna is a passive device designed for transmitting and receiving radio frequency signal which need to transceiving a high degree of performance over the entire potential environment range [4]. For these project's environment, antenna need to be designed in small size, low losses, high performance and not sensitive to the metal. A small patch antenna design by CST Microwave studio simulation is proposed to achieved all those characteristic as well as will be attached to the detector device.

Other focuses in these project is the detection part where a detector will be develop to sensing the current temperature and instantly been monitored by the loco pilot (train pilot) using installed android phone application. The detector device and the application are connected wirelessly to an access point or wireless fidelity.

2. Antenna Design

2.1. Initial design

Initial design is made by calculating the dimension of the antenna based on the frequency, wavelength and material's relative permittivity. The initial design uses equation (1) for the patch width. It was then followed up with equation (2) and (3) which is needed to find the patch length using equation (4).
An initial guess at the patch width:

$$\text{Width (mm)}=\frac{c}{2fr}\sqrt{\frac{2}{\varepsilon_r+1}} \tag{1}$$

Effective parameters:

$$\varepsilon_{reff}=\frac{\varepsilon_r+1}{2}+\frac{\frac{\varepsilon_r-1}{2}[1+12\frac{h}{W}]^{-\frac{1}{2}}}{2} \tag{2}$$

$$\frac{\Delta L}{h}=0.412\frac{(\varepsilon_{reff}+0.3)(\frac{W}{h}0.264)}{(\varepsilon_{reff}-0.258)(\frac{W}{h}+0.8)} \tag{3}$$

Patch length:

$$\text{Length (mm)}=\frac{c}{2f_r\sqrt{\varepsilon_{reff}}}-2\Delta L \tag{4}$$

The parameters calculated by equation (1-4) are applied in the antenna as seen in Figure 1.

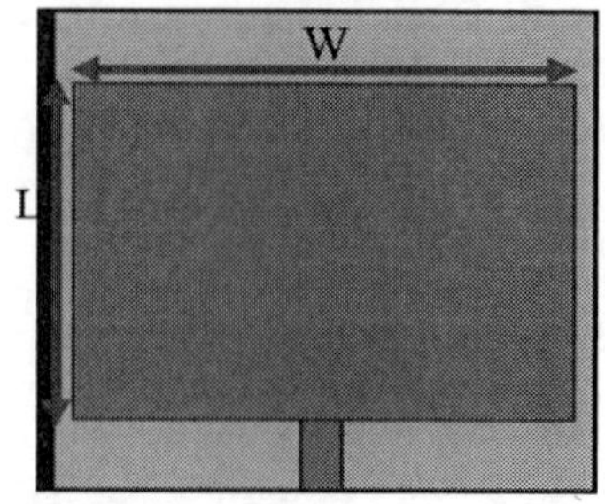

Figure 1. Antenna dimension.

The FR-4 material has a relative permittivity value of 4.8 and the initial height is 1.6mm. The result of the calculation is as follows:

Table 1. Calculated value of antenna parameter.

Parameters	Calculated value
Effective permittivity, ε_{reff}	3.494
ΔL	0.75 mm
Patch width, W	36.7 mm
Patch length, L	28.2 mm

To obtain the S11 graph that falls on the desired frequency, all the calculated values obtain in Table 1 are adjusted. For antenna simulation, a transmission line with the width of 1.3 mm from the edge of the substrate to the patch was added. The resulting antenna is shown in Figure 2.

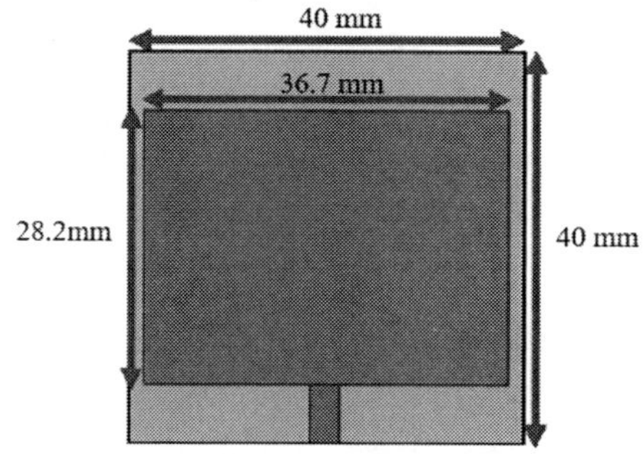

Figure 2. Initial design of antenna

2.2. Design modification

In order to achieve the main objective of this project, modifications are needed. The method used is by altering the design of the antenna in CST Software and simulate them to observe the reading. The modification is made based on the previous result. In each modification, several parameter sweeps were made. However, the overall parameter sweep results are not included here. Only the best parameters are taken based on the criteria stated in objectives of this thesis. Chronologically, the steps of the antenna's modification are as follows:

i. Adding slot in the top of patch as shown in figure 3.
ii. Adding inset feed at the connection of the patch and transmission line shown in figure 4.
iii. Adding stub in transmission line as shown in figure 5.

iv. Final adjustment to all parameters shown in figure 6.

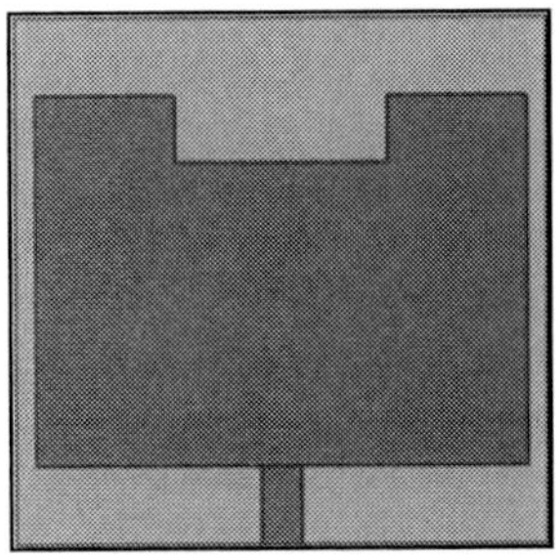

Figure 3. Slot added in the top of patch.

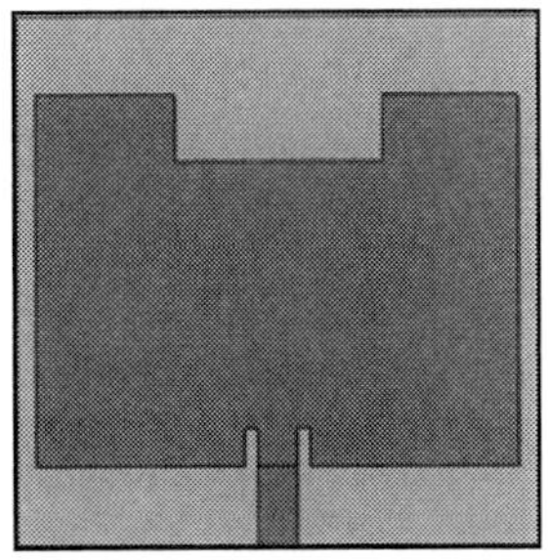

Figure 4. Inset feed added at the connection of the patch and transmission line.

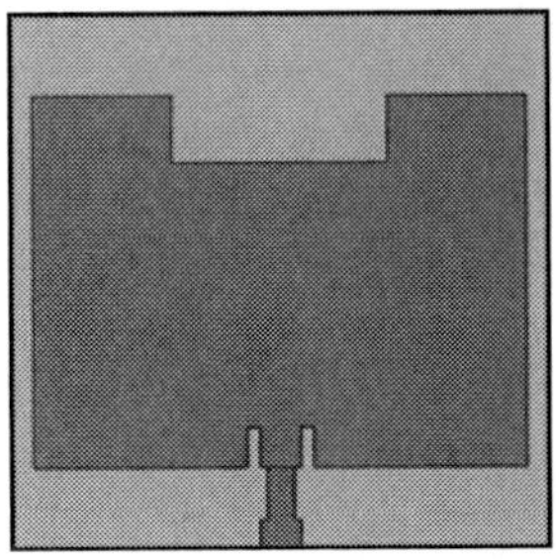

Figure 5. Stub added in transmission line.

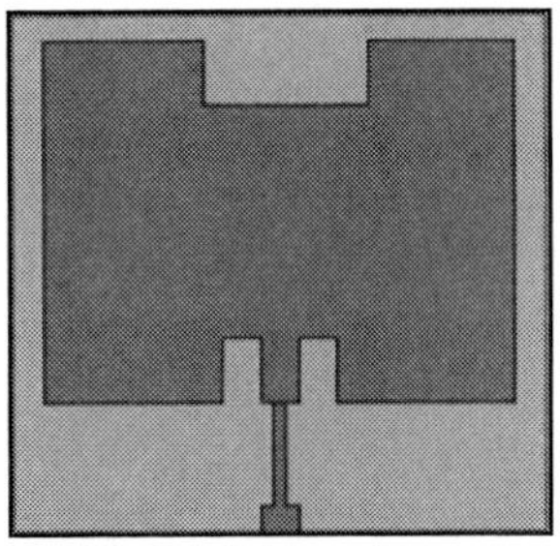

Figure 6. Height of transmission line and inset were increased

2.3. Antenna fabrication process

The designed antenna in CST is then fabricated into an actual antenna as seen in Figure 8. The antenna is then analyzed using the network analyzer equipment. The flow of fabrication process is shown in Figure 7.

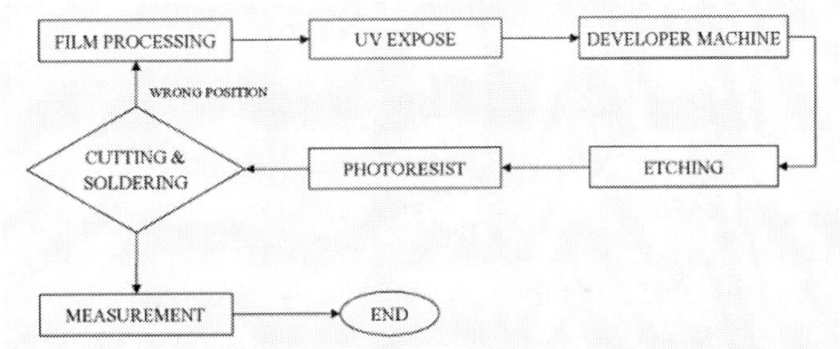

Figure 7. Flow of fabrication process

Figure 8. Fabrication antenna

2.4. Antenna results and discussion

After fabricating the antenna, the antenna is being analyzed by using network analyzer. S11 graph, VSWR and smith chart are the parameters analyzed by using network analyzer. Figure 9 shows the S_{11} result of simulation and measurement antenna. It can be seen that both simulation and measurement return loss of the antenna in good agreement. Both results manage to obtain less that -10dB at their resonance frequency. Based on the result, it shows that the measurement result is slightly different with simulation results. The simulation result drop exactly at 2.45GHz, while the measurement result drop at 2.46GHz with -29.81.

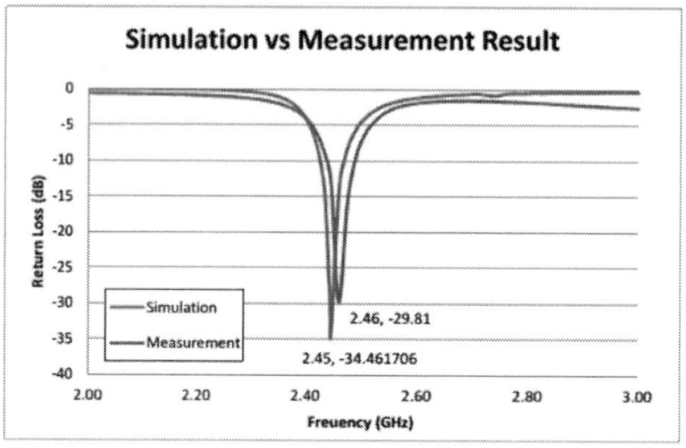

Figure 9. Simulation vs measurement of S_{11} result.

For VSWR, the simulation result VSWR is 1.0385692 while for the measurement is 1.07 as shown in Figure 7. From the value of VSWR, it can be conclude that there is no power reflected from antenna. The quality of an antenna can be defined by the value of VSWR, lower the value of VSWR more quality of antenna performance is matched to the Tx line - Transmission line and the high power is delivered to the antenna.

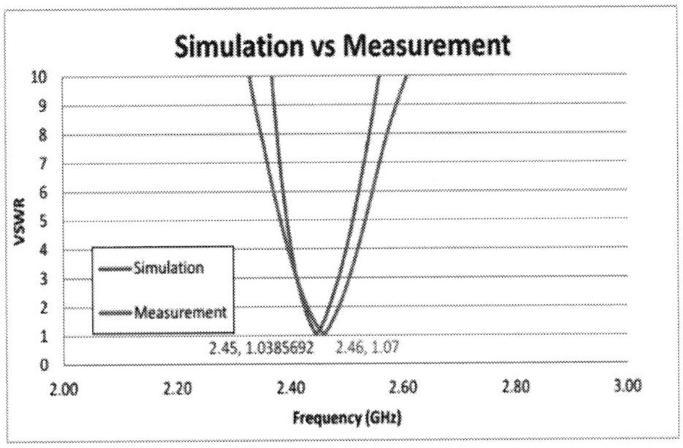

Figure 10. Simulation vs measurement of VSWR.

Lastly for the smith chart, the simulation results shows the impedance and transmission line is almost match with 50 Ω from the port. The result of the simulation is shown in figure 11 and measurement shown in figure 12.

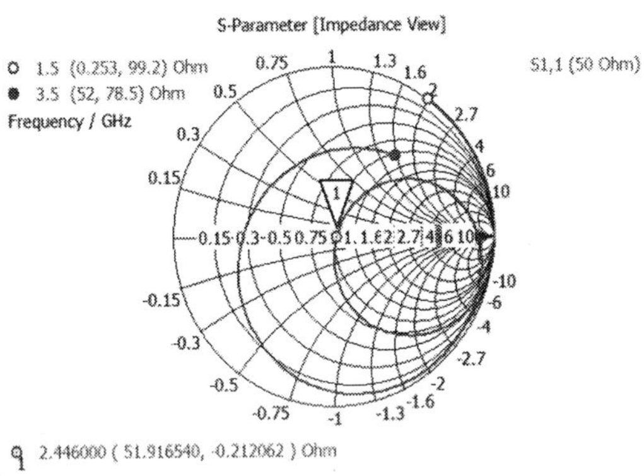

Figure 11. Simulation result of smith chart.

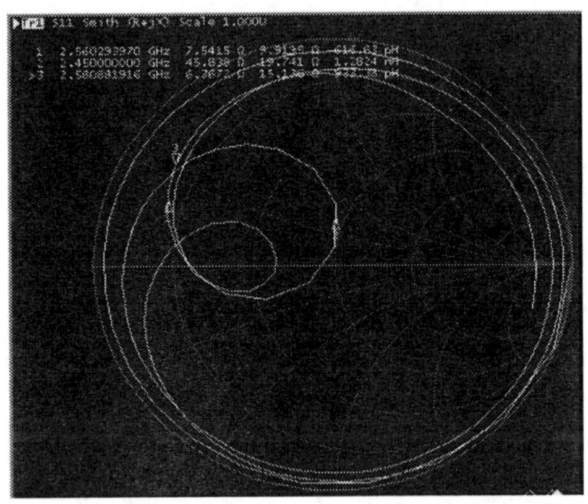

Figure 12. Measurement result of smith chart.

Table 2 shows the summary of radiation pattern for simulation and measurement result. The radiation pattern of the antenna is shown in figure 13 and figure 14.

Table 2. The result summary

	Simulation	Fabricated
Gain	4.95 dB	4.8 dB
Directivity	5.7 dBi	5.1 dBi
VSWR	1.0385692	1.07
Impedance matching	51.92 Ω	45.85 Ω

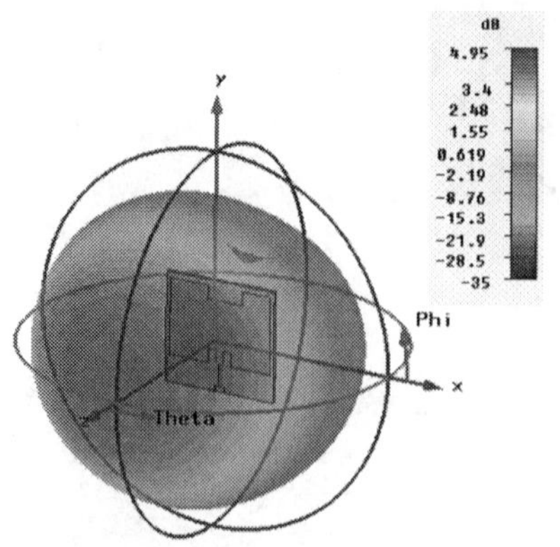

Figure 13. Farfield gain of simulation antenna.

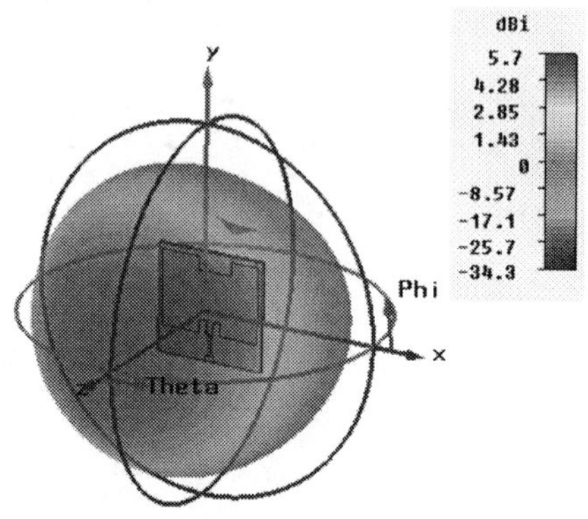

Figure 14. Farfield directivity of simulation antenna.

There are three main characteristic about of radiation pattern. There are:

- Efficiency
  The antenna efficiency is the ratio of the directivity to gain. For this antenna, the total efficiency is 84.12%.
- Gain
  The gain of the antenna is closely related to the directivity. For this antenna, the gain is higher than 3dB which is 4.97dB. Thus, this antenna capable to transmit their signal in long distance.
- Directivity
  The directivity is a measure which describes the directional of transmitting properties of the antenna. The directivity of this antenna is 5.701dBi. The directivity of the antenna is the capability of antenna to focus the signal in particular direction as well as transmitting and receiving signal. This antenna radiated in one direction to penetrate more power and obtain maximum performance of the antenna.

3. Development of monitoring system

Monitoring the wheels train current temperature is main objective of this project. There are few hardware need to be used where there are Arduino Nano as microcontroller, ESP8266-07 as Wi-Fi Module, LM35 as temperature sensor and light-emitting diode (LED) for the threshold indicator. The reading of the current temperature will be monitored in Android Application. The full system overview is shown in Figure 15.

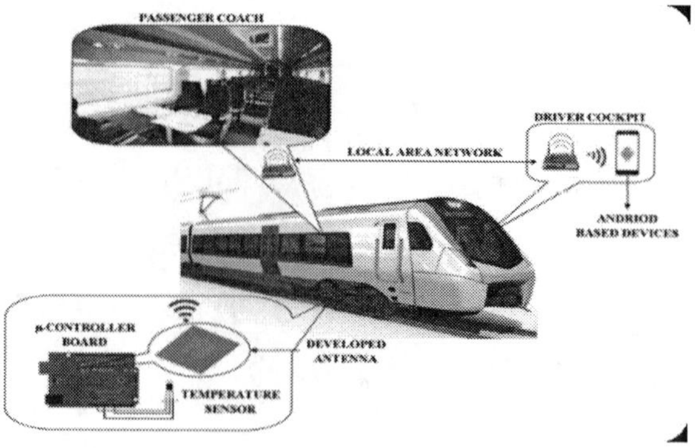

Figure 15. Overview of full system.

3.1. Data acquisition and processing

Figure 16(a) shows the connection of Arduino nano, LM35 and LED. The LM35 will acquired the reading of current temperature and the Arduino nano was programmed in Arduino IDE software using C language where to read instantly the reading as well as gain the output of LM35. For Figure 16(b), it shows the connection between Arduino nano and ESP8266-07, with AT-command and Hypertext Transfer Protocol (HTTP), the reading will be available in the ThinkSpeak.com cloud as mentioned in Figure 16(c). The ESP8266-07 was flashed with the latest firmware and has been configured as a station mode to connects to the train's WiFi. The LED used to indicate the threshold level of the reading temperature where it will be turned ON if the reading was exceed 75°C, if below than 75°C, the LED will be turned OFF. Figure 13 shows the temperature reading in two condition where in Figure 17(a), the temperature was under GOOD condition and in Figure 17(b), the temperature was exceed the limit of GOOD condition.

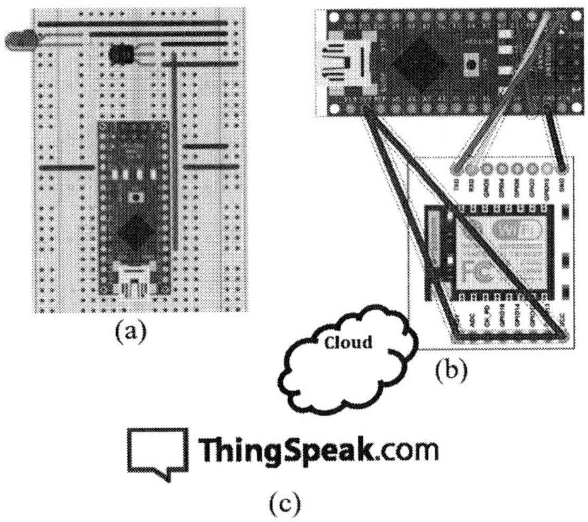

Figure 16. Arduino nano, LM35 and LED connection (a) Arduino nano and ESP8266-07 connection (b) The cloud server used (c).

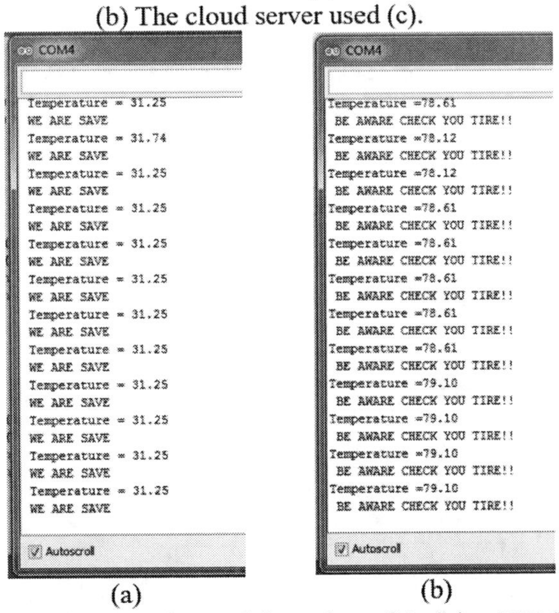

Figure 17. The temperature reading test using serial monitor of Arduino IDE (a) temperature was in GOOD conditions which is ≤ 70°C (b) temperature was exceed the limit of GOOD condition ≥ 71°C

3.2. Andriod application development

The reading that available in the cloud will be read by the android application where the application was developed in MIT App Inventor that using graphical programming as explain in figure 18. Same protocol with the microcontroller is a must where if not, the connection would not be success. The protocol used is HTTP and the transferred data of temperature reading will be programmed with generate a graph as shown in figure 19(b).

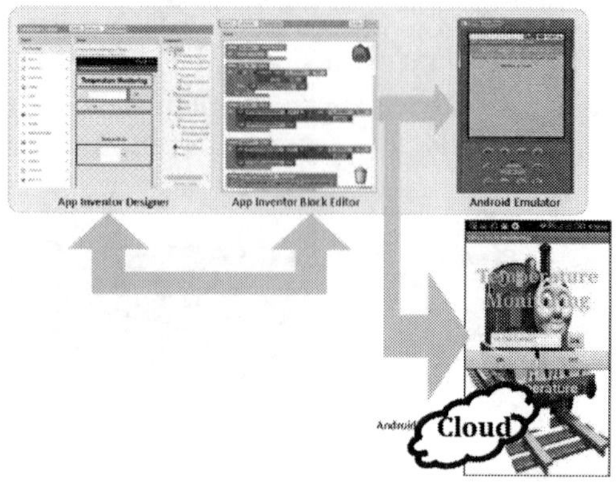

(a)

ThingSpeak.com

(b)

Figure 18. Flow of android application development.

This Android application will be installed in Android-based device and be placed in the train cockpit as well as been monitored by the train's pilot. The device must connected to the Train's WiFi where if not, it will not capable to read the temperature reading from the cloud server.

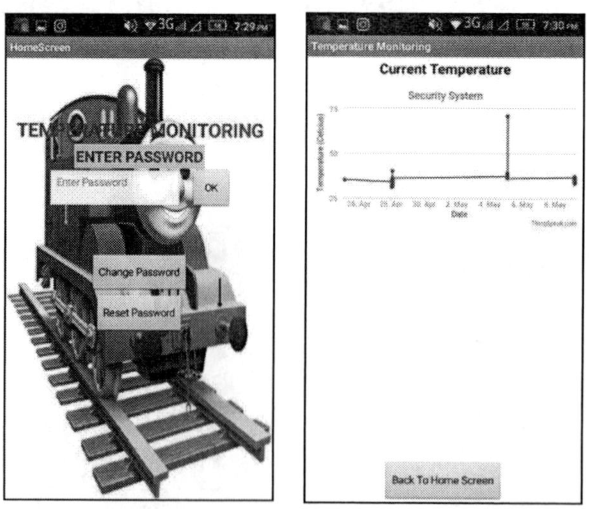

(a) (b)

Figure 19. (a) Home interface of andriod application (b) The monitoring interface of android application

As illustrated before in figure 15, the location of the device will be placed at the wheel train and to be exact is at the center bearing of the wheel train. In figure 20, shows how the electronic components, microcontroller, WiFi adapter and antenna are connected on the breadboard. For the future work, all that hardware will be fit into a customized casing which suits with the wheel train frame.

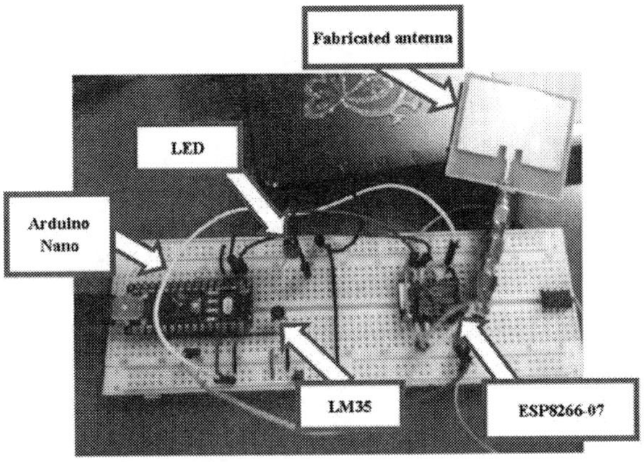

Figure 20. All the hardware connected to finalize the reading and measurement.

4. Conclusion

An antenna for a wireless sensor network for early wheel trains damage detection has successfully developed and presented. The new antenna was design to operate in single band frequency at 2.45GHz. The antenna is totally achieved the specification and requirements in term of gain, directivity, radiation pattern and efficiency. This antenna is very special because it is compact in their size. Based on the result of S11 or return loss graph, the antenna manages to obtain -34.46dB for simulation and - 29.81dB for measurement result at the resonance frequency. For the radiation pattern of this antenna are directional which capable to focus on power towards the transmitter or receiver. Besides that, the VSWR shows the almost an ideal value which is 1.04 for simulation and 1.07 for measurement. Whereas for the ideal case, the VSWR should be equivalent to 1 in order to make sure there is no power reflected back to the antenna. Furthermore, this project had achieved to create a detector device that capable to sensing the current temperature as well as sending the reading data into cloud wirelessly through Wi-Fi network. Thus this can ease the loco pilot to monitor the current temperature via android apps application. Last but not least, this project had been successfully compared the reading of current temperature between the detector and the android application.

References

[1] C. S. Mo, S. G. Kim, Y. J. Kwon, and K. S. Kang, "Improvement Directions through the Analysis of the Current States of Response Systems against Railway Casualties and Accidents of Urban Railways," Journal of The Korean Society of Civil Engineers, vol. 35, pp. 969-975, 2015.

[2] M. Macucci, S. D. Pascoli, P. Marconcini, and B. Tellini, "Wireless sensor network for derailment detection in freight trains powered from vibrations," in 2015 IEEE International Workshop on Measurements & Networking (M&N), 2015, pp.1-6.

[3] M. Hinnemo, "Design of an Antenna for a Wireless Sensor Network for Trains," 2011.

[4] J. Singh, T. Singh, and B. S. Sohi, "Design of slit loaded rectangular microstrip patch antenna," in 2015 2nd International Conference on Recent Advances in Engineering & Computational Sciences (RAECS), 2015, pp. 1-5.

IOP Conf. Series: Materials Science and Engineering **318** (2018) 012035 doi:10.1088/1757-899X/318/1/012035

The Design of Operational Amplifier for Low Voltage and Low Current Sound Energy Harvesting System

Liew Hui Fang[1,*], Rosemizi Bin Abd Rahim[2], Muzamir Isa[1], Syed Idris Syed Hassan[1] and Baharuddin Bin Ismail[3]
[1]School of Electrical System Engineering, University Malaysia Perlis, Pauh Putra Campus, Arau, Perlis, Malaysia
[2]School of Computer and Communication Engineering, University Malaysia Perlis, Pauh Putra Campus, Arau, Perlis, Malaysia
[3]Center of Excellence for Renewable Energy (CERE) Universiti Malaysia Perlis, Pauh Putra Campus, Arau, Perlis, Malaysia.

E-mail: g1330110940@studentmail.unimap.edu.my

Abstract. The objective of this paper is to design a combination of an operational amplifier (op-amp) with a rectifier used in an alternate current (ac) to direct current (dc) power conversion. The op-amp was designed to specifically work at low voltage and low current for a sound energy harvesting system. The goal of the op-amp design with adjustable gain was to control output voltage based on the objectives of the experiment conducted. The op-amp was designed for minimum power dissipation performance, with the means of increasing the output current when receiving a large amount of load. The harvesting circuits which designed further improved the power output efficiency by shortening the fully charged time needed by a supercapacitor bank. It can fulfil the long-time power demands for low power device. Typically, a small amount of energy sources were converted to electricity and stored in the supercapacitor bank, which was built by 10 pieces of capacitors with 0.22 F each, arranged in parallel connection. The highest capacitance was chosen based on the characteristic that have the longest discharging time to support the applications of a supercapacitor bank. Testing results show that the op-amp can boost the low input ac voltage (~3.89 V) to high output dc voltage (5.0 V) with output current of 30 mA and stored the electrical energy in a big supercapacitor bank having a total of 2.2 F, effectively. The measured results agree well with the calculated results.

1. Introduction

The field of power harvesting has experienced significant evolution over the past few years due to the growth of ever-increasing demand to produce portable and wireless electronic devices with prolonged lifespan [1]. The limited time span of a battery limits the reliability of a device. An alternative to conventional batteries for the energy source can be considered in providing maximum energy to prolong the lifespan of an electronic device.

The purpose of energy harvesting systems is as a simple, inexpensive alternative source for charging, towards improving a self-powered devices by reducing the dependency on the conventional copper cables and troublesome battery. Energy harvesting is a process where the ambient waste energy was captured and converted into usable electricity. This process was proposed previously and related research has been highly developed. A variety of ambient sources such as solar [2], thermal [3], acoustic noise [4], acoustic energy [5], nuclear power [6], wind [7], and mechanical vibration [8] have been studied as an additional energy supplies for the last decades.

With the increasing of serious polluted environment and resources shortage problems, the renewable energy attracts so much attention, especially for sound wave energy harvesting system, due to its wide availability. However, the power, the voltage, and the current produced by piezoelectric materials are far too small and lower than the power of most electronics applications. In addition, the piezoelectric harvesting system can produce a high level of output energy only when it generates resonance with external vibration excitation. This usually requires energy harvesting circuit to modulate the output power, which leads to a relatively large energy loss due to the switch control circuit. Eventually, it greatly affects the competence of energy harvesting system. Accordingly, the methods of collecting and storing parasitic energy are also the key to acquiring self-powered systems. Subsequently, op-amp was introduced alternatively as ac-dc converter circuit to increase the amount of energy generated by the energy harvesting device.

The recent research cited above focused on developing optimal energy harvesting structures. Nevertheless, the electrical outputs of these devices in many cases were too small to supply power to any electrical devices directly. Jamal et al [9] investigated sound energy harvesting devices and the methods of accumulating energy by using multiple piezoelectric transducers stored in multiple supercapacitors. It was then summed up and amplified through adder LM 324 op-amp and voltage multiplier circuits. The resultant electric power was used to charge a rechargeable 9 volts dc battery and charged within half an hour. Ali Far [10] developed a highly efficient ultra-low current and low voltage rail-to-rail input-output class of an AB amplifier for energy harvesting system. A rail-to-rail input output amplifier with new floating minimum current source was also proposed, which operates under low Vdd (drain) as low as ~0.6 V, amplifier consumes 330 nA current, gain of 78 dB and improves amplifier's dc and ac response. Kyriaki et al [11] proposed the possibility of harvesting the wasted heat in a power amplifier (PA) by using thermoelectric generator (TEG) from 1 Watt, 2.45 GHz class-E power amplifier circuits, and converting them to DC electrical power. The PA circuit was built using microstrip technology and the generated heat and temperature distribution on the PA circuit board were evaluated. Consequently, the harvested output power of 1.015 mW, 1.31 V and 3.1 mA with 10dB gain was measured, respectively. Fahmi et al. [12] reported that a class-D RF power amplifier has been cultivated into wireless energy transfer module for solar energy harvesting system. The class-D radio frequency (RF) power amplifier has been used as the driving circuit for transmitted coil switching to obtain the desired frequency level of power source with nearly 99% efficiency, theoretically. Madhuri et al. [13] suggested that direct ac-to-dc power converter was offered for efficient energy harvesting from a low voltage inertial micro generators. The main target was signal generator, followed by a high-current buffer, used to measure the micro generator output voltage. The buffer can be examined by using a power op-amp in voltage follower mode. The pulse signal produced by the pulse width modulation (PWM) was fed to the two buffers that can be enabled by external signals. The comparators in the polarity detector unit enabled the appropriate buffers to produce a gate pulses that control the MOSFETs of the boost and buck–boost converter during appropriate half cycles. The converter was successfully operated to directly increase the low ac voltage (400 mV) to a high dc voltage (3.3 V).

The general idea of this paper was to develop an op-amp specifically to harvest low voltage and low current produced by sound wave energy. Difficulties may arise if the energy output from a single strip is too low to turn on the converter section. This may be resulted from a threshold voltage cut-off, or may also be resulted from the very low current generated from the piezoelectric material. Thus, effective solution by using the op-amp as an active harvesting circuits is necessary. The system was experimentally studied by using single piezoelectric strips in a sound energy harvesting system. The raw energy will then be converted and conditioned to something more usable. The general block diagram for this process is shown in Fig. 1. The material used to harvest wind energy for this paper were the piezoelectric strips. The specific piezo-strip used was a Q220-A4-503YB, with dimensions of 31.8 x 66.7 mm. The strip was chosen for its convenience more than for specific performance, under the consideration that piezoelectric materials are capable in providing high voltages, but generally consumed very low currents (often 10's – 100V and in the range of nA-μA).

There are many types of energy storage systems. Currently, batteries which are an example of mature technology, are widely used for its reliable performance and availability. However, it has a few drawbacks, such as short charging/discharging, time providing only small amount of power density, and short life cycle [14]. On the other hand, supercapacitors have larger power density, smaller capacity and long life cycle [15]. Therefore, in this research, sound energy harvesting storage system combined the characteristics of the op-amp and the supercapacitor bank was introduced. The storage system was controlled by power electrical converters to charge or discharge in harmony, which not only guarantees the quality of the sound energy harvesting system, but also improves the overall performance of the energy storage system, extending the life cycle of the energy storage system and reduces the costs incurred. The calculated output voltage and power of the piezoelectric plates match well with the experimental data. The simulation results were then compared with the experimental data for validation purposes. The comparison agrees well with each other and the deviations were acceptably small for design purposes.

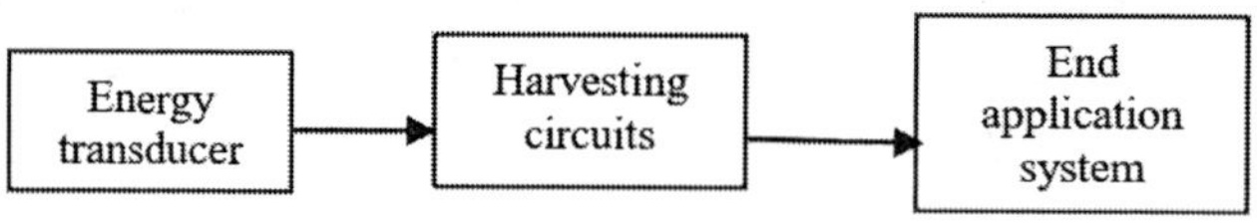

Figure 1. Energy Harvesting System.

2. Methodology

The methods used for the energy harvesting system was categorised into few parts, as follows. Firstly, the experiment started with the development of the op-amp which was to amplified the low voltage and low current sound energy harvester. The op-amp was designed with variable gain to be able to switch based on output voltage required. Secondly, for better gain of op-amp in this research, the single piezoelectric was selected to run the op-amp as a harvesting circuit. Then, the minimum gain will be chosen based on the target output of supercapacitor bank which is 5.0 V. Thirdly, the minimum gain of the op-amp with single piezoelectric was measured and analysed with single piece supercapacitor charging time with difference capacitance and finally, the highest capacitance of a supercapacitor will be considered into building the supercapacitor bank due to the longest discharging time or vice versa.

In this experiment, a transducer model Q220-A4-503YB was used to harvest and convert the sound waves into a useful electrical energy. The sound wave energy system was accumulated from a loud speaker. Piezoelectric, which was used as an energy transducer, was placed at various distance from the speaker and was connected to the harvesting circuitry. The specifications of loudspeaker are shown in Table 1. The piezoelectric transducer model Q220-A4-503YB was operated at 68 Hz and accomplished a maximum power response of 33.133 dBuW at sound level of 96 dB. This level is in the range of the sound level for ambience environment; which was between 50-100 dB.

Figure 2 shows the experimental setup used for the characterization of sound energy harvesting. The loud speaker was installed in a wooden box and connected to a function generator. During in-lab testing, instead of energy harvester, a function generator was used to provide input voltage waveforms with various amplitudes and frequencies. Oscilloscope was used to observe the input and output voltage signals for different frequencies and amplitudes. Sinusoidal waveform was provided as an input throughout the in-lab characterization.

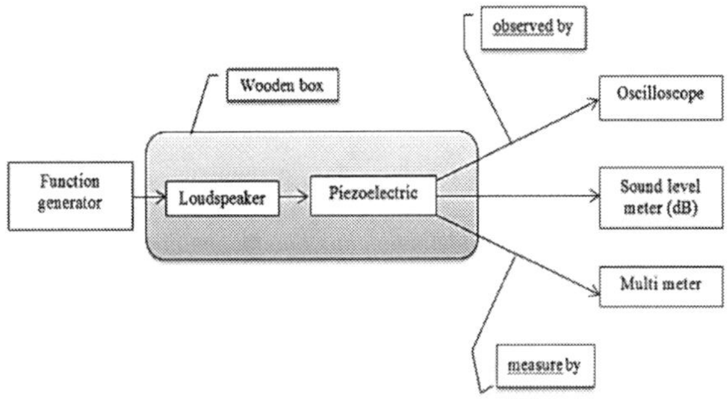

Figure 2. The experimental setup for the characteristics of the piezoelectric transducer harvester on sound wave.

Table 1. Specifications of 6.5" 2-way loudspeaker.

Nominal Impedance	4 ohms
Speakers Sensitivity	92 dB
Frequency Response	40 Hz – 20 KHz
Nominal Music Power	55 Watts
Maximum Music Power	250 Watts

2.1 Energy Harvesting System/Harvesting Interface Circuit

The block diagram for the harvesting circuit concept as shown in Fig. 3 can be categorised into a few parts, operational amplifier (op-amp), rectifier circuit, and energy storage devices. Piezoelectric (PE) transducers are commonly utilised to produce high output voltage and low output current. The utilisation of the op-amp instead of comparator provides lower power dissipation and improves power extraction capability of the piezoelectric harvesters. Energy conversion efficiency varies according to the op-amp and bridge rectifier. Performance optimisation of the op-amp and bridge rectifier is required to maximize the energy conversion efficiency.

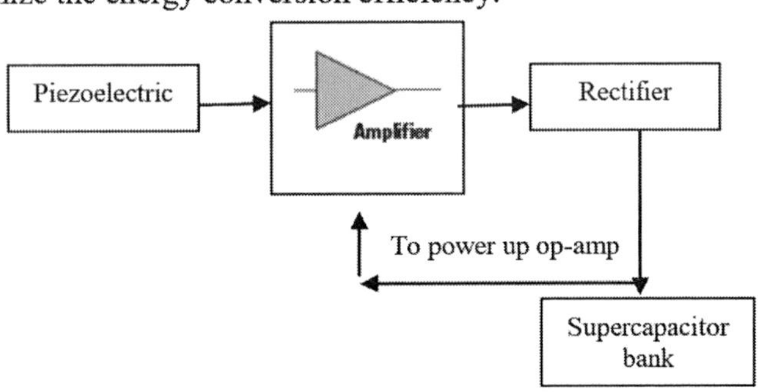

Figure 3. Schematic representation of the op-amp.

2.1.1 Operational amplifier (non-inverting)

The feature of non-inverting amplifier is an op-amp circuit configuration which produces an amplified output signal. The output signal of a non-inverting op amp was in-phase with the input signal applied. In other words, a non-inverting amplifier behaved like a voltage follower circuit. A non-inverting amplifier also used negative feedback connection, but instead of supplying the entire output signal back to the input, only a small part of the output signal voltage was fed back as an input to the inverting input terminal of the op-amp. The high input impedance and low output impedance of the non-inverting amplifier makes the circuit ideal for impedance buffering applications. Fig. 4 illustrates the LM 358N pin configuration.

The LM 358N is a duel single supply op-amp. As a single supply, it eliminates the need for a duel power supply, thus simplifying the design and basic application usage. LM 358N consists of two independent, high gain, internally frequency compensated op-amp which were designed specifically to operate from a single power supply over a wide range of voltage. The LM 358N is a duel op-amp, which means there are two op amps in one integrated circuit (IC). One op amp consists of a non-inverting input, and another one consists of voltage follower. LM 358N is also one of the types of op-amp. The LM 358N has around 30 mA current out.

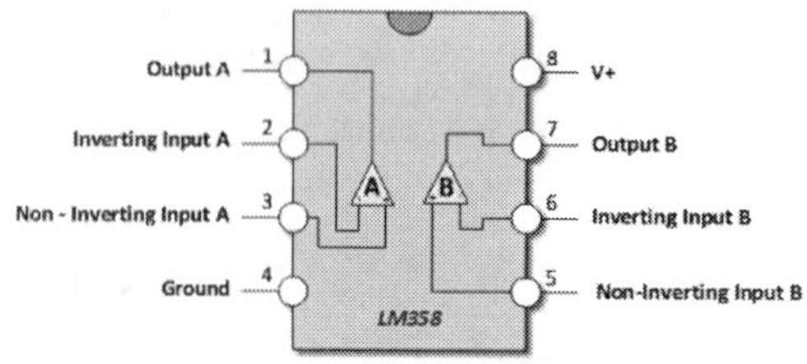

Figure 4. LM 358N pin configuration.

2.1.2 Operational amplifier (voltage follower)

The main purposes of a voltage follower are:

1. To provide a high impedance input which isolates the amplifier's output from the circuit it was attached to.
2. To provide a low impedance output to provide high current output and/or high bandwidth for a high capacitance load.

Voltage follower, also known as buffer amplifier, is a device which connects high impedance circuit to low impedance circuit. The gain of voltage follower is unity i.e. Av= 1. Voltage follower = Unity gain.

In voltage follower, the output signal is the same as the input signal. It is a device having high input impedance and low output impedance. Thus, voltage follower acts as a unity gain buffer to avoid loading effects in the circuit driving it.

The main reason the voltage followers were used was that they draw small amount of current, without disturbing the original circuit, and provide the same voltage signal as the output. They act as isolation buffers, isolating the circuit providing the least disturbance to the power of the circuit. When the circuit has a very high input impedance, only small amount of current is drawn from the circuit. According to ohm's law, current, I=V/R. Thus, the greater the resistance, the less current is drawn from a power source. As a result, the power of the circuit was unable to affect the supplied current on a high impedance load.

2.1.3 Rectifier as ac-dc converter signal

The rectifier is unique because its formation allows the negative portion of the input signal to follow the same current path as when it does during the positive input signal. The intension of using rectifier is to convert the AC signal from the output op-amp into dc power. The energy was then stored in the energy storage such as battery or supercapacitor bank. The full wave bridge has the same function as the half wave in the sense that only the positive portion of the ambient signal was used. To improve the extraction capacity, schottky diodes 1N5711 were widely used in rectifier due to their low forward conduction voltage ~0.45 V, large saturation current, low junction capacitance, low substrate leakage, very fast switching and small series resistance which was in good agreement with the theoretical values. The purpose of small voltage drop will allow as minimum as possible the voltage of the sound wave to be harvested.

2.1.4 Supercapacitor as energy storage device

The supercapacitor can be charged and discharged virtually with unlimited number of times. Aside from the fact that the supercapacitors can be charged very quickly due to their low internal resistance, which is known as ESR, they can also quickly be discharged. Table 2 organised a few different types of ESR and Farad value of a supercapacitor, which has been tested.

To develop the charging of supercapacitor from sound wave energy, an approximate calculation was based on the following parameters [14-16]:

Total charge stored, Q in the capacitor was calculated using;

$$Q = CV \quad (1)$$

Total energy stored, E in the capacitor was calculated using;

$$E = QV \quad (2)$$

Where, Q is charge (in coulombs), V is voltage required and C is capacitance. The time needed to charge the supercapacitor to a voltage of V_{sup} using the diode charger can be approximated based on Equation (3).

$$t_{diode} \approx \frac{C_{sup}V_{sup}}{I_{sc}} \quad (3)$$

Where C_{sup} represents capacitance of supercapacitor, V_{sup} represents the voltage of supercapacitor to be store and I_{sc} represents the short circuit current.

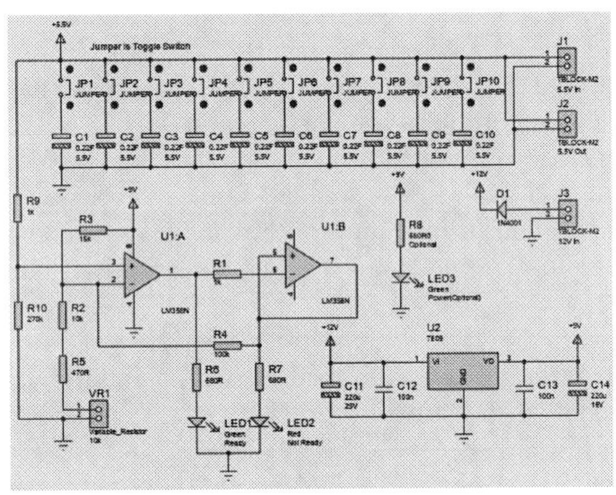

Figure 5. The schematic circuit of supercapacitor bank.

Supercapacitors have many advantages. For instance, they maintain a long life cycle, which can be cycled hundreds of thousands of times with minimal change in performance and are offset by their low energy density. A supercapacitor's lifetime spans 10 to 20 years, and the capacity might reduce from 100 % to 80 % after 10 or so years. On the other hand, supercapacitors with their low equivalent series resistance (ESR), provide high power density and high load currents to achieve almost instant charge in seconds.

The requirement for this paper is the efficiency of energy storage device. It needs to be able to accept a wide range of voltages instead of a fixed voltage. Storing the energy and accumulating it to a usable energy level had become a problem that needed to be solved. One end of the circuit was connected to a supercapacitor bank. The supercapacitor bank in Fig. 5 was charged completely so that it can recycle the electrical power by converting them into mechanical energy and vice versa. Hence, it can be used in case of power failure and in situations where reduction in electricity (by switching off the supply) expenditures is needed. The entire process mentioned in the design occurs only after the supercapacitor bank was fully charged. The supply was then cut off automatically from the device and finally the device operates by the used of supercapacitor bank.

Table 2. Specifications of supercapacitor value and ESR.

Capacitance	0.01 F	0.022 F	0.047 F	0.1 F	0.22 F
ESR (ohm)	300	20	200	8	6
Voltage rating	5.5 V	11 V	5.5 V	11 V	11 V
Capacitance tolerance	+80%, -20%	+80%, -20%	+80%, -20%	+80%, -20%	+80%, -20%
Lead spacing	5.08 mm	15 mm	5.08 mm	10.2 mm	15 mm
Capacitor case style	Coin	Coin	Coin	Coin	Coin

2.2 Principal of operational op-amp

In this study, the LM 358N was used to design the op-amp and also as comparator to the supercapacitor bank system. In Fig. 6, the completed schematic circuit of op-amp with bridge is presented. An op-amp was used as harvesting circuit for sound energy harvester and was categorised into two stages. First stage is voltage follower and the second stage is non-inverting op-amp. In Fig. 6, there was no gain for the first stage (voltage follower), in fact consists of two resistor function as voltage divider where, there are R4= 1 kohm connected between terminal input to op-amp (+) input and R3 = 100 kohm connected between op-amp (+) to ground on the input of (+) op-amp. This construction allows the op-amp do drop the voltage by about 0.01 %.

The gain of the non-inverting circuit for the op-amp was easy to identify. The calculation hinges around the fact that the voltage at both inputs was the same. This arises from the fact that the gain of the amplifier was exceedingly high. If the output of the circuit remains within the supply range of the amplifier, then the output voltage divided by the gain means that there is virtually no difference between the two inputs.

As the input to the op-amp draws no current, the current flowing in the resistors R4 and R3 is the same. The voltage at the inverting input was formed from a potential divider consisting of R4 and R3, and as the voltage at both inputs is the same, the voltage at the inverting input must be the same as the non-inverting input. This means that $V_{in} = V_{out}$ x R4 / (R4 + R3). Thus, the op-amp gain is available to adjust based on experiment persistence and adjusting the maximum and minimum by shifting the resistance value on RV2.

The purpose of building up the voltage divider is to protect the op-amp (+) input. Therefore, resistor R4 functioned as to provide high impedance between terminal input and op-amp input (+). Hence, there will be no damage to op-amp in case terminal input is accidentally connected to higher voltage than the op-amp can handled.

Moreover, the important of resistor R3 is to make the op-amp input less sensitive to noise and pull down the op-amp input to ground if the terminal input is left floated (not connected to anything). Overall, the op-amp output will remain off even if the terminal input is left floated. If R3 was not used, the op-amp input will be very sensitive, and the op-amp input will floated (not pulled down to ground) when the terminal input is left floated, resulting in the undesired output produced by the op-amp output.

Initially, this application uses only one op-amp as non-inverting op-amp and is sufficient instead of using voltage follower. However, for this research, LM 358N op-amp was used, and this configuration pin contains two op-amp built in inside one IC in Fig. 4. In IC pin, let assume if one of the op-amp left unused (floated), the overall circuit output performance will be affected. To overcome this problem, another op-amp acted as voltage follower was constructed with the advantage to amplify the current of the input for the output.

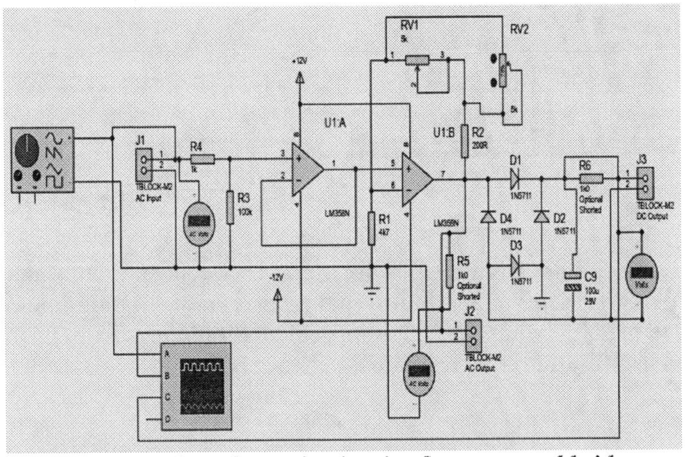

Figure 6. The schematic circuit of op-amp and bridge.

3. Results and Discussion

3.1 Designing op-amp with adjustable gain value

A single piezoelectric was tested with an op-amp to identify the most suitable performance by adjusting the op-amp variable gain in order to examine better gain. The minimum (min) gain of op-amp was chosen as the most suitable gain to achieve the experiment output of 5.0 V. To validate this method, the op-amp circuit was designed by using Proteus, where the prototype circuit was measured by using oscilloscope and implemented ac and dc voltage was measured by using multimeter. An observation on oscilloscope shows that CH1 represents AC voltage at signal input, CH2 represents AC voltage at signal output, and CH3 represents DC voltage at signal output from bridge. Fig. 7 until Fig. 10 depicted a comparison between simulation and experimental results for op-amp circuit voltage after amplified and rectification using conventional bridge circuit. Fig. 7 and Fig. 8 shows the simulation and experimental result, respectively for waveform signal from ac and dc voltage at minimum gain of op-amp. Whereas, Fig. 9 and Fig. 10 depicted the simulation and experimental result, respectively for waveform signal from ac and dc voltage at maximum (max) gain of op-amp. Overall, the results of experimental measurement and simulation are in good agreement with one another.

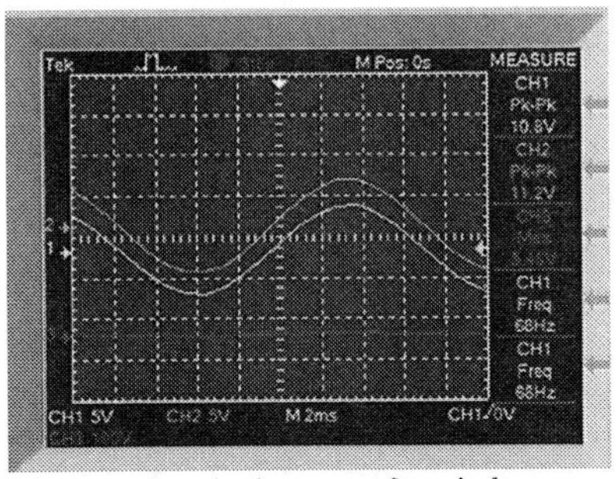

Figure 7. Simulation result minimum gain of op-amp.

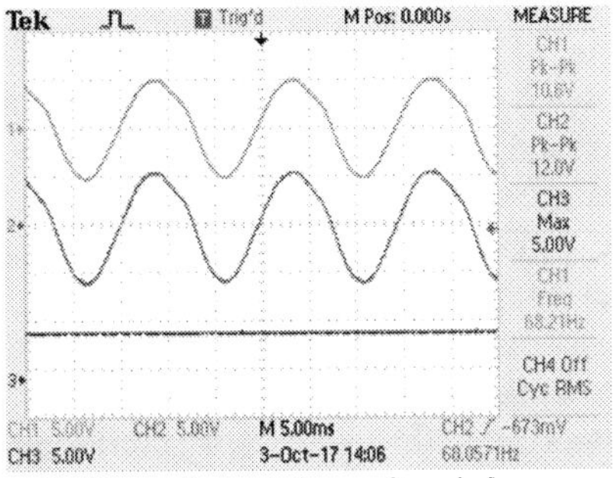

Figure 8. Experimental result minimum gain of op-amp.

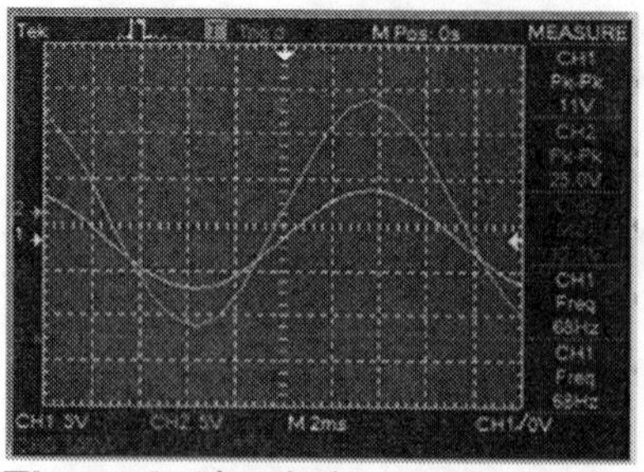

Figure 9. Simulation result maximum gain of op-amp.

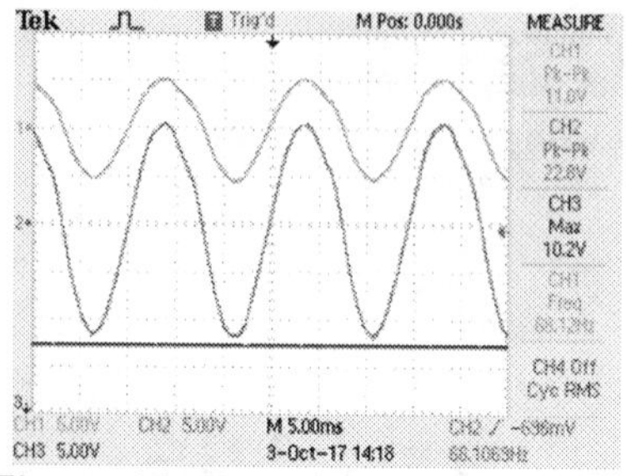

Figure 10. Experiment result of maximum gain of op-amp.

This paper provides comparative results of calculations between minimum and maximum gain on a non-inverting op-amp and gain of voltage follower. The maximum and minimum gain of op-amp is by considering the RV1 value of 5 kohm. Based on the calculation, the minimum and maximum gain of a non-inverting op-amp is 1.05 and 2.3 as explained in detail below.

Gain of non- inverting operational,

$$\frac{V_{out}}{V_{In}} = A_v = 1 + \frac{R_2}{R_1}$$

$$\frac{V_{out}}{V_{In}} = A_v = 1 + \frac{200}{4k}$$
$$A_v = 1.05 \text{ (Min gain)}$$

$$\frac{V_{out}}{V_{In}} = A_v = 1 + \frac{200 + (5k)}{4k}$$
$$A_v = 2.3 \text{ (Max gain)}$$

Gain of Voltage follower,

$$\frac{V_{out}}{V_{In}} = A_v = 1$$
$$V_{Out} = V_{In} = 1$$

3.2 The testing of min and max value of op-amp with single piezoelectric

Table 3 tabulated the result of ac voltage at both signal input and output, dc voltage at signal output, as well as min and max gain of op-amp. The overall result obtained from theoretical calculation, simulation and experimental measurement were compared, ensuring a reasonable agreement to one another. The higher values of experimental data compared to both calculated or simulated were due to current losses or error in system setup during experiment. Furthermore, the theoretical calculation was more accurate compared to simulated research work, since more assumptions were involved in theoretical calculation as compared to simulated ideal ones. The min gain of op-amp was selected to fulfil the supercapacitor bank with total output voltage of 5.0 V. All experimental results were comparatively similar with the simulation results. As tabulated in Table 3, the experimental results of output voltage were much closer to the theoretical values, as predicted.

Table 3. Overall result of min and max gain value of op-amp in energy harvesting system.

Single piezo	Calculation		Simulation		Experimental	
	Min gain	Max gain	Min gain	Max gain	Min gain	Max gain
ac input			3.889 V	3.889 V	3.90 V	3.801 V
ac output			4.043 V	8.855 V	4.084 V	8.00 V
dc output			5.203 V	12.068 V	5.01 V	12.76V
Gain	1.05	2.3	1.040	2.277	1.047	2.105

ac input*- ac voltage at signal input
ac output*- ac voltage at signal output
dc output*- dc voltage at signal output

3.3 The testing of charging time on difference supercapacitance with single piezoelectric and min gain value of op-amp

Supercapacitors has been increasingly used recently to enhance the source with low voltage and low current as in sound energy harvesting. The main advantages of a supercapacitors are higher power density, longer cycle period and higher charging/discharging efficiency compared with the conventional battery. Supercapacitors are able to charge and discharge at a faster rate compared to batteries and have significantly larger charge capacity compared to conventional capacitors.

Initially, the experiment was carried out with single supercapacitor of different capacitance (Farad) to investigate which type of supercapacitor are suitable to be chosen to be utilised into a supercapacitor bank. From Table 4, it was observed that both theoretical calculation and experimental charging time for the 0.22 F capacitor was longer compared to the 0.01 F capacitor. However, the experimental charging time was higher compared to theoretical calculation due to the small losses between connections of components which lowered the charging current supplied. As a result, longer charging time is required due to small amount of current supplied.

Hence, the highest capacitance of 0.22 F for charge storage was chosen and emerged as an alternative to batteries in applications where the importance of power deliver surpass the total energy storage. Supercapacitors, on the other hand, were based on a different technology since they are electrostatic devices. The supercapacitors discharge performance is the result of a compromise with its power capability. They were classified into a virtually unlimited life cycle, a high power density due to the low internal resistance and a fast recharging phase. Whereas, the fast discharging criteria can be replaced by building the supercapacitor bank in which each supercapacitors were connected in parallel connection.

Table 4. Comparison of both experimental and theoretical calculation of the time charging required for different value of supercapacitance

Supercapacitance (F)	Time charging (second) single supercapacitor	
	Calculation	Experimental
0.01	1.67 s	15 s
0.022	3.67 s	30 s
0.047	7.83 s	60 s
0.1	16.67 s	80 s
0.22	36.67 s	110 s

3.4 The testing of charging time on different number of supercapacitors with single piezoelectric and min value gain of op-amp

The highest capacitor with value of 0.22 F was selected to construct the supercapacitor bank. The smaller the resistance or the capacitance, the smaller the time constant, the faster the charging and the discharging rate of the capacitor, and vice versa. Supercapacitor with the highest capacitance value has longer charging and discharging time required and vice versa. Due to slow discharging characteristic, it can hold the stored electrical energy for longer duration than the usual supercapacitor. Consequently, the output from multiple supercapacitor bank can be added easily. Table 5 presents the time needed to charge a supercapacitor bank having different pieces of 0.22 F supercapacitor. The results show that the number of supercapacitors was directly proportional to charging time needed. The more number of supercapacitors involved, the longer the duration needed in charging the supercapacitor bank.

With a continuous supply current of 30 mA, the supercapacitor bank (5.0 V) will be fully charged in less than 2 hours for each module. The total numbers of modules for supercapacitors used in this system were 10 modules, resulting in a total of 2.2 F connected parallel to one another. The system with supercapacitor shows significant enhancements compared with the used of battery. In Fig. 5, the LM 358N was used in supercapacitor bank as comparator to control both the green and the red LED condition. The green light in the supercapacitor bank will lit when it was fully charged. Whereas, the red light will lit when the total voltage was remain less than 5.0 V.

Supercapacitor with highest capacitance values per unit volume has the highest energy density compared to other capacitors. With their high capacitance values, supercapacitor can fill in the gap between capacitor and batteries. One important feature of supercapacitors compared to batteries is that it can be charged very quickly and discharge slowly due to the large value of Farad compared to conventional capacitor.

Table 5. Comparison on both experimental and theoretical calculations of the time charging required for different number of 0.22 F supercapacitor

Number of 0.22 F supercapacitor (pieces)	Supercapacitor bank time charging (second)	
	Calculation	Experimental
2	73.34 s	230 s
4	146.68 s	458 s
6	220.02 s	686 s
8	293.36 s	900 s
10	366.7 s	1350 s

4. Conclusions

The topology combines an op-amp and a bridge converter to condition the positive input cycles, respectively. The topology successfully boosts the 3.89 V, 68 Hz ac to 5 V dc. The output voltage amplified to 5.0 V through an ac-dc converter and an op-amp voltage control. In comparison to rectifiers, this study employed the minimum number of passive energy storage components and achieved the maximum conversion efficiency. The output power and the output voltage are compared between all these topologies. The simulation results and experimental results were also compared. It was certified that the simulation results are in a good agreement with the experimental results. It was also proven that the op-amp was able to perform better than other ac-dc circuit, due to the simplicity and cost reduction criteria. An op-amp also capable to increase the output current to speed up the load charging time and improved the overall system efficiency up to 90 %.

Experimental and simulation results displayed that the supercapacitor bank was able to reduce the stress on the conventional battery power sources which has some drawbacks due to its higher volume and a limited life cycle. Furthermore, using the op-amp, maximum energy harvesting can be implemented proficiency. Analysis of the op-amp was carried out. The recommended control scheme with the op-amp circuit was implemented and the bridge converter was successfully operated to directly step-up the low ac voltage to a high dc voltage as well. Future research can focused on investigating and utilized much more voltage for various other applications and the distortion on the input current can be reduced by designing a suitable circuit.

Acknowledgement
The authors would like to acknowledge University Malaysia Perlis and the Malaysian Ministry of Higher Education for providing the Fundamental Research Grant Scheme (FRGS Grant No: 9003-00561), which made this study possible to be conducted and successfully published.

References

1. Deepti and S. Sharma. Energy harvesting using piezoelectric for Wireless Sensor Network. *2016 IEEE 1st International Conference on Power Electronics, Intelligent Control and Energy Systems (ICPEICES)*, Delhi, India, July (2016), 1-3.
2. R. A. Roseline, and P. Sumathi. Solar power for Wireless Sensor Networks in Environment Monitoring Applications-A Review. *International Journal of Scientific and Research Publications*, **4**, 8 (2014), 1-3.
3. X. Lu and S. H. Yang. Thermal energy harvesting for WSNs. *2010 IEEE International Conference on System, Man and Cybernetics*, Istanbul, Oct. (2010), 3045-3052.
4. N. Salim, M. F. M. Idros, S. A. M. Al-Junid and A. H. A. Razak. Study on the capability of Acoustic Energy Harvesting for low power device application. *2016 IEEE Student Conference on Research and Development (SCOReD)*, Kuala Lumpur, (2016), 1-4.
5. K. K. Selim, A. Haggag, H. M. Yehia, F. Z. Amer and A. M. El-Garhy. Acoustic energy conversion into useful electrical energy from Disk Jockey by using piezoelectric transducer. *2016 Eighteenth International Middle East Power Systems Conference (MEPCON)*, Cairo, Dec. (2016), 871-876.
6. W. Turner, C. M. Tang, B. Beck, T. H. Mareci and R. Bashirullah. Nuclear magnetic resonance energy harvesting for ultra-low power biomedical implants. *WAMICON 2011 Conference Proceedings*, Clearwater Beach, FL, (2011), 1-4.
7. A. Jusho, A. Pegatoquet and T. N. Le. Wind Energy Harvesting for Autonomous Wireless Sensor Network. *2016 IEEE Euromicro Conference on Digital System Design (DSD)*, Limassol, Aug-Sept. (2016), 301-308.
8. Y. J. Hu, I. C. Chen and T. H. Tsai. A piezoelectric vibration energy harvesting system with improved power extraction capability. *2016 IEEE Asian Solid-State Circuits Conference (A-SSCC)*, Toyama, Nov. (2016), 305-308.
9. G. R. A. Jamal, H. Hassan, A. Das, J. Ferdous and S. A. Lisa. Generation of usable electric power from available random sound energy. *2013 International Conference on Informatics, Electronics and Vision (ICIEV)*, Dhaka, (2013), 1-4.
10. A. Far. Amplifier for energy harvesting: Low voltage, ultra low current, rail-to-rail input-output, high speed. *2016 IEEE International Autumn Meeting on Power, Electronics and Computing (ROPEC)*, Ixtapa, (2016), 1-6.
11. K. Niotaki, A. Georgiadis and A. Collado. Thermal energy harvesting for power amplifiers. *2013 IEEE Radio and Wireless Symposium*, Austin, TX, (2013), 196-198.
12. I. Fahmi Zambari, C. Yi Hui and R. Mohamed. Development of Wireless Energy Transfer Module for Solar Energy Harvesting. *Procedia Technology*, 11, (2013), 882-894.

13. K. Madhuri and Dr.A Srujana. Low Voltage Energy Harvesting by an Efficient AC-DC Step-Up Converter, *IOSR Journal of Electrical and Electronics Engineering (IOSR-JEEE)*, **7**, 3(2013), 32-47.
14. A. Kailas, D. Brunelli and M. A. Ingram. A simple energy model for the harvesting and leakage in a supercapacitor. *2012 IEEE International Conference on Communications (ICC)*, Ottawa, ON, (2012), 6278-6282.
15. S. Saggini, F. Ongaro, C. Galperti and P. Mattavelli. Supercapacitor-based hybrid storage systems for energy harvesting in wireless sensor networks. *2010 Twenty-Fifth Annual IEEE Applied Power Electronics Conference and Exposition (APEC)*, Palm Springs, CA, (2010), 2281-2287.
16. D. Torregrossa, S. Karkar, E. Rivet, H. Lissek and M. Paolone. Acoustic energy harvesting using Electrochemical Double Layer Capacitors: Technical feasibility and performance assessment. *IECON 2016- 42nd Annual Conference of the IEEE Industrial Electronics Society*, Florence, Oct. (2016), 5495-5500.

IOP Conf. Series: Materials Science and Engineering **318** (2018) 012036 doi:10.1088/1757-899X/318/1/012036

Development and Analysis of Hybrid Thermoelectric Refrigerator Systems

M. Saifizi[1], M. S. Zakaria[1], Sazali Yaacob[2] and Khairunizam WAN[3]
[1]Faculty of Engineering Technology, Universiti Malaysia Perlis, 02100 Padang Besar, Perlis, Malaysia.
[2]Malaysian Spanish Institute, University of Kuala Lumpur, 09000 Kulim, Kedah, Malaysia.
[3]School of Mechatronic Engineering, Universiti Malaysia Perlis, 02600 Arau, Perlis, Malaysia.

E-mail: saifizi@unimap.edu.my

Abstract. Thermoelectric module (TEM) is a type of solid-state devices which has the capability to maintain the accuracy of small temperature variation application. In this study, a hybrid thermoelectric refrigerator system is introduced by utilizing TEMs; direct and air to air thermoelectric heat pump to cool down and maintain low temperature for vaccines storage. Two different materials which are aluminum and stainless steel are used as container in hybrid thermoelectric refrigerator (HTER) configuration to investigate the response of every system in transient and steady state mode. A proper temperature sensor calibration technique is implemented to make certain real time data acquisition of the systems are not affected very much from the noise generated. From step response analysis, it is indicated that HTER I (aluminum) has rapid settling time from transient to steady state than HTER II (stainless steel) since aluminum has better thermal conductivity as compared to stainless steel. It is found that HTER I is better in cooling capability with the same input current instead of HTER II which required a longer time to achieve steady state mode. Besides, in Pseudo Random Binary Sequence (PRBS) response analysis injected to both systems shows HTER I is very sensitive to current input as the sequence length of HTER I is shorter than HTER II. However both systems depict the varying temperature in the range of 4 oC due to differences in thermal conductivity of container.

IOP Conf. Series: Materials Science and Engineering **318** (2018) 012036 doi:10.1088/1757-899X/318/1/012036

1. Introduction

Thermoelectric module (TEM) is a type of solid-state devices. It functions to cool and heat without any movement of mechanical parts. The combination ability to cool down and heat up, TEM is used for heating and cooling on a small-scale application and electric saving. The performance of thermoelectric cooling rely on a variety of variables, such as operating temperature, voltage or current supplies[1]. The efficiency of a heat sink on the hot side will help to heat pumped on the cold side surface. Normally, a thermoelectric cooling system does not operate at a same condition level. It is affected by the ambient environment that probably changing and the power dispersed at different level[2]. The capabilities of a thermoelectric cooling can be analyzed by referring to the transient and steady state condition. The transient condition is affected by thermal capacitance and thermal time constant. At the time of surface being cooled down isothermally, the temperature gradient of the cooling area becomes importance for designing a controller.

A single thermoelectric cooler will not be able to reach its optimum performance in all conditions [1]. Therefore, in order to suit with the desired condition, it is necessary to know the specific application. In normal cases, the most adequate model for large and complex analysis of thermoelectric systems is to choose a lumped model [3], whereas a distributed model is used for detailed description in order to determine the optimum number of thermocouples for the TEM, or when analyzing the performance of heat sinks. In this chapter, development of a prototype hybrid thermoelectric refrigerator using TEMs as plant is presented. The behavior of the system is also discussed in this chapter.

If your paper is intended for a conference, please contact your conference editor concerning acceptable word processor formats for your particular conference.

2. Hybrid thermoelectric refrigerator system

Developing and designing a proper controller of a thermoelectric refrigerator is the main focus of this work, which will help to keep at least 50 ml of vaccines in the indispensable temperature (from 2°C to 8°C) with ambient temperature of 30°C [4]. The Hybrid Thermoelectric Refrigerator systems (H-TER) is designed and developed based on the two parts; namely mechanical and electrical part as shown in Figure 1. The assembly technique adopted in the thermoelectric heat pump is crucial to minimize the thermal resistance which will cause a significant improvement in its performance.

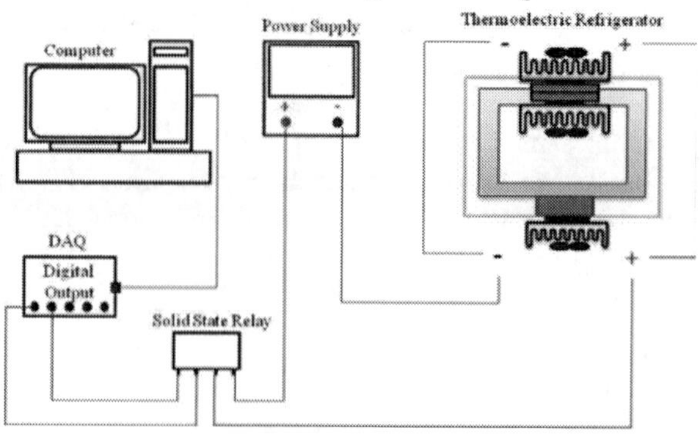

Figure 1. Hybrid Thermoelectric Refrigerator (H-TER) control system configuration

The primary structure of the proposed system consists of aluminium/stainless steel container (12 cm width x 12 cm length x 9 cm height with thickness of 3.5 mm) and insulated box with an isothermal material (polystyrene) to minimize loss of heat to the ambient. The insulation material also separates the container and acrylic casing as shown in Figure 2. Two thermoelectric heat pumps are utilized in this method, namely; direct and air-to-air thermoelectric heat pumps. Both thermoelectric heat pumps are connected in parallel configuration with 44 Watt input power at maximum input condition. The cold side heat exchanger will absorb the heat and the TEM immediately pump out. Whereas, air-to-air thermoelectric heat pump installation engages by cooling objects through convection. This is done via the hot side of the heat exchanger which is equipped with fans to dissipate the heat to its surrounding. Meanwhile, the cold surface of air-to-air thermoelectric heat pump is fixed to another heat sink, which is used to bring out the cold air and a compulsion of air convection through the length of the heat sink. In order to obtain better performance of cooling in the container, a fan is used. Furthermore, the internal air circulation will assist to maintain and keep the temperature gradient within the enclosure to a minimum level. Direct thermoelectric heat pump offer performance by cooling an object via conduction. A space block is also put to the direct thermoelectric heat pump which yields to a maximum heat transfer to the container.

The heat sink disperses heat from both pumps which are generated by the TEM as a result from the Joule effect. This indicates that the lower the thermal resistance of the heat sink, the better indication of performance of the TEM. The heat sink attached to a fan will change the performance of the heat sink by decreasing the thermal resistivity of dissipating heat via forced convection. The fans are operated at a constant speed with 12V direct current (DC) and 0.18A input.

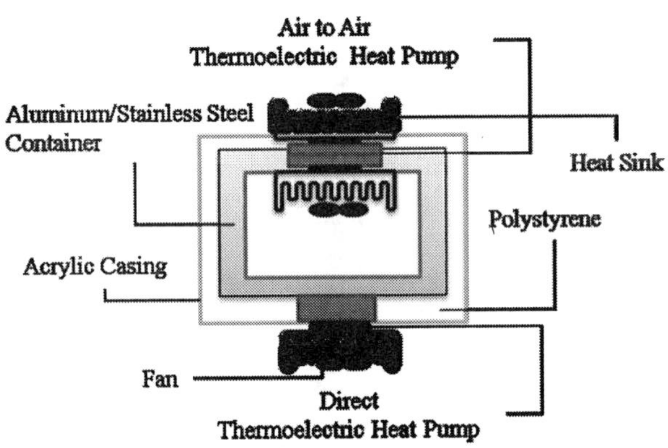

Figure 2. Hybrid Thermoelectric Refrigerator with insulation assembled and integrated with air-to-air and direct thermoelectric heat pumps

3. Experimental setup

3.1 Temperature measurement

In order to accomplish a successful system identification procedure and the implementation of on-line control strategies, a suitable computer program and interfacing device is very important. The system interface should be able to handle the implementation of real-time measurement capabilities as shown in Fig. 3. It is not only the ability to read and send data between computer and the plant, the computer memory and data acquisition (DAQ) should be adequately realized the online system identification and control strategies.

The setup and installation of the interfacing system should be easy in a very simple procedure. Both hardware and software fitted with the system is reliable in handling long-run experiment and analysis. The data collected by this system is useful and easily processed and retrieved without any issues or problems. Moreover, the input-output programming should be easy and simple by adopting reliable programming software. The following measurement instruments in Fig. 3 are used in the identification and control of H-TER systems.

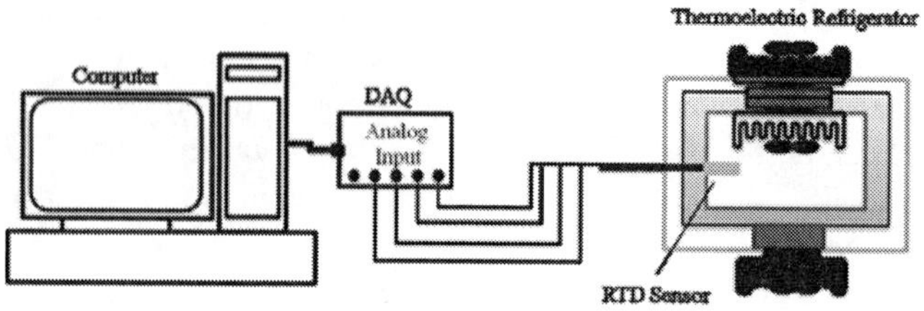

Figure 3. Temperature measurement instrumentation of H-TER systems using RTD sensor and DAQ

NI-9219 is plugged to NI-9174 as a USB modular instrument chassis. Such connection will allow synchronization and triggering capabilities as well as flexibility that will help to obtain a proper measurement. Fig. 4 described the connection of 4-Wire Resistance and 4-Wire resistance temperature detector (RTD) sensor, where the NI 9219 is generating current that depends on the load of resistance between the EX+ and EX– terminals. The measured resistance is computed by NI-9219 from voltage reading of the sensor. However, the wire resistance is not affecting this sensor since high input impedances of analog-to-digital converter (ADC) allow a slight amount of current flows across the HI and LO terminals.

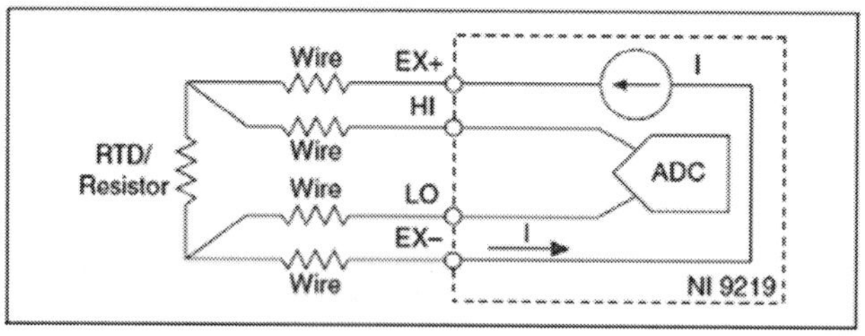

Figure 4. Connection in 4-wire resistance and 4-wire RTD sensor circuit to measurement DAQ instrument

The RTD is known as precision sensor and being used in industrial and laboratory's applications for low temperature measurement. The RTD element is more repeatable and accurate than the thermocouple. In fact, the RTD sensor produces stable readings longer than thermocouple. RTD also characterizes as more repeatable readings sensor, which is the same temperature readings produce same results over multiple trial. RTD sensor employs the property that the electrical resistance of metal varies with temperature. The resistance of the sensor is changed linearly according to the temperature change. This resistance is measured by DAQ and later converts into a temperature reading. The resistance reading can be carried out by passing a small current through RTD's element.

NI-6008 digital analog input-output (IO) data acquisition (DAQ) is a control unit that interfaces computer and H-TER systems. This DAQ supplies digital analog channel that allows for closed loop control through connections to solid-state relay (SSR) in order to control the input actuator that is, the thermoelectric heat pumps. The NI USB-6008 DAQ is a USB supported as a standard control device using analog as well digital IO. Pulse width modulation (PWM) is generated by digitally controlling current using NI-6008 DAQ by digital encoding analog signal levels. The duty cycle is able to encode a specific signal of current level by modulating the square wave. However, the signals of PWM are always in digital form since the supply of DC is either fully on/off, or intermediate. The current output is applied to the H-TER systems through series of recurring of on and off pulses.

The thermoelectric heat pump is used for refrigerator application that will function directly from DC power. The input current will be controlled by a digital computer programming, interfaced to NI-6008 DAQ output to maintain temperature inside the container that produces by SSR. It automatically corrects the temperature using a feedback loop. The MOSFET transistor junction in SSR is a component to control the current input through photodiodes that generates electric insulation. This technology allows NI-6008 DAQ devices to control the flow of PWM input current of thermoelectric heat pumps in H-TER systems applications. In addition, the SSR enables for a lower current controlling signal like NI-6008 DAQ. The SSR is connected in series to 12V DC power supply, thermoelectric heat pumps and DAQ digital output. The SSR is also wired with digital input DAQ to generate PWM input current from power supply to control the inner temperature of H-TER systems as shown in Fig. 5.

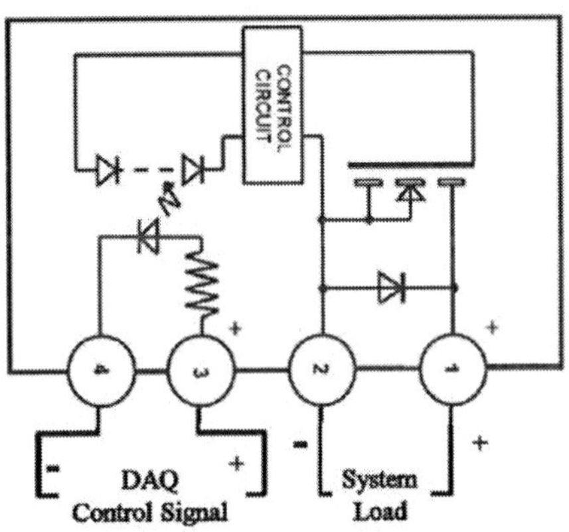

Figure 5. Control circuit of SSR configuration for controlling input and generate current output signal for system load

However, the SSR performance to generate the output current is depending on the ambient operating temperature of the device as shown in Fig. 3.9. The H-TER systems data is recorded by a DAQ for every single second. It is very useful in closed loop feedback application to control and correct any temperature variation. Fig. 6 shows configuration components within the system that dynamically do not intervene and affect its performance. There is an aluminium block spacer that connects the direct thermoelectric heat pump cold side to the aluminium/stainless steel vaccine's container.

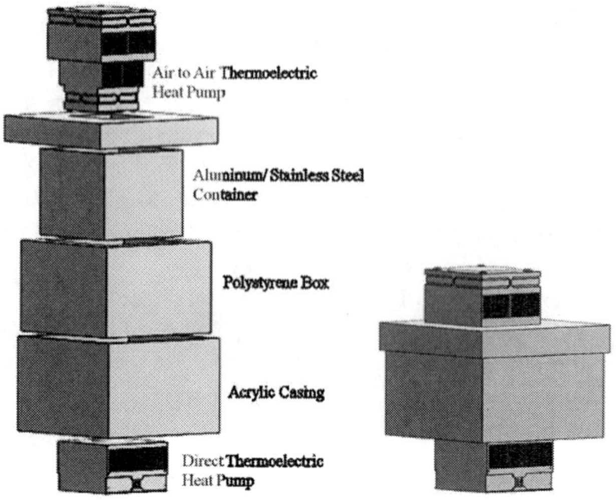

Figure 6. Hybrid Thermoelectric Refrigerator integration, explode and collapse views

This includes the material use in insulation, which made of polystyrene with thermal conductivity of 0.033 W/(m•K). The container is placed in acrylic casing as a holder. The vaccines will be kept and stored in a container and insulated by the polystyrene to avoid ambient or any other heat source affecting the inner temperature.

3.2 Sensor Calibration

The aim of performing sensor calibration is to make certain on the consistency aspect in measuring a temperature range under specified environmental condition. In this work, the sensor calibrated online to reduce the effect of noise and bias, which is affected by the instrument and transmission cables from the data logging devices and plant as well. The main aspect of calibration is to determine the relationship between the measured variables with the electrical signal such as resistance, voltage, and current. Platinum RTD is used with a typical resistance of 100Ω at 0°C. In addition, the calibration is done to get accuracy measurement for other than 0°C and 100°C, and able to measure up to 200°C. RTD can determine its resistance and hence the temperature by measuring the voltage across the element. The resistance and temperature have a linear relationship. The RTD is categorized at nominal resistance of 0°C with the sensor of 100Ω. The following Equation (1) describes linear relationship between resistance and temperature:

$$R_T = R_0\left[1 + aT + bT^2\right] \quad (1)$$

RT is resistance at temperature of T; R0 is nominal resistance; a and b are constant that applied in scaling the RTD. The calibration test of temperature in a specific range is very important to perform accuracy measurement. In this work, the sensor will be calibrated at 0°C to 8°C as this temperature range is considered safe storage for vaccine. The average resistance or temperature curve for this 100Ω pure platinum RTD is shown in Fig. 7. A specific calibration technique, i.e. in-situ technique is implemented for this work in order to obtain its accuracy. The calibration is conducted at Calibration and Measurement Laboratory of SIRIM Sdn. Bhd. at Penang, Malaysia. RTD sensor is calibrated using water bath at 28°C -30°C ambient condition to determine scaling parameter by direct measurement.

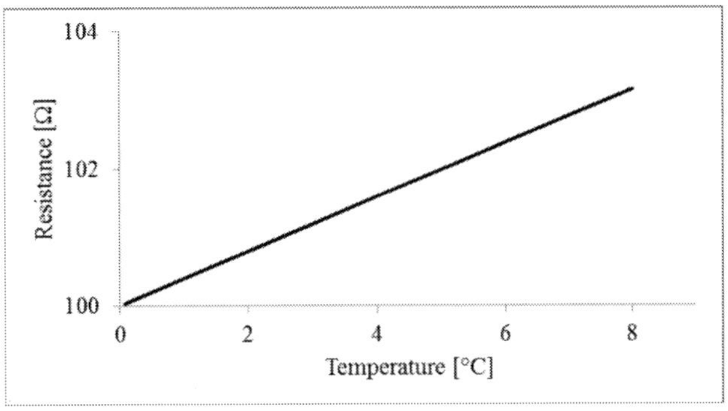

Figure 7. The average resistance/temperature curve for RTD sensor calibration graph using ice water bath

The relationship between resistance and temperature is found to be relatively linear. The fitted curve is used to ensure an accurate RTD measurement. The constant values of a and b are extracted as and respectively by the curve fitting.

3.3 Filtering Operating Data

A first order low-pass filter is designed for noise minimization of the sensor. The transfer function for the filter can be written in as:

$$\frac{Y(s)}{U(s)} = \frac{1}{T_c s + 1} \tag{2}$$

where Tc is the time constant, U(s) is the input and Y(s) is the output. The above equation is discretized Euler Backward finite difference method as:

$$T_c \frac{y(k) - y(k-1)}{T_s} + y(k) = u(k) \tag{3}$$

$$y(k) = \frac{T_c}{T_c + T_s} y(k-1) + \frac{T_s}{T_c + T_s} u(k) \tag{4}$$

where Ts is sampling time of the system. From the Equation (5), F_a is defined as filter parameter:

$$F_a = \frac{T_s}{T_c + T_s} \tag{5}$$

Haugen suggested the following rules, which is the sampling time should be smaller than time constant of the system to avoid elimination of system dynamic information [5]:

$$T_a \leq \frac{T_c}{5} \tag{6}$$

The filter is fitted into the H-TER systems with appropriate specification and significant frequency range. Signal output temperature that has been measured to be filtered using a computer program such as the equations above to remove noise. This is important in order to obtain better parameter estimation.

4. H-TER Systems Response Performance

4.1 Step Response Analysis

The objective of this section is to analyze the dynamic behavior of thermoelectric refrigerator under the effect of heat conduction and convection. In practical sense, this refrigerator always work under transient operating conditions due to the time change in the current and voltage supply, hot and cold side of TEM surface temperatures, and electronic devices performance. A study of transient behavior of the thermoelectric refrigerator rather than the steady state behavior is very important for two reasons: (a) to investigate the H-TER behavior during the startup and shutdown period and (b) to understand the system behavior when all or some of the operating condition are varying with time. Examples of these operating conditions are the cold temperature, hot temperature, and the imposed electric field. The transient behavior of the open loop H-TER systems due to a sudden imposed electric current is illustrated in Fig 8. Initially, the input current is set to 85% duty cycle, which is the current input varies linearly with duty cycle. By using this input current, the air temperature of refrigerator is 8.4°C and 8.7°C for H-TER I and H-TER II systems respectively. At 1000 seconds, a constant 90% duty cycle of input current is imposed suddenly, in the form of a unit step function, on the system causing the air temperature inside the both containers stainless steel and aluminum are decreased drastically and maintained at around 3.93°C and 4.19°C for H-TER I and H-TER II systems respectively. After 2000 seconds, the temperature continues to decrease at a slower rate response.

The response for both H-TER I and H-TER II is given in Fig. 8. It shows the response towards the experiment data is recorded at 20 second sampling time. It also indicates that the H-TER systems have slow settling time of transient response with time delay performance. However, the system did not provide any overshoot performance. The H-TER II system take longer time to reach steady state condition than H-TER I system because of thermal resistance of stainless steel container is higher than aluminium container.

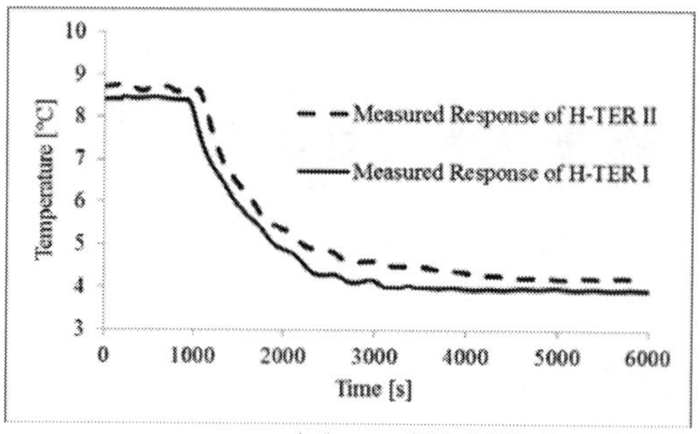

Figure 8. Step response measured data of H-TER I and H-TER II systems

Table 1 shows the characteristics step response performance for H-TER1 and H-TER II systems. It shows that H-TER I transient response is faster than H-TER II system. The H-TER I found to be better in cooling efficient with the same step input current. It can be deduced that any increase of thermal resistance of the container will cause air temperature respond more slowly and increase the time required to reach steady state condition.

Table 1. Characteristics measured step response performance of H-TER 1 and H-TER II

System performance	H-TER I system	H-TER II system
Time constant [s]	660	710
Settling time [s]	2490	2960
Time delay [s]	20	40

The generated step response data of the experiment in Fig. 8 is compared with simulated data that is generated by transfer function in Equation 7 and Equation 8. The transfer functions are developed base on characteristic of H-TER systems in Table 1. The validation of a simulated first order model with a measured step response of H-TER I and H-TER II systems are shown in Fig. 9 and 10 respectively. The estimation model of the system is digitized by implementing zero order hold method. Based on the converted model to digital form in Fig. 9 and 10, it indicates that the simulated first order model provides a close transient and steady state characteristics as compare to the measured data of H-TER systems. It can modeled in the discrete time transfer function of H-TER I and H-TER II as Equation (7) and Equation (8) respectively,

$$\frac{T_{H-TER1}(z)}{U_{H-TER1}(z)} = z^{-1}\frac{0.1334z^{-1}}{1-0.9702z^{-1}} \quad (7)$$

$$\frac{Y_{H-TERII}(z)}{U_{H-TERII}(z)} = z^{-2}\frac{0.1217z^{-1}}{1-0.9722z^{-1}} \quad (8)$$

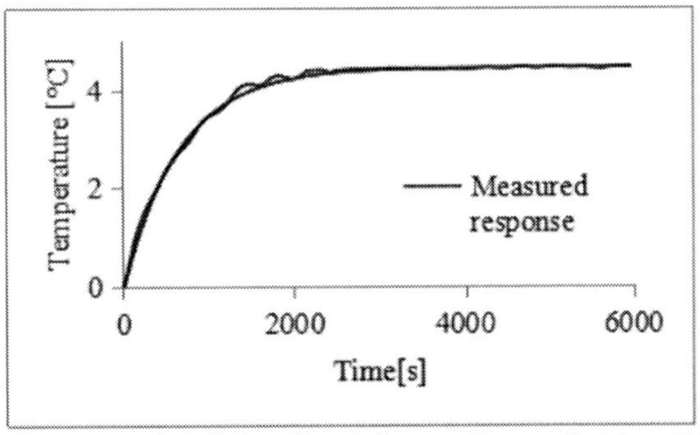

Figure 9. Step response of measured and simulated first order model for H-TER I system

IOP Conf. Series: Materials Science and Engineering **318** (2018) 012036 doi:10.1088/1757-899X/318/1/012036

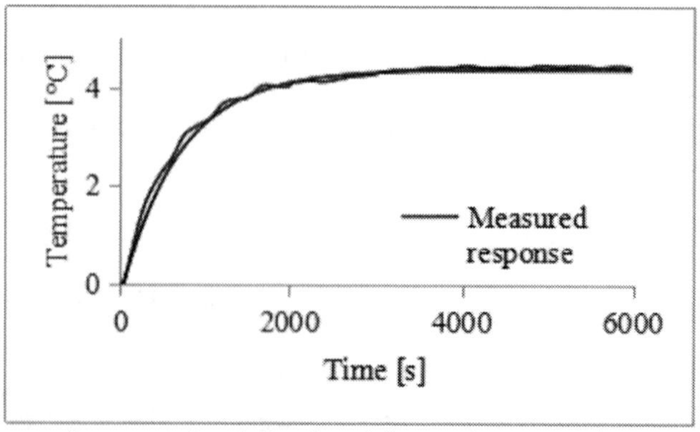

Figure 10. Step response of measured and simulated first order model for H-TER II system

Another way to understand about system's properties is by referring to the poles and zeros in z-plane plot of the system in Fig. 11. The pole-zero diagrams are generated based on Equation 7 and Equation 8 discrete transfer function. Particularly, the poles area has a direct effect upon the properties of a dynamic system which is noted its stability. The Fig. 11 also indicates that the system is found to be stable as the pole located within the unit circle of z-plane.

The location of the poles is relatively near to the unit circle's circumference of z-plane. The transient properties will depict an oscillatory response. Thus, a first order model is maybe not adequate representation of the H-TER systems due to oscillatory characteristic of step response in Fig. 11. This called for higher order model to represent the H-TER systems. A second order system may be appropriate since the response of the first order model indicates poor relative stability as a result from the pole location. This supports the second approach of modeling by which it is required to optimize the identification with the model.

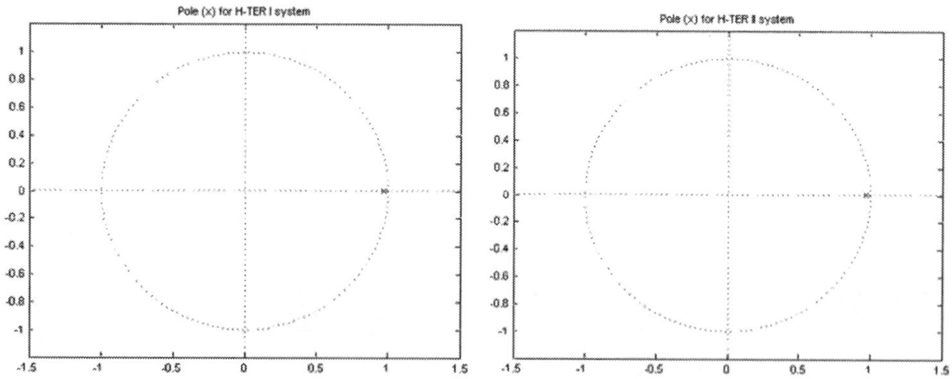

(a) (b)

Figure 11. First order model Pole-zero diagrams of (a) H-TER 1 and (a) H-TER II systems showing the poles located relatively near the unit circle onto z-plane

4.2 PRBS Response Analysis

Pseudo Random Binary Sequence (PRBS) signal is injected to the H-TER systems interfaced with computer. During the data collection the noise is filtered using a low pass filter of 200 seconds time constant. Starting from steady state condition, a sequence of input signal is injected to excite the H-TER systems. Two set of real-time data are collected from the H-TER systems having different container material. Each set of measurement consists of 250 output discrete data with 20 seconds sampling time. PRBS sequence length and magnitude are required and have to be selected carefully especially in real-time parameter estimation. In parameter estimation the choice of input magnitude is critical in the sense that if the magnitude input is too small, system cannot be simulated enough dynamic then the response to be improperly represented. However, if the magnitude is too large, then the system will have nonlinearity characteristics and the stability will be affected.

A PRBS is a periodic, deterministic signal, which approximate a discrete-time white noise-like properties may switch between two constant levels only at certain equally spaced time intervals, at time t=0, Δt, $2\Delta t$, $3\Delta t$..., where Δt is clock period. In this experiment, the useful rule of thumb is used to calculate clock period according to [3] in Equation (9):

$$\Delta t = (0.2 \;\; to \;\; 0.5) \times T_C \qquad (9)$$

where T_C is time constant of the system.

Furthermore, since there is a time delay in the H-TER systems response, the time period chosen in designing the PRBS is then become:

$$\Delta t = (0.2 \;\; to \;\; 0.5) \times T_C + time \;\; delay \qquad (10)$$

In order to identify correctly sequence of length of PRBS input, the total duration time of the signal to exhibit a complete the steady state gain of the plant dynamic model should be considered. Therefore, the settling time of the H-TER systems is approximated equal to the sequence period T of the PRBS signal, the minimum PRBS sequence length can be determined by:

$$T = \Delta t \times N \approx T_s \qquad (11)$$

where T_S is settling time and N is sequence length. The time constant, settling time and time delay is referred to Table 1 to calculate sequence length and clock period. The maximum of sequence length in which n is the number of cells of the shift register as in following equation:

$$N = 2^n - 1 \qquad (12)$$

where n is the number of registers. The input signal sequence repeats itself every 2^n-1 bit interval for a n-bit shift register. A PRBS signal with maximum length sequence of 6 to 16 is used during data collection for H-TER systems. However, the sequence length form did not produce significant changing on output of H-TER systems, which is not varying at 4°C. By using trial and error procedure, a PRBS signal input with maximum length sequence of 3 and 7 are used during data collection for H-TER I and H-TER II respectively. The sequence length of H-TER I is smaller than H-TER II is maybe due to H-TER I is sensitive to changing input current compared to H-TER II. The input magnitude of the PRBS signal is chosen to be ±5% of steady state value of the H-TER systems temperature, which is 85% to 90% duty cycle of current input switching up and down continuously. Fig. 12 and Fig. 13 are the measured temperature response of H-TER systems for filtered and unfiltered condition. The result shows that the temperature responses of H-TER I and H-TER II systems slightly vary in the range of 4°C as optimum operating point. It may be due to differences in conduction heat transfer rate and thermal conductivity of container material.

(a)

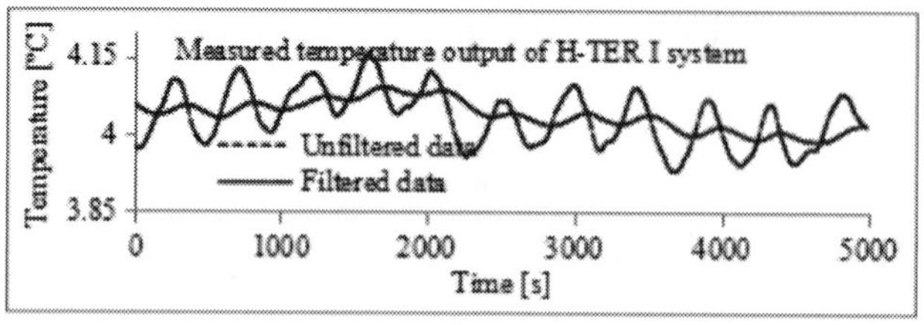

(b)

Figure 12. Filtered and unfiltered of measured temperature response of H-TER I system (a) PRBS input to SSR, (b) output

(a)

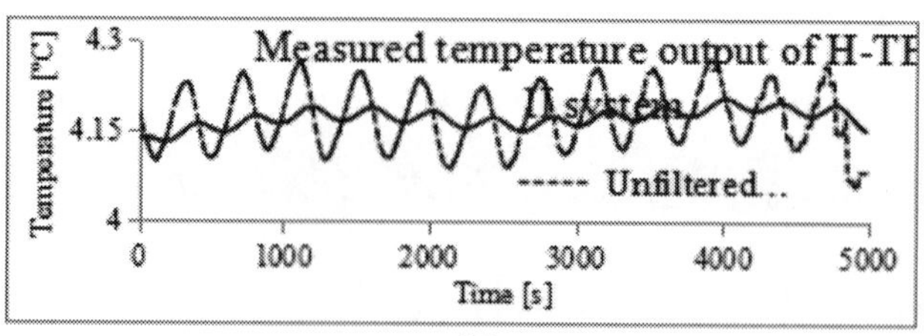

(b)

Figure 13. Filtered and unfiltered of measured temperature response of H-TER II system (a) PRBS input to SSR, (b) output

5. Conclusion

In this work, development of a refrigerator system called H-TER systems based on thermoelectric cooling technology is presented. The experimental setup is designed for modeling and controlling the system in real-time application. In addition, the importance of calibration and data conditioning process using first order low pass filter in work activities is implemented. This is crucial to confirm on the accuracy aspect of low temperature measurements for critical application, which are always influenced the dynamic ambient condition. As the nature of thermoelectric devices is nonlinear, therefore, the selection of optimum operating point for a thermoelectric refrigerator in given application requires an iterative process. The operating point of this work application is 85% to 90% duty cycle input current to achieve temperature range of vaccines storage. A step and PRBS input response experiments are conducted to analyze the dynamic characteristics of H-TER systems. The system is sensitive to the varied and fluctuation of current input especially the aluminium container material properties. A first order dynamic model of H-TER systems has been developed, to validate the system and to work the system properties based on pole and zero plots.

References

[1] Andrea Montecucco, Jonathan Siviter, Adrew R. Knox, "The effect of temperature mismatch on thermoelectric generators electrically connected in series and parallel ," in *Applied Energy*, vol. 123, pp. 47-54, Jun. 2014.
[2] C. A. Gould, N.Y.A. Shammas, S. Grainger, I. Taylor, "Thermoelectric cooling of microelectronics circuits and waste heat electrical power generation in a desktop personal computer," in Material Science and Engineering B, vol. 17, pp. 316-325, 2011.
[3] D. Mitrani, J. Salazar, A. Turo', M. J. Garcia, A. Chavez, "Lumped and Distributed Paramaeter SPICE Model of TE Devices Considering Temperature Dependent Material Properties," in Thermal Investigation of ICs and Systems, pp. 202-207, 2007.
[4] Zaffran M. Vaccine transport and storage: environmental challenges. Dev Biol Stand 1996;87:9–17.
[5] Finn Haugen, "Advanced Dynamic and Control," Tech Teach, 2012.
[6] A.J. Fairweather, M.P. Foster, D.A. Stone, " Battery Parameter Identification with Pseudo Random Binary Sequence Excitation (PRBS)," in Journal of Power Sources, pp. 9398-9406, 2011.

IOP Conf. Series: Materials Science and Engineering **318** (2018) 012037 doi:10.1088/1757-899X/318/1/012037

Lower Side Switching Modification of SHEPWM for Single H-Bridge Unipolar Inverter

M.Z. Aihsan
University Malaysia Perlis

Abstract. Selective Harmonic Elimination Pulse Width Modulation (SHEPWM) is a famous fundamental frequency method for both single stage H-bridge inverter and cascaded multilevel inverters. The main function of SHEPWM is to eliminate the selective lower order of odd harmonic such 3rd, 5th 7th and 9th of the output voltage of the inverter but maintain the fundamental component. In this paper, the 5kHz of the unipolar SHEPWM switching scheme of the inverter is developed and later will be compared to the modified SHEPWM switching scheme. The performance of this inverter is measured through the final total harmonic distortion (THD), the efficiency of the whole system and the natural shape of the output after LC filter.

Email: zaid@unimap.edu.my

1. Introduction

The full bridge inverter is a circuit consists of four units of power semiconductor switches that performs the switching operation between ON and OFF in a dedicated sequence with respect to the switch gate pattern and convert the DC input voltage into a symmetric AC output voltage with desired magnitude and frequency [1]. In the default switching scheme, the first phase of inverter will conduct from $0-\pi$, while the second phase will be in $\pi-2\pi$ and both phases are in high switching state. It is noticeable that the switching can be modified by changing the lower side switching of the inverter with a 50 Hz ON-state for both phases. Table 1 shows the desired modification scheme, Figure 1 illustrate the conventional switching strategy and Figure. 2 the proposed modified switching scheme.

Table 1. Switching strategy distribution

Gate Signal	Conventional		Modified	
	$0-\pi$	$\pi-2\pi$	$0-\pi$	$\pi-2\pi$
S1	High Frequency	**OFF**	High Frequency	**OFF**
S2	High Frequency	**OFF**	50 Hz ON-state	**OFF**
S3	**OFF**	High Frequency	**OFF**	High Frequency
S4	**OFF**	High Frequency	**OFF**	50 Hz ON-state

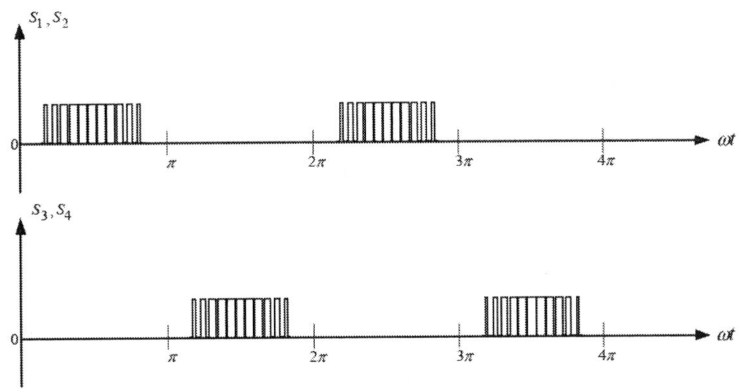

Figure 1. The conventional switching scheme

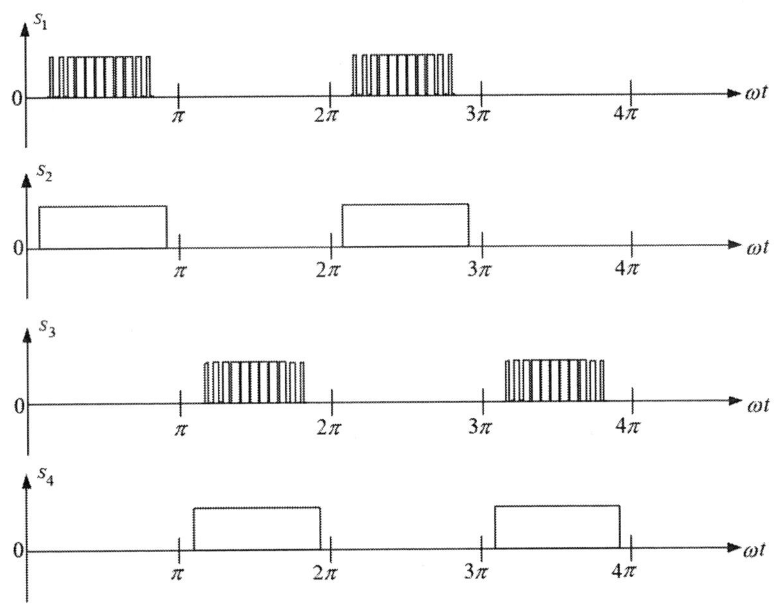

Figure 2. The proposed modified switching scheme

In inverter studies, the measurement of THD and efficiency of the whole system are important. These two criteria can be set and tuned at the switching scheme of the inverter itself. As an example, the higher switching frequency will lead to a bigger switching loss which later could impose to a low efficiency of the inverter. Improper switching angles also will lead to appearance of harmonic at odd orders resulting the increment percentage of final THD value. As seen in **Fig. 2.**, the switching scheme is modified by combining the high switching state with a single 50 Hz ON-State at the lower side MOSFET of the inverter.

By modifying this switching scheme will lead to lesser switching state as can be seen in **Fig. 6.** below. By right when the switching state in reduce, the efficiency of the inverter should increase and also the value of THD will be decreased. This matter will be discussed more in result section below.

2. Formulation for Unipolar SHEPWM

The SHEPWM switching angles are determined by using mathematical formulation. The waveform of the inverter for unipolar is given by the following Fourier series and the waveform has odd quarter-wave symmetry.

$$f(x)=\frac{a_o}{2}+\sum_{n=1}^{\infty}(a_n\cos(n\theta)+b_n\sin(n\theta)) \quad (1)$$

a_o, a_n and b_n are constant coefficient

This equation is used to solve the unknown angles that hold the desired harmonics order to be eliminated. By means this equation can calculate the right angles for not generating the harmonic at the desired harmonic order to be eliminated with respect following the condition of equation (4) below [4]. Usually for multilevel inverter, certain harmonic order will be eliminated as the nature shape of multilevel inverter output voltage is in symmetrical AC wave shape while for unipolar inverter natural shape before the filter is in square wave shaped. For this project, the f_{sw} selected is 5 kHz and the elimination is made up to 99th harmonic order cancelations by using the following equations:

$$F=[(\cos\theta_1)-(\cos\theta_2)+(\cos\theta_3)-....+(\cos\theta_n) \quad (2)$$

$$F=[(\cos\theta_1)-(\cos\theta_2)+(\cos\theta_3)-....+(\cos\theta_{99}) \quad (3)$$

$$\theta_1<\theta_2<\theta_3...<\theta_{99}<90° \quad (4)$$

2.1. MATLAB code generation

In order to solve the equation, MATLAB software is used. The equation need a proper initial guess of triggering angles and it will re-calculate the correct initial guess with MATLAB M-File [2-5]. Equation (5) shows that the transcendental equation and the value of output is set to zero except the fundamental equation. To solve this equation, Newton-Raphson method is used. MATLAB M-file code is used as this Newton-Raphson requires heavy computation strain [3-5].

$$\cos(1\times t1)-\cos(1\times t2)+\ldots+\cos(1\times t(n-2))=\frac{A\pi}{4} \quad (5)$$

$$\cos(3\times t1)-\cos(3\times t2)+\ldots+\cos(3\times t(n-2))=0 \quad (6)$$

$$\cos(m\times t1)-\cos(m\times t2)+\ldots+\cos(m\times t(n-2))=0 \quad (7)$$

Where,

Odd order: t1, t3, t5 = Turned ON state
Even order: t2, t4, t6 = Turned OFF state
m = harmonic order
m = n – 1
n = 2 × number of pulses per quadrant
m = (2 × 25) – 1 = 49 harmonic order (harmonic 3rd – 99th)

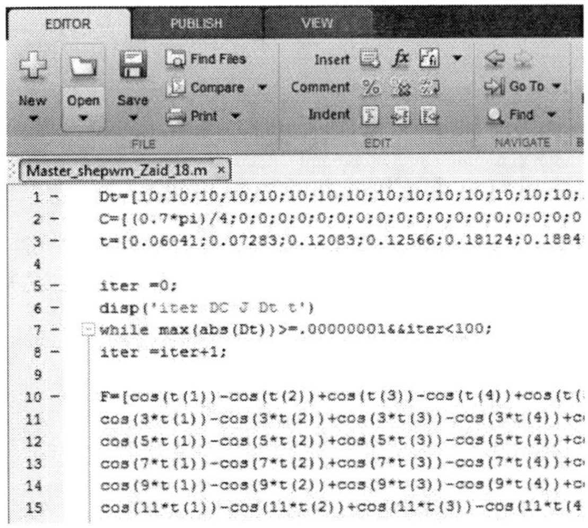

Figure 3. MATLAB code for angle generation

It has 4 major part in this MATLAB code generation where are to insert the proper initial guess, the iteration of the numerical method, the deviation of the actual answer and the final part is the Jacobian matrix as to differentiated the value of F matrix from the previous section as in equation (8):

$$J = \begin{bmatrix} -1\,\sin(1\times t1) & 1\sin(1\times t2)\ldots & 1\sin(1\times t50) \\ -3\,\sin(3\times t1) & 3\sin(3\times t2)\ldots & 3\sin(3\times t50) \\ & . & \\ & . & \\ -99\,\sin(99\times t1) & 99\sin(99\times t2)\ldots & 99\sin(99\times t50) \end{bmatrix} \tag{8}$$

The iteration process continued by finding a new angle by using the final equation (9):

$$t = t + D_t \tag{9}$$

After the iteration process is done, the calculated angles will be presented as follows in the MATLAB workspace as in Table 2 The iteration process usually repeats several times as to get the degree of accuracy [6].

Table 2. Example of calculated angles using MATLAB

State	Angles	Radian	Degree	Times
ton	t1	0.0583	3.3405	185.58
toff	t2	0.0619	3.5472	197.07

ton	t3	0.1166	6.6816	371.20
toff	t4	0.1238	7.0943	394.13
ton	t5	0.1750	10.0239	556.88
toff	t46	1.4208	81.4087	4522.70
ton	t47	1.4215	81.4483	4524.90
toff	t48	1.4837	85.0071	4722.62
ton	t49	1.4839	85.0209	4723.38
toff	t50	1.5708	89.9990	4999.95

3. Simulation Design using Simulink

At this time, this project is tests with simulation using MATLAB Simulink as to test the performance of the inverter and the THD behaviour using FFT analysis module in the Simulink. The circuits are constructed together with its controller as in Figure 4 and Figure 5. The circuit parameter for this project as shown in Table 3.

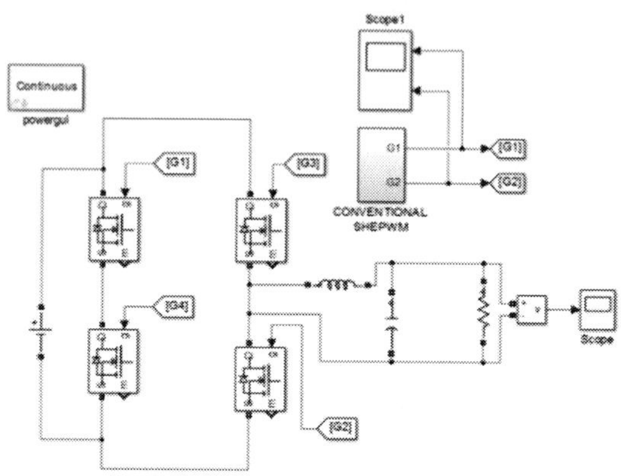

Figure 4. The conventional SHEPWM controller

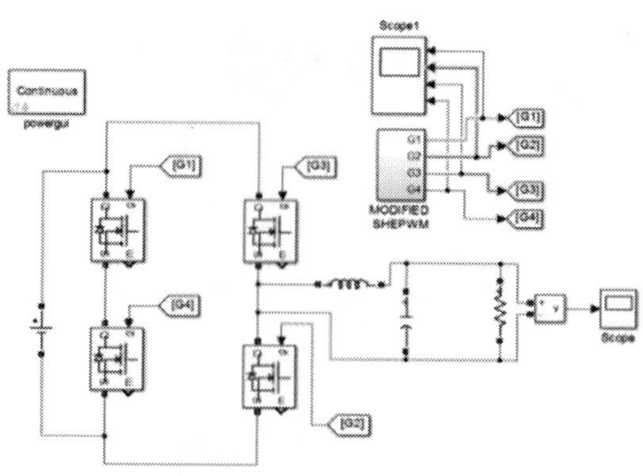

Figure 5. The Modified SHEPWM controller

Table 3. Example of calculated angles using MATLAB

Criteria	Conventional	Modified
Fundamental frequency, F	50 Hz	50 Hz
Switching frequency, F_{sw}	5 kHz	5 kHz
Input Voltage	100 V	100 V
Resistive load	25 ohm	25 ohm
Inductor	2.5 mH	2.5 mH
Capacitor	60 uF	60 uF

As shown above, the same circuit used for both cases, only the controller are different. Conventional SHEPWM share the same signal for both MOSFET 1 and MOSFET 2 while for modified SHEPWM, MOSFET 1 received the high switching state from G1 signal and MOSFET 2 received 50 Hz On-state from G2 signal.

4. Verification

4.1 Simulation results

To verify the results, both conventional and modified SHEPWM inverter is connected to normal R load under unfiltered and filtered condition. A low pass LC filter is used as filter to change the square wave shaped AC into a smooth symmetrical AC output voltage. Usually, the performance of SHEPWM is first tested without LC filter as to see whether the decided harmonic elimination is eliminated. Connecting an inverter to a LC filter is the purpose of to power up the normal AC load application. The modulation index set for this project is 0.8 the performance of inverter working fine under condition of under modulated [8].

4.1.1 Unfiltered inverter

A) Inverter output voltage

The circuit is first tested without LC filter and compared between the conventional and modified SHEPWM as in Figure 6.

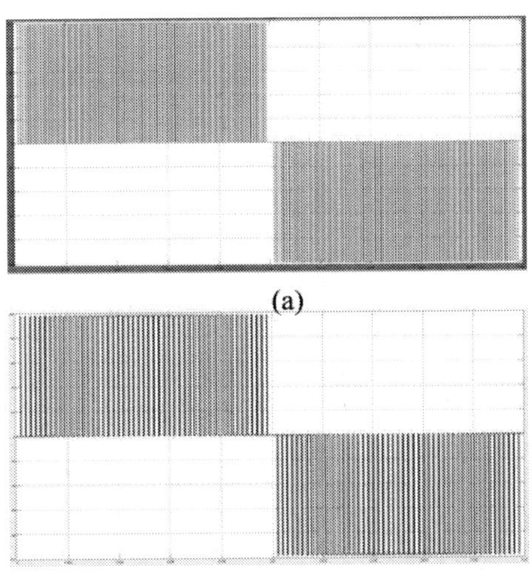

(a)

(b)

Fig 6. a) The conventional SHEPWM b) The modified SHEPWM

As shown above the comparison between two-switching scheme and the modified switching scheme has a lesser switching state even though it has the same switching frequency. A lesser switching state will result a better performance of efficiency as the higher switching state will lead to a huge switching loss.

B) Total harmonic distortion

As presented above on Figure 6 the inverter is tested with a normal resistive load. Usually for harmonic elimination resistive load is used at the beginning before performing for inductive load as resistive load nature behaviour will not have criteria such lagging and leading for rise current of the inverter. The rise and fall will lead the output voltage not align in a right phase in one cycle time of 50 Hz. If with resistive load, the cancelation is perfectly eliminated, an inductive load test is ready to be tested. Figure. 7 shows the fast Fourier transform, FFT of THD output voltage of the developed system.

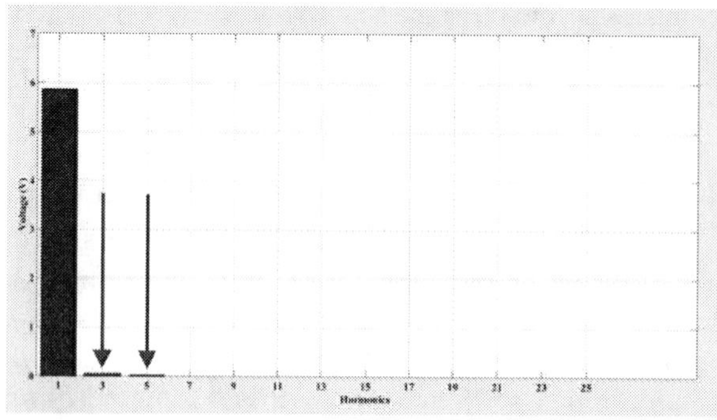

a)

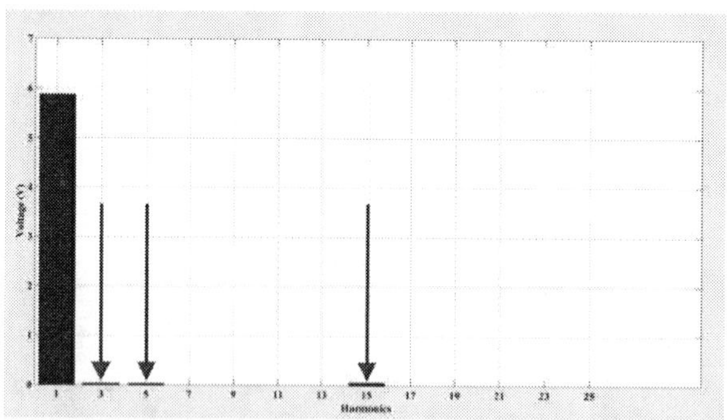

b)

Figure 7. a) THD result for conventional SHEPWM b) THD result for modified SHEPWM

Table 4. THD performance for unfiltered inverter

Switching strategy	Overall percentage
Conventional	37.2 %
Modified SHEPWM	38.5 %

In Figure 7a above, it can be observed on 3rd and 5th harmonic order is still appeared. On 3rd harmonic appear to be slightly higher than 5th harmonic and the rest of the harmonic is eliminated. While for Figure 7b, 3rd and 5th harmonic has the same magnitude and appearance of harmonic on 15th order. The possible reason of this phenomenon is during the combination of high switching state and 50 Hz ON-state reducing some of the switching state and it may not suitable.

4.1.2 Filtered inverter

A) Inverter output voltage

In this section, inverter will be tested with the low pass active filter with the value of inductor and capacitor mention in Table 3 earlier.

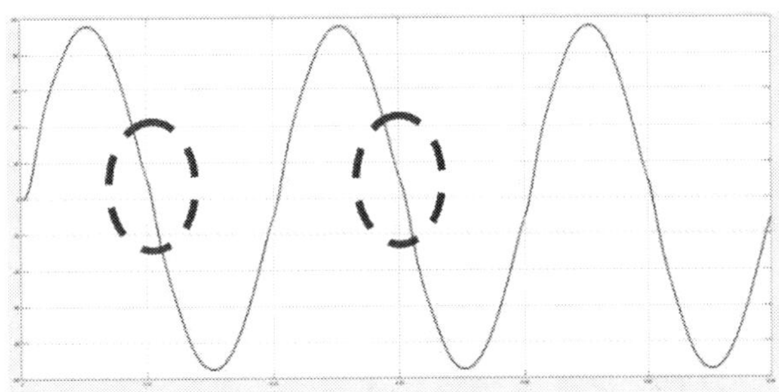

a)

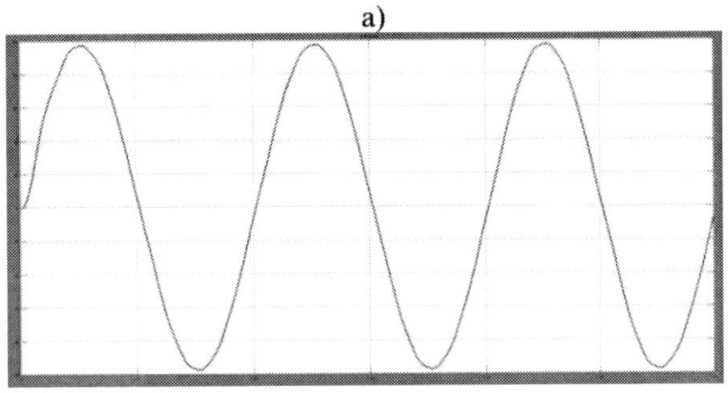

b)

Figure 8. a) The conventional SHEPWM b) The modified SHEPWM

For filtered cases, both inverter with different switching scheme are tested with the same LC filter design as to study the performance of the filter on handling higher switching scheme of the inverter and the other one with lesser switching. It can be observed that on **Fig. 8.** a), there is some disturbance during the switching is changing from positive side to negative side and the graph is not that smooth. This phenomenon is due to the high switching state undergo the filtering process where the rise and fall current need to be considered as well as sometime it need a longer time for filtering the square wave inverter into a better symmetrical AC voltage. The distortion occurs at 0.01 second and repeat the same problem at the second phase and third phase of the inverter.

B) Total harmonic distortion

Using an active low pass LC filter usually will clear all the small noise occurred in the inverter output simultaneously will result a better performance of THD of the inverter. Using filter also will clear some cases like smoothing the side band harmonic which usually appears at high order harmonic [7].

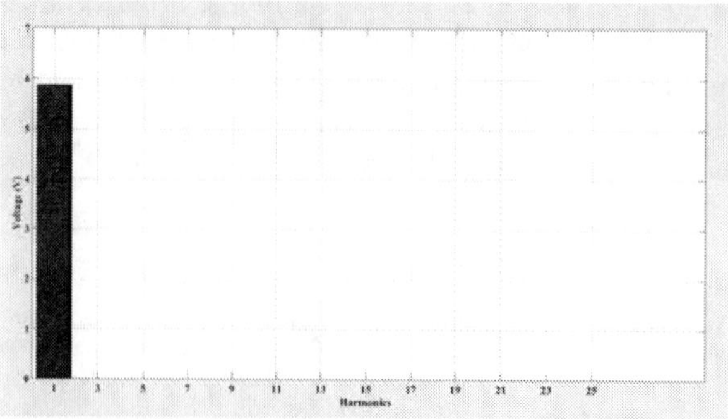

a)

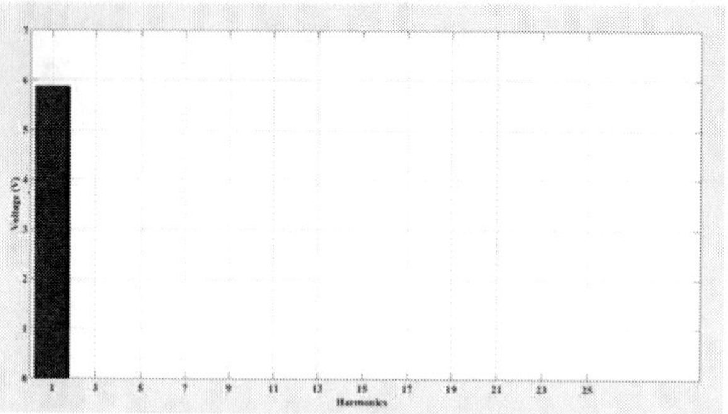

b)

Figure 9. a) THD result for conventional SHEPWM b) THD reult for modified SHEPWM

Table 5. THD performance for filtered inverter

Switching strategy	Overall percentage
Conventional	4.7 %
Modified SHEPWM	3.9 %

The amount of harmonic reduction after the filter is around 30% ~ 35% and it reduced to a value lesser than 5% as shown in table 5. It can be observed that the modified switching scheme has a lower THD percentage compared to the conventional one even before filter the result, the conventional THD is slightly lesser than the modified SHEPWM. For inverter efficiency, the modified SHEPWM gives out 87.4% which are higher 6.2% of the conventional SHEPWM inverter.

5. Conclusion

This paper presented a modification made to a normal unipolar SHEPWM inverter. The main objective is to study the performance of the modified switching scheme related to THD performance for both unfiltered and filtered cases. It can be concluded that the modified SHEPWM are way better than the conventional SHEPWM as the measurement of THD is slightly lower and the efficiency of the system also improve by 6.2%.

References

[1] W.C. Lin, "Comparative Study On Leakage Current Phenomena In Single Phase Transformerless Grid Connected PV Inverter Topologies – Unipolar PWM Switching H-Bridge Inverter, H5, HERIC And Modified HB-ZVR," M.Sc. Thesis, School of Electrical System Engineering, University Malaysia Perlis, 2012

[2] N. Veeramuthulingam, S. Sivajanani, Thebinaa Venugopal, N. Arunkumar, "GA Based Performance Enchancement SHE Methods for Nine-level Inverter". The International Journal of Emerging Trends & Technology in Computer Science (IJETTCS), vol. 2, pp. 146-151, Sept – Oct 2013.

[3] Keerthivasan, K. Narayanan, M. Deve, V.S. Jerome, J., "Estimation of optimal single phase inverter's switching angles using Pattern Search Method," Computer Communication and Informatics (ICCCI), 2012 International Conference on , vol., no., pp.1,6, 10-12 Jan. 2012.

[4] J.Baskaran, S.Thamizharasan, R.Rajtilak, "GA Based Optimization and Critical Evaluation SHE Methods for Three-level Inverter," The International Journal of Soft Computing and Engineering (IJSCE), vol.2, pp.321-326, July 2012.

[5] J.A. Mohammed, "Optimum Solving SHEPWM Equations for Single Phase Inverter Using Resultant Method," Engineering & Technology, vol.26, pp. 660-670, 2008.

[6] S.Siriroj, "Optimized Harmonic Stepped-Waveform for Multilevel Inverter,"M.Sc. thesis Dept. Elect. Eng., Virginia Polytechnic Institute and State University, U.S.,1999

[7] H. Kim, S. Sul, "A Novel Filter Design for Output LC Filter of PWM Inverters," Journal of Power Electronics, vol. 11, pp 74-81, 2011.

[8] Sine PWM and its Realization, 2nd ed., Indian Institute of Technology Kanpur, 2012.

IOP Conf. Series: Materials Science and Engineering **318** (2018) 012038 doi:10.1088/1757-899X/318/1/012038

Ethylene Gas Sensing Properties of Tin Oxide Nanowires Synthesized via CVD Method

Maisara A. M. Akhir[1*], Khairudin Mohamed[2], Sheikh A. Rezan[3], M. M. Arafat[4], A. S. M. A. Haseeb[4], M. N. A. Uda[1] and Nuradibah, M. A.[1]
[1]School of Bioprocess Engineering, Universiti Malaysia Perlis, 02600 Arau, Perlis, Malaysia
[2]School of Mechanical Engineering, Universiti Sains Malaysia, 14300 Nibong Tebal, Penang, Malaysia
[3]School of Materials & Mineral Resources Engineering, Universiti Sains Malaysia, 14300 Nibong Tebal Penang, Malaysia
[4]Department of Mechanical Engineering, Faculty of Engineering, University of Malaya, 50603 Kuala Lumpur, Malaysia

E-mail: maisaraazad@unimap.edu.my

Abstract. This paper studies ethylene gas sensing performance of tin oxide (SnO_2) nanowires (NWs) as sensing material synthesized using chemical vapor deposition (CVD) technique. The effect of NWs diameter on ethylene gas sensing characteristics were investigated. SnO_2 NWs with diameter of ~40 and ~240 nm were deposited onto the alumina substrate with printed gold electrodes and tested for sensing characteristic toward ethylene gas. From the finding, the smallest diameter of NWs (42 nm) exhibit fast response and recovery time and higher sensitivity compared to largest diameter of NWs (~240 nm). Both sensor show good reversibility features for ethylene gas sensor.

1. Introduction

Ethylene, C_2H_4 gas is a vegetal hormone and produced in most plant tissues and is known to be an influential factor in starting off the ripening of fruits. Furthermore, ethylene gas is an important substance in fresh argo-products marketing because it can be used commercially for artificial plant hormone and to monitor the ripening process of climacteric fruits. This has made it possible for tropical fruits such as mangoes and bananas to be harvested green and shipped to distant markets, where they are ripened under controlled conditions. However, natural ethylene production of fruits can cause problems in storage facilities. Ethylene can destroy the green color of plants.

Realizing the importance of detecting the ethylene gas, many researchers are focusing on the investigation of the most suitable sensing material for the ethylene gas sensor. Various sensing materials had been used to detect low concentration of ethylene including titanium dioxide (TiO_2) thin films [1], porous indium oxide (In_3O_3) nanospheres [2], tungsten trioxide (WO_3) thick film [3] and SnO_2 nanoparticles [4, 5]. Most of the materials mentioned above are used to model chemiresistive-based devices, where the conductivity of these materials either increases or decreases based on the exposure to different gas concentrations.

Among these materials, SnO_2 nanoparticles are the most popular material for ethylene gas sensor due to its wide band gap n-type semiconductor properties, which leads to low electrical resistance and better electrical conductivity for gas sensing properties. Furthermore, low operating temperature, high sensitivity, mechanical simplicity of sensor design and low manufacturing cost have made SnO_2 the preferred choice for gas sensor applications.

Besides, SnO_2 nanostructures has the ability to change its electrical and micro structural properties depending on its size and shape [6]. For example, one dimensional (1-D) nanostructured SnO_2 in the form of wires, tubes, belts with high aspect ratio offer unique surface properties. This includes photoluminescence and gas sensitive resistivity with better crystallinity, higher density, and lower power consumption [7, 8]. Properties of the materials such as melting point, electrical conductivity, chemical reactivity, fluorescence and magnetic permeability are size-dependent in the nano-scale range and these properties change as a function of the size of the particles. Recently, 1-D nanostructures have attracted much attention because of their potential applications in gas sensors due to its high surface to volume ratio, while maintaining good chemical and thermal stabilities with minimal power consumption and low weight.

In our previous paper, detail discussion on CVD synthesis and characterization method had been discussed [9]. In this paper, gas sensing performance of 1-D SnO_2 nanostructures toward ethylene gas were investigated in terms of sensitivity, response and recovery times, optimum operating temperature and reversibility features. SnO_2 NWs were synthesized using CVD technique.

2. Materials and Methods

2.1. Synthesis and Characterization

The details for the synthesis and characterization of SnO_2 NWs are described in our previous work [9]. The sample with the largest, C8 (~250 nm) and smallest, C7 (42 nm) diameter of NWs were selected for gas sensing properties test. Figure 1 shows the SEM images and X-ray diffraction spectroscopy (XRD) for the studied samples. XRD results confirm the formation of the tetragonal rutile structure of SnO_2 (ICSD # 98-005-8477).

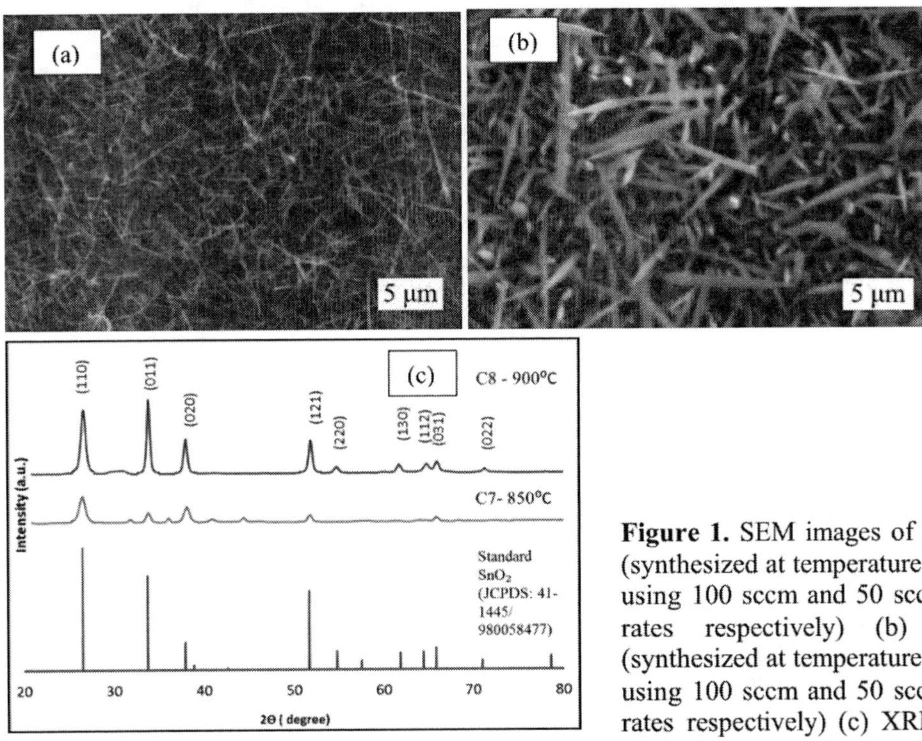

Figure 1. SEM images of (a) smallest NWs, C7 (synthesized at temperature 850°C for 60 minutes using 100 sccm and 50 sccm of Ar and O_2 flow rates respectively) (b) largest NWs, C8 (synthesized at temperature 900°C for 60 minutes using 100 sccm and 50 sccm of Ar and O_2 flow rates respectively) (c) XRD spectra for C7 and C8.

2.2. Gas sensing performance test

The as-prepared SnO_2 NWs were mixed with sensor ink [10] and drop coated onto the alumina substrate with interdigitated printed Au electrodes on the top of its surfaces for gas sensing element fabrication. The sensing element was sintered in a furnace at temperature of 500 °C for 1 hour under nitrogen flow rate of 200 sccm at which the phase contents were stabilized.

In studying the gas sensing performances of SnO_2 nanostructures, several parameters had been selected, which were sensitivity, optimum operating temperature, response time and recovery time. Response time is the time gap needed by the sensing material to reach 90 % of its steady-state resistance value after the introduction of ethylene gas. On the other hand, the recovery time is the time gap of gas sensing material to be within 10 % of its resistance value before exposure to the ethylene gas. These two responses were obtained by measuring the time interval of resistance versus time graph. The sensitivity of the sensor is defined as Ra/Rg, where Ra and Rg is the resistance of the sensor in air and in the target gas, respectively. The sensing materials had been tested under different working temperatures. The temperature with the highest sensitivity value will be considered as the optimum operating temperature and was used to measure the gas response at different ethylene concentrations.

The electrodes of gas sensing element were then connected to the electrometer by using gold wires and the electrometer was connected to a PC. Then, the elements were placed inside the quartz tube and inserted in the furnace. N_2 air was allowed to flow inside the test chamber at the pre-set temperature until the stable resistance was obtained. After that, an ethylene gas pulse was discharged for a certain period of time, followed by the nitrogen gas pulse and the process was repeated for different working temperature. The change in resistance in the nitrogen gas and ethylene gas over time was recorded. The dilution of target gas concentrations ranging from 50 to 100 ppm was done using digital mass flow controller (C100L-CM-NR-2-0V1-SV1-PV2-V1, Sierra) and the total flow rate of gases was maintained at 500 sccm throughout the experiments. Figure 2 illustrates the schematic diagram of gas sensing measurement system.

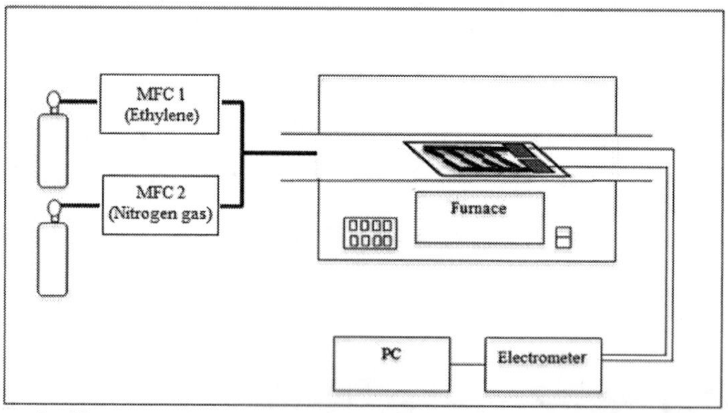

Figure 2. The schematic diagram of the gas sensing measurement system. This system was connected to a computer via the specific interface board. Data were collected in the form of resistance and temperature in real-time.

3. Results and Discussions

Figure 3 illustrates graph sensitivity versus temperature for both of the CVD samples. From the figure 3, the sensitivities of SnO_2 NWs with the smallest (C7) and the largest diameter (C8) to ethylene gas had increased when the temperature was increased. Furthermore, both sensors reached the maximum values when the temperature was at 450°C. Besides this, it can be seen that the sensor with the smallest diameter (C7) NWs had higher sensitivity of 10.55 than the sensor with the largest diameter by a magnitude of 6.54. The sensitivity difference between the smallest and largest diameter NW was about 38 %.

Table 1 presents the response and recovery times of SnO_2 NWs after exposure to 1000 ppm concentration of ethylene gas at different temperatures. The response time of the sensor with the largest diameter (C8) was much longer than the sensor with the smallest diameter (C7) at temperature 375 °C. On the other hand, between temperature 400 °C – 450 °C for CVD sample C8, the response time was similar to C7 which was less than 10 s. It should be noted that the scan time of the resistance measurement in this system was every 10 s. So, the sensors probably took less than 10 s to response to the changes in resistivity.

Besides this, the smallest diameter of SnO_2 NWs took less than 1 minute to recover to 90 % of the baseline resistance in comparison to the largest NWs, which took more than 1 minute to recover. The recovery times for both samples increased with operating temperature. From the results, it can be concluded that the sensor with the smallest diameter of NWs (C7) had better response and recovery times compared to the sensor with the largest diameter (C8).

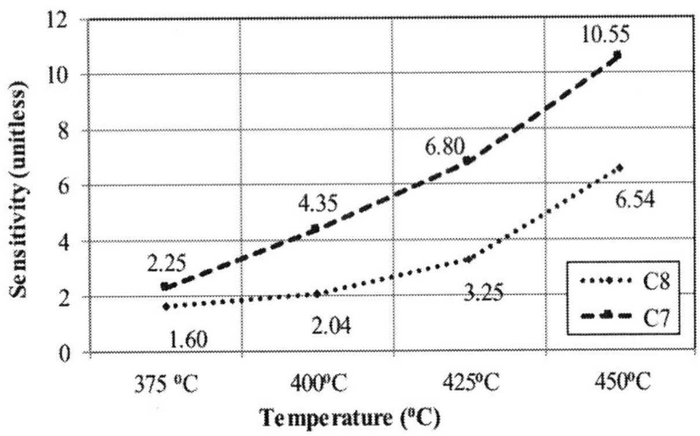

Figure 3. Sensitivity change with respect to temperature for CVD samples C7 and C8.

Table 1. Response and recovery time of SnO_2 NWs towards 1000 ppm concentration of ethylene.

Temperature (°C)	Response Time (s)				Recovery Time (s)			
	375	400	425	450	375	400	425	450
C7	≤10	≤10	≤10	≤10	20	30	50	50
C8	20	≤10	≤10	≤10	60	70	80	90

* ≤ 10 meant the response time may be equal to or less than 10s since the system measured resistance values every 10s.

The sensor's response characteristics toward various ethylene gas concentrations at the optimum temperature of 450 °C were investigated. This operating temperature of 450 °C was chosen based on sensitivity results in figure 3. Figure 4 (a)-(c) shows the sensor response characteristic at various ethylene gas concentrations. From the figure, it can be observed that the resistance of both sensors decreases steeply upon the exposure to ethylene gas and gradually increases after a specific recovery time. Moreover, based on figure 4 (c), the magnitude of the gas sensitivities for both of the sensors improved dramatically with the increase in ethylene gas concentrations. SnO_2 NWs with the smallest diameter (C7) had a higher gas sensitivity value response at all concentration compared to the largest diameter of NWs's (C8) sensor.

In terms of reversibility, both sensors showed excellent ability to recover to its initial value after exposure to the analyte gas was removed. As seen in figure 4 (a) and (b), the resistances in sample C7 and C8 had reached their baseline value upon removal of ethylene gas. It was noticeable that the baseline resistance for C7 was higher than C8 by ~96.8 % at 1000 ppm of ethylene. This difference may be due to the smaller NWs diameter sample having a larger electron depleted region. As a result of this, a higher resistance was obtained with the smaller diameter NWs. Similar observation was reported by Yamazaki [11] and the explanation for this was centered on the depleted region.

Furthermore, a shorter response and recovery time of the SnO_2 NWs was found at high ethylene gas concentration (1000 ppm) and this response and recovery time had increased with decreasing gas concentration. The response and recovery time for C7 were between 10 s – 20 s (1000 ppm) and 50 s - 80 at lower concentration (50 ppm). For sample C8, it was 10 s – 20 s at high concentration and 1 minutes - 2 minutes at lower concentration.

1D nanostructures offer an advantage in terms of high aspect ratio (L/D) ratio, which allows a larger active site surface for gas adsorption and desorption in gas sensor applications. Typical range for aspect ratio was between 5 – 1000 [12]. The high aspect ratio for SnO_2 NWs was expected to provide high gas sensitivity. Based on the findings in the present work, the surface reaction of the SnO_2 NWs gas sensor greatly depended on the working temperature. It was noticed that the resistance of SnO_2 NWs decreases with increase of temperature and followed closely to Ohm's law. During the increase in operating temperature while gas sensing, the adsorption and desorption processes of the gas molecules will be accelerated. The captured electrons by the adsorbed oxygen atom in SnO_2 are thermally excited to the conduction band as the temperature rises. Hence, this results in a decrease in resistivity.

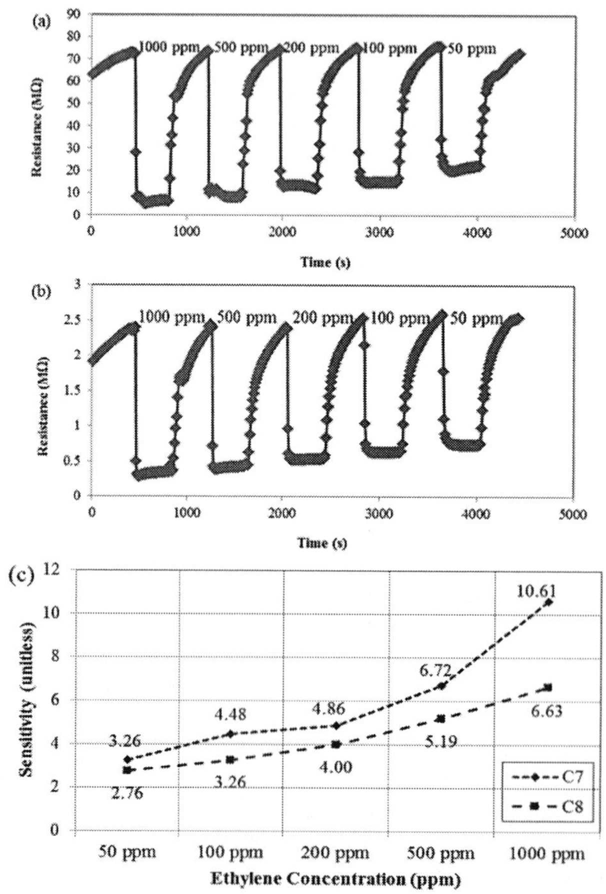

Figure 4. Gas sensitivity response with respect to ethylene gas (a) C7, (b) C8 and (c) varying ethylene gas concentration at temperature of 450°C.

Furthermore, at low temperature (<150 °C), the adsorbed oxygen species at the surface of sensing material will extract the electrons from the conduction band of the material to produce O_2^- by equation (2) [13]. At the same time, O^- and O^{2-} are produced when the temperature was between 150 °C to 450 °C (Equation 3 and 4) and O^- becomes dominant at this temperature [13,14]. In the latter case, the highest sensitivity of 10.61 was measured for SnO_2 NWs via CVD at 450 °C. Therefore, O^- atoms absorbed onto the NWs surface was believed to be the predominant oxygen species at this temperature based on studies by Ahsan [13].

The expected reactions between the analyte gas and oxide surface of the sensing material at elevated temperatures are written in equation (3) and (5). The sensing mechanism of SnO_2 material towards ethylene gas involves the adsorption of oxygen molecules from the atmosphere onto the SnO_2 surface. In the present study gas sensing setup, the origin of the oxygen molecule was from the premixed ethylene and air gas mixture. This chemisorbed oxygen species was formed by trapping of electrons from the conduction band of SnO_2. As a result of this trapping, a depletion layer consists of O^- ionic species was formed. This process is expressed in equation (3).

The depletion region size increases and the conduction band decreases as more and more oxygen molecules are adsorbed onto the sensing surface. This in turn causes a low electrical conductivity or high surface resistivity on the SnO_2 NWs. When the ethylene gas is exposed to SnO_2 sensing layer at an elevated temperature, the molecules will react with the adsorbed species and this process will revert back the electrons to the SnO_2 conduction band as given in equation (5). The high electron concentrations result in higher surface conductivity and thus, the resistance will decrease [5]. The decrease in resistivity will signal the presence of ethylene gas. Figure 5 illustrates the schematic diagram of the depletion and conduction region before and after exposure to ethylene gas.

$$O_2\,(g) \rightarrow O_2\,(ads) \quad (1)$$
$$e^- + O_2\,(ads) \rightarrow O_2^-\,(ads) \quad (2)$$
$$2e^- + O_2\,(ads) \rightarrow 2O^-\,(ads) \quad (3)$$
$$e^- + O^-\,(ads) \rightarrow O^{2-}\,(ads) \quad (4)$$
$$C_2H_4 + O^{2-} \rightarrow C_2H_4O + 2e^- \quad (5)$$

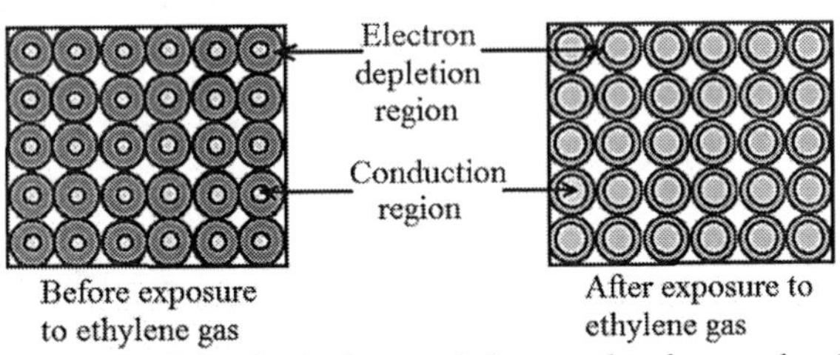

Figure 5. The schematic diagram of the depletion and the conduction region before and after exposure to ethylene gas.

With reference to the table 1 and figure 4, both SnO_2 NWs exhibit fast response and recovery time, which was 10 s to 20 s (C7) and 50 s to 120 s (C8) respectively upon exposure to ethylene gas. This response and recovery behavior was due to the high surface area of SnO_2 NWs structure that were randomly oriented. This randomization allows rapid diffusion between the NWs gaps or voids with a shorter time required to return to baseline resistance. Furthermore, gas sensitivity behavior for C7 and C8 show that it has the ability to recover to its initial baseline resistance when ethylene gas was purged with nitrogen from the test chamber. These findings implied a good reversibility property of the sensor with NWs like structure synthesized using CVD methods.

Besides this, SnO_2 NWs with the smallest diameter (C7) had higher sensitivity values than the largest diameter NWs (C8) at all relative ethylene gas concentrations at 450 °C. Wire-like structure of C7 and C8 allowed the gas molecules to absorbed on the NWs with ease. Figure 6 presents the illustration of the potential difference between the intra-particles connection of SnO_2 NWs.

From the figure, it can be seen that the geometrical aspect (L/D) of the NWs affects the resistance of the sensor by changes in the conduction (E_C) and Fermi energy levels (E_F). These changes are mostly governed by the bulk interface and interface between the electrode and sensing NWs rather than the grain boundaries [11].

The higher gas sensitivity of the smallest NW diameter (C7) compared to largest (C8) was due to a wider conductive bulk region in the smaller diameter sample (refer to figure 6). The smaller NWs have more inter-grain contacts and in turn enhances the gas sensitivity than the larger NWs. Therefore, the conductance of the bulk region for the NWs was influenced by its exposure to ethylene gas. This exposure led to a large change in surface resistivity and higher gas sensitivity for sample C7 than C8.

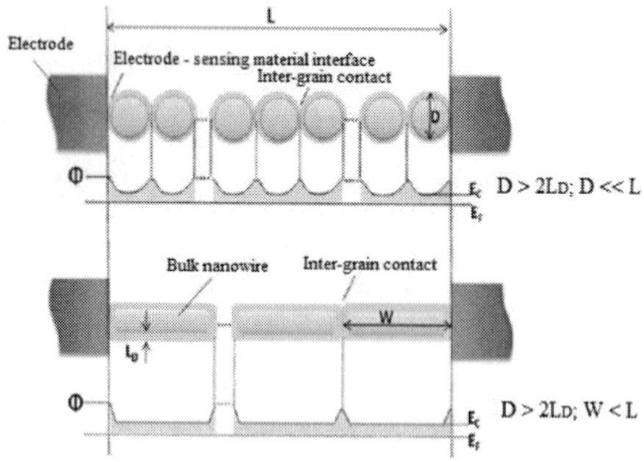

Figure 6. The illustration of the potential intra-particles connection SnO_2 NWs [11]

4. Conclusions

CVD sample with the smallest NWs diameter (C7) exhibit the quickest response and recovery time towards ethylene gas than the largest NWs diameter (C8) samples. Moreover, the gas sensitivity was higher for smallest diameter of NWs than the largest diameter. In addition, the 1-D NWs shows a good reversibility characteristic with the lowest recovery time among the tested samples (50 s). This work shows successful fabrication of 1-D SnO_2 NWs gas sensing material to detect low concentrations of ethylene gas.

Acknowledgments

The author is grateful to School of Bioprocess Engineering, UniMAP, Nanofabrication and Functional Materials Lab, School of Mechanical Engineering, USM and Department of Mechanical Engineering, Faculty of Engineering, UM, for providing the facilities to carry out the research. The research is funded by UniMAP/STG/2016/9001-00553.

References

[1] R. Zhang, M. I. Tejedor, M. A. Anderson, M. Paulose, and C. A. Grimes, "Ethylene Detection Using Nanoporous $PtTiO_2$ Coatings Applied to Magnetoelastic Thick Films," *Sensors*, **2002**, 331 (2002).

[2] Y. Cao *et al.*, "Synthesis of porous In_2O_3 microspheres as a sensitive material for early warning of hydrocarbon explosions," *RSC Adv.*, **5**, 5424 (2015)

[3] O. Nimittrakoolchai and S. Supothina, "High-yield precipitation synthesis of tungsten oxide platelet particle and its ethylene gas-sensing characteristic," *Mater. Chem. Phys.*, **112**, 270 (2008).

[4] W. Chen, Q. Zhou, T. Gao, X. Su, and F. Wan, "Pd-Doped SnO_2-Based Sensor Detecting Characteristic Fault Hydrocarbon Gases in Transformer Oil," *J. Nanomater.*, **2013**, 1 (2013).

[5] M. Agarwal, M. D. Balachandran, S. Shrestha, and K. Varahramyan, "SnO_2 Nanoparticle-Based Passive Capacitive Sensor for Ethylene Detection," *J. Nanomater.*, **2012**, 1 (2012).
[6] C. H. Drake, "Understanding the Low Temperature Electrical Properties," (2007).
[7] Y. H. Y. and G. W. Y. B Wang, "Growth mechanisms of SnO_2/Sn nanocables," *Nanotechnology*, **17**, 7 (2006).
[8] H.-C. Z. Hang Chi Hong-Jun Xu, Xu-Dong Shan, Zhi-Min Liao, and Da-Peng Yu, "One-Step Growth and Field Emission Properties of SnO_2-Capped Silicon Nanowires: A Sn-Catalyzed Approach," *J. Phys. Chem.*, **113**, 4 (2009).
[9] M. A. M. Akhir, K. Mohamed, S. A. Rezan, H. L. Lee, and S. S. M. Izah, "An assessment of chemical vapor deposition synthesis of SnO_2 nanowires by statistical design," *Key Engineering Materials*, **701**, 52 (2016).
[10] M. Arafat, A. Haseeb, and S. Akbar, "A Selective Ultrahigh Responding High Temperature Ethanol Sensor Using TiO_2 Nanoparticles," *Sensors*, **14**, 13613 (2014).
[11] T. Yamazaki, "Formation of oxide nanowires by thermal evaporation and their application to gas sensors," *Nanowires - Implementations Appl.*, **7**, 117 (2012).
[12] J. Huang and Q. Wan, "Gas sensors based on semiconducting metal oxide one-dimensional nanostructures," *Sensors*, **9**, 9903 (2009).
[13] M. Ahsan, "Thermally Evaporated Tungsten Oxide (WO_3) Thin Films for Gas Sensing Applications," (2012).
[14] C. Wang, L. Yin, L. Zhang, D. Xiang, and R. Gao, "Metal oxide gas sensors: sensitivity and influencing factors.," *Sensors (Basel).*, **10**, 2088 (2010).

Evaluation of Short-Term Cepstral Based Features for Detection of Parkinson's Disease Severity Levels through Speech signals

Qi Wei Oung[1],*, Shafriza Nisha Basah[1], Hariharan Muthusamy[2], Vikneswaran Vijean[1], Hoileong Lee[1]
[1]School of Mechatronic Engineering, Universiti Malaysia Perlis (UniMAP), Campus Pauh Putra, 02600, Arau, Perlis, Malaysia
[2]Department of Biomedical Engineering, Faculty of Engineering and Technology, Kattankulathur Campus, SRM University, SRM Nagar, Kattankulathur - 603 203 Kancheepuram District, Tamil Nadu, India

Email: aliciaoung@gmail.com

Abstract. Parkinson's disease (PD) is one type of progressive neurodegenerative disease known as motor system syndrome, which is due to the death of dopamine-generating cells, a region of the human midbrain. PD normally affects people over 60 years of age, which at present has influenced a huge part of worldwide population. Lately, many researches have shown interest into the connection between PD and speech disorders. Researches have revealed that speech signals may be a suitable biomarker for distinguishing between people with Parkinson's (PWP) from healthy subjects. Therefore, early diagnosis of PD through the speech signals can be considered for this aim. In this research, the speech data are acquired based on speech behaviour as the biomarker for differentiating PD severity levels (mild and moderate) from healthy subjects. Feature extraction algorithms applied are Mel Frequency Cepstral Coefficients (MFCC), Linear Predictive Coefficients (LPC), Linear Prediction Cepstral Coefficients (LPCC), and Weighted Linear Prediction Cepstral Coefficients (WLPCC). For classification, two types of classifiers are used: k-Nearest Neighbour (KNN) and Probabilistic Neural Network (PNN). The experimental results demonstrated that PNN classifier and KNN classifier achieve the best average classification performance of 92.63% and 88.56% respectively through 10-fold cross-validation measures. Favourably, the suggested techniques have the possibilities of becoming a new choice of promising tools for the PD detection with tremendous performance.

1. Introduction

Speech is a complicated task that involves parallel and sequential control of numerous mechanisms and systems in a highly detailed and refined ways. In the human anatomy of the speech system, lungs are the primary source of speech production, whereby they produce sufficient airflow through the glottis in order for the vibration of the vocal folds. When the vocal folds vibrates, they generate a source excitation signal holding the properties of pressure wave expelled from the lungs. Next, the source signal passes through the vocal tract to be filtered by the spectral envelope to form a speech signal [1, 2].

For a population that is moving towards increased number of elderly people, Parkinson's disease (PD) is placed in the second place for the most common progressive neurodegenerative disorder in the world, which affects people above 60 years old. According to the statistics revealed by the World Health Organization (WHO), it was expected that the world is currently having seven to ten million people with Parkinson (PWP). For the forthcoming years, this amount is predicted to upsurge by two-fold as the age is the leading risk feature for the onset of PD [3-6]. At present, the cause of PD is still unidentified and there is no cure for PD, although it has been managed by some types of drugs known as levodopa. The characteristic motor features of PD include various kinds of motor and non-motor deficits such as behaviour, resting tremor, bradykinesia, sensation, akinesia, postural instabilities, and hypokinetic dysarthria, which is speech impairment [7]. Current existing techniques for diagnosis of PD are according to the patient's previous health report and neurological inspection conducted either through observation by the neurologist or some technological tools such as electromyography (EMG), electrocardiography (EEG) or brain imaging modalities, which are magnetic resonance imaging (MRI) and computer tomography (CT) scans [8-10].

Vocal impairment is one of the initial indicators for the onset of PD, where it is predicted that nearly 90% of people with Parkinson (PWP) show speech deficiencies. Typical symptoms of speech impairment include the inability in pronouncing words, decreased voice tone, dysphonia (failure to produce normal vocal sounds) and dysarthria (inability to articulate speech normally). Speech recording is an uncomplicated and non-invasive method, which is the core benefits of introducing it as one of the notable biomarkers for assessment of PD. Detection of voice changes in PWP would make the possibility of earlier intervention before the onset of disabling physical symptoms. Changes in voice might also function as an easily assessable and objective proxy to determine PD severity and monitor early trajectories of PD progression. Little et al. have proposed different dysphonia measures for the aim of differentiating speech signals of PWP from healthy subjects. These measures have the potentials for enhancing the current technique via speech signals [11-13]. Several works conducted by previous researchers were presented in Table 1 that includes voice data acquisition, feature extraction, classification and performance evaluation [14-22]. From these studies, it can be observed that there have been initiatives involving the application of speech signals that had the ability to differentiate between PWP and healthy subjects. However, evaluation of signal processing algorithms for speech analysis platform that focuses on distinguishing between PD patients with different severity levels from healthy subjects is still at its infancy. Therefore, with more in-depth researches are encouraged to analyse speech abnormalities in PWP, this study is proposing a computational framework for the ultimate objective analysis of differentiating PWP with mild or moderate severity level from healthy subjects.

2. Methodology

The overall block diagram of the proposed approach is demonstrated in Figure 1.

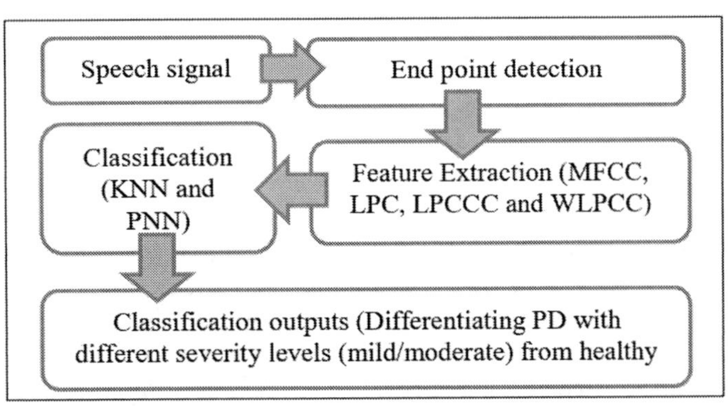

Figure 1. Overall procedure of the proposed approach.

Table 1. Summary of researches conducted on classification of PWP from healthy subjects.

Author (Year)	Number of subjects	Methods of data acquisition	Feature extraction	Classifier	Accuracy
Hamid Karimi *et al.*(2011) [16]	31 subjects (23 PWP and 8 healthy subjects)	Sustained vowel phonations	*7 dysphonia features	Support Vector Machine (SVM), K-Nearest Neighbor (KNN), Discrimination-Function-based (DFB)	SVM (91.04%), KNN (93.82%) and DFB (82.20%).
Athanasios Tsanas (2012) [12]	43 subjects (33 PWP and 10 healthy subjects)	Sustained vowel phonations	*10 dysphonia features	Random Forests (RF), SVM	SVM (98.6%) and RF (93.5%)
Hui-Ling Chen *et al.* (2013) [17]	31 subjects (23 PWP and 9 healthy subjects)	Sustained vowel phonations	*22 dysphonia features	Fuzzy K-Nearest Neighbor (FKNN), SVM	FKNN (96.07%) SVM (94.62%)
Yahia Alemami *et al.*(2014) [15]	40 subjects (20 PWP and 20 healthy subjects)	Sustained vowel phonations	**5 frequency features **5 amplitude features **3 harmonicity features **5 pitch features **5 pulse features **3 voicing features	Naïve Bayesian Classifier, KNN	Naïve Bayesian Classifier (93.3%), KNN (80.00%)
Mohammad Shahbakhi *et al.* (2014) [20]	31 subjects (23 PWP and 9 healthy subjects)	Sustained vowel phonations	*22 dysphonia features	SVM	94.50% (4 optimized features), 93.66% (7 optimized features), 94.22% (9 optimized features)
Saloni *et al.*(2015) [22]	31 subjects (23 PWP and 9 healthy subjects)	Sustained vowel phonations	*22 dysphonia features	Artificial Neural Network (ANN), SVM	ANN (87.70%) SVM (95.90%)
Haydar Ozkan *et al.*(2016) [14]	31 subjects (23 PWP and 9 healthy subjects)	Sustained vowel phonations	*22 dysphonia features	KNN	95.02% (2-fold cross-validation 96.72% (5-fold cross-validation) 99.10% (10-fold cross-validation)

**Brief description of these features are explained in Tsanas et al. [11-13]; **Brief description of these features are described in Betul et al.[23].*

2.1. Database

From earlier research, several vocal tests have been conducted for evaluating PWP vocal impairments. This includes (a) sustained vowel phonations where the speaker is asked to constantly maintain the vowel to a maximum period, and (b) running speech where the speaker is asked to read simple standard sentences [12]. Although some of the vocal deficiencies in running speech, for instance, the combinations of consonants and vowels can be captured and considered as a more convincing assessment in daily routines, the analysis of running speech is more complicated compared to sustained phonations. On the other hand, the assessment through sustained vowel phonations had the ability to provoke speech impairment symptoms, where identification of dysphonia is best carried out with absence of confounding effects from articulatory or linguistic components of running speech.

Both clinical practice and extensive studies showed that the assessment through sustained vowel phonation is adequate for many speech test applications, particularly for PD speech assessment. Generally, sustained vowel phonation test is easy to conduct and also it can give satisfying results in discriminating PWP from healthy subjects [11-13]. Therefore, in this research, we will focus on sustained phonation tests.

The speech data collection had been carried out from 30 subjects (14 males and 16 females with an average age of 65 years). 20 participants (8 males and 12 females) are PWP and 10 participants are healthy subjects. Each subject was rated by a neurologist using the Hoehn and Yahr scale where the symptom ratings have ranged from 0 to 4, where 0 = normal, 1 = mild, 2 = moderate, 3 = severe and 4 = unintelligible). Out of 30 subjects, 10 subjects were rated '0', 10 subjects were rated '1', and 10 subjects were rated '2'. The voice phonations of the subjects were recorded using a headset (DW Pro2 series) where the speech files were recorded with 44.1 kHz sampling rate, 16-bit resolution and stored as '.wav' file. Audio speech files were recorded using Matlab Simulink Tools. The experiment was conducted in an auditory room of the Neurology Department, Penang General Hospital. A small briefing about the test was given to each subject before starting the experiment to ensure that the subject can give full co-operation and fully understand the overall recording process. The headset (DW Pro2 series) was fitted on the subject's head; approximately 5cm from their mouth, recording the voice while the subject is conducting their speaking task. The speech database comprised of sustained vowels /a/ and /o/ phonations, where the speaker is requested to produce /a/ phonation for three time and /o/ phonation for three times in a random sequence. For every assessment, the speaker is asked to sustain the phonation to a maximum period, attempting to maintain steady frequency and amplitude.

2.2. Pre-processing (End-Point Detection Technique)

In speech signals, some of the segments such as the silent segments would have a much lower amplitude compared to the spoken segments, causing these silent segments with less energy than their spoken energy. Thus, this technique can be applied for differentiating between the speech segments and the silence segments [24]. For this research, segmentation for each voice signal is conducted by segmenting the stable portion from the overall signals. As these signals would normally be stable in the middle of the whole signals, segmenting the central portion is to eliminate the silent periods at the beginning and the ending of the signals. 2s segments from the middle, stable portion of the speech signals were chosen for the subsequent acoustic analysis in order to avoid problems during the onset and offset of the phonations.

2.3. Feature Extraction

One of the important tasks to produce produce a better recognition result is extracting the greatest representation of the parameter from the audio signals. The speech signals were first down sampled from a sampling frequency of 44.1 kHz to 16 kHz. Then, the signals were pre-emphasized with the value of the coefficient, α equal to 0.97. This process will upsurge the energy of the signals at a greater frequency. Equation (1) is used for building a pre-emphasis filter.

$$C(n) = B(n) - \alpha * B(n-1) \quad (1)$$

where C(n) symbolised the pre-emphasized signal, B(n) symbolised the actual signal, and α is the pre-emphasis coefficient. After pre-emphasis, the process of framing was conducted to divide the speech signals into smaller segments and overlapping is required to capture the subject's specific features in the speech data. Windowing is performed on the framed signal to smooth the abrupt frequencies at the end points of the frames and undesirable frequencies in the speech frames. In this research, the pre-processed speech signal is divided into fixed frames of 20ms with an overlap of 50%. For windowing, Hamming window is chosen as the window shape by taking consideration of the next block in the feature extraction processing chain and integrates all the closest frequency line [25].

In this study, Mel Frequency Cepstral Coefficient (MFCC), Linear Predictive Coefficients (LPC), Linear Prediction Cepstral Coefficients (LPCC) and Weighted Linear Prediction Cepstral Coefficients (WLPCC) are chosen as the feature extraction algorithms.

2.3.1. Mel Frequency Cepstral Coefficients (MFCC).
The MFCC working principle is based on human hearing perceptions whereby it cannot perceive frequencies over 1kHz. MFCC consists of two types of filters that are spaced linearly in frequency below 1000Hz and spaced logarithmically in frequency above 1000Hz. A subjective pitch is presented on Mel Frequency Scale to extract the significant characteristic in speech signals. MFCC is widely used in audio feature extraction due to its high sensitivity properties and the total spectra slope of the high-order cepstral coefficients. After the windowing step, Discrete Fourier Transform (DFT) is performed for converting each frame of time domain N samples into the frequency domain. This conversion is performed to obtain the desired frequency resolution on a Mel scale. After the computation of DFT on the windowed signals, the resultant signals were squared to get the Mel-scaled filtered energies. Each filter's bank has a triangular bandpass frequency response, whereby the filter output is the weighted sum of its filtered spectral components and the Mel for given frequency f in Hz is computed. Equation (2) is used for converting the frequency in term of Hz into Mel scale. Lastly, the log Mel spectrum is transformed back into time domain using Discrete Cosine Transform (DCT). The result after DCT conversion is called MFCC [25-27]. For this research, MFCCs with 40 coefficients were generated to study their impact on sustained vowel phonations classification accuracy.

$$m = 2595 \log_{10}(1+f) * 700 \tag{2}$$

2.3.2. Linear Predictive Coefficients (LPC).
In addition, the second chosen technique for extracting the useful features is a linear predictive analysis. It is one of the important features that reflect the differences in the biological structure of the human vocal tract. It also provides an accurate, reliable and robust method for parameter estimation that characterises the linear time-varying system representing vocal tract. The essential concept of linear predictive is each voice signal is predictable as the linear combination of past p samples, whereby the autocorrelation of p denotes the order of LPC analysis and p is fixed as 13 [28-30]. Next, the autocorrelation coefficients will be then be converted into Linear Prediction Cepstral Coefficients (LPCC). This conversion is performed by LPC analysis implemented based on the Levinson-Durbin recursive algorithms. LPCC are the cepstral coefficients of the Fourier transform representation of the logarithmic magnitude spectrum, where LPCC have been proven to be more robust to noise and reliable compared to LPC. LPCC are the recursion of LPC parameters to LPC spectrum. In this work, the LPCC features that represent each frame was determined based on this equation, $Q = (3/2)p$. Next, a standard method to weigh the cepstral coefficients is performed for reducing the sensitivity of low-order cepstral coefficients and high-order coeficients to total spectral slope and noise respectively. Weighted Linear Prediction Cepstral Coefficients (WLPCC) can be simply obtained by multiplying LPCC with the weighted formula [28, 30]. Finally, this coefficient will be normalised between 0 and 1 before proceeding to the classification stage.

2.4. Classification
Two techniques chosen as classifiers for this research are Probabilistic Neural Network (PNN) and k-Nearest Neighbour (KNN). PNN is a feed-forward neural network, which was derived from the Bayesian network. PNN architectures. This network is described as an application of statistical algorithm called Kernel discriminate analysis where the procedures consist of 4 layers: an input layer,

hidden layer, summation layer and the output layer. The advantages of PNN are fast training procedure, permanently parallel structure and confirmed to converge to an optimal classifier, where the increased representative training set and sample size can be added or removed, without any additional retraining. PNN classifier learns more quickly than other neural network model and had favourable results in many applications. With these reasons and advantages, PNN can be considered as a supervised neural network that had the capability to be used in this PD classification system [31, 32]. The accuracy of PNN highly depends on suitable smoothing parameter or spread factor (ŋ). Appropriate ŋ value was found between 0.01 and 0.1 through experimental investigations.

Next, KNN is a supervised learning algorithm where the outcome of the new instance query will be grouped according to the majority of the value of k in this algorithm. It is a very popular and one of the most fundamental classification algorithms that demonstrate excellent performance characteristics with a short period of training time. The classification is using majority vote theory among the class of the k objects. With a given query point, the k number of training points that represents the number of nearest neighbour points closest to the query point will be discovered. The KNN algorithm used neighbourhood classification as the prediction value of the new query instance. The label of a class is determined by the KNN category using majority voting. The effect of the different neighbourhood in the classification results was studied by varying the k values from 1 to 10 [33].

3. Results and Discussion

There was a total of four features used in this research that are 40 MFCC features (MFCC-40), 13 LPC features (LPC-13), 19 LPCC features (LPCC-19) and 13 WLPCC features (WLPCC-13). In this research, 10–fold cross-validation was performed on the features extracted where the data is randomly divided into 10 equal pieces. Each selected piece is chosen as the test set with training done on the remaining of the data. The cross-validation is then repeated for 10 times (folds) with each subsample used exactly once as the validation data. The advantage of this technique compared to conventional validation is that all observations are used for both training and validation, and each observation is used for validation exactly once. Herein, 10 subjects per group with 3 trials and 2 sustained vowels per features which resulted in a total feature vectors of 60 X 40 MFCCs, 60 X 13 LPCs, 60 X 19 LPCCs and 60 X 13 WLPCCs were analysed. The mean classification results for 3 severity levels, whereby '0' representing healthy, '1' representing mild and '2' representing moderate together with an average of the 3 classes are tabulated in Table 2 and Table 3.

Table 2. Mean classification accuracy results for KNN

	KNN (%)			
	'0'	'1'	'2'	Mean
MFCC-40	88.08± 0.58	87.66± 0.50	86.71± 0.40	87.48± 0.27
LPC-13	85.67± 0.57	84.90± 0.65	85.20± 0.76	85.26± 0.40
LPCC-19	86.47± 0.38	85.26± 0.33	85.57± 0.33	85.77± 0.21
WLPCC-13	86.98± 0.43	85.97± 0.53	85.81± 0.58	86.25± 0.30
LPC 13+ LPCC-19+ WLPCC-13	89.80± 0.44	88.81± 0.57	87.07± 0.75	88.56± 0.39

Table 3. Mean classification accuracy results for PNN

	PNN (ŋ =0.09) (%)			
	'0'	'1'	'2'	Mean
MFCC-40	92.46± 0.32	90.94± 0.52	89.04± 0.63	90.81± 0.24
LPC-13	89.00± 0.67	88.07± 0.77	86.91± 0.64	88.00± 0.31
LPCC-19	90.13± 0.46	89.53± 0.56	87.01± 0.53	88.89± 0.20
WLPCC-13	91.94± 0.63	93.11± 0.64	90.29± 0.75	91.78± 0.33
LPC 13+ LPCC-19+ WLPCC-13	93.97± 0.41	92.81± 0.48	91.10± 0.53	92.63± 0.28

From Table 2 and Table 3, it is clearly shown that PNN with a smoothing parameter, η equal to 0.09 showed slightly better performance measure for all the features extracted compared to KNN. As seen in Table 2, the mean accuracy achieved for LPC-13 features using KNN classifier for class '0' is 85.67%, class '1' is 84.90%, class '2' is 85.20%, and average among the 3 classes is 85.26%. On the other hand, as seen in Table 3, when LPC-13 features are used in PNN classifier, the mean precision for class '0' obtained is 89%, class '1' is 88.07%, class '2' is 86.91%, and an average of the 3 classes is 88%. For LPCC-19 features, as seen in Table 2, when LPCC-19 features are used in KNN classifier, the average accuracy obtained for the healthy class is 86.47%, the mild class is 85.26%, the moderate class is 85.57%, and an average of the three classes is 85.77%. On the other hand, as seen in Table 3, when LPCC-19 features are used in PNN classifier, the mean precision achieved for the healthy class is 90.13%, the mild class is 89.53%, the moderate class is 87.01%, and average among the 3 classes is 88.89%. For WLPCC-13 features, as seen in Table 2, when the WLPCC-13 features are used in KNN classifier, the average accuracy obtained for class '0' is 86.98%, class '1' is 85.57%, class '2' is 85.81%, and an average of the three classes is 86.25%. However, as seen in Table 3, when WLPCC-13 features are applied in PNN classifier, the highest average accuracy obtained is 93.11% from class '1' followed by 91.94% from class '0' and 90.27% from class '2'. The average among 3 classes for PNN applying WLPCC features is 91.78%. It is undoubtedly seen that PNN classifier achieved better performances compared to KNN classifier when using LPC based features that consist of LPC, LPCC, and WLPCC features.

In addition, for MFCC-40 features, PNN classifier also performed well compared to KNN classifier with 92.46%, 90.94%, and 89.04% of average accuracy for class '0', '1' and '2' respectively. These results go the same for the combination of all LPC based features, where these features showed the mean accuracy of 93.97% for class '0', 92.81% for class '1' and 91.90% for class '2'. Among all the features, the combination of all LPC features presented the highest mean classification results for average among the 3 classes of 92.63% using PNN classifier and 88.56% using KNN classifier. Although it is difficult to make direct comparison between present approach with our proposed techniques, as all the approaches are unique in term of signal analysis, selecting feature extraction and classification framework, we have put the summary of performance from various approaches conducted for the diagnosis of PD in Table 1. As shown in Table 1, previous work done on speech signals has only presented results on distinguishing between healthy controls and PWP through some dysphonia features. However, for classification of healthy subjects from PWP with the mild and moderate level of PD, to the best of authors' knowledge, there has been very little analogous published methodological framework conducted. In our present study, we have applied two widely used short-term cepstral-based features in audio feature extraction, which is MFCC and LPC based features (LPC, LPCC, and WLPCC) to differentiate PWP with mild and moderate severity level of PD from healthy subjects, where KNN classifier achieved highest mean accuracy of 89.80% and PNN classifier achieved highest mean accuracy of 93.97%.

4. Conclusion

PD is grouped as the second commonest neurological illness after Alzheimer. Researches have shown that voice signal may be useful for PD diagnosis based on the sources of clinical evidence, which suggested the majority of PWP usually reveal some form of vocal disorder. In this study, techniques using two types of classifier, KNN and PNN for distinguishing PWP with mild and moderate severity level from healthy subjects through voice signals by extracting a number of features were conducted. The experimental results presented the best mean performance was reached when extracting the combination of LPC based features (LPC, LPCC and WLPCC) of 93.97% for PNN classifier and 89.80% for KNN classifier. Therefore, it can be summarized that this proposed approach is a good starting point to present the performance for efficiently classify PD severity levels (mild and moderate) from healthy subjects through speech features.

The existing works can be further enhanced by increasing the number of PD severity levels into mild, moderate and severe and numbers of subjects in each class and through improvements in the existing signal processing and classification techniques.

5. Acknowledgment

All the authors would like to acknowledge the neurologists from the Penang General Hospital for their support and assistance in providing the PD patients and healthy subjects throughout the research.

REFERENCES

[1] Hariharan M, Yaacob S, Hasrul M and Oung Q W 2006 SPEECH EMOTION RECOGNITION USING STATIONARY WAVELET TRANSFORM AND TIMBRAL TEXTURE FEATURES *ARPN Journal of Engineering and Applied Sciences* **9** 1316-22

[2] Khan T, Westin J and Dougherty M 2014 Classification of speech intelligibility in Parkinson's disease *Biocybernetics and Biomedical Engineering* **34** 35-45

[3] Lee H L, Shahriman A B, Sazali Y, Zuradzman M R, Khairunizam W, Ahmad W, Zunaidi I, Cheng E M, Khadijah S and Nisha S 2014 In vitro evaluation of finger's hemodynamics for vein graft surveillance using electrical bio-impedance method *Australian Journal of Basic and Applied Sciences* **8** 350-9

[4] Lee H L, Shahriman A B, Wan K, Roohi S A and Zuradzman M R 2014 Upper extremity vein graft monitoring device after surgery procedure: a preliminary study. In: *Advanced Materials Research*: Trans Tech Publ) pp 656-60

[5] Organization W H 2006 *Neurological disorders: public health challenges*: World Health Organization)

[6] Sim O F 2001 Ageing in Malaysia: National policy and future direction *University of Malaya, Kuala Lumpur*

[7] Elbaz A, Bower J H, Maraganore D M, McDonnell S K, Peterson B J, Ahlskog J E, Schaid D J and Rocca W A 2002 Risk tables for parkinsonism and Parkinson's disease *Journal of clinical epidemiology* **55** 25-31

[8] Kupryjanow A, Kunka B and Kostek B 2010 Updrs tests for diagnosis of parkinson's disease employing virtual-touchpad. In: *IEEE 2010 Workshops on Database and Expert Systems Applications* pp 132-6

[9] Oung Q W, Hariharan M, Basah S N, Yaacob S, Sarillee M and Lee H L 2014 Use of technological tools for Parkinson's disease early detection: A review. *2014 IEEE International Conference on Control System, Computing and Engineering (ICCSCE)* pp 343-8

[10] Oung Q W, Hariharan M, Lee H L, Basah S N, Sarillee M and Lee C H 2015 Wearable multimodal sensors for evaluation of patients with Parkinson disease. *2015 IEEE International Conference on Control System, Computing and Engineering (ICCSCE)* pp 269-74

[11] Tsanas A, Little M A, McSharry P E and Ramig L O 2011 Nonlinear speech analysis algorithms mapped to a standard metric achieve clinically useful quantification of average Parkinson's disease symptom severity *Journal of the Royal Society Interface* **8** 842-55

[12] Tsanas A, Little M A, McSharry P E, Spielman J and Ramig L O 2012 Novel speech signal processing algorithms for high-accuracy classification of Parkinson's disease *IEEE Transactions on Biomedical Engineering* **59** 1264-71

[13] Little M A, McSharry P E, Hunter E J, Spielman J and Ramig L O 2009 Suitability of dysphonia measurements for telemonitoring of Parkinson's disease *IEEE transactions on biomedical engineering* **56** 1015-22

[14] Ozkan H 2016 A comparison of classification methods for telediagnosis of Parkinson's disease *Entropy* **18** 115

[15] Alemami Y and Almazaydeh L 2014 Detection of Parkinson disease through voice signal features *The Journal of American Science* **10** 44-7

[16] Karimi Rouzbahani H and Daliri M R 2011 Diagnosis of Parkinson's disease in human using voice signals *Basic and Clinical Neuroscience* **2** 12-20

[17] Chen H-L, Huang C-C, Yu X-G, Xu X, Sun X, Wang G and Wang S-J 2013 An efficient diagnosis system for detection of Parkinson's disease using fuzzy k-nearest neighbor approach *Expert systems with applications* **40** 263-71

[18] Hariharan M, Polat K and Sindhu R 2014 A new hybrid intelligent system for accurate detection of Parkinson's disease *Computer methods and programs in biomedicine* **113** 904-13

[19] Oung Q W, Hariharan M, Lee H L, Basah S, Yaacob S, Sarillee M and Lee C 2015 Objective assessment of Parkinson's disease symptoms severity: A review. *2015 2nd International Conference on Biomedical Engineering (ICoBE)* pp 1-6

[20] Shahbakhi M, Far D T and Tahami E 2014 Speech analysis for diagnosis of parkinson's disease using genetic algorithm and support vector machine *Journal of Biomedical Science and Engineering* **7** 147-56

[21] Oung Q W, Muthusamy H, Lee H L, Basah S N, Yaacob S, Sarillee M and Lee C H 2015 Technologies for assessment of motor disorders in Parkinson's disease: a review *Sensors* **15** 21710-45

[22] Sharma R and Gupta A K 2015 Voice Analysis for Telediagnosis of Parkinson Disease Using Artificial Neural Networks and Support Vector Machines *International Journal of Intelligent Systems and Applications* **7** 41

[23] Sakar B E, Isenkul M E, Sakar C O, Sertbas A, Gurgen F, Delil S, Apaydin H and Kursun O 2013 Collection and analysis of a Parkinson speech dataset with multiple types of sound recordings *IEEE Journal of Biomedical and Health Informatics* **17** 828-34

[24] Li Q, Zheng J, Tsai A and Zhou Q 2002 Robust endpoint detection and energy normalization for real-time speech and speaker recognition *IEEE Transactions on Speech and Audio Processing* **10** 146-57

[25] Shah S and Bhise A 2013 Fast Speaker Recognition using Efficient Feature Extraction Technique 1 *International Journal of Computer Science and Network* **2** 6-12

[26] Singh P P and Rani P 2014 An approach to extract feature using mfcc *IOSR Journal of Engineering* **4** 21-5

[27] Holi M S 2013 Automatic detection of neurological disordered voices using mel cepstral coefficients and neural networks. *2013 IEEE Point-of-Care Healthcare Technologies (PHT)* pp 76-9

[28] Nehe N S and Holambe R S 2012 DWT and LPC based feature extraction methods for isolated word recognition *EURASIP Journal on Audio, Speech, and Music Processing* 1-7

[29] Antoniol G, Rollo V F and Venturi G 2005 Linear predictive coding and cepstrum coefficients for mining time variant information from software repositories. *ACM SIGSOFT Software Engineering Notes* pp 1-5

[30] Schroeder M 1984 Linear prediction, entropy and signal analysis *IEEE ASSP Magazine* **1** 3-11

[31] El Emary I M and Ramakrishnan S 2008 On the application of various probabilistic neural networks in solving different pattern classification problems *World Applied Sciences Journal* **4** 772-80

[32] Specht D F 1990 Probabilistic neural networks *Neural networks* **3** 109-18

[33] Wu X, Kumar V, Quinlan J R, Ghosh J, Yang Q, Motoda H, McLachlan G J, Ng A, Liu B and Philip S Y 2008 Top 10 algorithms in data mining *Knowledge and information systems* **14** 1-37

IOP Conf. Series: Materials Science and Engineering **318** (2018) 012040 doi:10.1088/1757-899X/318/1/012040

Wideband Circularly Polarized Printed Ring Slot Antenna for 5 GHz – 6 GHz

Mohamed Nasrun Osman[1], Mohamad Helmi A. Rahim[1], Muzammil Jusoh[1], Thennarasan Sabapathy[1], Mohamad Kamal A. Rahim[2], and Saidatul Norlyana Azemi[3]

[1]Bioelectromagnetics Research Group (BioEM), School of Computer and Communication Engineering, Universiti Malaysia Perlis, Pauh Putra Main Campus, 02600 Arau, Perlis.
[2]Advanced RF and Microwave Research Group (ARFMRG), Communication Engineering Department, Faculty of Electrical Engineering, Universiti Teknologi Malaysia, 81310 Skudai, Johor.
[3]Advanced Communication Engineering Centre (ACE), School of Computer and Communication Engineering, Universiti Malaysia Perlis, Pauh Putra, 02600 Arau, Perlis.

E-mail: nasrun@unimap.edu.my

Abstract. This paper presents the design of circularly polarized printed slot antenna operating at 5 – 6 GHz. The proposed antenna consists of L-shaped feedline on the top of structure and circular ring slot positioned at the ground plane underneath the substrate as a radiator. A radial and narrow slot in the ground plane provides coupling between the L-shaped feedline and circular ring slot. The circular polarization is realized by implementing the slits perturbation located diagonally to perturb the current flow on the slot structure. The antenna prototype is fabricated on FR4 substrate. The simulated and measured results are compared and analyzed to demonstrate the performance of the antenna. Good measured of simulated results are obtained at the targeted operating frequency. The simulated -10dB reflection coefficient bandwidths and axial ratio are 750 MHz and 165 MHz, respectively. The investigation on the affect of the important parameters towards the reflection coefficient and axial are also presented. The proposed antenna is highly potential to be used for wireless local area network (WLAN) and wireless power transfer (WPT).

1. Introduction

Wireless communication undergoes rapid changed in recent years. More peoples are using modern communication services, thus increases the demand on the used of the antenna that can covers a lot of communication services [1]. The utilization of the linear polarization (LP) is only capable to detect signals in one direction and will facing polarization mismatch loss - loss in signal strength when the signal is curving from different angles. Therefore, the used of circular polarization (CP) instead of the LP has overcome this problem by constantly receive a component of the signal due to the angular variation regardless of receiver orientation [2], thus making the CP antennas capable to send and receive signals in all angles.

The CP has the capability to enhance the system gain, cross polar discrimination and dealing with the multipath fading environment, consequently can enhance the system performances [3]. The polarization of the signal might change to orthogonal type of polarization when reflected with metallic objects, hence effect the polarization to be reverses and conversely will produced mostly orthogonal in polarized [4]. In addition, the CP is also preferable on penetrating the bend around the obstruction and have better resistance to the degradation of signals in bad weather conditions. Hence, CP is effective to be deployed for maintaining and establishing in communication systems, especially for unwanted and unpredicted propagation scenario. The ability of CP to reduce the mismatch polarization loss has made CP more attractive for some applications such as radia frequency identification, navigation and radar [5].

The CP antenna can be developed by different methods as reported in the literature. The CP antenna is successfully realized by arranging the LP antenna with controlling the feeding network [6] and the used of assymetric bowtie dipoles [7]. However, this approach taken will increased the complexity and the size of the structure. The conventional technique to produce a CP is by creating a perturbation segment on the radiator. Work done in [8-10] implemented perturbation segment on the rectangular patch either chamfering corner, L-slits or orthogonal assymetric slots. Still, the perturbation technique on the patch has a drawback in term of narrow S_{11} and axial ratio (AR) bandwidth. Due to this drawback, slot antenna is gaining popular to fulfill the aim of wideband operation. Paper presented in [11, 12] shows the used of slot antenna to obtain wideband AR of more than 1 GHz. Therefore, the slot antenna is chosen due to its wideband characteristics and less sensitive to manufacturing tolerances than the conventional microstrip antennas.

In this paper, a circularly polarized circular ring slot antenna is proposed, which is capable to produce left-hand CP (LHCP) with wideband impedance and AR bandwidth. CP characteristics is achieved by implementing slits perturbation method located specifically on the circular ring slot antenna. Simulated and measured results indicate that the proposed antenna has successfully achieved the CP capability at 5-6 GHz frequency band with a maximum gain of 5 dBi.

2. Antennas Structure and Mechanism

The optimized structure of proposed CP slot antenna is indicated in Figure 1. It is composed of two parts : a feeding line with energy coupling and a circular ring slot as a radiator. The printed slot antenna is fabricated on a FR4 substrate with relative permittivity of 4.7 and a thickness of 1.6 mm. The dimension of the structure is L_s x W_s. The front side of the antenna consists of L-shaped feedline with width of W_f and length of $L_{f1}+L_{f2}$.

The circular ring slot with a inner radius of a and outer radius of r_1 is etched on the opposite side of the substrate. Radial slot with diameter of r_2 and narrow slot with length of 5 mm are located at the y-axis below the ring slot. These slots are attached to the edge of the ring slot, and is positioned to be underneath and across by the feedline on the top. This is to ensure the good energy coupled between feedline and the ring slot. Two slits (length of L_{slit} and width of W_{slit}) are inserted diagonally -45°, connecting towards the inner radius. The slits and inner radius are separated with a gap of 0.3 mm. To obtain CP, the slits and the narrow feeding slot have to be 45° or 135° apart. The photograph of the fabricated CP slot antenna is shown in Figure 2.

The approach and mechanism of the proposed antenna is as follows. To achieve circular polarization, the antennas have to be excited by two orthogonal modes of the same magnitude and in phase quadrature. The existing of the slit will perturbed the flow of the current at the edge of the ring slot. This will lengthened the path of the current and drive it to travel along the slit. At the specific slit length, the proposed antenna fulfill the conditions to excite CP. Therefore, the length and width of the slits mostly determines the the response of AR and have to be optimized to ensure the AR value drop below than 3dB (CP is produced).

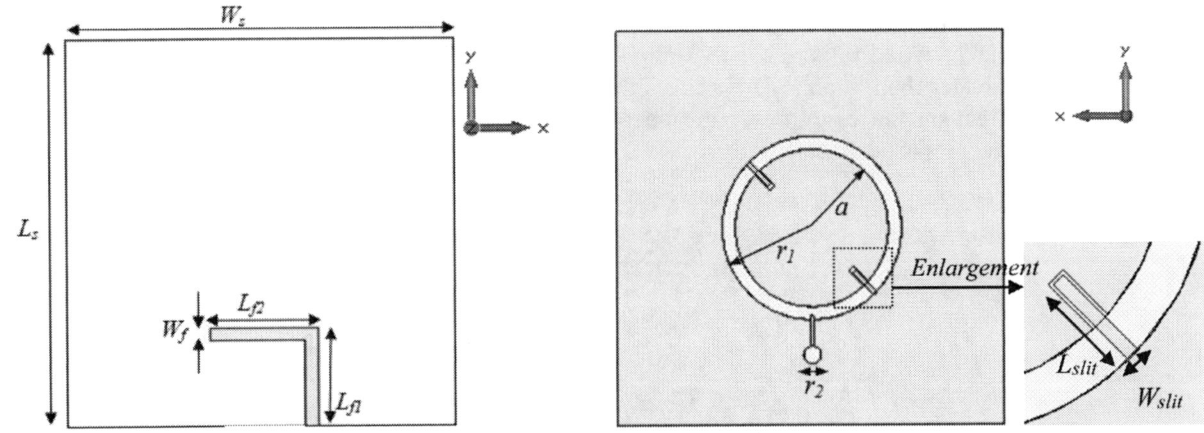

(a) (b)

Figure 1. Geometry of the proposed antenna. (a) front view and (b) back view. (Dimensions in mm: W_s=90, L_s=90, L_{f2}=25, W_f=3, L_{f1}=22, L_{sl}=7, a=17, r_1=20, r_2=2, L_{slit}=8 , W_{slit}=1)

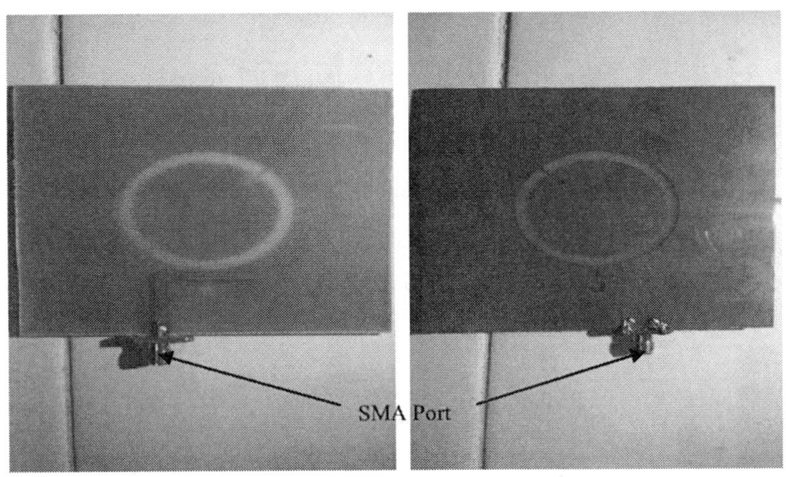

(a) (b)

Figure 2. Fabricated CP printed slot antenna. (a) front structure and (b) back structure.

3. Parametric Studies

Figure 3 depicts the step of designing the proposed antenna. At first, the feedline and the circular ring slot is created, followed with radial slot at second stage. Here, the dimension of the feedline, radial slot and narrow slit is optimized to achieve good impedance matching when the energy is coupled to the circular ring slot. This have to be matched well with the width of the ring slot. At this condition, the antenna is excited with linear polarization. To achieve CP, the slit is introduced at the diagonal position (third step). Then, to obtain desired and optimized design, the parametric studies of the important parameters are conducted and investigated. This studies helped to identify the effect of the important parameters and provide useful informations to obtain optimized design. Thus, the affect towards S_{11} reflection coefficient and AR are presented.

Investigation began with the study on the effect of the radial slot. Fig. 4(a) shows the effect of size of the radial, r_2 to the S_{11} result when it is varies from 1mm to 5mm. Another important parameter that affect the impedance matching is the width of the feedline, W_f, as shown in Fig. 4(b). The parameter W_f is varied from 2 mm to 6 mm. From the graph, it can be seen that good matching (S_{11} < -10 dB) is achieved when W_f is less than 3 mm. The S_{11} value is shift above the threshold -10dB when the width is greater than 3 mm. This proves that the dimension of the feedline and the radius of the radial slot affect on the coupling of the energy, thus give significant influence to the impedance matching of the proposed antenna.

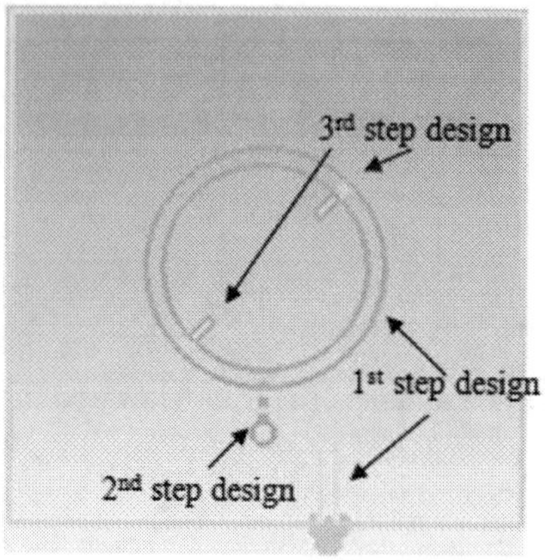

Figure 3. The step of designing CP printed slot antenna.

The investigation also has been carried out on the length of the slit, L_{slit} to the S_{11} and AR result, as shown in Figure 5(a) and Figure 5(b), respectively. For this analysis, the slit length is lengthened from 2 mm to 6 mm. From these figures, it proves that slit length is greatly affecting the CP characteristic and slightly influence the S_{11} response. When this length is further extended until 6 mm, the AR values is begun to fall. Therefore, based on this analysis, the value of AR is dropped below than threshold 3 dB when the slit length is between 3 mm to 4.2 mm. When the length is more than 4.2 mm, the AR value starts to increase, indicate CP is no longer excited.

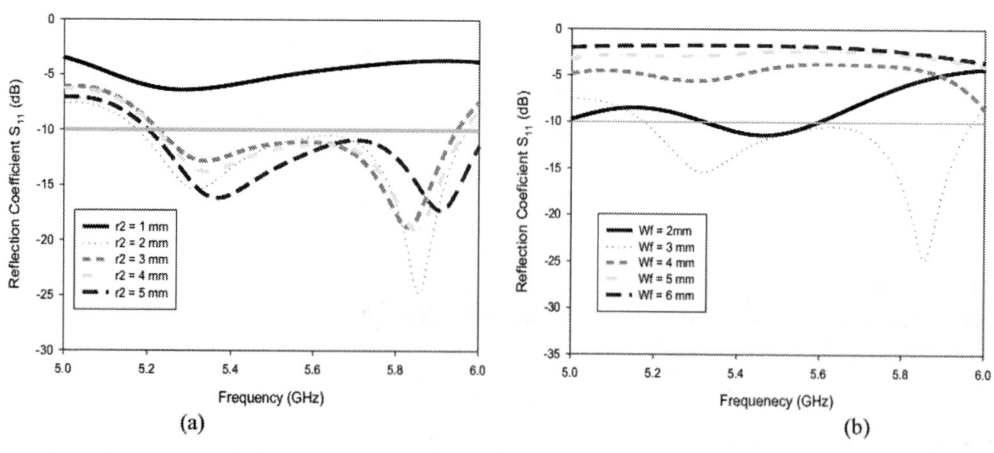

Figure 4. Effect towards S_{11} result for (a)small circular slot, r_2 and (b) width of the feedline, W_f

IOP Conf. Series: Materials Science and Engineering **318** (2018) 012040 doi:10.1088/1757-899X/318/1/012040

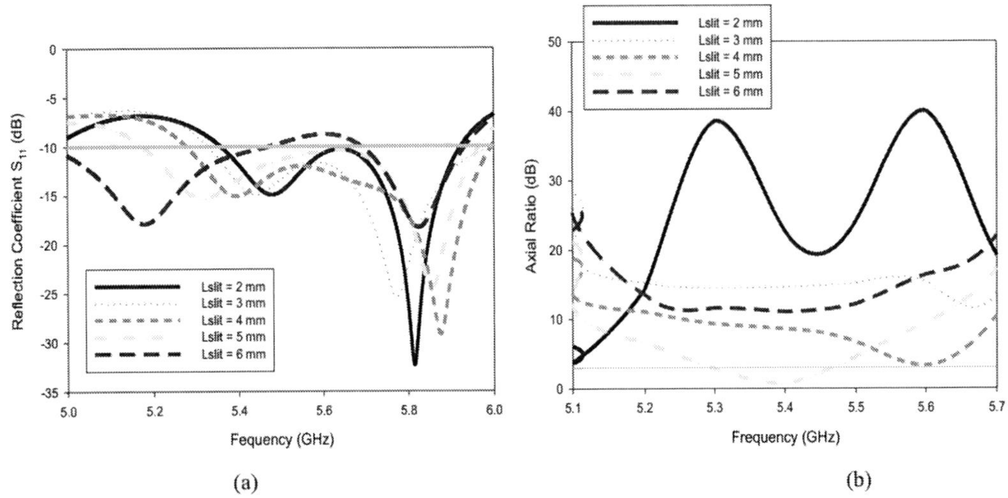

Figure 5. Effect of the slit length, L_{slit} to the (a) S_{11} and (b) axial ratio results

4. Result and discussion

Simulated and measured input reflection coefficient for CP printed slot antenna is shown in Figure 6. As shown in the figure, good agreement has been achieved between the simulated and measured. The measured impedance bandwidth of reflection coefficient, as referred to -10 dB is 514.9 MHz (5.5643 - 6.072 GHz). Slight discrepancies were occurred, which might be due to the fabrication tolerance. The etching process was done manually, which could make the slits slightly bigger or smaller than the actual size, thus affecting the antenna effective surface and resonance frequency.

Figure 7 plot the simulated AR curves over frequency. The simulated 3-dB AR bandwidth, BW_{AR} is 164.9 MHz (5.2952 – 5.4601 GHz). It can be noticed that the bandwidth of the AR lies within the targeted frequency band. Meanwhile, the simulated electric field at frequency 5.4 GHz is illustrated in Figure 8 and shown the proposed antenna produce LHCP when the slits are inserted at the -45° diagonal. Another type of orthogonal CP can be achieved if the position of the slits are changed to 45° diagonal (mirror at *y*-axis). By doing this, right hand CP can be produced.

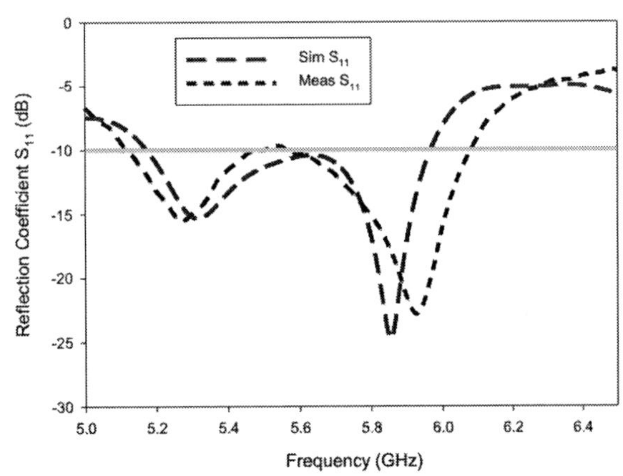

Figure 6. Comparison of simulated and measured reflection coefficient result.

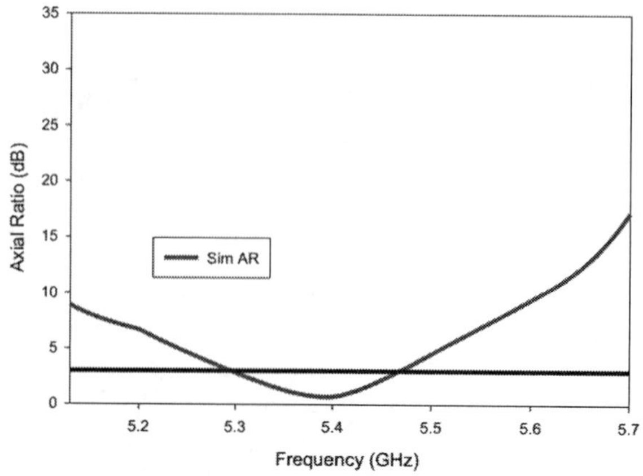

Figure 7. Simulated AR (dB) of the proposed antenna.

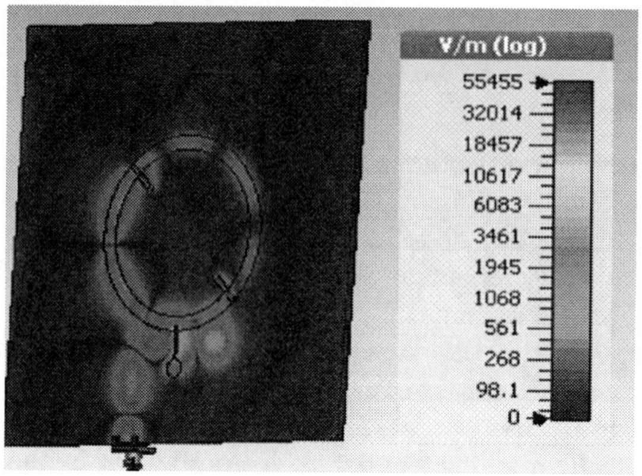

Figure 8. Simulated E-field for LHCP at 5.4 GHz.

5. Conclusion

A CP ring circular slot antenna having a wideband impedance and AR bandwidth has been presented. A circular ring slot antenna is used as a radiator. In this work, CP characteristic is developed using a simple method, which is by using slits perturbation. From the parametric studies conducted, the position and the length of the slits mainly influence the response of the AR. The simulated and measured results of the reflection coefficient and AR are presented. The results illustrate the proposed antenna is succesfully achieved according to the desired specification. Compact in dimension, the proposed antenna is designed to be operated in the 5-6 GHz and potential suitable to be used in wireless power sensors application.

The author thanks Universiti Malaysia Perlis and Research Management and Innovation Centre for the support of the research work under grant 9001-00569.

References

[1] X. Qian, G. Zhang, X. Lei, L. Hou, and S. Ma, "Design of A Broadband Circularly Polarized UHF RFID Tag Antenna for Metallic Objects," IEEE Antennas Wirel. Propag., vol. 2, pp. 1–3, 2015.

[2] B. Zhan, W. Ding, and S. Liu, "Design Parameter Optimization of a Dual-band Right-handed,"

IEEE Antennas Wirel. Propag., pp. 8–11, 2016.
[3] S. Latha, "Circular Polarized Microstrip Patch Array Antenna for Wireless LAN Applications," IEEE Antennas Wirel. Propag., pp. 1197–1200, 2016.
[4] K. George Thomas and G. Praveen, "A novel wideband circularly polarized printed antenna," IEEE Trans. Antennas Propag., vol. 60, no. 12, pp. 5564–5570, 2012.
[5] A. Goyal, A. Gupta, and L. Agarwal, "A Review Paper on Circularly Polarized Microstrip Patch Antenna," IEEE Trans. Antennas Propag., no. 3, pp. 138–142, 2016.
[6] W. Cao, B. Zhang, A. Liu, T. Yu, D. Guo, and K. Pan, "A reconfigurable microstrip antenna with radiation pattern selectivity and polarization diversity," IEEE Antennas Wirel. Propag. Lett., vol. 11, pp. 453–456, 2012.
[7] H. H. Tran and I. Park, "Wideband Circularly Polarized Cavity-Backed," IEEE Trans. Antennas Propag., vol. 15, pp. 358–361, 2016.
[8] Y. J. Sung, T. U. Jang, and Y.-S. Kim, "A reconfigurable microstrip antenna for switchable polarization," IEEE Microw. Wirel. Components Lett., vol. 14, no. 11, pp. 534–536, 2004.
[9] Y. J. Sung, "Reconfigurable patch antenna for polarization diversity," IEEE Trans. Antennas Propag., vol. 56, no. 9, pp. 3053–3054, 2008.
[10] M. S. Nishamol et. al, "An electronically reconfigurable microstrip antenna with switchable slots for polarization diversity," IEEE Trans. Antennas Propag., vol. 59, no. 9, pp. 3424–3427, 2011.
[11] M. Shokri et.al, "Tiny circularly polarized printed slot antenna for UWB usage," Life Science Journal, vol. 9, no. 3, pp. 2288–2291, 2012.
[12] R. K. Saini and S. Dwari, "A broadband dual circularly polarized square slot antenna," IEEE Trans. Antennas Propag., vol. 64, no. 1, pp. 290-294, 2016.

IOP Conf. Series: Materials Science and Engineering **318** (2018) 012041 doi:10.1088/1757-899X/318/1/012041

Multiple Vehicle Detection and Segmentation in Malaysia Traffic Flow

Ahmad Fariz Hasan[1], Mohd Fikri Che Husin[1], Khairul Affendi Rosli[1], Mohd Norhafiz Hashim[1] and Amar Faiz Zainal Abidin[2]

[1]Department of Electronic Engineering Technology, Faculty of Engineering Technology, Universiti Malaysia Perlis, Malaysia.
[2]Faculty of Engineering Technology, Universiti Teknikal Malaysia Melaka, 76100 Durian Tunggal Melaka

E-mail: farizhasan@unimap.edu.my, fikri@unimap.edu.my, khairulaffendi@unimap.edu.my, mnorhafiz@unimap.edu.my, amarfaiz@utem.edu.my

Abstract. Vision based system are widely used in the field of Intelligent Transportation System (ITS) to extract a large amount of information to analyze traffic scenes. By rapid number of vehicles on the road as well as significant increase on cameras dictated the need for traffic surveillance systems. This system can take over the burdensome task was performed by human operator in traffic monitoring centre. The main technique proposed by this paper is concentrated on developing a multiple vehicle detection and segmentation focusing on monitoring through Closed Circuit Television (CCTV) video. The system is able to automatically segment vehicle extracted from heavy traffic scene by optical flow estimation alongside with blob analysis technique in order to detect the moving vehicle. Prior to segmentation, blob analysis technique will compute the area of interest region corresponding to moving vehicle which will be used to create bounding box on that particular vehicle. Experimental validation on the proposed system was performed and the algorithm is demonstrated on various set of traffic scene.

1. Introduction

Developments of Intelligent Transportation Systems (ITS) have received a lot of attention in recent years. ITS refers to a variety of tools, such as traffic engineering concepts, software, hardware, and communications technologies, that can be applied in an integrated fashion to the transportation system to improve services in transportation systems operations, such as traffic management, commercial vehicle operations, transit management, and information to travelers [1].

Rapid development of technologies and the emergence of a new information age offer a new dimension in the operation and management of transport systems and facilities. ITS involved the application and integration of advance communication, microprocessor and information technologies into transport systems to achieve efficient utilization of infrastructure and energy resources, to improve safety and reduce the environmental impact of traffic.

One of the most significant efforts in ITS is developing visual surveillance systems that help reduce numbers of traffic incidents and traffic jams in urban and highway scenario. There are many types of devices that has been used in traffic monitoring systems such as loop detectors, sensors and cameras but vision based analysis technique has only become popular recently in transportation management due to its capability to extract rich information of road traffic compared to other sensor based systems. Normally, the detection of vehicles based on traffic surveillance in highways is performed by human operator at control centers.

Vision based system is a promising alternative since it requires no pavement adjustments and has more potential advantages such as larger detection areas, more flexibility and affordable. At the same time the system performs better and provides good quality results.

However traffic flow raises interesting but difficult problems for image processing. The various light conditions and different weather circumstance places a strong need on the robust algorithms, which require a great amount of computational power to meet the real-time operations of the traffic monitoring system [2]. This work focused on developing multiple vehicle detection and segmentation to automatically detect moving vehicle based on automated vision system. This is initial step before it can be advance to another system such as travel speed estimation, vehicle tracking and classification and so on so forth.

2. Related Works

Recently the use of image processing in Intelligent Transportation Systems (ITS) has increased with tremendous usage in traffic monitoring. This situation led to increasing growth in research and development in this field. Most of the research and development are strongly motivated by significant increase in the number of vehicles, change in population density and as well as urbanization.

Common techniques used for vehicle detections and segmentation can be summarized as follows: Background subtraction and Frame differencing. A. Elgammal et al. [3] introduces background subtraction where technique involves is subtracting the observed image from the estimated image and threshold the result to generate the detect vehicles. This technique can be used to detect and segment both static and moving vehicles but it depends on the background modeling technique. If the background is updated periodically then only vehicle in motion can be detected because static vehicle will become part of the background. Shuguang Zhao et al. [4] introduce frame differencing method which is almost similar to background subtraction technique. The only different in this technique is, it does not use specific background frame (without vehicles) for subtraction. This technique is robust to environmental change, but unable to detect stationary vehicles.

Several studies exist in the literature on automatic video analysis for vehicle detection and segmentation. Techniques explained earlier alone are insufficient to extract traffic flow measurements, thus, most of the work incorporates additional algorithms such as tracking, occlusion removal and etc which are essential for extracting traffic flow parameters. Goo Jun et al. [5] presented a method where the occlude vehicles were divided into many small patches and group the patches according to the clusters of motion vector found by tracking features point. Their proposed system is constructed in several steps which are: background subtraction, occlusion detection, motion analysis and vehicle segmentation.

In [6] vehicle counting method based on blob analysis and feature based tracking is presented. The proposed algorithm is composed of three steps, moving object segmentation (background subtraction), blob analysis, and tracking. A vehicle is modeled as a rectangular patch and classified via blob analysis. Through blob analysis feature such as texture, color and shape are extracted. Those features used to discriminate different objects at the same frame and to recognize the same object at different time which is important for counting. Tracking moving targets is achieved by comparing the extracted features and measuring the minimal distance between two temporal images.

3. Research Methodologies

The basic idea of the vehicle segmentation and tracking was to use three major techniques which are optical flow, morphology and Blob Analysis.

3.1 Grayscale

Processing a color image is three times slower as it has to be performed on three different layers. Besides that, a color image also requires three times the memory space compared to a gray level image. These drawbacks are definitely undesirable for real-time application especially when the color information in not required. For an input video, the first step is to generate its gray scale value from the video itself. Since some cameras capture color images, this process is significant because it removes 24-bit color value and converts them into 8-bit gray level image. This is necessary since gray scale contains the value of each pixel in a single sample which means it carries only intensity information and makes it easier for further analysis.

3.2 Optical Flow Estimation

In this paper, optical flow is used to estimate the velocities of vehicle prior for segmentation. There are two methods widely used to compute optical flow estimation which are Lucas Kanade method and Horn Schunck method. For this work, Lucas Kanade method has been choosen. Optical flow is a pattern of apparent motion of object in a visual scene caused by the relative motion between an observer and the scene. The sequences ordered of image allow the estimation of motion image velocities. The optical flow methods try to calculate the motion between two image frames which are taken at time 't' and 't+1' at every pixel position].

A common starting point, for optical flow estimation is to assume that pixel intensities are translated from one frame to the next,

$$I(\vec{x}, t) = I(\vec{x} + \vec{u}, t+1) \quad (1)$$

Where $I(\vec{x}, t)$ is image intensities in the function of space $\vec{x} = (x, y)^T$ and time t, and $\vec{u} = (u_1, u_2)^T$ is the image velocities. Given a pixel P, all the pixels in the neighbourhood of P have the same velocity as P. Consider a $m \times m$ window $m > 1$, centred at pixel P, the following equation can be formulated as,

$$I_{x1}V_x + I_{y1}V_y = -I_{t1} \quad (2)$$
$$I_{x2}V_x + I_{y2}V_y = -I_{t2} \quad (3)$$
$$I_{xn}V_x + I_{yn}V_y = -I_{tn} \quad (4)$$

Where Vx and Vy are the components of the velocity of optical flow associated with the considered voxel, respectively laying on the X axis and on the Y axis. From the equation above, there are more than two equations; therefore it can be represented in matrix form. Hence,

$$\begin{bmatrix} I_{x1} & I_{y1} \\ I_{x2} & I_{y2} \\ \vdots & \vdots \\ I_{xn} & I_{yn} \end{bmatrix} \begin{bmatrix} V_x \\ V_y \end{bmatrix} = \begin{bmatrix} -I_{t1} \\ -I_{t2} \\ \vdots \\ -I_{tn} \end{bmatrix} \quad (5)$$

Or equivalently it can be written as;

$$A\vec{v} = -b \quad (6)$$

Using least square method;

$$A^T A\vec{x} = A^T(-b) \text{ or } \vec{v} = (A^T A)^{-1} A^T(-b) \tag{7}$$

Or equivalently it can be written as;

$$\begin{bmatrix} V_x \\ V_y \end{bmatrix} = \begin{bmatrix} \Sigma I_{xi}^2 & \Sigma I_{xi} I_{yi} \\ \Sigma I_{xi} I_{yi} & \Sigma I_{yi}^2 \end{bmatrix}^{-1} \begin{bmatrix} -\Sigma I_{xi} I_{ti} \\ -\Sigma I_{yi} I_{ti} \end{bmatrix} \tag{8}$$

With the summation running from 1 to n. Therefore the value of velocity between frame to frame can be computed here.

In another word, equation 1 can be writen as follows;

$$f(u, v) = \iint (I_x u + I_y v + I_t)^2 \, dxdy) \tag{9}$$

Equation 7 is integrated with respect to the dimension of W. Assuming the velocity is the same all over W, that is u and v do not depend upon dx nor dy, therefore it can be writen as the following system,

$$u \iint I_x^2 dxdy + v \iint I_x I_y dxdy + \iint I_x I_t dxdy = 0 \tag{10}$$

$$u \iint I_x I_y dxdy + v \iint I_y^2 dxdy + \iint I_y I_t dxdy = 0 \tag{11}$$

This technique will estimate the motion of vehicles velocities based on image sequences. The motion of vehicles in a visual scene is caused by the relative motion between an observer and the scene.

3.3 Threshold

Thresholding is required to perform segmentation on the basis of the different intensities or colors in the foreground and background regions of an image. Thresholding is a pixel point operation that performs a comparison on each pixel value to determine if the intensity value is above or below an input threshold variable Tval. The goal of thresholding is to partition an image into regions that separates objects from the background. In this project, global thresholding will be used. Global thresholding uses a fixed threshold for all pixels in the image and therefore works well only if the gray-level distribution histogram contains distinctively separated peaks corresponding to the objects and background. It will creates a binary image from optical flow estimation by turning all pixels below a certain threshold value to 0 and value above threshold to 1 whereby 0 will represent black image while 1 will refer to white image.

Morphological dilation and erosion are performed to obtain a clear and filled region upon the obtained result from threshold technique. Square structuring element is used for one time. Median filter was been used in filtering process in order to remove some of the unwanted noise from threshold result.

Once the threshold is cleared and the region is filled from morphological process, blob analysis is used to compute statistic for connected regions in a binary image. In this work, blob analysis technique has been set to calculate the area of connected region. It will scan through the entire pixel to determine the connected region based on threshold in order to determine whether the interested value is belong to moving vehicle or not. As the result from the blob analysis is figure out, the segmentation process will take place to detect the vehicle.

3.4 Segmentation

Segmentation process is the last part of the algorithm which is to distinguish the objects of interest from the background and unwanted objects. This part of processing is responsible to identify the moving vehicle from the static objects or the background. Segmentation is preferred based on optical flow estimation as discussed earlier. It involves, partitioning an image into groups of pixels which are homogeneous with respect to some criterion. Different groups must not intersect each other and adjacent groups must be heterogeneous. The bounding boxes were determined based on the ratio of interested region when performing blob analysis.

4. Experiments and Results

The video is captured using a camera video with a resolution 384×288 pixel from a fixed distance. The distance between camera and the target object is an important factor as the ratio of object size in pixels over the real size varies. The algorithm was tested on two different video sequences. Table 1 shows video specification on different frame.

Table 1. Video specification of test samples

Video	Length	Size	Frame Number	No. of Vehicle
1	29 sec	384 x 288	166	17
			212	15
			221	14
2	16 sec	384 x 288	48	26
			77	28
			141	24

4.1 Analysis from video 1

Figure 1 shows the result obtained for video 1. It can be summarized in Table 2. Some of the vehicles are not detected and segmented due to occlusion. This is due to the fact that the vehicles are moving together. This is one of the problems using optical flow that need further research in the future.

Table 2. Comparison for analysis 1

Frame Number	Manual Inspection	Proposed Algorithm	Correctness (%)
166	17	6	35.0
212	15	7	46.6
221	14	7	50.0

4.2 Analysis from video 2

Figure 2 shows the results obtained for video 2. It can be summarized in Table 3. Some of the vehicles are not detected and segmented due to some of the vehicles are far at the back, and are just about to start moving. The optical flow technique is only able to segregate out the moving vehicles only. This technique is not robust when applied on stationary vehicles.

Table 3. Comparison for analysis 2

Frame Number	Manual Inspection	Proposed Algorithm	Correctness (%)
48	26	9	36.6
77	28	14	50.0
141	24	14	58.3

4.3 Comparison between Proposed Algorithm and other technique when tested with different resolution.

As can be seen, the proposed technique show the promising result compare to other technique where in this work, other technique developed is used foreground detection technique [8]. Table 4 and 5 shows the evaluation on the different technique while running on the different video resolution. Furthermore, while running on high resolution video, the result is much better because the video resolution is much more clearer compare to the one that has been compressed but the trade of is, it required huge computation time in order to complete for 1 video since it consist up to millions of pixel in one frame. For video number one, it requires 15 minutes while for video two is up to 10 minutes to complete the task. Figure 3 and Figure 4 shows the result obtained for video 1 and video 2 respectively.

Table 4. Comparison using different technique on different resolution video for video 1

Case Study	Evaluation of vehicle detection			
Manual Inspection	Proposed Algorithm (Optical Flow Technique)		Other Algorithm (Foreground Subtraction Technique)	
	CCTV	HD	CCTV	HD
28	15	22	8	6
Frame Number	257	257	257	257
Percentage of Inspection (%)	53.57	78.57	28.57	21.43

Table 5. Comparison using different technique on different resolution video for video 2

Case Study	Evaluation of vehicle detection			
Manual Inspection	Proposed Algorithm (Optical Flow Technique)		Other Algorithm (Foreground Subtraction Technique)	
	CCTV	HD	CCTV	HD
28	15	22	8	6
Frame Number	316	316	316	316
Percentage of Inspection (%)	53.57	78.57	28.57	21.43

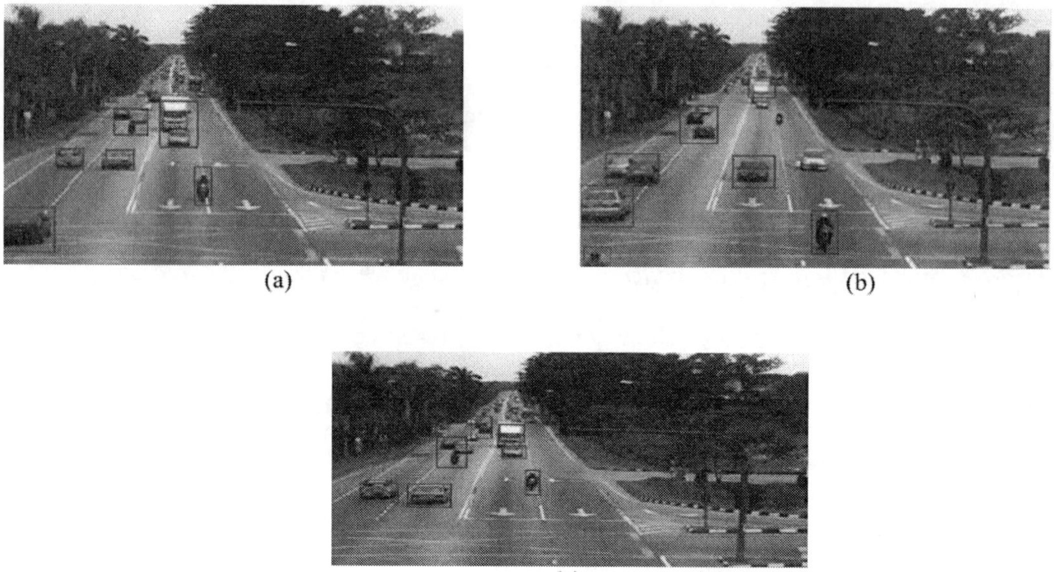

(a) (b) (c)

Figure 1: Result from video 1 (a) Frame 166 (b) Frame 212 (c) Frame 221

(a) (b) (c)

Figure 2: Result from video 2 (a) At frame 48 (b) At frame 77 (c) At frame 141

IOP Conf. Series: Materials Science and Engineering **318** (2018) 012041 doi:10.1088/1757-899X/318/1/012041

Figure 3: Result for video 1, (a) using optical flow on HD resolution, (b) using optical flow on CCTV video, (c) using foreground detection on HD resolution, (d) using foreground detection on CCTV video

Figure 4. Result for video 2, (a) using optical flow on HD resolution, (b) using optical flow on CCTV video, (c) using foreground detection on HD resolution, (d) using foreground detection on CCTV video.

IOP Conf. Series: Materials Science and Engineering **318** (2018) 012041 doi:10.1088/1757-899X/318/1/012041

5. Conclusion

In this study, the algorithm has been has successfully implemented using MATLAB 2010. As a result, the algorithm is able to detect and segmented moving vehicles, as long as the targeted objects appear fully within the camera view range. Although the algorithm has a reasonable success rate, the algorithm has various limitations and does not perform well on stationary vehicles and occlusion. Thus the performance can be improved and the present algorithm can be further developed for better reliability and effectiveness. Even the HD resolution gives better result but in order to achieve real-time performance, the algorithm was tested on CCTV video which greatly reduces processing time.

References

[1] C. Jotin Khisty and B. Kent Lall. Transportation engineering. 2nd Edition. Upper Saddle River, NJ: Prentice- Hall Inc. 1998.

[2] K. Toyama, J. Krumm, B. Brumitt, and B. Meyers, "Wallower: Principles and practice of background maintenance," In Proc. 7th Intl. Conf. on Computer Vision. Kerkyra, Greece, 255-261, 1999.

[3] A. Elgammal, D. Harwood, and L. Davis, "Non-parametric model for background subtraction," In Proceedings of IEEE ICCV 1999, 751-767. 68, Sept 1999.

[4] Shuguang Zhao, Jun Zhao , Yuan Wang and Xinlin Fu, "Moving Object detecting using gradient Information, three frame differencing and connectivity testing," Advance in Artificial Intelligence. Berlin; Springer, 510-518, 2006.

[5] Goo Jun and J. K. Aggarawal, "Tracking and Segmenting of Highway Vehicles in Cluttered and Crowded Scenes," Workshops on Applications of Computer Vision Copper IEEE 2008, Colorado, Jan 2008.

[6] Thou-Ho (Chao-Ho) Chen, Yu-Feng Lin, and Tsong-Yi Chen, "Intelligent Vehicle Counting Method Based on Blob Analysis in Traffic Surveillance," Proceedings of the Second International Conference on Innovative Computing, Information and Control, 238, 2007.

IOP Conf. Series: Materials Science and Engineering **318** (2018) 012042 doi:10.1088/1757-899X/318/1/012042

A GUI Based Software for Sizing Stand Alone AC Coupled Hybrid PV-Diesel Power System under Malaysia Climate

Nor Syafiqah Syahirah Mohamed, Noor Amalina Banu Mohamat Adek , Nurul Farhana Abd Hamid.
School of Electrical System Engineering,Unimap Perlis, Malaysia.

E-mail: norsyafiqah.unimap@edu.my

Abstract. This paper presents the development of Graphical User Interface (GUI) software for sizing main component in AC coupled photovoltaic (PV) hybrid power system based on Malaysia climate. This software provides guideline for PV system integrator to design effectively the size of components and system configuration to match the system and load requirement with geographical condition. The concept of the proposed software is balancing the annual average renewable energy generation and load demand. In this study, the PV to diesel generator (DG) ratio is introduced by considering the hybrid system energy contribution. The GUI software is able to size the main components in the PV hybrid system to meet with the set target of energy contribution ratio. The rated powers of the components to be defined are PV array, grid-tie inverter, bi-directional inverter, battery storage and DG. GUI is used to perform all the system sizing procedures to make it user friendly interface as a sizing tool for AC coupled PV hybrid system. The GUI will be done by using Visual Studio 2015 based on the real data under Malaysia Climate.

1. Introduction

In Malaysia, most of the rural area lacked of electricity from the grid utility power supply especially in Sabah and Sarawak. They have to rely on decentralized DG as the main source[1] which it does not guarantee 24-hour electricity supply. To overcome this problem the Ministry Education of Malaysia has launched a large electrification program using PV hybrid program for rural schools in Sabah [1]. This program aimed to solve the electricity problem in Sabah using Stand-Alone PV Hybrid System [2]. The Stand-Alone PV hybrid (SPVH) system is system that combine PV with the another power generating source as a backup such as batteries and DG. The main advantage of SPVH is to produce as much as possible energy from renewable energy sources to ensure that it is sufficient energy to meet the load demands [3].

PV hybrid system installed in rural schools in Sabah can be classified in two types Direct Current (DC) and Alternate Current (AC) coupled topology [4]. The DC coupled system is widely accepted to provide power to small load applications. However, AC coupled system offers more advantages for large system [5-7]. Generally, the PV hybrid system involves five components which are PV array, grid-tie inverter, bi-directional inverter, battery storage and backup source. Figure 1 shows the block diagram of hybrid PV system.

However, AC coupled PV hybrid system is a relatively new system in Malaysia. Therefore, the sizing issue becomes a major challenge and there is still no detailed procedures that are based on

Malaysia's climate [8]. Usually, most of AC coupling system in Malaysia was designed by system provider from country origin [9, 10].

Based on study [11] ,the performance of the actual system is presented by using AC coupled PV hybrid power system installed at one of the rural school electrification in Sabah. It shows the low performance ratio of the system and indicates the PV system is oversized and the unbalance PV energy generated with the energy consumption by user.

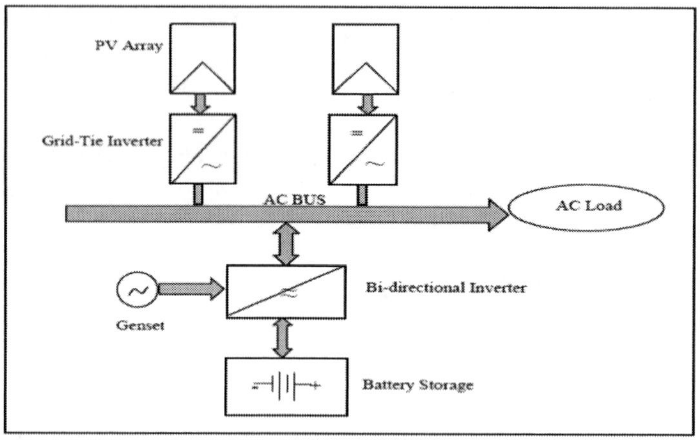

Figure 1: Block Diagram of AC Coupled Hybrid PV Power System.

Therefore, this software is developed as a tool to design AC Coupled Hybrid PV-Diesel Power System by using Graphical User Interface (GUI). Besides that, each components in AC coupled PV hybrid system can be sized accurately to meet the load demand and prevent the un-utilized power in the system.

2. Software Design Methodology

The concept of this sizing tool acts as a guideline for PV system integrator to design effectively the size of AC coupled PV hybrid system components. This sizing tool is developed by using GUI platform based on Visual Studio 2015.The overall procedures in this sizing tool are summarized in the flow chart in Figure 2 begins with main menu interface and ends with summary of PV hybrid configuration. Main menu is the first interface of the system. For the first windows, the user can analyze the stages involved in sizing a stand-alone solar hybrid system.

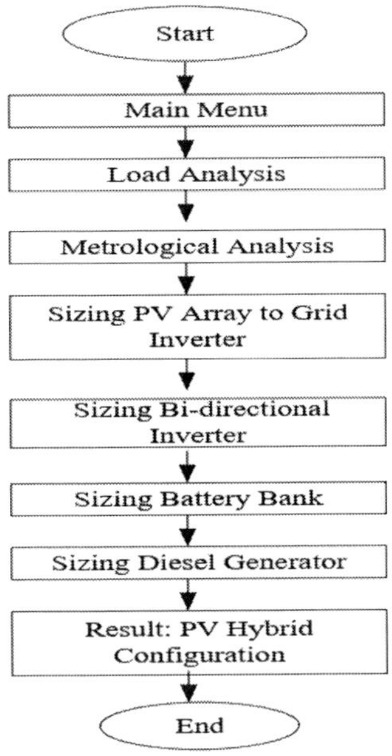

Figure 2. Procedure in sizing an AC Coupled Hybrid PV System.

3. System Sizing

Proper system sizing is essential for sustainability and reliability. The system sizing will be used to decide on size of each components involved in the PV hybrid system.

3.1 Sizing PV Array to Grid Inverter

Using this software, the PV to grid inverter sizing for the hybrid system gives the optimum size of PV array that matched with the grid inverter based on the energy requirement. In this step, the configuration of PV array that matches to the grid inverter is to be defined.

3.1.1 PV Array Power. The PV array as the main source of renewable energy used to supply the consumer with required amount of energy demand based on their load. This amount can be defined by the PV to DG ratio, n of the system. There are several losses affected the output of PV array power. The following formula is proposed to estimate the required PV array power at Standard Test Condition (STC) by using equation (1):

$$P_{STC} = \frac{n \times E_d}{PSH \times f_{tem} \times f_{mm} \times f_{dirt} \times f_{ag} \times \eta_{inv} \times \eta_{cab} \times \eta_{bd} \times \eta_{batt}} \tag{1}$$

P_{stc} is power of PV array at STC.
n is fraction of energy contributed by PV (PV to DG ratio) .
E_d is energy demand per annum.
PSH is peak sun hour per annum.
f_{dirt} is power de-rating factor due to dirt.
f_{mm} is power de-rating factor due to the module mis-match.
f_{tem} is power de-rating factor of power due to average daily maximum ambient temperature.
f_{ag} is power de-rating factor of aging.

η_{inv} is maximum efficiency of inverter.
η_{cab} is efficiency of DC cable from PV to inverter
η_{bd} is maximum efficiency of bi-directional inverter.
η_{batt} is efficiency of battery.

3.1.2 Total Number of PV Module Required.
By using the value of PV array power, the estimated total number of PV modules required for the system based on the selected module is determined by equation (2)[12].

$$N_{tot} = \text{round\_up}\left[\frac{P_{stc}}{P_{mod\_stc}}\right] \quad (2)$$

P_{stc} is power of PV array at STC.
N_{tot} is total number of PV modules required.
P_{mod_stc} is maximum power rating of PV module at STC

3.1.3 Optimum Range Size of Grid Inverter.
The optimum size of grid inverter for the system is calculated based on the PV array capacity. The design factors, f_{d1} and f_{d2} must be selected appropriately based on the types technology of module used and how they are mounted at the site [12]. In this study, the values for f_{d1} and f_{d2} are 0.9 and 1.0 respectively. The technology of the module used is mono-crystalline with free standing structure. The optimum range of size of grid inverter can define by equation (3) as follows:

$$f_{d1} \times (N_{tot} \times P_{mod\_stc}) \leq P_{nom\_inv} \leq f_{d2} \times (N_{tot} \times P_{mod\_stc}) \quad (3)$$

f_{d1}, f_{d2} are design factors.
P_{nom_inv} is nominal AC power output of the selected inverter.

3.1.4 Sizing of PV Array to Grid-Inverter selected.
The range of estimated total number of PV modules that matches with a selected inverter calculated according to the nominal output power of grid tie inverter selected and also the design factor for the system. To determine the range of number of PV modules that matched a specific inverter[12] can define by equation (4) as follows:

$$N_{pv\_inv} = N_{min} \ldots N_{max} \quad (4)$$

N_{pv_inv} is range of number of PV modules matched to the selected inverter
N_{min} is minimum number of PV modules.
N_{max} is maximum number of PV modules.

$$N_{min} = \text{round\_up}\left[\frac{P_{nom\_inv}}{P_{stc\_mod} \times f_{d1}}\right] \quad (5)$$

$$N_{max} = \text{round\_down}\left[\frac{P_{nom\_inv}}{P_{stc\_mod} \times f_{d2}}\right] \quad (6)$$

P_{nom_inv} is nominal AC output power of the inverter.

3.1.5 Number of Modules in Series.
The range number of PV modules connected in series in a string.

$$N_S = N_{S\_min} \ldots\ldots N_{S\_max} \tag{7}$$

N_S is number of PV modules in series.
N_{S_min} is minimum number of PV modules in series.
N_{S_max} is maximum number of PV modules in series.

$$N_{S\_min} = \text{round\_up}\left[\frac{V_{min\_inv\_mppt} \times f_{s2}}{V_{mp\_min} \times \eta_{cab}}\right] \tag{8}$$

$$N_{S\_max} = \text{round\_down}\left[\frac{V_{max\_inv\_input} \times f_{s1}}{V_{oc\_max}}\right] \tag{9}$$

$V_{min\_inv\_mppt}$ is minimum MPPT input of voltage of inverter.
f_{s1}, f_{s2} are safety margins .

The safety margin f_{s1} and f_{s2} are valued 0.95 and 1.1. These values are introduced for a safety purpose. The values of safety margin are depending on 5% less and 10% over from the window voltage of grid-inverter selected.

3.1.6 Number of Modules in Parallel.
The maximum possible number of string in parallel is calculated according to the input DC current of grid-inverter selected, the current short circuit at STC of module and the safety margin. In general, grid-inverter of PV system categorized base on power rating have single MPPT and multiple MPPT which is take into account during calculate the number of PV string connected in parallel. It is determined by equation (10) as [12]:

$$N_p = \text{round\_down}\left[\frac{I_{dc\_inv} \times No\_of\_MPPT}{I_{sc\_stc} \times f_{s3}}\right] \tag{10}$$

N_P is maximum number of string in parallel.
I_{dc_inv} is maximum input current of the inverter.
I_{sc_stc} is short circuit current of PV module.
f_{s3} is safety margin.(varies from 1.1 to 1.25).

3.1.7 Array Configuration.
The array configuration is the optimum numbers of PV module which are connected in series and parallel for the system. The total number of PV module is determined by equation (11) as[12]:

$$N_t = N_s \times N_p \tag{11}$$

N_t is the total number of PV modules in the array.

3.2 Number of Grid-Tie Inverter.

IOP Conf. Series: Materials Science and Engineering **318** (2018) 012042 doi:10.1088/1757-899X/318/1/012042

The total number of grid inverter can calculated according to the actual power array and nominal power of inverter selected in equation (3). The total number of grid inverter can define using equation (12) as [12] :

$$N_{grid\_inv} = roud\_up\left[\frac{N_t \times P_{mod\_stc} \times f_{de-rate}}{P_{nom\_inv}}\right] \quad (12)$$

N_{grid_inv} is the number of grid inverter.

3.3 Bi-directional inverter Sizing.

To select the power rating of bi-directional inverter suitable for the system, the power peak demand need to considered. The total number of bi-directional selected can defined by equation (13) :

$$N_{bd\_inv} = round\_up\left[\frac{PPD}{P_{ac\_bd}}\right] \quad (13)$$

N_{bd_inv} is the number of bi-directional inverter.
PPD is power peak demand.
P_{ac_bd} is nominal output power of the selected bi-directional inverter.

3.4 Battery Sizing.

The suggested sizing process of battery bank is valid for different types of batteries available. The several steps involved to identify the size of battery bank are as follows:

3.4.1 Total Capacity Required Daily.

The total capacity required to satisfy the total energy required daily. It can be calculated by equation (14) as [12]:

$$C_{req} = \frac{E_{d\_daily}}{SV} \quad (14)$$

C_{req} is total battery capacity required daily.
E_{d_daily} is daily energy demand.
SV is system voltage.

3.4.2 Battery Bank Capacity Required.

Battery bank capacity required is determined as a function of several parameters such as autonomy day, maximum of DOD, temperature correction factor of battery and efficiency of the bi-directional inverter. The autonomy days is defined as the number of days that the battery able to fully support the load demands without recharging by PV or the DG. The autonomy days can be set smaller if the system has DG. The value of temperature correction factor can be identified by defining the lowest daily temperature for the site and correlate with the value given by the manufacture. The amount of battery bank capacity is determine by equation (15) as [12]:

$$C_{bat\_req} = C_{req} \times \frac{T_{aut}}{DOD_{max}} \times \frac{1}{f_{bat\_tem} \times \eta_{bd}} \quad (15)$$

C_{bat_req} is actual amount of battery capacity required.
T_{aut} is the days of autonomy (day/s).
DOD_{max} is the maximum depth of discharge of battery.

f_{bat_tem} is temperature correction factor of battery.

3.4.3 Total Bank Discharge Current.

The total battery bank discharge current is the maximum amount of current needed to satisfy the load demand. The total bank discharge current from the battery is equal to the total load current. This is calculated by equation (16) as[12]:

$$I_{bat\_dis} = I_{tot} = \frac{1}{SV} \times \left[\sum_{i=1}^{n} P_{dc} + \sum_{i=1}^{n} \left(\frac{\frac{P_{ac}}{pf}}{\eta_{bd}} \right) \right] \quad (16)$$

I_{bat_dis} is discharge current from the battery bank
P_{dc} is total power of DC load.
P_{ac} is total power of AC load.
pf is power factor of the AC load.

3.4.4 Number of Batteries in Series String.

The total number of batteries in series can be calculated according to the system voltage and by the chosen nominal voltage of battery. The number of batteries in string can be determined using equation in (17) as [12]:

$$N_{s\_bat} = round\_up\left(\frac{SV}{V_{nom\_bat}} \right) \quad (17)$$

N_{s_bat} is number of batteries/cells in series.
V_{nom_bat} is nominal voltage of battery/cell.

3.4.5 Number of Batteries String in Parallel.

The total number of battery strings can be calculated according to the capacity of battery bank required and the capacity of each battery selected. Selection of the next large standard size of battery is recommended The number of batteries in parallel can be calculated by equation (18) as[12]:

$$N_{p\_bat} = round\_up\left(\frac{C_{bat\_req}}{C_{bat\_selected}} \right) \quad (18)$$

N_{p_bat} is number of battery string in parallel.
$C_{bat\_selected}$ is capacity of each battery selected.

3.5 Diesel Generator Sizing

The DG rating is based on the maximum load demand during battery charging. The power from DG is used to charge the battery and provide AC power to the load. When selecting the size of DG, need to be making sure that the DG can meet all the power requirement of the load. The oversize factor of DG is also considered in this study. To ensure a long service life and also good diesel utilization, the DG need to be operated at optimum load [18].The DG rating is calculated by equation (19) as follows [12]:

$$S_{dg} = \frac{S_{max\_dem} \times f_{over}}{f_{dg\_derate}} \quad (19)$$

S_{dg} is the size of a DG required.

S_{max_dem} is the maximum load demand during battery charging.
f_{over} is oversize factor .
f_{dg} is the total de-rating factor of DG (eg:temperature,humidity,altitude).

4. Result

The results of this research consist of the rated powers and configuration of main components in AC coupled PV hybrid power system depends on the load demand. In this research, an AC coupled PV hybrid power system installed at one of the rural school in Sabah is chosen as a case study case to validate the feasibility of developed software. The system size is 15kWp. The AC coupled PV hybrid system is designed to supply electricity to the whole school building included office computer lab, teacher's quarter, and guard house. The actual load demand for 7 months used and analyze to calculate the size of algorithm for the system components.

Figure 3 shows the main interface of the PV system design tool for AC Coupled PV hybrid system. The main interface shows the list of the steps involved in sizing AC Coupled PV hybrid system. It allow for easy navigation of the system. It includes the load analysis, and sizing for each of the parameter such as PV Array, Grid – Tie Inverter, Bi-Directional Inverter, Battery Bank, and DG.

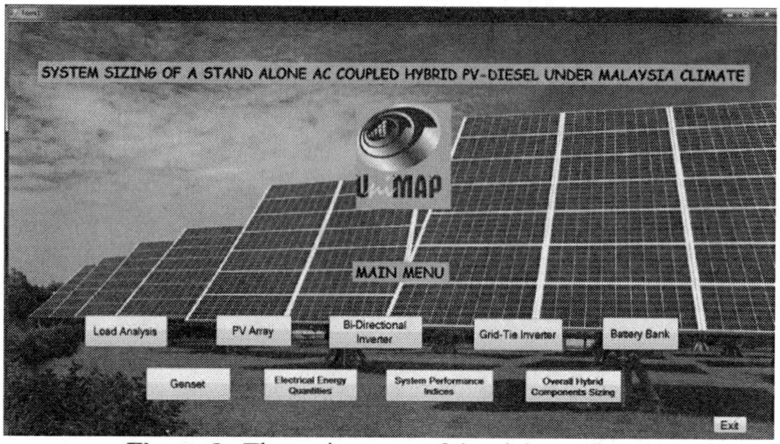

Figure 3. The main menu of the sizing software.

In this research, the weather profile was studied under Malaysia's climate. It is very important to understand the geographical distribution of the solar irradiance and the operating temperature condition at the site before the PV system installation. The daily solar irradiance is graphically represented in Figure 4. From that, the amount of PSH will be obtained. This value is important to determine the amount of PV array power of the system. The system is designed to supply electricity to the whole school building included office computer lab, teacher's quarter, and guard house. The types of connected loads are ceiling fans, computers, projectors, photocopy machine, water pump and etc. The entire appliances are operating using AC power. This system supply power continuously to the respective building with backup by DG. Figure 4 also shows the load analysis interface. In this section the user can view the actual load graph plotted for daily by selecting the month and day buttons. In addition, from this part the user can also know the values of power peak demand and the amount of energy demand. These values are important to determine the size of bi-directional inverter [8, 10] and also the size of PV array respectively later on [8].

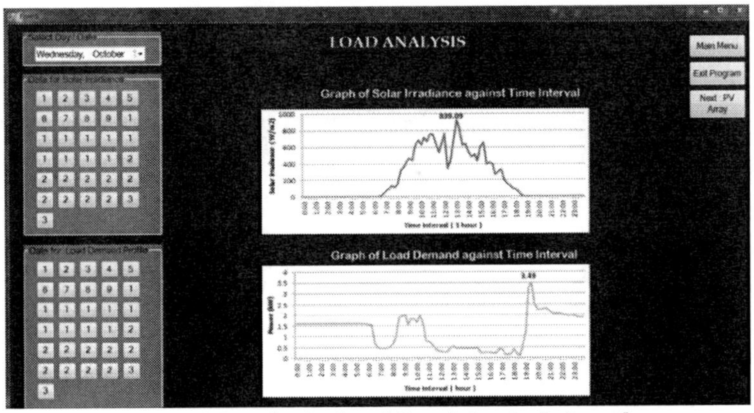

Figure 4. Metrological and Load Analysis Interface.

The first step in sizing an AC coupled PV hybrid system is sizing the Photovoltaic Array to Grid Inverter. The PV to grid inverter sizing gives the optimum size of PV array that matched with the grid inverter based on the energy requirement.

From that, the optimum size of grid inverter for the system is calculated based on the actual final number of PV array capacity and the optimum range size of inverter is determined. There are 6 types of grid inverter available in database and the new data specification can be added manually. From the optimum range of grid inverter above, the actual size, the total number and the type of grid inverter available as in data base can be selected shown in Figure 5.

The second step in sizing an AC coupled PV hybrid system is to determine the PV array configuration. The array configuration is the optimum numbers of PV module which are connected in series and parallel for the system. Figure 6 shows an example of a PV array which consists of 12 PV modules connected in series and 2 PV modules connected in parallel.

In the same interface shows in Figure 6, the third step in sizing an AC coupled PV hybrid system which is calculated the total number of grid inverter according to the actual power array and nominal power of Grid inverter selected.

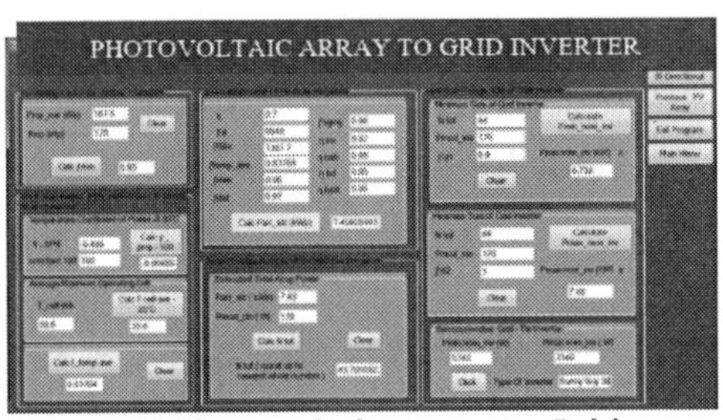

Figure 5. Photovoltaic Array to Grid Inverter Interface.

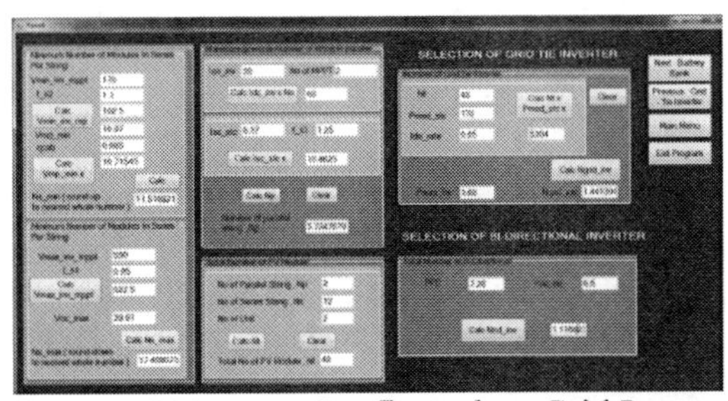

Figure 6. PV Array Configuration, Grid Inverter and Bi-directional Interface.

The fourth step in sizing AC coupled PV hybrid system is sizing bi-directional inverter. As mentioned, the value of power peak demand from the load analysis interface will be used to calculate the size and number of bi-directional as shown in Figure 6. The fifth step is sizing the capacity of the battery bank. The first step of sizing the battery bank is on energy requirement and capacity required daily. The tool will display the possible configuration in series and parallel of batteries that can match the required capacity as shown in Figure 7.

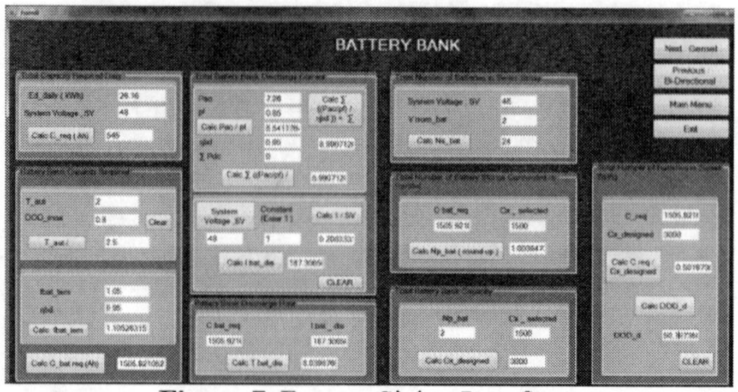

Figure 7. Battery Sizing Interface.

The DG will be used for supplying electrical power to the loads when the PV array cannot produce electricity due to no sunlight. The DG must be able to supply at least the same amount of power as the total AC power demand obtained in load analysis and it must have the same output voltage with the AC system voltage. Then, the user can directly key in the value of de-rating factor involved base on site of case study in the sizing window in Figure 8. After that, the actual size of DG will appear on the tools. After complete all the steps, the user can review the detail recommended configuration of the PV hybrid system as shown in Figure 9.

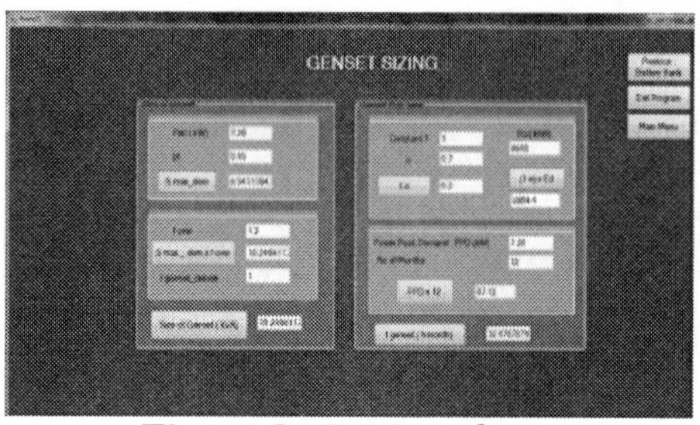

Figure 8. DG Interface.

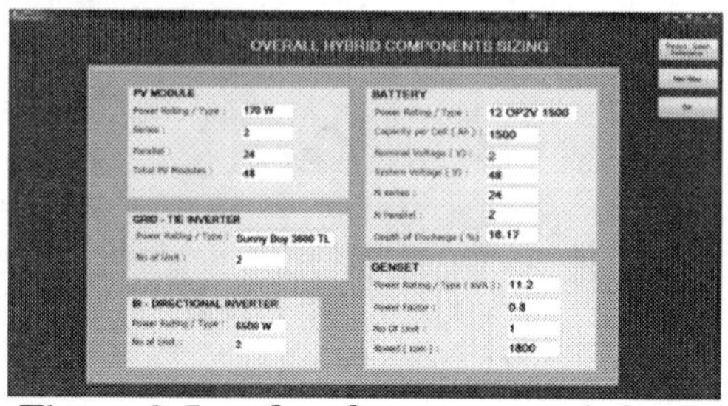

Figure 9. Interface for recommended AC coupled PV hybrid system

5. Conclusion

The development of software tool by using GUI interface has been developed in this study. This software tool is used to find an algorithm for sizing of AC coupling PV hybrid system components. The developed algorithm consists of PV array, grid-tie inverter, bi-directional inverter, battery storage and DG as a backup system for the electrification in rural area. In addition, this software is important to assist the user with minimal knowledge and enhance the user's understanding on design the AC coupled PV hybrid system; conveniently they can use this tool to develop their own PV hybrid system.

References

[1] A. M. Mahmud, "Evaluation of the Solar Hybrid System for Rural Schools in Sabah, Malaysia," presented at the World Renewable Energy Congress Sweden, May 2011.

[2] A. M. Mahmud, "Evaluation of the Solar Hybrid Systems for Rural Schools in Sabah Malaysia," University Oldenburg Germany, January 2010.

[3] S. B. D. Rekioua, and N. Bettar., "Development of Hybrid Photovoltaic-fuel Cell System for Stand-Alone Application," International Journal of Hydrogen Energy, vol. 39, pp. 1604-1611, 2014.

[4] A. M. B. Mahmud, "Analysis of the Performance and impact of the Rural Electrification using Solar Hybrid System for Rural Schools in Sabah, Malaysia – Case Study," presented at the 6th European Conference on PV-Hybrid & Mini Grid Chambery,France.

[5] M. R. J.R.G Cramer, M.Wolly, "The Future of Village Electrification -More than two years of Experiance with AC-Coupled Hybrid System," presented at the 2nd European PV -Hybrid and Mini-Grid conference, Kassel,Germanny, 2003.
[6] M. Meinhardt, M. Rothert, M. Wollny, and A. Engler, "Pure AC-Coupling -The Concept for Simplified Design of Sclable PV Hybrid Sytem using Voltage/Frequency Statics Controlled Battery Inverters " SMA Regelsystem GmbH 2005.
[7] G. e. a. Cramer, "PV System Technologies -State of-the-Art and Trends in Decentetralized Electrification," REFOCUS Journal, pp. 38-42, Febuary 2004.
[8] N. S. syahirah, "Sizing Algorithm for AC Coupled Hybrid PV-Diesel Power System Under Malaysia Climate," Master, Electrical Engineering, Universiti Teknologi MARA (UiTM), 2014.
[9] Leonics. (9/10/2012). How to Design Solar PV System. http://www.leonics.com/support/article2_12
[10] SMA. (2010). Solar Stand-Alone Power and Backup Power Supply. Available: www.SMA.de
[11] A. M. O. Nor Syafiqah Syahirah Mohamed, Faranadia.A.H., presented at the IEEE 5th Control and System Graduate Research Colloquium UiTM, Shah Alam, Malaysia, 2014.
[12] S. Shaari, A. M. Omar, A. H. Haris, and S. I. Sulaiman, Photovoltaic Energy System-Design Principles: Malaysian Green Technology Corporation, 2013.

IOP Conf. Series: Materials Science and Engineering **318** (2018) 012043 doi:10.1088/1757-899X/318/1/012043

Morphology and Structural Characterization of Carbon Nanowalls Grown via VHF-PECVD

M Akmal Hasanudin[1], Y Wahab[1], A K Ismail[2] and Z A Zahid Jamal[1]

[1]Advanced Multidisciplinary MEMS-Based Integrated Electronic NCER Centre of Excellence (AMBIENCE), School of Microelectronic Engineering, Universiti Malaysia Perlis, 02000 Kuala Perlis, Perlis, Malaysia.
[2]Department of Physics, Faculty of Science, Universiti Teknologi Malaysia, 43010 UTM Skudai, Johor, Malaysia.

E-mail: akmalhasanudin@studentmail.unimap.edu.my

Abstract. A 150 MHz very high frequency plasma enhanced chemical vapor deposition (150 MHz VHF-PECVD) system was utilized to fabricate two-dimensional carbon nanostructure from the mixture of methane and hydrogen. Morphology and structural properties of the grown nanostructure were investigated by FESEM imaging and Raman spectroscopy. Carbon nanowalls (CNW) with dense and wavy-like structure were successfully synthesized. The wavy-like morphology of CNW was found to be more distinct during growth at small electrode spacing and denser with increasing deposition time due to better flux of hydrocarbon radicals to the substrate and higher rate of reaction, respectively. Typical characteristics of CNW were observed from strong D band, narrow bandwidth of G band and single broad peak of 2D band of Raman spectra indicating the presence of disordered nanocrystalline graphite structure with high degree of graphitization.

1. Introduction

Since the breakthrough made by Novoselov *et al.* [1], graphene-based carbon nanostructures also known as two-dimensional carbon nanostructures (2DCN) have been extensively studied due to their excellent intrinsic properties. Recently, 2DCN have found their application in variety of fields including biosensor, field emission, supercapacitor, blackbody absorber and hydrophobic coating [2-6]. The versatility of 2DCN has offered many development and improvement in engineering field as the next generation superior material in nanoelectronics, energy storage and solar device [7,8].

In principal, 2DCN are a self-organized, stacked flat graphene sheets forming graphitic building block of walls or petal-like structure and have been known by several names such as carbon nanosheets (CNS), carbon nanoflakes (CNF) and carbon nanowalls (CNW). Their structure exhibits thin sharp edges with high aspect ratio that is typically in the range of few nanometers to tens of nanometers [9-11]. For example, CNW is composed of combination of planar graphene layers, stacking towards each other to form almost vertical graphitic walls, with thin sharp edges of the wall determine its aspect ratio.

There have been numerous reports regarding the growth method of CNW found in the literature. In general, chemical vapor deposition (CVD) method such as hot filament CVD (HFCVD), microwave plasma enhanced CVD (MPCVD), MPCVD with hydrogen radical injection and radio frequency plasma enhanced CVD (RFPCVD) have been adopted to synthesize CNW [3,11-14]. Carbon precursors such as methane, acetylene and tetrafluorocarbon have been widely employed. Each of these methods owned different advantages in terms of film density, wall size and film controllability.

To date very few reports have managed to produce CNW with controllable film morphology. This feature is crucial since its application is highly dependent on film density and structural arrangement of the walls. Current researches on the fabrication of CNW often employed a highly modified PECVD system such as MPCVD and RFPCVD with hydrogen radical injection. Their system configuration is more complex and expensive compared to conventional PECVD 13.56 MHz RF plasma due to presence of magnetron component, and requires two-stage plasma source to operate. In addition, the synthesis of CNW employing a VHF-PECVD system is less to be found in the literature. In most reports, the plasma regions of either direct or remote plasma setup are not stated.

In this paper, the synthesis of CNW by using a capacitively coupled VHF-PECVD with working frequency at 150 MHz has been reported. The studied parameters were electrode separation (anode to sample stage), deposition time and type of substrates. This work focused on the investigation of surface morphology and structural properties of the grown CNW. The potential of capacitively coupled PECVD employing VHF radio frequency with remote plasma setup was elucidated in this report.

2. Experimental works

2.1. Sample preparation and materials

Fused silica (quartz) glass and Si crystal (University Wafer Inc.) were used as substrate throughout this study. All substrates were cut into a square dimension of 10 mm x 10 mm. The substrates were cleaned by immersion in acetone (Merck Company), ethanol (HACH Company) and deionized water with ultrasonic bath for 15 minutes to eliminate oil, dirt and debris originated from substrate cutting. High purity methane (CH_4) and hydrogen (H_2) of 99.999% were utilized as the carbon source and carrier gas for the nucleation of the CNW film.

2.2. Nickel catalyst deposition

A RF magnetron sputtering system (Penta Vacuum Inc.) was utilized to deposit nickel (Ni) thin film on quartz substrate. The sputter pressure for Ni deposition was fixed at 6.0×10^{-3} Torr. 50 W of RF power was set and the RF impedance matching were adjusted until a purple glow-discharge of plasma can be seen through the chamber viewport. Sputtering parameters were fixed at 50 W of RF power, 5.0 sccm of argon supply, $6.0\text{-}6.5 \times 10^{-3}$ Torr of deposition pressure and base pressure of 1.0×10^{-5} Torr. Sputtering time was set to 105 seconds. Sputtering parameters were maintained at low power and low argon flow rate in purpose to induce low deposition rate and high uniformity of Ni thin film.

2.3. Synthesis of carbon nanowalls with varying electrode separation

A 150 MHz VHF-PECVD system was equipped with a linear motion feed through attached to the anode, thus enabling the adjustment of substrate position relative to the anode. The gas flow rate was set at 17.0 sccm (CH_4) and 5.0 sccm (H_2) with deposition time of 3000 s. RF power and temperature were kept constant at 20 W and 750 °C, respectively. The electrode separation was varied from 25 mm to 55 mm; with a certain electrode distance was set for each run. After the introduction of CH_4 and H_2 into the reactor, the linear motion feed through knob was adjusted to the desired substrate position before the growth process was initialized.

2.4. Synthesis of carbon nanowalls with varying deposition time

Flow rate of CH_4 and H_2 were maintained at 17.0 and 5.0 sccm, respectively. For this experiment, the growth time was varied from 20 minutes to 50 minutes with each run was set at a specific growth pressure. Reactor pressure was maintained at 7.7×10^{-2} Torr, with substrate temperature of 750 °C, RF power of 20 W and electrode distance at 30 mm.

2.5. Investigation of surface morphology and structural properties of carbon nanowalls

FESEM from Zeiss (SUPRA 35VP) was utilized to gather information on topographical (in 2D image), surface area coverage (CNW density) and size of the deposited CNW films. The CNW length and wall width were estimated by using the built-in software from the FESEM controller. Raman spectrometer (Renishaw inVia Raman Microscope) was utilized to investigate CNW film structural properties. This model is equipped with He-Cd laser for the monochromatic light source. In this study, a 514 nm laser source was utilized as light excitation of the sample. All samples were scanned in the range of 1000 cm^{-1} to 3500 cm^{-1} at five different spots with three repetitions to minimize error in signal count.

3. Results and discussion

3.1. Influence of electrode separation and deposition time on CNW surface morphology

Figure 1(a-c), provides the aerial view of SEM images of grown CNW at various electrode separations. Figure 1(a) and 1(b) shows that the CNW have been successfully grown with a wavy-like structure. Denser CNW is achieved as electrode distance increases from 25 mm to 45 mm with more distinct shape at smaller spacing. However, the morphology changed drastically during growth at 55 mm.

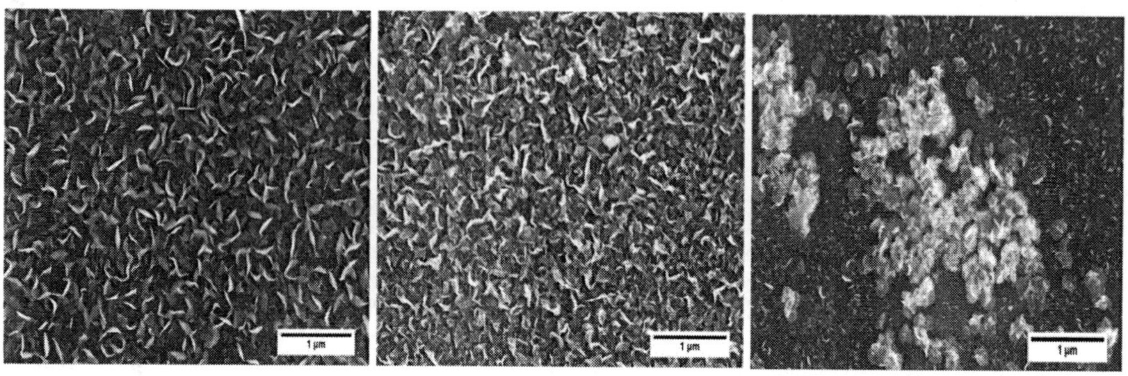

(a) **(b)** **(c)**

Figure 1. SEM images of CNW grown at different electrode separation: (a) 25 mm, (b) 45 mm, and (c) 55 mm with growth condition of 17:5 (CH_4:H_2) sccm; 20 W RF power; temperature of 750°C and growth time of 3000 s. All images were captured at 20000 times magnification.

Figure 2 (top) verifies the average wall density and size of CNW grown at different electrode separation. They are found to be ranging from 11 wall/μm^2 to 24 wall/μm^2 and 0.46 µm to 0.48 µm, respectively, with highest density is obtained during separation of 45 mm. It is clear that wall size does not affected during growth at small spacing.

In contrast, during growth at largest electrode separation, 55 mm, dramatic change in morphology was observed thus causing the wall size to increase 115.52 µm. Figure 1(c) shows agglomeration of graphene flakes characterized by its round shape attributes and absence of vertical nanowalls. The growth of CNW is directly influenced by the supply of carbon species to the surface [17]. Thus, in a VHF-PECVD that employs remote plasma setup, the directed flux of C_2 hydrocarbon species or radicals towards the substrate surface may be varied by distance.

Figure 3(a-d) presents the top view SEM images of CNW morphology grown at different deposition time. It can be noted that CNW density increases with deposition time. It also can be noticed that wall spacing becomes smaller as the deposition time increases.

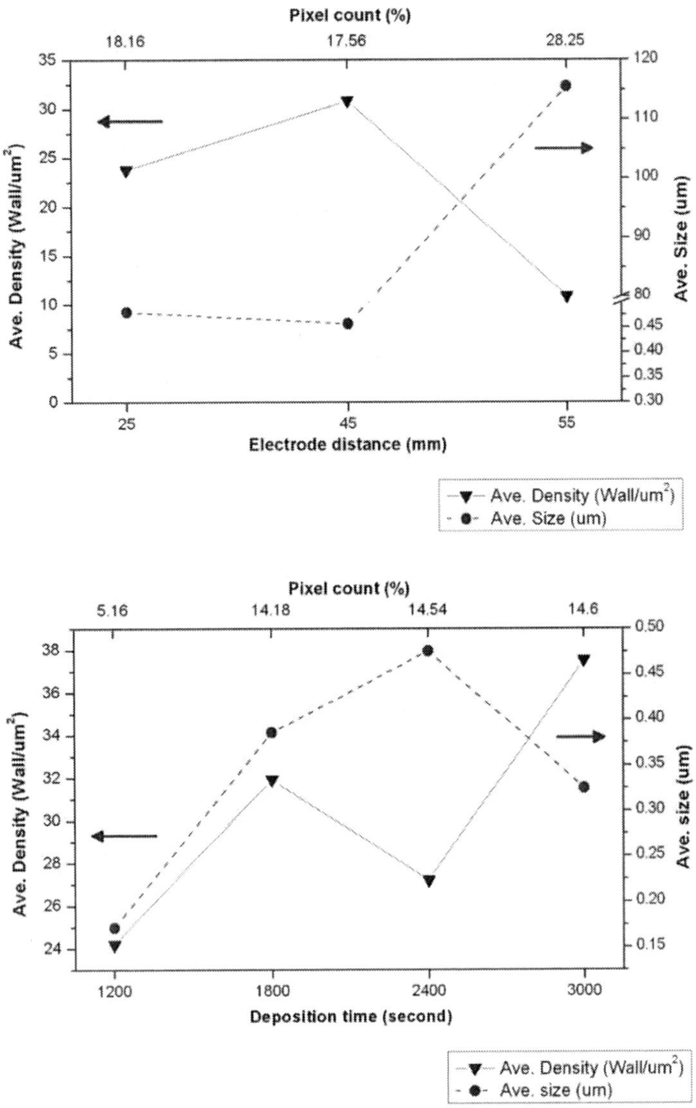

Figure 2. Relationship of wall density and size at varying electrode separation (top) and deposition time (bottom).

Figure 2 (bottom) shows the average wall density and sizes were found to be ranging from 24 wall/μm² to 38 wall/μm² and 0.17 μm to 0.48 μm, respectively. The highest average density is obtained during growth at 3000 s with value of 38 wall/μm². In contrast, largest average wall size is observed during growth at 2400 s with value of 0.48 μm. This distinct change in morphology is due to higher rate of surface reaction of hydrocarbon radical species at an elevated temperature [16].

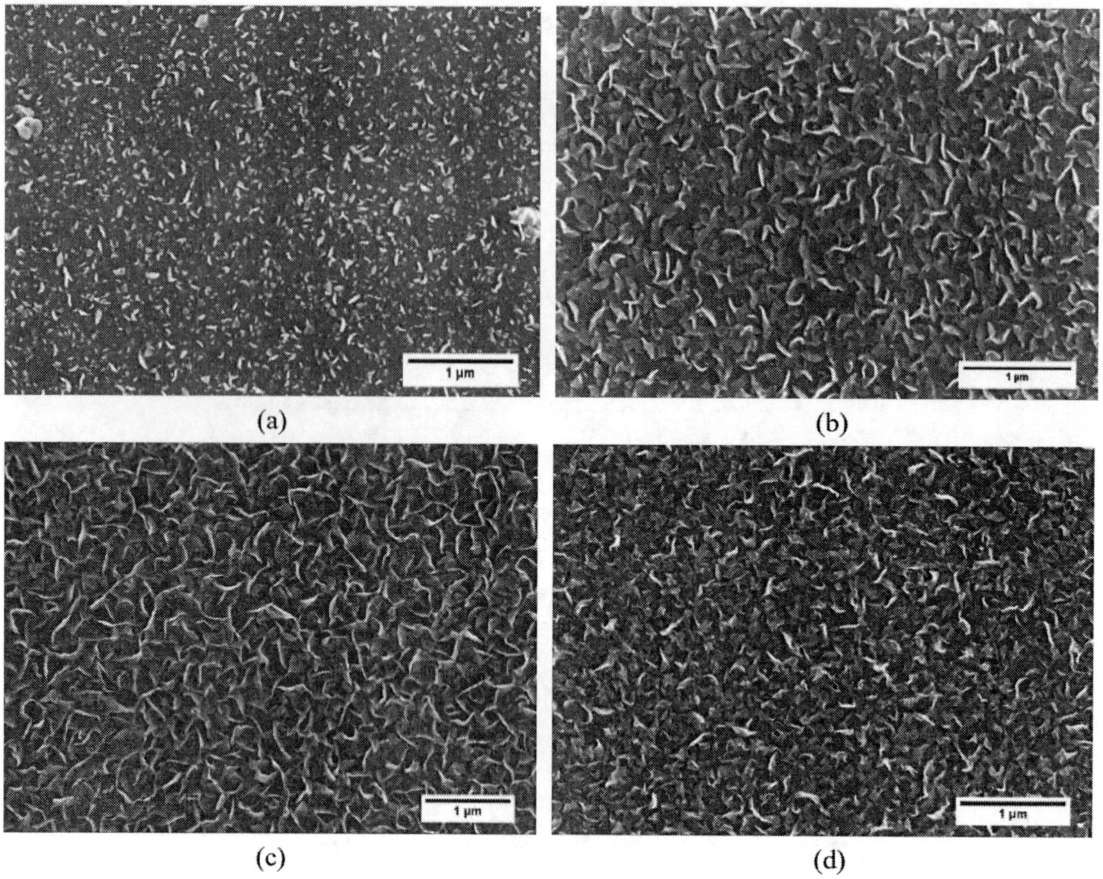

Figure 3. SEM images of CNW grown at varying deposition time: (a) 1200 s, (b) 1800 s, (c) 2400 s and (d) 3000 s with growth condition of 17:5 (CH_4:H_2) sccm; 20 W RF power; and temperature of 750°C. All images were captured at 20000 times magnification.

3.2. Influence of nickel catalyst and substrate type on CNW surface morphology

Figure 4 shows a comparative top view SEM image of grown CNW on Ni-catalyzed Si and silica substrate. It was found that the CNW is denser during growth on single crystal (sc)-Si substrate compared to silica. The average length and width of CNW were also drastically reduced. It also can be noted that, the morphology of CNW grown on sc-Si was composed of higher number of vertically standing walls compared to silica where a wavy and slightly inclined structure is dominant. The calculated average density of CNW grown on silica and sc-Si is 24 wall/µm^2 and 37 wall/µm^2, respectively (results not shown here). No noticeable change in average wall size as the value slightly changed from 0.63 to 0.64 µm. This result suggests that the density of CNW has increased approximately 37 % during growth on sc-Si while maintaining its size. Presence of Ni catalyst catalyst (top) and single crystal p-type Si coated with Ni catalyst (bottom) somehow did not influence the growth morphology of CNW. Still, it can be claimed that the growth rate of CNW is higher compared to silica.

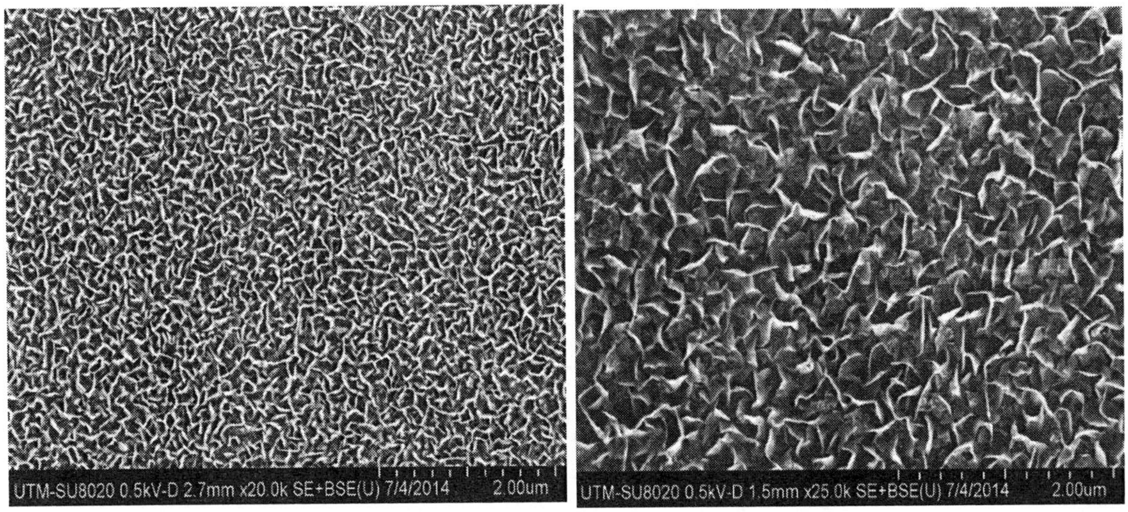

Figure 4. SEM images of CNW grown on different type of substrates; monocrystalline Si (sc-Si) (left) and on silica glass coated with Ni (right).

3.3. Raman scattering of grown CNW

Raman spectra of grown CWN are represented in figure 5(a-c). It was found that all experimental sets inherit similar bands located at peak position of acceptable range as has been reported in many literatures [18,19]. These bands are D (at ~ 1347 cm^{-1}), G (at ~ 1575 to 1577 cm^{-1}), D' (at ~ 1609 cm^{-1}), 2D (at ~ 2683 to 2687 cm^{-1}) and D+G (at~2925 cm^{-1}). The existence of D and G band can be considered as the molecular fingerprint of carbon-based materials and were observed in all poly-aromatic hydrocarbons.

The occurrence of D band in CNW is mainly due to disorder that corresponds to crystallographic or lattice defects of the structure [8]. It also can be related to the disorder, induced by finite crystallite size in nanocrystalline graphite (nc-graphite) [15]. Furthermore, the D band is always absence in the Raman spectra of graphene and defect-free graphite [11]. Thus, the high intensity of D band obtained in every experimental set was originated from the disorder activated by lattice defects, small crystallite size and density of edges of the grown CNW.

For the G band or graphite band, its bandwidth, W_G is prone to crystal imperfections of the graphite regions and has been a measure of graphitization of graphite-based structure [8,11]. A narrow bandwidth of the G band can be considered as a graphitic structure that possesses high degree of graphitization or in other words, low disorder nc-graphite.

The occurrence of D' peak is also due to the imperfection or disorder of nc-graphite regions in the CNW film. Interestingly, the D' peak is only revealed in a low disorder graphite structure and is absent in highly ordered graphite. In the case of disordered graphite, the peak broadening of G and D' band often represents disorder where broader G band will be accompanied by the elimination of the D' band and shifts into a broad single line G peak. Cho *et al.* [14] suggest that the D' band is originated from the breaking of symmetrical bonding due to finite crystalline size of the graphite regions representing a highly ordered nc-graphite region constituting the CNW.

Second order D peak known as 2D band was observed in every sets of experiment. This 2D band is due to stacking behavior of sp^2 orbitals in planar periodic arrangement in graphite, a misaligned stacking behavior is known as turbostratic (without AB stacking). For CNW the 2D peak is a single line with broader bandwidth approximately nearly two times compared to a sharp single line obtained in Raman spectra of a monolayer graphene [20]. This indicates that CNW is composed of 2D graphite structure.

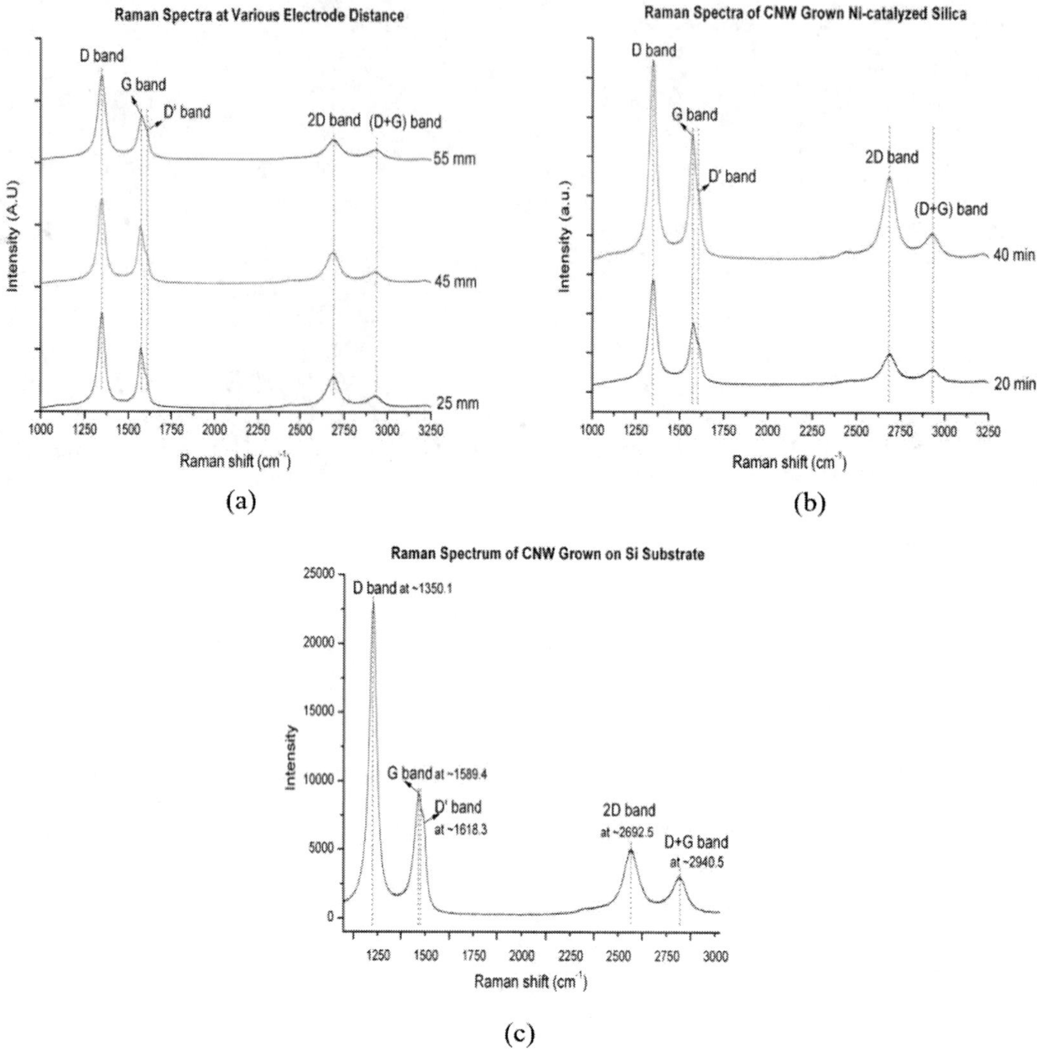

Figure 5. Raman spectra of CNW grown at (a) various electrode separation. Raman spectra of grown CNW (b) with presence of Ni-catalyst on silica and (c) on Si substrate without catalyst

3.4. CNW film quality during growth at varying electrode separation and different substrate type

For growth at various electrode separation, no significant change can be observed in value of I_D/I_G.(refer Table 1). Still, higher crystallinity of CNW can be achieved during growth at 45 mm electrode separation due to its lower defect density. This also can be supported its $I_{D'}/I_G$ value however, too large electrode separation may promote growth of CNW with high crystallographic disorder. This may be due to the limited supply of radicals flux and higher collision between radical species as the space or distance between anode and cathode become larger thus inducing low quality of nanocrystalline graphite. It also can be derived that as the electrode distance grew larger, the W_G slightly increases. This indicates that at larger electrode distance, CNW with lower degree of graphitization are produced.

Raman spectra obtained from the synthesis of CNW on different substrate with the presence of Ni catalyst. Still, it was found that all five main bands (D, G, D', 2D and D+G) were detected for all samples. Strong D band is collected during growth on sc-Si compared to silica with a value approximately twice as much. This may be related to the high thermal conductivity of Si and the number of vertically grown CNW on sc-Si [16], where higher fractions of edges were involved thus inducing stronger D band signal. In addition, the I_D/I_G (refer Table 2) ratio during growth on sc-Si revealed that even though the structure is high in defect, it exhibits better crystallinity due to smaller grain size. Based on the W_G value, it can be stated that the degree of graphitization is comparable between growth on sc-Si and silica at an equal amount of deposition time. Thus, it can be claimed that CNW grown on both sc-Si and Silica are constitute of highly graphitized nanocrystalline regions as has been suggested by [15]. However, the nanographite regions will be smaller during growth on sc-Si. It is also important to note that responses obtained during growth on Ni-catalyzed silica were in line with growth without catalyst. Thus, this revealed that the presence of Ni catalyst had no influence on the structural characteristics of CNW. This claim was also suggested by [3].

Table 1. Intensity ratio and bandwidth of Raman signal during growth at varying electrode separation

Electrode separation (mm)	I_D/I_G	$I_{D'}/I_G$	I_{2D}/I_G	W_G (cm^{-1})	W_{2D} (cm^{-1})
25	1.66	0.29	0.53	38	83
45	1.55	0.28	0.56	42	86
55	2.04	0.31	0.48	50	104

Table 2. Intensity ratio and bandwidth of Raman signal during growth of CNW with the presence of Ni catalyst.

Type of Substrate	I_D (cts)	I_D/I_G	$I_{D'}/I_G$	W_G (cm^{-1})	W_{2D} (cm^{-1})
sc-Si G_t=20 min	22652	2.77	0.41	52	93
Silica, G_t=20 min	13040	1.81	0.32	50	91
Silica, G_t=40 min	25160	1.67	0.26	42	85

4. Conclusion

A capacitively coupled 150 MHz VHF-PECVD with remote plasma setup is a promising technique to synthesize CNW structure. From this method, a dense wavy-like CNW structure was successfully grown. Morphology of CNW is possible to be controlled by varying the electrode distance and deposition time. Apart from that, the quality of the CNW structure can be claimed as highly graphitized which composed of finite size of nanocrystalline graphite regions. In summary, CNW can be considered as a unique 2D carbon nanostructure due to its composition of less ordered graphite made up of small nanocrystalline graphite domain with a high degree of graphitization.

Acknowledgements

The authors would like to thank CS Nano UTM for the VHF-PECVD equipment setup. This setup is financially sponsored by Ministry of Higher Education (MOHE) Malaysia under FRGS grant scheme (Vot: 4F847). Also, the authors are indebted to the Department of Physics, University Malaya for the assistance during Raman spectroscopy. The author is supported by the Graduate Assistance scheme from Universiti Malaysia Perlis.

References

[1] Novoselov K S, Geim A K, Morozov S V, Jiang D, Zhang Y, Dubonos S V, Grigorieva I V and Firsov A A 2004 *Science* **306** (5696) 666

[2] Kurbanoglu S and Ozkan S A, *J. Pharm. Biomed. Anal.* 2017 (in press)

[3] Cui L, Chen J, Yang B, Sun D and Jiao T 2015 *Appl. Surf. Sci.* **357** 1-7

[4] Yen H F, Horng Y Y, Hu M S, Yang W H, Wen J R, Ganguly A, Tai Y, Chen K H and Chen L C 2015 *Carbon* **82** 124-134

[5] Davami K, Cortes J, Hong N and Bargatin I 2016 *Mater. Res. Bull.* **74** 226-233

[6] Gao J M, Song X F, Hu J, Guo S C, Fang L, Wu F and Wei D P 2016 *Mater. Lett.* **184**, 273-277

[7] Russo P, Xiao M and Zhou N Y 2017 *Carbon* **120** 54-62

[8] Krivchenko V A, Evlashin S A, Mironovich K V, Verbitskiy N I, Nefedov A, Wöll C, Kozmenkova A Y, Suetin N V, Svyakhovskiy S E, Vyalikh D V, Rakhimov A T, Egorov A V and Yashina L V 2013 *Scientific Reports* **3** 3328

[9] Wang Z, Shoji M, Baba K, Ito T and Ogata H 2014 *Carbon* **67** 326-335

[10] Shih W C, Jeng J M, Huang C T and Lo J T 2010 *Vacuum* **84** (12) 1452-1456

[11] Davami K, Shaygan M, Kheirabi N, Zhao J, Kovalenko D A, Rümmeli M H, Opitz J, Cuniberti G, Lee J S and Meyyappan M 2014 *Carbon* **72** 372-380

[12] Itoh T, Nakanishi Y, Ito T, Vetushka A, Ledinský M, Fejfar A, Kočka J and Nonomura S, *J. Non-Cryst. Solids* 2012 **358** (17) 2548-2551

[13] Kar R, Patel N N, Chand N, Shilpa R K, Dusane R O, Patil D S and Sinha S 2016 *Carbon* **106** 233-242

[14] Cho H J, Kondo H, Ishikawa K, Sekine M, Hiramatsu H and Hori M 2014 *Carbon* **68** 380-388

[15] Kurita S, Yoshimura A, Kawamoto H, Uchida T, Kojima K, Tachibana M, Molina-Morales P and Nakai H 2005 *J. Appl. Phys.* **97** (10) 104320

[16] Seiya S, Anindita C, Chia-Liang C and Masamichi Y 2011 *Japan. J. Appl. Phys.* **50** (1S1) 01AF08

[17] Banerjee D, Mukherjee S and Chattopadhyay K K 2011 *Appl. Surf. Sci.* **257** (8) 3717-3722

[18] Cott D J, Verheijen M, Richard O, Radu I, Gendt S D, Elshocht S V and Vereecken P M 2013 *Carbon* **58** 59-65

[19] Kim S Y, Choi W S, Lee J H and Hong B 2014 *Mater. Res. Bull.* **58**, 112-116

[20] Tzeng Y, Chen W L, Wu C, Lo J Y and Li C Y 2013 *Carbon* **53** 120-129

Orthogonal Array Testing for Transmit Precoding based Codebooks in Space Shift Keying Systems

Mohammed Al-Ansi[1,*], Syed Alwee Aljunid[1], Essam Sourour[2], Anuar Mat Safar[1], and C B M Rashidi[1]
[1]School of Computer and Communication Engineering, University Malaysia Perlis (UniMAP), Perlis, Malaysia.
[2]Department of Electrical Engineering, College of Engineering, Prince Sattam Bin Abdulaziz University, 11991, Wadi Addwasir, Saudi Arabia.

E-mail: ma.ahmed@studentmail.unimap.edu.my

Abstract. In Space Shift Keying (SSK) systems, transmit precoding based codebook approaches have been proposed to improve the performance in limited feedback channels. The receiver performs an exhaustive search in a predefined Full-Combination (FC) codebook to select the optimal codeword that maximizes the Minimum Euclidean Distance (MED) between the received constellations. This research aims to reduce the codebook size with the purpose of minimizing the selection time and the number of feedback bits. Therefore, we propose to construct the codebooks based on Orthogonal Array Testing (OAT) methods due to their powerful inherent properties. These methods allow to acquire a short codebook where the codewords are sufficient to cover almost all the possible effects included in the FC codebook. Numerical results show the effectiveness of the proposed OAT codebooks in terms of the system performance and complexity.

1. Introduction

Spatial Modulation (SM) is a hybrid Multiple Input Multiple Output (MIMO) and digital modulation techniques proposed as a potential technology to satisfy both requirements of low complexity and Energy Efficiency (EE) in the next 5G communication systems [1, 2]. In each instant of SM transmission, the conventional Amplitude and Phase Modulation (APM) constellation symbol is transmitted through a single Transmit Antenna (TA) while the other TAs are idle where the index of active TA is specified randomly according to the user data. Therefore, the first part of the user data is mapped into the conventional APM constellation symbol while the second part is mapped to a spatial constellation. One special form of SM is the Space Shift Keying (SSK) where the user information bits are mapped only to the spatial constellation and unmodulated data, such as +1 symbol is transmitted through the active TA. This elimination of constellation symbol in SSK decreases the data rate but it reduces the detection complexity in the receiver side. SM/SSK methods are then generalized to increase the Spectrum Efficiency (SE) through activating more than one antenna at a time as in [3] and [4].

The novel idea of SM/SSK systems offers several benefits in comparison to the conventional MIMO. One of them is the low receiver complexity since the optimal Maximum Likelihood (ML) detector is applied at a single-stream decoding complexity. SM systems also consume less power because they require single Radio Frequency (RF) Chain which is one of the most consuming power in the transceiver.

Moreover, the inter channel interference is avoided and no need for antenna synchronization due to transmit from single TA at each transmission time [5].

One limitation of SM/SSK methods is the lack of diversity gain, thus gets low Bit Error Rate (BER) system performance due to activating a less number of TA's. Transmit Precoding (TPC) techniques are recently proposed to weight the channel matrix and accordingly improve the system BER performance provided that the instantaneous Channel State Information (CSI) is known. However, the challenge is how to find the best precoder that maximize the Minimum Euclidean Distance (MED) between the received constellation points since the BER is predominated by the MED [6]. The precoder is employed to construct a TPC diagonal matrix which basically consists of two matrixes, real numbers represent the Power Allocation Precoding (PAP) matrix and a complex numbers represent the Phase Rotation Precoding (PRP) matrix. In PAP, the values are used to scale the magnitude of the channels while the phase can be scaled through the PRP matrix.

In the literature, the authors of [7, 8] proposed iterative algorithms with the aim of maximizing the MED through joint design of PAP and PRP matrixes and consequently get minimum BER. Other iterative schemes are introduced in [9, 10] based on designing a PAP matrix, while the TPC matrix constructed in [11, 12] is designed to change the phase of the TAs through PRP matrix. These iterative methods provide a global minimum of BER, since they designed to match the instantaneous CSI. However, they cannot be considered in practice due to their high complexity, especially for the high number of TAs. Further, full knowledge of CSI is essential to apply the algorithms and this necessity an infinite feedback channel.

On the other hand, most of the restrictions found in the iterative algorithms can be tackled when TPC based codebook precoding is applied. In codebook precoding, the CSI is known only to the receiver and a codebook is predefined to both sides of the transceiver. Given the instantaneous CSI, the receiver selects one of codewords in the codebook that satisfies maximizing the MED and then fed back the codeword index to the transmitter. In [13], the researchers apply the TPC based codebook precoding and noteworthy performance was achieved. However, the selected codewod is not exactly matching the instantaneous CSI and hence the MED is not globally minimized. Accordingly, the performance is less than what accomplished in the case of iterative schemes. One solution proposed for this problem is to design the codebook through matching the CSI on the average as in [14]. Though, this method requires infinite rate of feedback channel where the transmitter should be updated with the whole codebook whenever the CSI changes.

In this research, we consider constructing a Full-Combinations (FC) codebook with systematic structure based PRP precoding [15]. Given the CSI to the receiver, it tests all the combinations in FC codebook and selects the most desirable codeword that satisfy the selection criteria to maximize the MED. Then, the index of this codeword is fed back to the transmitter in order to weight the channel matrix. However, this exhaustive search increase the time of the whole precoding process and it becomes more challenging, particularly in the case of high number of TAs. Moreover, we noticed that there are several combinations (i.e. codewords) can satisfy the maximum MED criteria. Our initial objective of this work is to design a codebook contains only the most significant and concrete codewords which have the capacity to cover all the possible phase changes in order to precode the TAs. Thus, this research proposes to take advantage of the well-known Orthogonal Array Testing (OAT) methods for constructing the TPC codebooks. OAT properties allow designing codebooks with very short codebook length and consequently attain a considerable reduction of time during the selection process. Furthermore, these OAT codebooks have the capability to provide near to the high BER performance obtained by FC codebooks.

The remainder of this paper is organized as follows: in section 2, SSK system model and FC codebook precoding are introduced. The proposed OAT codebook is explained in section 3. Numerical results are presented in section 4. Finally, section 5 concludes the paper.

2. SSK System Model with Codebook Precoding

Consider SSK system with N_t antennas at the transmitter and N_r antennas at the receiver as shown in Figure 1. In each transmission instant, a length of $\log_2(N_t)$ bits is taken from the user data bits and mapped to one of the spatial constellations to get the $(N_t \times 1)$ transmission vector $\mathbf{x}_k$ given by:

$$x_k = [0, \ldots, \underbrace{1}_{kth}, \ldots, 0]^T \tag{1}$$

this indicates that only the kth transmit antenna is active. By applying the TPC Codebook predefined for both sides of the transceiver, the $N_c \times N_t$ codebook $\mathbf{C}$ is comprised of N_c codewords as

$$\mathbf{C} = [\mathbf{p}_0, \mathbf{p}_1, \ldots, \mathbf{p}_b, \ldots, \mathbf{p}_{N_c-1}]^T \tag{2}$$

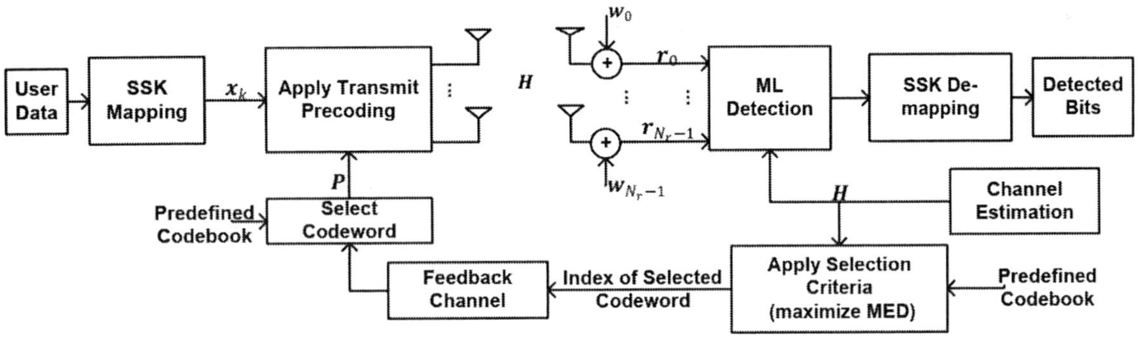

Figure 1. SSK system model based TPC codebook with finite rate feedback channel.

each codeword $\mathbf{p}_b$ in the codebook contains N_t elements, where $\mathbf{p}_b = \left[p_{b,0}, p_{b,1}, \ldots, p_{b,N_t-1}\right]^T$, for $0 \le b \le N_c - 1$.

In the receiver, the complex received signal can be expressed as:

$$\mathbf{r} = \mathbf{H}\,\mathbf{P}\,x_k + w \tag{3}$$

where $\mathbf{H} = [\mathbf{h}_0, \mathbf{h}_1, \ldots, \mathbf{h}_{N_t-1}]$ is the $(N_r \text{ x } N_t)$ complex channel matrix known to the receiver only with entries are assumed to be identical and independently distributed (i.i.d.) complex Gaussian random variables with zero mean and unit variance. The precoding matrix $\mathbf{P}$ is formed in a diagonal structure to satisfy the SSK requirement of single TA active at a time as $\mathbf{P} = diag(\mathbf{p}_b)$. The $(N_r \text{ x } 1)$ noise vector $\mathbf{w}$ is the white noise whose elements are i.i.d. complex Gaussian random variables with zero mean and variance $1/\gamma$, where γ is the Signal to Noise Ratio (SNR) per receive antenna.

Among all the codewords in the codebook $\mathbf{C}$, the codeword $\mathbf{p}_b$ is selected as the best one to maximize the MED between the received constellation points where the upper bound BER of SSK is dominated by the MED [6]. Therefore, the precoding codeword $\mathbf{p}_b$ is selected as follows:

$$\mathbf{p}_b = \arg\max_{\mathbf{p}_k \in \mathbf{C}} \left(\min_{i \neq j} \left(\left\| p_{k,i}\mathbf{h}_i - p_{k,j}\mathbf{h}_j \right\|^2 \right) \right) \tag{4}$$

then, the receiver feeds-back the index b of this codeword to the transmitter using N_{fb} bits.

2.1. *Full-Combinations (FC) Codebook*

Full-Combination (FC) Codebook contains all the combinations cover the possible L phase changes for N_t antennas. Each combination represents one codeword and its elements can be obtained through $p_{b,i} = \exp\left(j\,2\pi l_{b,i}/L\right)$. This allows the FC codebook to have a systematic structure. We assume the first column elements are equal to one (i.e. $p_{b,0} = 1$) which means no phase change, so the weight of the first antenna is not scaled and becomes as a reference. Then, the other elements in each codeword

IOP Conf. Series: Materials Science and Engineering **318** (2018) 012044 doi:10.1088/1757-899X/318/1/012044

$\left[l_{b,1}, l_{b,2}, \cdots \quad , \right]^T$ for the remaining antennas are obtained through the conversion of the codeword index b into a Base-L number with $N_t - 1$ digits. Hence, the codebook length is $N_c = L^{(N_t-1)}$ codewords, each with N_t elements.

All the N_c codewords go through the selection criteria (i.e. eq. (4)) at the receiver to find the best one provides the maximum MED. Then, the index of this codeword is fed back to the transmitter and the number of feedback bits is $\lceil N_{fb} = (N_t - 1)\log_2 L \rceil$. Therefore, FC codebook is considered as a reference since it covers all the L possible phase changes to weight the channel matrix and it can guarantee reaching the minimum BER performance achieved by any codebook based PRP. Even though, the exhaustive search among all combinations becomes more challenging when N_t is high. Moreover, we noticed from the numerical simulation that there are several combinations provide the same maximum MED and the first one tested is being selected. This motivates us to look for a low complexity method or TPC codebook maintain less BER with acceptable complexity.

3. Orthogonal Array Testing (OAT) Codebook

Orthogonal Array Testing matrix (OAT) is a statistical method mainly used for experimental testing in industrial engineering. It's useful to avoid the difficulty of testing all possible combinations of experimental factors when each factor has various possible levels [16]. The Methodology of the OAT is based on the selection of a concise and effective set from the full-combinations such that this set is enough to consider all effects of each factor [35]. This selection guarantees to test all the pairwise combinations in the set. Typically, OAT is described with the notation $O_{Runs}\left(Levels^{Factors}\right)$, where *Runs*, *Levels*, and *Factors* are integers. The parameter *Runs* is the total number of chosen combinations to test (number of rows in the matrix). The parameter *Factors* is the number of factors to test in each combination (number of columns in the matrix). Hence, the matrix size is $(Runs \times Factors)$ with entries from 0 to *Levels*-1, where the *Levels* is the number of levels to test for each factor. For efficient testing, the integer *Runs* typically satisfies $Runs << Levels^{Factors}$. More details about the description of the properties and implementation of OAT are available in [16-19].

3.1. OAT Codebook Design

While OAT is invented for experimental testing of an industrial design with sufficient test coverage, it found applications in communication systems [20]. In this paper, the concept of the OAT is utilized to design the precoding codebook where the integers *Factors*, *Levels*, *Runs*, represent the number of TAs (N_t), the number of phase rotations (L), and the number of codewords in the codebook (N_c), respectively. Therefore, the OAT codebook dimension is ($N_c \times N_t$). The codebook entries are integers $l_{b,i}$ which take values from 0 to $L-1$. Similar to the FC codebook, these entries are converted into the L phasors $p_{b,i}$.

As an illustration, the OAT codebook $O_8(2^4)$ is shown in Table 1 for $N_t = 4$ and $L = 2$. This codebook involves eight codewords to test the effect of four factors and two levels for each factor. However, the same situation requires 16 codewords in the FC codebook.

Constructing OAT codebook for certain factors and levels can be done by several schemes, but the simplest way is to select the appropriate table from the available libraries [17, 21] and in the case if it is not available in the library, it can be designed with taking into consideration the rules and properties of the OATs.

Table 1. OAT Codebook $O_8(2^4)$

Antenna1	Antenna2	Antenna3	Antenna4
-1	-1	-1	-1
-1	-1	-1	1
-1	1	1	-1
-1	1	1	1
1	-1	1	-1
1	-1	1	1
1	1	-1	-1
1	1	-1	1

The size of OAT codebooks implemented in this paper is the typical size $N_c = L \times N_t$. Hence, the number of feedback bits is $N_{fb} = \log_2 L + \log_2 N_t$, which is much smaller than the FC codebook for large N_t. The construction of the OAT codebooks for $L = 2$ with $N_t = 8$ and $N_t = 16$ is based on the tables available in [22]. The codebooks for $L = 4$ with $N_t = 8$ and $N_t = 16$ are based on the tables available in [23] and [24] respectively.

4. Numerical Results

This section presents the BER performance of SSK system using the proposed OAT codebook precoding. Then, we compare between the OAT, FC codebook [15] and DFT codebook [13]. Perfect channel estimation is assumed at the receiver and finite feedback channel is applied in the system. Figure 2 shows the SSK BER performance with $N_t = 8$ and $N_r = 2$. It's observed that almost 4 dB gain at BER of 10^{-3} is achieved applying OAT codebook with $L = 2$ phases compared to the conventional SSK without precoding. Additional of 1 dB gained when $L = 4$ phase changes are employed. Besides, this figure shows the capability of OAT codebooks to provide a close BER to the FC codebook [15]. From the system complexity side, the OAT codebook length $N_c = 16$ and $N_c = 32$ codewords for $L = 2$ and 4 phases, respectively. While the FC codebooks contain $N_c = 128$ and $N_c = 16384$ codewords for $L = 2$ and 4 phases, respectively.

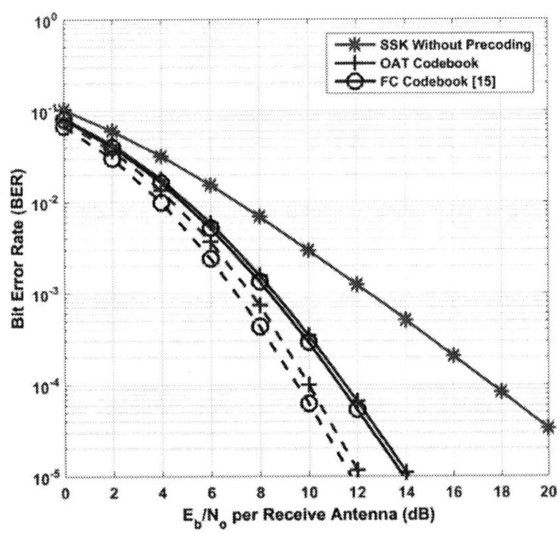

Figure 2. BER performance comparison between OAT and FC codebooks for SSK systems with $N_t = 8$ and $N_r = 2$. Solid lines indicate employing $L = 2$ phases, while dashed lines indicate $L = 4$ phases.

Tables 2 and 3 illustrate the considerable gain accomplished by reducing the codebook size where the precoding time and the number of feedback bits are decreased when OAT codebook implemented. For example, OAT codebook contains 64 codewords and 6 bits are required to feed-back the index for $N_t = 16$ and $L = 4$ phases. In contrast, FC codebook comprises more than 10^9 codewords and needs more than 29 feedback bits for the same number of TA and phases.

Table 2. Comparisons between the length of OAT and FC Codebooks for different cases of N_t with $L = 2$ phases.

Codebook Type	$N_t = 8$	$N_t = 16$	$N_t = 32$
FC	128	32768	2.14E+9
OAT	16	32	64

Table 3. Comparisons between the length of OAT and FC Codebooks for different cases of N_t with $L = 4$ phases.

Codebook Type	$N_t = 8$	$N_t = 16$	$N_t = 32$
FC	16384	1.07E+9	4.6E+18
OAT	32	64	128

Figure 3 compares the OAT codebook to the related method in the literature DFT codebook [13] with $N_t = 16$ and $N_r = 2$. It shows that applying the OAT codebook with $L = 4$ phases provides lower BER than the case of codebook generated using DFT matrix.

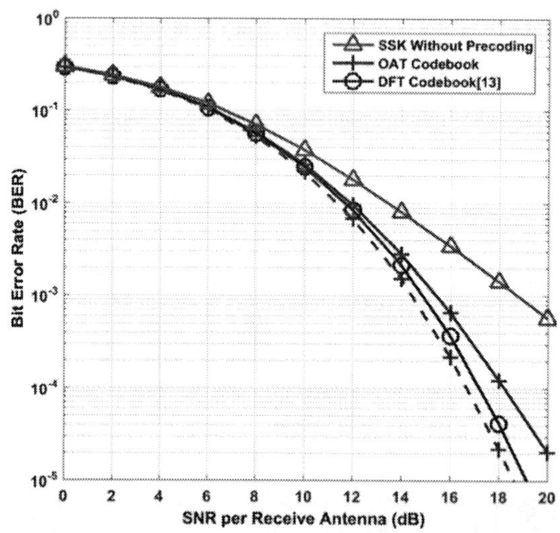

Figure 3. BER performance comparison between OAT and DFT codebook precoding for SSK systems with $N_t = 16$ and $N_r = 2$. Solid lines indicate employing $L = 2$ phases, while dashed lines indicate $L = 4$ phases.

5. Conclusions

This research proposed Orthogonal Array Testing (OAT) methods to construct a codebook based Phase Rotation Precoding (PRP) to weight the transmit channel vectors of SSK systems. In the Full-Combinations (FC) codebook, all possible combinations of the phase changes should be verified to catch the optimal codeword that maximizes the MED between the received constellation points. However, this extensive search becomes more challenging, particularly in the case of high number of transmit antennas. In contrast, the proposed OAT methods demonstrate that it has a robust inherent properties allow to design a codebook contains less number of combinations. These combinations are effective and enough to cover all needed effects of the reference FC codebook. Thus, the numerical results show the great reduction achieved when OAT codebook applied. Accordingly, the implementation time is considerably minimized and consequently the number of feedback bits is reduced.

References

[1] E. Basar, Index modulation techniques for 5G wireless networks, IEEE Commun. Mag., **54**, pp. 168-175, (2016).
[2] C. X. Wang, F. Haider, X. Gao, X. H. You, Y. Yang, D. Yuan, *et al.*, Cellular architecture and key technologies for 5G wireless communication networks, IEEE Commun. Mag., **52**, pp. 122-130, (2014).
[3] J. Jeganathan and A. Ghrayeb, *Generalized space shift keying modulation for MIMO channels*, presented at the Proc. IEEE 19th PIMRC'8, Cannes, (Sep. 2008).
[4] A. Younis, N. Serafimovski, R. Mesleh, and H. Haas, *Generalised spatial modulation*, in Proc. IEEE Signals Syst. Comput., Nov. 2010, pp. 1498-1502.
[5] R. Mesleh, H. Haas, S. Sinanovi´c, C. Wook Ahn, and S. Yun, Spatial modulation, IEEE Trans. Veh. Tech. , **57**, pp. 2228-2241, (Jul. 2008).
[6] J. Jeganathan, A. Ghrayeb, L. Szczecinski, and A. Ceron, Space shift keying modulation for MIMO channels, Trans. Wireless Commun., **8**, pp. 3692-3703, (Jul. 2009).
[7] M. C. Lee, W. H. Chung, and T. S. Lee, Generalized precoder design formulation and iterative algorithm for spatial modulation in MIMO systems with CSIT, IEEE Trans. Commun., **63**, pp. 1230-1244, (Apr. 2015).
[8] P. Yang, Y. L. Guan, Y. Xiao, M. D. Renzo, S. Li, and L. Hanzo, Transmit precoded spatial modulation: Maximizing the minimum Euclidean distance versus minimizing the bit error ratio, IEEE Trans. Wireless Commun., **15**, pp. 2054-2068, (Mar. 2016).
[9] P. Yang, Y. Xiao, S. Li, and L. Hanzo, A Low-Complexity Power Allocation Algorithm for Multiple-Input Multiple-Output Spatial Modulation System, IEEE Trans. Veh. Technol., **65**, pp. 1819-1825, (2016).
[10] P. Yang, Y. Xiao, B. Zhang, S. Li, M. El-Hajjar, and L. Hanzo, Power allocation-aided spatial modulation for limited-feedback MIMO systems, IEEE Trans. Veh. Techn., **64**, pp. 2198-2204, (May 2015).
[11] C. Masouros, Improving the diversity of spatial modulation in MISO channels by phase alignment, IEEE Commun. Lett., **18**, pp. 729-732, (May 2014).
[12] P. Yang, Y. Xiao, B. Zhang, M. El-Hajjar, S. Li, and L. Hanzo, Phase rotation-based precoding for spatial modulation systems, IET Commun., **9**, pp. 1315-1323, (Feb. 2015).
[13] M. C. Lee, W. H. Chung, and T. S. Lee, *Precoder design for space shift keying in MIMO systems with limited feedback*, in Proc. IEEE 19th PIMRC'8, Sep. 2014, pp. 6-10.
[14] M. C. Lee, W. H. Chung, and T. S. Lee, Limited feedback precoder design for spatial modulation in MIMO systems, IEEE Commun. Lett., **19**, pp. 1909-1912, (Nov. 2015).
[15] M. Al-Ansi, S. A. Aljunid, and E. Sourour, *Phase Rotation Codebook Precoding for Space Shift Keying MIMO Systems*, in *Mobile and Wireless Technologies 2017: ICMWT 2017*, K. J. Kim and N. Joukov, Eds., ed Singapore: Springer Singapore, (2018), pp. 94-101.
[16] A. Hedayat, N. Sloane, and J. Stufken, *Orthogonal Arrays: Theory and Applications*. New York, USA: Springer, (Jul. 1999).
[17] W. F. Kuhfeld. (2009). *SAS Technical Support Notes*. N. J. A. Sloane. (2005). *A Library of Orthogonal Arrays*.
[18] C. Weng, F. Yang, and A. Z. Elsherbeni, Linear antenna array synthesis using Taguchi's method: A novel optimization technique in electromagnetics, IEEE Trans. Ant. and Propag., **55**, pp. 723-730, (Mar. 2007).
[19] C. Ying and L. Derong, Multiuser detection using the Taguchi method for DS-CDMA systems, IEEE Trans. Wireless Commun., **4**, pp. 1594-1607, (July 2005).
[20] W. F. Kuhfeld. *TS-723 Orthogonal Arrays*.
[21] W. F. Kuhfeld. (2009). *A library of Strength-Two Orthogonal arrays*. Available: N. J. A. Sloane. (2005). *A Library of Orthogonal Arrays with 64 Runs and 21 Factors*.
[22] N. J. A. Sloane. (2005). *A Library of Orthogonal Arrays with 32 Runs and 9 Fcators*.

IOP Conf. Series: Materials Science and Engineering **318** (2018) 012045 doi:10.1088/1757-899X/318/1/012045

The Effect of Aptamer Concetration towards Reduced Graphene Oxide-Field Effect Transistor Surface Channel for Biosensor Application

Azrul Syafiq Zainol Abidin[1], Ruslinda Abdul Rahim[1], Chow Yong Huan[2], Nur Nasyifa Mohd Maidin[1], Nurul Atiqah Ahmad[1], Saeed S. Ba Hashwan[1] , Fatin Nabilah Mohd Faudzi[1], and Voon Chun Hong[1].
[1]Institute of Nano Electronic Engineering
[2]School of microelectronic Engineering
Universiti Malaysia Perlis (UniMAP), 01000 Kangar Perlis, Malaysia

E-mail: azrulzainol@gmail.com, ruslindarahim@unimap.edu.my

Abstract. Aptamer are artificially produce bioreceptor that has been developed to bind with various target biomolecules such as ion, cells, protein and small molecules. In this research, an aptamer concentration of 0.5 nM, 1 nM, 5 nM, 10 nM, and 50 nM were immobilized on reduced graphene oxide (rGO) integrated with field effect transistor (FET) respectively to study the effect of aptamer concentration toward rGO surface for stable biosensing platform. The 0.5 nM concentration of aptamer shows the highest current result of 84.3 μA at 1 V applied through the source and drain. After immobilized with aminated aptamer, the conductivity shows significant reduction due to the formation of amide bond on rGO surface between aminated aptamer and carboxyl group on rGO.The electrical performance of FET integrated with rGO shows stable electrical performance suitable to be used in the biosensing application.

1. Introduction

The previous technology of detecting biomolecules has faces various challenges. The detection of biomolecules usually involves complexity of device structure, long response time and low sensitivity [1].For example, the chromatographic techniques have issues relating to the formation of assay, complexity of the system, and multi-step purification limits [2]. To solve these issues, biosensor technology is introduced in the arena. Aptasensor and immunosensor are the biosensor that commonly used in detection of biomolecules. The immunosensor technology such as enzyme linked immunorsorbent assay (ELISA) technology has an advantage of lower detection limit than aptasensor but it faces problem of long time-consuming labeling process and complicated features [3]. Besides, immunosensor is being replaced by aptasensor due to the cost constrained and undesirable immune response [4]. Thus, an aptasensor based reduced graphene oxide field effect transducer (rGO-FET). to detect biomolecules has been established.

Aptamers are single stranded DNA or RNA molecules that made of from artificially in vitro evolution process called systematic evolution of ligands by exponential enrichment (SELEX) [5]. In the SELEX process, the binding targets are first to immobilize, then the aptamers are applied in the pool and the binding process will carry on. After that, the weak binders are washed away while the bound aptamers are eluted and magnified. It was then reapplied to the binding target again and repeated several times to get a strong stringency. Aptamer has properties that could bind to various target from small organic and inorganic molecules to cells with high sensitivity. This is due to their tendency to form single stranded and helices loop. The advantage of using aptamer instead of antibody is aptamer has a smaller size compare to antibody and will result in a higher sensitivity and lower limits of detection in biosensor [6]. It can be developed to detect almost any analyte or biomolecule. Up until now, there are no report studies on aptamer concentration influence the biosensor performance.

In this research, we presented a solid platform for the aptamer as bioreceptor immobilized on rGO-FET. Different aptamer concentration were used to investigate the response of the aptamer towards rGO-FET. The optimum aptamer concentration are needed to be applied to get reliable biosensor output. In FET biosensor technology history, graphene field effect transistor (GFET) has been commonly developed and researched [4] but then, graphene is transferred during the micromechanical cleavage cause complicated and limited in implantation [7-8]. Hence, rGO that are prepared by reducing graphene oxide was introduced to the FET system to detect the biomolecules [8]. rGO can be developed with low-cost technique and high yield production can be archieved and can be easily applied to wide range of substrates [9]. Besides, rGO has a better conductivity compare to graphene oxide due to the reduction of the oxygen group which makes the ratio of carbon higher. However, it also has shortcoming compare to graphene oxide which is it has a lower binding rate than graphene oxide during addition of the linker. Nevertheless, this type of FET has advantages of detecting biomolecules with ultra-sensitivity, label-free, and real-time monitoring system [11-12].

2. Methodology

2.1. Fabrication of FET

A thick oxide layer with 3000 Å above was developed on a <111> dimension wafer to isolate the device by acting as a dielectric layer and insulator in bio-FET. Aluminum was then deposited on the SiO_2 by using Physical Vapor Deposition. The wafer was put in the spin-coater and a positive photoresist was dropped on it. 3 steps of spin rpm parameter were used which are 700 rpm for 10 s, 3000 rpm for 25 s and 0 rpm for 5 s. Then, the wafer was soft bake by heating it with 100℃ for 90 s before photolithography process. A mask with 200 μm channel was used to pattern the transducing channel of the wafer. Resist developer (RD6) was mix with diluted water in ratio of 1:1 to etch the exposed photoresist area. After that, the aluminum was etched by using aluminum etchant. Figure 1 shows the fabrication process of FET.

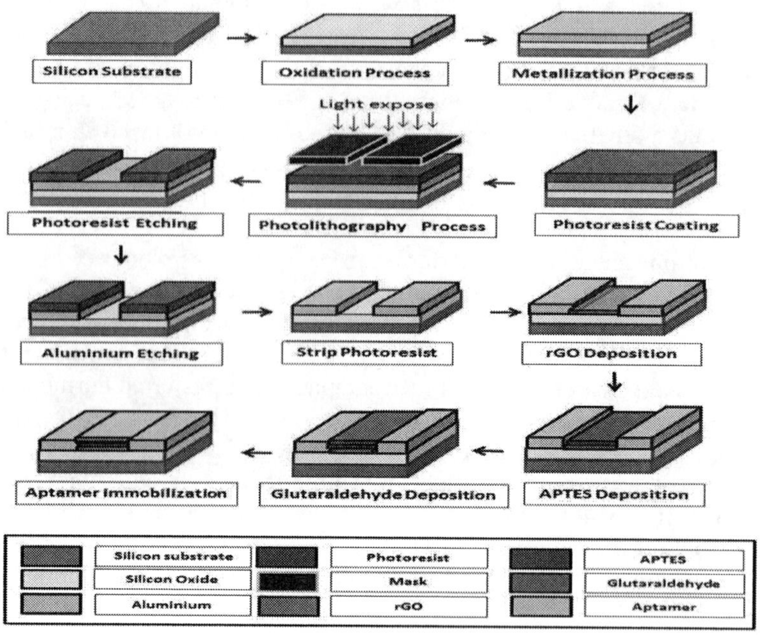

Figure 1. Overview of fabrication process aptamer integrated rGO-FET.

2.2. Deposition of rGO

Before deposit rGO on the transducing channel surface of FET, the rGO solution was sonicated for 1 hour. Next, by using a micropipette, a total volume of 30 μL of rGO solution is drop-casting onto the transducing channel of FET device. It is then was dry at 70℃ for 3 minutes on a hot plate to dry up the rGO.

2.3. Surface functionalization

3-aminopropyltriethoxysilane (APTES) was mixed with ethanol and diluted in deionized water in ratio of 2:50:50 to increase the surface tension. 30 μL of APTES was then dropped onto rGO surface with micropipette and incubated at room temperature inside a dry chamber for 1 hour. APTES was attached to the hydroxyl group of rGO and provide amine functional group for further binding process. After that, glutaraldehyde (GA) was mixed with diluted water in a ratio of 1:39 and drop onto the rGO by micropipette with a volume of 30 μL to make sure the GA cover the whole surface area. It was incubated in a dry chamber at room temperature for 1 hour.

2.4. Aptamer immobilization

5 different concentration of aptamer was used to detect the effect of the aptamer towards the FET surface. The concentration used are 50 nM, 10 nM, 5 nM, 1 nM and 0.5 nM. The aptamers are first mix with phosphate buffer saline (PBS). Then 30 μl of the mixture was deposited on the FET transducing channel and incubated at 38℃ with 85% humidity for 1 hour. The surface was then washed three times with 50 μL of PBS to clean out the unbound aptamer. The process was repeated for different concentration by using same procedure. Figure 2 shows the schematic diagram of surface functionalization on the transducing channel.

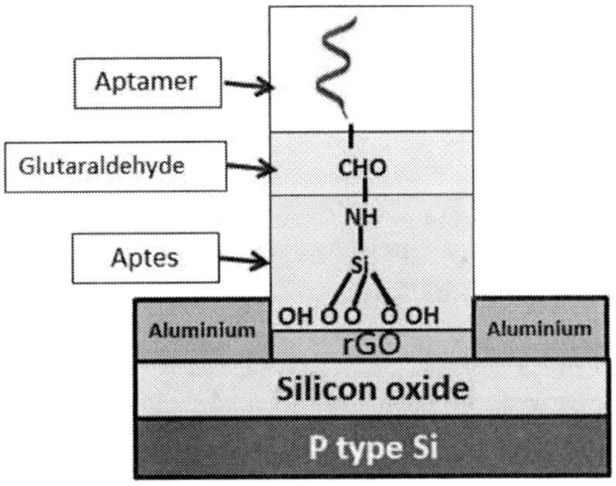

Figure 2. Schematic diagram of linkage on the transducing channel surface

3. Result and discussion

3.1. Scanning electron microscope (SEM)

Figure 3 and Figure 4 shows image of rGO and GO in a working distance of 20 kV with spot size of 10 mm and magnification of 50k, respectively. By comparing the image, we can clearly see that rGO in Figure 3 has wrinkle structure with corrugation and scrolling that are typical structure in graphene. Besides, it is also has a more separated flake compare to the GO structure in Figure 4. GO has a sponge-like structure and multilayer disorder sheet morphology. This is due to rGO has a lower ratio of hydroxyl group and hence shows the properties of graphene clearer than GO.

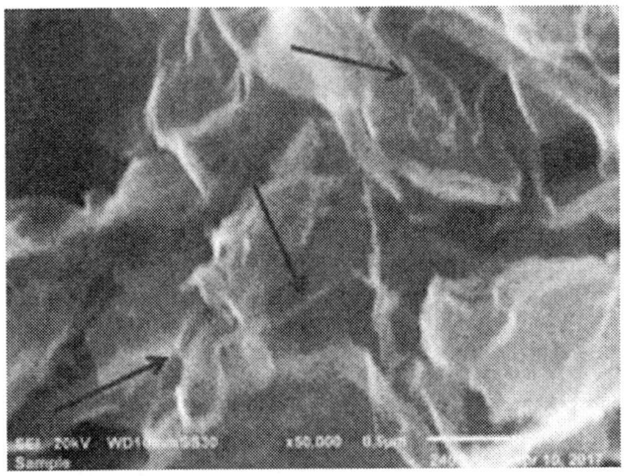

Figure 3. SEM image of rGO

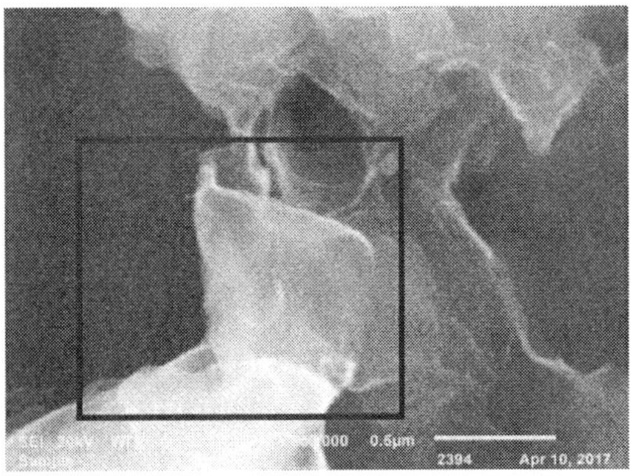

Figure 4. SEM image of graphene oxide

3.2. Fourier transform infrared spectroscopy

Figure 5 shows FTIR results of rGO and it shows several peaks which represent the functional group contained in the solution. rGO contains hydroxyl group and carbonyl group in its chemical structure. The peak shows at 3290cm^{-1} represent hydroxyl group exists in the rGO solution used. Whereas the carbon double bond oxygen has a small peak at 1720cm^{-1} reveals the carbonyl group in the solution.

IOP Conf. Series: Materials Science and Engineering **318** (2018) 012045 doi:10.1088/1757-899X/318/1/012045

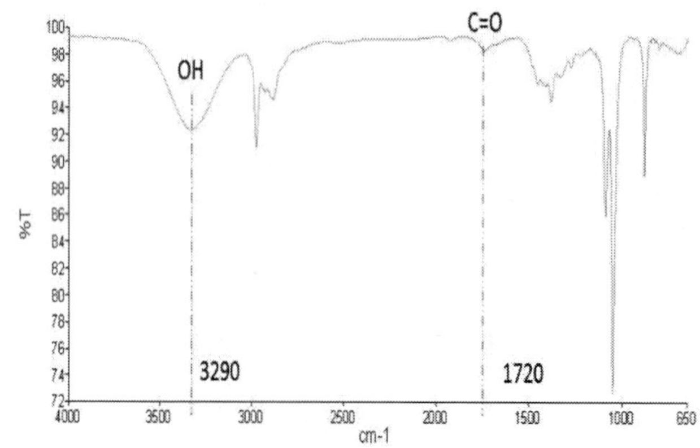

Figure 5. FTIR results of rGO

3.3. Electrical characterization

Figure 6 shows the conductivity performance of the device. After deposition of rGO, steep increase of conductivity can be seen. Before the rGO deposition, electron could not pass through the transducing channel area since the oxide layer act as dielectric layer. rGO is a conductive material hence when it is deposited on the transducing channel, the conductivity of the device will increase. The conductivity of the device decrease when the biochemical linker APTES and GA were added. This is due to the cross linker layer which formed on the surface of the transducing channel. The conductivity after immobilization of aptamer, the electrical reading decrease due to DNA aptamer contains negative charge ion thus inhibit the electrical flow from drain to source.

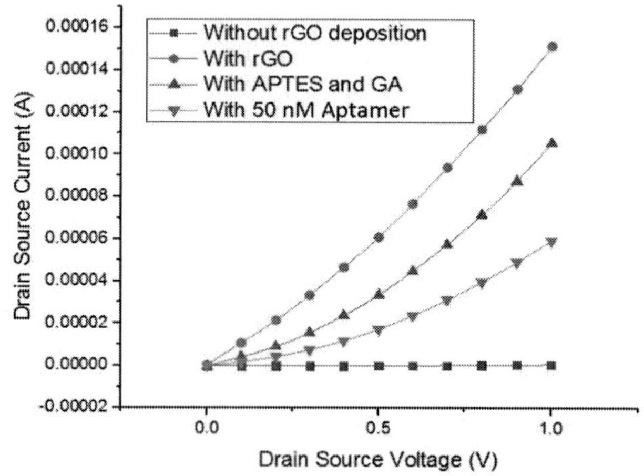

Figure 6. Electrical measurement of aptamer-rGO FET

3.4. Stability of rGO deposition

rGO material was deposited on the surface of FET by dropping a constant amount of volume towards the transducing channel which is SiO_2 layer. rGO will act as a conducting layer and immobilization site for the probe which is aptamer. Figure 7 shows a slight decrease of conductivity each time rGO surface being washed and thus proves that rGO can be deposited on the SiO_2 of FET channel without need any modification.

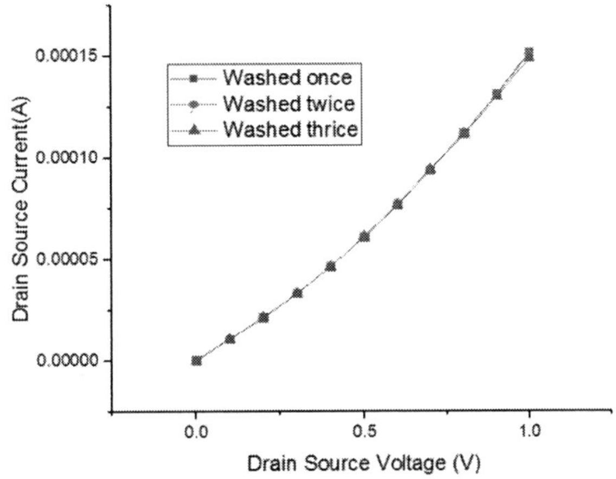

Figure 7. Stability of rGO deposition

3.5. Response of aptamer concentration toward rGO-FET

Figure 8 shows the effect of aptamer concentration on rGO-FET in electrical measurement. The voltage-current (I-V) was measured from 0 V to 1 V. The lowest concentration of aptamer shows highest electrical conductivity which is at 84.3 μA by using 0.5 nM of aptamer. When 0.5 nM, 1 nM and 5 nM of aptamer were immobilized, the device shows decrease of conductivity but the decrease level are small. However, during 10 nM and 50 nM concentration of aptamer were immobilized on the rGO-FET, the conductivity of the device shows a significant reduction in conductivity. Aptamer used are modified with amine, decreased in current was observed due to the formation of amide bond on rGO surface between aminated aptamer and carboxyl group on rGO. Higher concentration of aptamer will have more negative charge[13] that binds to the linker thus, prove the properties of aptamer binding with linker on rGO surface.

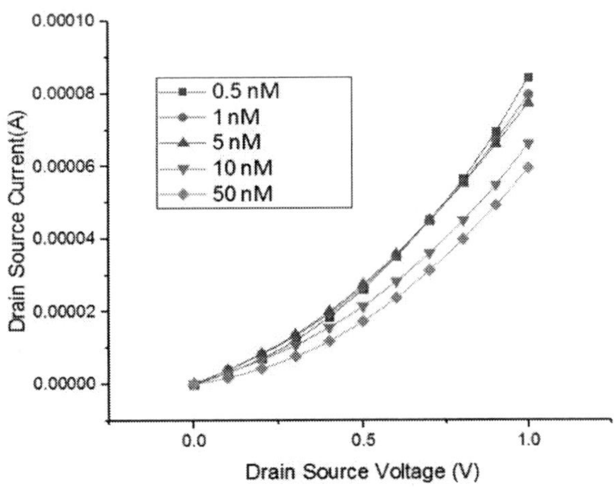

Figure 8. Electrical measurement of Aptamer concentration towards FET surface channel

4. Conclusion

In a nutshell, the fabricated device shows high sensitivity in application for detection of biomolecule. rGO-FET can be a promising detection system with integration of aptamer with concentration range from 0.5 nM to 50 nM. Among these concentrations, 0.5 nM concentration of aptamer shows highest drain-source current at 1V whereas 50 nM concentration of aptamer has lowest current pass through source-drain at 1 V. This is due to rGO contains negative charge that formed by oxygen group in their structure repels with the electron contain in aptamer and creates depletion on the FET surface. Higher concentration of aptamer will have a higher amount of electron and thus decrease the rate of current flow through the gate. Hence, lower concentration ranging from 0.5 nM to 5 nM of aptamer suitable to be integrated with rGO-FET to get excellent conductivity between source and drain. The stability of rGO verify that rGO suitable to be used as a transducing material and immobilization site for aptamer probe.

Acknowledgement
This work is funded by Fundamental Research Grant Scheme (9003-00596) under the Ministry of Higher Education (MOHE), Malaysia. The authors fully acknowledged Graduate Assistance Scholar from Universiti Malaysia Perlis and Institute of Nano Electronic Engineering (INEE), Universiti Malaysia Perlis for the approved fund which makes this important research viable and effective.

References
[1] Van Baelen, H., Beck, M., & De Moor, P. A cortisol- induced charge difference in desialylated human transcortin detected by isoelectric focusing. *Journal of Biological Chemistry*, *247*(**9**), 2699-2703 (1972)

[2] Schöning, M. J., & Poghossian. A. Recent advances in biologically sensitive field-effect transistors (BioFETs). *Analyst*, *127*(**9**), 1137-1151 (2002)

[3] Cheng, S., Hotani, K., Hideshima, S., Kuroiwa, S., Nakanishi, T., Hashimoto, M., & Osaka, T. Field effect transistor biosensor using antigen binding fragment for detecting tumor marker in human serum. *Materials*, *7*(**4**), 2490-2500 (2014)

[4] Monaghan, P. J., Owen, L. J., Trainer, P. J., Brabant, G., Keevil, B. G., & Darby, D. Comparison of serum cortisol measurement by immunoassay and liquid chromatography-tandem mass spectrometry in patients receiving the 11 β-hydroxylase inhibitor metyrapone. *Annals of Clinical Biochemistry*, *48*(**5**), 441-446 (2011)

[5] Taghdisi, S. M., Danesh, N. M., Lavaee, P., Ramezani, M., & Abnous, K. An electrochemical aptasensor based on gold nanoparticles, thionine and hairpin structure of complementary strand of aptamer for ultrasensitive detection of lead. *Sensors and Actuators B: Chemical*, ***234***, 462-469 (2016)

[6] Crivianu-Gaita, V., & Thompson, M. Aptamers, antibody scFv, and antibody Fab'fragments: an overview and comparison of three of the most versatile biosensor biorecognition elements. *Biosensors and Bioelectronics*, ***85***, 32-45 (2016)

[7] Gómez-Navarro, C., Meyer, J. C., Sundaram, R. S., Chuvilin, A., Kurasch, S., Burghard, M., & Kaiser, U. Atomic structure of reduced graphene oxide. *Nano letters*, *10*(**4**), 1144-1148 (2010)

[8] Yuan, H., Jiao, Q., Liu, J., Liu, X., Yang, H., Zhao, Y., & Li, H. Ultrathin-walled Co 9 S 8 nanotube/reduced graphene oxide composite as an efficient electrocatalyst for the reduction of triiodide. *Journal of Power Sources*, ***336***, 132-142. (2016).

[9] He, Q., Sudibya, H. G., Yin, Z., Wu, S., Li, H., Boey, F., & Zhang, H. Centimeter-long and large-scale micropatterns of reduced graphene oxide films: fabrication and sensing applications. *Acs Nano*, *4*(**6**), 3201-3208 (2010)

[10] Truong, T. K., Nguyen, T. N. T., Trung, T. Q., Sohn, I. Y., Kim, D. J., Jung, J. H., & Lee, N. E. Reduced graphene oxide field-effect transistor with indium tin oxide extended gate for proton sensing. *Current Applied Physics*, *14*(**5**), 738-743 (2014)

[11] Chang, J., Zhou, G., Gao, X., Mao, S., Cui, S., Ocola, L. E., & Chen, J. Real-time detection of mercury ions in water using a reduced graphene oxide/DNA field-effect transistor with assistance of a passivation layer. *Sensing and Bio- Sensing Research*, **5**, 97-104 (2015)

[12]Sohn, I. Y., Kim, D. J., Jung, J. H., Yoon, O. J., Thanh, T. N., Quang, T. T., & Lee, N. E. pH sensing characteristics and biosensing application of solution-gated reduced graphene oxide field-effect transistors. *Biosensors and Bioelectronics*, ***45***, 70-76 (2013)

[13]Ruscito, A., & DeRosa, M. C. Small-Molecule Binding Aptamers: Selection Strategies, Characterization, and Applications. *Frontiers in Chemistry*, *4*, **14** (2016)

IOP Conf. Series: Materials Science and Engineering **318** (2018) 012046 doi:10.1088/1757-899X/318/1/012046

An Improvement To The k-Nearest Neighbor Classifier For ECG Database

Haryati Jaafar[1,*], Nur Hidayah Ramli[1], Aimi Salihah Abdul Nasir[1]

[1]Faculty of Engineering Technology, Universiti Malaysia Perlis, UniCITI Alam Campus, Sg. Chuchuh, 02100 Padang Besar, Perlis, Malaysia

[*]haryati@unimap.edu.my

Abstract. The k nearest neighbor (kNN) is a non-parametric classifier and has been widely used for pattern classification. However, in practice, the performance of kNN often tends to fail due to the lack of information on how the samples are distributed among them. Moreover, kNN is no longer optimal when the training samples are limited. Another problem observed in kNN is regarding the weighting issues in assigning the class label before classification. Thus, to solve these limitations, a new classifier called Mahalanobis fuzzy k-nearest centroid neighbor (MFkNCN) is proposed in this study. Here, a Mahalanobis distance is applied to avoid the imbalance of samples distribition. Then, a surrounding rule is employed to obtain the nearest centroid neighbor based on the distributions of training samples and its distance to the query point. Consequently, the fuzzy membership function is employed to assign the query point to the class label which is frequently represented by the nearest centroid neighbor Experimental studies from electrocardiogram (ECG) signal is applied in this study. The classification performances are evaluated in two experimental steps i.e. different values of k and different sizes of feature dimensions. Subsequently, a comparative study of kNN, kNCN, FkNN and MFkCNN classifier is conducted to evaluate the performances of the proposed classifier. The results show that the performance of MFkNCN consistently exceeds the kNN, kNCN and FkNN with the best classification rates of 96.5%.

1. Introduction

Classification processes have been extensively studied in the machine learning literature, such as face classification, speech classification, handwriting character classification and others. These processes are divided into unsupervised and supervised classification [1, 2]. In unsupervised classification, a clustering technique is applied to find the similarity of samples before the samples are classified into a small number of homogenous or same groups. For a supervised classification, the classification is tackled through two classifiers, namely parametric and nonparametric classifiers. The parametric classifier is based on the assumption of given samples that are normally distributed. Some parametric classifiers include the support vector machine (SVM), naïve Bayes and decision tree. In contrast, the non-parametric classifiers do not make any assumptions on the samples' distribution. These classifiers include the k-nearest neighbor (kNN), multilayer perceptron (MLP), artificial neural network (ANN) and Euclidean distance (ED) [3].

This research focuses on the kNN which is under the supervised and non-parametric category. This classifier was introduced by Cover and Hart in 1976 [4]. Consequently, it is regularly used in practice since this classifier is simple to implement and quite straightforward in the classification process.

According to the kNN rule, the classification is based on a majority voting process. In this process, a testing sample or query point is assigned to the class label which has been represented by the majority label of its k-nearest neighbors. The k-nearest neighbors are determined from the minimum distance between the query point and training samples. Nevertheless, the drawback of kNN is that the distributions of samples are not taken into account in this classifier. There are different types of samples distribution either around or aside from the query point. Hence, the nearest neighbors may not have the symmetry around the query point if the samples are not homogenous or of the same class in its neighborhood [5]. Another major problem is that the kNN rule is no longer optimal if the number of training samples is limited. This is because the limited number of training samples can affect the region of neighbourhood to be smaller. Thus, the kNN rule will select the nearest training samples within the region. However, the nearest neighbour classes in the region may be unreliable which misclassify the query point [6].

In order to overcome the problems occurring in kNN, studies on the concept of neighborhood are proposed. According to Chaudhuri [5], the concept of neighborhood should be such that (a) the neighbors are as near to the query point as possible and (b) the neighbors sit as symmetrically around the query point as possible. To tackle the problem, Chaudhuri [5] proposes a simple and intuitively appealing definition of neighborhood called the nearest centroid neighborhood (NCN) and nearest median neighborhood (NMN). In the NCN concept, the idea of the neighbourhood is based on two criteria. Firstly, by the distance criterion where the k neighbours of a query point must be as near as possible. Secondly, by the symmetry criterion, their centroid must also be as close to the query point as possible. For the second approach which is the NMN, the median point is used instead of centroid to define the neighbors. Following the earlier work, Sanchez et al., [3] introduce a surrounding neighborhood (SN) to employ the distances and distributions of training samples before the classification process.

On the other hand, an extension of the kNN using the fuzzy rule has been proposed by Keller et al., [7]. In the fuzzy rule-based system, the fuzzy kNN (FkNN) assigns a fuzzy membership for the query point rather than assigning the query point to the class label. The fuzzy membership is determined based on the class weighted. The class weighted is inversely proportional to the distance between training samples and the query point [13]. The highest fuzzy membership value is considered as a winner for that query point.

Motivated by the capability of the NCN and FkNN to handle the kNN problems, an extensive improvement classifier of kNN called a Mahalanobis fuzzy k-nearest centroid neighbor (MFkNCN) classifier is proposed in this paper. This classifier is divided into two parts. In the first part, an adaptive technique by using Fuzzy Inference System (FIS) is introduced to reduce the size of the training set aiming for both reducing complexity and avoiding the outliers when using the entire training set. A Mahalanobis distance is employed in these algorithms.

Subsequently, in second part, a surrounding-fuzzy based rule is employed. This rule is adapted from the SN rule and fuzzy rule applied in the NCN and FkNN, respectively. The surrounding rule is used to ensure that the training samples are distributed sufficiently in the region of neighbourhood with the nearest neighbours located around the query point. Consequently, the fuzzy-based rule is employed to solve the ambiguity of the weighting distance between the query point and its nearest neighbors.

In order to determine the effectiveness of propsed classifier, experimental studies from an ECG signal database is employed in this study. This database is obtained freely from public heart sound database, assembled for an international competition, the PhysioNet/Computing in Cardiology (CinC) Challenge 2016. The archive comprises nine different heart sound databases sourced from multiple research groups around the world.

This paper is outlined as follows. The proposed classifier, MMFkNCN and the rationale to use FIS and Mahalanobis distance measure is presented in Section 2. The surrounding-fuzzy based rule is explained in Section 3 and Section 4 discusses the experimental results and discussion. Finally, conclusions are summarized in Section 5.

2. Part 1: Fuzzy Inference System

The goal in this section is to investigate an iterative way to reduce the storage requirement and select the best candidate from the training samples without degrading performance of the MMFkNCN. They are achieved by employing the outlier detection based on fuzzy-rule. Here, two algorithms which are the Weighted Similarity Function (WSF) and triangle inequality are considered. A weighted similarity function is used to calculate the distance (or similarity) of a query point to the training samples while triangle inequality is implemented to search for the threshold for the outlier detection. The training samples that fell within a threshold are defined as the candidate samples. The unselected training sample is called as the outlier where it is defined as a noisy sample that is excluded during the second part operation.

2.1. Mahalanobis distance

In most of the nearest neighbor classifier, the Euclidean distance is conventionally used to measure the distance between the query point and training samples. It is appropriate for all numeric training samples with similar length. However, the class distribution of the data in the training samples is imbalanced when one of the classes is more under-represented than the other classes [3]. Hence, a severe bias can occur when using the Euclidean distance as the training samples are assumed as balanced distribution in most cases.

Therefore, the Mahalanobis distance is introduced in this study instead of measuring the training samples using the Euclidean distance. This distance is a well-known criterion which depends on estimated parameters of the multivariate distribution. Given a query point y, a set of training samples $T=\{x_j\}_{j=1}^{N}$ with the label patterns of $\{y_i \mid i=1,2,\ldots,n\}$ where N is the number of training samples, x_j is the training sample and n is the number of class. Then, the sample mean vector by $\overline{x}_N$ and the sample covariance matrix by C_N,

$$C_N = \frac{1}{N-1}\sum_{j=1}^{n}(y-\overline{x}_N)(y-\overline{x}_N)^T \quad (1)$$

The Mahalanobis distance for each multivariate data point $\{y_i \mid i=1,2,\ldots,n\}$ is given as :

$$M(y,x_j)=\left(\sum_{j=1}^{n}(y-\overline{x}_N)^T C_N^{-1}(y-\overline{x}_N)\right)^{1/2} \quad (2)$$

2.2. Weighted similarity function

The WSF is applied to efficiently find the similarity of the query point and the training samples. This is done by assigning the weight to each query point in the training set. The similarity is important to estimate and check the similarity and dissimilarity values between them. Technically, the similarity values would increase and dissimilarity values would decrease when both the query point and training sample became more similar. In the WSF, the computation of similarity depended on weight coefficients $w(y,x_j)$. As the training sample has larger $w(y,x_j)$, it became more similar and closer to the query point [8].

In order to apply the WSF, the information of a query point y and a set of training sets $T=\{x_j\}_{j=1}^{N}$ are firstly determined. Then the Mahalanobis distance, $M(y,x_j)$ between the query point and the training samples in Equation (2) is computed. Subsequently, the WSF based on weight coefficients, $w(y,x_j)$ is determined by modifying the distance function in Equation (2) to form the following equation:

$$w(y,x_j)=\frac{1-M(y,x_j)}{M_{\max}} \quad (3)$$

IOP Conf. Series: Materials Science and Engineering **318** (2018) 012046 doi:10.1088/1757-899X/318/1/012046

where $w(y,x_j)$ is in the range [0,1], M_{max} is the maximum possible distance between two training samples in the feature space.

2.3. *Triangle inequality*

To define whether the sample is either a candidate sample or outlier, the triangle inequality is employed in this study. Considering x represented the training sample, y as a query point and z is the reference point, the reference point is the furthest training sample from the query point that is obtained from Equation (2). The distances are sorted in ascending order before determining the reference point. The condition for applying the triangle inequality technique is that the distance between two objects should not be less than the difference in the distances to any objects [9]. Specifically, if the distance between the query point and the training samples satisfies Equation (3), the following inequality would always true.

$$M(y,x_j) \le M(y,z) + M(x_j,z) \tag{4}$$

By comparing the training sample and query point to a reference point, an upper bound is obtained as:

$$|M(y,z) - M(y,x_j)| \le M(x_j,z) \le M(y,z) + M(y,x_j) \tag{5}$$

According to Equations (4) and (5), the triangle inequality and upper bound depended on the distance between the training sample to the reference point, $d(x_j,z)$. In order to avoid calculation of $d(x_j,z)$, a lower bound is needed such that $M(x_j,z) \ge M(y,x_j)$.

Let:

$$M(y,x_j) \le \frac{1}{2} M(y,z) \tag{6}$$

Hence, Equation (4) turned to be:

$$2M(y,x_j) \le M(y,z) + M(x_j,z) \tag{7}$$

Thus, Equation (7) proved that:

$$M(y,x_j) \le M(x_j,z) \tag{8}$$

Equation (6) is rearranged and as a result:

$$M(x_j,z) \ge M(y,x_j) - M(y,z) \tag{9}$$

Assuming $M(x_j,z) \ge 0$, the lower bound is obtained as:

$$M(x_j,z) \ge \max\{0, M(y,z) - M(y,x_j)\} \tag{10}$$

and the distance $M(x_j,z)$ can be avoided. Since $M(y,x_j) \le M(x_j,z)$, the training sample is selected as the candidate sample after the Equation (6) is rewritten as:

$$\frac{M(y,x_j)}{M(y,z)} \le \frac{1}{2} \tag{11}$$

The process of traversing a list of the candidate samples continued until it met the lower and upper bound criterion.

2.4. Design on FIS

There are three main stages to compute the output of the FIS i.e. fuzzification process, rule extraction and defuzzification process. The Mamdani model was also adopted in this study since it effectively modelled the common-sense rules on fuzzy variables.

The fuzzification is the process of finding the degree of membership of a value in a fuzzy set. In this process, the input variables are converted from crisp values to fuzzy values by using a set of input Membership Functions (MF) In this study, the Gaussian curve is chosen to determine the membership values. In this process, the weight coefficients, $w(y,x_j)$ from Equation (3) and the ratio of triangle inequality from Equation (11) are employed as the input variables and the MFs and the outlierness is assigned as an output variable. The quantization level [10] for weight coefficients, $w(y,x_j)$, $\mu_A(x)$, triangle inequality, $\mu_B(x)$ and outlierness, $\mu_C(x)$ are summarized in Table 1.

Table 1. Quantization level of A and B fuzzy sets

Quantization level	A (weight coefficients)	B (ratio of triangle inequality)	C (outlierness)
V1=0	Small	Narrow	High
V2 = 0.5	Medium	Moderate	Intermediate
V3 = 1	Large	Wide	Low

As the MFs of input and output variables are defined in the FIS, the rule base is then evaluated. The input membership values such as the weighting factors are used by the rules to determine their influence on the set of fuzzy outputs in the final output conclusion. In this study, nine rules are used to characterize fuzzy rules and they are shown in Table 2.

Table 2. Rule extracted from A and B for outlier detection

Rules	Inputs		Output
	A (weight coefficients)	*B* (ratio of triangle inequality)	*C* (outlierness)
Rule 1	Small	Narrow	Intermediate
Rule 2	Small	Moderate	Not high
Rule 3	Small	Wide	High
Rule 4	Medium	Narrow	Not low
Rule 5	Medium	Moderate	Intermediate
Rule 6	Medium	Wide	Not high
Rule 7	Large	Narrow	Low
Rule 8	Large	Moderate	Not low
Rule 9	Large	Wide	Intermediate

In the defuzzification process, the fuzzy output values are converted into a single crisp value or final decision. The COG was applied and the crisp value obtained is:

$$y_i = \frac{\int_y y.\mu_C(y_i)dy}{\int_y \mu_C(y_i)dy} \quad (12)$$

where *Yi* is a centroid of fuzzy set *C* or outlierness.

The defuzzified output of the fuzzy procedure is influenced by the value of weight coefficient and the ratio of the triangle inequality. For example, if the value of weight coefficient is large and ratio of the triangle inequality is narrow, the outlierness was low hence the training sample is accepted as a candidate training sample.

3. Part 2: Surrounding-fuzzy Based Rule

In the second part, the surrounding-fuzzy based rule is proposed where this rule is modified from the surrounding rule and fuzzy rule that is applied in the kNCN and FkNN, respectively. The main objective of this classifier is to optimize the performance results while considering the surrounding-fuzzy based rule which are

i) the k centroid nearest neighbors should be close to the query point as possible and located symmetrically around the query point.

ii) the query point is classified by taking the fuzzy membership values into account.

In this rule, the candidate samples that obtained from Part 1 is used to define the k neighbors. A set of candidate samples $T_s = \{x_s\}_{s=1}^{Q}$ with the label pattern of $c_1, c_2, \ldots c_n$, is given where Q is the number of training samples, x_s is the candidate sample and n is the number of class. The distance between the query point and the candidate samples is found by using Equation (13).

$$d(y, x_s) = \|y - x_s\|_{L2} \quad (13)$$

A set of k nearest centroid neighbor $X^{NCN} = \{x_r^{NCN}\}_{r=1}^{k}$ is considered where k is the number of nearest centroid neighbors. By applying Equation (13), the first centroid neighbor is determined by selecting the training sample that is located the nearest to the query point. The first nearest centroid neighbor is set as x_1^{NCN}. The subsequent nearest centroid neighbors for r=2,…,k are determined by calculating the centroid position of all the training samples except the training sample that has been selected as the nearest centroid neighbor. The centroid of a set of training samples, x_{rj}^{c} can be defined as:

$$x_{rj}^{c} = \frac{\left(\sum_{r=1}^{r-1} x_r^{NCN}\right) + x_j}{r} \quad (14)$$

The k nearest centroid neighbor is determined by computing the shortest distance between the centroid of a set of a training samples and the query point. It is given as:

$$x_r^{NCN} = \min \|(y - x_{rj}^{c})\|_{L2} \quad (15)$$

The process is repeated by determining the centroid between the training samples and previous nearest centroid neighbors for the value of r=3.

After the k nearest centroid neighbors has been determined, the next step is to assign a fuzzy membership for the k nearest centroid neighbors to class labelled as y rather than voting the k nearest centroid neighbors as a majority. By assigning the fuzzy membership, the MFkNCN classifier become more advantageous as it is able to provide the next potential secondary structural class if the first one is wrongly predicted. This process is done by determining the weight that is inversely proportional to the distance between x_r^{NCN} and y. The fuzzy membership found is:

$$u_i^{NCN}(y) = \frac{\sum_{r=1}^{k} u_{ir}\left(1/\left\|y - x_r^{NCN}\right\|^{2/(m-1)}\right)}{\sum_{r=1}^{k} 1/\left\|y - x_r^{NCN}\right\|^{2/(m-1)}} \tag{16}$$

where $i = 1,2,...,C$, with C as the number of classes and u_{ir} as the membership degree of k nearest centroid neighbors.

The value of the fuzzy strength parameter is set at 2. This was because the fuzzy membership values are proportional to the inverse square of the distance. When it is set at 2, it leads to an optimal result in the classification process. The constraint fuzzy membership is given as [7]:

$$u_{ir}(x_k) = \begin{cases} 0.51 + 0.49(n_r / k) & r = i \\ 0.49(n_r / k) & r \neq i \end{cases} \tag{17}$$

where n_r denoted the number of rth nearest centroid neighbors.

After calculating all memberships for a query point, it is assigned to the class of the highest membership value as shown in Equation (17).

$$C(y) = \arg\max(u_i^{NCN}(y)) \tag{18}$$

The process of the MFkNCN classifier during the searching stage is then repeated for a new query point until all query points are successfully classified [11].

4. Experimental results

4.1. Data acquisition

In this study, the Shiraz University adult heart sounds database (SUAHSDB) is employed. This database was constructed using recordings made from 79 healthy subjects and 33 patients (total 69 female and 43 male, aged from 16 to 88 years). During the recording, the subjects were asked to relax and breathe normally during the recording session. The database consists of 114 recordings (81 normal recordings and 33 pathological recordings). The recording length varied from approximately 30s–60s. The sampling rate was 8000 Hz with 16 bit quantization except for three recordings at 44 100 Hz and one at 384000 Hz. The data were recorded in wideband mode of the digital stethoscope, with a frequency response of 20 Hz–1 kHz [12].

4.2. Experimental setup

A selected signal from 79 healthy subjects with 79 recordings is applied for experiment. The principle component analysis with autocorrelation method is selected as a feature extraction due to the fact that the feature is more robust to noise compared to other feature extractions [12]. Subsequently, a comparative study of kNN, kNCN, FkNN and MFkNCN classifiers is conducted to evaluate the performances of the classifiers. Two experiments were conducted in this study, where in the first experiment, performance evaluations with different values of k were observed. The values of k ranging was set from 1 to 10 while the size of feature dimension was fixed to 64. In this experiment, the training samples were set to 30 and the testing data set to 49. In the second experiment which is performance evaluation based on different sizes of feature dimensions, four sizes of feature dimensions i.e. 10, 16, 32 and 64 are computed.

The proposed methods have been implemented in Matlab R2010 (b) and have been tested in Intel Core i5, 2.1GHz CPU, 6G RAM and Windows 7 operating system. In all experiments, the performance was evaluated based on the classification accuracy (CA) which is calculated as;

$$CA = \frac{N_C}{N_T} \times 100\% \tag{19}$$

where Nc is the number of syllables which is recognized correctly and N_T is the total number of test syllables.

4.3. Performances evaluation

Figure 1 illustrates the performance of four classifiers in term of different values of k. It shows that the kNN gives the poorest classification performance as compares to kNCN, FkNN and MFkNCN. In addition, the classification gradually decreases as the values of k increases. Although the highest classification accuracy of kNN is 94.59% at k=1 and 2, the performance of the kNN is still poor compared to the kNCN, FkNN and MFkNCN. For the kNCN, it is observed that the performance of the kNCN increases as the number values of k varies from 1 to 7 with the highest classification accuracy 96.49% at k=7. Moreover, this classifier is able to perform better than the kNN and FkNN in most of the values of k. On the other hand, the highest accuracy for the FkNN is 95.88% at k=8. The results show that the FkNN outperforms the kNN at every values of k. Nevertheless, this classifier performs poorly than the kNCN and MFkNCN in most of the values of k. For the MFkNCN, the results show that the performance of the MFkNCN consistently exceeds that of the kNN and FkNN. Although the performance of MFkNCN is lower than kNCN at k=7 and 10, the MFkNCN is able to achieve the better performance than kNCN almost every values of k with the best classification performance is obtained at k=9 with 96.55%. However, all of the classifiers show an instable trend when the number of k is increased. This is because some of the training samples from different classes have a very similar characters and these training samples are called overlapping samples. Thus, the misclassification often occurs near class boundaries where overlapping usually occurs as well [10].

For the second experiment, the value of k was fixed to 9 based on the highest accuracy that achieved in MFkNCN from the previous experiment. Figure 2 shows the classification performance at different sizes of feature dimensions for each classifier i.e. kNN, kNCN, FkNN and MFkNCN. As it can see in Figure 2, the performance of kNN provides worst performance, yielding 90.8% and 92.09% classification accuracy for the highest and lowest accuracy at 10 and 64 feature dimensions, respectively. The results show the kNCN outperforms the kNN and FkNN where the highest average accuracy achieved is 96.19%. For the FkNN, the performance of this classifier is better than kNN with the highest classification accuracy is 95.29%. It can be observed that the performance of the MFkNCN consistently exceeds of the kNN and FkNN at all sizes of feature dimensions. In 16 feature dimension, the kNCN slightly outperform the MFkNCN on this dataset. However, the performance of MFkNCN varies more in other feature dimensions. The best classification performance is achieved at 64 size of feature dimension with 96.55%.

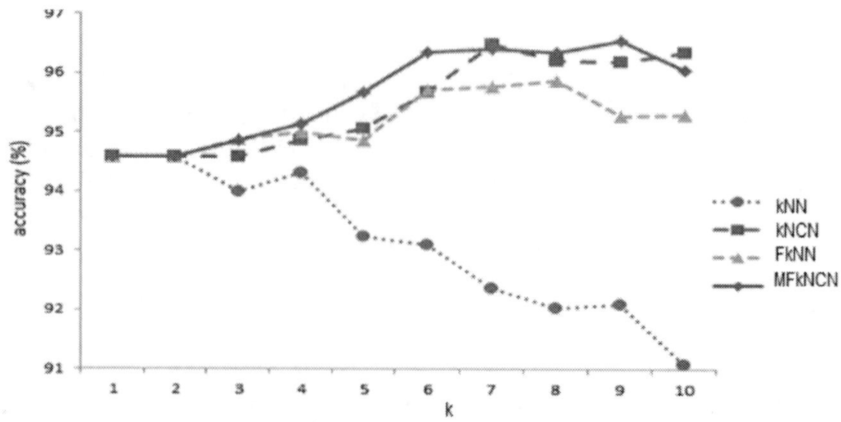

Figure 1. Performances of kNN, kNCN, FkNN and MFkNCN based on different values of k

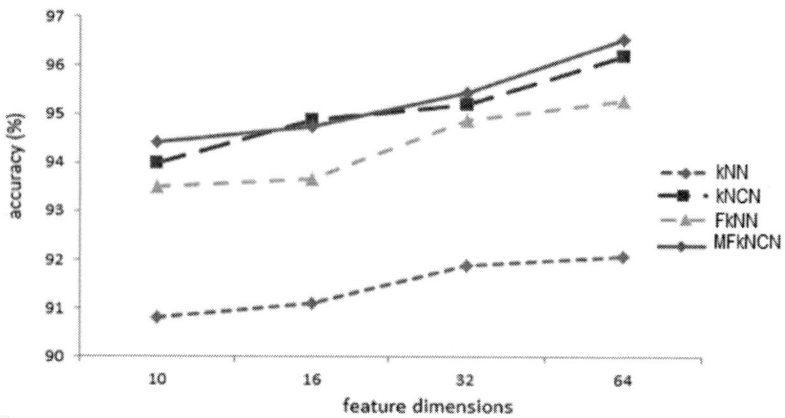

Figure2. Performances of kNN, kNCN, FkNN and MFkNCN based on different sizes of feature dimensions

5. Conclusion

In this paper, an empirical work of the centroid neighborhood and fuzzy-rule based system with Mahalanobis distance called a a Mahalanobis fuzzy k-nearest centroid neighbor (MFkNCN) has been proposed and successfully implemented. The proposed classifier aims at exploiting the strength of the centroid neighbourhood while solving the ambiguity of the weighting distance between the query point and its nearest neighbours by using Mahalanobis distance. The proposed classifier firstly computed the k nearest centroid neighbourhood for each class separately. Then, the fuzzy membership is constructed to assign the query point to the right class label. In this paper, the proposed classifier has been compared with the kNN, kNCN and FkNN to validate the effectiveness of the MFkNCN. A series of experiments, based on different values of k and different sizes of feature dimensions have been performed to determine the competence of the proposed classifier. It was found that the classification accuracies were up to 92.09%, 96.19%, 95.29% and 96.55% for kNN, kNCN, FkNN and MFkNCN, respectively. Meanwhile, for the palm prints dataset, the kNN, kNCN, Results indicate that the MFkNCN provides the best classification among the classifiers.

Acknowledgement
The authors would like to express their gratitude for the financial support provided by Universiti Malaysia Perlis Short Term Grant (9001-00500).

References

[1] Sheskin D J 2011 *Handbook of Parametric and Nonparametric Statistical Procedures,* (Chapman & Hall)
[2] Kumar Y and Sahoo G 2012 Analysis of Parametric & Non Parametric Classifiers for Classification Technique using WEKA *International Journal Information Technology and Computer Science* Volume **7** pp 43-49
[3] Sanchez J S, Pla F and Ferri F J 1997 On the use of neighbourhood-based non- parametric classifiers *Pattern Recognition Letters* Volume 18 pp 1179–1186
[4] Cover T and Hart P 1967 Nearest neighbour pattern classification *IEEE Trans on Information Theory* Volume 13 Number 1pp 21-27
[5] Chaudhuri B B 1996 A new definition of neighbourhood of a point in multi-dimensional space *Pattern Recognition Letters* Volume 17 Number 1 pp 11-17
[6] Imandoust S B and Bolandraftar M 2013 Application of K-Nearest Neighbor (KNN) Approach for Predicting Economic Events: Theoretical Background. *Int. Journal of Engineering Research and Application* Volume 3 Number 5 pp 605-610
[7] Keller J M Gray, M R and Givens, J A 1985 A fuzzy k-nearest neighbor *algorithm IEEE Trans.Syst., Man, Cybern. SMC* Volume 15 Number 4 pp 580-585
[8] Triguero I, Derrac J, Garcia S and Herrera F 2012 A taxonomy and experimental study on prototype generation for nearest neighbor classification *IEEE Trans. on Sys, Man, and Cybernetics, Part C: Apps & Reviews* Volume **42** Number 1 pp 86-100
[9] Berman A and Shapiro L G 1998 Selecting good keys for triangle-inequality-based pruning algorithms *IEEE International Workshop on Content-Based Access of Image and Video Database* pp 12-19
[10] Jaafar H , Ibrahim S and Ramli D A 2015 Robust and fast computation touch-less palm print recognition system using LHEAT and IFkNCN classifier Computational *Intelligence and Neuroscience* pp 1-17
[11] Jaafar H, Mukahar N and Ramli D A 2016 A methodology of nearest neighbor: Design and comparison of biometric image database *IEEE Student Conference on Research and Development (SCOReD)* pp 1-6
[12] Liu C, Springer D, Li Q, Moody B, Juan R A, Chorro F J, Castells F, Roig J M, Silva I, Johnson A E and Syed Z 2016 An open access database for the evaluation of heart sound algorithms *Physiological Measurement* Volume 37 Number 12 pp 2181

IOP Conf. Series: Materials Science and Engineering **318** (2018) 012047 doi:10.1088/1757-899X/318/1/012047

Numerical Investigation of a Chip Printed Antenna Performances for Wireless Implantable Body Area Network Applications

N.H.Ramli[1], H.Jaafar[2], Y.S.Lee[1]
[1]Department of Electronic Engineering Technology, Faculty of Engineering Technology, Universiti Malaysia Perlis, Malaysia
[2]Department of Electric Engineering Technology, Faculty of Engineering Technology, Universiti Malaysia Perlis, Malaysia

Email:hidayahramli@unimap.edu.my

Abstract. Recently, wireless implantable body area network (WiBAN) system become an active area of research due to their various applications such as healthcare, support systems for specialized occupations and personal communications. Biomedical sensors networks mounted in the human body have drawn greater attention for healthcare monitoring systems. The implantable chip printed antenna for WiBAN applications is designed and the antenna performances is investigated in term of gain, efficiency, return loss, operating bandwidth and radiation pattern at different environments. This paper is presents the performances of implantable chip printed antenna in selected part of human body (hand, chest, leg, heart and skull). The numerical investigation is done by using human voxel model in built in the CST Microwave Studio Software. Results proved that the chip printed antenna is suitable to implant in the human hand model. The human hand model has less complex structure as it consists of skin, fat, muscle, blood and bone. Moreover, the antenna is implanted under the skin. Therefore the signal propagation path length to the base station at free space environment is considerably short. The antenna's gain, efficiency and Specific Absorption Rate (SAR) are -13.62dBi, 1.50 % and 0.12 W/kg respectively; which confirms the safety of the antenna usage. The results of the investigations can be used as guidance while designing chip implantable antenna in future.

1. Introduction

Recently, the wireless communication through human body is catching the attention for researcher. The system is applied to enhance and support the quality of human lives. There are lots of implantable devices for medical systems are already used such as pacemakers and cardioverter defibrillators, temperature monitors, blood glucose sensor, cochlear and retinal implants and functional electrical stimulator [1].

The development of the technology gives high impact to the size of the antenna. Consequently, there are necessities to design small size and low power consumption of antenna for wireless implantable body area network application [2-4]. Designing an implantable antenna is very challenging compared to the antenna at free space condition. The antenna efficiency is reduced and faces very strong effect of multipath losses since the lossy environment (such as human body) has a different layer of boundaries [5-14]. Attenuation is occurred due to the weakly conductive tissue and reflection at each of the boundaries of dissimilar tissue. There are also difficulties for human body to act as a medium for signal to propagate due to dissimilar tissue and organ. The value of power absorption by human body is necessity to follow the standard in order to confirm the safety of user.

Previous work is presented an implantable chip antenna for WiBAN application at 4.8 GHz. In this work, the antenna has been tested in selected parts of human body (hand, chest, leg, heart and brain). The antenna performances have been investigated in terms of gain, efficiency, return loss and SAR. The human voxel model in CST Microwave Studio software has been used throughout in this work for all the simulations.

2. Antenna configurations

The implantable antenna is composed of a single metallic layer which is copper with 0.036 mm thickness and it is printed on both side surfaces of the FR4 substrate. The dielectric constant of the substrates is ε_r=4.7 with loss tangent 0.019 and 1.6 mm thickness. Figure 1 shows the optimum configuration of the proposed antenna. In general, the size of the antenna is 5 mm × 5 mm × 1.6 mm dimension. The radiating elements of the antenna are composed of rectangular shape with multiple E-slots at the center and feeding line. The radiator is center fed by microstrip line with 4.1 mm length and 3 mm width. The optimized dimensions of the antenna are: a = 5 mm, b = 5 mm, c= 0.8 mm, d = 0.8 mm, e= 0.5 mm, f= 0.7 mm and g= 4.5 mm.

While designing the antenna, some techniques are applied to reduce the size where the slots are added in radiating element, and then the antenna design is folded. There is no ground pin is added as the antenna is too small to create a hole.

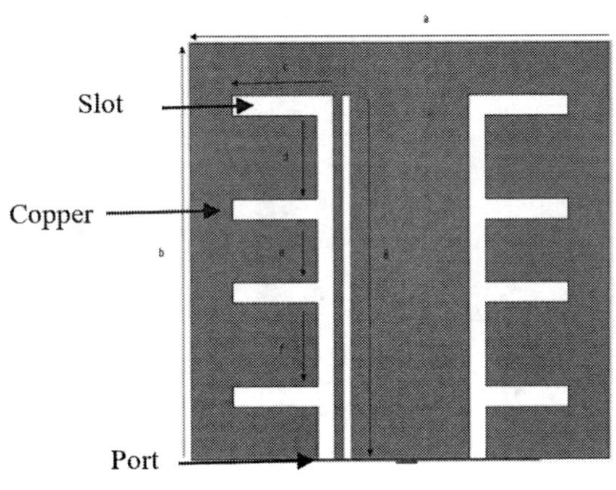

(a)

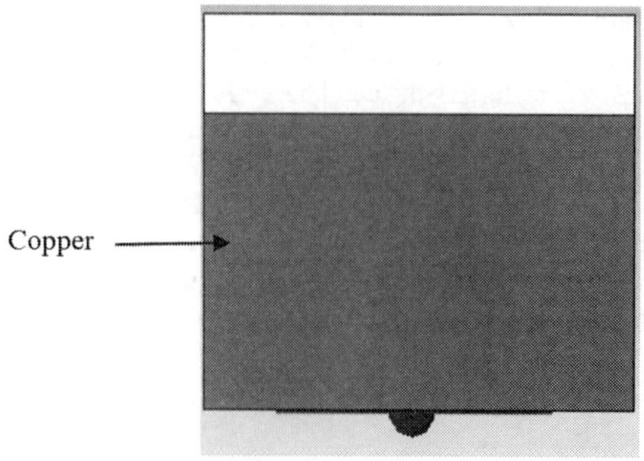

(b)

Figure 1. The configuration of the antenna (a) Front view (b) Back view

3. Experiment setup

In this section, the antenna is implanted in the human voxel model by using the CST Microwave Studio 2012. The antenna is implanted in five selected parts of the human body which are hand, chest, leg, heart and brain. The antenna performances are evaluated in term of operating frequency, bandwidth, gain, efficiency and SAR. The main purpose of this study is to analyse the antenna performances at different part of the human body and to determine the best location for antenna to operate. The input reference power of an implantable antenna is 0.1W. The implantations of antenna in some selected parts of the human voxel model are shown in Figure 2.

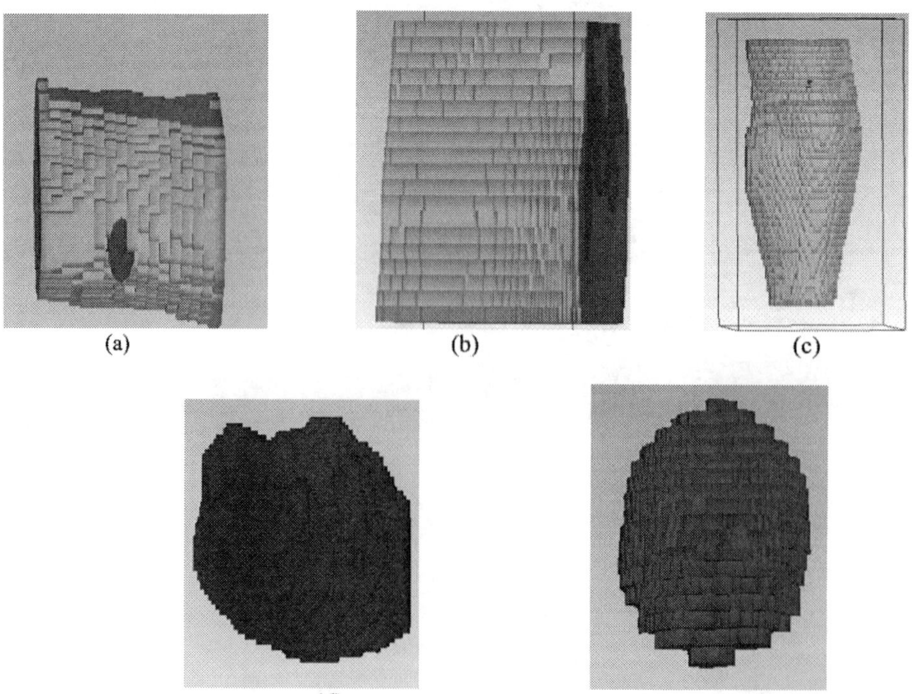

(a) (b) (c) (d) (e)

Figure. 2. Antenna implanted in the human voxel model (a) hand (b) chest (c) leg (d) heart (e) brain

The antenna is coated with a thin layer of plastic in order to preserve the antenna's biocompatibility. Subsequently, the antenna is implanted under the skin in the cases of hand, leg and chest part. For the cases of heart and brain, the antenna is implanted approximately 3 mm from the surfaces of the organs.

4. Results and discussions

The reflection coefficient magnitudes of antenna for several parts in the human body are presented in Figure 3. The observed results have revealed that the antenna offers the highest impedance bandwidth and lowest reflection coefficient magnitude when implanted in the hand model, compared to other models. The results demonstrate that the antenna is tuned at frequency 5.0 GHz with a -10 dB impedance bandwidth of 900 MHz (4.6-5.5 GHz).

However, when the level of reflection coefficient magnitude is remarked below -6 dB (useful for certain application), the antenna implanted in the chest and leg model is acceptable to apply for communication. In the chest model, the illustrated results have shown that the antenna operates at 4.2 GHz with impedance BW of 300 MHz (4.1-4.4) GHz while the antenna implanted in the leg model is tuned at 5.0 GHz with impedance BW 500 MHz (4.5-5.5) GHz. The antenna implanted in heart and brain presents less stability as the antenna cannot perform well in these complex structures. Based on the results, it can be seen that the operating frequency of the antenna is detuned for each part of the human body due to the effect of different electrical properties thus mismatched the input impedance.

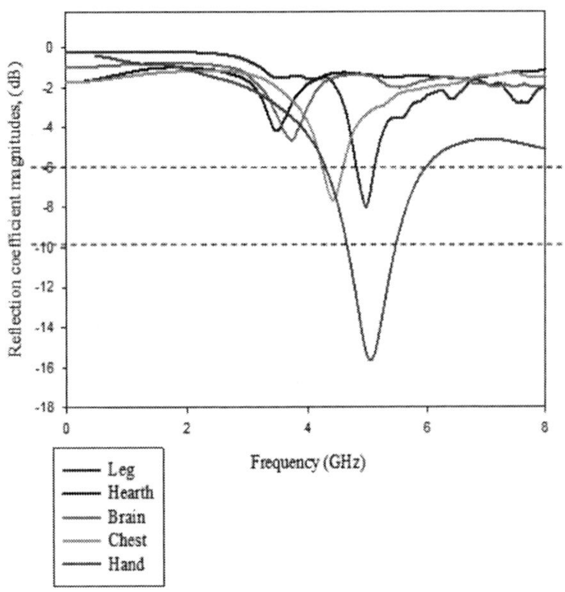

Figure 3. Comparisons of resonant frequency of each part of human body model

The electrical properties of material may affect the input impedance of the antenna as its value is depending on electrical properties thus tuned the resonant frequency. According to the equation 1, input impedance is defined as

$$Z_A = R_A + jX_A \tag{1}$$

Where:

Z_A= antenna impedance
R_A= antenna resistance
X_A= antenna reactance

In general, the resistive part of the equation 2 consists of two components which are

$$R_A = R_r + R_L \tag{2}$$

Where:
Rr = radiation resistance of an antenna
R_L= loss resistance of the antenna

According to the equation 3, the loss resistance of the antenna is represents the conductivity and dielectric losses associated with the antenna structure. Hence, this indicates that the different electrical properties of the human body model provide different input impedance, thus detuned the operating frequency of the antenna.

$$R_L = \frac{l}{P}\sqrt{\frac{\omega\mu_0}{2\sigma}} \tag{3}$$

Since the antenna is used for implantation in the human body, the value of specific absorption rate (SAR) is very important in order to protect human from any harm. Up to now, there is no specific SAR standard applied to any specific parts of the human body, thus the SAR is based on the SAR standard for mobile application. According to the standard from the European Union, the SAR limit is 2 W/kg averaged over 10 g of tissue. Table 1 shows the gain, efficiency and SAR for antenna implanted in hand, leg, chest, heart and brain.

Table 1. The gain, efficiency and SAR of antenna for several parts of human body model (input power = 0.1 W)

Human part	Gain (dB)	Efficiency	SAR (W/kg)
Hand	-13.62	1.50%	0.12
Chest	-15.86	-23.16%	0.33
Leg	-13.35	-21.47%	0.31
Heart	-16.66	-26.08%	0.21
Brain	-17.78	-25.78%	0.17

Based on the data in Table 1, an antenna implanted in the human hand voxel model provides the most reasonable gain and efficiency compared to others organ. This factor may happen since the human hand voxel has less complexity of structure (skin, blood, fat and bone) thus providing lower attenuation. Besides, the antenna is placed under the skin, hence the distance for a signal to propagate to the base station is short. Although the value is lower, the gain and efficiency of the antenna are still in reasonable rate since the antenna is placed in lossy environment.

5. Conclusion

The implantable chip printed antenna with multiple E- slots operating at frequency 4.8 GHz is presented. Generally, the size of the antenna is 5 mm x 5 mm x 1.6 mm and it's composed by FR4 substrates and copper as radiating element. The antenna is implanted in five selected parts of the human body which are hand, chest, leg, heart and brain. The antenna performances are evaluated in term of operating frequency, bandwidth, gain, efficiency and SAR. The main idea of this investigation is to evaluate the antenna performance at different part of the human body and to verify the best location for antenna to operate. The observed results have proven that the antenna offers the highest impedance BW when implanted in hand, compared to other parts. This factor could be happened as the impedance matching of the antenna is changed from each part of the human body due to the material that composed of human body part. The gain and efficiency of implantable antenna are also high in human hand model. This is because the human hand voxel has less complexity of structure (skin, blood, fat and bone) and antenna is placed under the skin hence the distance for a signal to propagate is short. The results of the investigations can be use as guidance while designing chip implantable antenna in future.

Acknowledgment

The authors wish to acknowledge the Short Term Grant (STG) provided by Universiti Malaysia Perlis (no. of project: 9001-00516) for funding this work. The members of Department of Electronic Engineering Technology, Faculty of Engineering Technology, UniMAP and Wireless Communication Centre (WCC), Faculty of Electrical Engineering, Universiti Teknologi Malaysia are also specially thanked for their helps and kindness.

References

[1] P.Soontornpipit, C.M,Furse, Y.Chung,. Design ofImplantable Microstrip Antenna for Communication with Medical Implants.IEEE Trans.MicrowaveTheory Tech. vol.52, pp.1944-1951. (2004).

[2] J.Kim, Y.R.Samii. Implanted Antenna inside a Human Body: Simulation , Designs and Characterizations. IEEETrans.Microwave Theory Tech., vol.52, pp.1934- 1943, (2004).

[3] W.Xia, K.Saito, M.Takahashi, K.Ito .Performances of an Implanted Cavity Slot Antenna Embedded in the Human Arm.*IEEE* Trans. On Antenna and Propag,vol.57, pp.894-899, (2009).

[4] D.Panescu. Wireless Communication Systems for Implantable Medical Devices. IEEE Engineering In Medicine And Biology Magazine, pp. 96-101, (2008).

[5] P.S Hall and Y.Hao, Antennas and Propogation for Body Centric Wireless Communications.Artech House Boston/London, (2006)

[6] P.S.Hall, Z.H.Hu, Y.I Nechayev.Characterisation of Narrowband Communication Channels on the Human Body at 2.45GHz.IET Microw. Antennas Propag, vol.4,pp. 722-732, (2010).

[7] K.Y.Yazdandoost.A 2.4GHz Antenna for Medical Implanted Communications.AsiaPacific Microwave Conference ,pp. 1775-1778,(2009).

[8] M.R.Kamarudin,.Y.I.Nechayev, P.S.Hall, , Performance of antennas in the on-body environment .Antenna and Propagation Society International Symposium,pp. 475-478,(2005).

[9] K.Ito, ; Ida, I.; Men-Shien Wu, Body effect on characteristics of small loop antenna in pager systems. Antenna and Propagation Society International Symposium,pp. 1081-1084 (1992).

[10] N.H.Ramli, M.R.Kamarudin, N.A. Samsudin, E.N.Ahyat. Investigation on a Compact Ring Printed Monopole Antenna for WiBAN Application Systems, Microwave and Optical Technology Letters, (2013), **55**(5): 1053-1058.

[11] Y.Gao, B. L. Ooi, et al. Band-notched ultra-wideband ring-monopole antenna. Microwave and Optical Technology Letters ,(2006),**48**(1): 125-126.

[12] J.X.Xiao, M. F. Wang, et al. "A printed circular-ring monopole antenna with band-notched feature for uwb application." Microwave and Optical Technology Letters (2010) **52**(4):'827-830

[13] N.H.Ramli., M.R.Kamarudin, N.A,Samsuri, E.N.Ahyat. A 6.0 GHz Chip Printed Monopole Antenna For Wireless Implantable Body Area Network (Wiban). Progress In Electromagnetics Research C– Volume 41. Pages 189-200, (2013).
[14] N.H.Ramli,,M.R.Kamarudin.N.A.Samsuri, Comparison of performance of compact ring printed monopole and loop antennas for wireless-implantable body area network applications. *Microwave and Optical Technology Letters.* – Volume 56. Issue 6. Pages 1270-1074. Impact Factor : 0.585.(2014)

An Adaptive Insertion and Promotion Policy for Partitioned Shared Caches

Norfadila Mahrom[1,*], **Michael Liebelt**[2], **Rafikha Aliana A. Raof**[1], **Shuhaizar Daud**[1] and **Nur Hafizah Ghazali**[1]
[1]School of Computer and Communication Engineering, Universiti Malaysia Perlis, Arau, 02600 Perlis, Malaysia
[2]School of Electrical and Electronic Engineering, Faculty of Engineering, Computer and Mathematical Sciences, The University of Adelaide, South Australia, 5005 Australia

E-mail: norfadila@unimap.edu.my

Abstract. Cache replacement policies in chip multiprocessors (CMP) have been investigated extensively and proven able to enhance shared cache management. However, competition among multiple processors executing different threads that require simultaneous access to a shared memory may cause cache contention and memory coherence problems on the chip. These issues also exist due to some drawbacks of the commonly used Least Recently Used (LRU) policy employed in multiprocessor systems, which are because of the cache lines residing in the cache longer than required. In image processing analysis of for example extra pulmonary tuberculosis (TB), an accurate diagnosis for tissue specimen is required. Therefore, a fast and reliable shared memory management system to execute algorithms for processing vast amount of specimen image is needed. In this paper, the effects of the cache replacement policy in a partitioned shared cache are investigated. The goal is to quantify whether better performance can be achieved by using less complex replacement strategies. This paper proposes a Middle Insertion 2 Positions Promotion (MI2PP) policy to eliminate cache misses that could adversely affect the access patterns and the throughput of the processors in the system. The policy employs a static predefined insertion point, near distance promotion, and the concept of ownership in the eviction policy to effectively improve cache thrashing and to avoid resource stealing among the processors.

1. Introduction

With the increasing number of applications simultaneously sharing the last level cache, cache contention and memory coherence issues have become one of the major focuses of processor architects. Competition among multiple applications for the last level cache (LLC) without efficient cache management may lead to an under-utilized system due to various forms of interference between the applications. However, cache replacement policy has been identified as one of the contributors to the overall improvement of system performance and has received much attention from researchers and computer architects. As the conventional LRU replacement policy is widely accepted as the standard replacement policy, many attempts have been made to improve the policy in respect of the three component policies of *insertion decision*, *promotion* and *victim selection*.

The commonly used LRU replacement policy employed in a multiprocessor system is inherited from uniprocessor systems. For that reason, some drawbacks of the LRU replacement policy are identified, which are generally due to lines or data that reside too long in the cache. The insertion of new lines at the highest priority position (i.e., the Most Recently Used (MRU) position) upon a miss has in turn caused zero-reuse lines and lines without high locality to spend a large amount of time occupying the cache [1][2]. On the other hand, the promotion strategy of the LRU replacement policy, which automatically moves hit lines to the MRU position, has also resulted in cache waste because it maximizes the time taken for *dead lines* to traverse from the MRU position to the lowest priority position (i.e., the LRU position) and eventually be evicted. A dead line refers to a cache line that is not reused from the time of its last access until it is evicted from the cache [1][8]. Even though the LRU policy is good at handling lines with high utility and temporal locality, other cores in the system have to wait longer before they can make use of the cache capacity while dead lines remain in the cache [8]. Nevertheless, the insertion policy and the associated promotion policy of LRU give each line a good chance of obtaining a hit while it traverses from the MRU to the LRU position, but may be inefficient for workloads in which the working set size is greater than the available cache space. This is because there are lines that are re-referenced immediately after they are replaced and this may lead to some problems such as cache thrashing, which can be improved by retaining some fraction of the working set in the cache long enough to generate more cache hits [3][4][11]. Therefore, many researchers have proposed a variety of techniques intended to improve the replacement policy so that better performance of the shared cache can be obtained.

In this paper, all three components of the replacement policy for a partitioned shared cache are investigated. The aim is to provide an adaptive LRU replacement policy that is targeted for a system where the shared cache has been carefully partitioned and distributed among multiple cores by a well-accepted cache partitioning scheme. It is expected that the new adaptive dynamic cache replacement policy for a partitioned shared cache could achieve better performance to support image processing analysis, with smaller hardware overhead and power consumption. It is also expected that the result could be used to assist pathologists to diagnose TB slides in a shorter time.

2. Motivation and Background

Earlier studies have proposed to insert a new line either at the MRU position or the LRU position. That is, when the new line is inserted at the MRU position, frequent accesses of the line while it traverses towards LRU position can derive additional hits, but it will also waste cache capacity in the case of *dead-on-arrival lines*. The *dead-on-arrival lines* is defined as lines that are inserted and used only once in the cache and will never be reused at all after they get evicted. This is as opposed to the *zero-reuse* lines, which will not be reused between insertion and eviction, but may be reused again at a greater distance (after the lines have been evicted from the cache). Therefore, by comparison, insertion at the LRU position has shown substantial benefits on the zero-reuse lines as the lines can be evicted immediately after they are used [3][7]. However, on every cache hit, both strategies automatically move all the hit lines to the MRU position. This may be good for cache lines with multiple-reuse, but not for single-reuse lines or lines with many references at greater distance (particularly after these lines become dead). Therefore, it would be desirable if the cache spaces that these lines are occupying are used by other lines to generate additional hits.

The *pseudo insertion/promotion policy* (PIPP) [1] is one of the cache replacement techniques that have been augmented from the traditional LRU policy, particularly in the context of insertion and promotion policies. The main differences between PIPP and the LRU policy are: (1) PIPP does not always insert incoming cache lines at the highest priority position, and (2) upon a cache hit, PIPP will not automatically promote the hit line to the highest priority position. Note that the *utility-based cache partitioning* (UCP) scheme [6] is used by PIPP to determine the allocation of cache ways for competing applications in the system. However, PIPP does not explicitly partition the shared cache as UCP does. Instead it uses the cache allocation information to implicitly partition the cache. That is, the number of ways allocated to an application is used to specify the priority position for the application to install all new incoming lines. The sequence order of the insertion position employed by PIPP starts from the lowest priority position to the highest priority position. Upon cache hits, PIPP promotes hit lines by a single priority position towards the highest priority position with a certain probability, ensuring that the priority increment remains unchanged. The promotion probability is used to prevent streaming applications from hurting the performance of non-streaming applications that are running in the system, as the streaming accesses into the cache are very unlikely to be reused. Therefore, the value of the promotion probability of a streaming application is set to be much smaller (e.g. P_{stream}=1/64) than the promotion probability of a non-streaming application (e.g. P_{prom}=3/4), so that a better chance can be given to the lines of the non-streaming applications to reside longer in the cache. The value of the promotion probability can be any reasonable values as [1] has proven that a wide range changes to the probability values has resulted in small overall performance improvement to the system.

Figure 1 illustrates the operation of the insertion policy of PIPP, in which the UCP has allocated 5 ways to Core1 and 3 ways to Core2. Thus, all new lines of Core1 are inserted at priority position 5 while all new lines of Core2 are installed at position 3. In PIPP, evictions always choose the line at the lowest priority position. Therefore, on every cache miss, the miss-causing core can evict cache lines that belong to another core in the system. This allows cache capacity stealing among the cores, which may result in an inefficient strategy in maintaining the cache ways that have been allocated to the cores.

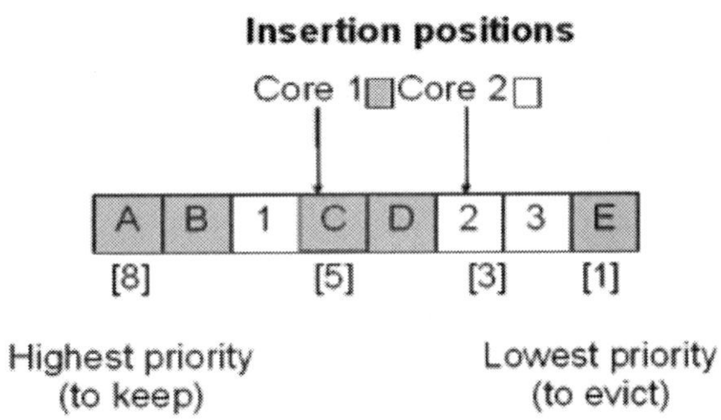

Figure 1. Example of insertion positions in PIPP. Consider Core 1 is allocated with 5 ways and Core 2 receives 3 ways from UCP scheme.

It is observed that the insertion policy of PIPP seems to treat zero-reuse lines more effectively than the LRU policy because the lifetime of the zero-reuse lines can be reduced due to a shorter distance to traverse towards the lowest priority position for eviction. Additionally, the single priority position promotion of PIPP enables the single-reuse lines and the lines with a burst of references to occupy the cache long enough before they become dead, but will quickly become a potential victim once they are ready to be replaced. However, PIPP's pseudo insertion/promotion policy does not explicitly enforce the cache partitioning and may lead to a large number of lines being taken away from any core(s) in the system to be used by another core(s). This is because PIPP only stimulates the system to create results similar to the hard/strict partition (i.e. the number of ways allocated to each core for every execution intervals will remain unchanged), so that the cache capacity stealing allowed by PIPP among all the applications may produce benefits in the utilization of the available cache resources. Hence, it is concluded that uncontrolled line stealing among cores by PIPP may severely degrade the system performance.

Therefore, this paper proposes algorithms that are intended to strictly partition the shared cache and explicitly enforce the original way-partitioning decisions. One may say that some of the cache capacity allocated to the cores may be under-utilized, but such a situation is not likely to happen if the cache is properly partitioned by the cache partitioning scheme. Moreover, if all components of the cache replacement policy are improved accordingly, the occupancy of the partitioned shared cache can be controlled thereby improving the overall performance of the system.

Since UCP is widely accepted, it is used in this paper with the implementation of an adaptive LRU replacement policy that has been adopted from the concept of PIPP. The adaptive insertion and promotion strategies of the new policy are augmented from the concept introduced by PIPP in order to maintain the benefits received by zero-reuse lines, single-reuse lines and multiple-reuse lines at a short distance. Additionally, the eviction policy of the traditional LRU policy will be improved so that it better satisfies the requirements of a multiprocessor system, such that the cache capacity stealing can be controlled among the cores.

3. Adaptive Insertion and Promotion Policy

In this section, the concept of adaptive insertion and promotion policy of the new replacement policy is presented. Then, an example to detail how the new policy works in a partitioned shared cache is provided.

3.1 Middle Insertion 2 Positions Promotion Policy (MI2PP)

The MI2PP policy is proposed with the objective of providing better thrashing resistance to a partitioned shared cache by maintaining the number of cache ways received by all cores equal to the number of ways that have been allocated originally. The policy is targeted to improve the insertion and promotion strategies of a cache replacement policy so that the cache occupancy of zero-reuse and single-reuse lines is not wasted, while multiple-reuse lines could be retained in the cache long enough to incur additional cache hits before eviction. In other words, the MI2PP policy tries to retain as many of the beneficial cache lines as possible, so that they could manifest more cache hits in the allocated cache space and minimize the residency of zero-reuse and single-reuse lines in the allocated space without yielding significant hardware overhead. Therefore, the MI2PP policy chooses to insert the new lines at the middle position of the priority stack based on the associativity of the shared cache and the number of cores in the system.

Thus:

$$InsertionPosition = \frac{SetAssociativity}{NumberOfCores} \tag{1}$$

The MI2PP insertion may not be as good as LRU insertion in minimizing the residence time of the zero-reuse lines in the cache. However, it could reduce the time spent by the zero-reuse lines by approximately half compared to those installed at the MRU position. MI2PP could also reduce the number of misses incurred due to dead-on-arrival lines. In addition, for the lines with single-reuse and multiple-reuse, they can be protected and have more chances for further hits while they traverse from the middle position to the LRU position. This is in contrast to the immediate disadvantage if they are inserted at the LRU position. However, if the next references are not in the near-future, LRU insertion will remain a better policy than MI2PP.

On a cache hit, MI2PP implements a different promotion policy from that of LRU. The hit line will automatically be moved by two priority positions towards the highest priority position. The reason for this change is that if the lines were automatically shifted to the MRU position, as in the LRU policy, single-reuse and multiple-reuse lines at a greater distance may consume cache capacity without providing any benefit. Hence, MI2PP will shorten the distance for the dead lines to traverse towards the LRU position. This means that the lines will have potential to be evicted sooner than in the conventional MRU promotion policy.

The MI2PP promotion strategy will also slow down the movement of the lines with multiple-reuse towards the MRU position if the next accesses of other cache lines happen at priority positions lower than the positions of the multiple-reuse lines, providing the multiple-reuse lines with more opportunity to derive additional hits along the way. Given that all new lines are installed at the middle priority position and get promoted by only two positions upon cache hits, this approach does not cause any disadvantage to the many frequent access lines. This is because these lines are preserved at priority positions which are high enough for them to be kept in the cache long enough to generate more hits before eviction.

Since the target of the MI2PP policy is to ensure that the number of cache ways received by all cores is equal to the number of ways they have been allocated, victim selection for the cache line replacement is modified by taking into account the total number of ways belonging to the miss-causing application/core. If the miss-causing core gets more ways than it is allocated at the time of the cache miss, the LRU line of the core will be selected as an evictee. Otherwise, the LRU line in the set that does not belong to the miss-causing core will be replaced. This technique is used to enforce the cache partition and helps to preserve the benefits of partitioning the cache by sustaining adequate cache resources among all the cores.

In summary, the goal of MI2PP policy was to provide an effective yet not so complex cache replacement policy that works well in a partitioned shared cache, since an optimal cache partitioning scheme used in the system may cost significant hardware overhead. Therefore, MI2PP policy uses information already available from the baseline LRU replacement policy and exploits the information to make better use of the partitioned shared cache resources to gain better overall system performance.

3.2 Implementation Example

Since MI2PP policy was adapted from PIPP, a simple example used in [1] is presented in Figure 2 to deliver a detailed explanation of MI2PP. Figure 2(a) illustrates a simple operation of PIPP, while Figure 2(b) demonstrates the operation of MI2PP policy on a dual-core system.

The two cores, namely core0 and core1 are allocated with five and three cache ways, respectively. To differentiate the lines occupied by the cores, the cache lines of core0 are represented by numerals in squares, while letters in black circles represent cache lines of core1. For both of the policies, the insertion position of each core is shown in the figure. For PIPP, when core1 makes a first request for line D which is not in the cache, the line is installed at position 3. Similarly with core0, when references to line ζ and η during the second and third requests incurred cache misses, both lines are inserted at position 5. Conversely, MI2PP always inserts all the new lines requested by any core at the same location, which is at position 4. When the next reference of line D (i.e., the fourth request) by core1 creates a hit, PIPP will automatically move the line by a single position. On the other hand, the line will be shifted slightly higher by MI2PP, which is by two priority positions.

One of the significant differences between the two policies that can be observed is in the context of cache capacity management. In some execution intervals, it can be seen that PIPP demonstrates cache allocation deviation. This is due to the highlighted fact that the policy does not explicitly enforce the cache partitions, but only pseudo-partitions the cache. On the contrary, MI2PP strictly enforces the cache partitions and always maintains the target allocation of 5 lines to core0 and 3 lines to core1 in every execution interval.

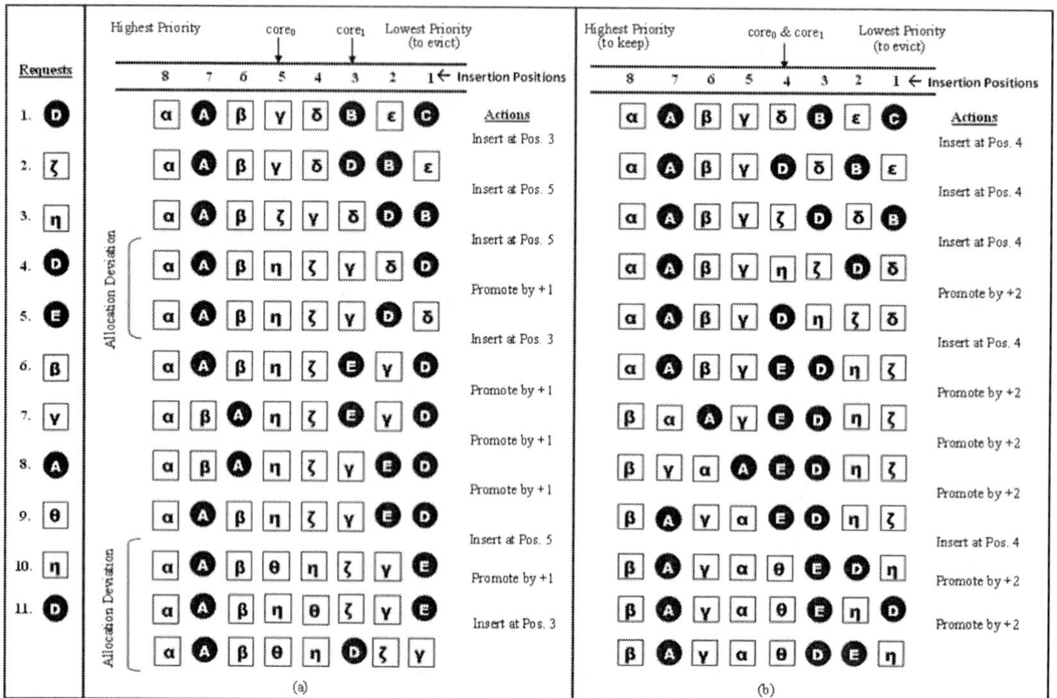

Figure 2. Example operation of (a) PIPP from Xie and Loh [2009], and (b) MI2PP policy for a variety of insertions and promotions activities. Assume evictions for both policies always choose the lowest-priority cache line in the set.

To better understand the similarities and differences between PIPP and MI2PP policy, this discussion is narrow down to the operation of both policies towards zero-reuse, single-reuse and multiple-reuse lines. Assume that line ζ that belongs to core0 does not exhibit any reuse. PIPP inserts the line at position 5, while MI2PP installs at position 4. After several hits and misses, MI2PP eliminates line ζ from the cache more quickly than PIPP. This shows the insertion at priority position 4 (at the middle of the priority stack) can reduce the lifetime of dead lines in the cache. However, if MI2PP is compared with the LRU insertion such as implemented in thread-aware dynamic insertion policy (TADIP) [3], MI2PP is not efficient in minimizing the age of the dead lines although it has been shown in Figure 2 to perform better than PIPP. This is because the LRU insertion in TADIP will immediately evict zero-reuse lines from the cache upon next cache access as all new incoming lines will be inserted at the LRU position.

Next, consider line η as a single-reuse line which is reused while it is still in the cache (i.e., the tenth request). MI2PP will promote the line by two priority positions, which is slightly higher than PIPP. However, the lower insertion at position 4 by MI2PP helps to push the line downwards to the end of the priority order sooner than PIPP. If line E (i.e., the fifth request) is also considered as a single-reuse line which is installed by PIPP at a lower priority position than MI2PP, a cache miss would be incurred in PIPP if the next reuse happens at the final interval. This is because the line is already replaced by another line, while MI2PP retains the line in the cache. Thus, the promotion by 2 positions in MI2PP is beneficial in keeping the single-reuse line long enough for the next reuse, while insertion at the middle position is an advantage to shorten the lifetime of the single-reuse line after the line has been reused.

Finally, after line D is brought into the cache, MI2PP keeps the multiple-reuse line long enough to be exposed to cache hits on every next reference. On the other hand, PIPP eliminates the line more quickly than MI2PP after the line has been first reused, which costs a cache miss upon second reference to the line. The treatment of line D illustrates how the promotion by two priority positions of MI2PP can be better than PIPP in maximising the amount of time for multiple-reuse lines residing in the cache. Overall, the example demonstrates that MI2PP can save more cache misses compared to PIPP.

4. Conclusion

The implementation of MI2PP policy will be done using Qsys System Integration Tool. Then, the multiprocessor system will be evaluated by using the Altera DE1-SoC development board that is equipped with the Cyclone V FPGA with ARM Cortex-A9. For all of the MI2PP evaluations, the performance metrics that will be used to quantify the performance of the system are the *IPC throughput*, *weighted speedup* which indicates reduction in execution time, and the *harmonic mean* of weighted speedup which accounts for and balances both fairness and performance of the system [5][12].

Let IPC_i denote the IPC of the ith application, $SingleIPC_i$ the stand alone IPC of the same application if it is executed in isolation, and N the number of processes the system executes concurrently. The formulae for the aforementioned metrics are:

$$\text{IPC throughput} = \sum_{i=0}^{N} IPC_i \tag{2}$$

$$\text{Weighted speedup} = \sum_{i=0}^{N} \frac{IPC_i}{SingleIPC_i} \tag{3}$$

$$\text{Harmonic mean} = \frac{N}{\sum_{i=0}^{N} \left(SingleIPC_i / IPC_i\right)} \tag{4}$$

It is expected that the new adaptive dynamic cache replacement policy for a partitioned shared cache in system will increase the hit rate of the shared cache and the instructions per cycle executed by the multiple processors in the system. Hence, the replacement policy could support the image processing analysis that required massive memory utilization, as well as reducing hardware overhead and saving power consumed by the shared cache. In future, more review will be done to further strengthen this research. Quantitative analysis of different techniques and policies used to achieve better performance of the multiprocessors also will be done.

Acknowledgements

The authors wish to thanks Universiti Malaysia Perlis for funding this research project under the Short Term Grant Scheme (Grant No. 9001-00568).

References

1. Y. Xie and G.H. Loh. (2009). PIPP: promotion/insertion pseudo-partitioning of multi-core shared caches. *ACM SIGARCH Computer Architecture News, 37(3),* pp. 174-183.

2. C. Li, D. Wang, Y. Xue, H. Wang and X. Zhang, "Enhanced adaptive insertion policy for shared caches," in *Advanced Parallel Processing Technologies*, 2011, pp. 16-30.
3. A. Jaleel, W. Hasenplaugh, M. Qureshi, J. Sebot, S. Steely Jr and J. Emer, "Adaptive insertion policies for managing shared caches," in *Proceedings of the 17th International Conference on Parallel Architectures and Compilation Techniques*, 2008, pp. 208-219.
4. M. K. Qureshi, A. Jaleel, Y. N. Patt, S. C. Steely and J. Emer. (2007). Adaptive insertion policies for high performance caching. *ACM SIGARCH Computer Architecture News, 35(2),* pp. 381-391.
5. K. Luo, J. Gummaraju and M. Franklin, "Balancing throughput and fairness in SMT processors," in *IEEE International Symposium on Performance Analysis of Systems and Software (ISPASS),* 2001, pp. 164-171.
6. M. K. Qureshi and Y. N. Patt, "Utility-based cache partitioning: A low-overhead, high-performance, runtime mechanism to partition shared caches," in *Proceedings of the 39th Annual IEEE/ACM International Symposium on Microarchitecture,* 2006, pp. 423-432.
7. V. Mekkat, A. Holey, P. C. Yew and A. Zhai, "Managing shared last-level cache in a heterogeneous multicore processor," in *Proceedings of the 22nd International Conference on Parallel Architectures and Compilation Techniques*, Edinburgh, 2013, pp. 225-234.
8. Cong Thuan Do, Hong-Jun Choi, Jong Myon Kim, Cheol Hong Kim. A new cache replacement algorithm for last-level caches by exploiting tag-distance correlation of cache lines. *Microprocessors and Microsystems, Volume 39, Issue 4,* 2015, pp. 286-295.
9. Terasic Technologies Inc. (2016, August 24). DE1-SoC User Manual. Terasic Technologies Inc.
10. K. Veropoulos, C. Campbell, Learmonth, G., B. Knight, and J.Simpson (1998), "The Automated Identification of Tubercle Bacilliusing Image Processing and Neural Computing Techniques," in *Proceeding of the 8th International Conference on Artificial Neural Networks*, Skövde, Sweden, pp. 797-802.
11. J. Wu, X. Sui, Y. Tang, X. Zhu, J. Wang and G. Chen, "Cache management with partitioning-aware eviction and thread-aware insertion/promotion policy." in *International Symposium on Parallel and Distributed Processing with Applications (ISPA),* 2010, pp. 374-381.
12. A. Jaleel, H. H. Najaf Abadi, S. Subramaniam, S. C. Steely and J. Emer. (2012, March). "CRUISE: cache replacement and utility aware scheduling," in *ACM SIGARCH Computer Architecture News* (Vol. 40, No. 1, pp. 249-260).
13. G. Jia, G. Han, J. Jiang and L. Liu, "Dynamic Adaptive Replacement Policy in Shared Last-Level Cache of DRAM/PCM Hybrid Memory for Big Data Storage," in *IEEE Transactions on Industrial Informatics*, vol. 13, no. 4, pp. 1951-1960, Aug. 2017.

IOP Conf. Series: Materials Science and Engineering **318** (2018) 012049 doi:10.1088/1757-899X/318/1/012049

Gas Source Localization via Behaviour Based Mobile Robot and Weighted Arithmetic Mean

Ahmad Shakaff Ali Yeon[1], Kamarulzaman Kamarudin[1,2], Retnam Visvanathan[1], Syed Muhammad Mamduh Syed Zakaria[1], Ammar Zakaria[1,2], Latifah Munirah Kamarudin[1]
[1]CEASTech, Universiti Malaysia Perlis (UniMAP), Malaysia
[2]School of Mechatronic, Universiti Malaysia Perlis (UniMAP), Malaysia

E-mail: kamarulzaman@unimap.edu.my

Abstract. This work is concerned with the localization of gas source in dynamic indoor environment using a single mobile robot system. Algorithms such as Braitenberg, Zig-Zag and the combination of the two were implemented on the mobile robot as gas plume searching and tracing behaviours. To calculate the gas source location, a weighted arithmetic mean strategy was used. All experiments were done on an experimental testbed consisting of a large gas sensor array (LGSA) to monitor real-time gas concentration within the testbed. Ethanol gas was released within the testbed and the source location was marked using a pattern that can be tracked by a pattern tracking system. A pattern template was also mounted on the mobile robot to track the trajectory of the mobile robot. Measurements taken by the mobile robot and the LGSA were then compared to verify the experiments. A combined total of 36.5 hours of real time experimental runs were done and the typical results from such experiments were presented in this paper. From the results, we obtained gas source localization errors between 0.4m to 1.2m from the real source location.

1. Introduction

Work on mobile olfaction began by mimicking the behaviour of organisms based on the studies of chemotaxis on bacteria being attracted and repelled by chemicals. Combining our understanding of chemotaxis with the recent advancements in mobile robotics, the world now have a mature and popular research field which consists of using mobile robots and gas sensors to find, trace and localize chemical gas plumes. While the early works on mobile olfaction only began in the early 1990s, it was believed that Larcombe and Halsall [1] was the first time where chemically sensitive robot devices were seriously being discussed to be deployed in the hazardous nuclear industry's environment. In general, mobile olfaction can be divided into three equally important sub problems that need to be solved which are gas plume searching, gas plume tracing and gas source declaration.

1.1. Gas Plume Searching

Commonly used technique or algorithm for gas plume searching is by searching upwind [2-5] because gasses will always travel in the air along with the wind in the same direction. The mobile robot might still miss the gas plume as it travels upwind if the implemented searching technique is not able to cover a large dynamic environment. There are also other factors that are detrimental to the success of solving this step such as gas sensor position, gas sensor sensitivity, gas sensor response time and others.

In an indoor environment or where the physical state of the space is known, wind information is less likely needed. A carefully planned searching algorithm can manage to traverse through the whole area without problems [6] but the time to cover the whole area might differ based on the mobile robot's speed and the size of the area.

1.2. Gas Plume Tracing

After finding the gas plume, the mobile robot will need to solve the second problem of mobile olfaction which is gas plume tracing or in other words, move reactively along or within the gas plume. The most common technique in plume tracing is by taking measurements of the gas concentration at different points and calculates its differences to gain information about the gas plume gradient which then can be exploited to navigate the mobile robot to the gas source.

1.3. Gas Source Declaration

The definition of gas source declaration varies from research group to research group. Gas source declaration is the process of determining the certainty that the gas source is in the immediate vicinity thus ending the robot's run [7].

Up till now, there are only a few techniques to solve the gas source declaration problem and most of them involved the authors of past works to declare an area surrounding the known gas source location as the goal which means that if the mobile robot enters the area, it is count as a successful experimental run. Other techniques are machine learning, neural network algorithms and gas distribution mapping.

1.4. Problem Statement

Earlier works focused on making the mobile robots to arrive within a certain radius from the actual gas source location. This method while valid is biased with the assumption that the mobile robot knows that it has arrived near the actual gas source. Furthermore, artificial neural network (ANN) [7, 27-28] was utilized to make the mobile robot recognize that it has found the actual gas source but this technique requires extensive development time for training.

Recent advancements in solving the gas source localization problem are more concentrated in developing gas distribution map building techniques [32-34]. In general, the mobile robot will construct a map of the environment and overlaying it with gas distribution map. The downside of gas distribution mapping is that the approach is complex and requires high computing power for analysis and computation [35]. Researchers often neglect autonomous path planning of the mobile robot as a compromise.

Finally, there is no standard external system or method to verify the gas sensor measurements from the mobile robot. Researchers have been using cameras to verify mobile robot's trajectory, pose, and odometry but very few have tried to verify the gas sensor measurements taken by the mobile robot.

In this work, a large gas sensor array was utilized to verify the gas sensor measurements taken by the mobile robot. This was done by comparing the mobile robot's measurement with the measurement take by the sensor arrays [30]. A combination of low level algorithms was also implemented on the mobile robot to navigate the mobile robot autonomously while the mobile robot transmits its data to a central workstation for calculating the gas source location. These steps were taken to keep the mobile robot a low monetary and low computation cost and its data verifiable while also able to perform gas source localization tasks autonomously.

2. Related Works

2.1. Chemotaxis

Chemotaxis is a strategy used by researchers to solve the gas source localization problem by exploiting gas concentration measurements from gas sensors.

Rozas algorithm [9] has been the kickstarter for chemotaxis. It emphasized on gradient following strategy by taking multiple spatial measurements at different times. After taking the gas sensor measurements, the robot will turn to the direction of the sensor with highest gradient of chemical concentration then move forward into the direction for m meters.

Pre-programmed moveset like the Rozas were implemented in a few other works. Kuwana [10] worked on mimicking the Silkworm Moth behaviour to move the robot towards a gas source. Pheromone sensors were used and have been found to be fast degrading compared to chemical sensors. The work was improved at the cost of sensor's longevity [11]. The mobile robot was successful in mimicking the Silkworm Moth's zig-zag movements towards the gas source. Kazadi [12] improved the works of Kuwana with polymer sensors which are sensitive to wind. By using polymer sensors, the mobile robot managed to turn faster towards the gas source.

Furthermore, a popular method of navigation strategy which is the Braitenberg strategy. A pure Braitenberg vehicle has two sensors sufficiently separated and placed opposite of each other with each of them controlling the wheel on the same side or cross-coupling to control the wheel on the opposite side [13]. Sandini [14] executed the strategy with his mobile robot. From the work, the mobile robot showed reactive potential towards gas concentration. The Braitenberg has been studied quite thoroughly. A reactive Braitenberg or in other words, continuous sampling of gas concentration has been studied and was found to be far more successful [15]. Two Braitenberg strategies (Attract and Repel) were also studied and comparison work has been done [16]. Repel strategy often avoid the gas source and the avoided areas can be deemed as the gas source location.

Moreover, other biologically inspired algorithms have been utilized in mobile olfaction. A popular one is the biased random walk of E.coli [17-19]. Cleverly planned navigation strategy such as Sweeping [6], Hex-path [18], Spiral [19] and Surge [20] were also implemented in past works which can be utilized to solve the navigation part of gas source localization problem.

2.2. Gas Source Declaration

The common technique used to declare a gas source is by manually declaring an area surrounding the actual gas source location as the gas source. When a part of the mobile robot reaches said area, it usually counted as a 'hit' and the experiment deemed successful [11,19,21-23].

Buscemi [24] developed a heuristic approach based on gas sensor saturation for gas source declaration. Then, Grasso and Atema [25] tried a similar approach with the addition of a threshold over a sampling period. Due to the volatile nature of the gasses, Cowen [26] proposed another heuristic method comparing concentration measurements at different heights to help solving the gas source declaration in a case of gasses that are lighter than air and was verified [5]. In addition to heuristics, machine learning, support vector machine and neural network approaches were also taken by researchers to tackle this problem [7, 27-28]. Although these works showed positive results in declaring the gas source location, neural network algorithms and machine learning requires extensive training period.

Particle filter based odour source declaration algorithm was also implemented using mobile robot [20]. The usage of the particle filter algorithm was extended to gas source declaration using multiple static gas sensor nodes via wireless sensor network [29].

Gas distribution mapping technique by combining gas sensor data with location estimations can also be classified as a gas source declaration technique by some research groups where the declaration is performed by the researcher by interpreting the map generated by the robots [8].

3. System Overview

Figure 1 shows the overall architecture of the system for solving gas source localization problem. Within the system, there are three smaller subsystems namely the integrated mobile robot, large gas sensor array (LGSA) for data verification [30], and the robot tracking system [31] working simultaneously to gather data and transmit the data through WSN and intranet to a remotely located base station PC for processing.

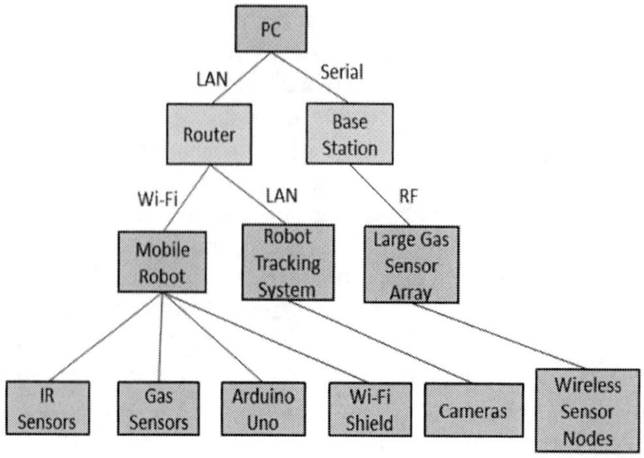

Figure 1. System block diagram of the integrated system.

3.1. Mobile Robot

Arduino Mobile Robot was chosen to be the mobile robot platform for this work. The Arduino robot is a very low-cost consumer grade hobby kit which is suitable for educational purposes. It consist two processors and two microcontrollers with one each on its motor board and control board. It is easily programmable and highly modifiable but limited amount of digital and analog I/O pins available on the two boards. The robot was modified to meet our standards to perform gas source localization task and programmed to move its wheels based on certain conditions which will be explained further in Section 4. Figure 2 shows the mobile robot system used throughout this work.

Wi-Fi shield is needed to establish communication with the base station wirelessly. An additional microcontroller which is the Arduino Uno is mounted and connected to the mobile robot to accommodate a Wi-Fi shield. The data collected are raw voltages of the gas sensors in form of 10-bit data. Conversion of the raw voltage values from digital, V_d to analog raw voltage values, V_a is shown in Equation (1) where V_{ref} is the reference voltage of the board. For Arduino Uno and Arduino robot, V_{ref} is 5V. The analog digital converter of the Arduino Uno is similar to the analog digital converter of the Arduino Robot.

$$V_a = (V_d / 1024) \times V_{ref} \quad (1)$$

3.2. Gas Sensor

Two Figaro TGS 2600 MOX gas sensor were mounted on the robot with one on each side with a 18cm of separation between each other. Two electrical circuit boards consisting of a single 10kΩ resistor, RL each were made to mount the sensors and connect it to the robot's and the Arduino UNO's digital I/O channels. 10-bit data of raw voltages, V_d were obtained and converted into analog values, V_a as shown in Equation (1). From the analog raw voltage, the resistance of the gas sensor, R_s can be calculated as shown in Equation (2).

$$R_s = [(V_{ref} \times R_L) / V_a] - R_L \quad (2)$$

From Equation (2), sensor signal value s can be obtained as shown in Equation (3)

$$s = \Delta R / R_0 = [(R_0 - R_s) / R_0] = 1 - (R_s / R_0) \quad (3)$$

where R_0 is the baseline resistance taken at the start of an experiment. The sensor reading before gas is released is taken to be the baseline reading, R_0. The sensor signal values were used throughout this work mainly to control the mobile robot's motor in the Braitenberg algorithm and for the generation of gas distribution map of the testbed for verification.

Figure 2. The modified Arduino Mobile Robot platform with a pattern template on top for robot tracking purposes.

4. Algorithms

4.1. Zig-Zag

In this work, the the traditional Zig-Zag was modified and will primarily act as the searching technique to cover the area within the testbed with the aid of on board compass and IR sensors. The mobile robot will perform the programmed evasive manoeuvre when the IR sensors detect an obstacle which in this case is the testbed's wall.

```
Zig-Zag Algorithm
Initialize parameters
Wait 90 seconds
while(robot power on)
        record sensors measurement
        set direction
        read compass value
        if(obstacles detected)
                evasive manuever
        else if(compass value > direction)
                set speedLeft >  speedRight
        else if(compass value < direction)
                set speedLeft < speedRight
        else if(compass value =direction)
                set speedLeft = speedRight
        end
end
```

Figure 3. Pseudocode of the Zig-Zag algorithm.

IOP Conf. Series: Materials Science and Engineering **318** (2018) 012049 doi:10.1088/1757-899X/318/1/012049

4.2. Braitenberg

Variety of Braitenberg strategy has been used by past works to perform olfaction research with mobile robots. In this work, the 'attract' and 'repel' behaviour were implemented. During the 'attract' behaviour, the motor will turn to the side of the higher gas sensor measurement while the opposite happen during the 'repel' behaviour.

```
Braitenberg Algorithm
Initialize parameters
wait 90 seconds
record initial sensor measurement
calculate initial sensor resistance, R_0l, R_0r
while(robot power on)
    record sensors measurement
    calculate current sensor resistance, R_sl,R_sr
    calculate sensor signal, s_l,s_r
    if(obstacles detected)
        evasive maneuver
    else
        calculate speedRight
        calculate speedLeft
        set speedRight and speedLeft
    end
end
```

Figure 4. Pseudocode of the Braitenberg algorithm.

4.3. B+Z

Both the Braitenberg and Zig-Zag algorithm were then combined into one to form a multi-phase algorithm which consists of a searching phase utilizing the Zig-Zag algorithm and a tracing phase using the Braitenberg algorithm.

The algorithm works with by executing one of the two algorithms within certain conditions. The mobile robot will always calculate the values of sensor signals s_l and s_r while always searching for obstacles with its IR sensors.

4.4. Gas Source Declaration Strategy

For gas source declaration (GSD), we would like to distance ourselves from the traditional way of declaring an experimental run as successful or a failure depending on the ability of the mobile robot to get into the near vicinity of the known actual gas source location or also known as 'hits'. Strategies such as machine learning and neural network were utilized in past works but it proved to be too costly in terms of memory and computing power to be implemented on our Arduino robot system.

In this work, weighted arithmetic mean is proposed. The centroid of the gas plume position were calculated by exploiting the information gained throughout the experimental run such as the sensor voltages and sensor signal values by applying them to act as weights to the robot's position.

The simplified mathematical definition can be explained as follows. A non-empty set of data consists of x-axis and y-axis coordinates denoted as $\{x_1,x_2,x_3,\ldots,x_n\}$ and $\{y_1,y_2,y_3,\ldots,y_n\}$. A set of sensor measurements taken at the particular coordinate such as sensor voltages or signal values act as weights can be denoted as $\{w_1,w_2,w_3,\ldots,w_n\}$. Equation (4) and (5) denotes the calculation of the weighted mean where xp and yp is the predicted x-axis and y-axis position of the gas source.

$$x_p = (x_1w_1 + x_2w_2 + \cdots) / (w_1 + w_2 + \cdots) \quad (4)$$

$$y_p = (y_1w_1 + y_2w_2 + \cdots) / (w_1 + w_2 + \cdots) \quad (5)$$

In this work, the variables used as weight are the raw gas sensors voltages and sensor signals. All the variations are as follows:

1. Type A: GSD using raw voltage measurement.
2. Type B: GSD using all sensor signal values.
3. Type C: GSD using all positive sensor signal values.
4. Type D: GSD using all sensor signal values over 0.3

Note that the GSD performed is not in real-time. The predicted gas source location was calculated immediately after the experimental run was completed using MATLAB on the base station PC as all recorded data from the LGSA, mobile robot tracking system and the gas sensor measurements from the mobile robot were sent to and stored in the base station PC.

```
B+Z Algorithm
Initialize parameters
wait 90 seconds
record initial sensor measurement
calculate initial sensor resistance, R0l, R0r
while(robot power on)
    record sensors measurement
    read compass value
    calculate current sensor resistance, Rsl,Rsr
    calculate sensor signal, sl,sr
    if(obstacles detected)
        evasive manuever
    else if(Rsl/R0l ≤0.3 and Rsr/R0r≤0.3)
        set direction
        if(compass value > direction)
            set speedLeft > speedRight
        else if(compass value < direction)
            set speedLeft < speedRight
        else if(compass value =direction)
            set speedLeft = speedRight
        end
    else if(Rsl/R0l >0.3 or Rsr/R0r>0.3)
        calculate speedRight
        calculate speedLeft
        set speedRight and speedLeft
    end
end
```

Figure 5. Pseudocode of the B+Z algorithm.

5. Experimental Setup

The setups for the experiments are as follows:

1) Robot starting position at $(x,y) = (30,550)$.
2) Ethanol gas released position at $(x,y) = (150,220)$, at the height of 20cm from the ground and at actual robot start time.
3) Number of runs for each algorithm:
    a. Zig-Zag: 5 runs.
    b. Braitenberg (Attract & Repel): 10 runs.
    c. B+Z (Attract & Repel): 20 runs.
5) Stop conditions:
    a. 10 minutes of run time.

b. Mobile robot reached the other end of the testbed after completing either searching or tracing phase.

Before the mobile robot perform its task, it will first sit through a wait time period of 90 seconds where the robot is turned on but not doing anything of significance. This is so that the gas sensors on board the mobile robot have enough time to heat up and normalize its measurements.

6. Experiments & Results

The results for GSD errors for all types are presented in Table 1. Examples of the robot's trajectory when implementing the Zig-Zag, Braitenberg and B+Z algorithms are illustrated in Figure 6 to Figure 8.

Table 1. Average GSD Error for all types of algorithm.

Algorithm	Average GSD Error (*m*)			
	A	B	C	D
Zig-Zag	0.68	0.78	0.75	1.05
Braitenberg (Attract)	0.74	0.57	0.39	0.79
Braitenberg (Repel)	0.63	0.62	0.62	0.72
B+Z (Attract)	0.82	1.22	0.65	0.68
B+Z (Repel)	0.82	0.91	0.51	0.66

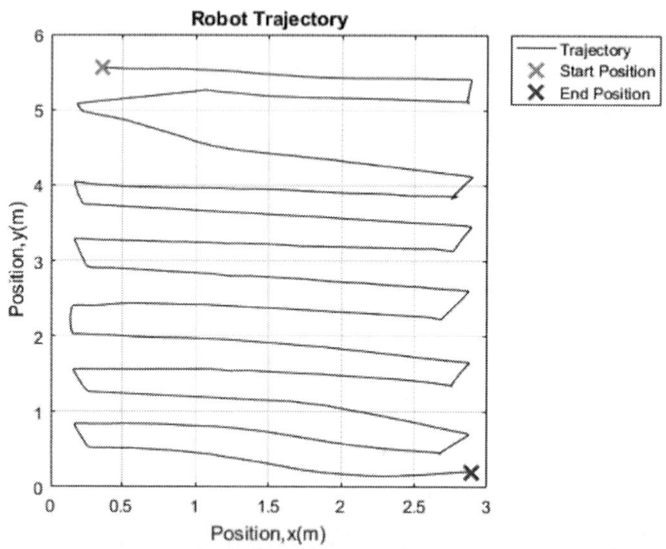

Figure 6. The mobile robot's trajectories after implementing the Zig-Zag algorithm Run #3.

From Table 1, it was found that localization errors by calculating the source location using the sensor's raw voltage data (Type A) are very reasonable ranging from 0.63m to 0.82m. The errors are worse when calculating the source location by using raw sensor signal values (Type B). It is believed that the baseline sensor resistance value R_0 taken at the beginning of every run to calculate the sensor signal values is the main reason for the worse results as it regularly dropped to negative values of s. Apart from circuitry noises skewing the values, a longer wait time for the sensor to normalize is needed. By removing only the negative values (Type C), the errors were reduced greatly.

Moreover, from the results, it can be seen that by calculating the gas source location by only using $s \geq 0.3$ (Type D), or in other words, taking only the higher end of the measurements, the calculated source location got worse. This confirms the idea that the areas of high gas concentration values are not always at the source or near its vicinity [16] due to the turbulent nature of the gasses in the environment.

(a) (b)

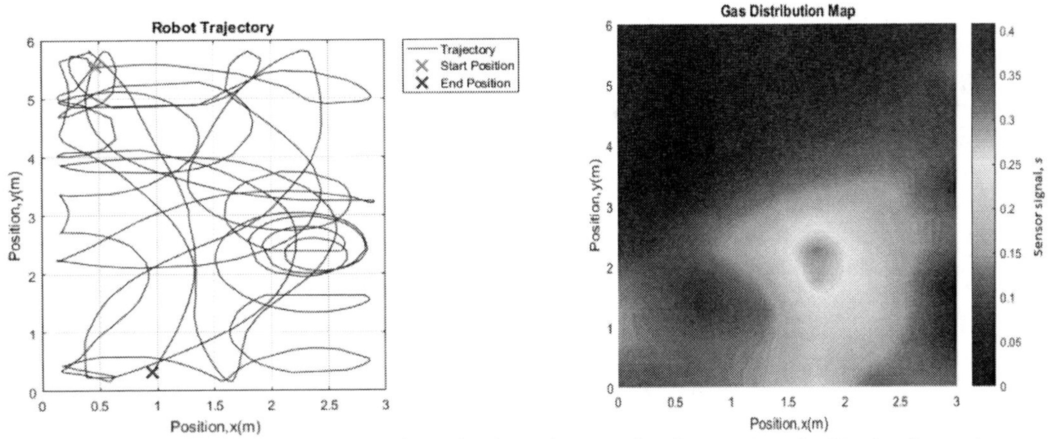

Figure 7. The figure (a) illustrates the robot's trajectory implementing the Braitenberg Attract algorithm Run #6 while figure (b) is the average gas concentration map for the run.

(a) (b)

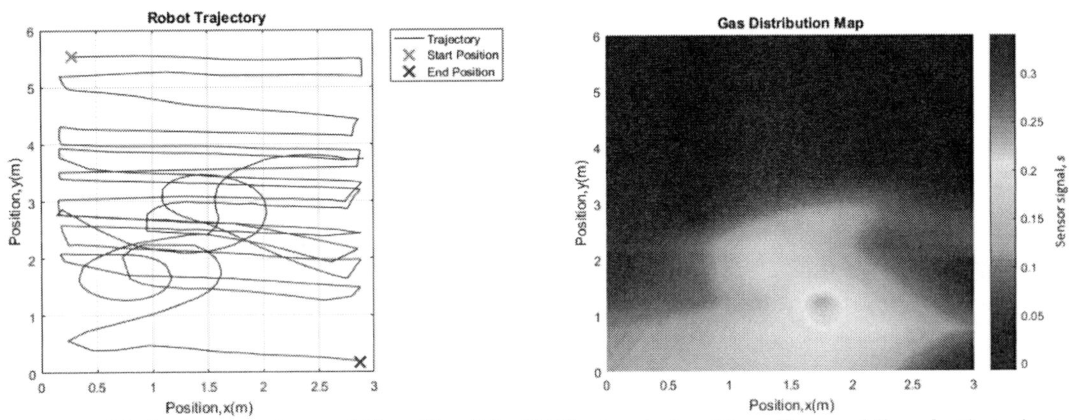

Figure 8. This is the experimental Run #7 of the B+Z attract algorithm. The mobile robot's trajectory (a) after completing a B+Z (Attract) algorithm clearly depicts the combination of the Zig-Zag and the Braitenberg. While (b) shows the average gas concentration map for B+Z (Attract) Run #7.

7. Conclusion & Future Work

A weighted arithmetic mean to calculate and predict the gas source location was successfully implemented. From the results, it was found that even with the rigid Zig-Zag moveset, a reasonable average error of 0.68m from the actual gas source location can be achieved. The error can be further lowered by the Braitenberg and the B+Z due to its tracing capability. The lowest average error (0.39m) achieved in this research is by the Braitenberg (Attract) but at the cost of a longer run time due to continuous roaming of the mobile robot while performing the Braitenberg. A balance of time and error is possible to achieve by the B+Z algorithms.

To be a fully independent system, the external mobile tracking system needs to be removed. A GPS may be integrated with the mobile robot to give precise positioning information. A mobile robot platform with more memory is also desirable so that the robot can process the information itself and self-calculate the gas source location thus removing the need of a base PC.

References

[1] Larcombe, M.H.E., 1984. Robotics in nuclear engineering: Computer assisted teleoperation in hazardous environments with particular reference to radiation fields.

[2] Kazadi, S., Goodman, R., Tsikata, D., Green, D. and Lin, H., 2000. An autonomous water vapor plume tracking robot using passive resistive polymer sensors. *Autonomous robots*, *9*(2), pp.175-188.

[3] Russell, R.A., Thiel, D., Deveza, R. and Mackay-Sim, A., 1995, May. A robotic system to locate hazardous chemical leaks. In *Robotics and Automation, 1995. Proceedings., 1995 IEEE International Conference on* (Vol. 1, pp. 556-561). IEEE.

[4] Ishida, H., Nakayama, G., Nakamoto, T. and Moriizumi, T., 2005. Controlling a gas/odor plume-tracking robot based on transient responses of gas sensors. *IEEE Sensors Journal*, *5*(3), pp.537-545.

[5] Marques, L., Almeida, N. and De Almeida, A.T., 2003, October. Olfactory sensory system for odour-plume tracking and localization. In *Sensors, 2003. Proceedings of IEEE* (Vol. 1, pp. 418-423). IEEE.

[6] Neumann, P.P., Hernandez Bennetts, V., Lilienthal, A.J., Bartholmai, M. and Schiller, J.H., 2013. Gas source localization with a micro-drone using bio-inspired and particle filter-based algorithms. *Advanced Robotics*, *27*(9), pp.725-738.

[7] Lilienthal, A., Ulmer, H., Frohlich, H.A.A.S., Stutzle, A., Werner, F. and Zell, A., 2004, April. Gas source declaration with a mobile robot. In *Robotics and Automation, 2004. Proceedings. ICRA'04. 2004 IEEE International Conference on* (Vol. 2, pp. 1430-1435). IEEE.

[8] Cabrita, G. and Marques, L., 2013, June. Divergence-based odor source declaration. In *Control Conference (ASCC), 2013 9th Asian* (pp. 1-6). IEEE.

[9] Rozas, R., Morales, J. and Vega, D., 1991, June. Artificial smell detection for robotic navigation. In *Advanced Robotics, 1991.'Robots in Unstructured Environments', 91 ICAR., Fifth International Conference on* (pp. 1730-1733). IEEE.

[10] Kuwana, Y., Shimoyama, I. and Miura, H., 1995, August. Steering control of a mobile robot using insect antennae. In *Intelligent Robots and Systems 95.'Human Robot Interaction and Cooperative Robots', Proceedings. 1995 IEEE/RSJ International Conference on* (Vol. 2, pp. 530-535). IEEE.

[11] Kuwana, Y., Nagasawa, S., Shimoyama, I. and Kanzaki, R., 1999. Synthesis of the pheromone-oriented behaviour of silkworm moths by a mobile robot with moth antennae as pheromone sensors. *Biosensors and Bioelectronics*, *14*(2), pp.195-202.

[12] Kazadi, S., Goodman, R., Tsikata, D., Green, D. and Lin, H., 2000. An autonomous water vapor plume tracking robot using passive resistive polymer sensors. *Autonomous robots*, *9*(2), pp.175-188.

[13] Braitenberg, V., 1986. *Vehicles: Experiments in synthetic psychology*. MIT press.

[14] Sandini, G., Lucarini, G. and Varoli, M., 1993, July. Gradient driven self-organizing systems. In *Intelligent Robots and Systems' 93, IROS'93. Proceedings of the 1993 IEEE/RSJ International Conference on* (Vol. 1, pp. 429-432). IEEE.

[15] Farah, A.M. and Duckett, T., 2002. Reactive localisation of an odour source by a learning mobile robot. In *In Proceedings of the Second Swedish Workshop on Autonomous Robotics*.

[16] Lilienthal, A. and Duckett, T., 2003. Experimental analysis of smelling Braitenberg vehicles. *environment*, *5*(10).

[17] Marques, L., Nunes, U. and de Almeida, A.T., 2002. Olfaction-based mobile robot navigation. *Thin solid films*, *418*(1), pp.51-58.

[18] Russell, R.A., 2003. Chemical source location and the robomole project. In *Proceedings Australian Conference on Robotics and Automation*.

[19] Ferri, G., Caselli, E., Mattoli, V., Mondini, A., Mazzolai, B. and Dario, P., 2009. SPIRAL: A novel biologically-inspired algorithm for gas/odor source localization in an indoor environment with no strong airflow. *Robotics and Autonomous Systems*, *57*(4), pp.393-402.

[20] Li, J.G., Meng, Q.H., Wang, Y. and Zeng, M., 2011. Odor source localization using a mobile robot in outdoor airflow environments with a particle filter algorithm. *Autonomous Robots*, *30*(3), pp.281-292.
[21] Lilienthal, A. and Duckett, T., 2003. Approaches to gas source tracing and declaration by pure chemo-tropotaxis. In *Autonome Mobile Systeme 2003* (pp. 161-171). Springer, Berlin, Heidelberg.
[22] Villarreal, B.L., Olague, G. and Gordillo, J.L., 2016. Synthesis of odor tracking algorithms with genetic programming. *Neurocomputing*, *175*, pp.1019-1032.
[23] Ferri, G., Caselli, E., Mattoli, V., Mondini, A., Mazzolai, B. and Dario, P., 2006, February. A biologically-inspired algorithm implemented on a new highly flexible multi-agent platform for gas source localization. In *Biomedical Robotics and Biomechatronics, 2006. BioRob 2006. The First IEEE/RAS-EMBS International Conference on* (pp. 573-578). IEEE.
[24] Buscemi, L., Prati, M. and Sandini, G., 1994. Cellular robotics: behaviour in polluted environments.
[25] Grasso, F.W. and Atema, J., 2002. Integration of flow and chemical sensing for guidance of autonomous marine robots in turbulent flows. *Environmental Fluid Mechanics*, *2*(1), pp.95-114.
[26] Cowen, E.A. and Ward, K.B., 2002. Chemical plume tracing. *Environmental Fluid Mechanics*, *2*(1), pp.1-7.
[27] Duckett, T., Axelsson, M. and Saffiotti, A., 2001. Learning to locate an odour source with a mobile robot. In *Robotics and Automation, 2001. Proceedings 2001 ICRA. IEEE International Conference on*(Vol. 4, pp. 4017-4022). IEEE.
[28] Weissburg, M.J., Dusenbery, D.B., Ishida, H., Janata, J., Keller, T., Roberts, P.J.W. and Webster, D.R., 2002. A multidisciplinary study of spatial and temporal scales containing information in turbulent chemical plume tracking. *Environmental Fluid Mechanics*, *2*(1-2), pp.65-94.
[29] Li, Q., Liu, Z. and Xiao, X., 2015. A gas source localization algorithm based on particle filter in wireless sensor network. *International Journal of Distributed Sensor Networks*, *11*(11), p.874532.
[30] Syed Zakaria, S.M.M., Visvanathan, R., Kamarudin, K., Ali Yeon, A.S., Md Shakaff, A.Y., Zakaria, A. and Kamarudin, L.M., 2015. Development of a Scalable Testbed for Mobile Olfaction Verification. *Sensors*, *15*(12), pp.30894-30912.
[31] Visvanathan, R., Mamduh, S.M., Kamarudin, K., Yeon, A.S.A., Zakaria, A., Shakaff, A.Y.M., Kamarudin, L.M. and Saad, F.S.A., 2015, November. Mobile robot localization system using multiple ceiling mounted cameras. In *SENSORS, 2015 IEEE* (pp. 1-4). IEEE.
[32] Ishida, H., Wada, Y. and Matsukura, H., 2012. Chemical sensing in robotic applications: A review. *IEEE Sensors Journal*, *12*(11), pp.3163-3173.
[33] Hernandez Bennetts, V., Schaffernicht, E., Pomareda, V., Lilienthal, A.J., Marco, S. and Trincavelli, M., 2014. Combining non selective gas sensors on a mobile robot for identification and mapping of multiple chemical compounds. *Sensors*, *14*(9), pp.17331-17352.
[34] Turduev, M., Cabrita, G., Kırtay, M., Gazi, V. and Marques, L., 2014. Experimental studies on chemical concentration map building by a multi-robot system using bio-inspired algorithms. *Autonomous agents and multi-agent systems*, *28*(1), pp.72-100.
[35] Lochmatter, T., 2010. Bio-inspired and probabilistic algorithms for distributed odor source localization using mobile robots. *EPFL Distributed Intelligent Systems and Algorithms Laboratory, Lausanne.*

IOP Conf. Series: Materials Science and Engineering **318** (2018) 012050 doi:10.1088/1757-899X/318/1/012050

Wearable Inset-Fed FR4 Microstrip Patch Antenna Design

S. R. Mohd Zaini and K. N. Abdul Rani*
Bioelectromagnetics Research Group (BioEM), Department of Electronics Engineering Technology, Faculty of Engineering Technology, Universiti Malaysia Perlis (UniMAP)

E-mail: khairulnajmy@unimap.edu.my

Abstract. This project proposes the design of a wireless body area network (WBAN) microstrip patch antenna covered by the jeans fabric as the outer layer operating at the center frequency, f_c of 2.40 GHz. Precisely, the microstrip patch antenna with the inset-fed edge technique is designed and simulated systematically by using the Keysight Advanced Design System (ADS) software where the FR4 board with the dielectric constant, ε_r of 4.70, dissipation factor or loss tangent, tan δ of 0.02 and height, h of 1.60 mm is the chosen dielectric substrate. The wearable microstrip patch antenna design is then fabricated using the FR4 printed circuit board (PCB) material, hidden inside the jeans fabric, and attached to clothing, such as a jacket accordingly. Simulation and fabrication measurement results show that the designed microstrip patch antenna characteristics can be applied significantly within the industrial, scientific, and medical (ISM) radio band, which is at f_c = 2.40 GHz.

1. Introduction

In a communications system, antenna is the most important part, which operates as a transducer for sending and receiving the electromagnetic waves. In this case, antenna becomes as an electrical device or conductive element, which helps convert the radio frequency (RF) or microwave to electrical power and vice versa. Recently, there is a rapid growth of wearable antenna development. Wearable antennas are becoming more and more lightweight that can be integrated into or hidden inside clothing to improve communication links and support a so-called wireless body area network (WBAN) application [1-2].

The WBAN antenna can be used as clothing used for communication purposes, which includes tracking and navigation, mobile computing, public safety and wireless communication [3]. Moreover, the wearable WBAN antennas need to be integrated within everyday clothing, be low profile, light weight, small size and be hidden as much as possible [4]. Several frequency bands have been assigned for WBAN systems, which are the Medical Implant Communication System band (MICS: 400 MHz), the Industrial Scientific Medical band (ISM: 2.40 GHz and 5.80 GHz), and the Ultra-wideband (UWB: 3.00-10.00 GHz), respectively [5].

This project aims to design a wearable microstrip patch antenna design. The main requirement is to design and fabricate the microstrip patch antenna using the FR4 substrate material covered by the jeans fabric as the outer layer. The FR4 microstrip patch antenna is hidden inside the jeans because of its small size and lightweight.

In order to design a microstrip patch antenna, some of basic requirements need to be considered, such as radiation, bandwidth and also the parameter of patch antenna [6]. There are two major problems of the microstrip patch antenna, which are narrow bandwidth and low efficiency [7]. Narrow bandwidth means having limited bandwidth and difficulties to capture the signals whereas low efficiency means that the designed antenna cannot operate properly. Antenna size and ground partial location is related to the operating bandwidth and antenna efficiency. The larger antenna and ground partial on substrate will help to improve higher operating bandwidth and higher antenna efficiency.

To overcome the narrow bandwidth constraint, the antenna is designed with the inset-fed edge technique at the operating bandwidth greater than 200 megahertz (MHz). The inset-fed microstrip patch antenna can capture more signals with the improved accuracy. In order to solve for the low efficiency limitation, the antenna is designed and fabricated with the aim of producing both simulated and measured return loss, S_{11} lower than -10.000 decibel (dB) ideally to obtain a good performance [2].

2. Antenna Structure and Design

In this project, a microstrip patch antenna is chosen because it has some advantages compared to other microwave antennas, which includes lightweight, thin profile configuration, cheap cost of fabrication, and can operate at a broad range of 100 MHz to 100 GHz frequency. Normally, the microstrip antenna consists of a conductive material, which radiates and absorbs electromagnetic wave, dielectric substrate, and ground plane. The conductive material and ground plane are basically made up of a thin copper, which is a good material to conduct electrical energy. The conductive material is embedded on the intermediate layer to be exposed to free air. The performance of the conductive material in converting electromagnetic wave to electrical energy and vice versa depends directly on feeding method, shape and dimension (e.g. width, w and length, l) applied [8]. Figure 1 shows the methodology applied in designing and fabricating the wearable microstrip patch antenna in this study.

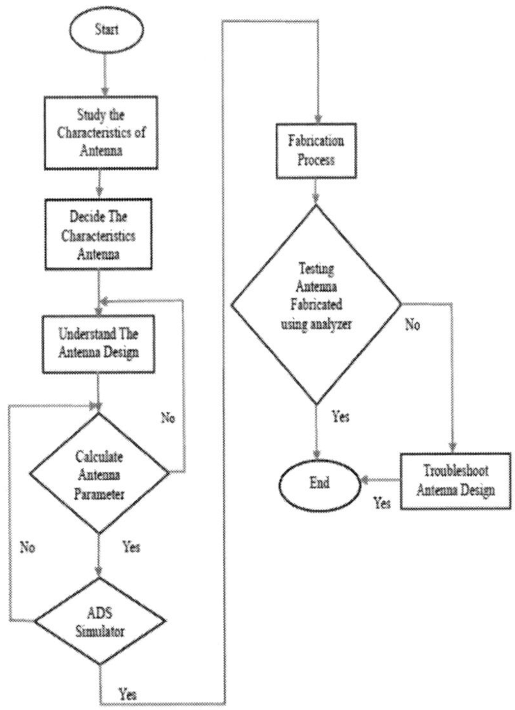

Figure 1. Methodology Flow Chart

IOP Conf. Series: Materials Science and Engineering **318** (2018) 012050 doi:10.1088/1757-899X/318/1/012050

2.1 Patch Antenna Design

2.1.1 Calculation of Microstrip Patch Antenna

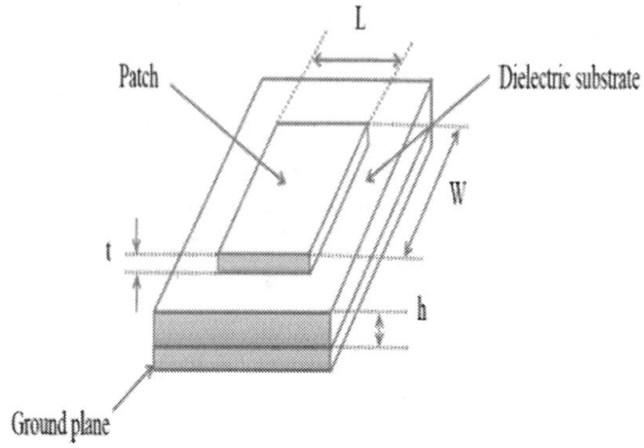

Figure 2. Microstrip Patch Antenna Structure

Figure 2 shows the rectangular microstrip patch antenna structure consists of conductive, substrate and ground planes. The microstrip patch antenna with the inset-fed design depends upon three critical parameters, which are the center frequency, f_c or resonant frequency, f_r, dielectric constant, $\mathcal{E}_r$, and height of dielectric substrate, *h*, respectively [9]. The f_r selected for this design is 2.40 GHz, which is in the ISM band. The dielectric material used is the FR4 board with the $\mathcal{E}_r$ of 4.70, dissipation factor or loss tangent, tan δ of 0.02 and *h* of 1.60 mm, respectively. The width of the patch, *W* is calculated using (1):

$$W = \frac{c}{2f_r\left(\sqrt{\frac{\varepsilon_r + 1}{2}}\right)} \tag{1}$$

According to (1), *c* is the speed of light where the increment of ε_r will decrease the size of the antenna patch. The actual length of the patch is computed as in (2):

$$L = \frac{c}{2f_r\sqrt{\varepsilon_{reff}}} - 2\Delta L \tag{2}$$

Based on (2), the effective dielectric constant, ε_{reff} is defined as in (3):

$$\varepsilon_{reff} = \left(\frac{\varepsilon_r + 1}{2}\right) + \left(\frac{\varepsilon_r - 1}{2}\right) + \left(1 + \left(\frac{12h}{W}\right)\right)^{-\frac{1}{2}} \tag{3}$$

Moreover, the length extension, ΔL is calculated as shown in (4) below:

$$\Delta L = 0.412h\left[\frac{(\varepsilon_{reff} + 0.3)}{(\varepsilon_{reff} - 0.258)}\right]\left[\frac{\left(\frac{W}{h} + 0.264\right)}{\left(\frac{W}{h} + 0.8\right)}\right] \tag{4}$$

2.1.2 Calculation of Microstrip Line Feed

Then, the microstrip synthesis, *H* with characteristic impedance, Z_0 equals to 50 Ω is generated using (5):

$$H = \left[\frac{Z_0\sqrt{2(\varepsilon_r + 1)}}{119.9}\right] + \frac{1}{2}\left[\frac{\varepsilon_r - 1}{\varepsilon_r + 1}\right]\left[\ln\left(\frac{\pi}{2}\right) + \frac{1}{\varepsilon_r}\ln\left(\frac{4}{\pi}\right)\right] \tag{5}$$

Based on (5), the width of microstrip line feed, W_f is computed as in the following (6):

$$W_f = \left[\left(\frac{e^H}{8} - \frac{1}{4e^H}\right)^{-1}\right] \times 1.60 \text{ mm} \tag{6}$$

Moreover, the length of microstrip line feed, L_f is obtained through (7) below:

$$L_f = \theta \times \frac{\lambda_g}{360°} \tag{7}$$

where,

$$\lambda_g = \frac{c}{f\sqrt{\varepsilon_{reff}}} \tag{8}$$

2.1.3 Calculation of the Antenna Ground Dimension

The length of the ground plane is calculated using the following (9):

$$L_g = L + 6h \tag{9}$$

Moreover, the width of the ground plane is computed as shown below (10):

$$W_g = W + 6h \tag{10}$$

After finding the microstrip patch antenna, line feed and ground dimension values through calculations using (1) - (10), the design of the wearable microstrip patch antenna schematic and layout is designed and modeled using the Keysight Advanced Design System (ADS) software. Deploying the ADS built-in controllers, optimization of some key parameters are also performed to obtain the optimum simulated antenna performance measurements.

2.1.4 Calculation of the Inset-Fed Line

This is the last step of design process, which is to calculate the parameter of inset-fed line. It assumes that the input impedance is 50.000 Ω, the width of microstrip feed line, W_f, the length of inset, F_i and the gap between the patch and the inset-fed line, G_{pf}, respectively. Figure 3 shows the designed inset-fed microstrip patch antenna parameters, which are recalculated after some troubleshooting processes using the ADS software to achieve the resonant at f_c = 2.40 GHz and simulated return loss, S_{11} less than -10.000 decibel (dB), respectively. Figure 4 displays the layout of the wearable microstrip patch antenna using the inset-fed edge technique.

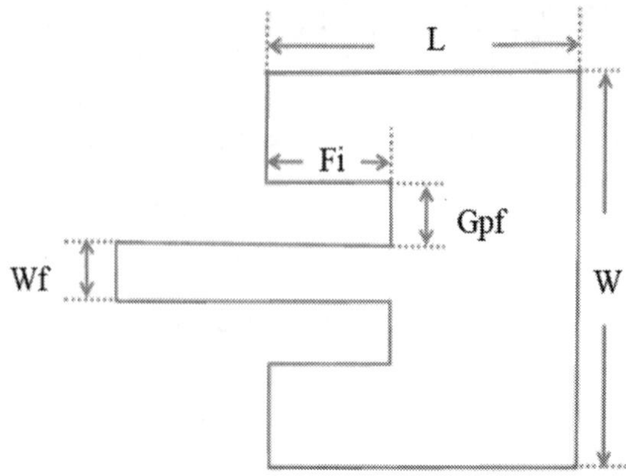

Figure 3. Inset-Fed Line Parameters

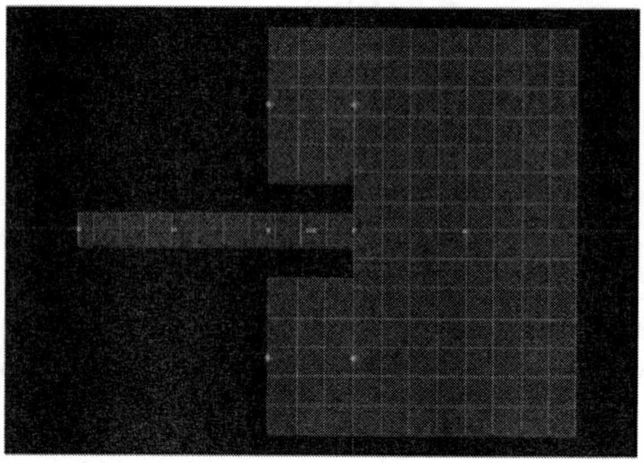

Figure 4. Edge Inset-Feeding Technique Layout

In this project, the wearable microstrip patch is inset-fed at an edge. Edge inset-feeding technique on a microstrip patch has few advantages compared to other direct contact and non-contact microstrip feeding technique. One of the advantages is power distribution scheme, which is easy to be fabricated. In this design, the inset-fed network and radiating patches will be etched on the same board. Table 1 shows the list of microstrip patch antenna parameters before and after undergoing optimization using the ADS software. Fig. 5 shows the schematic of the microstrip patch antenna with the optimal parameters of microstrip line (MLIN), microstrip line open-circuited stub (MLOC), and microstrip asymmetric coupled line (MACLIN), respectively. The MACLIN is added in the design to create the inset-fed line. In this case, width 1, W_1 and width 3, W_3 are equal and calculated based on the width of the substrate, W as in (11):

$$W_1 = W_3 = \frac{W - 9}{2} \tag{11}$$

On the other hand, the width 2, W_2 is equal to the width of microstrip line feed, W_f as in (6). Moreover, S_1 and S_2 are the gaps of the inset-fed where both are equal to W_f.

Table 1. Optimal Antenna Parameter

Parameter	Non-Optimal Value	Optimal Value
Length of Patch Antenna, L	30.00 mm	21.00 mm
Width of Patch Antenna, W	37.00 mm	37.00 mm
Length of Line Feed, L_f	18.00 mm	18.21 mm
Width of Line Feed, W_f	3.00 mm	3.00 mm

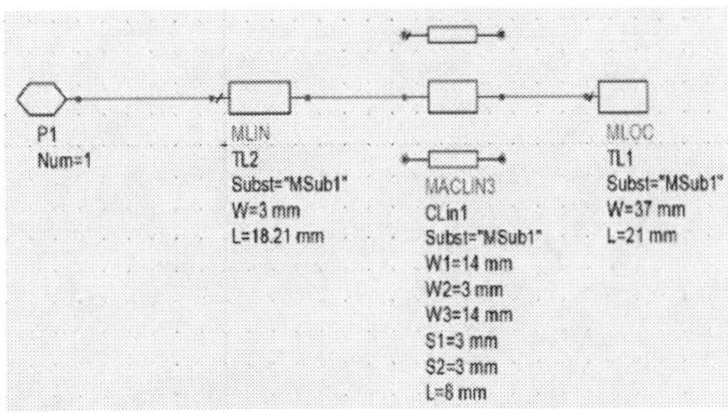

Figure 5. Microstrip Patch Antenna Design in Schematic

Figure 6. Fabricated Inset-Fed Microstrip Patch Antenna

Figure 6 shows the fabricated microstrip patch antenna after being soldered with a 50.000 Ω-impedance subminiature version A (SMA) probe connector. The purpose is to connect the antenna with a vector network analyzer (VNA) via a radio frequency (RF) coaxial cable for performance measurements. Figure 7 (a) and (b) below display the fabricated microstrip patch antenna is covered by the jeans fabric and then is attached to the clothing, such as a jacket. In this case, the inset-fed microstrip patch antenna design is considered as the wearable antenna that can support various WBAN applications operating at f_c = 2.40 GHz.

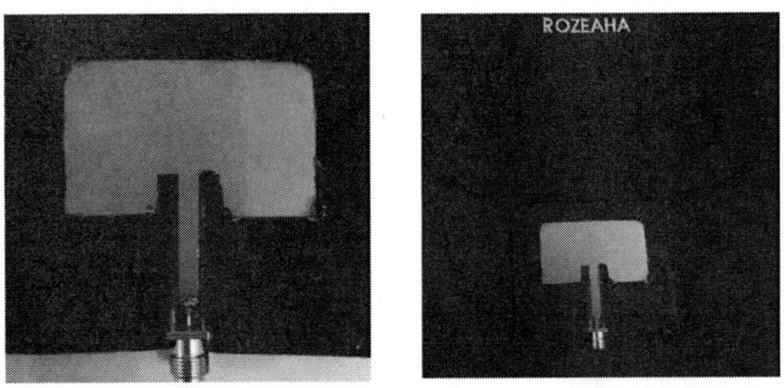

(a) (b)

Figure 7. (a) FR4 Covered with the Jeans Fabric (b) Microstrip Antenna Attached on Clothing

3. Result and Discussion

In this section, both simulation and actual measurements of the inset-fed microstrip patch antenna performance in the free space is discussed. The simulation is performed by using the Keysight ADS software whereas the measurement is prepared using the Agilent E5071C Vector Network Analyzer (VNA). The antenna performance is simulated and measured precisely in terms of return loss (S_{11}), impedance matching, three-dimensional (3D) radiation pattern, and gain.

S_{11} in an antenna is a parameter that states the total of power that is lost to the load and does not return as a reflection. The designed antenna should ideally achieve the return loss, S_{11} lower than -10 dB to be considered as a good performance [2]. Figure 8 depicts the simulated S_{11} magnitude in dB of the designed inset-fed microstrip FR4 patch antenna. It is clearly shows that the microstrip patch antenna located on the free space has the S_{11} parameter of -15.284 dB at the f_c = 2.40 GHz.

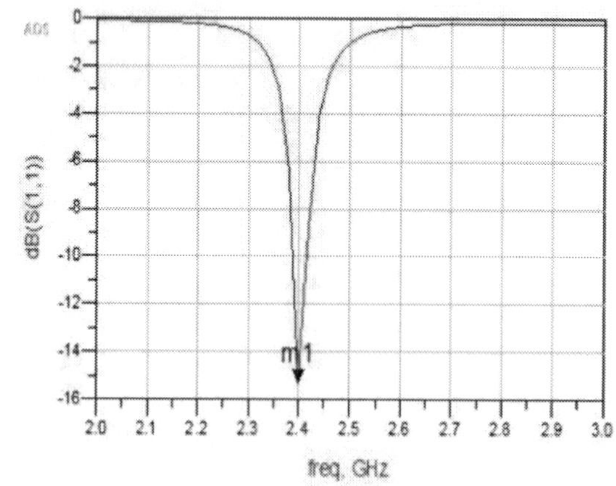

Figure 8. Simulated Return Loss, S_{11}

Figure 9 below shows the S_{11} measurement on the fabricated FR4 microstrip patch antenna, which is equal to -30.601 dB at the shifted f_r of 2.33 GHz.

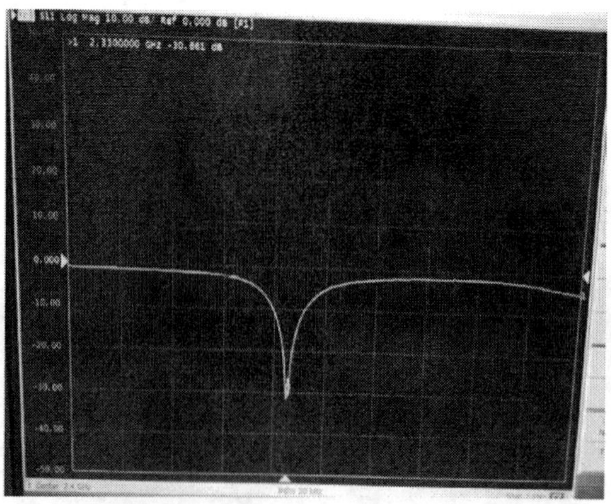

Figure 9. Measured Return Loss, S_{11}

Impedance matching is important in antenna design where the input impedance, Z_{in} of an antenna should be practically near to the characteristic impedance, Z_0 of 50.000 Ω in this study. If Z_{in} does not match with Z_0, the signal will be reflected back to the amplifier, hence, will not be radiated by the antenna. By using a marker function, the normalized impedance simulation using the ADS software is equal to 1.000 - j0.349 Ω as shown in Figure 10. So, multiply by 50 the impedance simulation magnitude is equal to 52.958 Ω. Figure 11 shows the Smith chart where the measured impedance matching using the Agilent E5071C VNA is equal to 49.815 Ω at the frequency of 2.33 GHz. In general, the magnitude impedance matching measurement is also acceptable and almost equal to the simulation, respectively.

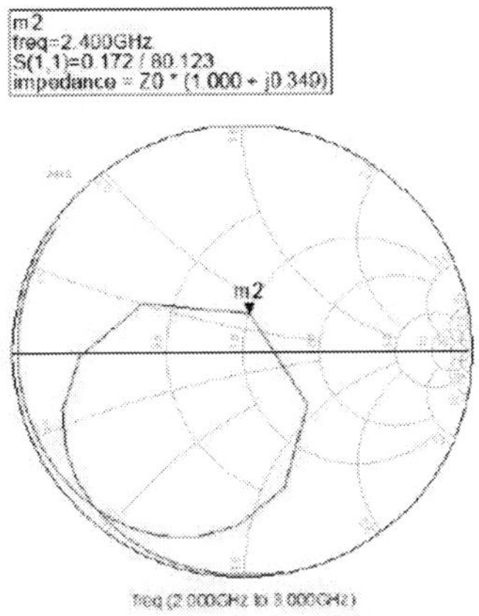

Figure 10. Simulated Impedance Matching

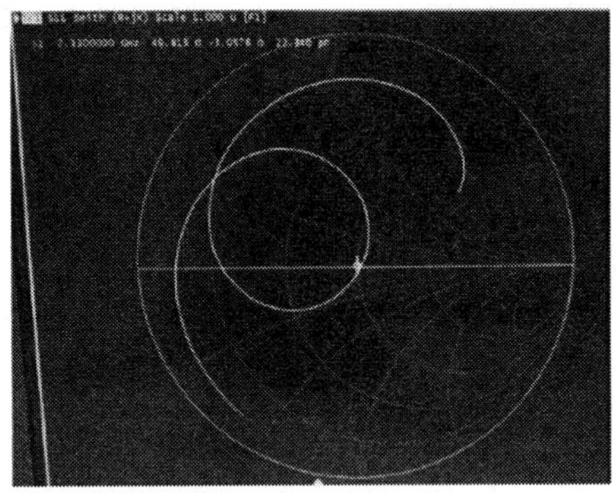

Figure 11. Measured Impedance Matching

Gain, directivity and radiated power are also antenna performance indicators. Figure 11 shows the gain, directivity, efficiency, and power radiated simulated measurements versus frequency. Precisely, the designed FR4 microstrip antenna has the simulated gain of 5.20905 decibel isotropic (dBi), directivity of 6.28795 dBi, input power of 2.42596 milliwatt (mW), and radiated power of 1.89231 mW at f_c = 2.40 GHz, respectively. In this case, the directivity of the designed microstrip antenna is theoretically in the normal range, which is between 5.00000 dBi and 8.00000 dBi. The expression dBi basically defines the gain of an antenna system relative to an isotropic radiator at the ISM radio band. Moreover, the antenna has acceptable radiation efficiency. The radiation efficiency is based the percentage of the radiated power-to-input power ratio, which is 78%. In summary, all the simulated and actual measurements indicate that the microstrip patch antenna has a good and adequate performance.

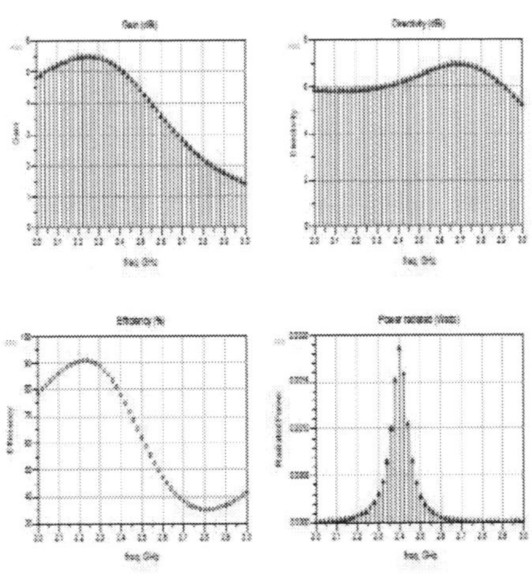

Figure 12. Antenna Performance Parameter vs. Frequency

4. Conclusions

In sum, the designed FR4 microstrip patch antenna covered with the jeans fabric as the outer layer and then attached to clothing, such as a jacket proven can have a satisfactory performance operating at the ISM band, which is f_c = 2.40 GHz. However, there is a deviation between simulated and actual S_{11} measurements where the f_r shifts from the desired 2.40 GHz to 2.33 GHz. This might be due to the manual antenna fabrication process deficiency, SMA connector intermittent, coaxial cable loss, and radiated power loss to surrounding, respectively. However, the actual S_{11} measurement, which is -30.601 dB is far below than -10.000 dB threshold. Similarly, the actual impedance matching magnitude deviates from the simulated impedance matching magnitude from 52.958 Ω to 49.815 Ω. The result for actual impedance matching magnitude is still acceptable, which is near to the predefined characteristic impedance, Z_0 of 50.000 Ω. Furthermore, substrate relative permittivity, substrate thickness, and conductive conductivity are critical in affecting the dimension values of microstrip patch and inset-fed line.

For future work, different substrates as the dielectric materials and fabrics as the outer layers can be possibly deployed and compared in terms of antenna radiation performance. Furthermore, this design can also be extended where many microstrip patch antennas will be positioned in certain geometry configuration of array. Besides, the design can also be applied with sensor to construct a portable Internet of Things (IoT) sensing device or rectifier circuit to develop a rectenna, which can convert harvested radio frequency (RF) energy into the direct current (DC).

Acknowledgement

This work is partially supported by the Universiti Malaysia Perlis (UniMAP) Short Term Grant (STG) scheme, under the grant 9001-00525. The authors would like to express their highest gratitude to UniMAP.

References

1. S. Park & S. Jayaraman, "Wearable Biomedical Systems: Research to Reality", IEEE International Conference (2007).
2. C. Parmar & S. Joshi, "Wearable Textile Patch Antenna for On-Body Communications", Global Conference on Communications Technology (2015).
3. N. Singh, et al., "Design & Performance of Wearable Ultra Wide Band Textile Antenna for Medical Applications", Open Engineering (2015).
4. M. I. Ahmed, et al., "Investigation and Comparison of 2.4 GHz Wearable Antennas on Three Textile Substrates and Its Performance Characteristics", Open Journal of Antennas and Propagation (2017).
5. E. G. Lim, et al., "Compact Size of Textile Wearable Antenna" International Multiconferences of Engineering and Computer Sciences (2014).
6. S. Sankaralingam & B. Gupta, "Effects of Bending on Impedance and Radiation Characteristics of Rectangular Wearable Antenna utilizing Smart Clothes", Microwave and Optical Technology Letters (2012).
7. A. Tsolis, et al., "Embroidery and Related Manufacturing Techniques for Wearable Antennas: Challenges and Opportunities," Electronics (2014).
8. J. Q. Howell, "Microstrip Antennas", International Symposium on Digital Antennas and Propagation Society (1972).
9. S. D. Murali, et al., "Development of Wearable Antennas with Different Cotton Textiles", International Journal of Engineering Research and Application (2014).

Design and Development of Multi-Transceiver Lorafi Board consisting LoRa and ESP8266-Wifi Communication Module

Noraini Azmi[1,2], Sukhairi Sudin[1], Latifah Munirah Kamarudin[1,2,*], Ammar Zakaria[1], Retnam Visvanathan[1], Goh Chew Cheik[1,2], Syed Muhammad Mamduh Syed Zakaria[1] and Khudhur Abdullah Alfarhan[1], R Badlishah Ahmad[3]
[1]Centre of Excellence for Advanced Sensor Technology (CEASTech), Malaysia.
[2]School of Computer and Communication Engineering, Universiti Malaysia Perlis (UniMAP), Perlis, Malaysia.
[3]Universiti Sultan Zainal Abidin (UniSZA), Terengganu, Malaysia.

E-mail: latifiahmunirah@unimap.edu.my

Abstract. The advancement of Micro-Electro-Mechanical-Systems (MEMS), microcontroller technologies and the idea of Internet of Things (IoT) motivates the development of wireless modules (e.g. WiFi, Bluetooth, Zigbee, and LoRa) that are small and affordable. This paper aims to provide detailed information on the development of the LoRaFi board. The LoRaFi 1.0 is a multi-protocol communication board developed by Centre of Excellence for Advanced Sensor Technology (CEASTech). The board was developed for but not limited to monitor the indoor air quality. The board comprises two different wireless communication modules namely, Long-range technology (LoRa) and WiFi (using ESP8266). The board can be configured to communicate either using LoRa or WiFi or both. The board has been tested and the wireless communication operates successfully. Apart from LoRa, WiFi enables data to be forwarded to the cloud/server where the data can be stored for further data analysis. This helps provide users with real-time information on their smartphones or other applications. In the future, researchers will conduct tests to investigate the communication link quality. Newer version with reduced board size and additional wireless communication module will be developed in the future as to increase board flexibility and widen the range of applications that can use the board.

1. Introduction

The increasing demand for better quality of life has inspired the academician and industrial communities known as Internet of Things (IoT). The possible interaction between living (e.g. human, livestock) and non-living things (e.g. cars, house, and building) through wireless network communication opens a door to the development of new and advance applications. By integrating sensing and actuation features to the developed device or system with artificial intelligent enable the creation of smart environments that is applicable for various application domains such as in healthcare, environment[1], structural monitoring, communication networking, and military applications. The advancement of Micro-Electro-Mechanical-Systems (MEMS) enable the miniaturization of mechanical and electro-mechanical component that motivates the development of microcontroller and wireless modules technologies that are small and affordable. Presently, there are a number of different wireless technologies available in the market which includes but not limited to WiFi, Bluetooth, Zigbee, GSM, GPRS and RFID. These wireless technologies differ from each other in terms of radio frequency band, modulation scheme, transmission power, transmission rate, and power consumption. Among the available wireless module, WiFi provides higher data rate and packet size compared to other wireless technologies such as Bluetooth and Zigbee. The popular frequency band often used by these wireless technologies is the free and unlicensed 2.4GHz band that is specifically allocated for Industrial, Scientific and Medical (ISM) applications. However, the increasing number of devices occupying the 2.4GHz ISM band leads to the increasing network traffic and spectrum congestion. The collisions and interference of the wireless data packet in this heterogeneous wireless environment reduce the Quality of Service (QoS) of each wireless technology.

Recently, the introduction of LoRa module provides an alternative for longer transmission range up to 22km (line of sight) and up to 2km (non-line of sight) according to test conducted by Libelium Incorporated[2]. Unlike WiFi, Zigbee and Bluetooth, LoRa module operate at other ISM frequency band which is in 433MHz, 868MHz or 915MHz. In Malaysia the allowed frequency band is the 433MHz and 915MHz. Studies have been conducted to understand the characteristics LoRa communication network behavior. This is because LoRa provides a long distance communication at low power consumption. Besides that, LoRa also operates at frequency other than the congested 2.4GHz band[3]. In this era of IoT, the increasing attention towards multi-protocol wireless communication board has led Centre of Excellence for Advanced Sensor Technology (CEASTech) to seize the opportunity to contribute to the society by developing a multi-protocol communication board named LoRaFi. The LoRaFi Board 1.0 as shown in figure 1 was developed by CEASTech. The board comprises of two different wireless technologies which is WiFi (using the ESP8266 module) and Long-range technology (LoRa). LoRa that operates in the lower frequency band of 433MHz, 868MHz and 915MHz captures developer interest as it able to serve as an alternative to the congested 2.4GHz band. Apart from being able to avoid the congested 2.4GHz spectrum, LoRa also capable of transmitting data over long distance that is approximately 22km[2]. On the other hand, the inclusion of WiFi on the board enables the collected data to be transmitted to the cloud. Data stored in the cloud can be utilized for further data analysis and prediction. In addition, the data also can be utilized and used as a real-time monitoring which can be displayed on smart phones.

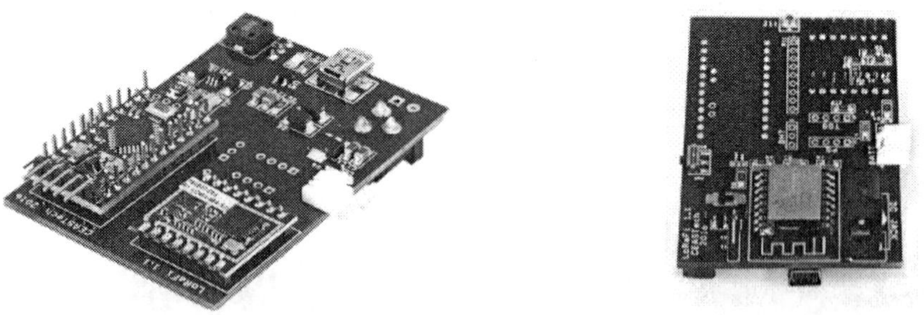

(a) Top view (b) Bottom view

Figure 1. The top and bottom view of LoRaFi board 1.0. LoRa module is at the top while WiFi is at the bottom.

This paper aims to provide detail information on the design and development of the LoRaFi board starting from the design concept until the fabrication of the final product. The rest of this paper was arranged as follows: Section 2 provides the LoRaFi development stage. Section 3 and 4 elaborates the LoRaFi board architecture and specification. Section 5 provides result obtained from simple test conducted to investigate the operation of the wireless communication. Section 6 describes the development of database and mobile application. Lastly, the future works is included in Section 7.

2. LoRaFi Design and Development

The summary of design and development of LoRaFi board is provided in the flow chart (refer figure 2). At first, block diagram of components required for the board from brainstorming session was roughly drawn. Next, the circuitry design was done using circuit drawing software (OrCAD PSpice). Upon successful compilation, the circuit was built on a prototyping board (breadboard). The board was tested per units (power, sensors, wireless module, and communication) and also as a complete board.

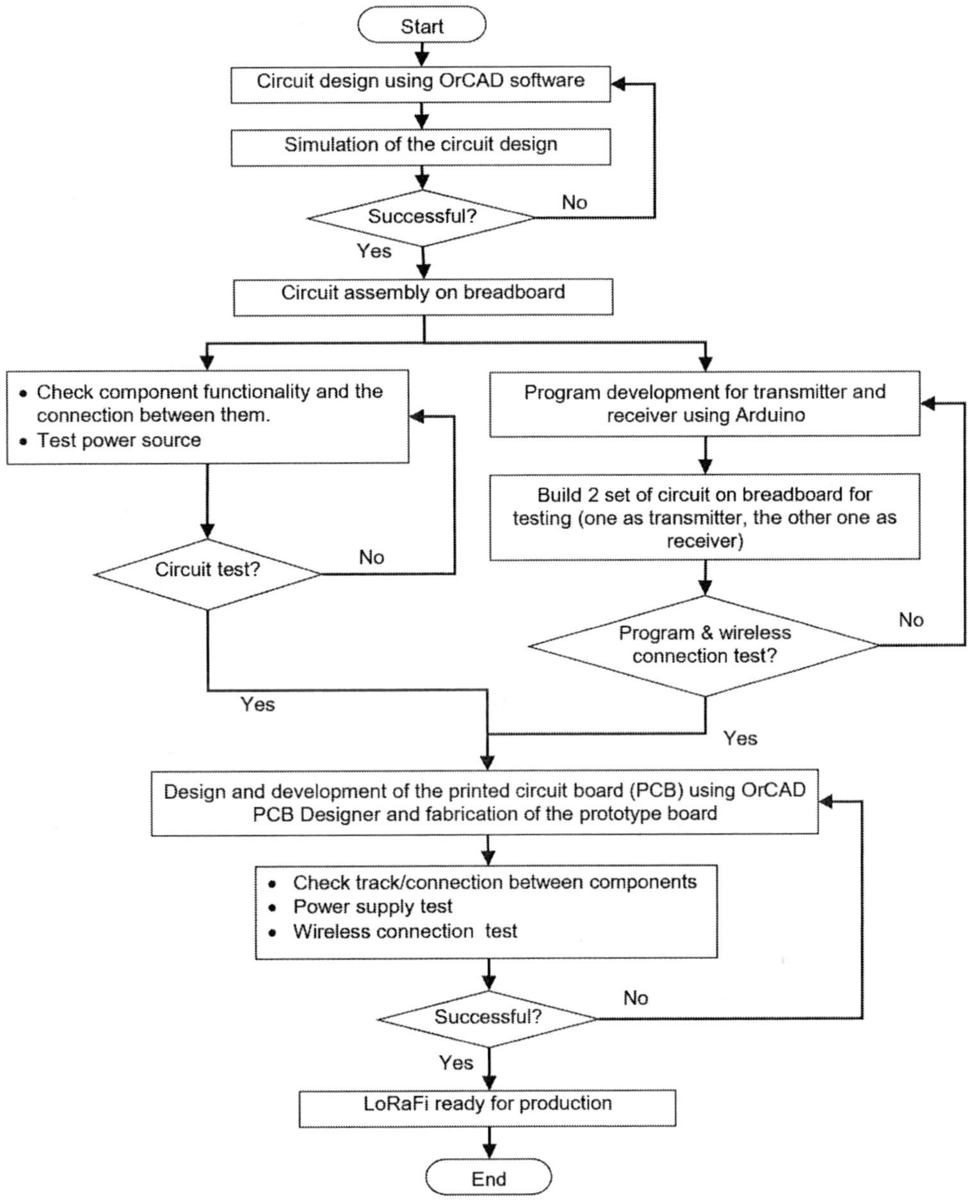

Figure 2. Flow chart of design and development of LoRaFi board.

When the board works successfully, the next step is to transform the circuitry from drawing to printed circuit board (PCB). There is a number of software available for PCB drawing such as OrCAD PCB Designer, Altium and others. In this research, OrCAD PCB Designer was chosen because it is software that is used by the etching department. Apart from that, there are various features that need to be noted such as board size/dimension, number of PCB layer, track size, through hole or surface mount component and others. Therefore, using the same software will help reduced the editing and synchronization time. The LoRaFi board was designed as two layers PCB with dimension of 5×7cm. The board contains combinations of through-hole and surface mount components. Therefore, components and tracks need to be carefully placed as to not overlap with each other.

3. LoRaFi Board Architecture

The LoRaFi board developed for but not limited to monitor the indoor air quality. The LoRaFi architecture shown in figure 3 can be divided into several parts that are interconnected and related to each other. The LoRaFi architecture includes Central Processing Unit (CPU), wireless communication module, communication with computer or other microcontroller, power source, and sensors and actuators. The components on the LoRaFi board is shown and labelled as in figure 4.

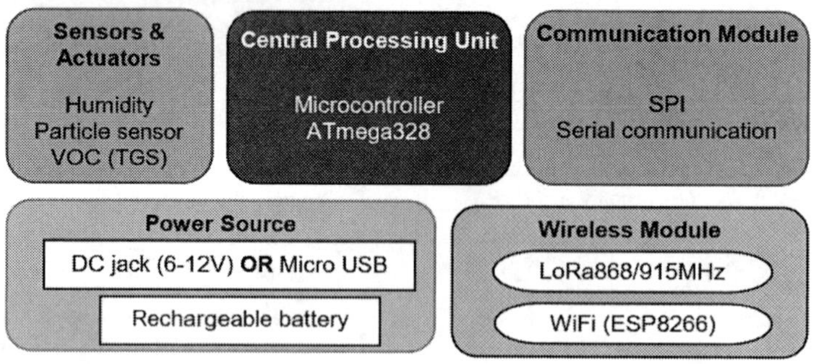

Figure 3. LoRaFi board design concept or architecture.

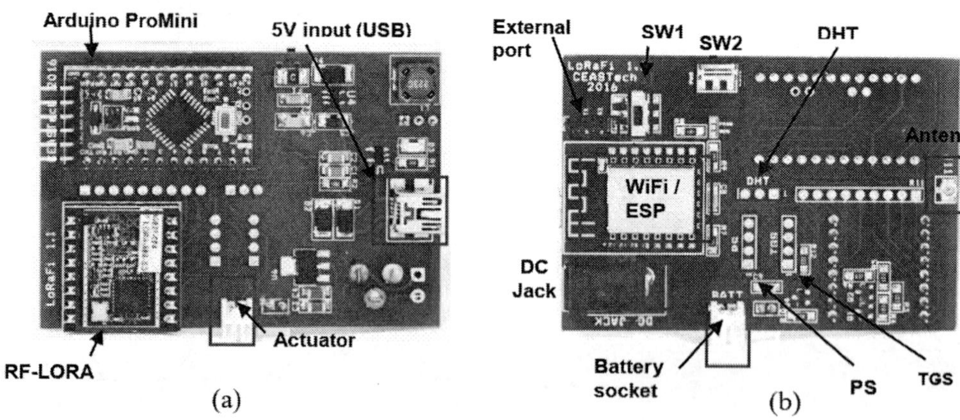

Figure 4. LoRaFi board where (a) Top view and (b) Bottom view.

3.1. Power Source

The LoRaFi board can be powered up by an external power supply or rechargeable 3.7V LiPo battery. The external power supply can be supplied from mini USB (plug into the mini USB header) or AC-to-DC adapter (plug into the DC jack). However, only one external power supply (either the mini USB or the DC jack) can be connected at a time. Meanwhile, the rechargeable battery can be used together with either DC jack or mini USB. A power switch (SW2) is provided to select between the external and the battery power for the board. In case of both external power and battery connected, the battery will be charged through the built-in battery charger. The LoRa and WiFi operate at 3.3V while the Arduino ProMini (a.k.a. ProMini) circuit operating voltage ranges from 3.3V up to 5V. Therefore, a level shifter is required in order to enable the communication between ProMini with LoRa and WiFi. The sub-section 3.2.1 provides further information on the level shifter on the LoRaFi board.

3.2. Microcontroller
Microcontroller is required especially for smart artificial system. There are various microcontrollers available in the market that varies in terms of resources, processing capabilities, speed and storage. For LoRaFi board, a microcontroller from Arduino known as ProMini[4] was used. The microcontroller on the ProMini is a microcontroller based on the ATmega328. The ATmega328 has 32kB of flash memory for storing code (of which 0.5kB is used for the bootloader). Besides that, the microcontroller also has 2kB of SRAM and 1kBs of EEPROM that can be read and written with the EEPROM library.

3.2.1. Level Shifter TTL. A level shifter is needed to step down the voltage from 5V to 3.3V when ProMini wants to send data to LoRa via Serial Peripheral Interface (SPI) communication link. Likewise, the ESP module also needs TTL when it wants to communicate with the ProMini via serial communication. However, there is no TTL required when LoRa or ESP wants to send data to ProMini.

3.3. Sensors and Actuators
The LoRaFi board has ports designated for different sensors which is humidity and temperature sensor, particle sensor and volatile organic compound (VOC). The following sub-sections described each of the sensors. Besides that, the LoRaFi board is also equipped with a port for an actuator. The pins configuration for the DHT11, particle sensor and VOC sensor is provided in Table 1. The port dedicated for the sensors was previously depicted in figure 4.

Table 1. Pins Configuration on LoRaFi Board.

Sensors	Pin 1	Pin 2	Pin 3	Pin 4
Humidity and temperature	3.3V	D2	GND	-
Particle sensor	GND	D4	D5	5V
Volatile organic compound	GND	A3	3.3V	5V

3.3.1. Humidity and Temperature Sensor. The LoRaFi board was designed with ports ready for plug-in sensors. The sensor that is used to measure the humidity and temperature is DHT11 sensor. The DHT11 sensor is displayed in figure 5(a). Other sensors that have similar specification to DHT11 can be used alternatively.

3.3.2. Particle Sensor. The particle sensor tested with the LoRaFi board is a particle sensor developed by SHINYEI Corporation (PPD42NS model[5]). The port for the particle sensor was labelled as PS in figure 4. The particle sensor shown in figure 5(b) creates digital (Lo pulse) output to Particulate Matters (PM). The Lo Pulse Occupancy (LPO) time is in proportion to PM concentration. The output from "P1" is for minimum particle size of 1 micro meter and larger.

3.3.3. Volatile Organic Compound (VOC). Volatile organic compounds (VOCs) are chemicals that can easily evaporate into the air such as ethanol, benzene and etc. TGS sensor (from Figaro) can be used to measure VOCs in the air. The TGS is a 3.3V analog sensor that has 4 pins as shown in figure 5(c).

3.3.4. Actuator. Actuator receives instruction from the microcontroller to perform a specific mechanical action. The LoRaFi board is equipped with a port for actuator, labelled as ACT (refer figure 4). The actuator is supplied with 5V power supply and connected to ProMini through digital pin 6 (D6). Some examples of actuator that can be used are fan, lamp and speaker/alarm. The remaining pins on the ProMini can be accessed directly from connectors over the board. Apart from the dedicated ports for the sensors and actuators, the LoRaFi boards also have analog pins that can be used with analog sensors.

(a) (b) (c)

Figure 5. Type of Sensors: (a) DHT11, (b) PS, and (c) TGS.

3.4. Communication with Computer or other Microcontroller
The ProMini can be programmed using Arduino IDE. The ESP module also can be externally programmed via external port (refer figure 4).

3.4.1. Serial Peripheral Interface (SPI). The Long-range technology (LoRa) module uses Serial Peripheral Interface (SPI) protocol to communicate with other microcontroller (ATmega328 on ProMini). Thus, several input/output (I/O) pins of the ProMini are dedicated specifically for the LoRa module. The pins are D13 (SCK), D12 (MISO), D11 (MOSI), D10 (chip select, SS), D9 (RST), D8 (Receiver Switch), D7 (Transmitter Switch), and D3 (LoRa DIO0). These pins are dedicated specifically for LoRa thus these pins cannot be used for other application. The advantage of the SPI communication is that additional devices that support SPI communication such as Secure Digital (SD) card can be added (perhaps for the next development board).

3.4.2. Serial Communication. Serial pins (TX0 and RX1) of the ProMini are shared between FTDI programmer and ESP module. A selector switch (SW1) is included on the board to disconnect the ESP module from the ProMini while uploading program to ProMini via FTDI programmer. Before uploading the program, the selector switch (SW1) needs to be toggled to "PROG". When done uploading, the SW1 must be toggled to "RUN" so that ESP can operate correctly. In addition, the ESP module can be configured as Access Point (AP), client or as both. It also supports basic TCP/IP, UDP, GET, POST, and Webserver applications.

3.5. Wireless Communication
There are two different wireless communication modules onboard which are LoRa and WiFi as shown in figure 6. LoRaFi board can be configured as either transmitter, receiver or as both interchangeably. In addition, UFL connector (antenna) is also provided on the board so external antenna can be attached for the LoRa module. ESP7S have port for antenna unlike ESP12E. The following sections will further elaborate on the communication protocols used for the LoRaFi board.

(a) (b) (c)

Figure 6. (a) LoRa 868/915MHz with SX1272 chip (b) ESP12E and (c) ESP7S.

4. Wireless Communications

The LoRaFi board comprised of two different wireless communication modules namely LoRa and WiFi. The LoRa communication was achieved by using LoRa module while WiFi communication was achieved by using either ESP12E or ESP7S module. The board can be configured to communicate using only LoRa or WiFi or both. Figure 7 shows the example of LoRaFi network.

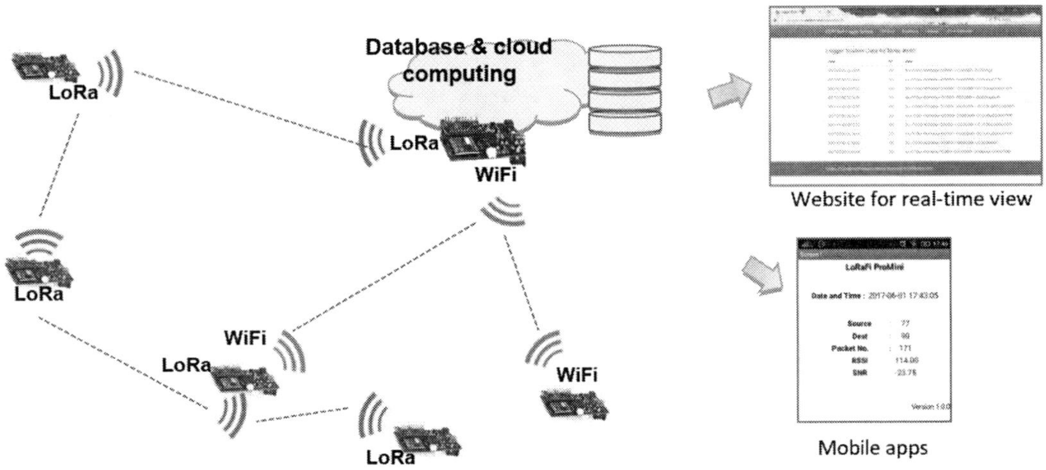

Figure 7. LoRaFi network.

The WiFi allows data to be sent to the cloud/server where data can be stored for further data analysis and/or providing users with real-time information on their smart phones. In this research, for testing purposes, the transmitter (also known as end device) is configured to operate using LoRa. The receiver, on the other hand, receives data from the transmitter via LoRa and forwarded the received data to the cloud via WiFi. The data can be viewed using internet browsers.

4.1. LoRa from Semtech

The LoRa 868/915MHz module used for the LoRaFi board is shown in figure 6(a). The LoRa module was integrated with SX1272 chip that supports different modulations such as FSK, GFSK, MSK, GMSK, LoRa and OOK modulation. By using the SEMTECH's patented LoRa modulation technique, the SX1272 able to achieve a sensitivity of over -137dBm using a low cost crystal [3]. Besides that, by using the LoRa modulation scheme, the transceiver is capable of achieving significantly longer range than existing systems based on FSK or OOK modulation [3]. LoRa employs spread spectrum. The spread spectrum modulation bandwidth (BW), spreading factor (SF) and error correction rate (CR) can be manipulated to achieve maximum flexibility.

Table 2 shows some of the available transmission quality configuration for LoRa. Throughput and range depend on the BW, SF and CR. BW is the physical bandwidth for RF modulation and there are three main bandwidth value specified by SEMTECH which is 125kHz, 250kHz and 500kHz. Referring to Table 2, larger signal bandwidth permits higher effective data rate, therefore the transmission time reduced at the tradeoff for reduced sensitivity. Another benefit of the spread modulation is that each spreading factor is orthogonal, thus multiple transmitted signals can occupy the same channel without interfering. This also permits simple coexistence with existing FSK based systems [3].

Table 2. Different combination of SF, BW and CR.

LoRa mode	SF	BW	CR	Reach	Time on air
1	12	125	4/5	Better	Medium
2	12	250	4/5	Medium	Less
3	10	125	4/5	Worst	Less
4	12	500	4/5	Better	Low
5	10	250	4/5	Better	Medium
6	11	500	4/5	Better	Worst
7	9	250	4/5	Medium-High	Medium-Low
8	9	500	4/5	Medium	Medium
9	8	500	4/5	Medium-Low	Medium-High
10	7	500	4/5	Worst	Least

Equation (1) and (2) shows the RSSI value calculation in FSK/OOK mode. Similarly, Equation (3) is the formula for RSSI in LoRa mode. Apart from that, the formula in Equation (4) can be used to calculate the signal strength of the received packet [3].

$$\text{RssiValue} = -2 \cdot \text{RFlevel [dBm]} + \text{RssiOffset [dB]} \quad (1)$$

$$\text{Rssi [dBm]} = -\frac{\text{RssiValue}}{2} \quad (2)$$

$$\text{Rssi [dBm]} = -139 + \text{Rssi, (with LnaBoostOn)} \quad (3)$$

$$\text{Packet Strength [dBm]} = -139 + \text{PacketRssi} \times 0.25, \text{(with LnaBoost on the SNT} \geq 0) \quad (4)$$

4.2. WiFi-ESP8266

The ESP module was developed by Ai-Thinker and two different processors available namely ESP8266[6] and ESP32. ESP8266 is a popular platform for IoT applications relying on WiFi, while ESP32 is the newly developed processor that adds a CPU core, faster WiFi, Bluetooth 4.2, and various I/O. The comparison between ESP-12E, -2F and -07S is given in Table 3.

Table 3. Comparison between ESP8266 with ESP32.

Feature	ESP 12E	ESP 12F	ESP-07S
WiFi chip/module		ESP8266	
Microcontroller		Tensilica 32-bit micro MCU	
Operating voltage		3.0 – 3.6V	
WiFi standard		IEEE802.11b/g/n	
Frequency		2.4GHz	
GPIO Pins	11	11	9
Pins	22	22	16
Antenna	PCB	PCB	External
Pitch		2mm	
Dimension (mm)	24 x 16	24 x 16	21.2 x 16

The series available for ESP8266-based module are ESP-01S, -07S, -12E, -12F and -12S, whereas the series available for ESP32-based module is ESP-32S. The ESP8266 core processor module is integrated with Tensilica L106 ultra-low power 32-bit microcomputer (MCU), with 16-bit compact mode, supports 80MHz frequency and 160MHz, and the RTOS support. It supports standard IEEE802.11b/g/n protocol and a complete TCP/IP protocol stack. There are two different version of ESP that can be used for the LoRaFi board which is ESP12E or ESP7S as previously shown in figure 6(b) and figure 6(c).

5. Wireless Communication Test

Simple test was conducted to investigate the wireless communication between two LoRa modules. In order to perform this test, two LoRaFi nodes are required. One node is set as transmitter while the other node is set as receiver. Both nodes are set to operate using the configuration shown in figure 8. The transmitter programmed to transmit string (e.g. `Hello`). On the other hand, the receiver was connected to computer via serial communication. Thus, data received by the receiver can be viewed via serial monitor available in Arduino IDE as shown in figure 9.

```
LoRaFi.TXpower(MAX_POWER);
LoRaFi.ChannelFrequency(915E6);
LoRaFi.Bandwidth(125E3);
LoRaFi.SpreadingFactor(12);
LoRaFi.CodingRate(8);
```

Figure 8. Code snippet.

```
Received Message: Hello
Received Message: Hello
Received Message: Hello
Received Message: Hello
Received Message: Hello
```

Figure 9. Data received by receiver and displayed on the serial monitor

6. Database and Mobile Application

The LoRaFi board able to receive real time data but the data need to be somehow stored for further processing. One way to do that is by having the gateway uploading the received data to the cloud. The data that is stored in the cloud can then be retrieved for data analytics, real-time display on website, alarm or notifications and for mobile application. Data fetched from the cloud by the developed mobile application will provide user with real time information. The mobile application in this research was developed using the open source Android MIT app 2 Inventor.

7. Conclusion and Future Works

The LoRaFi board developed can be used to measure ambient temperature, humidity, and volatile in the air. However, the sensors are replaceable with other sensor that have similar number of pins and power requirement. At present, the LoRaFi board can be purchased from Cytron Malaysia website using the following link: https://www.cytron.com.my/p-lorafi. In the future, developer will work on reducing the size of the board in order to widen the application and increase the flexibility of the board.

References

[1] M.H.M. Razali, A. Zakaria, A.Y.M. Shakaff, R. V.F.S.A. Saad, L.M. Kamarudin, and N.S.H. Abdullah, "Visualisation of acoustic entropy index for rainforest health monitoring system," *2015 IEEE SENSORS - Proc.*, pp. 5–8, 2015.
[2] Libelium Comunicaciones Distribuidas, "Extreme Range Links: LoRa 868 / 900MHz SX1272 LoRa module for Arduino Waspmote and Raspberry Pi." [Online]. Available: https://www.cooking-hacks.com/documentation/tutorials/extreme-range-lora-sx1272-module-shield-arduino-raspberry-pi-intel-galileo.
[3] Semtech, "Datasheet LoRa SX1272/73 -860 MHz to 1020 MHz Low Power Long Range Transceiver,". pp. 1–129, 2017.

IOP Conf. Series: Materials Science and Engineering **318** (2018) 012051 doi:10.1088/1757-899X/318/1/012051

[4] Arduino, "Arduino Pro Mini," *Arduino Products*, 2017. [Online]. Available: https://www.arduino.cc/en/Main/ArduinoBoardProMini. [Accessed: 20-Jun-2017].
[5] Shinyei Corporation, "Particle Sensor Model PPD42NS," vol. 2010. p. 3, 2010.
[6] Espressif Systems, "ESP8266EX Datasheet version 4.3," *Espressif Systems Datasheet*. pp. 1–31, 2015.

IOP Conf. Series: Materials Science and Engineering **318** (2018) 012052 doi:10.1088/1757-899X/318/1/012052

Design of UHF Antenna Sensor for Partial Discharge Detection in High Voltage S ubstation

Sharulnizam Mohd Mukhtar[1,*], Muzamir Isa[2,] and Azremi Abdullah Al-Hadi[3,***]**

[1] Faculty of Engineering Technology, Universiti Malaysia Perlis, Malaysia,
[2] School of Electrical System Engineering, Universiti Malaysia Perlis, Malaysia,
[3] School of Computer & Communication Engineering, Universiti Malaysia Perlis, Malaysia

E-mail: sharul@unimap.edu.my, *muzamir@unimap.edu.my, azremi@unimap.edu.my

Abstract. Partial discharge early detection is a major issue in preventative maintenance. This condition is necessary to prevent greater damage in the high voltage system. The PD detection method using UHF sensors has become an attractive method because of its high sensitivity and the ability to detect the PD location. The objective of this study was to investigate the performance of an antipodal Vivaldi antenna (AVA) as the UHF sensor for the partial discharge detection. The simulation was carried out using CST Design Studio™ to analyze the optimal parameters of the antenna before it was manufactured. The result revealed that the AVA has the good capability to be used as UHF sensor for PD detection with the high sensitivity and good directivity.

1. Introduction

Partial discharge (PD) is a localized dielectric breakdown of a small portion of an electrical insulation. PD could occur in the gaseous, liquid or solid form in the insulating medium. It often starts with a small gas voids, such as voids in solid epoxy insulation or bubbles in the liquid insulation medium such as transformer oils [1]. Voids also could occur in the insulation by two factors either manufacturing defects or by the aging of the equipment. Premature failure of an asset also could occur at the manufacturing stages that caused by poor workmanship. A high percentage of insulation failure of the asset can be observed in the first 3 years of its working life [2].

Early PD detection is very important and can prevent more serious damage to high voltage equipment. Once the PD is detected, faults can be expected and preventive maintenance can be done earlier. There are various methods that can be used to detect PD such as electrical methods, chemicals, acoustics, optics and electromagnetic. Among these, the ultra-high-frequency (UHF) method is widely used to detect PD on power transformers, gas-isolated substation (GIS) and air-insulated substation (AIS) [3][4].

In the early studies, increased time of the PD pulses generated in the high voltage equipment is very short, the pulse width was shown to be a nanosecond [5]. Therefore, an electromagnetic signal caused by the PD pulses may include a UHF band and lies in the range of 300 MHz to 3 GHz. UHF signal is extracted to reduce the effects of noise on the PD signal. Frequency band of common electromagnetic interference is not in the range of UHF band. Therefore, the interference is eliminated by UHF detection.

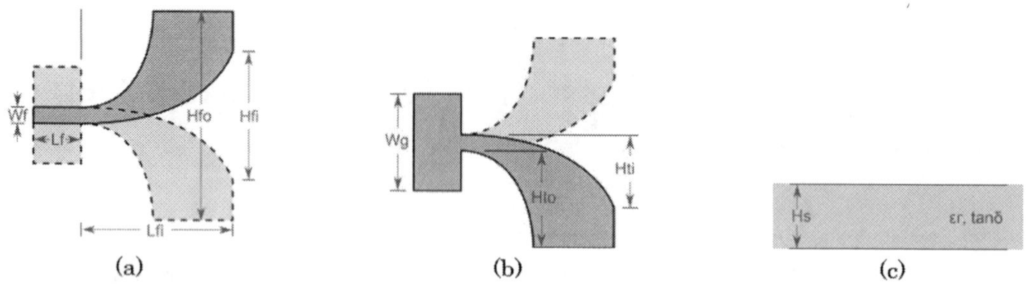

Figure 1. (a) Geometry of a top view of the AVA (b) Geometry of a bottom view of the AVA (c) Geometry of a side view of the AVA

Liu proposed a miniaturized equiangular spiral antenna (ESA) to overcome the disadvantage of most frequency independence antenna like the Archimedean spiral antenna and log-periodic dipole antenna [6]. In [7] it has been using a single-arm Archimedean spiral antenna to detect the UHF partial discharge signal emitted from the power transformer. While [8] has applied the use of the meander antenna for the same purpose.

2. Design of the UHF Antenna Sensor

The important factor that influence detection's efficiency of the sub-station UHF PD, is the performance of the sensor. Since the frequency component of the PD's signal can be as high as GHz, and there is attenuation during the transmission. The sensor must have enough wide band range and has a high sensitivity [9].

In UHF networks, digital radios, televisions, telecommunications signals, and periodic pulses from switching operations can affect the accuracy of PD detection. Hence the need for a UHF sensor that has a wide band and can reduce the band at certain frequency disturbances such as the GSM system that operates at 890 MHz to 960 MHz [10].

Therefore, based on wide band and high-gain antenna, Vivaldi was selected in this study as a detection antenna. The antipodal Vivaldi antenna (AVA), also known as the dual exponentially tapered slot antenna (DETSA), is part of a group of antennas of the end-fire tapered slot antennas (TSAs). End-fire tapered slot antenna includes constant width (CWSA), linear tapered (LTSA) and exponential tapered (ETSA), also known as Vivaldi antenna. Vivaldi antennas are widely used in wireless and radar applications due to their wide bandwidth, low cross- polarization, and highly directive patterns. The geometry of an AVA is shown in Figure 1.

2.1. Design of a Antipodal Vivaldi Antenna

To assess the performance of the AVA, the effects of antenna parameters such as dimensions and thickness of printed circuit boards are carefully investigated using commercially available software CST Design Studio^{TM}. With this parametric investigation, the antenna with the desired parameters can be expected for the future, after the necessary changes.

The AVAconsists of substrate and copper on both sides of the substrate and is divided into 2 parts, which are part of the balun and radiation area. The AVA is made of metalized dielectric substrate. Both flare, each exponentially tapered to its inner and outer profile on both sides of the substrate. To create an AVA, the equation used to describe the inner and the outer profile should be written as:

$$y = C_1 e^{\alpha x} + C_2 \tag{1}$$

where

$$C_1 = \frac{y_2 - y_1}{e^{\alpha x_2} - e^{\alpha x_1}} \tag{2}$$

Table 1. Design parameters of the AVA

Modeling parameters	Dimensions
Inner flare height, Hfi	106.5mm
Outer flare height, Hfo	245.0mm
Inner flare length, Lfi	210mm
Outer flare length, Lfo	120mm
Ground plane width, Wf	1.338mm
Feed line width , Wg	1.338mm

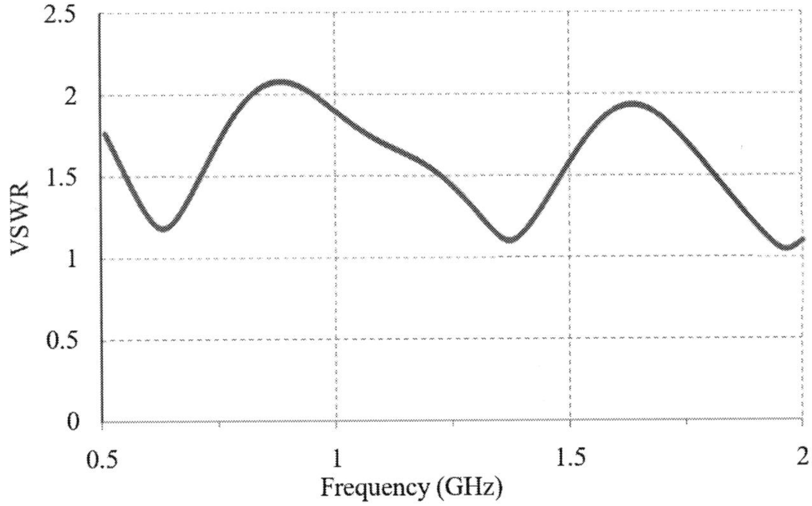

Figure 2. Simulation result of Voltage standing-wave ratio (VSWR) of the AVA

and

$$C_2 = \frac{y_1 e^{\alpha x_2} - y_2 e^{\alpha x_1}}{e^{\alpha x_2} - e^{\alpha x_1}} \quad (3)$$

where x_1, x_2, y_1, y_2 are the start and end x and y coordinates of the taper profiles. Determined parameter for the antenna are listed in Table 1.

3. Preliminary Result

Simulations are conducted using EM simulation software CST Design Studio™. The results of the designed AVA are shown in Figure 2 through Figure 5. The VSWR is a reflection coefficient function, which illustrates the power reflected by the antenna. An antenna must meet the needs of bandwidth provided in terms of VSWR is less than 2. Simulation result in Figure 2 shows that the VSWR value for whole frequency band is less than 2 except for the range from 825 Mhz to 950 Mhz. For other frequency, the VSWR value was fluctuate between 1 to 1.9.

S11 represents how much power is reflected from an antenna, therefore it's known as a reflection coefficient or return loss. S11 = 0 dB indicates that all the power is reflected from the antenna and nothing is transmitted. And if S11 = -10 dB, this indicates that if the 3 dB power is transmitted to the antenna, -7 dB is the reflected power. Figure 3 shows the S11 parameter for the designed AVA. As can see, the value of S11 parameter is less than -10 dB except for the frequency of 825 Mhz to 950 Mhz portion.

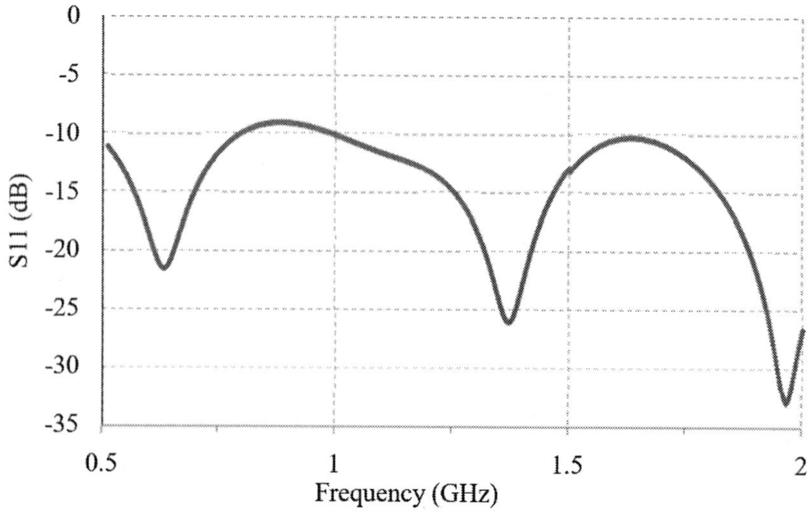

Figure 3. S11 parameter simulation result of the AVA

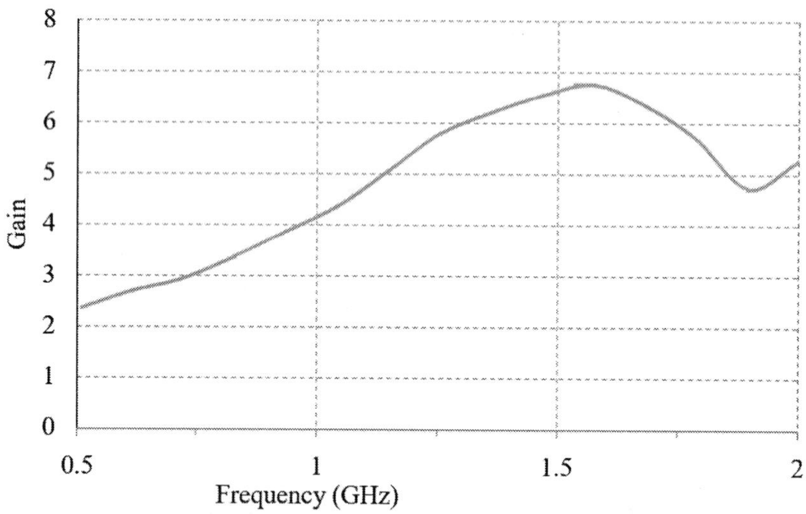

Figure 4. Simulation result of Gain versus frequency

Gain over frequency for this antipodal Vivaldi antenna is shown in Figure 4. As can be seen, gain increased with increasing frequency until it reached a maximum value at 1.575 GHz and thereafter decreased at a higher frequency. Although the gain at the lower frequency is low, about 2.3 dB it's indicates that this antenna is still able to radiate.

Antenna radiation patterns or antenna patterns are defined as "mathematical functions or graphical representations of antenna radiation characteristics as spatial coordinate functions" [11]. Figure 5 show the radiation pattern of the antipodal Vivaldi antenna for the different frequencies. H plane is marked with a solid line while E plane is marked with a dotted line. The result shows that these antenna has a good directivity at the radiation frequency of 2 GHz, 1.66 GHz, and 1.255 GHz. However, the radiation pattern shows it to be omnidirectional antenna when at a frequency of 0.5 GHz.

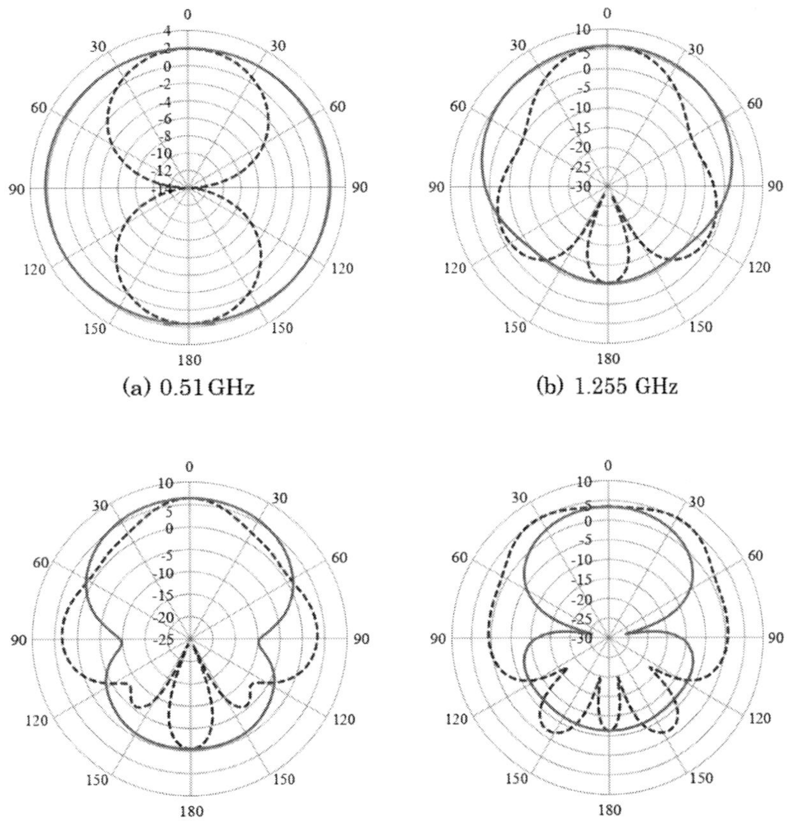

(a) 0.51 GHz (b) 1.255 GHz

(a) 1.68 GHz (b) 2 GHz

Figure 5. Simulation result of the radiation pattern of the AVA at different frequencies

4. Conclusion

In this work, an AVA antenna has been designed and analyzed using CST Design Studio™in order to assess the capability of antipodal Vivaldi antenna to be used as UHF sensor for partial discharge detection.

It was found the proposed AVA antenna satisfied the requirement for PD detection with the working frequency band of 510 Mhz to 825 Mhz and 950 Mhz to 2 Ghz. The denoised process to remove unnecessary signals before PD analysis is made will be easier with this AVA UHF sensor because the interference on the GSM band has been reduced.

Although the design of this UHF antenna was not completely accurate, it can serve as a foundation before carried out the manufacturing process of the sensor to which improvement can be made in the future.

References

[1] Chudnovsky B H 2013 Electrical power transmission and distribution: aging and life extension techniques (CRC Press)
[2] An introduction to partial discharge (pd) URL http://www.hvpd.co.uk/technical/
[3] Hikita M, Okabe S, Murase H and Okubo H 2008 IEEE Transactions on Dielectrics and Electrical Insulation **15** 505–518 ISSN 1070-9878
[4] Bell R, Charlson C, Halliday S P, Irwin T, Lopez-Roldan J and Nixon J 2003 IEEE Transactions on Power Delivery **18** 1187–1191 ISSN 0885-8977

[5] Li T, Rong M, Zheng C and Wang X 2012 IEEE Transactions on Dielectrics and Electrical Insulation **19** 1421–1430 ISSN 1070-9878
[6] Liu J, Zhang G, Dong J and Wang J 2015 Sensors **15** 29434–29451 ISSN 1424-8220 URL http://www.mdpi.com/1424-8220/15/11/29434
[7] Wang Y, Wu J, Chen W and Wang Y 2014 Journal of Sensors **2014** 8 URL http://dx.doi.org/10.1155/2014/839386
[8] Publishing I, Li M, Guo C and Peng Z 2014 Design of meander antenna for uhf partial discharge detection of transformers
[9] Judd M D, Yang L and Hunter I B B 2005 IEEE Electrical Insulation Magazine **21** 5–14 ISSN 0883-7554
[10] Global system for mobile 900 (gsm900) URL http://sppd.skmm.gov.my/skmmgovmy/files/attachments/GSM900.pdf
[11] Balanis C A 2016 Antenna Theory: Analysis and Design (John Wiley & Sons)

Performance Evaluation of Low Cost LoRa Modules in IoT Applications

Shuhaizar Daud[1], Teoh Shi Yang[1], Muhamad Asmi Romli[1], Zahari Awang Ahmad[1], Norfadila Mahrom[1], and Rafikha Aliana A. Raof[1]

[1]School of Computer & Communications Engineering, Universiti Malaysia Perlis, 02600 Arau, Perlis, Malaysia.

Abstract. LoRa is a low power long range wireless communication platform that is designed as an efficient communication platform for small, low powered devices. This makes it very suitable for battery powered devices and IoT implementation. This paper evaluates some low cost LoRa modules available on the market and their suitability, energy efficiency and performance during operation. Two low cost LoRa transceiver from Semtech Industries, the SX1272 and SX1278 were tested for their power consumption and maximum transmission range. This study have evaluated the two LoRa solutions and found that the SX1278 have a better transmission range and uses lower energy compared to the SX1272 thus making it more suitable for embedded implementation as a data gateway.

1. Introduction

An increasing number of radio technologies enabling low-power wireless communication over long distances has emerged in the past years. Ultra-narrowband technologies such as Sigfox (Labège, France) and Weightless-N, as well as spread-spectrum technologies such as LoRa allow for communicating up to few kilometers, and to build up low-power wide area networks (LPWANs) that do not require the construction and maintenance of complex multi-hop topologies [1].

A key characteristic of LPWAN technologies is indeed the ability to trade throughput for range and vice versa, i.e., one has the ability to fine-tune physical layer (PHY) settings to select a more sensitive (but slow) configuration that allows communication over a longer distance. This flexibility makes LPWAN technologies particularly appealing to developers of Internet of Things (IoT) applications requiring long-range communications with relatively low data rates.

LoRa technology is a new wireless protocol designed specifically for long-range, low-power communications. LoRa stands for Long Range Radio and is mainly targeted for Machine-to-Machine (M2M) and IoT networks. LoRa can be operated at the Industrial, Scientific and Medical (ISM) frequency band of 433 MHz, 868 MHz, and 915 MHz. LoRa is primarily intended for portable devices operating up to ten years on battery power alone in regional, national, or global deployment [2, 3]. This makes it very suitable for Internet of Things deployment since it uses very little power and could run for years without maintenance.

This paper tested 2 of the most popular LoRa solutions available on the market on an embedded platform for their effectiveness and efficiency as a data communication gateway.

IOP Conf. Series: Materials Science and Engineering **318** (2018) 012053 doi:10.1088/1757-899X/318/1/012053

1.1. LPWAN and LoRa
Low-power wide area networks complement short range wireless technologies such as Wi-Fi, Bluetooth Low Energy, and IEEE 802.15.4, and represent an interesting alternative to cellular technologies for urban-scale IoT applications. The success of LPWANs is due to their ability of providing long-range communication to thousands of devices at minimal cost and limited energy expenditure [4]. Longer communication ranges allow for drastically simplifying duty cycling and networking protocol, as LPWANs can form star topologies where the low-power end devices are able to directly communicate with a more powerful orchestrator [5]. This also allows for designing asymmetric communication schemes and to shift the load to the more powerful central device.

In order to increase the communication range, LPWAN technologies must improve the signal-to-noise ratio (SNR) at the receiver, either by narrowing down the receiver's bandwidth (reducing the receiver's noise-floor) or by spreading the energy of the signal over a wider frequency band (effectively reducing the spectral power density of the signal). NB-IoT and Weightless-P, for example, encode the signal in low bandwidth (<25 kHz) to reduce the noise level and keep the transceiver design as simple and cheap as possible. Sigfox an
d Weightless-N further narrow the signal into ultra-narrow bands as narrow as 100 Hz, further reducing the perceived noise.

1.2. LoRa Technology
LoRa is a proprietary LPWAN technology from Semtech (Camarillo, CA, USA) that recently attracted significant attention due to its ability to trade efficiently communication range against high data-rates, thus enabling IoT applications at an urban scale [6]. LoRa spreads the signal over a wider frequency band, and is more resilient to jamming and interference.

LoRa architecture defines Data Link (DL) layer above the physical layer or wireless modulation used to make long distance communication link [6]. There are many wireless systems that use modulation Frequency Shifting Keying (FSK) at the physical layer because it is a highly efficient modulation to achieve low power. In addition, FSK modulation facilitates demodulation process with the possibility of a small error rate.

The advantage of LoRa is that the ability of long-distance communication, depending on the surrounding environment or barrier at a particular location. The technology uses low power, high security transmission with AES encryption that allows for longer battery lifetime and better mobility [1]. LoRaWAN is a Low Power Wide Area Network (LPWAN) based on LoRa technology that allows LoRa devices communicate to each other and work as a wide area network. This is particularly beneficial for long distances between short range protocols and GSM [7]. LoRa and LoRaWAN™ both have a link budget that is better than the standard of other communications technology.

The core of LoRa technology is its Chirp Spread Spectrum (CSS) modulation: the carrier signal of LoRa consists of chirps, signals whose frequency increases or decreases over time [7]. LoRa's chirps allow the signal to travel long distances and to be demodulated even when its power is up to 20 dB lower than the noise floor. Because of this aspect, carrier sensing in LoRa is quite challenging: LoRa radios allow carrier detection via Channel Activity Detection (CAD) mode, a special reception state consuming half of the energy compared to the normal reception mode. However, the signals produced by different LoRa networks operating on different settings could create interference leading to false detections.

2. Test Setup
In this paper, we have tested two LoRa modules for their suitability to be used in Internet of Things (IoT) implementation. As IoT devices are usually powered using battery supply, we have tested both LoRa modules for their power consumption and maximum transmission range. Two low cost LoRa transceiver modules consisting of SX1272 and SX1278 from Semtech are tested. Using a microcontroller operating as the transmitter and receiver modules, system power consumption during operation are logged and maximum transmission range are tested.

2.1. Arduino Controller and Data Gateway

Arduino is a microcontroller ecosystem that consists of a processor platform and a software IDE. Because of its open-source nature, it is very popular in the embedded world because of its simplicity and ease of use. The platform supports Java and C/C++ programming language together with a multitude of other languages including assembly.

In this paper, Arduino Uno as shown in **Figure 1** is used as the controller platform to handle sensor input and connectivity to the LoRa modules and Internet communication.

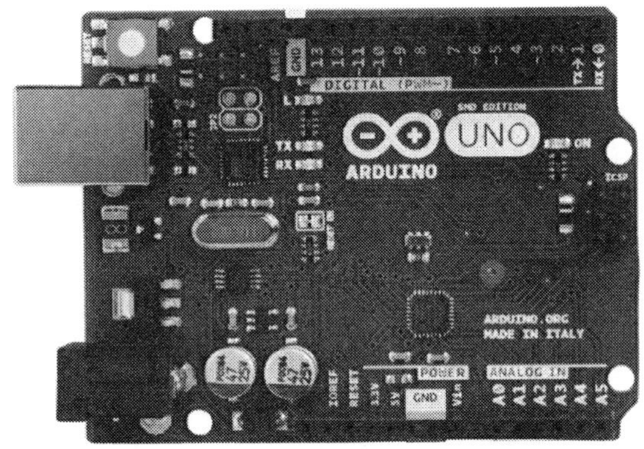

Figure 1. Arduino Uno microcontroller.

Arduino supports both battery and USB power and could be easily be programmed through the USB interface. The specification of the Arduino Nano used in this project is given in **Table 1**.

Table 1. Arduino Nano specifications.

Microcontroller	ATmega328p
Operating Voltage	5V
Input Voltage (recommended)	7-12V
Input Voltage (limit)	6-20V
Digital I/O Pins	14 (of which 6 provide PWM output)
PWM Digital I/O Pins	6
Analog Input Pins	6
DC Current per I/O Pin	20 mA
DC Current for 3.3V Pin	50 mA
Flash Memory	32 KB (ATmega328P) of which 0.5 KB used by bootloader
SRAM	2 KB (ATmega328P)
EEPROM	1 KB (ATmega328P)
Clock Speed	16 MHz
LED_BUILTIN	13
Length	68.6 mm
Width	53.4 mm
Weight	25 g

IOP Conf. Series: Materials Science and Engineering **318** (2018) 012053 doi:10.1088/1757-899X/318/1/012053

2.2. LoRa Modules

In this paper, two low cost LoRa modules utilizing LoRa chip from Semtech are used. The SX1272 and SX1278 are both capable of long range wireless transmission with the SX1278 a newer model with some improvement made to it.

2.2.1. SX1272.

The SX1272 (**Figure 2**) can act as transceivers for long range modem. The SX1272 have high interference immunity and provides long range spread spectrum communication.

Figure 2. SX1272 LoRa module.

It is a technology that can achieve over -137dBm by using precision crystal. This transceiver support high performance IEEE802.15g. It offers 3 bandwidths with is 125 kHz, 250 kHz and 500 kHz. The frequency set from 433 MHz to 1020 MHz.

2.2.2. SX1278.

The SX1278 (**Figure 3**) also can act as transceivers for long range modem. The SX1278 have high interference immunity and provides very long range spread spectrum communication

.

Figure 3. SX1278 LoRa module.

It is a technology can achieve of over -144 dBm. This transceiver support high performance IEEE802.15g. It offers 3 bandwidths with is 125 kHz, 250 kHz and 500 kHz. The frequency set from 137 MHz to 1020 MHz.

2.3. System Architecture

The system consists of a transmission node acting as the transmitter node and a data gateway as the receiver which provides network communication to the internet. By using this approach, it is possible to have multiple remote nodes connected to single data gateway and at the same time reduce cost and complexity to implement.

2.3.1. System Block Diagram
System architecture consists of two nodes acting as the transmission node and another one acting as the gateway. The system architecture is outlined in **Figure 4**. The entire system is divided into transmission node (sensor node) which will gather all sensor data and transmit it to the data gateway node.

The transmission node gathers data from input sensor ranging from humidity sensor, temperature sensor and luminance sensor. It is acquired by an ADC inside the Arduino microcontroller, digitized and averaged before being transmitted by the LoRa transceiver. The data gateway also consists of another Arduino microcontroller connected to a LoRa transceiver configured as a receiver. An Ethernet module provides internet access through RJ45 connection to the network.

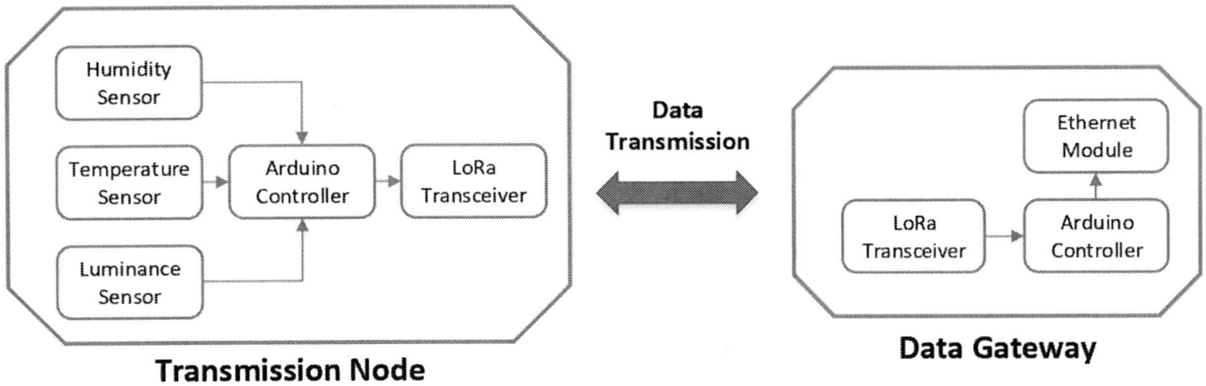

Figure 4. System architecture.

2.3.2. Test Sensors
In this system, 3 input parameters have been selected for implementation ranging from luminance sensor (LDR), humidity sensor and temperature sensor. All the sensors are integrated to the Arduino microcontroller for conversion process before being prepped for transmission to the gateway.

LDR Light Sensor
The Light Dependent Resistors (LDR) is a sensor that is used to detect light intensity. The resistance of the sensor varies with light intensity detected by the sensor with higher resistance in low intensity and lower resistance in higher intensity. The LDR specifications are given in **Table 2**.

Table 2. LDR sensor specifications.

Resistance at 1000 lux	400 ohms
Dark resistance	1M ohm
Maximum current	75 mA
Maximum power dissipation	250 mW
Maximum voltage	320 V

IOP Conf. Series: Materials Science and Engineering **318** (2018) 012053 doi:10.1088/1757-899X/318/1/012053

DHT11 Humidity Sensor

DHT11 is humidity sensor measures both temperature and humidity of surrounding air and sends the reading in digital format. The sensor is rated for an accuracy of 5% from 20%-90% humidity and is used as an input sensor in this system. The specification for the sensor is given in **Table 3**.

Figure 5. DHT11 sensor.

Table 3. DHT11 humidity sensor specifications.

Parameter	DHT11
Relative Humidity Range	20 ~ 90%
Relative Humidity Accuracy:	±5%RH
Temperature Accuracy:	±2°C
Resolution	1%RH,1°C
Repeatability	±1%
Long Term Stability	±1% per year
Operating Temperature Range	0 ~ 50°C
Power Supply:	3,3V ~ 5.5V
Supply Current:	0.5mA ~ 2.5mA
Idle Supply Current:	100uA ~ 150uA
Max sampling period (Max device update rate).	1 second

Sampling for all the sensors are done every 5 second and filtered using a moving window average before being transmitted to the gateway.

2.4. Experimental Test

To test the performance of the transceiver, two testing procedures have been carried out which is power consumption test and maximum distance test. Power consumption test are done by logging the current requirement of the nodes during transmission and calculating the power used by multiplying the voltage and current used during the operation.

Maximum transmission range test are carried out by checking the RSSI until the signal is lost. The test are done in a line of sight inside the university compound with the aid of GPS and Google Maps.

2.4.1. System Setup

In this system, the transmission node (sensor node) setup consists of the DHT11 sensor which gathers humidity and temperature reading and an LDR which gathers luminance reading as shown in Fig. 6. Both of the sensor are connected to an Arduino Uno that acquires and processes the sensor data. DHT11 are connected to the digital interface (DIO1) and the LDR is connected to the analog input (AI0) while the LoRa transceiver is connected through the SPI protocol at the ICSP header.

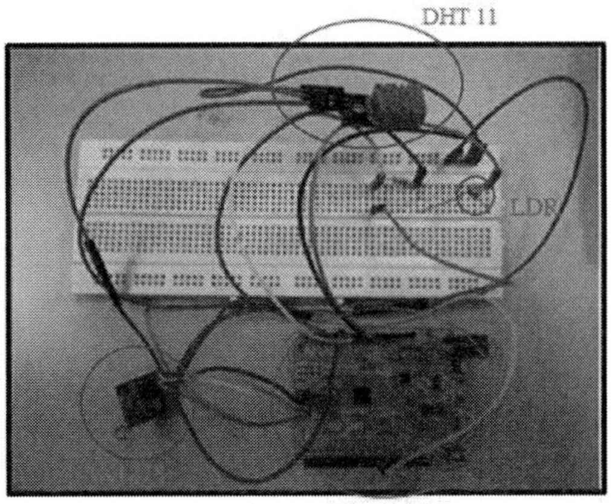

Figure 5. Actual transmit node setup.

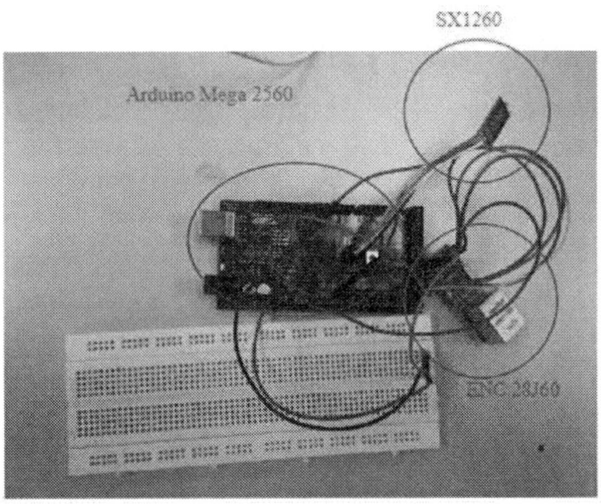

Figure 6. Actual data gateway setup.

The data gateway (**Figure 6**) does not have any sensor attached to it and merely transmits the incoming data from the sensor node and handles communication to the Internet and cloud service. An Arduino Mega2560 is used for as the main controller and an Ethernet module (ENC28J60) is used to provide internet access to the gateway. Data transmitted from the sensor node are processed here before being transmitted to the Internet to be stored in a cloud service.

IOP Conf. Series: Materials Science and Engineering **318** (2018) 012053 doi:10.1088/1757-899X/318/1/012053

3. Result and Discussion

In this study we have collected power consumption and transmission range data from both of the modules during transmission. Average power requirement of the system during data transmission are averaged across 10 samples. Range test are taken after 10 consecutive timeout that signifies out of range condition.

3.1. Power Consumption

Power consumption of the transmission node and gateway node are monitored using a current meter and voltmeter. Actual power requirement of the system are then calculated by multiplying the values to produce the power rating. An average reading of 10 values are taken and tabled in Table 4, Table 5 and Table 6. The transmission power settings are set inside the code and passed to the transmitter modules as configuration flags.

Table 4. SX1272 average power consumption during operation (average of 10 reading).

Power Setting	*Reading*	*Average Reading*	*Average Power, W*
Max	Voltage, V	5.047	0.2423
	Current, A	0.048	
High	Voltage, V	5.044	0.2219
	Current, A	0.044	
Intermediate	Voltage, V	5.042	0.2168
	Current, A	0.043	
Low	Voltage, V	5.038	0.2015
	Current, A	0.04	

Four transmission power settings are tested for the SX1272 and the average power requirement of the SX1272 shows an increase from the low power transmission to the maximum power. The same pattern is also visible for the SX1278 whereby the power requirement increased according to the transmission power.

IOP Conf. Series: Materials Science and Engineering **318** (2018) 012053 doi:10.1088/1757-899X/318/1/012053

Table 5. SX1278 average power consumption during operation (average of 10 reading).

Power Setting	*Reading*	*Average Reading*	*Average Power, W*
Max	Voltage, V	5.145	0.3499
	Current, A	0.068	
High	Voltage, V	5.142	0.2622
	Current, A	0.051	
Intermediate	Voltage, V	5.149	0.2523
	Current, A	0.049	
Low	Voltage, V	5.146	0.247
	Current, A	0.048	

Table 6. SX1272 and SX1278 power consumption difference.

	Average Power Requirement (W)		*Difference (%)*
Transmission Power	**SX1272**	**SX1278**	**%**
Low	0.2015	0.247	18.42
Intermediate	0.2168	0.2523	14.07
High	0.2219	0.2622	15.37
Max	0.2423	0.3499	30.75

Comparison of the SX1272 and SX1278 shows a higher power requirement of the SX1278 compared to the SX1272. An average difference of about 15% is expected except at the maximum transmission power where the power consumption differs by about 30% between the SX1272 and SX1278.

3.2. Transmission Range

Maximum transmission range test are carried out by moving the receiver away from the transmitter until the signal is lost. The transmitter location and test range are given in **Figure 7**.

Figure 7. Maximum range test setup.

The maximum range are tested for a number 10 times and averaged. The SX1272 library supports a total number of 3 transmission mode which varies the transmission sensitivity, bandwidth and speed. The SX1278 library supports all 10 different transmission modes and the results are logged in Table 7 below. The difference between transmission modes can be referred inside respective modules datasheets.

Table 7. SX1272 and SX1278 maximum transmission range.

	SX1272	*SX1278*
Mode	Range	Range
1	350 m	1km
2	250 m	1km
3	N/A	1km
4	220 m	900m
5	N/A	800m
6	N/A	700m
7	N/A	650m
8	N/A	600m
9	N/A	600m
10	N/A	500m

From the results gathered, it is found that for the SX1272, lower transmission mode which is optimized for low speed, maximum range does give the longest transmission range which is 350 meters. But for the SX1278, lowering the transmission speed and increasing the sensitivity using lower mode configuration does not improve the range after mode 3. The maximum range achievable could probably be improved with better antenna.

While both modules could transmit to relatively long range, the SX1278 have a much further transmission range compared to the SX1272 albeit at a higher power requirement.

4. Conclusion

The test carried out have shown that the SX1278 has a much further transmission range compared to the SX1272 albeit using more power during transmission operation. While the power requirement at the maximum transmission power varies by as much 30%, at lower settings the difference is much smaller ranging from 14-18%. This is still within acceptable values considering our test have shown that the SX1278 has a much further transmission range reaching a distance of 1km.

Acknowledgement

The project is partly funded by Universiti Malaysia Perlis's Short Term Grant (STG) program.

References

[1] W. Yang, M. Wang, J. Zhang, J. Zou, M. Hua, T. Xia, *et al.*, "Narrowband Wireless Access for Low-Power Massive Internet of Things: A Bandwidth Perspective," *IEEE Wireless Communications,* vol. 24, pp. 138-145, 2017.

[2] A. Rahman and M. Suryanegara, "The development of IoT LoRa: A performance evaluation on LoS and Non-LoS environment at 915 MHz ISM frequency," in *2017 International Conference on Signals and Systems (ICSigSys)*, 2017, pp. 163-167.

[3] T. Petrić, M. Goessens, L. Nuaymi, L. Toutain, and A. Pelov, "Measurements, performance and analysis of LoRa FABIAN, a real-world implementation of LPWAN," in *2016 IEEE 27th Annual International Symposium on Personal, Indoor, and Mobile Radio Communications (PIMRC)*, 2016, pp. 1-7.

[4] K. Zhang and A. Marchiori, "Demo Abstract: PlanIt and DQ-N for Low-Power Wide-Area Networks," in *2017 IEEE/ACM Second International Conference on Internet-of-Things Design and Implementation (IoTDI)*, 2017, pp. 291-292.

[5] K.-H. Ke, Q.-W. Liang, G.-J. Zeng, J.-H. Lin, and H.-C. Lee, "A LoRa wireless mesh networking module for campus-scale monitoring: demo abstract," presented at the Proceedings of the 16th ACM/IEEE International Conference on Information Processing in Sensor Networks, Pittsburgh, Pennsylvania, 2017.

[6] M. Rizzi, P. Ferrari, A. Flammini, E. Sisinni, and M. Gidlund, "Using LoRa for industrial wireless networks," in *2017 IEEE 13th International Workshop on Factory Communication Systems (WFCS)*, 2017, pp. 1-4.

[7] E. Aras, G. S. Ramachandran, P. Lawrence, and D. Hughes, "Exploring the Security Vulnerabilities of LoRa," in *2017 3rd IEEE International Conference on Cybernetics (CYBCON)*, 2017, pp. 1-6.

IOP Conf. Series: Materials Science and Engineering **318** (2018) 012054 doi:10.1088/1757-899X/318/1/012054

The Rogowski Coil Sensor in High Current Application: A Review

Ayob Nazmy Nanyan[1, a], Muzamir Isa[1, b], Haziah Abdul Hamid[1, c], Mohamad Nur Khairul Hafizi Rohani[1,d], Baharuddin Ismail[1,e]
[1]Centre of Excellence for Renewable Energy (CERE), School of Electrical System Engineering, Universiti Malaysia Perlis (UniMAP), Malaysia

Email: [a]ayob.nazmy@studentmail.unimap.edu.my; [b]muzamir@unimap.edu.my; [c]haziah@unimap.edu.my; [d]khairulhafizi89@yahoo.com; [e]baha@unimap.edu.my

Abstract. Rogowski coil is used for measuring the alternating current (AC) and high-speed current pulses. However, the technology makes the Rogowski coil (RC) come out with more improvement, modification and until today it's still being studied for the new application. The Rogowski coil has a few advantages compared to the high frequency current transformer (HFCT). A brief review on the basic theory and the application of Rogowski coil as a current sensor measurement that been done by previous researchers are presented and discussed in this paper. Additionally, the review also focused on the capability of Rogowski coil for high current sensor measurement and their application for fault detection, over voltage current sensor, lightning current sensor and high impulse current detection. The experimental set up, techniques and measurement parameters in models also been discussed. Finally, a brief review on the performance analysis of current sensor measurement of Rogowski coil likes sensitivity, the maximum and current detection which could be used as a guideline to another researcher in order to develop an advanced RC as high current sensor in future is presented. This review reveal that the RC has a very good performance in high current sensor detection in term of sensitivity which is up to a few nanosecond, higher bandwidth, excellent in detection of high fault and also could measuring lightning current up to 400kA and has many advantages compare to conventional current transformer(CT).

1. Introduction

Rogowski coil (RC) has been introduced by German physician Walter Rogowski [1] and it was proposed in 1912 [2]. The Rogowski coil is an air core coil that can measure both alternating current and high-speed impulse current based on Ampere's and Faraday's law [1]. However, the application of Rogowski coil is very limited due to the limitation of technology and source on that time. Nowadays, with the technology such as microprocessor –based, modern signal processing technique and others have made an improvement of the Rogowski coil and application of the Rogowski coil has become wider and more suitable for various applications [3]. There are many advantages of Rogowski coil [1] [4][5]; Cause no damage by large load, Capability to measure large currents (ranging from a few milliamperes and up to 1 MA without saturating because of a non-magnetic core, Compact and light weight, Wide bandwidth and typically between 0.1 Hz and 1 GHz, Low cost and inexpensive, Safe because there is no direct connection to the main circuit, Excellent transient response, Good linearity due to absence of magnetic material, Isolation and non-intrusive measurement between testing circuit and transducer.

Rogowski coil is commonly used in pulsed applications to make an indirect measurement of very large current [6]. This paper is a review of Rogowski coil as a current sensor by focusing on the over voltage measurement application or high impulse current that commonly occur in high voltage system such as lightning strikes [7], switching operation of circuit breakers [8], gas insulated switchgear switching and electromagnetic pulses. Additionally, Rogowski coil is recognized as primary current sensors and mostly suited to high voltage application [9]. Commonly, there are a few devices that are used for measuring impulse currents such as shunt device, current transformer (CTs) and Rogowski coil[10]. The sections of this paper are as follows. In Section 2, this paper presents the basic of Rogowski coil theory and follows by operating principles in Section 3. Meanwhile, in Section 4 Rogowski coil as high current sensor is reviewed and discussed according to the their specific applications. The various type of Rogowski coil also been highlight in this section. Finally, Section 5 provides a conclusion of this review paper.

2. Basic and Theory of High Current Rogowski Coil

Rogowski coil current sensors operate on the same principles as conventional iron-core current transformers (CT) and classified as low-power standalone current sensors since their secondary signal is different than the typical CT secondary signal [11]. Basically, the Rogowski coil consists of wire wound on a non-magnetic core [12]. For measuring the current, the Rogowski coil is placed around the conductor. A magnetic field is generated as current flows around the conductor. This field is converted into a signal according to Faraday's induction law, which is proportional to the current or proportional to the rate of change of the current flowing in the conductor [13]. The Rogowski coil consists of a helical coil of wire with the lead from one end returning through the center of the coil to the other end and making both of terminal are at the same end of the coil. The voltage that is induced in the coil is proportional to the rate of change (derivative) of current in the straight current conductor. The Rogowski coil is connected with integrator circuit (an electronic circuit device) which provides an output signal that is proportional to the output current as shown in

Figure *1*. A measurement current *i(t)* induces an AC magnetic flux *ϕ(t)*, which induces a voltage in the sensing winding of an RC. The voltage induced in the coil is defined[14] by Equation 1 :

$$Ui = N_{RC}\frac{d\phi(t)}{dt}\cdot\frac{N_{RC}N_1}{R_m}\cdot\frac{di(t)}{dt}\cdot M\frac{di(t)}{dt} \tag{1}$$

Where N_{RC} is the number of turns of the pick-up winding, N_I is the number of turns passed by the current.

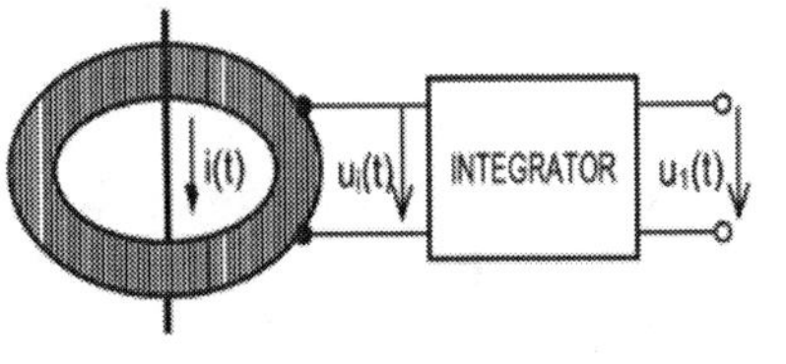

Figure 1. Rogowski coil with integrator [14]

i(t) and R_m are the reluctance of toroid with winding N_{RC} and M is indicates the mutual inductance between N_{RC} and N_1. Normally, one conductor (*N1=1*) will be used when measuring high currents. The time course of the measured current is obtained by integrating the induced voltage according to Equation 2 :

$$i(t) = \frac{1}{M}\int_0^1 ui(t)dt = \frac{ki}{M}u_i(t) \quad \textbf{(2)}$$

where *Ui* is the output voltage integrator determine by *Ui*, and k_1 is the integrator constant. There are harmonic waveforms, *I (ωt)* when measuring high current, and the induced voltage can be expressed as in Equation 3:

$$Ui = j\omega MI \quad (3)$$

Where $\omega = 2\pi f$ is the angular frequency of the measured current with frequency, *f*.

For measuring mains current by using RC with an integrator, the constant of the whole system is given as in Equation 4:

$$k_s(t) = \frac{u_1}{I} \quad (4)$$

where the integrator of output voltage is U_I and the current passing through the central conductor is *I*.

3. Types of Rogowski Coil

In this section, the advantages of Rogowski coil are briefly explained. Nowadays, there are many researchers have come out with the new current sensor technology by using Rogowski coil with various types, shapes and applications such as single- or multi-layer, rigid or flexible, the H-field sensor, Rogowski Coil designed with Discrete Coils made on Printed Circuit Boards (PCB) and others.

i. *Single or multi-layer Rogowski Coil* has coils that have lower values of mutual inductance, series self-inductance, series resistance, and distributed capacitance than multi-layer coils that consist of two windings wound over each other. These parameters should be taken into consideration when performing measurements when high-frequency and low phase-displacement is required. Single- layer-type Rogowski coils typically have several hundred turns to from the practical-sized of Rc. The mutual inductance that is usually obtainable ranges from about 0.1-1 μH [15].

ii. *Rigid Rogowski* Coils are wound over a non-magnetic core usually having a toroidal shape. This core is made of plastic, epoxy, or other insulating material. The main advantage of using non-ferromagnetic material is it can obtain the coil linearity since the non-ferromagnetic material cannot saturate at high currents [15][16][17] .

iii. *Flexible Rogowski Coils* may be wound over a silicone rubber core. The wire from the end of the coil winding returns along the length of the coil through a conductor along the axis of the core. Flexible Rogowski coils are convenient for measuring electric current in large or —awkwardly shaped conductors, where space around the conductor is restricted or where only a lightweight transducer can be suspended on the conductor. An Accuracy of flexible coils is not as high (can be improved by proper positioning of the primary conductor)[15].

iv. *H-field current sensor* suitable for applications in 10 kV, 200A. In comparison to classical RC, this discrete coil solution is smaller thus it is suitable for applications with small space available [15].
v. *The Printed Circuit Board Rogowski (PCB) coil* has three type; Rogowski coil designed with Discrete Coils made on Printed Circuit Boards, PCB design technique in a split-core, non-split-core PCB Rogowski coils. The output voltage of PCB Rogowski coils is smaller than the output voltage of Rogowski coils designed with conventional wire wound over an air-core since the core cross-section area is smaller. The use of more PCB coils results in higher accuracy, lower sensitivity to positioning the conductor, and better external field rejection performance. Another advantage of this solution is lower production costs [15] and can be placed on bus bars, bonding wires, and terminals [18].

4. Rogowski Coil for Transient, Overvoltage & High Current Sensor

This section discussed the Rogowski coil for high current measurements application focusing on various types of Rogowski coil shapes or types. High-Impulse currents commonly occur in high voltage power systems and may arise from different sources such as lightning strikes, switching operations of circuit breakers, switching manoeuvres in gas-insulated switchgear, and electromagnetic pulses [19]. High-impulse current and voltage measurements were originally associated almost entirely with short-circuit and insulation testing of high-voltage ac bulk power system equipment and with the recording of actual or simulated power system switching and lightning surges [20]. Additionally, the over voltages may go as high as six (6) times normal power frequency voltage [21]. In September 2010, a new standard IEC 62475 has been introduced as a horizontal standard to cover high current measurement techniques in general [22]. Basically, Rogowski coil can be considered as special current transformer [23]. The Rogowski coil can be used instead of CTs for fault detection in the transmission line and electrical machine as well [24].The advanced Rogowski coil is made or designed by using printed circuit board PCB, Rogowski Coil. Rogowski coil based on the technology of PCB has high accuracy in design and manufacture, and is flexible in wiring way and is easily mass produced [25]. Furthermore, PCB has high-temperature stability which is important for industrial applications [23]. The Rogowski coil also can be placed around high-voltage bushings, for on-line monitoring [26].

4.1. Rogowski coil Application for Fault Current & High Current Sensor

Fault current magnitude is determined by the source inductance which in effect, limits the maximum rate of change of the current [9].The Rogowski coil has a high reliability as a current sensor for protection and monitoring the power system and is the best choice for high current measurement in high-power laboratories [27]. The research of PCB with high-precision Rogowski coils has been carried out by L.Jubomir A. Kojovic which the applications are for protection, control, and metering in electric power systems and applied Rogowski coils as current sensors. The function of this sensor is to detect the fault current in different protection system for power transformers, power cables, and capacitor banks and connected with a relay. The relays communicate over fibre-optic cables connected to Ethernet [28]. The designed sensor has the capability with the following [29]; 1) Linearity, no saturation provides excellent protection security even at high fault currents exceeding 100 kA, 2) Light weight and compact size and 3) Increased personnel safety since opened secondary.

Figure *2* illustrates the Rogowski coils that are designed using two PCBs sandwiched together as a multi-layer PCB design and it provides high accuracy of the coils. Figure 3 presents the application of designed Rogowski coil as a current sensor for integrated protection system in power system which the sensor's installed for high-voltage power cable protection, capacitor bank protection, and power transformer protection.

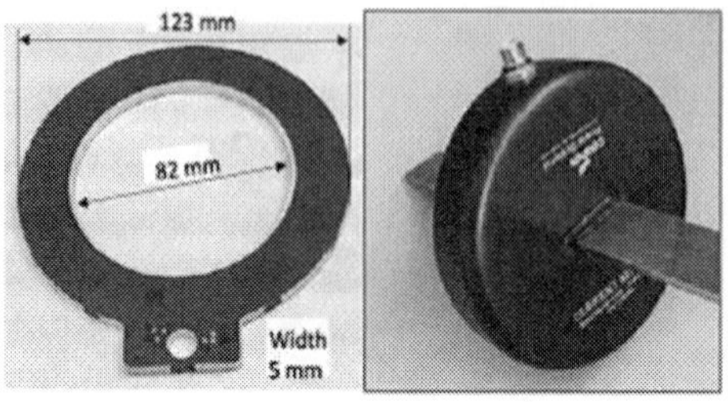

Figure 2. The PCB Rogowski coil Design [29]

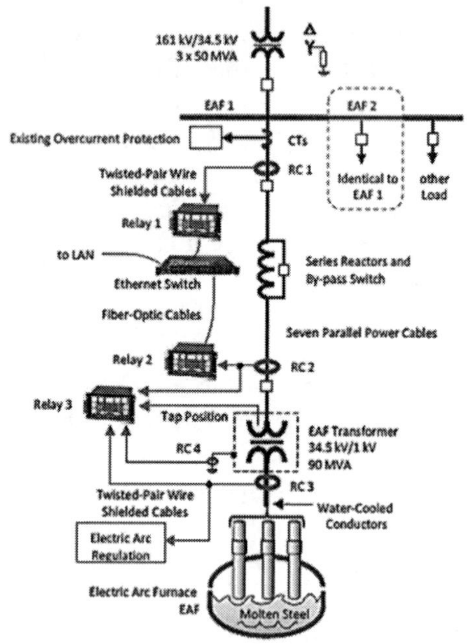

Figure 3. The Integrated Protection, Control, and Metering Functions[29]

The Rogowski coil sensor has a high possibility operating in secondary circuits fed by conventional current transformers (CTs). Hence, the Rogowski coil had been implemented inside the protective relays to replace the CTs with the same function but more reliable [9]. The Rogowski coil sensor worked very well which can detects current up to 2kA rms as well as operating from a 5A auxiliary CT secondary. The high sensitivity current sensor is needed to protect the transmission line system such as detection of the faulty current and normally the conventional CTs are often employed. The PCB Rogowski coil with high precision was designed to overcome this matter [26]. The sensor response time for impulse signals testing is less than 50ns, less transmission delay and fits to apply in traveling wave based faults detection and location [26]. Meanwhile, a hybrid current sensor for metering and protective relaying also been developed [30] which was a combination of Rogowski coil with a passive integrator and integrated-optic Pockels cell (IOPC). The sensor's had to be a highly linear and exceeding 0.2% linear standards and capable detect the current between 30 Ampere to 30 kA.

In power system, the bus bar protection system is one of the most important element and the sensor with high reliability, high speed, and stability are required. An overcurrent may damage or interrupt the whole system. With the CTs saturation problem, it may provide the abnormal situation to the CTs which errors happen due to the mismatch of CT turns ration and magnetization current resulting the CTs could not operate as required. Thus, the Rogowski coil was introduced to replace the CTs and as protection to the bus bar and combined with fuzzy reasoning system [31]. The advantage of this sensor is when fault is detected, the sensor will response very fast to clear the fault. The other test conducted shows that the Rogowski coil with the integrator performed the current measurement was up to 25kA at 50Hz frequencies [14]. Due to the limitation of CTs, if fault current occur to capacitor bank, high current flows which results in unbalance of CTs (may explode the CTs) and significantly damage the capacitor bank. Hence, to overcome this problem, the Rogowski coil current sensor has been introduced for capacitor bank protection system [28]. The indicated result shows that the installed Rogowski coils have performed very well and the circuit and equipment test set up as shown in Figure 4. Similarly, Relays is used to communicate over fiber-optic cables connected to an Ethernet switch [28] [29][11]. Meanwhile, the reliability of Rogowski coil for fault current detection was investigated by A.Dos Santos.et.al [11]. The low power stand-alone current sensor based on Rogowski coil was developed and this sensor had the capability of detecting high fault current up to 60kA.

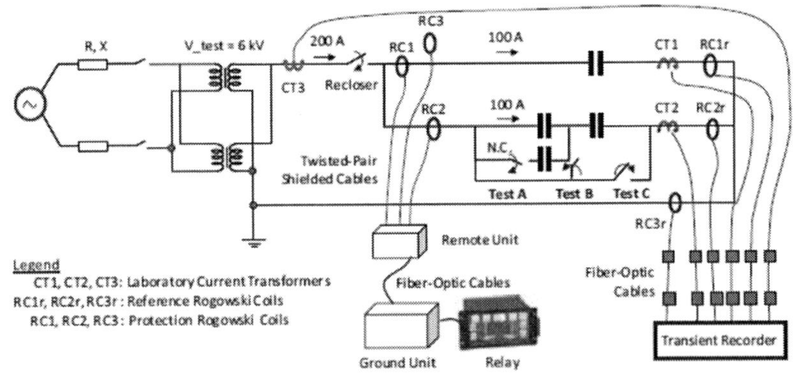

Figure 4. The high-power Test Setup [28]

4.2. Rogowski Coil Application for Another High Impulse Current Sensor

An accurate measurement for high-impulse current at ns/kA range is important [30][6]. In order to test different impulse current waveforms, a new design of a side-looking ''flat spiral'' self-integrating Rogowski coil that is wound by twin coaxial cable with individual sheath was designed by I.A.Metwally [32]. The improvement had been done to this sensor is it is designed without magnetic core and it gives high self-inductance and low self-resistance because of its winding structure. The advantages of the design are it gives higher bandwidth and lower plateau low corner frequency compared to the other designed coils, possesses a wide range of sensitivity, less susceptibility to measure other currents and stray magnetic fields in close proximity, and can be used in some applications when the conventional toroidal air-cored or PCB Rogowski coils are not suitable [32]. The measured results indicate that achieved bandwidth and sensitivity are up to 7.854 MHz and 3.623 V/kA, respectively. Additionally, sensitivities for the designed coil deviate from the corresponding measured ones by relative errors percentage within ±0.15 %. Figure 5 presents a new design of a side-looking "flat spiral" self- integrating Rogowski coil.

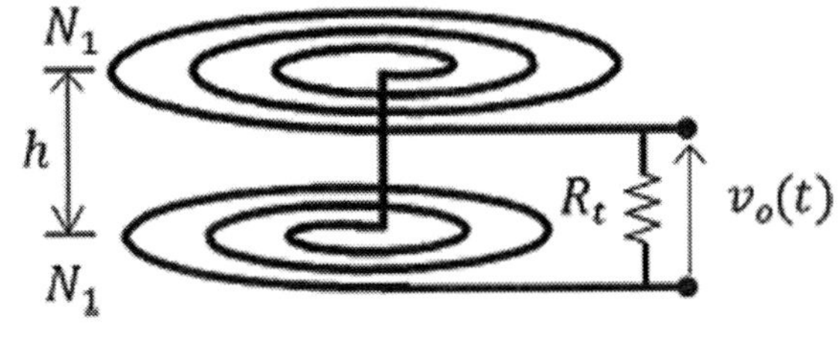

Figure 5. Connection of the two sub coils [32]

The development of Rogowski coil for impulse current measurements also has been carried out by Peerawut Yutthagowith and this sensor is very convenience in current measurements for the high-voltage test and based on the clamp on device type [33]. One of the characteristics of high impulse current is it has a fast rise time in a range of a few microseconds and a good sensor must be designed to overcome this matter. Thus, there are two (2) types of integrator circuits constructed to overcome this problem [32]; (a) inverting active integrator and (b) non-inverting active integrator. However, the Rogowski coil with the passive RC integrator fails in measurement of the impulse current of the duration time over 20 ms, with the inverting active integrator provides the deviated waveform with pre-shoot signal at the beginning time of the fast impulse current, and with the developed non-inverting integrator, it shows promising performance for measurement of all standard impulse currents in high-voltage tests. It can be concluded that Rogowski coil with the non-inverting active integrator shows very satisfied performance for measurement of the fast rise time current with a front time of 1 ms and the long-duration impulse current with the duration time of up to 2 ms.

Meanwhile, I. A. Metwally also proposed a method by using Rogowski coils for high impulse current measurement. Three (3) slow wave air-core Rogowski coils were developed with performance improved that enable plateau bandwidths up to 11 MHz and sensitivities of less than 0.25 V/kA. These three (3) Rogowski coil have different parameters, performance, and limitation. The designed coils are calibrated versus a commercial impulse current transformer and a commercial Rogowski coil for different impulse currents up to 10 kA. According to the analysis, the author found that these coils achieved plateau bandwidths up to 11 MHz and sensitivities of less than 0.25 V/kA [34]. All of these designed coils have a constant sensitivity which for RC1, RC2 and RC3 are 0.249 V/kA, 0.141 V/kA and 0.099 V/kA for the commonly optimized coil termination resistance of $R_t = 0.13\ \Omega$, respectively. Coil RC2 is the best because it has the lowest low corner frequency and the intermediate transit time in comparison to those of RC1 and RC3. Furthermore, the investigation of the peak current measurement error shows that RC2 is the lowest error recorded within ±0.7%, ±0.9%, and ±1.1% respectively, contrary to those of RC1 as it has the lowest inductance. The coil bandwidth decreases with the increase in the coil termination resistance leading to a nonlinear dependence of the coil output voltage peak on this value and sensitivity phase angle increases with the increase in the coil termination resistance resulting in distortion of the impulse waveform [34].

Additionally, an investigation of high accuracy measurement of the high impulse current sensor was conducted by Yu.Zhang et.al by using self-integrating with magnetic core [35]. By adding 50 Ω damping resistance, the amplitude of parasitic resonant signal has been damped by 5.7 times in contrast to the situation without any damping resistance [35]. The results indicate that this sensor coil has a response time is about 1.1 nanosecond and its sensitivity is 3.34 A/V. Moreover, the improvement that been done could accurately detect kA-range nanosecond pulse current in the laboratory and the capability of this sensor is proven. Meanwhile, the Rogowski coil was applied industrially in the early 2000's to Gas Insulated Substations (GIS) [23]. A PCB Rogowski for the current sensor in GIS system has been introduced [23]. There are many advantages of PCB Rogowski coil such as thermal stability, high linearity, high bandwidth, noise reduction due to digital integration, etc. The research trend for Rogowski coil as current transducer work has become very aggressive. Due to the upper limit of Rogowski-coil current transducer with passive integration is not clearly determined, the researcher had come out with the upper-frequency limit of the Rogowski coil current transducer with passive integration with 200 kA range [36]. Rogowski current transducers are frequently used for measuring very fast switching transients in power semiconductors [37].

4.3. Rogowski Coil Application for Lightning Current Detection Sensor

This section is specific to lightning current sensor application. Measuring lightning current is the most important criteria to study the lightning behaviour. Lightning current has a high amplitude and its rising time is short enough to cause over voltage in the power system. Lightning tests are necessary to measure the current flow along critical parts of the structure under test and Rogowski coils have been found very useful [38]. Thus, a good high current sensor is needed for this purpose. The Rogowski coil was tested for measuring the lightning current especially to the telecommunication tower [39] and wind turbine [40]. A shunt lightning current measuring system was introduced for measuring lightning current by using Rogowski coil [40]. The sensor consists of the integration circuit with a low-frequency amplifier, and low pass filter were designed and has a good accuracy for integration in the range of 100 Hz–1 MHz [40]. The range of measuring current was up to 1 kA and confirmed that the Rogowski coil could be used for measurement up to 2.4 MHz, which is higher than the maximum frequency of 1 MHz [40]. Lightning current contains high frequency, thus the B-Dot sensor was proposed in conjunction with Rogowski coil [39].

Similarly, the Rogowski coil has been tested for measuring lightning current up to several kilo Ampere (kA). This sensor was developed in order to measure and locate the lightning current on the transmission line based on the designed Rogowski coil with an external integrator and detected the lightning current up to 400 kA [40]. However, there is unwanted electromagnetic interference that will affect the performance of the sensor. Hence, an electrostatic shield which is made of copper or aluminium is added and it is often wrapped around the coil. Furthermore, to prevent the eddy current that induced by the magnetic flux, a slot along the inner circumference should be made to ensure the shielding box is not completely closed [41]. From their measurement result, this sensor has the capability of measuring lightning current in power grid up to 400 kA and the sensitivity is 0.1144 V/kA for output voltage below 50 V. The sensitivity of the coil linearly increases with the increase number of turns and sectional area. The sensitivity of the coil decreases exponentially with the increases of the effective magnetic path, integral resistance or capacitance [41]. Furthermore, the reliability PCB Rogowski to measure high amplitude of lightning current has been proved by integrating with passive external circuit which does not require the power [42]. The designed sensor has a good bandwidth and it measures the lightning current correctly.

5. Conclusion

In this paper, a high current sensor based on Rogowski coil with various types and applications were presented. The PCB Rogowski coil is a transformation of conventional Rogowski coil and it has very useful, high technology and has many advantages especially for high current measurement. From this review, it could say that Rogowski coils are widely used for protection, measuring and metering in power system and its performance is better than conventional CTs. It is proved that the Rogowski coil has a high capability for high current measuring or sensor such as fault current detection, high impulse current detection, and lightning current. Furthermore, Rogowski coil is very suitable to replace the CTs because it has more advantages such as it will not damage when a large load applied, no saturation has a wide bandwidth, able to measure large current, low cost (inexpensive), good transient response and much more. Additionally, all the high current test and measurement technique must follow the standard IEC 62475. However, the wireless Rogowski coil for high current measurement has not been developed yet, thus if this technology is applied to the Rogowski coil and it is believed that it will be more interesting.

References

[1] M. H. Samimi, A. Mahari, M. A. Farahnakian, and H. Mohseni, "The Rogowski Coil Principles and Applications: A Review," *IEEE Sens. J.*, vol. 15, no. 2, pp. 651–658, Feb. 2015.

[2] Yi Li, Yunchen Guo, Yi Long, Chenguo Yao, Yan Mi, and Junfeng Wu, "Novel lightning current sensor based on Printed Circuit Board Rogowski coil," in *2012 International Conference on High Voltage Engineering and Application*, 2012, pp. 334–338.

[3] C. C. Yii, M. N. K. H. Rohani, M. Isa, and S. I. S. Hassan, "Multi-end PD location algorithm using segmented correlation and trimmed mean data filtering techniques for MV Underground Cable," *IEEE Trans. Dielectr. Electr. Insul.*, vol. 24, no. 1, pp. 92–98, Feb. 2017.

[4] Chucheng Xiao, Lingyin Zhao, T. Asada, W. G. Odendaal, and J. D. van Wyk, "An overview of integratable current sensor technologies," in *38th IAS Annual Meeting on Conference Record of the Industry Applications Conference, 2003.*, vol. 2, pp. 1251–1258.

[5] I. A. Metwally, "Self-Integrating Rogowski Coil for High-Impulse Current Measurement," *IEEE Trans. Instrum. Meas.*, vol. 59, no. 2, pp. 353–360, Feb. 2010.
[6] J. Douglass, J. Greenly, D. Hammer, B. Kusse, R. McBride, and S. Pikuz, "Design and use of Small Rogowski Coils for use with Large, Fast Current Pulses," in *2005 IEEE Pulsed Power Conference*, 2005, pp. 717–720.
[7] I. A. Metwally, W. J. Zischank, and F. H. Heidler, "Measurement of Magnetic Fields Inside Single- and Double-Layer Reinforced Concrete Buildings During Simulated Lightning Currents," *IEEE Trans. Electromagn. Compat.*, vol. 46, no. 2, pp. 208–221, May 2004.
[8] A. Haddad *et al.*, "Direct voltage and trapped charge effects on the protective characteristic of ZnO surge arresters," *IEE Proc. - Sci. Meas. Technol.*, vol. 142, no. 6, pp. 442–448, Nov. 1995.
[9] V. Skendzic and B. Hughes, "Using Rogowski coils inside protective relays," in *2013 66th Annual Conference for Protective Relay Engineers*, 2013, pp. 1–10.
[10] "4-2013 IEEE Standard for HV Testing Techniques."
[11] A. dos Santos, J. F. Martins, P. Monteiro, L. j. A. Kojovic, M. T. Bishop, and T. R. Day, "Line differential protection systems based on low power stand alone current sensors," in *10th IET International Conference on Developments in Power System Protection (DPSP 2010). Managing the Change*, 2010, pp. 121–121.
[12] L. A. Kojovic, "Comparative Performance Characteristics of Current Transformers and Rogowski Coils used for Protective Relaying Purposes," in *2007 IEEE Power Engineering Society General Meeting*, 2007, pp. 1–6.
[13] N. Karrer and P. Hofer-Noser, "PCB Rogowski coils for high di/dt current measurement," in *2000 IEEE 31st Annual Power Electronics Specialists Conference. Conference Proceedings (Cat. No.00CH37018)*, vol. 3, pp. 1296–1301.
[14] K. Draxler and R. Styblikova, "Calibration of Rogowski coils at high currents," in *CPEM 2010*, 2010, pp. 537–538.
[15] L. a. Kojovic and R. Beresh, "Practical Aspects of Rogowski Coil Applications to Relaying," *IEEE PSRC Spec. Rep.*, no. September, pp. 1–72, 2010.
[16] M. N. K. H. Rohani *et al.*, "Evaluation of Rogowski coil sensor performance using EMTP-ATP software," in *2016 3rd International Conference on Electronic Design (ICED)*, 2016, pp. 446–451.
[17] M. N. K. H. Rohani *et al.*, "Geometrical Shapes Impact on the Performance of ABS-Based Coreless Inductive Sensors for PD Measurement in HV Power Cables," *IEEE Sensors Journal*, vol. 16, no. 17. pp. 6625–6632, 2016.
[18] M. Tsukuda, M. Koga, K. Nakashima, and I. Omura, "Micro PCB Rogowski coil for current monitoring and protection of high voltage power modules," *Microelectron. Reliab.*, vol. 64, pp. 479–483, 2016.
[19] I. A. Metwally, "Self-Integrating Rogowski Coil for High-Impulse Current Measurement," *IEEE Trans. Instrum. Meas.*, vol. 59, no. 2, pp. 353–360, Feb. 2010.
[20] R. J. Thomas, "High-Impulse Current and Voltage Measurement," *IEEE Trans. Instrum. Meas.*, vol. 19, no. 2, pp. 102–117, 1970.
[21] MS Naidu & V.Kamaraju, "Overvoltage Phenomenon & Insulation Coordination in Electric Power System," in *High Voltage Engineering*, Third Edit., Singapore: McGraw-Hill Education Publishing Company Limited, 2004, pp. 285–356.
[22] C.-H. Stuckenholz and M. Gamlin, "Overview of impulse current test standards and the impact on test equipment," in *2012 International Conference on Lightning Protection (ICLP)*, 2012, pp. 1–6.
[23] J. P. Dupraz, A. Fanget, W. Grieshaber, and G. F. Montillet, "Rogowski Coil: Exceptional Current Measurement Tool For Almost Any Application," in *2007 IEEE Power Engineering Society General Meeting*, 2007, pp. 1–8.
[24] O. Poncelas, J. A. Rosero, J. Cusido, J. A. Ortega, and L. Romeral, "Motor Fault Detection Using a Rogowski Sensor Without an Integrator," *IEEE Trans. Ind. Electron.*, vol. 56, no. 10,

pp. 4062–4070, Oct. 2009.

[25] Z. hua Li, S. hong Yan, W. zhong Hu, Z. xing Li, and Y. chun Xu, "High Accuracy On-Line Calibration System for Current Transformers Based on Clamp-Shape Rogowski Coil and Improved Digital Integrator," *Mapan - J. Metrol. Soc. India*, vol. 31, no. 2, pp. 119–127, Jun. 2016.

[26] C. Xianghu, Z. Xiangjun, D. Feng, and L. Ling, "Novel PCB sensor based on Rogowski coil for transmission lines fault detection," in *2009 IEEE Power & Energy Society General Meeting*, 2009, pp. 1–4.

[27] E.-P. Suomalainen and J. Hallstrom, "On-site calibration of a current transformer using a Rogowski coil," in *2008 Conference on Precision Electromagnetic Measurements Digest*, 2008, pp. 468–469.

[28] A. dos Santos, J. Lourenco, J. F. Martins, P. Monteiro, and L. A. Kojovic, "Current sensors for improved capacitor bank protection," in *11th IET International Conference on Developments in Power Systems Protection (DPSP 2012)*, 2012, pp. 153–153.

[29] L. A. Kojovic, "New Protection Schemes Based on Novel Current Sensors for Up-To-Date Grid," no. 1226, pp. 10–13, 2013.

[30] "A new hybrid current sensor for high-voltage applications," *IEEE Trans. Power Deliv.*, vol. 20, no. 1, pp. 32–38, Jan. 2005.

[31] A. H. Abdulwahid and Shaorong Wang, "A busbar differential protection based on fuzzy reasoning system and Rogowski-coil current sensor for microgrid," in *2016 IEEE PES Asia-Pacific Power and Energy Engineering Conference (APPEEC)*, 2016, pp. 194–199.

[32] I. A. Metwally, "New Side-Looking Rogowski Coil Sensor for Measuring Large-Magnitude Fast Impulse Currents," *Sens. Imaging*, vol. 16, no. 1, p. 100, Dec. 2015.

[33] P. Yutthagowith, "Rogowski coil with a non-inverting integrator used for impulse current measurement in high-voltage tests," *Electr. Power Syst. Res.*, vol. 139, pp. 101–108, 2016.

[34] I. A. Metwally, "Performance Improvement of Slow-Wave Rogowski Coils for High Impulse Current Measurement," *IEEE Sens. J.*, vol. 13, no. 2, pp. 538–547, Feb. 2013.

[35] Y. Zhang, J. Liu, G. Bai, and J. Feng, "Analysis of damping resistor's effects on pulse response of self-integrating Rogowski coil with magnetic core," *Measurement*, vol. 45, no. 5, pp. 1277–1285, 2012.

[36] T. Orosz, Z. A. Tamus, and I. Vajda, "Modeling the high frequency behavior of the Rogowski-coil passive L/r integrator current transducer with analytical and finite element method," in *2014 49th International Universities Power Engineering Conference (UPEC)*, 2014, pp. 1–4.

[37] W. F. Ray, C. R. Hewson, and J. M. Metcalfe, "High frequency effects in current measurement using Rogowski coils," in *2005 European Conference on Power Electronics and Applications*, 2005, p. 9 pp.-pp.P.9.

[38] D. A. Ward and J. L. T. Exon, "Using Rogowski coils for transient current measurements," *Eng. Sci. Educ. J.*, vol. 2, no. 3, p. 105, 1993.

[39] C. Romero *et al.*, "Measurement of lightning currents using a combination of Rogowski coils and B-dot sensors," in *2010 30th International Conference on Lightning Protection (ICLP)*, 2010, pp. 1–5.

[40] T. Kawabata, S. Yanagawa, H. Takahashi, and K. Yamamoto, "Development of a shunt lightning current measuring system using a Rogowski coil," *Electr. Power Syst. Res.*, vol. 118, pp. 110–113, 2015.

[41] Z. Li, Q. Zhang, L. Zhang, F. Liu, and X. Tan, "Design of Rogowski Coil with external integrator for measurement of lightning current up to 400kA," *PRZEGLĄD ELEKTROTECHNICZNY (Electrical Rev.*, vol. 87, no. 7, pp. 33–2097, 2011.

[42] Yi Li, Yunchen Guo, Yi Long, Chenguo Yao, Yan Mi, and Junfeng Wu, "Novel lightning current sensor based on Printed Circuit Board Rogowski coil," in *2012 International Conference on High Voltage Engineering and Application*, 2012, pp. 334–338.

IOP Conf. Series: Materials Science and Engineering **318** (2018) 012055 doi:10.1088/1757-899X/318/1/012055

Formation of Maximum Eddy Current Force by Non Ferrous Materials

M M A Kader, Z B Razali, N S M Yasin and M H Daud

Robotics and Automation Technology, Faculty of Engineering Technology, Universiti Malaysia Perlis, 02100 Padang Besar, PERLIS, MALAYSIA

E-mail: zolbahri@unimap.edu.my

Abstract. This project is concerned with the study of eddy current effects on various materials such as aluminum, copper and magnesium. Two types of magnets used in this study; magnetic ferrite (ZnFe+2O4) and magnetic neodymium (NdFeBN42). Eddy current force will be exerted to these materials due to current flows along the magnet. This force depends on the type of magnet, type of material and the gap between the magnet and the material or between the two magnets. The results show that at constant magnet to material gap, the eddy current force decreases as the magnet to magnet gap increases. Similarly, at constant magnet to magnet gap, the eddy current force decreases as the magnet to material gap increases. The minimum force was achieved when the gap of magnet to material is maximum, similarly to the gap of magnet to magnet. The weakest force was between Copper and Neodymium at a magnet to material gap of 20 mm and magnet to magnet gap of 40 mm; the eddy current force was 0.00048 N. The strongest force (maximum) was between Magnesium and Ferrite and 0.42273 N at a magnet to material gap of 3 mm and magnet to magnet gap of 5 mm.

1. Introduction

This project focused on the fundamental study and application of eddy current technique as a separator in separating non-ferrous waste. Conceptually, a variety of metal scrape waste are passed through a separating station on a conveyor system. In the station, the waste can be recognized, differentiated and separated between ferrous and non-ferrous materials, where the non-ferrous waste will be thrown away a bit far automatically from the conveyor, due to internal force created by the eddy current. Thus the research is focusing on the fundamental of eddy current on the non-ferrous materials and the relationship with the gap between magnets and materials to produce eddy current force.

This research consists of studying eddy current forces exerted by magnets (ferrite and neodymium) to aluminium, copper and magnesium at various gaps between magnet and material, and between magnet and magnet. The research is conducted through simulation using ANSYS.

IOP Conf. Series: Materials Science and Engineering **318** (2018) 012055 doi:10.1088/1757-899X/318/1/012055

2. Eddy Current Technique

Changing magnetic field in a conductor will induce electric currents within the conductors; the induced currents are called eddy currents or also known by another name as Foucault currents. Since these circulating eddy currents have inductance, they will induce magnetic fields [1].

Figure 1 shows a rectangular-shaped conductor with an AC flux ϕ. According to Faraday's law, an AC voltage E_1 is induced across its terminals. If the conductor is short-circuited, a substantial alternating current I_1 will flow. If more conductor is placed inside the first one as in Figure 2, a smaller voltage will be induced because it links a smaller flux [2]. There are a number of factors, apart from flaws, which will affect the eddy current response from a probe [3] and material conductivity, permeability, frequency, geometry and proximity/lift T off [4].

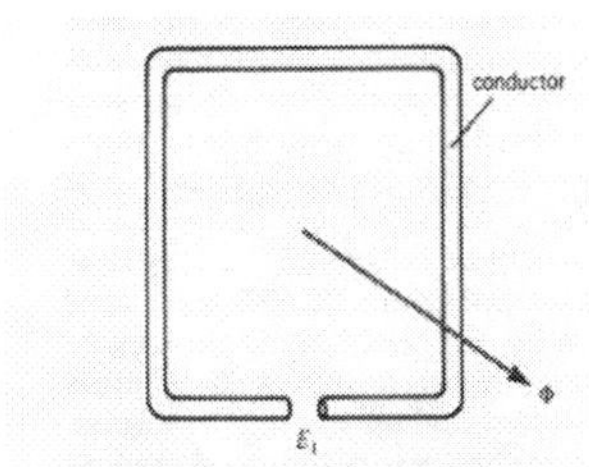

Figure 1. Conductor with an ac flux ϕ [2].

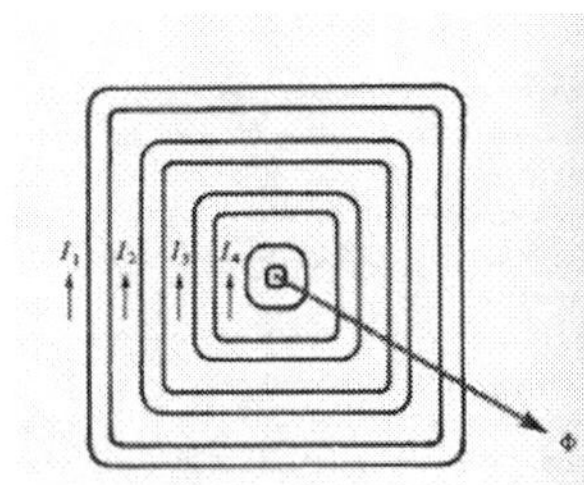

Figure 2. More conductors carry AC currents [2].

2.1 Force acting on particles due to eddy current

In order to calculate the forces acting on particles due to eddy current, a dipole model was developed. It was assumed that the magnetic field induces Eddy-currents within a particle, which in turn generates a magnetic moment in that particle. The interaction between the magnetic moment and the external magnetic field results in Eddy-current forces. The force can then be expressed as:

$$F_{LP} = \mu_M \nabla B \quad (1)$$

Since,

$$\mu_M = M_P V_P \quad (2)$$

Thus, the torque on the particle is given by:

$$T = V_P M_P B_0 \quad (3)$$

Where,

V_P – Volume of the particle
M_P – Magnetization of the particle
μ_M – Effective magnetic dipole moment of the particle
B_0 – External magnetic induction

In order to determine the force acting on a particle, given by eq. (3), a magnetic field generated by the magnetic system must be calculated. Shown that the components of the magnetic induction in the cylindrical coordinate system (r, Ø, z) relative to the drum axis can be written:

$$B_r = \sum_{n=0}^{\infty} b_n \left(\frac{r}{R_{drum}}\right)^{-2(n+1)k-1} \times \sin[(2n+1)k(\emptyset - \omega_{drum}t)] \quad (4)$$

$$B_r = \sum_{n=0}^{\infty} -b_n \left(\frac{r}{R_{drum}}\right)^{-2(n+1)k-1} \times \cos[(2n+1)k(\emptyset - \omega_{drum}t)] \quad (5)$$

Where b_n are the Fourier coefficients, R_{drum} the radius of the drum, k the number of pairs of the magnets used in the drum and ω_{drum} the angular velocity of the drum. The Fourier coefficients can be determined by measuring the magnetic field strength as a function of distance from the magnetic pole surface.

Based on equation (4) and (5), obtained the following expressions for the tangential and radial eddy-current forces acting on the particle:

$$F_{LPt} = C_1 \frac{C_2}{1+C_2^2} B^2 \quad (6)$$

$$F_{LPr} = C_1 \frac{C_2^2}{1+C_2^2} B^2 \quad (7)$$

Where,

$$C_1 = \frac{2\pi s V_P}{\pi_0 \omega} \quad (8)$$

$$C_2 = (k\omega_{drum} + \Omega)\tau \quad (9)$$

ω – Width of one pair of magnet
s – Shape factor of the particle
Ω – Angular velocity of particle rotation
τ – Characteristic time with induced magnetic field decays in the particle

$$\tau = \mu_0 \sigma_P s B^2 \quad (10)$$

Where,
B– Particle radius
The torque on the particle was found to be:

$$T = \frac{sV_P}{\mu_0} \frac{C_2}{1+C_2^2} B^2 \quad (11)$$

It can be seen that the eddy-current forces and the torque depend on the square of the magnetic field strength. By replacing ceramic ferrite permanent magnets by rare-earth magnets, a considerable increase in the separating force can be, therefore, achieved [5].

A non-ferrous metal is any metal either pure or alloy that does not contain iron in appreciable amounts. Non-ferrous metals are generally more expensive than ferrous metals. They are used because of their desirable properties such as low weight (e.g., aluminium), higher conductivity (e.g., copper), non-magnetic property or resistance to corrosion (e.g., zinc) [6].

Material selection mainly depends on how easily the can be shaped. Although the same operations are used with ferrous as well as nonferrous metals and alloys, the reaction of nonferrous metals to these forming processes is often more severe. Consequently, properties may differ considerably between the cast and wrought forms of the same metal or alloy [7].

2.1.1 Ferrite magnet.
A ferrite is a chemical compound of ceramic materials with iron (III) oxide (Fe2O3) as its main component. Yogoro Kato and Takeshi Takei of the Tokyo Institute of Technology invented magnet from ferrite in1930 [5][6]. The density of ferrite magnets is about 5 g/cm^3 [8][9].

2.1.2 Neodymium magnet.
A neodymium magnet (also known as NdFeB or Neo magnet) is the most widely used type of rare-earth magnet. It is a permanent magnet made from an alloy of neodymium, iron and boron to form the Nd2Fe14B tetragonal crystalline structure. Developed in 1982 by General Motors and Sumitomo Special Metals, neodymium magnets are the strongest type of permanent magnet commercially available [14]. The greater force exerted by rare-earth magnets creates hazards that are not found in other types of magnet. Neodymium magnets larger than a few cubic centimetres are strong enough to cause injuries

to body parts pinched between two magnets, or a magnet and a metal surface, even causing broken bones [11]. the greater strength of neodymium magnets has inspired new applications in areas where magnets were not used before. Neodymium magnets are used in the surgically placed Linx anti-reflux system [12].

2.2 Analysis of eddy current

This research consists of studying eddy current forces exerted by magnets (ferrite and neodymium) to aluminium, copper and magnesium at various gaps between magnet and material, and between magnet and magnet. The research is conducted through simulation using ANSYS. The ANSYS program uses Maxwell's equations as the basis for electromagnetic field analysis. Electromagnetic analysis may be coupled to circuit, heat transfer, mechanical, or fluid dynamics analysis [12, 14]. The type of formulation suitable for an analysis is determined by the way the current is introduced into the model [13].

3. Methods and Materials

3.1 Eddy current proximity sensors.

Alternating current and alternating magnetic field is produced if a coil passes a current. If there is a metal object in close proximity to this alternating magnetic field, then eddy currents are induced in it. The eddy currents themselves produce a magnetic field.

This distorts the magnetic field responsible for their production, which changes the impedance of the coil and the amplitude of the alternating current. At some pre-set level, these changes can be used to trigger a switch. The changes can be utilized to make a sensor which is used for the detection of non-magnetic but conductive materials; See Figure 3, the sensors have the advantages of being relatively inexpensive, small in size, with high reliability and can have high sensitivity to small displacements [14].

3.2 Simulation of the magnetized material.

The arrangement of the magnet and the material to be investigated is shown in Figure 4. The red rectangular is the magnet (either ferrite or neodymium) while the blue circle is the material (aluminum, copper or magnesium). When a current flows in the magnet, it will exert eddy current force to the material [14]. The force will depend on the gap between the magnet and the material, the amount of the current, the material and the type of the magnet. The gap is 3 mm, 5 mm, 10 mm, 15 mm and 20 mm.

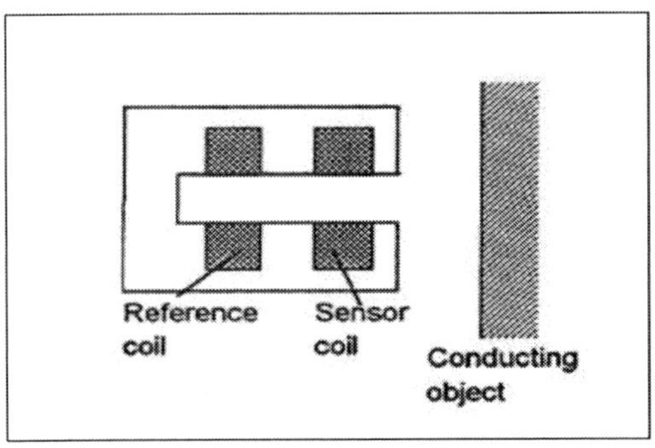

Figure 3. Eddy current sensor [3].

Figure 4. Arrangement of magnet and material.

3.3 Method of analysis of eddy current.

Analysis of eddy-current effect on aluminum material is done by ANSYS Software where Eddy force, magnetic flux density, magnetic field intensity and interaction force can be determine. The main reason of this analysis is to determine the force acting on non-ferrous material (aluminum) produced by the Neodymium magnets (NbFebN42). The analysis manipulated data is the distance between magnet to magnet and the distance between magnets to material [14].

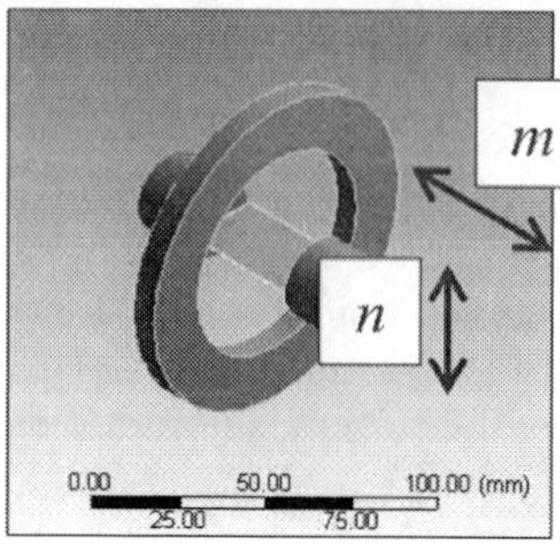

Figure 5. Analysis on effect of Eddy-Current on aluminum

Figure 5 shows the two distances between magnet to magnet and magnet to material. Here, *m* is representing magnet to magnet gap and *n* representing magnet to material gap. The gaps are measured in millimeter (mm).

3.4 Materials and gaps for the simulation.
In the simulation done, there are three materials to be chosen (aluminum, copper and magnesium). Those materials are chosen because in industry, there are over 60% of them which become waste. For the magnets, ferrite and neodymium magnets are used.

To analyze the eddy-current effects, two different kinds of gaps are used; they are magnet to magnet gap and magnet to material gap. Figure 6 shows the placement of magnet and material while Figure 7 shows Mesh of the analyzed parts.

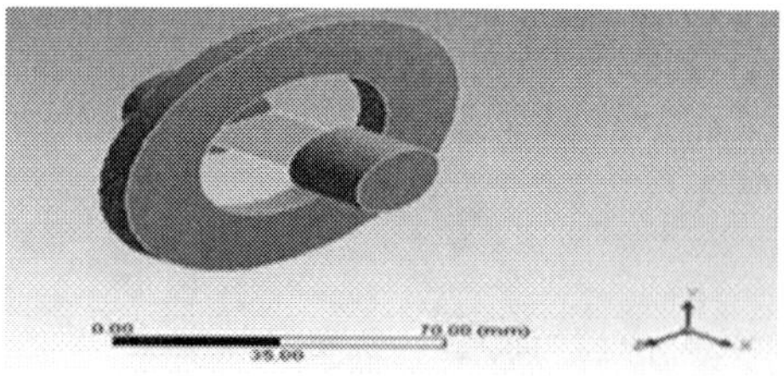

Figure 6. Placement of magnet and material.

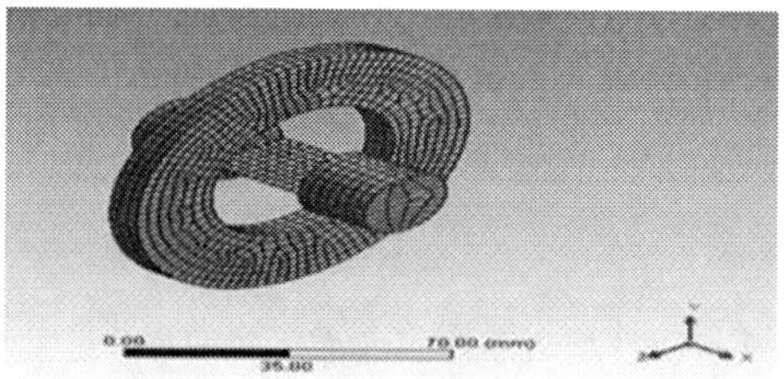

Figure 7. Mesh of the analyzed parts.

4. Results and Discussions
In the simulation done, it is decided to use smaller materials to reduce the force used to rotate the magnets. From the results of preliminary analysis, it can be seen that the critical or greater force is at the non-ferrous materials. This shows that the closer the gap between the magnet and the materials, the higher the eddy current force required to push the materials away from the magnet. The results described in the next part prove that the eddy current could be formed if there is a considerable gap established

between magnet to magnet and magnet to materials. Total maximum Eddy force analysis is shown in Figure 8.

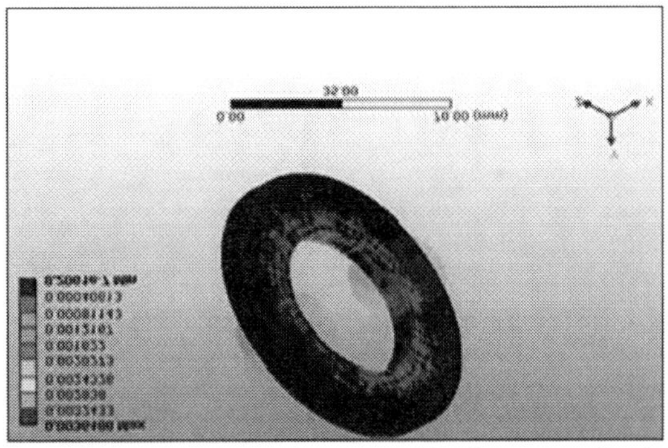

Figure 8. Total maximum Eddy force Analysis.

ANSYS software is used for analysis. Here, the simulation used the 5 mm Magnet to Magnet Gap and the 3mm Magnet to Material Gap. From the analysis for 5mm magnet to magnet gap and 3mm magnet to material gap, other result of analysis on different gap distance of magnet to magnet and magnet to material can be done. Figure 9 shows the respond of Ferrite Magnet to Magnesium, which results to the maximum eddy current force, while Figure 10 shows the respond of Neodymium Magnet to copper, which results to the minimum eddy current force.

The combination of overall result will show on the best result of maximum Eddy force acting on non-ferrous material. The highest maximum Eddy force acting on the material can be able to produce highest lift force to the material in the separation process of non-ferrous materials.

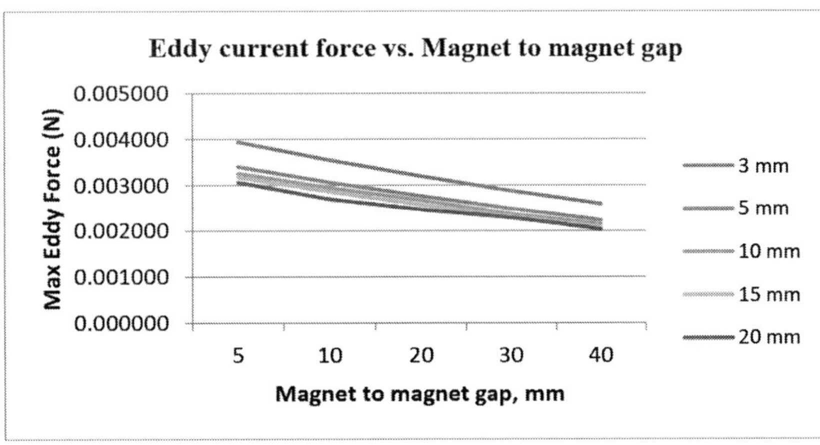

IOP Conf. Series: Materials Science and Engineering **318** (2018) 012055 doi:10.1088/1757-899X/318/1/012055

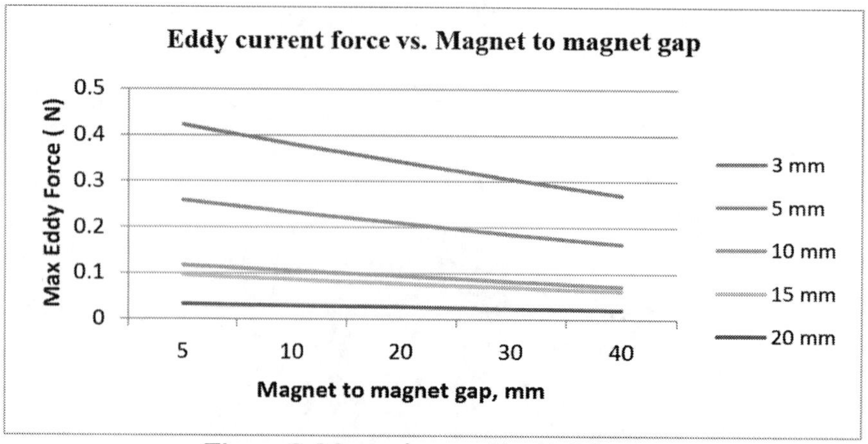

Figure 9. Magnesium vs. Ferrite Magnet

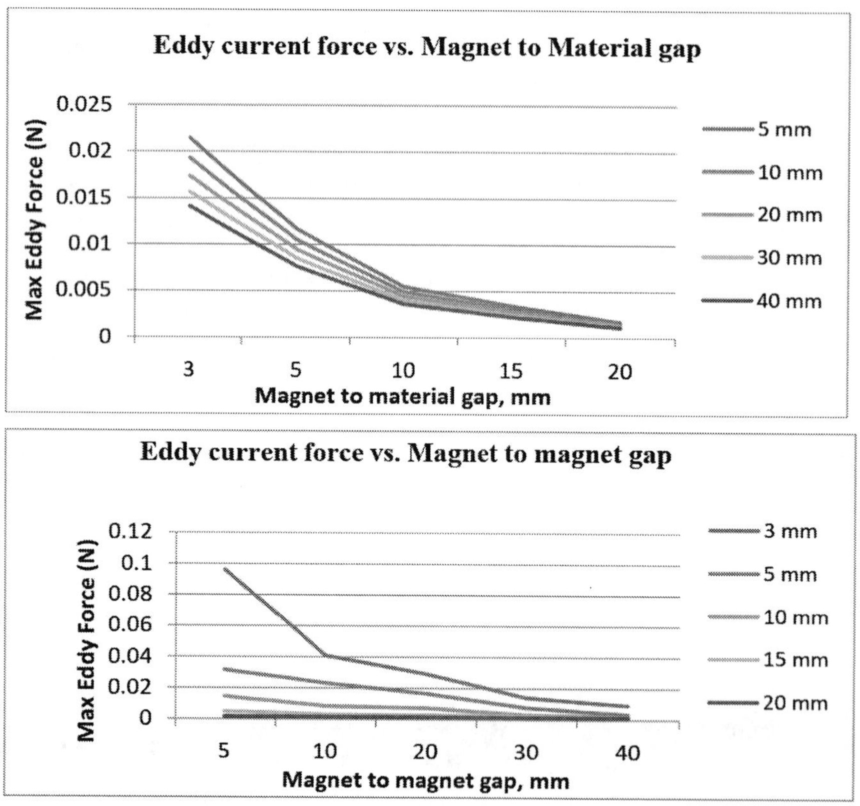

Figure 10. Copper vs. Neodymium Magnet

The pattern of eddy current force for other combination of magnet and materials is similar as has been explained for eddy current force for aluminum and ferrite [14]. The results of all the analysis are summarized in Table 1 and Table 2. From Table 1 and Table 2, it is seen that maximum eddy current force is achieved when the magnet is ferrite and the material is magnesium. Minimum eddy current force is achieved when the magnet is neodymium and the material is copper. X1=magnet to material (3 mm to 20 mm) and X2=magnet to magnet (5 mm to 40 mm)

Table 1: Maximum and minimum eddy current for Ferrite and the three materials.

	Max. eddy current at x1 = 3 mm and x2 = 5 mm	Min. eddy current at x1 = 20 mm and x2 = 40 mm
Magnesium-Ferrite	0.42273	0.02023
Aluminum-Ferrite	0.258140	0.002832
Copper-Ferrite	0.003940	0.002037

Table 2: Maximum and minimum eddy current for Neodymium and the three materials.

	Max. eddy current at x1 = 3 mm and x2 = 5 mm	Min. eddy current at x1 = 20 mm and x2 = 40 mm
Aluminum- Neodymium	0.096118	0.000926
Magnesium-Neodymium	0.02146	0.00114
Copper- Neodymium	0.001927	0.00048

5. Conclusion

From the simulation conducted, it is found that at constant magnet to material gap, the eddy current force decreases as the magnet to magnet gap increases. Similarly, at constant magnet to magnet gap, the eddy current force decreases as the magnet to material gap increases. It can be seen that the minimum force was achieved when the magnet to material gap is maximum and the magnet to magnet gap is also maximum.

The maximum force was achieved when the magnet to material gap is minimum and the magnet to magnet gap is also minimum. The results of the simulation are summarized in Tables 7 and 8. It is found that the weakest (minimum) force was between Copper and Neodymium at a magnet to material gap of 20 mm and magnet to magnet gap of 40 mm; the eddy current force was 0.00048 N. The strongest force (maximum) was between Magnesium and Ferrite and 0.42273 N at a magnet to material gap of 3 mm and magnet to magnet gap of 5 mm.

Acknowledgements

The author gratefully acknowledges the project is partial support by the Fundamental Research Grant Scheme (FRGS 9003-00442) Universiti Malaysia Perlis.

References

1. Schloemann E 1982 *Conserv. Recycl.* **5** (2/3) 149-162
2. Wildi T 2006 *Electrical Machines, Drives, And Power Systems*, 6th. Ed. Singapore: Pearson Prentice Hall
3. Bolton W 2003 *Mechatronics*, 3rd Ed, Pearson.: 28-29
4. Buckley JM 2009 *An introduction to eddy current theory and testing technology*. Retrieved from http://Tekintro.Doc
5. Robert T, Noguez ME, Salas G and Montejano S 2015 *Journal of material sciences*, **94** (3)

IOP Conf. Series: Materials Science and Engineering **318** (2018) 012055 doi:10.1088/1757-899X/318/1/012055

6. Hassan AM and Al-Bsharat AS 2009 *Improvements In Some Properties Of Non-Ferrous Metals By The Application Process*. Retrieved from http://Tekintro.Doc
7. Kwak JS 2009 *Enhanced magnetic abrasive polishing of non-ferrous metals utilizing a permanent magnet*. Retrieved from http://Tekintro.Doc.
8. Fiorillo F, Beatrice C, Bottauscio C, and Carmi E 2015 *Journal of Material Sciences*, **94** (3)
9. Moritz G 2004 *Eddy currents in accelerator magnets*. Pearson Prentice Hall
10. Sung DG, Kim DH, Cha H, Kim SH and ByunJK 2009 *Analysis of EMI Reduction Effect of a Magnetorheological Fluid-gap Inductor in DC/DC Converter*. Retrieved from http://Tekintro.Doc
11. Hamia R, Cordier C and Dolabdjian D. 2015 *Journal of material sciences*, **94** (3)
12. Zakaria Z, Yern LP and Abdullah AA 2015 *Journal of material sciences*, **94** (3)
13. Svoboda J 2004 *Magnetic Technique for the Treatment of Materials* Secaucus, NJ, USA: Kluwer Academic Publishers
14. Razali ZB and Nur Hazwan I 2015 *Applied Mechanics and Materials* Vols. **754-755** (2015) pp 576-580

IOP Conf. Series: Materials Science and Engineering **318** (2018) 012056 doi:10.1088/1757-899X/318/1/012056

Investigation on Convergence – Divergence Nozzle Shape for Microscale Channel in Harvesting Kinetic Energy

M S Zakaria [1], S Zairi [1], M N Misbah [1], M Saifizi [1] and Izzudin Rakawi [1]
[1] Faculty of Engineering Technology, Universiti Malaysia Perlis, Level 1, Block S2, UniCITI Alam Campus, Sungai Chuchuh, 02100 Padang Besar, Perlis, Malaysia

Email: syamil@unimap.edu.my

Abstract. This paper presents performance evaluation of nozzle shapes on microscale channel by employing different types of NACA airfoils profile and conventional profile. The deploying nozzle used are NACA 0012, NACA 0021 and NACA 0024 airfoils while for conventional convergence-divergence nozzle diameter ratio (d_2 /d_1) in the range from 1/4 to 3/4 are applied. These nozzles are assembled on rectangular cross sectional microscale channel which has designated constant fluid flow velocity at the channel inlet. This study revealed reduction on diameter ratio increased dramatically fluid velocity but further reduction on diameter ratio exposed fluid flow to fluctuate which slightly slowing down the fluid velocity. Nevertheless, curved NACA profiles are favourable for convergence – divergence nozzle in microscale channel as it significantly improved flow characteristics by enhancing fluid velocity and resultant kinetic energy as compared to conventional profile.

1. Introduction

The sustainable energy has been a point of interest for the researchers, due to the level of greenhouse effect and the fuel cost dilemma. Renewable energy is an efficient solution for the future to accomplish a perfect connection between renewable energy and economic improvement [1]. Small-scale hydro or micro-hydro power has been increasingly used as an alternative energy source which produces clean power resource, especially in remote areas where other power sources are not available [2]. Instream innovation is a small scale hydro-plant system and it depends on the hydrokinetic turbines in waterways or canal to produce energy. Usually, these systems do not require a huge infrastructure like dam or major water diversions, but by water wheels which reduce environmental impacts [2]. Harnessing kinetic energy from streaming water can optimize existing facilities [3]. For past years, researchers have been interested in the use of hydrokinetic turbines in stream of waterways, trench or channel to create power because of the greenhouse gas emissions and the fuel cost crisis. Numerous countries that are surrounded by watering system or rainy channels have a great potential for building up this innovation [4]. These countries have a primary issue for the advancement of hydrokinetic turbines which is low velocity of flow occurs in open flow microscale channel. An idea to increase the channels current flow is to set up nozzle in open channels.

The nozzle was considered to have the capability of intensifying the flow direction at the same time increasing the flow velocity. Based on research, it is found that the nozzle performance is important which feature the interactions of nozzle geometry and kinetic energy produce. Performance of nozzle can be explained as failure such as velocity of water loss and inconsistent flow of water produce occur on the nozzle which lead to incapability for the nozzle to generate energy. During harnessing energy from flowing water, there are a few problems that occur. The conventional nozzle that is commonly used in hydrokinetic plant is insufficient as in low velocity flowing water, the kinetic energy captured is limited. Moreover, the conventional nozzle loss high kinetic energy at high velocity due to turbulence flow in the stream. Nozzle geometry which are nozzle shapes and diameter ratio influence the performance and characteristics of the flow in the rectangular microcscale channel [5]. As nozzle geometry is vital factor in capturing energy in flowing stream and its alteration contribute to massive enhancement thus it is important to study on how the nozzle performance affected by the nozzle geometry. The focus of the study is the effect of the nozzle geometry in a rectangular microscale channel on nozzle performance in the aspect of kinetic energy.

2. Design of microscale channel

2.1. Configuration of microscale channel

The shape of the microscale nozzle channel design is shown in Figure 1 where L shows the overall length of channel, W is the width of the channel, T is the inlet water level, and h is the channel depth. The channel has rectangular cross section. The nozzle was deployed in the middle of the channels which Ln is the length of nozzle. NACA Four-Digit Series is the first family of NACA airfoils. NACA Four-Digit Series consists of 4 digits e.g. NACA 0012, which label the camber, position of the maximum camber and thickness [7]. For example, the NACA 0012 airfoil has a maximum thickness of 12% of chord length. The geometry of NACA 0012 is shown in Figure 2.

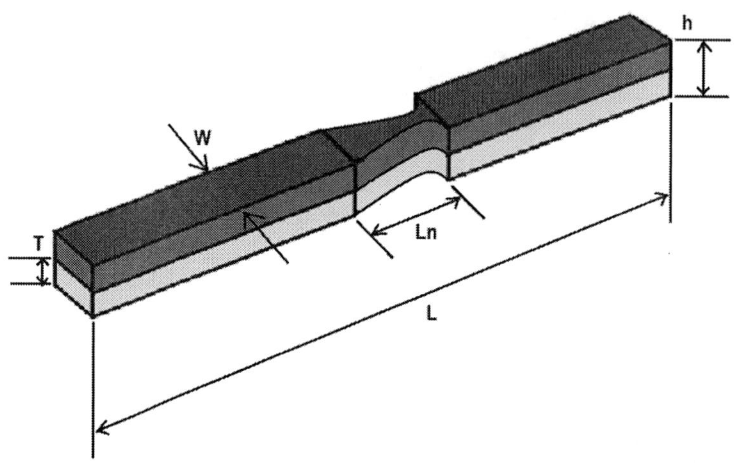

Figure 1. Microscale channel configuration

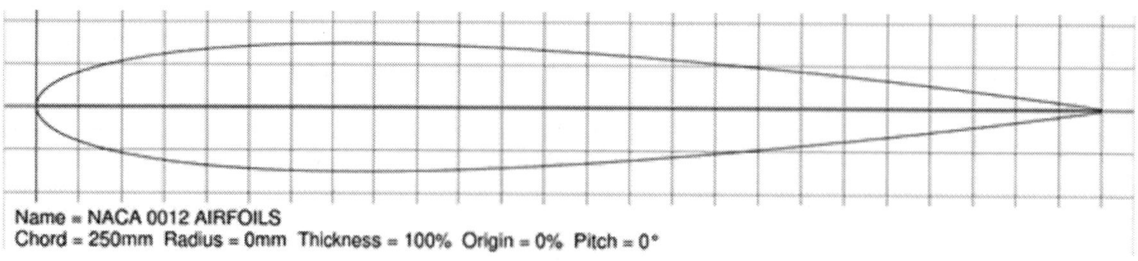

Figure 2. Cross Section of NACA 0012 [7].

Conventional nozzle used in this study for comparison to NACA nozzle is convergence - divergence system. In convergence – divergence nozzle, geometry difference between the conventional nozzles is the diameter ratio between the nozzles. Figure 3 shows dimension notation of d_2 / d_1 in convergence-divergence nozzle in which d_2 is diameter of nozzle and d_1 is width of channel.

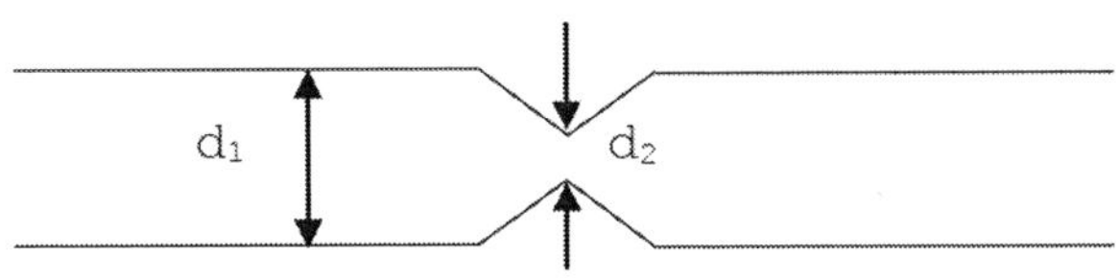

Figure 3. Conventional convergence - divergence nozzle

2.2. *Experimental work*

Microscale channel is fabricated internally which it can fit every nozzle employed in this study. The channel is made by aluminium thin sheet by forming it to rectangular cross section. The nozzles are placed at the slits on channel. For easy access, slits have been made at the bottom part of the nozzles to allow the nozzles be replaced with another nozzles with ease. NACA based nozzle are fabricated by using styrofoam. In order to harden the finished styrofoam, a polystyrene putty is used and sandpapers of different grade are used to smoothen the airfoils profile. Figure 4 exhibits a complete microscale channel used in conducting the experiment.

The inlet water velocity is set to constant velocity throughout the experiment. Velocity of the fluid stream is measured to determine kinetic energy produced consequent of accelerating fluid stream at the end of nozzle. In Table 1 it shows parameters for each nozzle applied in the experiment. The nozzles are assembled to microscale channel 40 cm from the channel inlet and a turbine is installed definitely at the end of nozzle to measure fluid stream velocity. Fluid stream velocity measurement is based on revolution of turbine rotation from the accelerating of fluid flow.

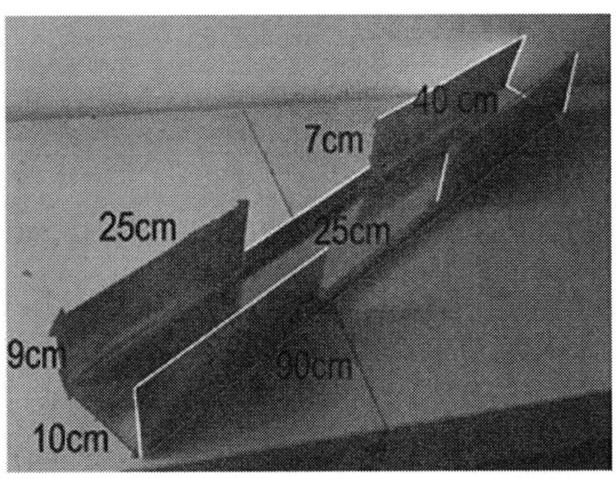

Figure. 4. Microscale channel

3. Mathematical equations

The following equation is used in calculating the fluid velocity.

$$v = \omega r \tag{1}$$

v = velocity, m/s
ω = angular velocity, rad/s
r = radius of turbine blade, m

The radius of the blade is 7.5cm (0.075m) and angular velocity is acquired directly from turbine speed. From the velocity measured, the kinetic energy produced by each configuration of respective nozzle is calculated accordingly.

$$K_e = \frac{1}{2} m v^2 \tag{2}$$

Continuity equation (3) is applied to calculate mass flow rate through the nozzle inlet.

$$m = \rho v A = \rho v_1 A_1 = \rho v_2 A_2 \tag{3}$$

where ρ is density of water and A is cross sectional area of the flow.

Table 1. Nozzle Parameters

Nozzle	**Profile**	**Parameter**
NACA 0012	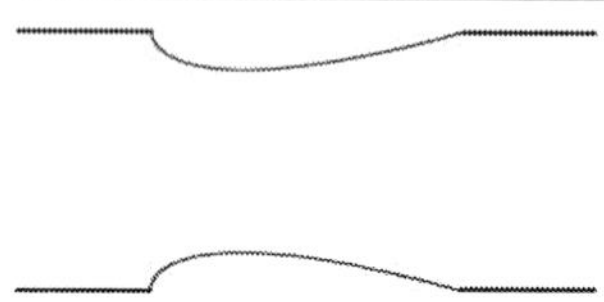	$d_2/d_1 = 0.76$ $L_n = 0.25$ m
NACA 0021	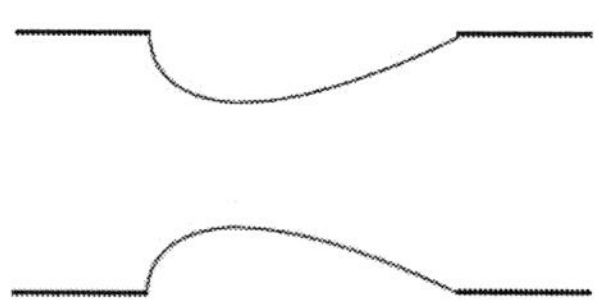	$d_2/d_1 = 0.58$ $L_n = 0.25$ m
NACA 0024	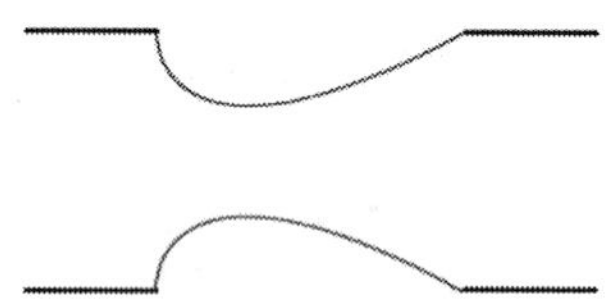	$d_2/d_1 = 0.52$ $L_n = 0.25$ m
Conventional convergence - divergence		$d_2/d_1 = 1/4$ $L_n = 0.25$ m $d_2/d_1 = 1/2$ $L_n = 0.25$ m $d_2/d_1 = 3/4$ $L_n = 0.25$ m

4. Result and discussion

Nozzle profile itself enable it to control the direction and characteristics of a fluid flow. This is due to the nozzle inner structure, especially during choking process at the throat. The flow is obstructs and compress during choking. The flow from the inlet section accelerates at the throat then decelerates through the outlet section. The concept of convergence-divergence nozzle employed both on NACA and conventional nozzle is proven offer good performance as it increased the mean velocity across nozzle [3] since this system capable to develop flow pattern properties behind the nozzle. Based on the graph in Figure 5, NACA 0024 produced the highest velocity which is 0.65 m/s while NACA 0012 has the least velocity 0.5 m/s produced as expected. By referring to continuity equation (3), velocity will increase as the area of contraction decreased. Since NACA 0024 has the maximum thickness, thus it has resulted in the lowest contraction area. Consequently NACA 0024 profile accelerated the fluid flow stream to the highest value.

Meanwhile for convergence – divergence nozzle, the maximum velocity produced is 0.554 m/s for diameter ratio 1/4 as depicted in Figure 6. When the diameter ratio increase by 25%, the resultant velocity reduced to 0.43 m/s. The principle lie on the nozzle concept which is effect of back pressure has contributed to this phenomena. Back pressure at the nozzle outlet is decreased when the diameter at choking area increase hence reducing the resultant velocity. The trend line for both NACA profile and convergence – divergence profile are proportionally linear which aligned with continuity equation. Nevertheless, both graphs in Figure 5 and Figure 6 exhibit slight reduction in velocity on further decreasing of diameter ratio specifically on NACA 0024 and ¼ diameter ratio respectively. It occur due to increasing of turbulence intensity behind the nozzle [5] which lead to friction loss and slowing down the fluid velocity. In Table 2 it presents comparison of diameter ratio to resultant velocity between NACA nozzle and convergence – divergence. It is obviously shows NACA profile nozzle capable of accelerating the fluid flow in microscale channel more than 30% as compared to conventional convergence – divergence nozzle.

Table 2. Velocity comparison for similar diameter ratio

Diameter ratio d_2/d_1	Velocity (m/s)
3/4	NACA 0012 = 0.493 Convergence - divergence = 0.241
1/2	NACA 0024 = 0.646 Convergence - divergence = 0.433

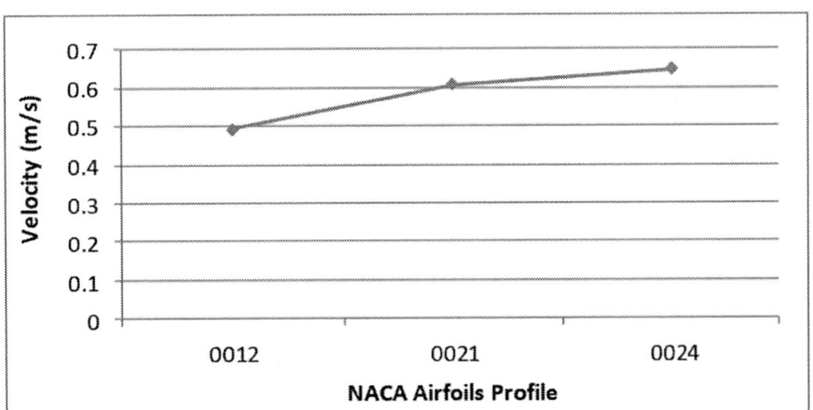

Figure. 5. NACA nozzle velocity produced

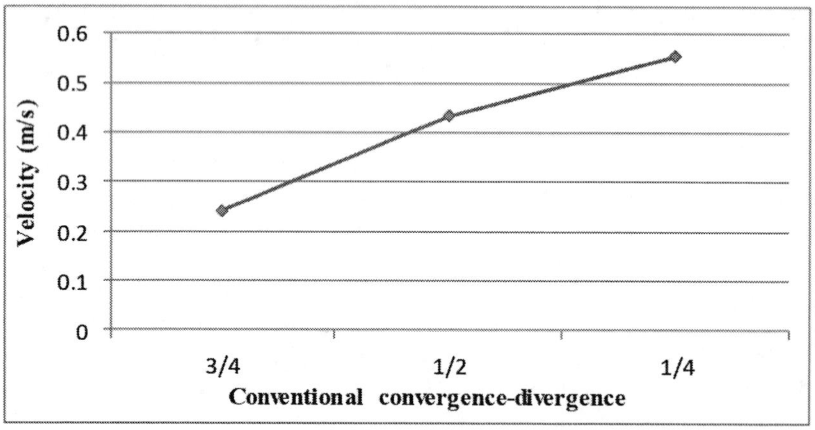

Figure. 6. Convergence - divergence nozzle velocity produced

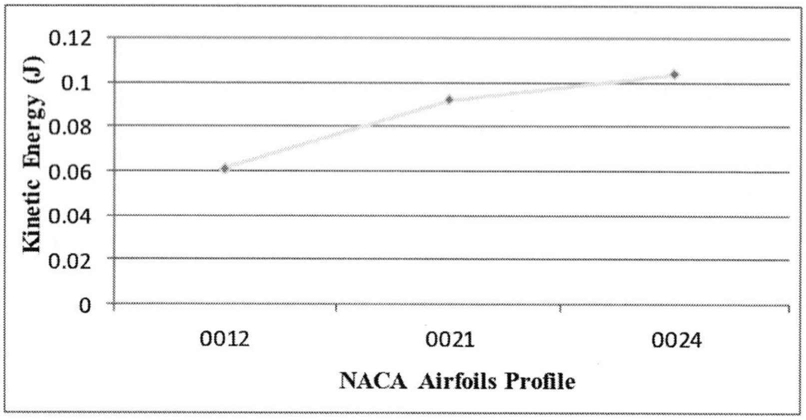

Figure. 7. NACA nozzle kinetic energy produced

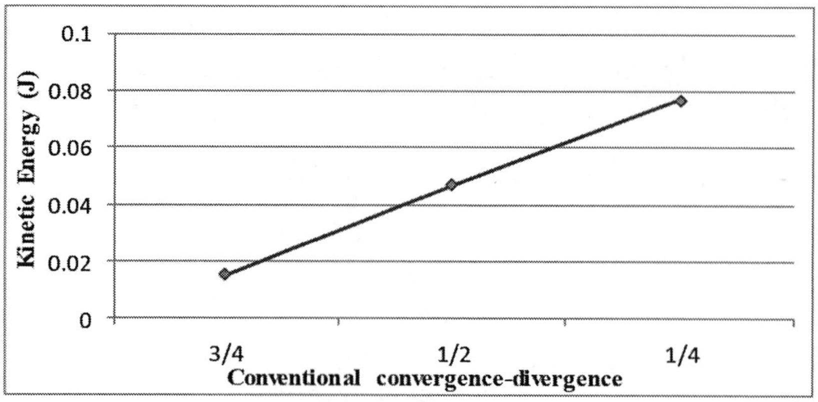

Figure. 8. Convergence-divergence kinetic energy produced

Figure 7 and 8 illustrate kinetic energy captured by respective nozzle. The potential of NACA nozzle in capturing kinetic energy in slow fluid stream is enlighten as it harnessed more energy than conventional convergence-divergence nozzle. Maximum kinetic energy produced by conventional convergence-divergence is only 0.08 J at ¼ diameter ratio while NACA 0012 is producing about 0.06 J even it is at the lowest among NACA profile.

Main factor that contribute to very significant gapping in velocity and kinetic energy relies on nozzle edge shapes. Besides of adequate diameter ratio that increase the velocity, nozzle edge profiles also have vital role which lead to reduction of turbulence flow. Even though inlet velocity of water is low which the flow is in laminar regime, the acceleration of flow across the nozzle profile caused disturbance in fluid flowing. Small fluctuations in the flow is sufficient enough to cause high frictional losses lead to high pressure drop [6]. In addition, it is found from this study curved nozzle profile is favourable in improving flow patterns as it relieves in delaying turbulence flow.

5. Conclusion

An experiment to investigate effect of nozzle profile to the flow velocity and potential kinetic energy in microscale channel were conducted in this study. By utilizing NACA profiles geometry and conventional convergence-divergence as nozzles, it indicated a significant finding on flow velocity and kinetic energy produced. This study revealed reduction of diameter ratio caused in accelerating the fluid flow but further reduction of contraction lead to frictional loss due to flow fluctuations. Furthermore, nozzles with NACA based profile enhanced flow characteristics by delaying turbulence therefore improve flow velocity and amount of kinetic energy.

Acknowledgments

The authors gratefully to acknowledge Universiti Malaysia Perlis for providing fund via Short Term Grant (STG) 9001-00524.

References

[1] Sopian, K., Ali, B., Asim, N. 2011, Strategies for renewable energy applications in the organization of Islamic conference (OIC) countries. *Renew. Sustain. Energy Rev.* **15**, 4706–4725.

[2] Khan, A. A., Khan, A. M., Zahid, M., & Rizwan, R. 2013, Flow acceleration by converging nozzles for power generation in existing canal system. *Renewable Energy*, **60**, 548–552.

[3] Khan, M.J., Bhuyan, G., Iqbal, M.T., Quaicoe, J.E. 2009: Hydrokinetic energy conversion systems and assessment of horizontal and vertical axis turbines for river and tidal applications: a technology status review. *Appl. Energy* **86**, 1823–1835.

[4] Elbatran, A.H., Yaakob, O.B., Ahmed, Y.M., Shabara, H.M.:Operation, 2015, Performance and economic analysis of low head micro-hydropower turbines for rural and remote areas: a review. *Renew. Sustain. Energy* Rev. **43**, 40–50

[5] A.H. Elbatran, O.B. Yaakob, Yasser M.Ahmed, H.M. Shabara, 2015, Numerical study for the use of different nozzle shapes in microscale channels for producing clean energy, *Int. J. Energy Eviron. Eng.* **6**: 137-146

[6] Munson, B.R., Young, D.F., Okishi, T.H. 2006, *Fundamentals of fluid mechanics*, Wiley, New York 5th edn, p. 576.

[7] Airfoil Tools, "NACA 4 digit airfoil generator" http://airfoiltools.com/plotter/index.

IOP Conf. Series: Materials Science and Engineering **318** (2018) 012057 doi:10.1088/1757-899X/318/1/012057

Biomechanical Influence of Implant Neck Designs on Stress Distribution over Adjacent Bone: A Three-Dimensional Non-Linear Finite Element Analysis

Muhammad Ikman Ishak[1], Aisyah Ahmad Shafi[2], Su Natasha Mohamad[3] and Noorlindawaty Md Jizat[4]
[1] Faculty of Engineering Technology, Universiti Malaysia Perlis, Level 1, Block S2, UniCITI Alam Campus, Sungai Chuchuh, 02100 Padang Besar, Perlis, Malaysia
[2] Faculty of Biosciences & Medical Engineering, Universiti Teknologi Malaysia, 81310 UTM Johor Bahru, Johor, Malaysia
[3] Faculty of Allied Health and Sciences, Cyberjaya University College of Medical Sciences (CUCMS), 3410, Jalan Teknokrat 3, Cyber 4, 63000 Cyberjaya, Selangor, Malaysia
[4] Faculty of Engineering, Multimedia University, Persiaran Multimedia, 63100 Cyberjaya, Selangor, Malaysia

Email: ikman@unimap.edu.my

Abstract. The design of dental implant body has a major influence on the stress dissipation over adjacent bone as numbers of implant failure cases reported in past clinical studies. Besides, the inappropriate implant features may cause excessive high or low stresses which could possibly contribute to pathologic bone resorption or atrophy. The aim of this study is to evaluate the effect of different configurations of implant neck on stress dispersion within the adjacent bone via three-dimensional (3-D) finite element analysis (FEA). A set of computed tomography (CT) images of craniofacial was used to reconstruct a 3-D model of mandible using an image-processing software. The selected region of interest was the left side covering the second premolar, first molar and second molar regions. The bone model consisted of both compact (cortical) and porous (cancellous) structures. Three dental implant sets (crown, implant body, and abutment) with different designs of implant neck – straight, tapered with 15°, and tapered with 30° were modelled using a computer-aided design (CAD) software and all models were then analysed via 3-D FEA software. Top surface of first molar crown was subjected to occlusal forces of 114.6 N, 17.2 N, and 23.4 N in the axial, lingual, and mesio-distal directions, respectively. All planes of the mandible model were rigidly constrained in all directions. The result has demonstrated that the straight implant body neck is superior in attributing to high stress generation over adjacent bone as compared to others. This may associate with lower frictional resistance produced than those of tapered designs to withstand the applied loads.

1. Introduction

The use of dental implants for the oral rehabilitation of fully and partially edentulous patients has greatly broadened the scope of clinical dentistry [1]. The success rate of dental implants in replacing the missing teeth has well been proven over the years through many studies. However, minor complication of implant post-surgery is still inevitable with common marginal bone loss occurrence found around oral implant locations. The marginal bone loss may significantly be associated with various factors such as gender, surgical trauma, plaque accumulation, smoking behaviour, biological bone width, bone quality, implant design and biomechanical factors [2].

Many previous clinical studies have been performed to review the biomechanical aspect of dental implant design, quality and strength of osseointegration, bone-implant interface, and their relationship to the long-term success of treatment. There is, however, no optimal design specifications of implant observed which could be considered for the best treatment outcome. The common design of dental implants can be improved to maximise their strength, interfacial stability, and load transfer by using appropriate material, surface treatment, and thread shape [1, 3].

According to a study conducted by Lee et al., the failures of implant were believed to be caused by poor biomechanics specifically poor implant stabilisation to anchor to the bones [3]. As far as stress within bones is concerned, the biomechanical design of dental implant, therefore, may not considerably be disregarded. The magnitude and dissipation of stress influence the rate of bone loss around implant body.

To date, many dental implant manufacturers have great interest in developing new implant design to improve the outcomes of treatment [4]. Stress concentration around implants and within bones are affected by various factors such as shape, geometry, and taper of the implant fixture. Most of past studies that investigate the effect of implant design have merely focused on the implant diameter, length, and thread. There is a few studies, on the other hand, aim to analyse the influence of the neck design of implant body on the stress distribution over adjacent bones. Previous findings indicated that the cortical bone areas surrounding implant neck sustained high stress concentration [5, 6]. This is also supported by a clinical and radiographic study by Tan et al. which investigated on the bone remodelling process of different implant neck configurations [7]. It reported that the implants with reduced height turned neck of 1.8 mm may, indeed, reduce the crestal bone resorption, and may maintain higher crestal bone levels than implants with a 2.8 mm turned, when sunk to the same depth.

There are numerous previous studies conducted to improve dental implant performance via computational analysis. Most of them have examined the distribution of forces in peri-implant bone using finite element analysis (FEA) to predict any possible failures. Finite element analysis plays a vital role in solving mathematical modelling problems in many fields of science and industry [8]. It has widely been used for quantitative evaluation of such stress distribution on implant and its surrounding bone since past three decades [9-14]. Among the advantages offered by FEA are including reliable stress and strain dispersal, simple model modification and accurate representation of complex geometries.

To the best of authors' knowledge, there is less study has been found, to date, to address the influence of different implant neck designs via meticulous finite element modelling. It is therefore a necessity for the present study to examine the effect of different implant neck designs – straight (ST), tapered with 15° (T15°), and tapered with 30° (T30°) towards the stability of implant body using three-dimensional (3-D) FEA. In this study, the stress dispersion within adjacent bones were investigated.

2. Materials and methods

2.1. Three-dimensional mandibular bone model construction

In this present study, a series of computed tomography (CT) image datasets of craniofacial was processed and analysed to reconstruct a 3-D model of mandibular bone using an image-processing software, Mimics 10.01 (Materialise, Leuven, Belgium). In order to differentiate between cortical and cancellous structures, the image datasets of CT scan were threshold using appropriate bone density scale.

For the bone region of interest, it covers the left side of mandibular bone which include the second premolar, first molar and second molar tooth regions as illustrated in figure 1. The finalised bone model has a thickness (cortical), length, height, and width of 1.5 – 4.5 mm, 37.5 mm, 28.5 mm, and 10.5 mm, respectively. As the porous structure or cancellous bone was taken into consideration in the

IOP Conf. Series: Materials Science and Engineering **318** (2018) 012057 doi:10.1088/1757-899X/318/1/012057

analysis, it was modelled as a solid continuum structure with spongy material properties assigned and it was surrounded by the dense cortical bone. The placement of implant into bones was simulated by removing the first molar, whilst the other two teeth were remained in place. As for the crown model, it was created by altering the coronal portion of first molar with the use of Boolean operation.

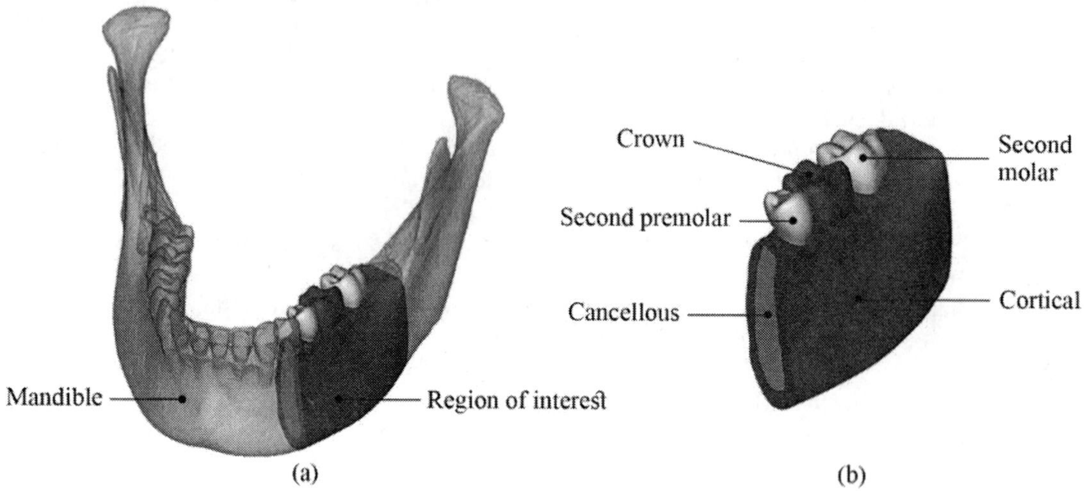

Figure 1. (a) Three-dimensional model of mandible with the selected region of interest. (b) The region of interest of analysis.

2.2. *Three-dimensional implant model construction*

A conventional computer-aided design (CAD) software (SolidWorks 2010, SolidWorks Corporation, Concord, Massachusetts, USA) was utilised to design the 3-D models of implant body and abutment. The length and diameter of implant body to be developed were 13 and 6 mm, respectively, which based on the existing dimensions obtained from a dental implant manufacturer, ANKYLOS®. The implant body was created as a one single body unit that would be attached to the abutment connection part.

In this study, there were three different types of common implant neck have been modelled – straight (ST), tapered with 15° (T15°), and tapered with 30° (T30°). The abutment-implant connection part had been designed as tapered integrated screw, whilst for the external implant thread profile, the buttress type was chosen. In other words, all the three cases investigated having an implant body with similar diameter, length, thread number, and thread profile as depicted in figure 2. As a result, the 3-D model of dental implant system in each case of ST, T15°, and T30° comprises three individual components which are crown, abutment and implant body.

2.3. *Finite element modelling*

The 3-D models of bone and implant system (implant body, abutment and crown) were converted to surface triangular mesh element with a size of 0.4 mm and saved in a stereolithographic file format (.stl). The element size considered was almost four times smaller than the one suggested by Lin et al [15]. For the validation of finite element models in this study, several comparisons had been made with the past related studies in terms of the methods of construction used, selection of parts to be analysed as well as complexity of model construction.

All 3-D models were then exported into a FEA software (Marc Mentat 2007, MSC Software, Santa Ana, California, USA) to convert surface triangular mesh into four nodes tetrahedral solid element with three degree of freedom. The total number of elements for each case of ST, T15°, and T30° was 184,988, 183,897, and 183,518, respectively.

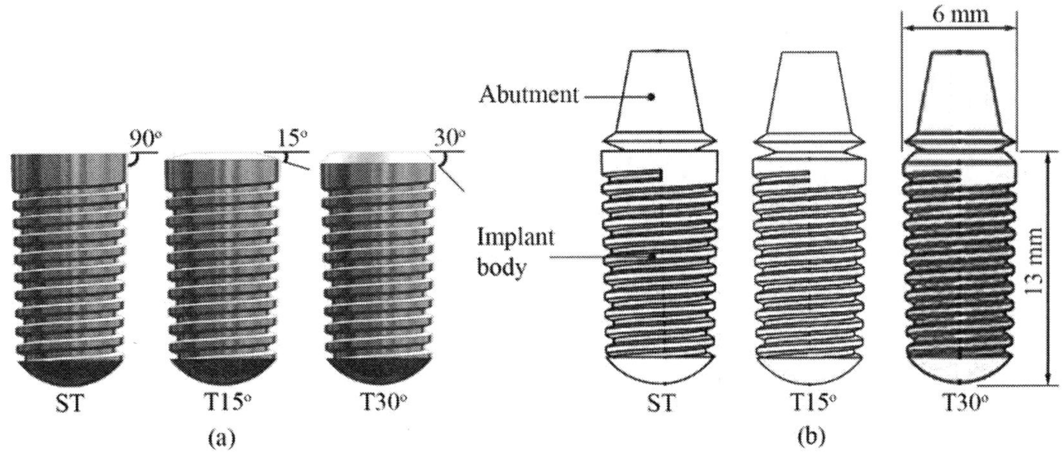

Figure 2. (a) Three-dimensional model of different designs of implant neck. (b) The placement of abutment in the implant body for all the three designs of implant neck.

2.4. Contact modelling

The implant body is rigidly anchored in the bone along its entire interface. This was therefore simulated via non-linear contact modelling in the pre-processing of analysis, accordingly.

A friction coefficient of 0.3 was assigned at the implant-bone, implant-abutment, implant-screw and abutment-screw interfaces to represent the immediate loading principle as well as to secure the transfer of force and prosthetic component deformation. Whilst, the contact surfaces between cortical-cancellous and cortical-mucosa soft tissue were assumed to be as perfectly bonded by merging the nodes between the two contacted models, therefore, no frictional contacts were assigned.

2.5. Material properties

All material properties of finite element model were assumed to be isotropic, homogeneous, static, and linearly elastic throughout the analysis. The crown was made of porcelain whilst the implant body and abutment were made of Ti-6Al-4V titanium alloy. The material properties of all models are shown in table 1.

Table 1. Young's modulus and Poisson's ratio for each finite element model.

Material	Young's Modulus, E (MPa)	Poisson's Ratio, v	References
Crown	68,900	0.28	[16]
Cortical bone	13,000	0.30	[17]
Cancellous bone	9,500	0.30	[18]
Implant body	110,000	0.35	[19]
Abutment	110,000	0.35	[19]
Teeth	18,600	0.31	[20]

2.6. Loading and boundary conditions

A resultant occlusal force of 118.2 N was applied obliquely at 75° respected to the occlusal plane on the top surface of the crown. This load, however, was represented by three different static occlusal load components. The load components with magnitude of –114.6 N, –23.4 N and –17.2 N were applied in the axial (z-axis), mesio-distal (x-axis), and bucco-lingual (y-axis) directions, respectively.

For the boundary condition, the bottom, mesial and distal surfaces of bone segment were fixed in the x, y and z directions to impede any movements or displacements. The loading configuration and boundary condition of the finite element model are illustrated in figure 3.

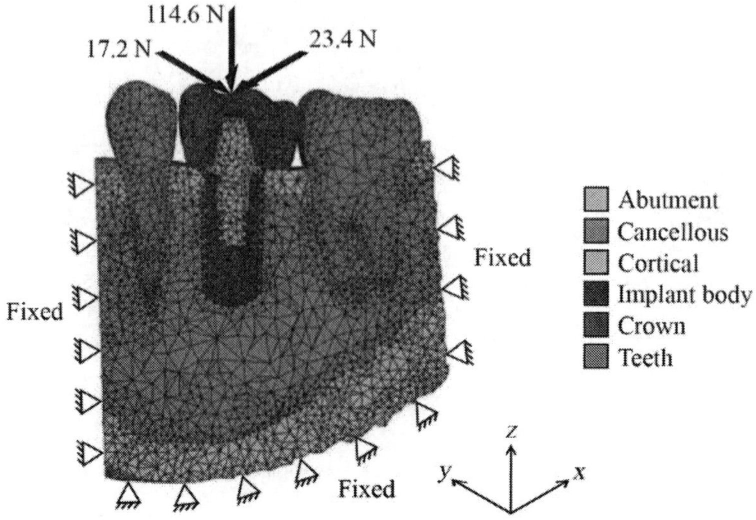

Figure 3. The boundary conditions showing rigid fixation at all planes of the bone segment and the static occlusal forces were applied on the top surface of crown.

2.7. Equivalent von Mises stress result index

The results of analysis were presented and discussed in terms of equivalent von Mises stress (EQV). The EQV index was analysed over adjacent bones for each design of implant necks.

Also, the results were presented in spectrum colouring scale with blue colour indicating low EQV magnitude while light grey colour indicating high EQV magnitude. The bone stress value and its distribution may provide significant information with regards to the influence of different designs of implant neck on mechanical understanding with low-value stress is favourable for a promising interpretation of the results.

3. Results

As far as stress parameter was concerned, our results showed that the greatest EQV value within the cortical bone was generated in ST with 2.8 MPa at the marginal region as exhibited in figure 4. A similar pattern of stress dissipation was found in T15° and T30° where the bone area towards apical portion seemed to sustain a lower stress level. The maximum EQV magnitude recorded in ST was observed to be approximately 2.8-fold higher than those in T15° (1 MPa) and T30° (1 MPa). In overall, the tapered implant neck designs (T15° and T30°) showed a more favourable stress dispersion within the marginal region of cortical bone. Figure 5(a) depicts the EQV distribution over the cortical bone for all designs of implant neck.

The stress results over the adjacent cancellous bone were parallel with the state of stress produced in the cortical bone where the implant neck through ST design recorded the highest value of EQV (2.1 MPa) as compared to others (T15°: 0.6 MPa and T30°: 0.5 MPa). There was less discrepancy of cancellous bone stress value between T15° and T30° which merely 16.7%. It seems that the high stresses were majorly concentrated at the edge region of second premolar-bone interface and gradually spread out towards the surrounding area as illustrated in figure 5(b). The level of stress developed in the cancellous was 1.3-fold lower than the one in the cortical bone for ST (2.8 MPa), whilst 1.7- and 2-fold lower than those in T15° (1 MPa) and T30° (1 MPa), respectively.

4. Discussions

The biomechanical factors of dental implant design such as size, shape, geometry, and neck type play a vital role in defining the success of teeth restoration treatment either for a short- or long-term evaluation. Based on past studies, the performance of dental implants is mainly dependent on the initial stability and osseointegration between implant body and bones. Implant neck or also known as collar is defined as the coronal feature of a dental implant. There are several different designs of implant neck may be found in practices, to date, namely straight, tapered, and filleted. The angle of those from tapered group is measured from the coronal plane to the slanted surface of collar.

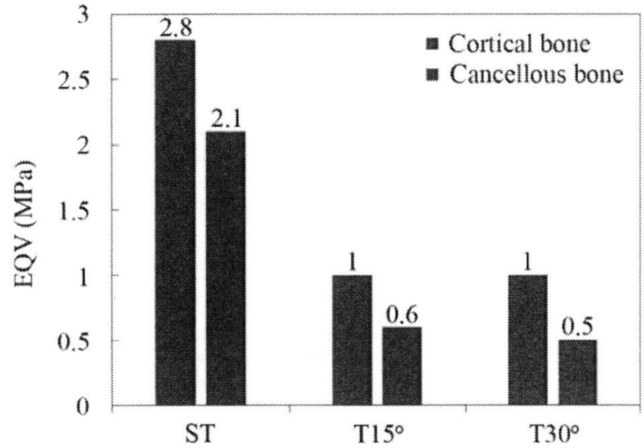

Figure 4. Comparison of EQV magnitude within the cortical and cancellous bone for all implant neck designs.

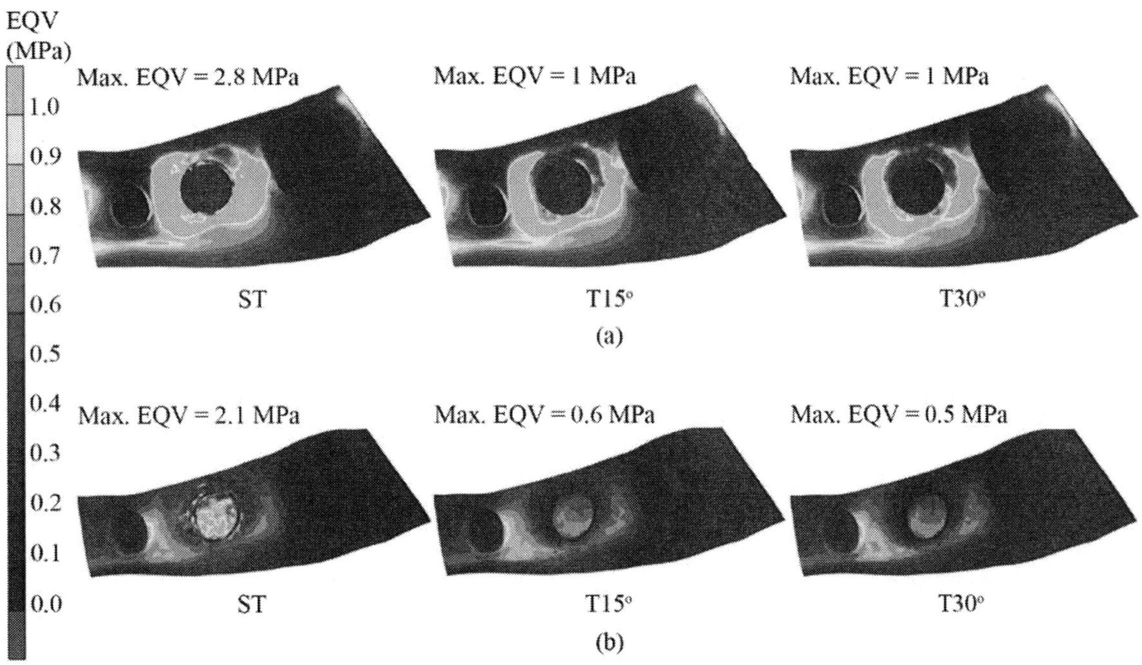

Figure 5. (a) Comparison of EQV magnitude within the cortical bone for all implant neck designs. (b) Comparison of EQV magnitude within the cancellous bone for all implant neck designs.

Crestal bone loss which occurred at the bone-implant interface as reported in previous studies could possibly be due to poor bone response towards mechanical factors of loading [21]. The findings from a study conducted by Karoussis et al., however, described that the hypothesised peri-implant marginal bone loss was also influenced by implant design [22].

The results of this study exhibited that the teeth restoration by the implant neck with ST design has increased the maximum cortical bone stress value approximately 2.8-fold greater than those by T15° and T30°. The stress magnitude difference between ST and T15°, and ST and T30° are similar with a percentage of 64.3%. The implant neck with tapered designs regardless of angle considered, depicted a more encouraging stress dispersion than the straight one. A possible explanation for this observation is that there may be a progressive transfer of occlusal load from the implant body to the cortical bone. As the straight and tapered implant necks have different configuration of contacting surface between implant body and bone tissue, they therefore leave a significant different effect on stress distribution. The straight design had smaller bone-to-implant contact area compared with the tapered due to the sharp edge of implant collar. The tapered design, on the other hand, is likely to have high contact area since the implant surface area increases due to the bevelled edge of implant neck. The slanted surface seems to tolerate better occlusal load than the straight neck due to high frictional resistance produced.

The result of cortical bone stress was in accordance with the cancellous wherein the tapered designs (T15° and T30°) promoted the most promising outcome. It showed that ST design was mostly responsible for the high stress of cancellous bone with about 3.5- and 4.2-fold higher than T15° and T30°, respectively. This was in agreement with the stress distribution developed where ST implant neck produced more highly-stressed regions that concentrated at the apical portion of bone-implant interface which support the applied occlusal load.

According to Hudieb et al., high strain concentration at the implant neck should be avoided as it may lead to crestal bone resorption [6]. In average, the crestal bone loss rate is more than 1 mm in the first year of treatment and about 0.1 mm for each of the following years. The potential zone for fracture when a dental implant system subjected to high bending forces is the neck of implant body. Past studies indicated that stress in the collar was higher than the other sections of dental implant [23]. Thus, it clearly shows that the implant neck plays an important role in stabilising the implant body during its final tightening especially in poor bone quality and marginal compact bone layer [24]. The findings of this study were also parallel with a study performed by Merdji et al. which specified that the mechanical stresses propagate in the areas of bone which are closer to the implant body and decrease towards the outer regions [5].

In all models tested, the greatest value of bone stress was recorded within the cortical bone specifically in ST design (2.8 MPa). Moreover, the maximum stress magnitudes recorded in the bones for all collar designs have no tendency to cause bone failures since the cortical bone is known to be able to sustain stress up to 69 MPa [25].

The limitations of the present study were as follow: (1) all material properties of finite element model had been assumed to be homogenous, isotropic, and linearly elastic; (2) the simulated occlusal load was only applied at one specific node; and (3) the scope of stress outcome was only presented in terms of equivalent von Mises in which may further be expanded in other types such as maximum principal stress.

5. Conclusions

The results of simulated static loadings and non-linear analysis support the following conclusions. The dental implant body with tapered implant neck was evident to be superior as depicted the most encouraging bone stress outcomes in comparison to straight design. The stress was fairly dispersed over the adjacent cortical and cancellous bones specifically at the bone-to-implant contact areas.

Acknowledgments

Appreciation is given to Faculty of Engineering Technology, Universiti Malaysia Perlis, Faculty of Biosciences and Medical Engineering, Universiti Teknologi Malaysia, and Faculty of Allied Health and Sciences, Cyberjaya University College of Medical Sciences. The study has been carried out with the support of Short-Term Grant (Project Number: 9001-00501) received from Research, Management and Innovation Centre, Universiti Malaysia Perlis. The authors reported no conflicts of interest related to this study.

References

[1] Steigenga J T, al-Shammari K F, Nociti F H, Misch C E and Wang H L 2003 *Implant Dent.* **12**

306-17
[2] Blanes R J, Bernard J P, Blanes Z M and Belser U C 2007 *Clin. Oral Implants Res.* **18** 707-14
[3] Lee J H, Frias V, Lee K W and Wright R F 2005 *J. Prosthet. Dent.* **94** 377-81
[4] Degerliyurt K, Simsek B, Erkmen E and Eser A 2010 *Oral Surg. Oral Med. Oral Pathol. Oral Radiol. Endod.* **110** e 1-11
[5] Merdji A, Bachir Bouiadjra B, Achour T, Serier B, Ould Chikh B and Feng Z O 2010 *Comput. Mater. Sci.* **49** 126-33
[6] Hudieb M and Kasugai S 2011 *Int. J. Oral Maxillofac. Surg.* **40** 200-06
[7] Tan W C, Lang N P, Schmidlin K, Zwahlen M and Pjetursson B E 2011 *Clin. Oral Implants Res.* **22** 14-9
[8] Ishak M I, Khor C Y, Jamalludin M R, Rosli M U, Shahrin S, Yeop Wasir N, Zakaria M S, Yamin A F M, Dahlan N D and Wan Draman W N A 2017 *Engineering Technology International Conference (ETIC 2016)* (Les Ulis Cedex A, France: MATEC Web of Conference) **97** 01040
[9] Ishak M I and Abdul Kadir M R 2012 *Biomechanics in Dentistry: Evaluation of Different Surgical Approaches to Treat Atrophic Maxilla Patients* (Springer Science & Business Media) p 1-89
[10] Ishak M I and Abdul Kadir M R 2013 *Finite Element Modelling: Biomechanics in Dentistry: Evaluation of Different Surgical Approaches to Treat Atrophic Maxilla Patients* (Berlin, Heidelberg: Springer) p 37-52
[11] Ishak M I, Abdul Kadir M R, Sulaiman E and Abu Kasim N H 2010 Effects of different zygomatic implant body surface roughness and implant length on stress distribution *IEEE EMBS Conference on Biomedical Engineering and Sciences (IECBES) (Kuala Lumpur, Malaysia, 30 November – 2 December 2010)* p 210-15
[12] Ishak M I, Abdul Kadir M R, Sulaiman E and Abu Kasim N H 2013 *Int. J. Oral Maxillofac. Implants* **28** e 151-60
[13] Ishak M I and Abdul Kadir M R 2013 *Treatment Options for Severely Atrophic Maxillae: Biomechanics in Dentistry: Evaluation of Different Surgical Approaches to Treat Atrophic Maxilla Patients* (Berlin, Heidelberg: Springer) p 9-26
[14] Ishak M I, Ahmad Shafi A, Abdul Kadir M R and Sulaiman E 2017 *J. Med. Biol. Eng.* **37** 336-44
[15] Lin C L, Chang C H, Cheng C S, Wang C H and Lee H E 1999 *Comput. Methods Programs Biomed.* **59** 187-95
[16] Alkan I, Sertgöz A and Ekici B 2004 *J. Prosthet. Dent.* **91** 319-25
[17] Li T, Hu K, Cheng L, Ding Y, Ding Y, Shao J and Kong L 2011 *Appl. Math. Model.* **35** 446-56
[18] Van Staden R C, Guan H and Loo Y C 2006 *Comput. Methods Biomech. Biomed. Engin.* **9** 257-70
[19] Sevimay M, Turhan F, Kilicarslan M A and Eskitascioglu G 2005 *J. Prosthet. Dent.* **93** 227-34
[20] Motoyoshi M, Ueno S, Okazaki K and Shimizu N 2009 *Int. J. Oral Maxillofac. Surg.* **38** 363-68
[21] Alexander H, Ricci J L and Hrico G J 2009 *J. Biomed. Mater. Res. B. Appl. Biomater.* **88** 306-11
[22] Karoussis I K, Bragger U, Salvi G E, Burgin W and Lang N P 2004 *Clin. Oral Implants Res.* **15** 8-17
[23] Akca K, Cehreli M C and Iplikcioglu H 2003 *Clin. Oral Implants Res.* **14** 444-54
[24] Ivanoff C J, Grondahl K, Sennerby L, Bergstrom C and Lekholm U 1999 *Int. J. Oral Maxillofac. Implants* **14** 173-80
[25] Kayabaşı O, Yüzbasıoğlu E and Erzincanlı F 2006 *Adv. Eng. Softw.* **37** 649-58

IOP Conf. Series: Materials Science and Engineering **318** (2018) 012058 doi:10.1088/1757-899X/318/1/012058

Manufacturing Process Selection of Composite Bicycle's Crank Arm using Analytical Hierarchy Process (AHP)

M. Luqman[1], M.U. Rosli[1], C.Y. Khor[1], Shayfull Zambree[2], H. Jahidi[1]
[1]Faculty of Engineering Technology, Universiti Malaysia Perlis (UniMAP), Level 1, Block S2, UniCITI Alam Campus, Sungai Chuchuh, 02100, Padang Besar, Perlis, Malaysia.
[2]Green Design and Manufacture Research Group, Center of Excellence Geopolymer and Green Technology (CEGeoGTech), School of Materials Engineering, Universiti Malaysia Perlis, 01000 Kangar, Perlis, Malaysia.

E-mail: uzair@unimap.edu.my

Abstract. Crank arm is one of the important parts in a bicycle that is an expensive product due to the high cost of material and production process. This research is aimed to investigate the potential type of manufacturing process to fabricate composite bicycle crank arm and to describe an approach based on analytical hierarchy process (AHP) that assists decision makers or manufacturing engineers in determining the most suitable process to be employed in manufacturing of composite bicycle crank arm at the early stage of the product development process to reduce the production cost. There are four types of processes were considered, namely resin transfer molding (RTM), compression molding (CM), vacuum bag molding and filament winding (FW). The analysis ranks these four types of process for its suitability in the manufacturing of bicycle crank arm based on five main selection factors and 10 sub factors. Determining the right manufacturing process was performed based on AHP process steps. Consistency test was performed to make sure the judgements are consistent during the comparison. The results indicated that the compression molding was the most appropriate manufacturing process because it has the highest value (33.6%) among the other manufacturing processes.

1. Introduction

Manufacturing process selection is a method to determine the most suitable process to fabricate a product. Many researchers have addressed in the literature on the importance of selection of an appropriate manufacturing process at the early stage of the product development process. Determining the most suitable process in the early stage can avoid the cost-penalty of making changes become large but it is a difficult task and a crucial decision because various factors has to be considered during the process selection [1].

Analytic Hierarchy Process (AHP) is a tool that employs at the conceptual design stage in the product development process [2, 3]. AHP is widely implanted on applications based on decision-making usually in the area of engineering, personal and social categories for the theme of selection and evaluation. AHP has been used to solve multi-criteria decision making in academic research and industrial training. Generally, the use of AHP is based on experience and skill of the users or experts to determine the factors that affecting the decision process. Furthermore, AHP can be integrated with other techniques such as quality function deployment (QFD) and data envelopment analysis (DEA) but in this research AHP will be used alone [4].

The automotive component is generally made from an expensive material such as aluminium alloy, titanium, chrome steel and carbon fibre [5, 6, 7]. In the manufacturing process, there are many processes with different amounts of cost and fabricating time. The manufacturing process also gives the manufacturers different types of problems. By doing this research, the problem faced by the manufacturers can be solved by using Analytical Hierarchy Process (AHP) to determine the most appropriate manufacturing process to fabricate a bicycle crank arm that will cut the unwanted cost [8].

Composite material exists in human technology for thousands of years, natural composite exist in animal and plant. The composite material is a mixture of two or more materials with different properties that produce unique composite properties. Nowadays, there are many types of composite material which can manipulate the cost and the properties such as polymer composites.

2. Analytical Hierarchy Process (AHP)

One of the multi-criteria decision-making support tools is the Analytic Hierarchy Process (AHP) and it was introduced by Saaty [9-10]. Many researchers getting interested in AHP mainly due to the nice mathematical properties of the method and it is easy to obtain the fact that required the input data. AHP can be used to solve the complex decision problem because it uses a multi-level of objectives, criteria, sub-criteria and alternatives. The related data are obtained by using a set of pairwise comparison. The comparisons are used to get the weights of the importance of the decision criteria and the relative performance measures of the alternatives in term of each individual decision criterion.

Many researchers applied AHP in their research, especially in task selection. Hambali employed AHP to select the most appropriate polymeric composite automotive bumper beam. The result revealed that the glass fibre epoxy is the most appropriate material because it has the highest value (25.7%, mass fraction) compared with other materials [2]. M.R Mansor integrated AHP, TRIZ and Morphological chart to select the best conceptual design of Kenaf fiber polymer composite automotive parking brake lever [11]. M.U Rosli applied AHP for Natural Fiber Composites Automotive Armrest Thermoset Matrix Selection [3].

However, there are some problems with the way pairwise comparisons are used and the way the AHP evaluates alternatives. Belton observed that AHP may reverse the ranking of the alternatives when an alternative identical to one of the already existing alternatives is introduced [12]. Belton and Gear proposed that each column of the AHP decision matrix to be divided by the maximum entry of that column in order to overcome the deficiency in AHP. Then, they introduced a variant of the original AHP, called the revised-AHP. Later, Saaty accepted the previous variant AHP and now called Ideal Mode AHP [9]. Besides the revised-AHP, other authors also introduced other variants of the original AHP. However AHP is the most widely accepted methods and is considered by many as the most reliable MCDM method.

3. Carbon Fiber Manufacturing Process

Composite manufacturing has been used for over 10,000 years ago when a composite material which is mud and straw used in the building, it is an early man-made manufactured composite [12]. There are many types of manufacturing process for polymer product but in this study only carbon fiber will be considered as a material to make bicycle crank arm. The suitable manufacturing processes for making carbon fiber bicycle crank arm are called hollow molding technology and these processes will be discussed.

3.1. Resin Transfer Molding (RTM)

RTM is a closed-mold, vacuum-assisted process that employs a flexible solid counter tool used for the B-side surface compression. This process yields excellent strength-to-weight characteristics, high glass-to-resin ratio and increased laminate compression. In this process, reinforcement mat or woven roving is placed in the mold cavity that has the shape of the desired part which is then closed and clamped [13]. Catalysed, low-viscosity resin is pumped in under pressure, displacing the air and venting it at the edges, until the mold is filled. After the filling cycle, the cure cycle starts during which the mold is heated, and the resin polymerizes to become rigid plastic. Gel coats may be used to provide a high-quality, durable finished product.

3.2. Compression Molding (CM)
Compression molding is a closed-mold composite manufacturing process that uses matched metal molds with the application of external pressure [14]. The mold is closed with a top force or plug member, pressure is applied to force the material into contact with all mold areas, while heat and pressure are maintained until the molding material has cured.

3.3. Vacuum Bag Molding (VBM)
Vacuum bagging is a process that uses a clamping method using atmospheric pressure to hold the adhesive or resin coated components of a lamination in place until the adhesive is cured. Vacuum bagging uses atmospheric pressure as clamps to hold together laminated plies that is sealed in an airtight envelope [15]. The pressure on the outside and inside of the envelope is equal to atmospheric pressure when the bag is sealed to the mold. As a vacuum pump evacuates air from the inside of the envelope, air pressure inside is reduce but air pressure outside remains at 14.7 psi [16]. Atmospheric pressure forces the side of the envelope and everything within it together putting equal pressure over the surface of the envelope. Pressurizing a composite lamination serves several functions. First, it removes trapped air between layers. Second, it compacts the fiber layers for efficient force transmission among fiber bundles and prevents shifting of fiber orientation during cure. Third, it reduces humidity. Finally, and most important, the vacuum bagging technique optimizes the fiber-to-resin ratio in the composite part.

3.4. Filament Winding (FW)
Filament winding is an automated method for creating composite structures by winding the filaments under tension over a rotating mandrel (tool). The fiber placement is guided by a machine with two or more axes of motion. This process is primarily used for hollow shapes of design [17]. There are two primary components involved in filament winding, first a stationary steel mandrel rotates, while a carriage arm travels horizontally up and down the length of the mandrel, second traveling arm includes a winding eye, which groups the roving and dispenses them to the mandrel. As the mandrel turns, the roving wrap around it to form a composite layer over the mandrel surface. The precise orientation of the composite matrix is determined by the rate of travel of the carriage and by the rotational speed of the mandrel, both of which are automated. Before encountering the mandrel, the fibers are impregnated with a resin, which later solidifies with the fiber to create the final composite material. After the filament winding lay-up pattern has been executed, the entire assembly (mandrel plus a layer of composite material) is cured in an oven.

4. Methodology
The methodologies of this research for an appropriate manufacturing process of composite bicycle's crank arm, according to the AHP selection method are as follows:

1. Define the objective of selection.
2. Develop a hierarchical framework.
3. Construct set of pairwise comparison matrix.
4. Perform judgement of pairwise comparison
5. Perform the consistency analysis. The consistency ratio (CR) must be less than 10% or 0.1.
6. Perform step 2-5 at all level of hierarchy.
7. Develop overall priority ranking.
8. Select the best alternatives.

5. Selection of manufacturing process by AHP method

Figure 1 shows the hierarchical framework of the research. The main goal is selecting the most suitable manufacturing process for bicycle's crank arm in order to produce a good quality product. The factors that considered in selection are structured in hierarchy form. The selected main criteria in level 2 are Production characteristic (PC), geometry of design (GD), cost consideration (CS), material (MT) and ease of maintenance (EM). The sub-criteria in level 3 are production quantity (PQ), rate of production (RP), processing time (PT), shape of design (SH), size (SZ), weight (WG), complex of design (CD), tolerance and surface finish (TS), tooling cost (TC) and equipment cost (EC).

Based on the gathered data from literature review as shown in Table 1, the pairwise comparison between each criteria and manufacturing process will be performed. Consistency analysis is performed in order to check the consistency of the pairwise comparison performed earlier. The pairwise comparison is accepted since that the result is 0.03 as shown is Table 2.

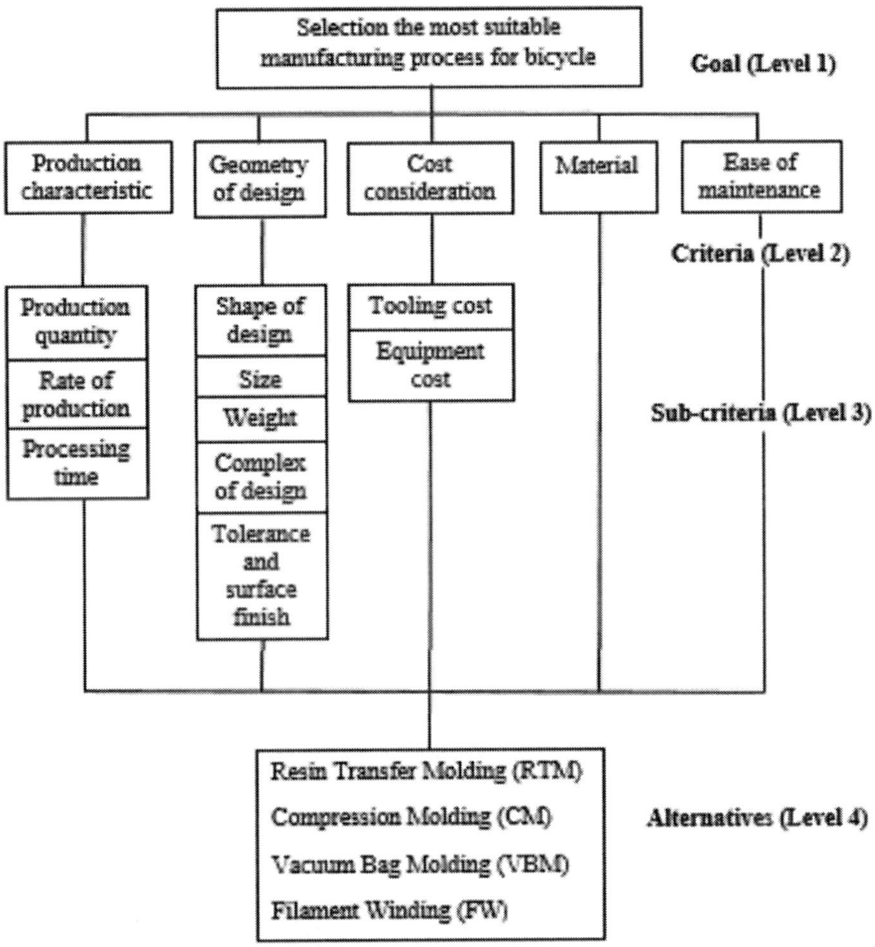

Figure 1. Hierarchical structure of AHP.

Table 1. Data used to perform pairwise comparison.

Criteria \ Process	RTM	CM	VBM	FW
PQ	Medium	High	Medium	High
RP	Medium	High	Low	High
PT	Average	Fast	Slow	Fast
SH	Highly possible	Highly possible	Possible	Least Possible
CD	Complex-simple	Complex-simple	Simple	Simple
SZ	Small-large	Small-large	Small-large	Small-large
WG	Light	Medium	Medium	Medium
TS	Good-excellent	Good-excellent	Good	Good
TC	Low-high	Medium-High	Low	Low-High
EC	Medium	High	Low	Low-High
MT	Carbon fiber, resin, polyester, epoxy	Carbon fiber, resin, polyester, epoxy, vinylester	Carbon fiber, resin	Carbon fiber, resin
EM	Easy	Hard	Easy	Easy

Table 2. Consistency test for criteria.

Goal	**PC**	**GD**	**CS**	**MT**	**EM**	**Priority vector (PV)**	**New Vector (NV)**	**NV/PV**	
PC	1	3	3	3	5	0.430	2.267	5.272	
GD	1/3	1	1	1	3	0.161	0.838	5.204	CI = 0.031
CS	1/3	1	1	3	3	0.210	1.114	5.305	
MT	1/3	1	1/3	1	3	0.138	0.698	5.058	CR = 0.03
EM	1/5	1/5	1/3	1/3	1	0.062	0.296	4.774	(acceptable)
						Total		25.613	
						Maximum eighenvalue (λ_{max})		5.123	

Next, all the pairwise comparison and consistency analysis for main criteria, sub-criteria and alternatives are performed as partly shown in Table 3. Consistency verification test is one of the most crucial steps and this is where the AHP method has the advantages. This step measures the degree of consistency among the pairwise comparisons by calculating the consistency ratio. Consistency ratio (CR) is the ratio of consistency index (CI) to random index (RI).

Table 3. Consistency test for alternatives.

	Priority vector											
	GOAL											
	PC			GD					CS		MT	EM
	PQ	RP	PT	SH	CD	SZ	WG	TS	TC	EC		
RTM	0.125	0.173	0.153	0.558	0.411	0.250	0.484	0.555	0.122	0.138	0.269	0.228
CM	0.375	0.372	0.389	0.263	0.380	0.250	0.224	0.252	0.057	0.058	0.564	0.097
VBM	0.125	0.092	0.068	0.122	0.147	0.250	0.161	0.097	0.558	0.385	0.091	0.384
FW	0.375	0.426	0.389	0.057	0.061	0.250	0.130	0.097	0.263	0.420	0.075	0.291
Consistency test												
λ_{max}	4.000	4.134	4.043	4.118	4.033	4.000	4.156	4.043	4.118	4.171	4.082	4.153
CI	0.000	0.045	0.043	0.039	0.011	0.000	0.052	0.014	0.039	0.057	0.027	0.051
RI	0.900											
CR	0.000	0.049	0.047	0.043	0.012	0.000	0.058	0.016	0.044	0.063	0.030	0.057

Table 4. Priority vectors.

	Priority vector											
	GOAL											
Criteria	PC			GD					CS		MT	EM
	0.430			0.161					0.210		0.191	0.062
Sub-criteria	PQ	RP	PT	SH	CD	SZ	WG	TS	TC	EC		
	0.287	0.140	0.574	0.415	0.251	0.056	0.127	0.151	0.750	0.250		
Alternatives												
RTM	0.125	0.173	0.153	0.558	0.411	0.250	0.484	0.555	0.122	0.138	0.269	0.228
CM	0.375	0.372	0.389	0.263	0.380	0.250	0.224	0.252	0.057	0.058	0.564	0.097
VBM	0.125	0.092	0.068	0.122	0.147	0.250	0.161	0.097	0.558	0.385	0.091	0.384
FW	0.375	0.426	0.389	0.057	0.061	0.250	0.130	0.097	0.263	0.420	0.075	0.291

Table 5. Overall priority vectors for sub-criteria with respect to criteria.

	Overall priority vector				
RTM	0.148	0.494	0.126	0.269	0.228
CM	0.383	0.285	0.057	0.564	0.097
VBM	0.088	0.137	0.515	0.091	0.384
FW	0.391	0.084	0.302	0.075	0.291

Table 5 summarizes the overall priority vector for four alternatives with respect to the sub-criteria. By multiplying the priority vector for process alternatives to the vector of sub-criteria, the overall priority criteria can be obtained. Table 6 shows the overall priority vector of the alternatives with respect to the criteria.

Table 6. Overall priority vectors for alternatives with respect to criteria.

	Priority vector					Overall priority vector
	PC	**GD**	**CS**	**MT**	**EM**	
	0.430	0.161	0.210	0.191	0.062	
RTM	0.148	0.494	0.126	0.269	0.228	0.235
CM	0.383	0.285	0.057	0.564	0.097	0.336
VBM	0.088	0.137	0.515	0.091	0.384	0.209
FW	0.391	0.084	0.302	0.075	0.291	0.227

Table 7 shows the compression molding process that has the highest value (0.336) among the other alternatives. The lowest value is the vacuum bag molding process with a value of 0.209.

Table 7. Result of selection.

No.	Best Selection	
1	Compression molding	0.336
2	Resin transfer molding	0.235
3	Filament winding	0.227
4	Vacuum bag molding	0.209

6. Conclusion

Nowadays, consumers are looking for vehicles more environmental friendly and lighter in weight. For this reason, the engineers are now focusing to substitute the metal parts on utilizing the natural fiber composites. Selecting the effective manufacturing process in product development is a crucial decision. The use of analytical hierarchy process (AHP) can assist manufacturers to evaluate and select the best manufacturing process based on the criteria and sub-criteria aspects of a decision. The analysis reveals that compression molding is the most suitable process for manufacturing carbon fiber bicycle crank arm as it has the highest value (0.336 or 33.6%) among the other manufacturing processes. It is proved that the analytical hierarchy process (AHP) is a useful method in solving the manufacturing process selection problem for the bicycle composite components during the conceptual design stage. Imprecise decision can cause the product to be remanufactured and not in optimized condition.

7. Acknowledgement
The research was financially supported by Universiti Malaysia Perlis (UniMAP) under Short Term Grant (STG) through project number 9001-00503.

References

[1] Hambali, A., Sapuan, S. M., Ismail, N., & Nukman, Y. (2009). Application of analytical hierarchy process in the design concept selection of automotive composite bumper beam during the conceptual design stage. *Scientific Research and Essays*, *4*(4), 198-211.

[2] Hambali, A., Sapuan, S. M., Ismail, N., & Nukman, Y. (2010). Material selection of polymeric composite automotive bumper beam using analytical hierarchy process. *Journal of Central South University of Technology*, *17*(2), 244-256.

[3] Rosli, M. U., Jamalludin, M. R., Khor, C. Y., Ishak, M. I., Jahidi, H., Wasir, N. Y., Ismail, R. I. (2017). Analytical Hierarchy Process for Natural Fiber Composites Automotive Armrest Thermoset Matrix Selection. In *MATEC Web of Conferences* (Vol. 97, p. 01039). EDP Sciences.

[4] Rosli, M. U., Ariffin, M. K. A., Sapuan, S. M., & Sulaiman, S. (2013). Integrating TRIZ and AHP: A MPV" s Utility Compartment Improvement Design Concepts. *International Journal of Materials, Mechanics and Manufacturing*, *1*(1), 32-35.

[5] Banhart, J., & Seeliger, H. W. (2008). Aluminium foam sandwich panels: manufacture, metallurgy and applications. *Advanced Engineering Materials*, *10*(9), 793-802.

[6] Smith, G. A., Kozuschek, M. A., & Nelson, D. A. (2001). *U.S. Patent No. 6,314,834*. Washington, DC: U.S. Patent and Trademark Office.

[7] Dweiri, F., & Al-Oqla, F. M. (2006). Material selection using analytical hierarchy process. *International journal of computer applications in technology*, *26*(4), 182-189.

[8] Rosli, M.U, Ariffin, M., Sapuan, S., & Sulaiman, S. (2014). Survey of Malaysian Car owner needs of a Car Interior. *International Journal of Mechanical & Mechatronics Engineering*, *14*(1).

[9] Saaty, T.L. (1977). A Scaling Method for Priorities in Hierarchical Structures, Journal of Mathematical Psychology, 15: 57-68

[10] Saaty, T.L. (1994). Fundamentals of Decision Making and Priority Theory with the AHP. RWS Publications, Pittsburgh, PA, U.S.A.

[11] Mansor, M. R., Sapuan, S. M., Zainudin, E. S., Nuraini, A. A., & Hambali, A. (2013). Hybrid natural and glass fibers reinforced polymer composites material selection using Analytical Hierarchy Process for automotive brake lever design. *Materials & Design*, *51*, 484-492.

[12] Belton, V. and Gear, T. (1983). On a Short-coming of Saaty's Method of Analytic Hierarchies. Omega, 228-230.

[13] Kalpakjian, S., & Schmid, S. R. (2014). *Manufacturing engineering and technology* (p. 913). K. V. Sekar (Ed.). Upper Saddle River, NJ, USA: Pearson.

[14] Wang, B. C., Zhou, X., & Ma, K. M. (2013). Fabrication and properties of CNTs/carbon fabric hybrid multiscale composites processed via resin transfer molding technique. *Composites Part B: Engineering*, *46*, 123-129.

[15] Suherman, H., Sulong, A. B., & Sahari, J. (2013). Effect of the compression molding parameters on the in-plane and through-plane conductivity of carbon nanotubes/graphite/epoxy nanocomposites as bipolar plate material for a polymer electrolyte membrane fuel cell. *Ceramics International*, *39*(2), 1277-1284.

[16] Matsuzaki, R., Kobayashi, S., Todoroki, A., & Mizutani, Y. (2013). Flow control by progressive forecasting using numerical simulation during vacuum-assisted resin transfer molding. *Composites Part A: Applied Science and Manufacturing*, *45*, 79-87.

[17] Xia, C., Shi, S. Q., Cai, L., & Hua, J. (2015). Property enhancement of kenaf fiber composites by means of vacuum-assisted resin transfer molding (VARTM). *Holzforschung*, *69*(3), 307-312.

IOP Conf. Series: Materials Science and Engineering **318** (2018) 012059 doi:10.1088/1757-899X/318/1/012059

Numerical Modelling and Simulation of Dynamic Parameters for Vibration Driven Mobile Robot: Preliminary Study

M E Baharudin[1,*], A M Nor[1], A R M Saad[2] and A M Yusof [2]
[1]School of Manufacturing Engineering, Universiti Malaysia Perlis
[2]Faculty of Technology, Universiti Malaysia Perlis

E-mail: mezral@unimap.edu.my

Abstract. The motion of vibration-driven robots is based on an internal oscillating mass which can move without legs or wheels. The oscillation of the unbalanced mass by a motor is translated into vibration which in turn produces vertical and horizontal forces. Both vertical and horizontal oscillations are of the same frequency but the phases are shifted. The vertical forces will deflect the bristles which cause the robot to move forward. In this paper, the horizontal motion direction caused by the vertically vibrated bristle is numerically simulated by tuning the frequency of their oscillatory actuation. As a preliminary work, basic equations for a simple off-centered vibration location on the robot platform and simulation model for vibration excitement are introduced. It involves both static and dynamic vibration analysis of robots and analysis of different type of parameters. In addition, the orientation of the bristles and oscillators are also analysed. Results from the numerical integration seem to be in good agreement with those achieved from the literature. The presented numerical integration modeling can be used for designing the bristles and controlling the speed and direction of the robot.

1. Introduction

For small and tiny crawling robots, vibration-driven mechanism is a suitable actuation system to be considered and implemented. Vibrational motion produced by periodic oscillations of internal masses is transformed into a directed locomotion. With increasing new applications in medical, manufacturing and maintenance, a greater demand for more efficient and reliable tiny robots are expected. This has created an interest to researchers to further enhance their understanding about vibration-driven robots for further improvement. For this purpose, numerical analyses and designs of vibration-driven robot for better motion and speed control were thoroughly studied [1-13].

Small vibration-driven robots are developed for different applications. The creation of different designs of mobile robots for movement through pipelines and similar technical systems can be found in [1]. An apedal locomotion system "Vibro-worm" to be used for inspection of pipelines was designed and tested. In another study [2], motion analysis, design and position control of a novel, low cost, sliding micro-robot which was actuated by centripetal forces generated by robot mounted vibration micro-motors was investigated. The micro-robot was able to perform translational and rotational sliding with sub-micrometer accuracy and develop velocities of up to 1.5 mm/s. interestingly, [3] studied the Setaria viridis spike's moving behaviour on vibrating cotton wire inspired from grass spike. The report claimed that the fabricated vibration-driven vehicle could move on a wide range of smooth and rough surfaces. In addition, research on the design, construction, performance and testing of a mechanical vibration exciter [4] concluded that the amplitude of vibration initially increases with increase in speed and reaches a maximum value (at resonance speed) and then decreases.

Analysis on the motion principle and behaviour of the vibration-driven robot also caught the attention of some researchers. For example, [5] investigated the performance of different friction models in microstick-slip motion systems, i.e. Coulomb, viscous, combined Coulomb and viscous model, Stribeck, Dahl, LuGre, and the elastoplastic friction models. The result showed that LuGre model had the best accuracy. Worked done by [6] proposed an efficient parameter estimation approach for LuGre friction model based on the chain-code method. In simulation studies, the locomotion behaviour of vibration-driven robot is frequently considered using a multibody system model. The aim is to obtain the influence of the system parameters on the motion behaviour [7], [8-9], and [10]. Research in [11] presented an analytical model and solved numerically for different configurations (direction of excitation) and parameters (length, bending stiffness, shape of the endpoint). The influence of vibration frequency on motion can be found in [12] and [13].

From the literature, it was found that the best parameters for vibrator position and the effect of bristle geometry are still vague. In this study, the changes in the motion behaviour when the parameter values are varied are observed numerically. As this study is still at the preliminary stage, the model has been simplified to ensure that the fundamentals of the model formulations are fully understood.

2. Numerical modeling

The numerical modeling of the vibration-driven robot is based on the prototype as in figure 1. The prototype consists of a rigid body with mass, m and length, L. The oscillator is attached onto the main body either at the centre of mass or any other position along the main body for experimental purposes. The main body is supported with two bristles where the bristles can be changed to different sizes, shapes, slope angles and stiffness.

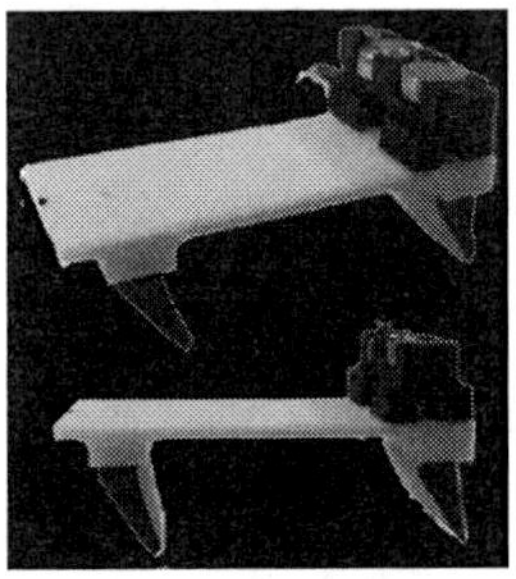

Figure 1. Design of prototype for experiment.

The oscillator is an eccentric rotating mass where the excitation force is created as in figure 2. The oscillator can be modelled as 1 DOF.

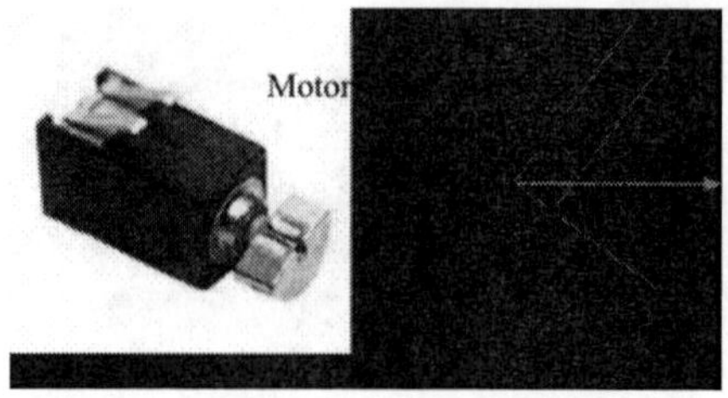

Figure 2. Eccentric rotating mass, m_e.

The rotation of the unbalanced mass creates the vertical, f_V and horizontal, f_H forces which can be derived as,

$$f_H = m_e e \omega_e^2 \sin\theta_e \tag{1}$$

$$f_V = -m_e g - m_e e \omega_e^2 \cos\theta_e \tag{2}$$

where m_e is the eccentric mass, e is the eccentricity of rotating mass, ω_e is the rotation velocity of the motor, g is the gravity acceleration and θ_e is the rotation angle of eccentric mass.

The vibration driven robot can be modelled as a 2-DOF platform as shown in figure 3.

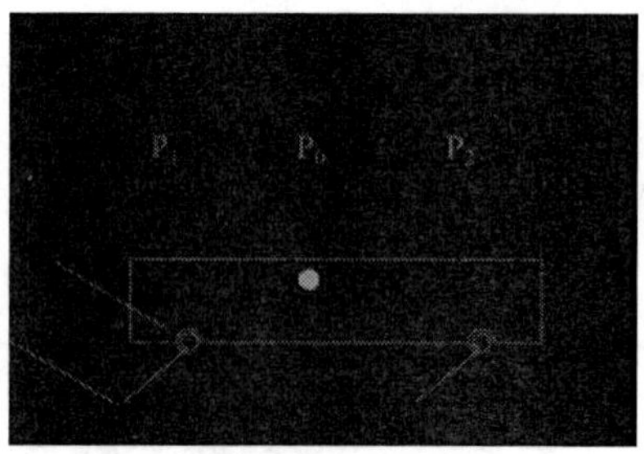

Figure 3. Model of vibration driven robot.

The excitation force, f_V can be positioned either at point P_1 or P_0. In this model, the damping factor is ignored and the centre of mass is assumed to be located at the centre of the platform. The equations of motion can be derived based on the energy method as,

$$T = \frac{1}{2} m \dot{} \quad \frac{1}{2} \tag{3}$$

$$V = \frac{1}{2} k_1 \left(y - L_1\theta_m\right)^2 + \frac{1}{2} k_2 \left(y - L_2\theta_m\right)^2 \tag{4}$$

where T is the total kinetic energy, V is the total potential energy, J_0 is the moment of inertia, θ_m is the angle of rotation at centre of mass, $\dot{\iota}_m$ is the first derivative of angle of rotation with respect to time, while k_1 and k_2 are the stiffness of spiral spring. From the Lagrangian, we know that,

$$L = T - V \tag{5}$$

and Lagrange's equations in the Cartesian coordinates can be derived as,

$$\frac{d}{dt}\left(\frac{\partial L}{\partial \dot{}}\right) - \frac{\partial L}{} = \mathbf{Q}_i \tag{6}$$

where **q** is the generalized coordinates and i is the number of the degree of freedom of the system. Therefore, equations (3) and (4) become as,

$$m\ddot{} \quad k_2)x + (k_2L_2 - k_1L_1)\theta_m = f_V \tag{7}$$

$$J_0\ddot{} \quad -k_1L_1)y + (k_1L_1^2 + k_2L_2^2)\theta_m = -f_VL_1 \tag{8}$$

Equation (7) and (8) can be written in the form of matrix as

$$\begin{bmatrix} m & 0 \\ 0 & J_0 \end{bmatrix}\begin{Bmatrix} \ddot{} \\ \ddot{} \end{Bmatrix} \begin{bmatrix} k_1 + k_2 & k_2L_2 - k_1L_1 \\ L_2 - k_1L_1 & k_2L_2^2 + k_1L_1^2 \end{bmatrix}\begin{Bmatrix} y \\ \theta_m \end{Bmatrix} = \begin{Bmatrix} f_V \\ -f_VL_1 \end{Bmatrix} \tag{9}$$

where equation (9) can be simplified into,

$$[\mathbf{M}]\{\ddot{}\} \quad]\{\mathbf{Y}\} = \{\mathbf{F}\} \tag{10}$$

The horizontal displacement, $x(t)$ can be calculated as

$$x(t) = d\sin\left(\cos^{-1}\left(\frac{y_0 - y(t)}{d}\right)\right) - d\sin\theta_B \tag{11}$$

where d is the length of the bristles and θ_B is the slope angle of the bristles. Equations (9) and (11) are integrated numerically based on the table 1 and table 2.

Table 1. Model parameters – Group 1.

Descriptions	Unit	Set A	Set B	Set C
Mass, m	g	40	40	40
Internal mass, m_e	g	2	2	2
Eccentric, e	mm	2	3	4
Bristle length, d	mm	2	4	6
Bristle angle, θ_B	degree	30	25	35
Stiffness, k_1=k_2	Nm	10	100	1000
Frequency, ω_e	Hertz	10	10	10

Table 2. Model parameters – Group 2.

Descriptions	Unit	Set A	Set B	Set C
Mass, m	g	40	40	40
Internal mass, m_e	g	2	2	2
Eccentric, e	mm	2	3	4
Bristle length, d	mm	2	4	6
Bristle angle, θ_B	degree	30	25	35
Stiffness, k_1,k_2	Nm	1,10	100,10	1000,10
Frequency, ω_e	Hertz	10	10	10

Table 3. Model parameters – Group 3.

Descriptions	Unit	Set A	Set B	Set C
Mass, m	g	40	40	40
Internal mass, m_e	g	4	6	8
Eccentric, e	mm	2	3	4
Bristle length, d	mm	2	4	6
Bristle angle, θ_B	degree	30	25	35
Stiffness, k_1,k_2	Nm	10,10	100,10	1000,1
Frequency, ω_e	Hertz	10	10	10

3. Results and discussion

Numerical parametric study is conducted to determine the dependency between the oscillator positions, the excitation force and bristles stiffness. The oscillator is placed at P_1 and the displacements at point P_1, P_0 and P_2 are observed. The results are given as in figure 4, figure 5 and figure 6.

From the results of figure 4, it is clearly shown that the oscillator position plays an important role to the robot motion. If the oscillator is placed at the centre of mass, the displacements for both bristles are the same. However, when the oscillator is placed at point P_1, the amount of displacement at each point are different and occurred at different phases for every time step. It shows that the displacement at the oscillator location is larger than at the other side of the robot platform.

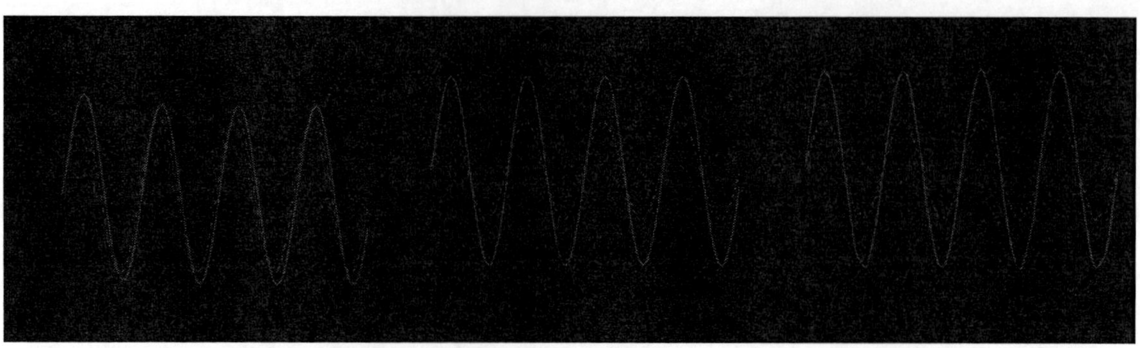

Figure 4. Bristles displacement at y-direction vs. time for table 1.

IOP Conf. Series: Materials Science and Engineering **318** (2018) 012059 doi:10.1088/1757-899X/318/1/012059

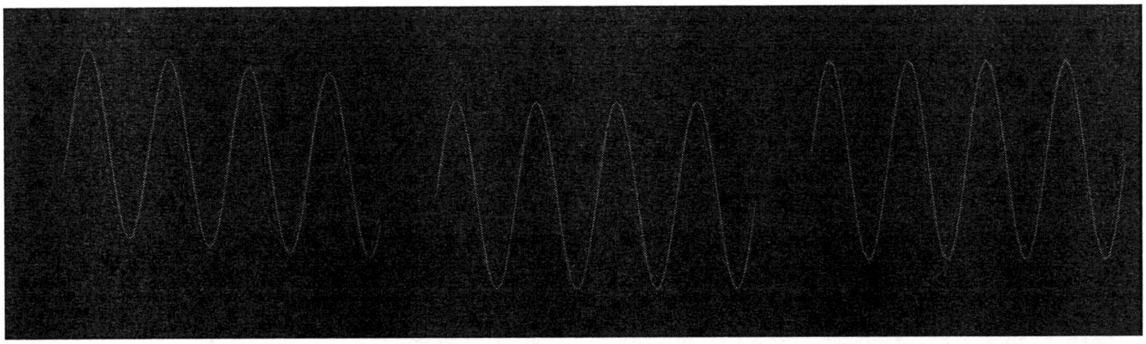

Figure 5. Bristles displacement at y-direction vs. time for table 2.

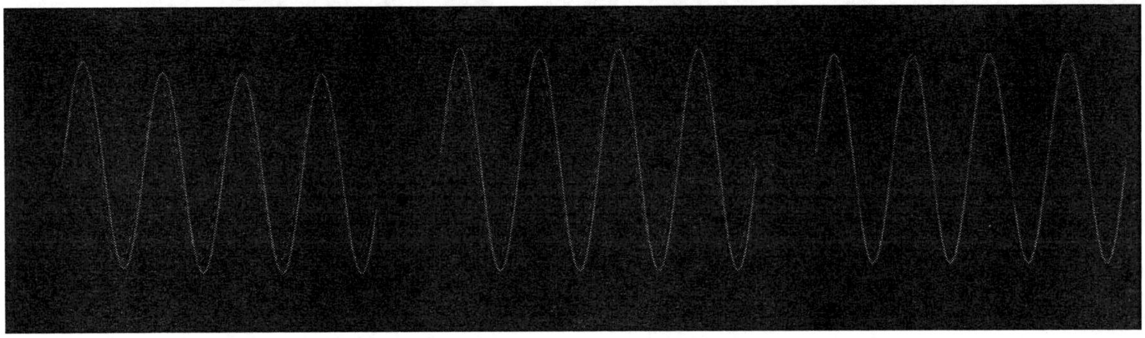

Figure 6. Bristles displacement at y-direction vs. time for table 3.

From figure 5, we can see the changes of spring stiffness, k_1 from 1 Nm to 1000Nm (while k_2 remain the same) did not show any significant effect at Point 2, P_2 as it only shows the displacement of the bristle almost similar at P_0 and P_1.

Form figure 6, the change of excitation forces and spring stiffness as in set A and B did not show significant effect to the displacement. However, when there is huge different of stiffness value between two springs and increasing of internal mass as in Table 3 – Set C, the phase of displacement sine wave for each point P_1, P_2 and P_3 started to decrease and close each other. Therefore, it can be said the time of force transmission between two points decrease when using different value of spring stiffness for each bristle.

In general, the stiffness of the bristles also affected the displacement of the bristles where high stiffness will reduce the bristle displacement and vice versa. On the other hand, low stiffness could reduce the spring response and hence will slow down the horizontal speed of the robot.

4. Conclusions

The numerical analysis conducted on the dynamic model of the vibration-driven robot shows that the displacements of the vibration-driven robot are affected by the position of the oscillator, the excitation force and the stiffness of the bristles. Adjusting the values of these crucial variables will cause different motion behaviours. Thus, it can be used as a consideration factor in designing an effective vibration-driven robot.

From the preliminary results, it can be said the best position for excitation force should be located at the end of the platform while the bristles stiffness did not produced significant influence to the displacement and velocity of the robot.

In conclusion, it can be said that to fully understand the motion behaviour of vibration-driven robots, the numerical dynamic model analysis should consider the effect of varying other parameters such as the damper characteristic of the bristles, geometric parameters of the bristles, and friction forces to make the motion more realistic.

References

1. V. Lysenko, K. Zimmermann, A. Chigarev, F. Becker, 2011, 56th International Scientific Colloquium, Ilmenau University of Technology.
2. P. Vartholomeos, E. Papadopoulos, 2006, IEEE International Conference on Robotics and Automation, 649-654.
3. S. Bai1, Q. Xul, Y. Qin, 2013, Sci. Rep. 3.
4. N. Anekar, V.V. Ruiwale, S. Nimbalkar, P. Rao, 2014, IJRET, **03**, Issue: 08.
5. Y. F. Liu, J. Li1, Z. M. Zhang, X. H. Hu, and W. J. Zhang, 2015, Mech. Sci., **6**, 15–28.
6. B.X. Liu, S.H. Nie, 2015, The 14th IFToMM World Congress.
7. F. Becker, 2015, IFAC, 48-1, 842–843.
8. L. Giomi, N. Hawley-Weld, L. Mahadevan, 2013, Proc. R. Soc. A 2013 469.
9. T. Al Ahmadi, A. Al Shamardi, N. Alotibi, 2016, ME 4773/5493 Fundamental of Robotics.
10. F. Becker, V. Lysenko, V. Minchenya, I. Zeidis, K. Zimmermann, 2013, Springer, 8-3, 299-306.
11. F. Becker1, S. Börner, V. Lysenko, I. Zeidis, K. Zimmermann, 2014, ISR Robotik, ISBN 978-3-8007-3601-0.
12. G. Cicconofri, F. Becker, G. Noselli, A. DeSimone, K. Zimmermann, 2016, ROMANSY 21 - Robot Design, Dynamics and Control, **569**, 225-232.
13. G. Cicconofri, A. DeSimone, 2015, IJNLM, **76**, 233-239.

IOP Conf. Series: Materials Science and Engineering **318** (2018) 012060 doi:10.1088/1757-899X/318/1/012060

Annealing Effects on the Formation of Copper Oxide Thin Films

Marina Marzuki[1], Mohd Zamzuri Mohamad Zain, Nurazhra Zarul Hisham, Nooraizedfiza Zainon, Azmi Harun and Rozie Nani Ahmad

School of Manufacturing Engineering, Universiti Malaysia Perlis, 02600 Perlis, Malaysia

E-mail: marina@unimap.edu.my

Abstract. This study approached the simple method of developing CuO thin films by thermal oxidation on pure Cu sheets. The effects of annealing temperature on the formation of CuO layers have been investigated. The oxide layers have been fabricated by annealing of Cu sheets for 5 hours at different temperatures of 980 ~ 1010 °C. The morphologies and optical properties of annealed Cu sheets were studied by using SEM and UV-Vis spectrophotometer respectively. It is revealed that the annealing temperature influence the grain growth and the grain size increases as the temperature increase. The highest grain size was observed on sample annealed at 1000 °C; with average area per grain size of 0.023 mm^2. Theoretically, larger grain size provides less barriers for electron mobility and increase the efficiency of solar devices. The optical absorption spectra of the oxide films was also measured. Interference pattern was noted at wavelength about 900 nm corresponding to the formation of CuO film. The interference noise observed could be due to the coarse surface and the presence of powdery oxide deposits that causes the scattering loses from the surface. CuO film obtained by this method may be further studied and exploited as low cost photovoltaic device.

1. Introduction

Thin film solar cells comprising of nontoxic, stable and abundant materials are widely studied in recent decades for renewable photovoltaic applications. The implementation of of new solar energy conversion device requires the use of less expensive materials that can make solar energy more competitive. Cuprous oxides (Cu_2O) are of interest because of their potential to reach 20% power conversion efficiency [1, 2]. It is said that this type of semiconductor could be utilized for fabrication of low cost solar cells. However, doping this material is proved to be challenging as Cu_2O has stability condition under limited range of temperature and pressure. Below 400°C, CuO is more stable than Cu_2O [3]. After a sufficient expose time to oxidation, CuO is formed [4]. This will result in poor mobility of minority carrier and limited light absorption. However, this could be improved by tuning the microstructure of Cu_2O/CuO layer through heat treatment.

Recently, the preparation of Cu_2O/CuO film by including annealing method in the process has been reported [4-6]. For example, Musa et. al. [4] studied the thermal oxidation conditions on the properties of plasma evaporation prefabricated cuprous oxide film. The mobility of minority carrier is reported improved for annealed samples. Johan et. al. [5] studied the annealing effects on the properties of copper oxide thin film pre-prepared by chemical deposition. The proportion of the two forms of copper oxides was reported varies with oxidation temperature. They also reported that photoluminescent intensity is greatly improved with the temperature increment. Sentra et. al. [6] studied the band gap energy of copper oxide film grown by plasma evaporation method. Band gap energy of Cu_2O is higher than CuO and both were indicated as p- type semiconductor.

This study approached the simple method of developing CuO thin films by thermal oxidation on pure Cu sheets. This is carried by annealing the samples in varied temperatures and the effect of the soaking temperature on the oxide layer is studied. The morphologies and optical properties of annealed Cu sheets were studied by using SEM and UV-Vis spectrophotometer respectively.

2. Experimental Methods

Experiment has been carried out using pure Cu as substrate (99.9%). Cu sheets (dimension 20 mm x20 mm x 0.2 mm) were cleaned and dried. To ensure the surface is free from contaminants, the sheets were immersed in acetone for 3 minutes. Annealing process was done for 5 hours at different temperatures; starting from 980 °C, 990 °C, 1000 °C, and 1010 °C. The heat treatment consists of heating up to destined temperature, then soaking the samples in predetermined temperature for 5 hours and cooling off within the furnace with controlled atmosphere. At cooling temperature of 500°C, the samples were taken out and were left in open air at ambient temperature. The formation of Cu_2O/CuO layer is then investigated.

Analysis is carried out by physical inspection, SEM micrograph and UV-VIS spectroscopy. The sample's surface topography and morphology as well as the composition information was performed with Scanning Electron Microscope (SEM) and Energy Disperse X-ray (EDS). ImageJ application was used to analyse the average measurement of length, perimeter, and surface area of grains structure. The optical absorption spectra were measured using a UV-Vis-near-infrared spectrophotometer with reference to the bare substrate of Cu sheet. The absorption scale start from 1500 nm to 300 nm with the scan rate 300 nm/min.

3. Results and Discussions

Experimental result has been analysed by characterizing the samples' physical appearance, morphology by Scanning Electron Microscopy (SEM), elemental properties (EDS) and light absorbance (UV-VIS)

3.1 Morphology

The initial appearance of the Cu sheets is standard pure copper colour shown in Fig. 1(a). By colour identification, it has been indicated that Cu_2O is formed first and at high temperature, dominant CuO is formed. Mixed oxides of Cu_2O and CuO are formed at temperatures below 1000 °C and at atmospheric pressure [7]. The nonstoichiometry of the oxides is formed from the reaction between the oxide crystal and O_2 surrounding the material [8].

Prior to annealing the samples from metallic brown changed to greyish black colour as depicted in **Figure 1** (**b,c,d** and **e**). This is standard identification of the formation of CuO layer and surface coarsening occurred.

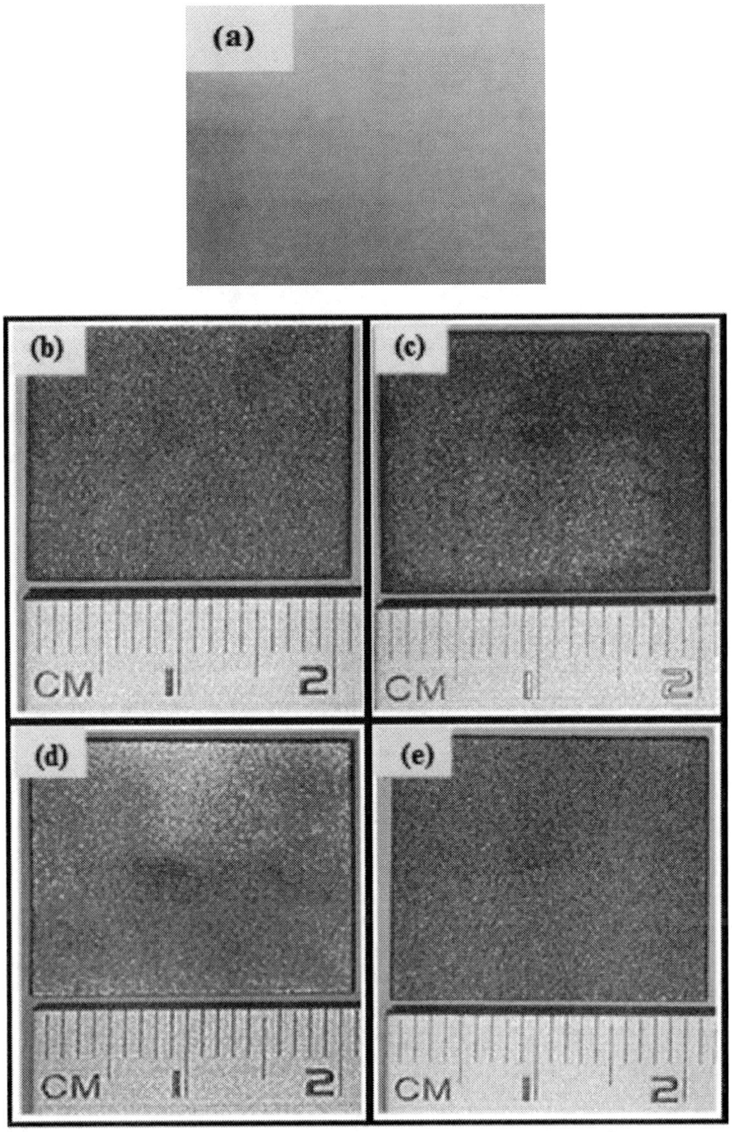

Figure 1. (a) Physical appearance of as-received Cu sheet; the annealed Cu sheets; (b) 980 °C, (c) 990 °C, (d) 1000 °C, and (e) 1010 °C.

Figure 2 (a,b,c,d) shows the microstructure of CuO layer images annealed at 980 °C, 990 °C, 1000 °C and 1010 °C by using SEM. Recrystallization and grain growth were observed on the microstructure. SEM morphology was taken before the specimens' oxidation took place at room temperature. Based on the micrographs, a considerable grain growth were observed on all the annealed specimens. Surface area per grain was obtained by triangulation method. There are discreet gaps difference on the grain boundaries, which attributed by inconsistent grain growth. The gaps between grains are the area full of defects such as dislocations and vacant spots [6].

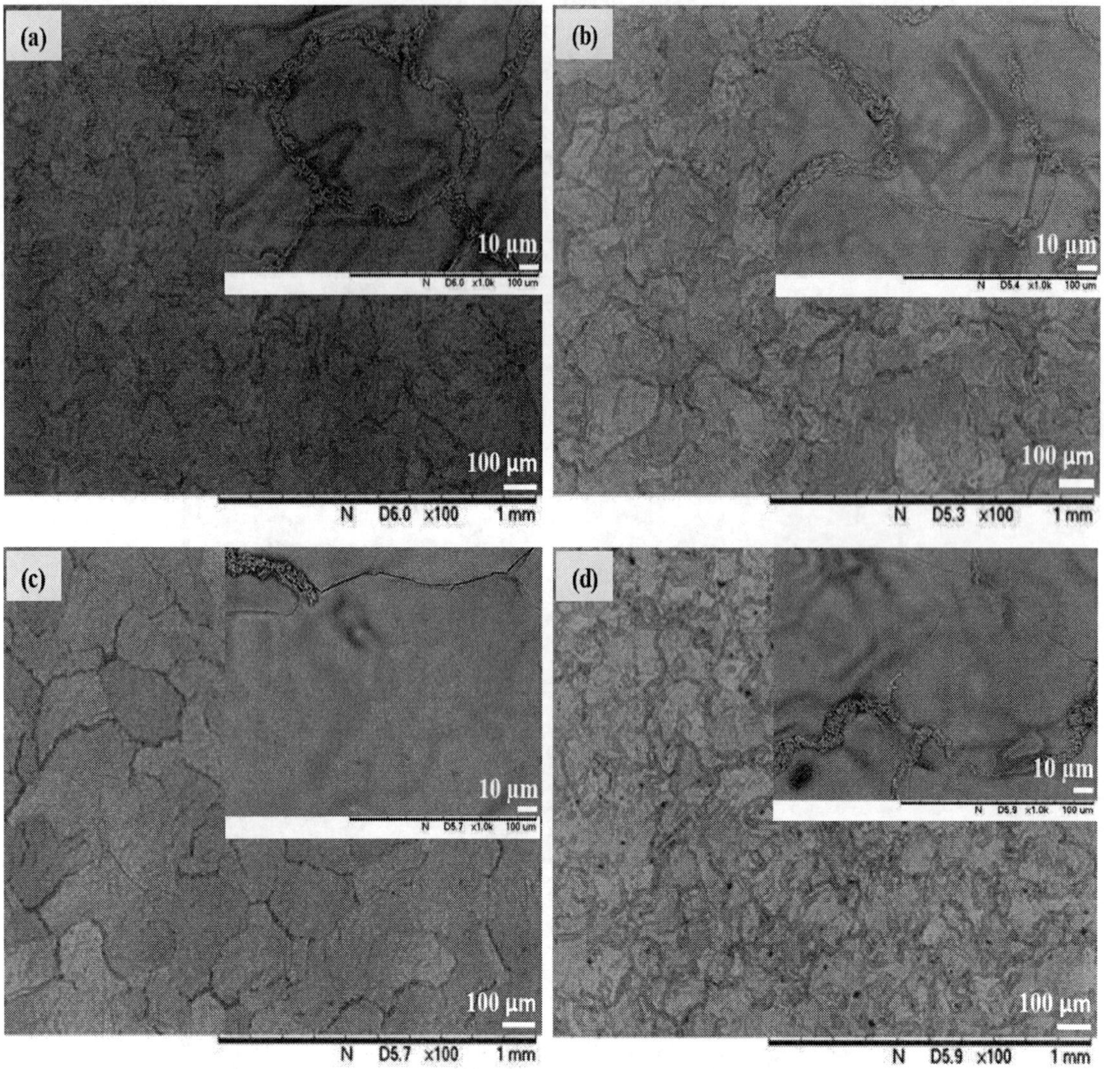

Figure 2. SEM images of CuO layer grown at (a) 980 °C, (b) 990 °C, (c) 1000 °C and (d) 1010 °C.

The growth of the CuO layer increases with the annealing temperature. Largest surface area per grain was observed on CuO layer of 1000 °C annealing temperature with average area per grain of 0.126 µm as which is evident in **Figure 3**. Theoretically, large grain size increase the mobility and reduce the resistivity for a carrier to move from one grain structure to the adjacent grain structure [9].

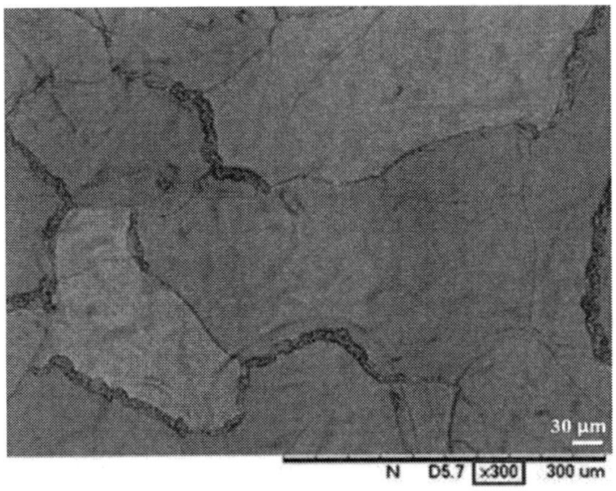

Figure 3. Magnified surface morphology of the CuO layer after annealed at temperature 1000 °C.

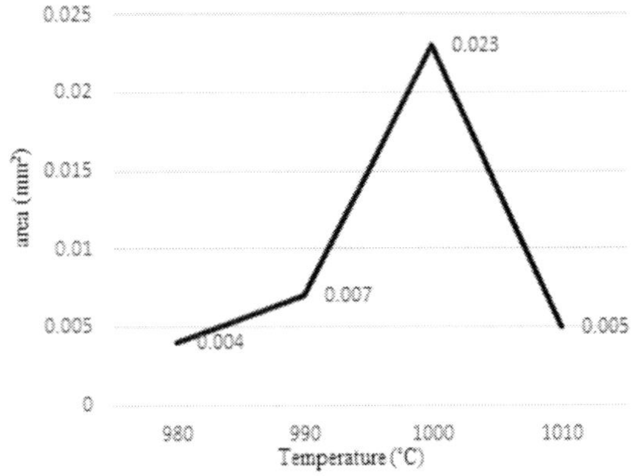

Figure 4. Grain growth over annealing time.

Figure 4 illustrates the average area of grain after annealing at different temperatures. The average area of grain for the crystal of annealed CuO sample were estimated to be 0.004 mm^2, 0.007 mm^2, 0.023 mm^2, and 0.005 mm^2 respectively to the annealed temperature. An increase in grain size tend to convert polycrystalline structure obtained after the oxidation process to single or nearly single crystal material.

Figure 5 is a magnified image of grain boundaries. Besides the noticeable grain growth, imperfections are evident along the boudaries. An ideal single orientation of crystalline structure of CuO layer could probably increase the mobility of electron inside CuO layer [10]. Because of perfect structure such as that does not exist, real crystalline structures includes several kinds of linear defects (Fig. 6) such as voids and uneven boudaries resulting from dislocations of grains [8].

Large grain size contributes to an increment in the mobility and reduce the resistivity of an electron movement from one crystal to another crystal structure. Thin boundary tends to let an electron to move smoothly with a low barrier across the grain boundaries. Several functional properties of metal oxides, particularly the photoreactivity and charge transport are related to defect disorders [8].

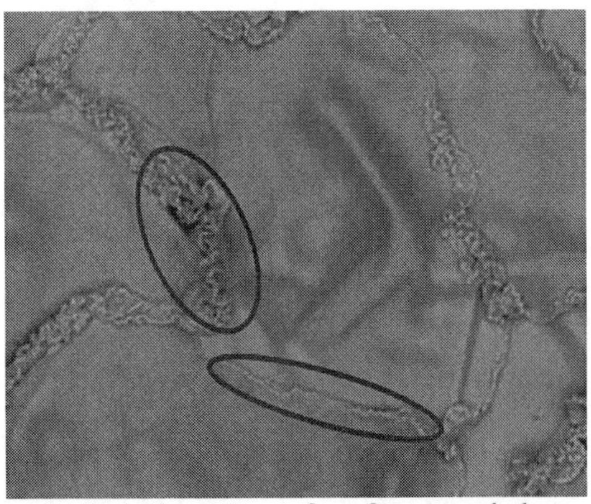

Figure 5. Grain boundary of surface morphology of CuO sample showing defects such as voids and uneven thickness of grain boundaries.

3.2. Elemental Properties

The CuO layer growth is in proportion with oxidation time. It was observed that the outermost layer has already developed the stable CuO oxide layer. This agrees with the EDS analysis exhibited that the highest content of oxygen was detected on sample annealed at 1010°C. Cu_2O is formed at lower temperature (~200°C) and followed by the formation of CuO layer as the temperature increase. The mixed oxides of Cu_2O and CuO could be found at temperatures below 1000 °C and at atmospheric pressure.

Figure 6 shows the images of distribution of element in annealed Cu sample at temperature of 980 °C, 990 °C, 1000 °C, and 1010 °C and Figure 7 summarizes the percentage of O_2 content. Cu and O_2 elements were detected in the CuO layer and thus, the composition of O_2 was focused. Based on the EDS analysis, an increment of amount of oxygen was increased along with the increase of annealing temperature. And the highest amount of O_2 of 18.8% after annealed at 1000 °C compared to other samples. While at temperature 980 °C and 990 °C O_2 was distributed uniformly and was estimated to be 15.4% and 15.7% respectively as per shown in lateral distribution of elements from Figure 6 (a) and Figure 6 (b). The percentage of O_2 slightly increased from low to the high annealing temperature. However, the O_2 concentration is slightly reduced to 14.5% at temperature 1010 °C as per shown in Figure 6 (d).

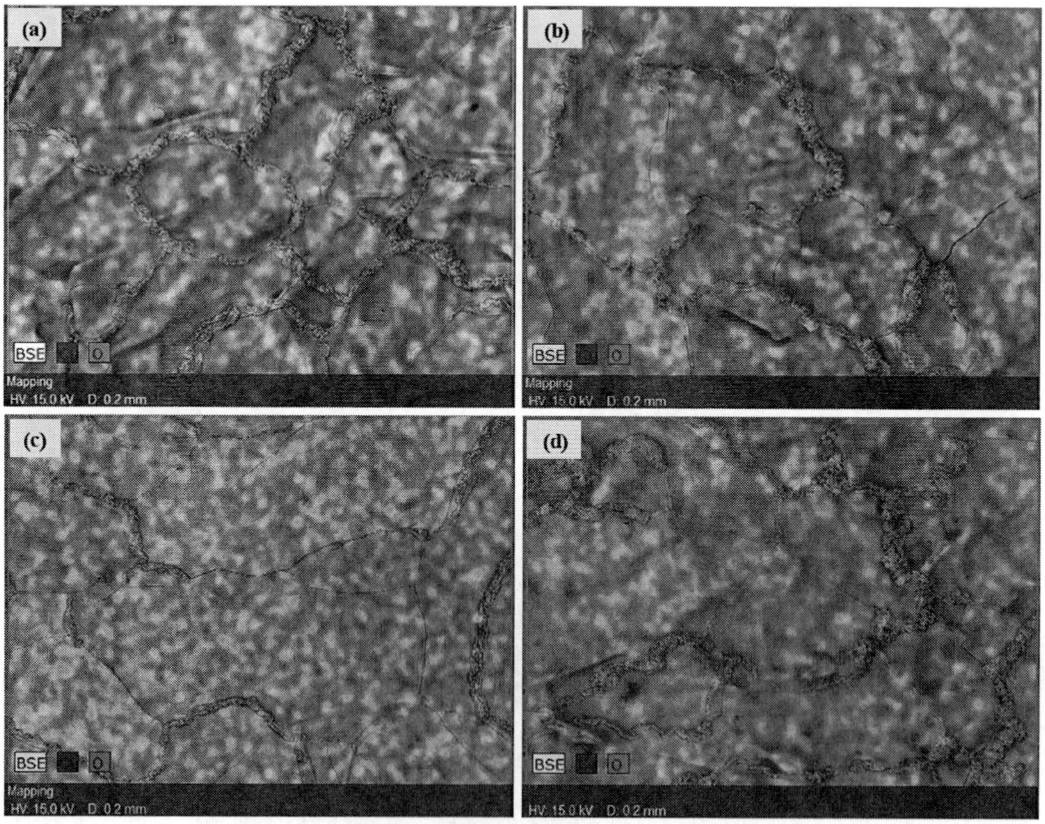

Figure 6. Percentage of element in annealed Cu sample at temperature (a) 980 °C, (b) 990 °C, (c) 1000 °C and (d) 1010 °C.

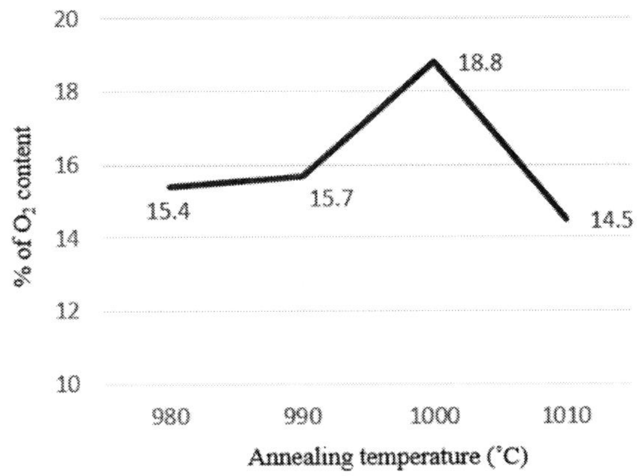

Figure 7. Percentage of O_2 content

3.3 Optical Properties

Optical analysis was carried out with UV VIS spectroscopy in wavelength range of 300 to 1000 nm. A typical recording of light absorption of the oxide layer is shown in **Figure 8.** The curve exhibits that the surface is slightly coarse at wavelength approximately 500 nm. The CuO layer showed an optical absorption threshold at 900 nm and the interference pattern was noted at wavelength at around 500 nm. The interferences may be due to the coarse surface and the presence of powdery oxide deposit that causes the scattering loses from the surface. This probably would continuously reduce the light transmittance of the material [5].

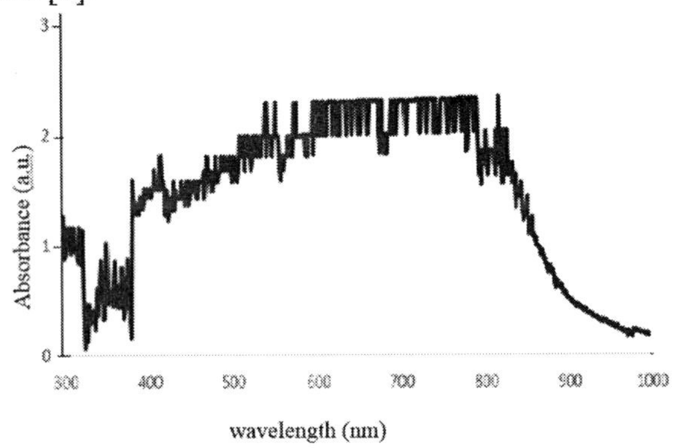

Figure 8. Absorption spectra for Cu_2O/CuO sheet annealed at 1000°C.

In theory, the conversion of Cu_2O into CuO can be testified by the determination of band gap. For this, the optical absorption coefficient, α was evaluated using the following **Equation (1):**

$$\alpha = (\ln T^{-1})/t \quad (1)$$

where t is the film thickness and T is the transmittance [9]. The equation is simplified by linear equation :

$$\alpha h\nu = A(h\nu - Eg)^{n/2} \quad (2)$$

where α is the absorption coefficient, A is a constant, n is the exponential integer depending upon the quantum selection rules for the particular material and hv is the photon energy obtained from **Equation (3)** [5] :

$$E = hv = hc/\lambda \tag{3}$$

where h is Plank's constant, c is the speed of light and λ is the wavelength.

Optical properties of a materials depend on physical characteristics of the films. Oxide layer consists of dominant CuO component has lower energy band gap (1.4 eV) compared to superior Cu_2O component (2.1eV) [5]. The lower energy band gap may be contributed by defects on the microstructure namely dislocations, vacancies, pits and pores.

4. Conclusion

In summary, all annealed samples were grown with CuO thin film. This was identified by its typical greyish black in colour. Though CuO has lower band gap energy than Cu_2O, the mobility was expected to improve with the recrystallization and grain growth after annealing. The result of CuO thin film formed as deduced from SEM, EDS and UV-VIS spectroscopy showed different growth behaviour with different annealing temperature. Grain growth has been evaluated in CuO thin film by measurement of its grain size at different temperatures. Highest grain growth was observed on samples annealed at 1000 °C. In general, CuO layer grown with increasing temperature. This was determined by the O_2 concentration by EDS analysis. The highest concentration of CuO layer was identified on the same samples with largest average area per grain. Poor mobility of an electron inside the oxide layer can be further improved by preparing a single orientation of CuO thin film. Larger grain size provides less barriers for electron mobility. The presence of defects on the boudaries also may contribute to the properties of the oxide thin film and could be potentially tailored for specific applications [8]. Hence, CuO film obtained by this method may be further studied and exploited as low cost photovoltaic device.

Acknowledgement

This work was supported with short term grant (STG) 9001-00563 granted by Universiti Malaysia Perlis

Reference

[1] B. K. Meyer, A. Polity, D. Reppin, M. Becker, P. Hering, P. J. Klar, T. Sander, C. Reindl, J. Benz, M. Eickhoff, C. Heiliger, M. Heinemann, J. Blasing, A. Krost, S. Shokovets, C. Muller and C. Ronning, *Phys. Status Solidi B*, **249**, 1487-1509 (2012).

[2] C. Xiang, G. M. Kimball, R. L. Grimm, B. S. Brunschwig, H. A. Atwater and N. S. Lewis, *Energy Environ. Sci.*, **4**, 1311-1318 (2011)

[3] T. Minami, T. Miyata and Y. Nishi, *Sol. Energy*, **105**, 206-217 (2014)

[4] A. C. Musa, T. Okomolafe and M. J. Carter, *Sol. Energ. Mat. Sol. C.*, 51, 305-306 (1998).

[5] M. R. Johan, M. S. Mohd Suan, N. L. Hawari and A. C. Hee, *Int. J. Electrochem. Sci.*, **6**, 6094-6104 (2011).

[6] K. Santra, C. K. Sarkar, M. K. Mukherjee and B. Gosh, *Thin Solid Films*, **213**, 226-229 (1992).

[7] F. De Lillo, F. Cecconi, G. Lacorata, A. Vulpiani, *EPL*, **84** (2008)

[8] N. Janusz.*Oxide Semiconductors for Solar Energy Conversion* (Taylor & Francis Group, 2012)

[9] T. Minami, Y. Nishi, Y. Miyata and J-L. Nomoto, *Appl. Phys. Exp.*, 4 (2011).

[10] C. R. Sekhar, *Solar Energy Mater.* 68, 307 (2001).

Optimization of Recycled Glass Fibre-Reinforced Plastics Gear via Integration of the Taguchi Method and Grey Relational Analysis

Nik Mizamzul Mehat[1,*], Noor Syuhada Zakarria[2], and Shahrul Kamaruddin[3]

[1]Department of Mechanical Engineering Technology, Faculty of Engineering Technology, Universiti Malaysia Perlis (UniMAP), Uniciti Alam Campus, Sungai Chuchuh, Padang Besar, Perlis, Malaysia.

[2] School of Mechanical Engineering, Universiti Sains Malaysia, Engineering Campus, 14300, Nibong Tebal, Penang, Malaysia

[2] Mechanical Engineering Department, Universiti Teknologi PETRONAS, 32610 Seri Iskandar, Perak Darul Ridzuan,

E-mail: nikmizamzul@unimap.edu.my

Abstract. The increase in demand for industrial gears has resulted in the increase in usage of plastic-matrix composites particularly glass fibre-reinforced plastics as the gear materials. The usage of these synthetic fibers is to enhance the mechanical strength and the thermal resistance of the plastic gears. Nevertheless, the production of large quantities of these synthetic fibre- reinforced composites poses a serious threat to the ecosystem. Comprehending to this fact, the present work aimed at investigating the effects of incorporating recycled glass fibre-reinforced plastics in various compositions particularly on dimensional stability and mechanical properties of gear produced with diverse injection moulding processing parameters setting. The integration of Grey relational analysis (GRA) and Taguchi method was adopted to evaluate the influence of recycled glass fibre-reinforced plastics and variation in processing parameters on gear quality. From the experimental results, the blending ratio was found as the most influential parameter of 56.0% contribution in both improving tensile properties as well as in minimizing shrinkage, followed by mould temperature of 24.1% contribution and cooling time of 10.6% contribution. The results obtained from the aforementioned work are expected to contribute to accessing the feasibility of using recycled glass fibre-reinforced plastics especially for gear application.

1. Introduction

To date, with advances in the development of plastic materials with wide spectrum of fiscal and practical properties available, plastics have become one of the most sought-after materials in the world particularly in the plastic gear industry. Plastic moulded gears offer some unique properties that are not achievable with metal based gears. The advantages include light weight, lower noise, higher modulus, self-lubrication, chemical resistance and, the most appealing of all, low cost. Because of their ability to run without lubrication and resist corrosion, plastic gears have been widely used in the automotive, food and textile industries, office machinery and household utensils as well as a host of other applications [1-3].

As mentioned the increase in demand for industrial gears has resulted in the increase in usage of plastic-matrix composites particularly glass fibre-reinforced plastics as the gear materials. This is to enhance the mechanical strength and the thermal resistance of the plastic gears. Nevertheless, the production of large quantities of these synthetic fibre- reinforced composites poses a serious threat to the ecosystem. These materials have become a major treat due to their non-biodegradability in the waste stream. Glass fibre filled plastic wastes can take decades and probably centuries to decompose. In Malaysia specifically, no viable recycling process appears to exist for these materials except being discarded to landfill and incineration. However, the limitation of landfill capacity and non-biodegradable properties of plastic makes the landfill method inevitable since the plastic waste consumes a large space and a lot of time to decompose in landfill sites On the other hand, incineration of plastic emits hazardous gasses during disposal of plastic waste which affects the environment and human health Comprehending to these limitations of landfill and incineration, recycling or reusing plastic waste as a single resin or in combination with other plastics were being developed and considered as a necessity [4-6].

However, plastic recycling is hindered by a wide range of barriers including mistrust of the quality of the recycled plastics as well as restrictions imposed by standards and specifications. Most plastic manufacturers and consumers were reluctant to use recycled plastics because of the perception that recycled plastics are inferior to the virgin resin and more difficult to handle [7]. Apart from that, improvements of mechanical and material properties of the recycled plastic products are still possible. From manufacturing perspective, optimization of injection moulding parameters is well known in improving the quality of the final product Optimization of injection moulding process is an appealing approach in improving the quality of the recycled plastic products. Injection moulding process is the most commonly used method in plastic processing industry [8-9]. Although injection moulding is one of the most widely used processes in plastic manufacturing, the quality of the products is easily affected by different injection moulding conditions. Material selection, part and mould design, and processing parameters play important roles in determining the quality of the final product [10-11]. In industrial practice, part and mould design are commonly assumed as fixed because any modification and alteration of part and mould design may lead to high production cost. Hence, processing parameters can be manipulated in order to improve the quality of the recycled product to a satisfactory level.

Comprehending to the facts stated, the proposed research is aimed at investigating the effects of incorporating recycled glass fibre-reinforced plastics in various compositions on dimensional stability and mechanical properties of gear produced with diverse injection moulding processing parameters setting. The integration of Grey relational analysis (GRA) and Taguchi method will be adopted to evaluate the influence of recycled glass fibre-reinforced plastics and variation in processing parameters on gear quality. The results obtained from the aforementioned work are expected to contribute to accessing the feasibility of using recycled glass fibre-reinforced plastics especially for gear application.

2. Methodology

In this study, polyamide 66 (PA66) filled with 33% glass fibre material was specified as the gear material to be studied, under the grade name of VITAMIDE AR17 supplied by Perrite. The general properties of the virgin PA66-GF were shown in Table 1. The recycled PA66-GF pellets were prepared with the same grade as the virgin material by crushing the existed products and post industrial wastes using a granulator. A spur gear design accordance with American Gears

Manufacturers Association (AGMA) was used. The details of geometry and specification of the gear were shown in Figure 1.

Table 1: General Properties of PA66-GF AR17

Properties	
Tensile strength (Mpa)	195
Strain at break (%)	3.5
Melting temperature (°C)	260
Density (kg/m^3)	1400
Flexural modulus (Mpa)	10000

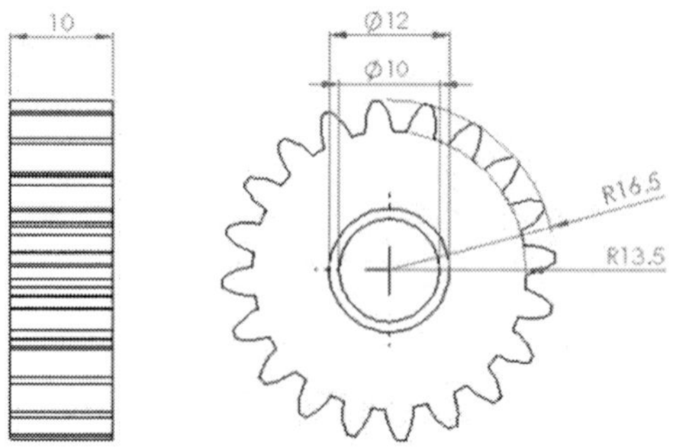

Figure 1. Geometry and specification of spur gear with module = 15; pressure angle = 20o; number of teeth = 20; face width = 10mm

2.2 Implementation of Taguchi method / GRA optimization procedures

2.2.1 Determination of quality characteristics

Regarding to the competitive and dynamics market of plastic gear in power and motion transmission application, simultaneous optimization of several quality characteristics is a major concern to quality engineer. There was numerous quality characteristics of injection moulded products such as warpage, shrinkage, sink index, weld line, shear stress, fibre orientation, etc. However, considering the functionality of the plastic gear, shrinkage and tensile properties were chosen as the quality characteristics to be studied in this research.

2.2.2 Selection of control parameters

There were four processing parameters including mould temperature, melting temperature, cooling time and injection time were selected for optimization with the aim of minimizing the shrinkage behaviour in tooth thickness, addendum as well as dedendum circles and enhancing the tensile ultimate strength, Young's modulus and elongation at break of the gears produced. In this study, the recycled blending compositions at various ratios was also considered as control parameters. The ratios were varied from 0, 5, 10, 15, 20, 25, 30, 35 to 40% by weight of recycled and virgin PA66-GF. Table 2 shows the control parameters that have been selected in this work.

Table 2. Controlled factors and their levels

Factors	Levels 1	2	3	4	5	6	7	8	9
A-Blending ratio (%)	0	5	10	15	20	25	30	35	40
B-mould temperature (°C)	50	60	70						
C-melting temperature (°C)	280	290	300						
D-cooling time (s)	10	20	30						
E-injection time (s)	0.6	0.8	1.0						

2.2.3 Selection of Taguchi's orthogonal array (OA)

Selection of an appropriate OA was based on the total degree of freedom (DOF) of the overall processing parameters. It was decided to study the interaction between the processing parameters because the processing parameters were dependent to each other in affecting the quality of a product. Hence, it was important to study the interaction among the processing parameters. In this study also, a mixed levels orthogonal array was decided to use since there was a nine levels parameter being considered in designing the experiment. Considering a nine levels parameter has eight DOF (DOF = number of levels – 1), four processing parameters each at three levels parameter has two DOF was equal to eight DOF and two interactions of three levels parameters contributes eight DOF (DOF interaction = DOF factor X DOF factor). Therefore, the total DOF required is 24 DOF. As discussed previously in screening experiment, the total DOF of selected OA must be greater than or equal to the total DOF required for the experiment. Hence, an L27 OA having 26 DOF was chosen for this study because it is the lowest order array that can accommodate all the parameters (Table 3)

Table 3. L27 array with interactions

FACTORS	COLUMNS	B	CXD	CXE	C				D	E		CXE	CXD
TRIAL NO.	1	2	3	4	5	6	7	8	9	10	11	12	13
1	1	1	1	1	1	1	1	1	1	1	1	1	1
2	1	1	1	1	2	2	2	2	2	2	2	2	2
3	1	1	1	1	3	3	3	3	3	3	3	3	3
4	1	2	2	2	1	1	1	2	2	2	3	3	3
5	1	2	2	2	2	2	2	3	3	3	1	1	1
6	1	2	2	2	3	3	3	1	1	1	2	2	2
7	1	3	3	3	1	1	1	3	3	3	2	2	2
8	1	3	3	3	2	2	2	1	1	1	3	3	3
9	1	3	3	3	3	3	3	2	2	2	1	1	1
10	2	1	2	3	1	2	3	1	2	3	1	2	3
11	2	1	2	3	2	3	1	2	3	1	2	3	1
12	2	1	2	3	3	1	2	3	1	2	3	1	2
13	2	2	3	1	1	2	3	2	3	1	3	1	2
14	2	2	3	1	2	3	1	3	1	2	1	2	3
15	2	2	3	1	3	1	2	1	2	3	2	3	1
16	2	3	1	2	1	2	3	3	1	2	2	3	1
17	2	3	1	2	2	3	1	1	2	3	3	1	2
18	2	3	1	2	3	1	2	2	3	1	1	2	3
19	3	1	3	2	1	3	2	1	3	2	1	3	2
20	3	1	3	2	2	1	3	2	1	3	2	1	3
21	3	1	3	2	3	2	1	3	2	1	3	2	1
22	3	2	1	3	1	3	2	2	1	3	3	2	1
23	3	2	1	3	2	1	3	3	2	1	1	3	2
24	3	2	1	3	3	2	1	1	3	2	2	1	3
25	3	3	2	1	1	3	2	3	2	1	2	1	3
26	3	3	2	1	2	1	3	1	3	2	3	2	1
27	3	3	2	1	3	2	1	2	1	3	1	3	2

IOP Conf. Series: Materials Science and Engineering **318** (2018) 012061 doi:10.1088/1757-899X/318/1/012061

3. Analysis and experimental results

After obtaining all the experimental results for shrinkage behaviour and tensile properties of all 27 trials as shown in Table 3, the analysis of the results was performed by computing the S/N ratio followed by data normalization of S/N results using GRA. Then, ANOVA was performed to determine the significance of each controlled parameters and interaction among the processing parameters. The procedures were described in details as follows:

3.1 Signal to noise (S/N) ratio

S/N ratio to measure the sensitivity of the quality characteristic being investigated in a controlled manner. Whenever an experiment involves with repeated specimens at each of the trial runs, the S/N ratio has been found to provide a practical way to measure and control the combined influence of deviation of the population mean from the target and the variation around the mean.

For shrinkage, the calculation of the smaller-the-better was selected meanwhile the higher-the-better was used to characterize the tensile properties. The results of shrinkage and tensile properties were converted into S/N ratio as shown in Table 4.

3.2 Data normalization using GRA

GRA is an effective way of analysing the relationship between data and can analyse many factors with complicated interrelationships among multiple performance characteristics which can overcome the disadvantages of statistical method. The computation of GRA was performed as follows:

Table 4. S/N ratio for shrinkage and tensile properties

Trial No.	Shrinkage			Tensile		
	Addendum	Dedendum	Tooth thickness	Maximum load	Young's Modulus	Elongation at break
1	-6.2419	-7.9824	7.6517	70.1427	39.6588	-1.6001
2	-7.2974	-9.3645	7.5852	68.1784	38.4801	-2.1279
3	-4.1664	-6.3677	5.3778	70.0858	39.9047	-1.8007
4	-6.3980	-8.4102	8.8147	69.0044	38.8287	-2.1229
5	-6.4782	-8.7043	13.8948	68.1497	38.7026	-1.9567
6	-7.6881	-9.5566	16.2873	67.5360	38.2935	-2.5066
7	-6.1101	-8.4782	10.8516	68.5553	39.7858	-2.1178
8	-6.9165	-9.0113	10.8516	69.4551	38.7428	-1.5358
9	-6.8158	-8.7548	8.4620	68.4894	38.6669	-1.8977
10	-7.3419	-9.6522	10.9437	68.3173	39.2959	-2.8893
11	-6.6602	-9.0399	8.9883	69.3218	38.9524	-1.2047
12	-7.4114	-9.3348	9.3743	68.1664	38.7455	-1.9824
13	-6.2067	-8.3169	8.6764	67.3651	39.1274	-3.2986
14	-6.5099	-8.3508	9.4606	69.2850	38.4389	-1.4825
15	-6.5486	-8.1286	9.9260	67.6051	37.6109	-1.9470
16	-7.2358	-8.8071	10.6718	68.2486	39.8158	-3.2112
17	-7.3889	-9.3751	11.5860	68.6876	39.3734	-2.7374
18	-6.2942	-8.2268	14.0472	67.7721	39.0182	-2.7212
19	-7.1774	-9.0490	11.7588	68.4630	39.3388	-2.7387
20	-6.2809	-8.4229	10.4117	67.7551	39.0730	-3.1859
21	-6.4580	-8.0619	9.9997	68.4037	38.4576	-1.3767
22	-7.1875	-8.7589	11.0920	68.8022	39.9728	-2.9000
23	-6.8050	-8.5466	12.1836	68.4736	39.4153	-2.6277
24	-6.4852	-8.2117	11.9438	68.3170	39.1067	-2.2354
25	-6.5085	-8.1574	8.7675	67.3709	39.2373	-2.9617
26	-7.1896	-9.0867	9.1474	68.0102	38.0747	-2.5266
27	-7.6208	-8.3342	6.2049	67.9104	38.1429	-2.2878

Step 1: Data pre-processing

Data pre-processing is a process of transforming the original data or also referred as an original sequence to a comparable sequence. This pre-process step also called as 'grey relational generation'. For this purpose, the shrinkage and tensile properties were normalized in the range between zero and one due to the different measurement units and scales of the raw data. The normalization can be done from three different approaches depending on desired quality characteristic

Lower the better

$$y_i^*(k) = \frac{max\ y_i^o(k) - y_i^o(k)}{\max y_i^o(k) - \min\ y_i^o(k)} \quad (1)$$

Higher the better

$$y_i^*(k) = \frac{y_i^o(k) - \min\ y_i^o(k)}{\max y_i^o\ (k) - \min\ y_i^o(k)} \quad (2)$$

Nominal the better

$$y_i^*(k) = 1 - \frac{|y_i^o(k) - y_i(k)|}{\max y_i^o\ (k) - y_i(k)} \quad (3)$$

where $y_i^*(k)$ is the value after the grey relational generation (data pre-processing), max $y_i^0(k)$ is the largest value of the original sequence, min $y_i^0(k)$ is the smallest value of the original sequence $y_i^0(k)$ and $y_i^*(k)$ is the desired value. Similar to S/N ratio characterization, the data pre-processing of shrinkage was done based on the smaller the better meanwhile the higher the better was used to characterize the tensile properties. The data pre-processing for the 27 trials were tabulated in Table 5.

Table 5. Data pre-processing of 27 trials

Trial No.	Data preprocessing					
	Shrinkage			Tensile		
	Addendum	Dedendum	Tooth thickness	Maximum load	Young's Modulus	Elongation at break
1	0.5893	0.4916	0.7916	1.0000	0.8670	0.8112
2	0.8891	0.9124	0.7977	0.2928	0.3680	0.5591
3	0.0000	0.0000	1.0000	0.9795	0.9712	0.7153
4	0.6337	0.6218	0.6850	0.5902	0.5156	0.5615
5	0.6565	0.7114	0.2193	0.2825	0.4622	0.6409
6	1.0000	0.9709	0.0000	0.0615	0.2890	0.3782
7	0.5519	0.6425	0.4983	0.4285	0.9208	0.5639
8	0.7809	0.8049	0.4983	0.7524	0.4792	0.8419
9	0.7523	0.7268	0.7173	0.4048	0.4471	0.6690
10	0.9017	1.0000	0.4898	0.3428	0.7134	0.1955
11	0.7081	0.8136	0.6690	0.7045	0.5679	1.0000
12	0.9214	0.9034	0.6337	0.2885	0.4804	0.6286
13	0.5794	0.5934	0.6976	0.0000	0.6421	0.0000
14	0.6654	0.6038	0.6258	0.6912	0.3505	0.8673
15	0.6764	0.5361	0.5831	0.0864	0.0000	0.6455
16	0.8716	0.7427	0.5147	0.3181	0.9335	0.0418
17	0.9150	0.9156	0.4309	0.4761	0.7462	0.2680
18	0.6042	0.5660	0.2053	0.1465	0.5958	0.2758
19	0.8550	0.8163	0.4151	0.3953	0.7315	0.2674
20	0.6004	0.6257	0.5386	0.1404	0.6190	0.0538
21	0.6507	0.5158	0.5763	0.3739	0.3585	0.9178
22	0.8579	0.7280	0.4762	0.5174	1.0000	0.1904
23	0.7493	0.6634	0.3762	0.3991	0.7639	0.3204
24	0.6584	0.5614	0.3981	0.3427	0.6333	0.5077
25	0.6651	0.5449	0.6893	0.0021	0.6886	0.1609
26	0.8585	0.8278	0.6545	0.2323	0.1964	0.3687
27	0.9809	0.5987	0.9242	0.1963	0.2253	0.4827

Step 2: Grey relational coefficient (GRC)
Following data pre-processing, a grey relational coefficient was calculated from the normalized data to express the relationship between the desired and actual shrinkage and tensile properties. The grey relational coefficient can be expressed as follows:

$$\xi_i(k) = \frac{\Delta_{min} + \xi . \Delta_{max}}{\Delta_{oi}(k) + \xi . \Delta_{max}} \quad (4)$$

where $\Delta_{oi}(k)$ is the deviation sequence of the reference sequence $y_o^*(k)$ and comparability sequence $y_i^*(k)$, namely:

$$\Delta_{oi}(k) = \| y_o^*(k) - y_i^*(k) \| \qquad (5)$$

$$\Delta_{max} = max_{j\epsilon i} max_k \| y_o^*(k) - y_j^*(k) \| \qquad (6)$$

$$\Delta_{min} = min_{j\epsilon i} min_k \| y_o^*(k) - y_j^*(k) \| \qquad (7)$$

ξ is distinguishing or identification coefficient which is defined in the range $0 \leq \xi \leq 1$. $\xi = 0.5$ is generally used.

Step 3: Grey relational grade (GRG)
After obtaining the GRC, the grey relational grade (GRG) was computed by averaging the grey relational coefficient. The grey relational grade can be defined as follows and tabulated in Table 6.

$$\gamma_i = \frac{1}{n} \sum_{k=1}^{n} \xi_i(k) \qquad (8)$$

The multiple quality characteristic performances of shrinkage and tensile properties were evaluated based on the grey relational grade. The complicated multiple performances have been converted into optimization of a single grey relational grade. The optimal level of the processing parameters was the level with the highest grey relational grade.

Table 6. Grey relational analysis results

Trial No.	Grey Relational Coefficient, GRC						Grey Relational Grade, GRG
	Shrinkage			Tensile			
	Addendum	Dedendum	Tooth thickness	Maximum load	Young's Modulus	Elongation at break	
1	0.5491	0.4958	0.7058	1.0000	0.7899	0.7259	0.7111
2	0.8184	0.8509	0.7119	0.4142	0.4417	0.5314	0.6281
3	0.3333	0.3333	1.0000	0.9607	0.9455	0.6372	0.7017
4	0.5772	0.5694	0.6135	0.5496	0.5079	0.5328	0.5584
5	0.5927	0.6340	0.3904	0.4107	0.4818	0.5820	0.5153
6	1.0000	0.9450	0.3333	0.3476	0.4129	0.4457	0.5808
7	0.5274	0.5831	0.4991	0.4666	0.8632	0.5341	0.5789
8	0.6953	0.7193	0.4991	0.6688	0.4898	0.7597	0.6387
9	0.6687	0.6466	0.6388	0.4565	0.4749	0.6017	0.5812
10	0.8357	1.0000	0.4950	0.4321	0.6357	0.3833	0.6303
11	0.6314	0.7284	0.6017	0.6285	0.5365	1.0000	0.6878
12	0.8642	0.8380	0.5771	0.4127	0.4904	0.5738	0.6260
13	0.5431	0.5515	0.6232	0.3333	0.5828	0.3333	0.4945
14	0.5991	0.5579	0.5719	0.6182	0.4350	0.7903	0.5954
15	0.6071	0.5187	0.5453	0.3537	0.3333	0.5851	0.4906
16	0.7956	0.6602	0.5075	0.4230	0.8826	0.3429	0.6020
17	0.8548	0.8556	0.4677	0.4884	0.6633	0.4059	0.6226
18	0.5582	0.5353	0.3862	0.3694	0.5530	0.4084	0.4684
19	0.7752	0.7314	0.4609	0.4526	0.6507	0.4057	0.5794
20	0.5558	0.5719	0.5201	0.3678	0.5676	0.3457	0.4881
21	0.5887	0.5080	0.5413	0.4440	0.4380	0.8588	0.5632
22	0.7786	0.6477	0.4884	0.5088	1.0000	0.3818	0.6342
23	0.6660	0.5976	0.4449	0.4542	0.6793	0.4239	0.5443
24	0.5941	0.5327	0.4538	0.4320	0.5769	0.5039	0.5156
25	0.5988	0.5235	0.6167	0.3338	0.6162	0.3734	0.5104
26	0.7794	0.7438	0.5913	0.3944	0.3835	0.4420	0.5557
27	0.9632	0.5548	0.8683	0.3835	0.3922	0.4915	0.6089

4. Discussion on results analysis

4.1 Main effects analysis

A main effect is a statistical term associated with experimental designs. To develop a clear understanding of how the optimum condition was selected, the main effects were further performed in this study and plotted in Figure 2.

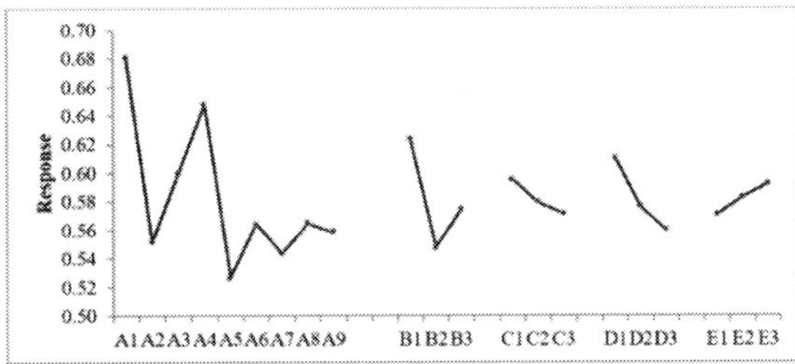

Figure 2. Main effects plot

Since the GRG represents the level of correlation between the reference and the comparability sequences, the larger GRG means the comparability sequence exhibits a stronger correlation with the reference sequence. In addition, the higher GRG also indicates the best multiple quality characteristic performances. Ignoring the interaction effects, notice that Fig. 2 shows the largest value of GRG for factors A, B, C, D and E are at level A1, B1, C1, D1 and E3 respectively. Hence, the optimal combination parameters achieving maximum tensile properties as well as minimum shrinkage were predicted to be A1 B1 C1 D1 E3, namely virgin PA66-GF, mould temperature of 50 °C, melting temperature of 280 °C, cooling time of 10s and injection time of 1s.

In studying the feasibility study of recycled PA66-GF blending composition, Fig. 2 illustrates that A4 of 15% wt recycled PA66-GF presents the most comparable performance with the virgin PA66-GF. Hence, another optimum parameter conditions to indicate the best multiple quality characteristic performances relative to the recycled PA66-GF blending composition were predicted to be A4 B1 C1 D1 E3, namely 15% wt PA66-GF, mould temperature of 50 °C, melting temperature of 280 °C, cooling time of 10 s and injection time of 1 s. Thus, these two sets of optimum conditions were verified in the confirmation test at the end of this study to assess the usability of the recycled blending material compared to its virgin.

4.2 Interaction Effects Analysis

To determine whether the interaction is present in this study, a proper interpretation of the results is necessary. The average effect of interactions C X D and C X E with respective factor level were performed and the results illustrated in Fig. 3.

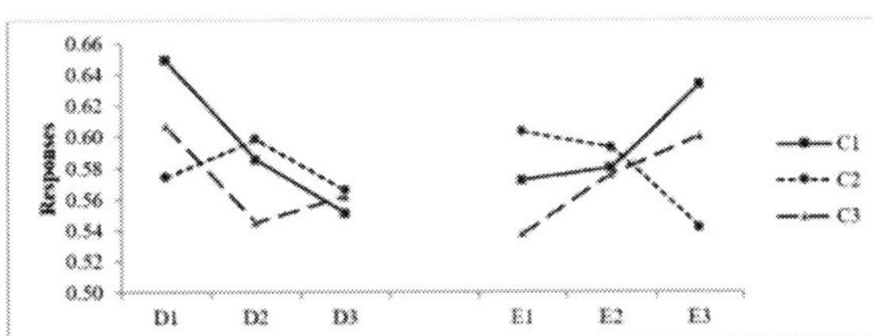

Figure 3. Interaction effects plot

The intersecting lines represent the interaction among parameters C, D and E. When re-examining the optimum condition obtained from the previous section of factors A1 B1 C1 D1 E3, Fig. 3 depicts that $\overline{C_1D_1}$ and $\overline{C_1E_3}$ have the higher values. Coincidentally, the optimum condition for interaction effects was the same as the optimum conditions obtained from main effects analysis in Fig. 2, hence no optimum condition modification is needed. Further analyses of the significance influence of interactions were made possible by the ANOVA.

4.3 Analysis of Variance (ANOVA)

Taguchi method does not only facilitate the response plot to visualize the impact of varying parameters on the multiple quality characteristic performance, but it can also perform ANOVA to estimate the relative contribution and significance of each parameter to the overall measured responses. The results of ANOVA were tabulated in Table 7.

Table 7. ANOVA results of the experiment

Symbol	Parameters/Factors	DOF, *f*	Sum of Squares, *S*	Variance, *V*	Variance Ratio, *F*	Pure Sum of Squares	Percent Contribution *P* (%)
A	Blending Ratio	8	0.0629	0.0079	2.73	0.0398	56.0%
B	Mold Temperature	2	0.0270	0.0135	4.69	0.0213	24.1%
C	Melting Temperature	2	0.0027	0.0013	0.47	-0.0031	2.4%
D	Cooling Time	2	0.0119	0.0059	2.06	0.0061	10.6%
E	Injection Time	2	0.0020	0.0010	0.35	-0.0037	1.8%
C X D		4	0.0065	0.0016	0.57	-0.0050	5.8%
C X E		4	0.0095	0.0024	0.82	-0.0020	8.5%
All other/ Error		2	0.0058	0.0029		0.07	5.1%

In view of the fact that the percentage contributions of all parameters and interactions on the multiple quality characteristic performance, the blending ratio was found as the most influential parameter in improving tensile properties as well as in minimizing shrinkage. The results match with the earlier findings of [12] where the fibre glass material exhibited a strong influence in reducing shrinkage compared to other parameters. This phenomenon was due to the glass fibres orientation effect along the direction of injection flow during the injection moulding process. Whereas, the injection time was found to have the least contribution in affecting the multiple quality characteristics performance with only 1.8% contribution.

4.4 Confirmation results

The final step of the design of experiment is the confirmation test, which verifies if the optimum conditions suggested by the matrix experiment do indeed give the improvement projected. The verification experiment was performed by conducting the test with predicted optimal parameter settings as highlighted in the previous section. In addition, the accuracy of Taguchi optimization approach with feasible recycled PA66-GF blending composition also can be proved via the confirmation test. In this study, the confirmation test was carried out by adopting the optimal processing parameter conditions of B1 C1 D1 E3 for virgin PA66-GF and 15% wt recycled PA66-GF to produce the plastic gear. The plastic gears undergo a tensile test and shrinkage test and the experimental performance of the plastic gears were tabulated in Table 8.

Table 8. Confirmation results of optimization prediction

	Shrinkage (%)			Tensile properties		
	Addendum	Dedendum	Tooth thickness	Maximum load (N)	Young's Modulus (Mpa)	Elongation at break(mm)
Virgin PA66-GF	2.1990	2.6757	0.1803	2569.5436	85.5000	0.8520
15% wt recycled PA66-GF	2.2213	2.7567	0.1811	3147.4450	79.8019	1.0194
Difference (%)	1.01	3.03	0.46	22.49	-6.66	19.65

The confirmation test indicates that the shrinkage of the plastic gear produced by 15% wt recycled PA66-GF blends has improved to a higher quality level which slightly better than the performance of the part produced by virgin PA66-GF. The reduction in molecular weight and fibre breakage during recycling process have indirectly minimized the shrinkage. Whereas for tensile properties, the plastic gear made of 15% wt recycled PA66-GF blend has also improved the strength of the plastic gear to withstand the maximum load and elongation at break at 22.49% and 19.65% respectively which better than virgin PA66-GF gear. The results match with the findings of Bernasconi et al. [13] where recycled products behaved in a stronger manner compared to its virgin because of fibre breakage during recycling led to a consequent decrement of fibres contribution in composite strength. However, the Young's modulus and melt flow index of virgin PA66-GF gear were still better than the gear made of 15% wt recycled PA66-GF blends since the properties of the material have been altered during the reprocessing of the material [14].

5. Conclusion

- From this research, the blending ratio is found as the most influential parameter of 56.0% contribution in both improving tensile properties as well as in minimizing shrinkage, followed by mould temperature of 24.1% contribution and cooling time of 10.6% contribution. In view of Taguchi's interaction effect, melting temperature and cooling time give 5.8% contribution and melting temperature and injection time give 8.5% contribution which is relatively a small contributions compared to parameter effect. Therefore, integration of both Taguchi method and grey relational analysis enhanced the efficiency of the optimization process in determining the influence of processing parameters and interaction among parameters subjected to desire multiple quality characteristics.
- The integrated methodology has proven its capabilities in determining the optimal conditions with multi responses quality characteristics. In this study, the optimal processing parameter conditions of B1 C1 D1 E3 (mould temperature of 50 °C, melting temperature of 280 °C, cooling time of 10 s and injection time of 1 s) were obtained for both virgin and 15% wt recycled PA66-GF to satisfy the multi performance quality characteristics of injection moulded plastic gear.
- The experimental results depicted that 15% of recycled PA66-GF blending composition behave almost similar to the virgin material in term of melt flow index. Comprehending to this fact, 15% recycled PA66-GF blend is considered reliable to satisfy quality requirements. This phenomenon is supported with improvement in shrinkage of 15% recycled PA66-GF blend gears compared to virgin PA66-GF gears. In addition, for tensile properties, the plastic gears made of 15% wt recycled PA66-GF blend have also improved the strength of the plastic gear to withstand the maximum load and elongation at break at 22.49% and 19.65% respectively better than virgin PA66-GF gears. However, the Young's modulus of virgin PA66-GF gear is still better than the gear made of 15% wt recycled PA66-GF blends. This phenomenon is due to the degradation on the molecular weight of recycled PA66-GF during recycling process. Apart from that, overall performance of 15% recycled PA66-GF is considered as a good candidate in substituting the virgin PA66-GF in plastic production.

The author acknowledges the STG grant (No. STG 9001-00523) provided by UniMAP for defraying the work that resulted in this manuscript.

References

1. H. İmrek, Trib. Int. **42**, 13 (2009)
2. M. Kalin, A. Kupec, Wear, **376–377**, Part B (2017)
3. K. Mao, Wear, **262**, 3-4 (2007).
4. N.M. Mehat, K. Shahrul, J. of Mat. Proces. Tech. 20 **211**, 5 (2011)
5. N.C. Fei, N.M Mehat, S. Kamaruddin, Z.M. Ariff, Int. J. of Manuf. Eng. **2013**, 1–9 (2013)
6. F. Gu, J. Guo, W. Zhang, P. A. Summers, P. Hall, Sci. of The Total Env. **601–602**, 1192-1207 (2017)
7. N.C. Fei, N.M Mehat, S. Kamaruddin, Z.M. Ariff, Int. J. of Res. in Mech. Eng. **1**,1 (2013)
8. S. Kitayama, S. Natsume, Sim. Model. Pract. and Theo. **48**, 35-44 (2014)
9. B. Ozcelik, Int. Comm. in Heat and Mass Trans. **38**, 8 (2011)
10. N.M. Mehat, S. Kamaruddin, Polym.-Plast. Tech. and Eng. **50**, 15 (2011)
11. C.P. Chen, M.T. Chuang, Y.H. Hsiao, Y.K. Yang, C.H. Tsai, Exp. Sys. with App. 36, 7 (2009)
12. E. Hakimian, A.B. Sulong, A.B. Mat. and Desig. 42, 4 (2012)
13. A. Bernasconi, D. Rossin, C. Armanni, C., Eng. Frac. Mech. 74, 4 (2007)
14. A.R. Breeds, S.N. Kukureka, K. Mao, D. Walton, C.J. Hookie, Wear, 166,.85–91 (1993)

The Effect of Copper Addition on the Properties of Sn-0.7Cu Solder Paste

R M Said[1], F H Mohamad Johari[1], M A A Mohd Salleh[1] and A V Sandu[2]
[1]Centre of Excellence Geopolymer and Green Technology, School of Materials Engineering, Universiti Malaysia Perlis (UniMAP), Taman Muhibbah, 02600, Jejawi, Arau Perlis, Malaysia.
[2]Faculty of Materials Science and Engineering, Gheorghe Asachi Technical University of Lasi, Romania

E-mail: ritamohdsaid15@gmail.com

Abstract. The effect of copper addition on the properties of Sn-Cu based solder paste were investigate through this study. The Sn-0.7Cu solder paste doped with different concentration of Cu were prepared using solder paste mixture. The bulk solder microstructure of as-solidified solder paste was studied. Besides that, intermetallic compound (IMC) formation on Cu substrate and hardness of all solder paste also being investigated. Results shows that increasing Cu concentration cause formation of large Cu_6Sn_5 IMC at bulk solder and the size of the IMC grew larger at high temperature. In addition, β-Sn area reduce when Cu concentration was high. The IMC morphology for all solder paste almost remain unchanged. However, there are large Cu_6Sn_5 IMC form near the interfacial IMC in Sn-Cu solder paste with high amount of Cu (Sn-10Cu). The hardness value was decrease when processing temperature at 250 °C due to present of small void in the microstructure while hardness of solder material increased at high temperature.

1. Introduction

Soldering technology plays an important role in electronic industry since it provided a metallurgical bonding between electronic component and substrate [1]. Lead (Pb) free solder has been highly demanded for usage in electronics packaging and assembly as a replacement for leaded solder [2]. The usage of leaded solder in electronic industry has been banned because of its toxicity that can harmful to individual health and also environment [3]. Since then, several studied have been made to developed new Pb free solder for replacing the existing Pb solder [4-8]. Nowadays, Sn-Ag-Cu based solder alloy was recognize as candidates for replacing Pb solder due to their good properties [9, 10]. On the other hand, less effort has been given to the development of much cheaper Sn-Cu solder alloys [11]. The properties of Sn-Cu solder could might be improve through micro-alloying [12] by addition of minor elements such as nickel [11, 13].

During liquid-solid reaction, solder paste will melt and wet the Cu substrate and hence resulting in the formation of intermetallic compound (IMC) layer. This IMC not only provide the joining but also the electrical and mechanical continuity in electronic assemblies. However, excessive IMC layer thickness prone to has deteriorate effect on solder properties due to high brittleness of the IMC [14]. Therefore, IMC formation of the solder joint in electronics packages become critical concern in the microelectronics industry [4, 15]. Basically, IMC that form from reaction between Sn-based solder with Cu substrate is scallop morphology of the Cu_6Sn_5 and Cu_3Sn when subjected to extensive soldering time or high temperature.

In this study, the main objectives are to investigate the effect of Cu addition in Sn-0.7Cu solder paste on the properties such as bulk microstructure, morphology of IMC layer and hardness during liquid solder/solid Cu substrate reaction for different soldering (processing) temperature.

2. Experimental procedure

Sn-0.7wt. % Cu solder paste (supplied by Nihon Superior Co. Ltd, Japan) was used as based material to produce new solder paste (Sn-4 wt. % Cu and Sn-10 wt. % Cu). The new solder pastes were prepared by mechanically mixing Sn-0.7wt. % Cu solder paste with copper particle using solder paste mixer. The mixing process was carried out for approximately 30 minutes to achieve homogeneity. To produce a solder joint, about 0.1 g of solder paste was then deposited on copper substrate and melted in furnace for temperature about 250 °C and soak for 2 minute. Besides that, solder paste also was place into crucible and melt for different temperature (250 and 500 °C).

The sample was then mounted using epoxy resin before metallography sample preparation (grinding, polishing and etching). The bulk microstructure and interfacial intermetallic compound (IMC) layer was observed under Optical Microscope (OM) and Scanning Electron Microscope (SEM). The mechanical properties of bulk solder was evaluated through vickers microhardness tester. The microhardness of the bulk solder alloy were tested using Micro Vickers testing machine with load of 9.81 N and dwell time of 10 s for each indentation. There are 15 indents were taken for each sample and the average value for the hardness value was then calculated.

3. Results and discussion

3.1. Bulk microstructure

Figure 1 show the optical microscope images of the Sn-Cu bulk solder microstructure with different Cu composition and also processing temperature. Two visible region of β-Sn phase and Sn-Cu eutectic phase of Cu_6Sn_5 can be observed on the microstructure of the solder paste. Generally, the microstructure of the Sn-0.7Cu solder after solidified consists of β-Sn phases (light colour), and SnCu phase which usually consist of needle-like Cu_6Sn_5 IMCs, eutectic Sn-Cu area [16].

Accordingly, it can be seen that there are changes in the bulk solder microstructures where the composition of Cu is different and the processing temperature varies. As can be seen in Figure 1(c and e), increasing amount of Cu reduce β-Sn area. Other than that, higher processing temperature results in larger grain size while lower temperature results in smaller grain size. In addition, Cu_3Sn IMC surrounded with Cu_6Sn_5 IMC was clearly observed at samples with high Cu concentration (Sn-10Cu).

Figure 2 shows the SEM image of Sn-4Cu solder after solidified. From the SEM/EDX analysis, it was clearly observed that the light area was β-Sn phase while the large particle is Sn-Cu eutectic phase of Cu_6Sn_5. Hung et al. [17] states that the amount of Cu composition in Sn-Cu solders influences the Cu-Sn intermetallic formation where increasing the Cu content increases the Cu-Sn intermetallic and decreases the area of the Sn-rich phase. According to Shen et al. [18] Cu has ability to refine the grain size of Bi-rich phase solder as Cu reacts with Sn to form Cu_6Sn_5 IMC particles and act as heterogeneous nucleation sites during solidification process.

Previous studies reported that adding some elements into solder alloy may alter the microstructure and thus improved the properties [19]. Zheng et al. reported that through microalloying, the microstructure of Sn-0.7Cu added with Ni and Zn were alter and a fully eutectic was obtained. Other than that, Nogita et al. found that Ni additions dramatically alter the nucleation patterns and solidification behaviour of the Sn–Cu_6Sn_5 eutectic in Sn-0.7Cu solder [20].

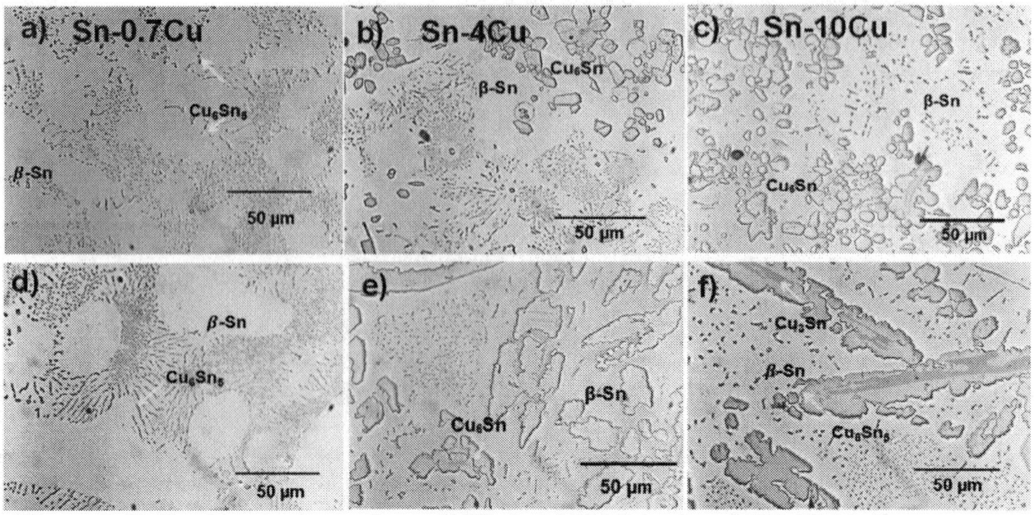

Figure 1. Bulk microstructure of as-solidified at 250 °C (a-c) and 500 °C (d-e) processing temperature.

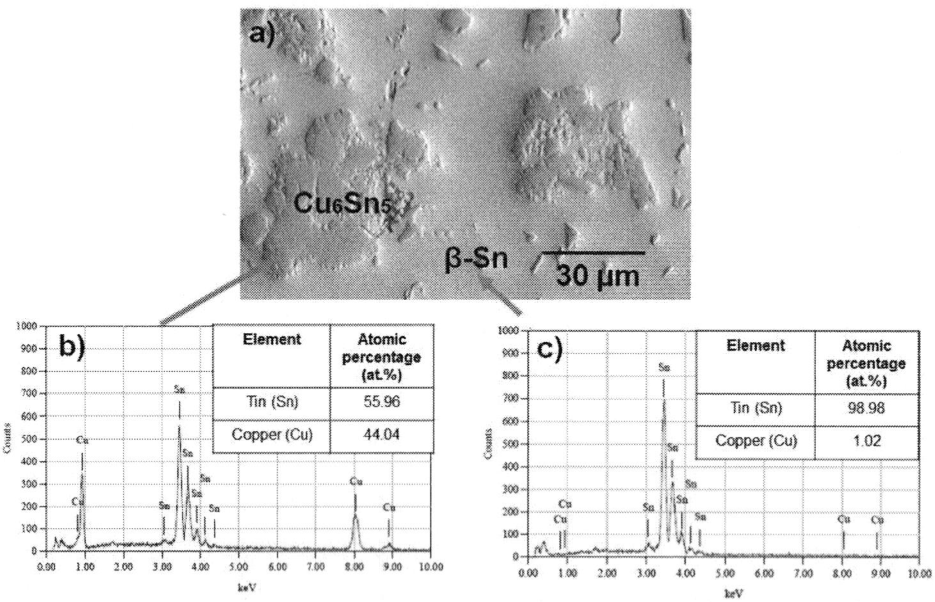

Figure 2. SEM/EDX Micrographs (a) bulk microstructure of Sn-4Cu, (b) Cu_6Sn_6 and (c) Tin (Sn) element detected.

3.2. Intermetallic compound formation

Figure 3 depict the SEM morphology of intermetallic compound (IMC) for Sn-Cu solder paste on Cu substrate. The scallop-type morphology was observed in all solder joints. Increasing amount of Cu into solder paste unchanged the morphology of the solder joint. In contrast, the formation of Cu_6Sn_5 IMC near the interfacial IMC results in larger size for solder paste with high Cu concentration (Figure 3c).

During soldering, solder paste will melt and wet Cu substrate resulting in the formation of IMC layer along the substrate. The diffusion of Cu atom from the Cu substrate and Sn atom from solder to the interface between melt solder and substrate will continue until Sn become supersaturated. Clearly, the Cu_6Sn_5 IMC was first precipitated from the supersaturated Sn liquid phase. Scanning Electron Microscope equip with Energy-Dispersive X-ray (SEM/EDX) result in Fig. 4 shows the IMC phase that form during early soldering was considered to be Cu_6Sn_5 by referring to the atomic ratio of Cu and Sn nearly to 6:5.

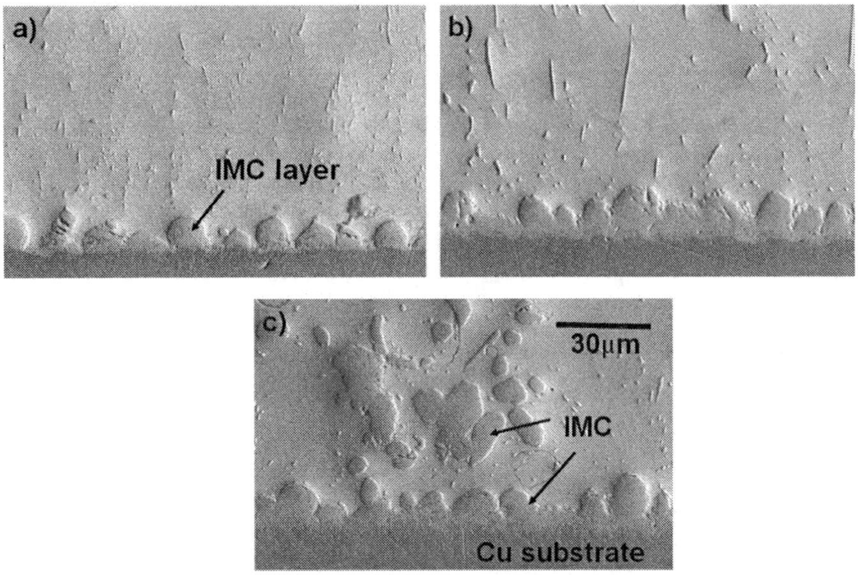

Figure 3. SEM image of intermetallic compound formation at 250 °C; (a) Sn-0.7Cu, (b) Sn-4Cu and (c) Sn-10Cu.

Cu_6Sn_5 IMCs are formed first because a large driving force for the metallurgical reaction between copper (Cu) and tin (Sn) atoms exists at the metastable composition. Scallop-type Cu_6Sn_5 first formed at the Sn/Cu interface during soldering, and its rate of formation is very fast. Cu_6Sn_5 was formed by the dissolution of Cu, followed by a metallurgical reaction. If contact with the molten solder is long enough, Cu_3Sn IMC formed between Cu_6Sn_5 and Cu. Cu_3Sn was formed by diffusion and by reaction type growth [21].

Formation of large Cu_6Sn_5 near the interfacial Cu_6Sn_5 IMC of Sn-10Cu might due to the high amount of Cu supplied in the solder compare to low Cu concentration of solder paste (Sn-0.7Cu and Sn-4Cu). Hence, this situation cause the formation of many Cu_6Sn_5 IMC located at near interfacial IMC also at bulk solder (Figure 1e).

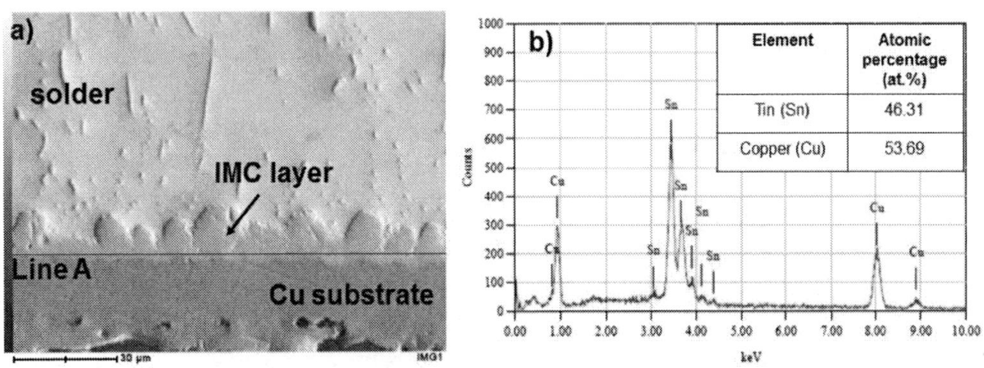

Figure 4. Line analysis of Sn-0.7Cu on Cu substrate.

IOP Conf. Series: Materials Science and Engineering **318** (2018) 012062 doi:10.1088/1757-899X/318/1/012062

3.3. Hardness

Figure 5 shows the bar chart for micro hardness value for Sn-Cu solder alloy with varying in Cu concentration at different processing temperature. Based on the bar chart (Figure 5a), the microhardness value of bulk solder alloys that melted at 250 °C has decreasing pattern as Cu composition in the solder alloys was increased.

Meanwhile, the value of microhardness for solder alloys melted at 500 °C were increased as the Cu composition in solder alloy increased as shown in Figure 5b. Rehim and Zahran have reported in their research that the increasing in microhardness is might be due to the presence of denser, harder and stiffer η-phase and ε-phase IMCs. This IMC act as a barrier for localized plastic deformation of the Sn-matrix during localized indentation (microhardness testing). The value of microhardness of the solder also may be affected by the microsturucture of the solder that is formed after solidified [11].

As for the solders melted at 250 °C, increasing Cu composition containing elevated number of porosity present in the bulk solder. Thus, this situation can lower hardness value of Sn-10%Cu. The mechanical properties always related to the microstructure of solder joint in which become very important to ensure prolong reliability during service.

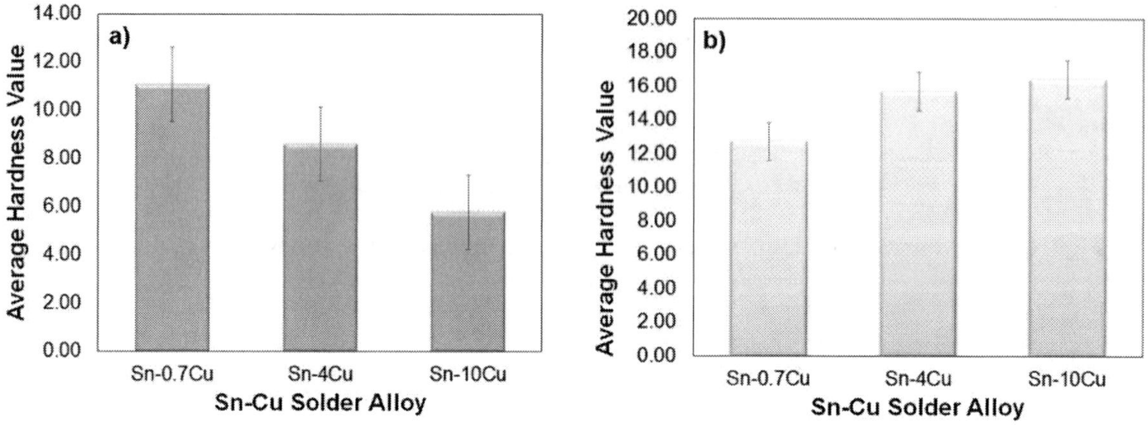

Figure 5. Microhardness value fo Sn-Cu solder alloys melted at (a) 250 °C and (b) 500 °C.

4. Conclusion

The effect of Cu addition on the microstructural and mechanical properties for Sn-Cu solder was examined through this project. These solder alloys were compared based on their properties after being melted in temperature of 250 °C and 500 °C with Cu composition of 0.7wt% Cu, 4wt%Cu and 10wt%Cu. The conclusions that can be drawn from this research are:

a) The microstructure and the morphology for bulk solder alloys were changed due to addition of Cu and increasing processing temperature. The size of Cu_6Sn_5 grains increases at higher Cu composition and higher melting temperature. Besides that, β-Sn area was reduce at higher Cu composition in bulk solder alloys.
b) As the Cu concentration increase, the Cu_6Sn_5 IMC grew larger near the interfacial IMC. Unfortunately, the interfacial IMC remain unchanged.
c) The solder paste that melted at 250 °C and solder with 10 % of Cu has the lowest hardness value compared to other two compositions. While for solder melted in 500 °C, pure Sn-0.7Cu solder has the lowest hardness value compared to other two compositions.

Acknowledgement
The authors gratefully acknowledge Nihon Superior Co. Ltd, Japan for supporting the research effort through materials.

References
[1] Musa S A, Mohd Salleh M A A and Saud N 2013 *Adv. Mater. Res.* **795** 518-521
[2] Mohd Salleh M A A, M.H.Z. Hazizi, Z.A. Ahmad, K. Hussin, and K.R. Ahmad 2011 *Adv. Mater. Res.*
[3] Mohd Salleh M A A, McDonalda S D, Terada Y, Yasuda H, and Nogita K 2015 *Mater. Des.* **82** 136–147
[4] Mohd Salleh M A A, McDonald S D, Gourlay C M, Yasuda H, and Nogita K 2016, Mater. Des. **108** 418–428
[5] Zhao Z, Liu L, Choi H S, Cai J, Wang Q, Wang Y, and Zou G 2016 *Microelectron. Relia.* **60** 126-134
[6] Zhao X, Wen Y, Li Y, Liu Y, and Wang Y 2016 *J. Alloys Compd.* **662** 272-282
[7] Mohd Salleh M A A, McDonald S D, and Nogita K 2013 *Appl. Mech. Mater.* 260-6
[8] Somidin F, Mohd Salleh M A A, and Ahmad K R 2013 *Adv. Mater. Res.* **620** 105-111
[9] Yahaya M Z, Ani F C, Samsudin Z, Sahin S, Abdullah MZ, and Mohamada A A 2016 *Mater. Sci. Eng., A* **669** 178–186
[10] Sun L and Zhang L 2015 *Adv. Mater. Sci. Eng.* 16
[11] Harcuba P and Janeček M 2008, *WDS'08 Proceedings of Contributed Papers* 220–224
[12] Zeng G, McDonald S D, Mu D, Terada Y, Yasuda H, Gu Q, Mohd Salleh M A A, and Nogita K 2016 *J. Alloys Compd.* **685** 471-482
[13] Mohd Salleh M A A, McDonald S D, Gourlay C M, Belyakov S A, Yasuda H, and Nogita K 2016 *J. Electron. Mater.* **45**(1)
[14] Ramli M I I, Saud N, Mohd Salleh M A A, Derman M N, and Said R M 2016 *Microelectron. Relia.* **65** 255–264
[15] Mohd Salleh M A A, McDonald S D, and Nogita K 2017 *Mater. Process. Technol.* **242** 235-245
[16] Abd El-Rehim A F and Zahran H Y 2017 *J. Alloys Compd.* **695** 3666-3673
[17] Hung F Y, T S Lui, Chen L H, and He N T 2008 *J. Alloys Compd.* **457**(1) 171-176
[18] Shen J, Pu Y, Yin H, Luo D, and Chen J 2014 *J. Alloys Compd.* **614** 63-70
[19] Zeng G, McDonald SD, Gu Q, Terada Y, Uesugi K, Yasudad H and Nogita K 2015 *Acta Mater.* **83** 2419-25
[20] Nogita K, Read J, Nishimura T, Sweatman K, Suenaga S and Dahle AK 2005 *Mater. Trans.* **46** 2419-25
[21] Lee LM and Mohamad AA 2013 *Adv. Mater. Sci. Eng.*

IOP Conf. Series: Materials Science and Engineering **318** (2018) 012063 doi:10.1088/1757-899X/318/1/012063

The Effects of Oxidation Layer, Temperature, and Stress on Tin Whisker Growth: A Short Review

Z Mahim[1], M A A Salleh[1], and C Y Khor[2]
[1]Center of Excellence Geopolymer & Green Technology (CEGeoGTech), School of Materials Engineering, Universiti Malaysia Perlis (UniMAP), Taman Muhibbah, 02600 Jejawi, Arau, Perlis, Malaysia.
[2]Faculty of Engineering Technology, Universiti Malaysia Perlis (UniMAP) Aras 1, Block S2 Kampus UniCITI Alam Sungai Chuchuh, Padang Besar, Perlis, Malaysia.

E-mail: zawawi870314@gmail.com.

Abstract. In order to reduce the Tin (Sn) whisker growth phenomenon in solder alloys, the researcher all the world has studied the factor of this behaviour. However, this phenomenon still hunted the electronic devices and industries. The whiskers growth were able to cause the electrical short, which would lead to the catastrophic such as plane crush, the failure of heart pacemaker, and the lost satellite connection. This article focuses on the three factors that influence the whiskers growth in solder alloys which is stress, oxidation layer and temperature. This findings were allowed the researchers to develop various method on how to reduce the growth of the Sn whiskers.

1. Introduction

Tin (Sn) whiskers have become a problem in electronic packing field since the beginning of this technology has been widely applied in all sectors [1-4]. Normally they have diameter 1-20 μm and their length is 5-500 μm [5]. Besides tin, another metal that popular for this issues were Cd and Zn but the Sn draw the most of attention compare to the others. Sn whiskers generally were good conductors, which this advantage were lead to electrical failures and even catastrophic disaster [6]. The growth of tin whisker was effectively controlled by alloying tin with Pb, and SnPb soldering and coating are until now widely used [7]. Yet, this practice alloying with Pb were strictly phased out by The Restriction of the Use of Certain Hazardous Substances (*RoHS*) due to the toxicity of lead that generally know [6]. To eliminate or reduce the Sn whisker growth, the growth phenomena of this problem were needed to be understood. Since 1940s, many theories, including dislocations-based theories [2-7], recrystallization theory [8], compressive stress-based theory [9-10] and many other [11-12] have been suggested. The latest factors that were studied by researcher related with compressive mechanical stresses, oxidation layer and temperature change.

2. Oxidation layer effect the whisker growth

There were several researcher such as Schroeder et al. [20] and Nakdaira et al. [21] studied the growth of whisker can be induced in the conditions at high humidity and temperature by oxidation or corrosion of the tin finish. Another researcher, Oberndorff et al. [22] have tested in high humidity inducing severe oxidation and corrosion by matte tin plated leadframe package. From the observation, the whisker growth causes by tin oxide that have high molar volume. Su et al. [23] have done a study of statistical for whisker growth and population. The mention that the last sample have a higher whisker population per component and have a longer maximum whisker length due to development corrosion stress

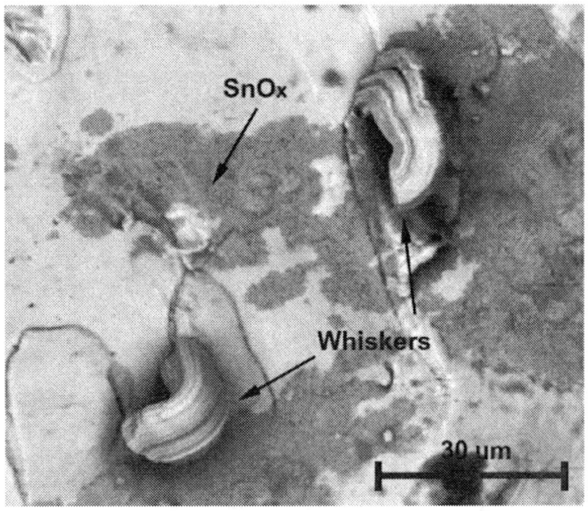

Figure 1. Corrosion spot on the surface area (darker spot). [23].

The observation by B. Horvath et al. [23] show the differences in trends of the whisker growth for tin-copper alloys. The whisker behaviour and corrosion resistance of the alloys was investigated by using polarisation measurements to reveal the relation between them. In figure 1, Scanning Electronic Microscope Backscattered Electron image mode was used to observe the condition of few corrosion spot for the sample that aging at 85°C/85% RH. It also can be observed at the corrosion spot, the whiskers were growth but the whisker's surface is free from dark corrosion. Another researcher, Osenbach et al. [22] found the area of corrosion crosses the non-corroded area of the surface film was parallel with no apparent height difference which shown that the surface of the film was pinned during the process of corrosion. With this method, the oxidation-induced excess tin atoms has been retarded in the original volume of the tin layer and create local stress and excess strain energy [22]. The corrosion may perform to spread only on the area near the top free surface but it may also spread deep inside the film putting mechanical pressure on the neighbouring tin grains. Nakadaira et al. [21] conclude as the corrosion increase, the corroded layer was in connection with the whisker base, retarding the growth of the whisker.

3. Temperature and humidity

European Union and Japan were restricted to ban the use of lead in electronic industry because of the hazardous to human and environment. This restriction force the researchers to improve the properties of solder by adding rare earth (RE) elements in to the solders. The result show the positive improvement mostly in all mechanical properties but the problem comes when tin whisker also found to grow faster compare to the lead solder [28-32].

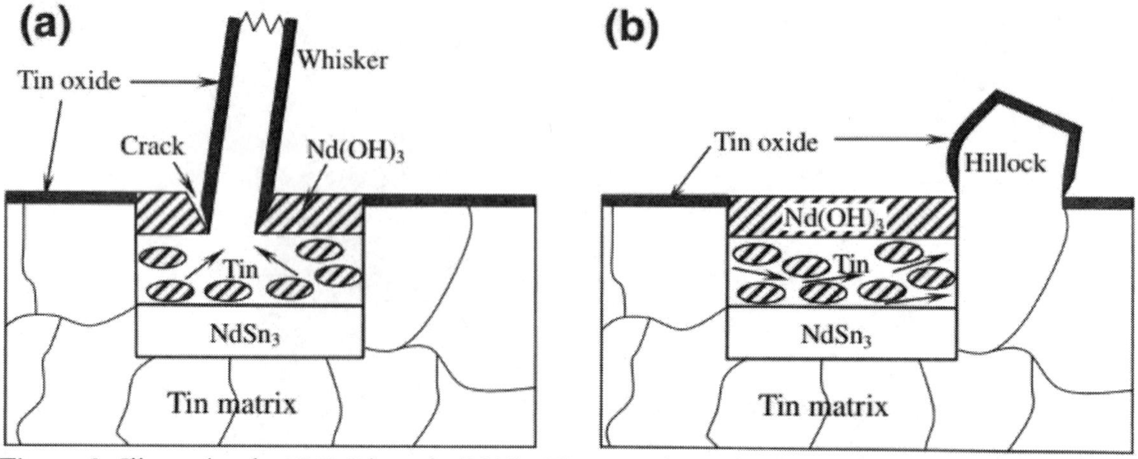

Figure 2. Illustration by C-F. Li et al. (2013). The growth mechanisms of (a) tin whisker and (b) hillock grown form Sn5Nd. [36]

The effect of temperature and humidity on tin whisker growth has been investigated since the early years of electronic application [1,25]. According to the opinion of Arnold and Osenbach, the higher temperature and humidity speed up tin whisker growth, which able to cause more atom mobility and in-depth oxidation of tin at higher temperature and/or higher humidity. Meanwhile the low temperature and low relative humidity could reduce whisker growth on tin coating, [25,26]. The oxidation of RE-Sn intermetallic compounds was caused the rapid growth of the tin whisker in lead-free solder [29,33,34]. At room temperature, thread-type of whiskers were found to growth, while at 150°C, hillock-type whiskers grew [29, 35]. Unfortunately, there are still limited study on the effects of temperature and humidity on tin whisker growth for RE-Sn alloys. Figure 2 shows the illustration for the mechanisms of tin whisker and hillock grown.

4. Stress effect on tin whisker growth

Generally, there are an amount of usually decided upon variables that effect whisker formation. Many researchers come to an understanding that compressive stress in tin is the essential factor that driving force behind whiskers growth [15]. A four-point bending technique was used by C-K. The growth trend of numerous whiskers was observed in specimen S1 by C-K. Lin.

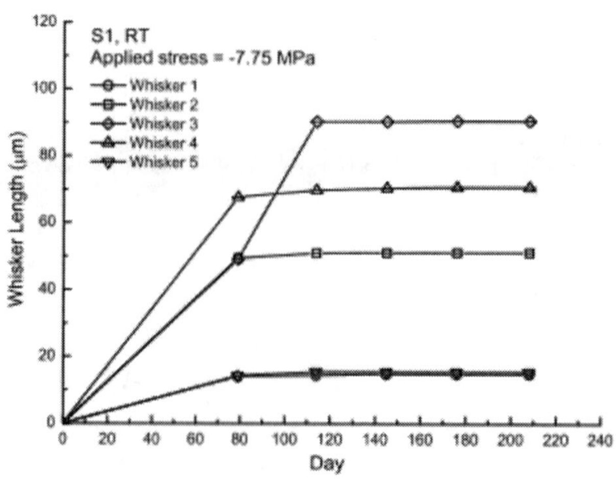

Figure 3. The trend of growth Specimen S1 for several tin whiskers [37].

The mechanical stress has been applied for 79 day and after that, observation the growth of whiskers was reported in figure 3. From the graph, they report the growth of whiskers almost stopped except whisker 3 that keep growing until day 114. The similar observation also was reported by the Sakuyama S et al. [39]. From the figure 4 and 5, tin whiskers were keep growing on whether mechanical stresses were applied or not. Unfortunately, the continuously applied the compressive stress able to increase the whisker density but decrease their length. When the tensile stress were applied continuously, the density and length of whisker were reduced by comparing with the case no stress applied.

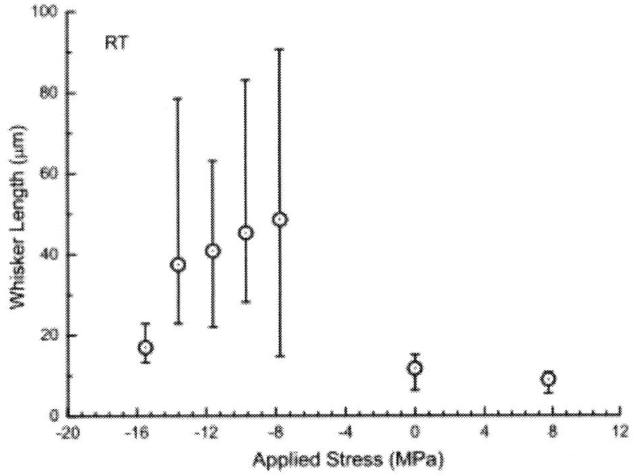

Figure 4. Interconnect between length and applied stress of whisker [37].

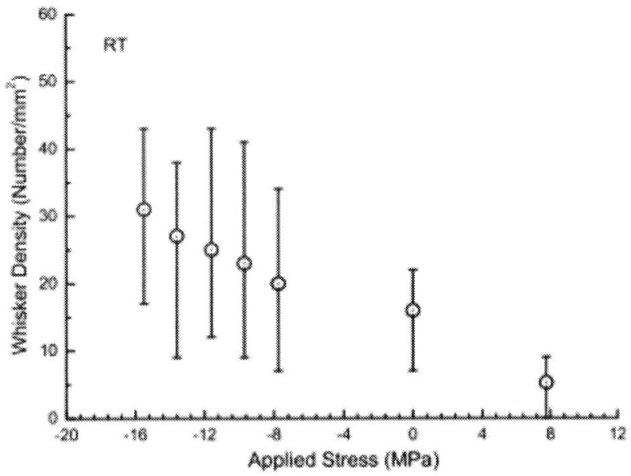

Figure 5. Interconnection between density and applied stress of whisker [37].

Tu et al. [38] explain that a compressive stress gradient was induced stress-free broken oxide spot for creep or the growth of a whisker to reduce the stress. The driving force was supplied a compressive stress gradient for the diffusion of tin from the stressed region to the root of a whisker by a diffusion type of creep mechanism [38].

5. Conclusion

The study of whisker phenomena is substantially improved knowledge about the factor effect the growth of tin whiskers. From this knowledge, it can allowed researcher to make demonstrable, efficient progress in identifying the key variable related with tin whisker. There are several special characteristics to approach the whisker studies;

i. As the corrosion increase, the corroded layer is in contact with the whisker base, disturbing the growth of the whisker

ii. The higher temperature and humidity speed up tin whisker growth and the low temperature and low relative humidity could reduce whisker growth

iii. Continuously applied the mechanical tensile stress were able to retarding the growth of tin whisker formation.

Acknowledgement
The author gratefully acknowledges the financial support of the Ministry of Higher Education of Malaysia through the Fundamental Research Grant Scheme (FRGS, No: 9003-00600).

References

[1] K G Compton A Mendizza, and S M Arnold. *Corros. Eng.*, **7** (1951), pp. 327-334.
[2] J.D Eshelby *J. Appl. Phys.*, **23** (12) (1952), pp. 1401-1403.
[3] J.D. Eshelby *Mater. Sci. Technol.*, **31** (7) (2015), pp. 675-698
[5] B Illés and B Horváth *J. Alloys Comp.* **616** (2014) 116–121.
[6] NASA, (https://nepp.nasa.gov/Whisker/background/index.htm#q3)
[7] ` S M Arnold *Proc. IEEE Elec Comp. Conf.* (1959), pp. 75-82.
[8] W C Ellis, D F Gibbons, and R G Treuting Cooperstown, New York, (1958) p. 102.
[9] M W Barsoum, E N Hoffman, R D Doherty, S Gupta, and A Zavaliangos *Phys. Rev. Lett.*, **93** (20) (2004).
[10] M A Dudek and N Chawla *Acta Mater.* **57** (15) (2009), pp. 4588-4599.
[11] K N Tu *Phys. Rev.* B, **49** (3) (1994), pp. 2030-2034.
[12] M Liu and A P Xian *J. Electron.Mater.*, **38** (11) (2009), pp. 2353-2361.
[13] E Chason and F Pei *JOM*, **67** (10) (2015), pp. 2416-2424.
[14] L Qiang and Z Huang *Microelectron. Reliab.* **54** (11) (2014), pp. 2494-2500.
[15] G T Galyon, C Xu S Lal, and B Notohardjono,A. Frye, *ECTC2005 iNEMI, IEEE Elec. Comp. & Tech. Conf.*, Lake Buena Vista, FL, (2005).
[16] B Z Lee and D N Lee *Acta Metar.* **46** (1998) 3701.
[17] S Madra *IPC/JEDEC Meeting*, Frankfurt, Germany, (2003) 115-120.
[18] D W Hoffman *J. Vac. Sci. Technol.* A **12** (1994) 953-961.
[19] M J Bozack, and S K Snipes *Surface Science.* **652** (2016) 355-366.
[20] V Schroeder, P Bush, M Williams, Nhat Vo, and H L Reynolds *IEEE Trans. Electron.Packag. Manuf.* **29** (4) (2006) 231–238.
[21] Y Nakadaira, S Jeong, J Shim, J Seo, S Min, T Cho, S Kang and S Oh *Microelectron. Reliab.* **48** (2008) 83–104.
[22] J W Osenbach, J M DeLucca, B D Potteiger, A Amin, R L Shook and F A Baiocchi *IEEE Trans. Electron. Packag. Manuf.* **30** (1) (2007) 23–35.
[23] P Su, J Howell and S Chopin *IEEE Trans. Electron. Packag. Manuf.* **29** (4) (2006) 246–251.
[24] B Hovath, T Shinohara and B Illes *J. Alloys Comp.* **577** (2013) 439-444
[25] S M Arnold *Proc. 43rd Am. Electroplat. Soc.* **43** (1956) 26–32.
[26] J W Osenbach, R L Shook, B T Vaccaro, B D Potteiger, A N Amin, K N Hooghan, P Suratkar and P Ruengsinsub *IEEE Trans. Electron. Packag. Manuf.* **28** (2005) 36–62.
[27] H Ye, S Xue, L Zhang, Z Xiao, Y Hu, Z Lai and H Zhu *J. Alloys Compd.* **509** (2011) L52–L55.
[28] C F Li, Z Q Liu, P J Shang, NS J K Shang *Scr. Mater.* **65** (2011) 1049–1052.
[29] T H Chuang *Scr. Mater.* **55** (2006) 983–986.

[30] A P Xian, M Liu *J. Mater. Res.* **24** (2009) 2775–2783.
[31] M Liu and A P Xian *J. Alloys Compd.* **486** (2009) 590–596.
[32] T H Chuang and C C Chi *J. Alloys Compd.* **480** (2009) 974–980.
[33] B Jiang and A P Xian *Philos. Mag. Lett.* **87** (2007) 657–662.
[34] M A Dudek and N Chawla *Acta Mater.* 57 (2009) 4588–4599.
[35] T H Chuang *Metallurgic. Mater. Trans. A* **38** (2007) 1048–1055.
[36] C-F Li, Z-Q Liu, and J-K Shang *J. Alloys Compd.* **550** (2013) 231-238.
[37] C-K Lin and T-H Lin *Micro. Reliability*, **48** (2008) 1737-1740.
[38] Tu K N, Chen C, and Wu A T *J. Mater Sci: Mater Electron*, **18** (2007) 269–281.
[39] Sakuyama S and Kutami M *Fujitsu Sci Tech J*, **41** (2005) 217–224.

IOP Conf. Series: Materials Science and Engineering **318** (2018) 012064 doi:10.1088/1757-899X/318/1/012064

Optimization of Polysulfone / Graphene Oxide / Polyethylene Glycol/ Triaminopyrimidine by Using Response Surface Methodology

Mohd Riduan Jamalludin[1,*], Zawati Harun[2], Siti Khadijah Hubadillah[3], Mohd Hafiz Dzarfan Othman[3], Noor Hasliza Kamarudin[2], Muhamad Zaini Yunos[2], Ras Izzati Ismail[1], Lailina N. M[1]

[1]Faculty of Engineering Technology, Universiti Malaysia Perlis (UniMAP), Kampus UniCITI Alam, Sungai Chuchuh, Padang Besar 02100 Perlis Malaysia.

[2]Integrated Material Process, Advanced Materials and Manufacturing Centre (AMMC), Faculty of Manufacturing and Mechanical Engineering, Universiti Tun Hussein Onn Malaysia, 86400 Parit Raja, Batu Pahat, Johor Darul Takzim, Malaysia.

[3]Advanced Membrane Technology Research Centre (AMTEC), Faculty of Chemical and Energy Engineering, Universiti Teknologi Malaysia, 81310 Skudai, Johor, Malaysia.

E-mail: riduanj@unimap.edu.my

Abstract. The addition of polyethylene glycol (PEG), graphene oxide (GO) and triaminopyrimidine (TAP) into polysulfone membranes was used to modify the membrane morphology and increase membrane performance. The central composite design of the response surface methodology was used to predict the maximum permeability and rejection of the PSf membrane. The parameter chosen for this study were PEG (7-14 wt%), GO (0-2.5wt%) and TAP (0-0.5 wt%) concentration. The flat sheet membrane was prepared via phase inversion technique where polysulfone (PSf) was used as base polymer. Various concentration of GO, PEG and TAP were added into the casting solution to produce different membrane composition. PEG was added as pore forming agent for the PSf membrane while inorganic additive such as GO was used to increase the hydrophilicity of the membrane. Besides that, the addition of TAP as a compatibilizer to enhance the physical interaction between GO and PEG. The characterization and surface morphology of produced membrane were analysed via scanning electron microscope, SEM and X-ray diffraction, XRD. The optimization of membrane performance was carried out by using response surface methodology (RSM). The performance of the membrane was analysed by using distilled water for pure water flux test and humic acid for rejection test. The optimized responses, membrane permeability and rejection obtained experimentally were 301.562 $Lm^{-2}h^{-1}$ and 91.562% respectively, with deviation from the predicted value of 7.884 and 0.4381 %, respectively.

1. Introduction

Membrane separation process for water treatment plays an important role to provide a clean and quality water system. Membrane technology are currently being applied to all sorts of fields including energy production, medical, industry, agriculture and food processing. In fact, membrane technology can be considered as the backbone in our industry and human life. As for example, the demand for high quality water has increased from time to time. To meet this demand, manufacturer and developer has manage to come out with a lot of techniques to produce high quality water. However, the most promising technique that has been studied by a lot of researcher is membrane technology process [1]. This process has found to be most suitable in water separation compared to other process such as halogen disinfection, radiation and chlorination. Membrane technology has become quite popular because of they are cheap and fast to fabricate, highly selective, flexible to be integrated with other process, required small space and its separation process will not change the water phase [1]. Briefly, membrane is a selective barrier that allows the passage of certain constituents and retains other constituents to pass through it. In other word, it acts as a filter. Separation is achieved when the membrane successfully permeating the passing of selective one or more components of a stream while retarding or restricted the passing of one or more other components. Polysulfone (PSf) are the most popular polymer to be used in membrane fabrication due to its good mechanical, thermal and chemical stability [2]. Even though it has good properties, PSf has a hydrophobic nature which make it prone to membrane fouling by the absorption of protein and other biomolecules from the feed stream [3]. Many researchers has studies a lot of alternatives to overcome this weakness. One alternative is via membrane modification. This technique require the incorporation of additive into the membrane formulation which may cause alteration in membrane morphology structure, pore size and its separation layer [1,2]. Additive has shown significant effect on the enhancement of protein rejection and antifouling properties of PSf membrane [4]. Even though a lot of additive studies and achievement in membrane properties is well published in various journals, however the selection of best additive for membrane has yet to be determine.

2. Experimental

2.1 Materials

Membrane were prepared from PSf was supplied from Udel P1700, N-methyl-2- pyrrolidone (NMP) was manufactured from Emplura®, Merck Schuchardt OHG. PEG was supplied from R&M Marketing, ESSEX, U.K, Graphene oxide was prepared via modified Hummers method and TAP was manufactured from Alfa Aesar, Johnson Mothey Company.

2.2 Preparation of graphene Oxide (GO)

Graphene oxide was synthesized from commercial graphite via modified Hummers method. 4g of commercial graphite powder was put into 230 mL concentrated of sulfuric acid (H_2SO_4) under ice bath. Then, 30g of potassium permanganate (KMnO4) was added slowly with stirring for 1 ½ hours. After that, the temperature of the reaction was increased to 40°C and the stirring continued for another 1 hour. Then 500mL of de-ionized water was added to the mixture and temperature were brought up to 100°C for 45 minutes. 7 mL of hydrogen pyroxide (H_2O_2) 30 wt % was slowly added to the mixture with additional of 500mL de-ionized water. The mixture was then diluted with 1:10 HCl solution (200mL) in order to remove metal ions by filter paper and vacuum pump. The suspension was washed several times with distilled water until the filtrate became neutral to remove remaining salt impurities. The final GO was sonicated for 1 hour and drying in vacuum oven at 40°C for 24 hours.

2.3 Membrane preparation

All membrane were prepared by using phased inversion method. PSf weighted at 18 wt% are stirred with 82 wt% solvent NMP for 4 hour with 500 rpm with 60 °C using a mechanical stirrer. After the solution has dissolved completely, additive such as Graphene Oxide (GO) and Polyethylene glycol 400 (PEG) were added into the solution and stirring continued for 30 minutes or until the additive has completely dissolved. The flat sheet membrane were fabricated by pouring dope solution on a glass plate with thickness of 120 μm. Distilled water was used as non-solvent in a coagulation bath. The produced flat sheet membranes were then left to dry for 24 hours in room temperature.

2.4 Characterization of membrane

2.4.1 Scanning electron microscope (SEM)

The surface and cross sectional structures of a few membranes were examine by SEM. The equipment used was a model JEOL JSM6380LA. The cross section of the membrane were prepared by immersed and fracturing the membranes at the temperature of liquid nitrogen manufactured by MOX. All the specimens were coated with a thin layer of gold before scanned with SEM.

2.4.2 X-ray diffraction (XRD)
XRD technique to determine the phase of crystalline material and provide the information on unit cell dimension. The machine used was XRD Bruker D8 Advance

2.4.3 Permeation flux (PWF)
The pure water flux is a measure of permeability of a membrane and from the knowledge of pure water flux can predict the permeability of the membrane. The formulas to calculate pure water flux are mention as below:

$$PWF = Q / [A.V.t] \tag{1}$$

2.4.4 Rejection (R)
In this study humic acid was prepared as water rejection. The percentage of humic acid are by using 0.2 g/l of humic acid, 2 liter distilled water and stir by using a magnetic stirrer The equation solute separation of the membrane:

$$R\ (\%) = [1 - (C_p / C_f)] \times 100 \tag{2}$$

2.4.5 Porosity
Membrane was immersed into distilled water for 24 hours at 25°C in order to determine the membrane surface porosity in a dimension of 2cm x2cm. The membrane in wet state was weight, W_w by using an electronic balance after carefully wiping the entire surface with clean tissue. Then the wet membrane dried in an oven for 50 to 60°C for 24 h. Then, membrane is weighed again in dry state. The membrane surface porosity was calculated using the following equation below:

2.4.6 Tensile strength
Tensile strength was determined by a universal electronic strength measurement (ABC machine). The standard tensile test was referred from American Society of Testing Method (ASTM) designation code of D882-12. In order to measure membrane tensile strength, each type were cut into strips 5 cm long and 1 cm wide. Five specimens were taken from each sample for the measurement and their average value are calculated.

3. Result and Discussion
3.1 X-ray Diffraction (XRD)
The characterization of GO was carried out by using XRD Bruker D8 Advance. Figure 3.1 shows the XRD pattern of natural graphite. From the pattern presented, natural graphite shows a peek at 26.22°. After the synthesization process, GO the obtained XRD pattern was showed the existence of GO as the peek has shifted to 10.9°. This pattern is similar to the GO obtained by Ganesh *et al.*

3.2 Scanning Electron Microscopy Analysis (SEM)
The Figure. 3.2 shows cross sections of flat sheet membrane containing PEG, GO and TAP. From the figure, it is found that all membranes are having asymmetric structure consisting of a dense top surface layer, porous sublayer and a small portion of sponge-like structure of bottom surface layer. Figure 2 demonstrated that the pore at the top and bottom layer increase with increasing of PEG concentration. The Figure 2 also shows the finger-like structure is shorten with the increasing of PEG concentration as the bigger porous started to growth This is due hydrophilicity of PEG properties that increased the inflow of water diffusion rate in the polymer solution during phase inversion process and leads to the formation of larger finger-like pores. Similar result was found by Ma *et al.* Meanwhile, the cross sectional morphology of prepared membrane at different GO concentration. The result shows that the number of finger-like pore at the top layer increases as the GO concentration increased. The membrane bottom layer was noticed to become more porous due to the increased of hydrophilic of GO

content. This condition might be due to the hydrophilic effect of GO that attracted more water and responsible for the fast exchange of solvent and non-solvent during the phase inversion phase which leads to this porous structure at both layers [4]. Figure 2 (c) shows the result of TAP with concentration of 0 wt%, 0.25 wt% and 0.67 wt% with the fixed value of PEG and GO. The cross-sectional and morphology analysis of the effect TAP crystalline characteristic revealed that the finger-like structure at top surface increases and macrovoid start to expand more that restrict the smaller finger-like formation as increases TAP concentration. Similar result was found by Basri *et al* that using TAP in PES-$AgNO_3$ [3]. This observation proved that TAP is functioned as a good compatibilizer between GO and PEG.

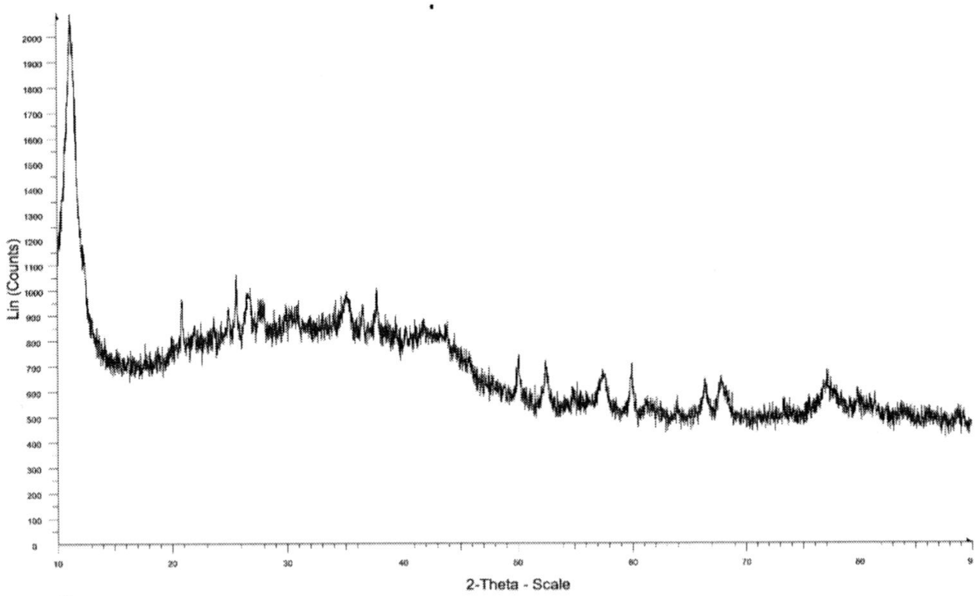

Figure 1. XRD pattern of graphene oxide

IOP Conf. Series: Materials Science and Engineering **318** (2018) 012064 doi:10.1088/1757-899X/318/1/012064

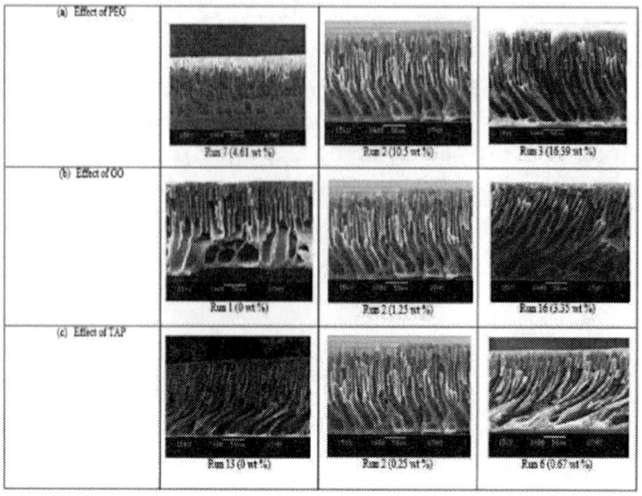

Figure 2. SEM image of membranes.

3.3 Response Surface Methodology (RSM) and Analysis of Variance (ANOVA)

3.3.1 Permeability

Figure 3 shows the three-dimensional graph for membrane performance in term of permeability response. Figure 3.3 (a) shows an effect of PEG and GO weight percentage with fixed TAP concentration of 0.25 wt% and Figure 3.3 (b) shows an effect of GO and TAP weight percentage with fixed PEG concentration at 10.50 wt% meanwhile Figure 3.3 (c) shows an effect of TAP and PEG at fixed GO concentration of 1.25 wt%. The result shows that PEG has significant effect to permeability compared to GO at fixed TAP value. The addition of TAP into membrane structure shows that when this concentration increases, the membrane permeability will also increase for both composition of fixed PEG and GO. This proved that TAP was a good compatibilizer of GO and PEG to enhance the physical interaction between polymer membrane and additives that would brings to good performance. Similar result was obtained by Basri *et al*, that using the additive and TAP as compatibilizer was increased the membrane performance [3,4].

3.3.2 Rejection

Figure 4 shows the three-dimensional graph of membrane rejection performance based on two selected factor. In this study, the effect of PEG and GO weight percentage with fixed TAP concentration of 0.25 wt% was shown in Figure 4 (a). Figure 4 (b) shows the effect of GO and TAP weight percentage with fixed PEG concentration at 10.50 wt% meanwhile Figure 4 (c) shows the effect of TAP and PEG with fixed GO concentration at 1.25 wt%. Both Figure 4 (a) and (c) that demonstrated by increasing PEG concentration reduced the rejection performance of membrane [5]. This is due to PEG function as pore forming agent that definitely increase the pore structure. Therefore, as the concentration of PEG increased, the number of pore and size also increased and thus promotes better permeation and reduce rejection. Meanwhile, in Figure 4 (a) and (b) the present of GO as additives promotes better rejection percentage for a certain range. As shown in Figure 3 (a) and (b), the increase of GO increase slightly the permeability value [6,7]. This result proved that GO has ability to increase the permeability with at the same time maintaining the rejection value which contradict to the PEG effect [8,9]. Thus, GO is a good additive to enhance the membrane performance. Figure 4 (b) and (c), shows the effect of increasing TAP concentration show an increase in rejection value. TAP as compatibilizer able to mix

GO and PEG and that result in good dispersion of GO. These will leads to formation of barrier on membrane and helps increase the rejection percentage [10,11,12].

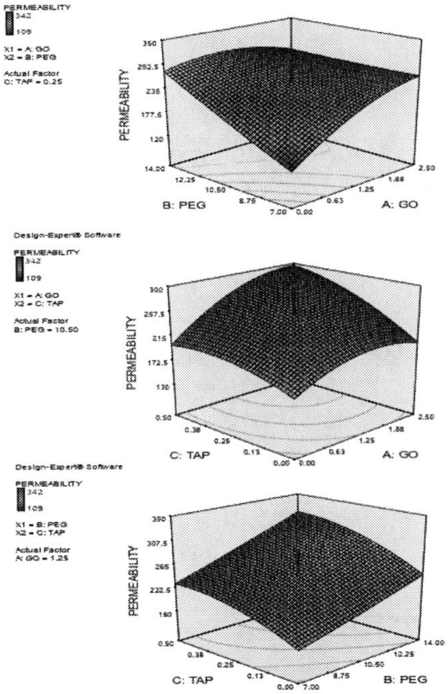

Figure 3. 3-D results of permeability

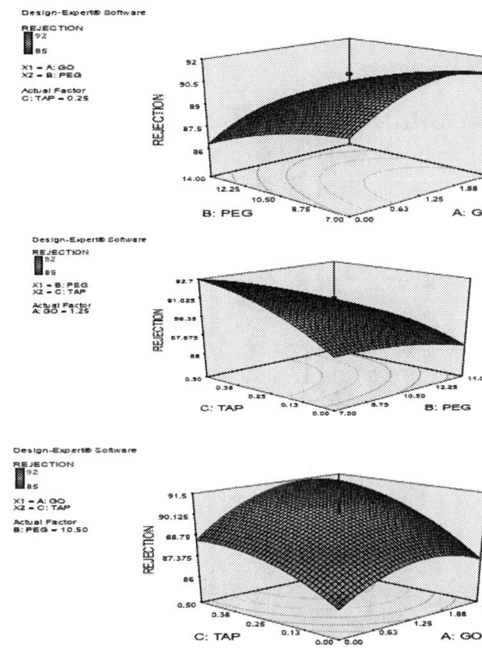

Figure 4. 3-D results of Rejection

3.3.3 Optimization of the Membrane Composition

The optimal prediction of the response obtained was at 0.890 desirability. Thus, it can be conducted with PEG (8.74 wt %), GO (2.50 wt %) and TAP (0.50 wt %). The result shows that the most desirable operating condition is 0.890. The desirability value that is closer to 1 is the ideal response [13]. Table 1 shows the prediction and actual obtained. The results presented that the predicted value for permeability was 293.678 and rejection 92.0001. Meanwhile, the actual membrane permeability was at 301.562 and the rejection was at 91.562. The error for permeability was 7.884% and for rejection 0.4381%.

Table 1. Prediction and actual obtained

	Permeability	**Rejection**
Predicted value	293.678	92.0001
Actual value	301.562	91.562
Error	7.884%	0.4381%

4. Conclusion

Flat sheet polysulfone membrane with containing Polyethtlene Glycol (PEG), Graphene Oxide (GO) and Triaminopyrimidine (TAP) was successfully prepared via phase inversion technique. The performance of PSf membrane were evaluated and optimized by using RSM approach. The effect of additive GO in membrane were carefully observed with interaction of other two parameter as it is the main factor in this study. The result found that the optimum composition of GO/PEG/TAP for membrane performance was 2.50 wt %, 8.74 wt %, and 0.50 wt % respectively with desirability at 0.89. The result also obtained that presences of GO/PEG/TAP in polysulfone membrane was changes the microstructure of membrane, enhanced hydrophilicity, water flux, and rejection property of membrane.

Acknowledgement
The authors would like to thank to Ministry of High Education Malaysia, Universiti Tun Hussein Onn Malaysia for their Research Grant Scheme (T001) for support in providing the grant implement and Universiti Malaysia Perlis for support of Short-Term Gerant (Project Number: 9001-00502).

References

[1] Arthanareeswaran, G., Mohan, D., and Raajenthiren, 2009 M. *Journal Of Membrane Science* **350** –130 – 138.
[2] Aryanti, P. T. P., Khoiruddin, Wenten, I. G., 2013 *Journal of Water Sustainability*, **3**, Issue 2, 85-96.
[3] Basri, H., Ismail, A.F., Aziz, M., 2012 *Desalination* **287**. 71-77.
[4] Cao, K., Jing, Z., Zhao, J., Zhao, C., Gao, C., Pan, F., Wang, B., Cao, X., Yang, J., 2014 *Journal of membrane science* **469**. 272-283.
[5] Mansourizadeh, A. and Ismail, A. F., 2009 *Journal Of Membrane Science* **348** – 260 – 267.
[6] Ganesh, B.M., Arun, M. I., Ismail, A. F., 2012 *Desalination* **313**.199-207.
[7] Jamalludin, M.R., Harun, Z., Basri, H., Yunos, M.Z., Shohur, M.F., Aminudin, N.N., Rosman, N., 2013, *e-ISSN* 2180-3722.
[8] Jamalludin, M.R., Harun, Z., Basri, H., Yunos, M.Z., Shohur, M.F. 2013 *Applied mechanics and materials* **372**, 8-12.
[9] Idris, A., Kormin, F., Noordin, M.Y., 2006 *Separation and Purification Technology* **49** 271–280.
[10] Hubadillah, S.K., Harun, Z., Othman, M.H.D., Ismail, A.F., Salleh, W.N.W., Basri, H., Yunos, M.Z., Gani, P., 2016 *Chemical Engineering Research and Design* **112**, 24-35.
[11] M. R. Jamalludin, Z. Harun, H. Basri, M. Z. Yunos, and M. F. Shohur, 2013 *Applied Mechanic and Materials*. **372**, 8–12.
[12] Jamalludin, M.R., Harun, Z., Hubadillah, S.K., H. Basri, Dzarfan M.H, A.F Ismail, Shohur, M.F., Yunos, M.Z. 2016 *Chemical Engineering Research and Design*, **114** 268–279.
[13] Yunos, M.Z., Harun, Z., Basri, H.,Hassan, S., Ismail, A.F et al. 2013 *Jurnal Teknologi* **65** (4), 111-115.

IOP Conf. Series: Materials Science and Engineering **318** (2018) 012065 doi:10.1088/1757-899X/318/1/012065

Surface Finish Effects Using Coating Method on 3D Printing (FDM) Parts

AHM Haidiezul[1], AF Aiman[2] and B Bakar[3]
[1] Faculty of Engineering Technology, Universiti Malaysia Perlis, Level 1, Block S2, UniCITI Alam Campus, Sungai Chuchuh, 02100 Padang Besar, Perlis, Malaysia
[2] School of Manufacturing Engineering,Universiti Malaysia Perlis,Kampus Alam Pauh Putra,02600 Arau,Perlis

Email: haidiezul@unimap.edu.my

Abstract. One of three-dimensional (3-D) printing economical processes is by using Fused Deposition Modelling (FDM). The 3-D printed object was built using layer-by-layer approach which caused "stair stepping" effects. This situation leads to uneven surface finish which mostly affect the objects appearance for product designers in presenting their models or prototypes. The objective of this paper is to examine the surface finish effects from the application of XTC-3D coating developed by Smooth-On, USA on the 3D printed parts. From the experimental works, this study shows the application of XTC-3D coating to the 3-D printed parts has improve the surface finish by reducing the gap between the layer

1. Introduction

In product design and development cycle, prototyping process is one of the important stages before it can go for manufacturing process. The quality of prototype is ranging from low to high fidelity. Vijay et al. [1] defined the prototypes deal with three aspects of interest:

- The implementation of the prototype from the entire product itself to its subassemblies and components.
- The form of the prototype from virtual to the physical prototype.
- The degree of the approximation of the prototype from a rough representation to an exact replication of the product.

Rapid prototyping (RP) is one of the emerging manufacturing techniques nowadays. It is a freeform manufacturing process that allows users to fabricate real physical part directly from a computer-aided design (CAD) model. Compared with other manufacturing processes, RP has several advantages such as it does not require any tooling to fabricate the part, less human intervention and ability to produce complex part. There are three common RP technologies - Fused Deposition Modelling (FDM), Strereolitography Apparatus (SLA), and Selective Laser Sintering (SLA).

Three-dimensional printing is a new advanced development of rapid prototyping technology and widely used by industrial designers and engineers. It is generally referred as a smaller system of RP and intended to be easily used and networked in an office environment [2]. Fused Deposition Modelling-based 3-D printing receives a significant demand around the world owing to several advantages offered by this technology such as wide variety of materials available, quick material change, low maintenance cost, quick production of thin parts, a tolerance equal to ±0.1 mm, no supervision, no toxic materials, very compact size, and low temperature operation [3]. Fused Deposition Modelling technique is prominent as an additive manufacturing technology which is commonly used in 3-D printing machines. It works by feeding molten thermoplastic material into the heated extruder to build physical form layer-by-layer. Figure 1 shows the schematic diagram of the FDM machine

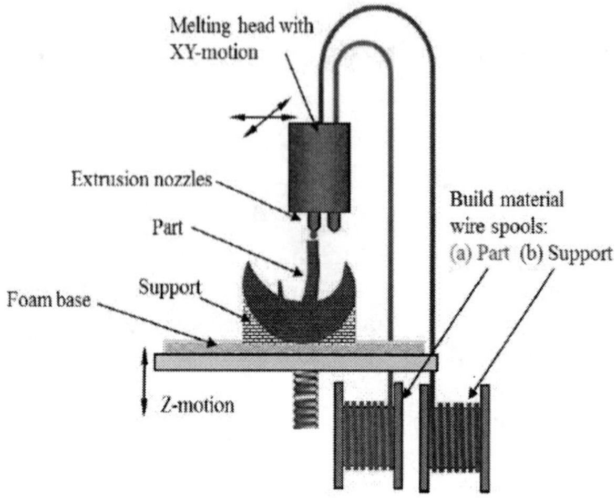

Figure 1. Schematic diagram of FDM [4]

Surface finishing of the FDM parts is very crucial in producing high fidelity prototypes and moulding.. For prototypes, low quality of surface finishing will affect the level of exactness compared with actual product. This situation will influence the decision making process because designers and engineers do not receive the right information when developing a new product. The common problem occurs in FDM technique is the layer remain visible on the surface after the fabricating process end. This problem is called stair stepping effect. This condition requires the FDM parts to undergo post-treatment process before it can be fully eliminated.

2.Previous works

To summarize the process of rapid prototyping, there are five steps that need to follow. Figure 2 exhibits the five main steps in rapid prototyping process.

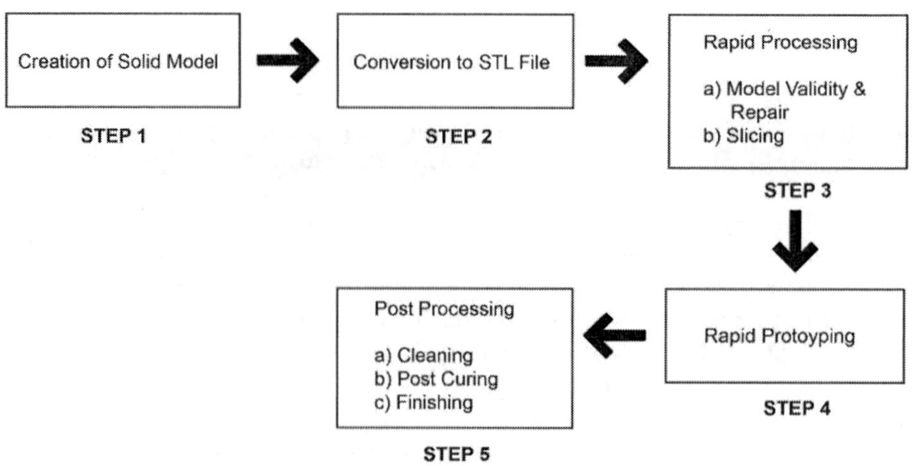

Figure 2. Five steps in the process of rapid prototyping [5]

Many past researchers have performed various types of study to improve the surface finish of FDM parts. Based on Figure 2, research on improving surface finish of FDM parts can be classified into rapid and post processings. Rapid processing is the step involved before the rapid prototyping process begins while post processing is after the process is completed. Gurpal and Parlad [6] had divided method of improving surface finish into four categories, namely:

- Optimization of build orientation.
- Slicing strategy (layer thickness)
- Fabricators parameterization optimization
- Post-treatment

Besides those parameters, Ongjan et al. [7] in their study had introduced two additional factors that can leave an impact on the surface quality:

- Extrusion speed
- Extrusion temperature

The primary goal of their research was to investigate the influence of extrusion speed and temperature to the surface quality. Implementation of rapid processing step does not fully remove layer visibility on the surface. However, it will reduce necessary post processing step. Reducing post processing step will reduce the manufacturing time and cost of the product.

Surface quality of the FDM parts can be enhanced by applying post processing step. Post processing step involves various methods such as sanding, chemical finishing, blasting and painting. Galantucci et al. [8] in their study used chemical finishing process to reduce the surface roughness of FDM parts. During chemical finishing process, specimen was immersed in dimethylketone (acetone). The author observed significant improvement in surface roughness of the treated specimen. A different approach has been proposed by Ivan et al. [3] to improve the quality of FDM parts surface finish. In their research, they had applied blasting technique where pre-treated sample was blasted with sodium-bicarbonate (SB) and glass beads (GB). Based on the results, this technique showed a minimal influence on the surface quality of the parts. This is in accordance to the aim of the present study which is to improve the surface finish of 3-D printing (FDM) parts by using surface coating method. Figure 3 illustrates the effect of various layer thicknesses on the stair stepping effect.

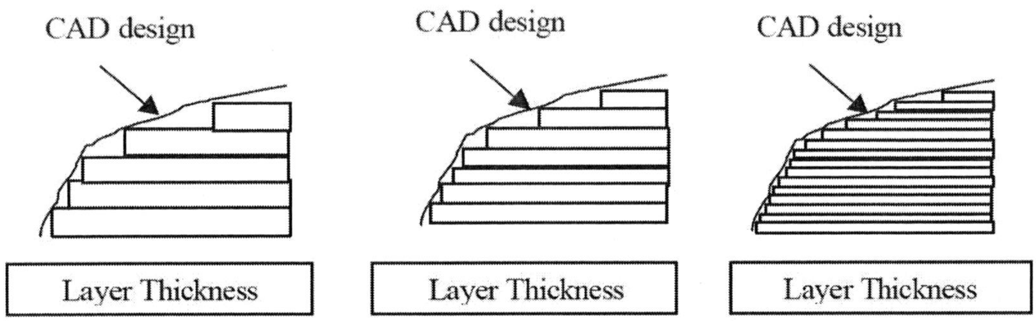

Figure 3. Effect of layer thickness to stair stepping effect [9]

Therefore, this study was intended to propose the application of surface coating method to remove the stair stepping effect. It works by fill the gap between the layer with XTC-3D.

IOP Conf. Series: Materials Science and Engineering **318** (2018) 012065 doi:10.1088/1757-899X/318/1/012065

3.Experimental work

3.1. Specimen preparation

To carry out the experimental work, a 3-D data of tested sample was prepared using CAD software. The dimensions of the tested sample – width, height and length are shown in Figure 4.

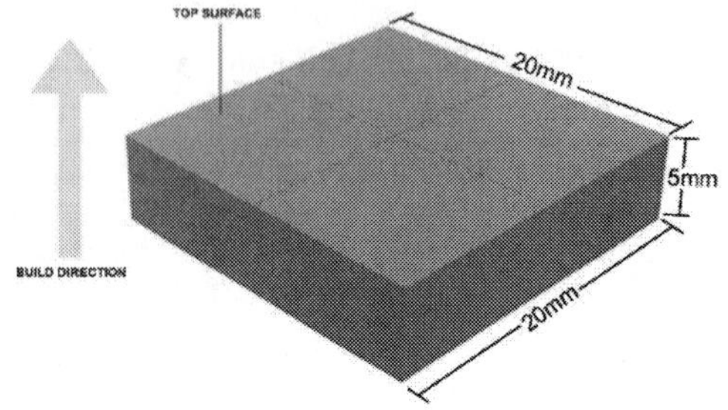

Figure 4. The dimensions of tested specimen

From the CAD software, the 3-D data was converted into Stereolitography format (STL). This data format is universal for 3-D printing software and it converts CAD data into triangulation mesh. Later, STL data was loaded into 3-D printing software, Flashprint. In this software, all printing paramaters were set according to the machine default setting that includes printing resolution, layer height, infill, shell, speed, and temperature. The parameters setting is shown in Table 1. After that, STL data was sliced into series of layer.

Table 1. Parameters setting

Parameters	Value
Infill	15%
No of Shells	2
Temperature	220°C
Layer Thickness	0.18 mm
Print Speed	60 mm/s

Three tested samples were produced using a 3-D printer from Flashforge Dreamer, USA. The printer has specified 230 x 150 x 140-mm build size and layer thickness ranging from 0.1 to 0.5 mm. It also has built precision ±0.2 mm. Material that has been used to produce this part is Acrylonitrile Butadiene Styrene (ABS). This material comes in the form of filament with 1.75-mm diameter. Figure 5 depicts the slicing process of STL file using flashprint software.

IOP Conf. Series: Materials Science and Engineering **318** (2018) 012065 doi:10.1088/1757-899X/318/1/012065

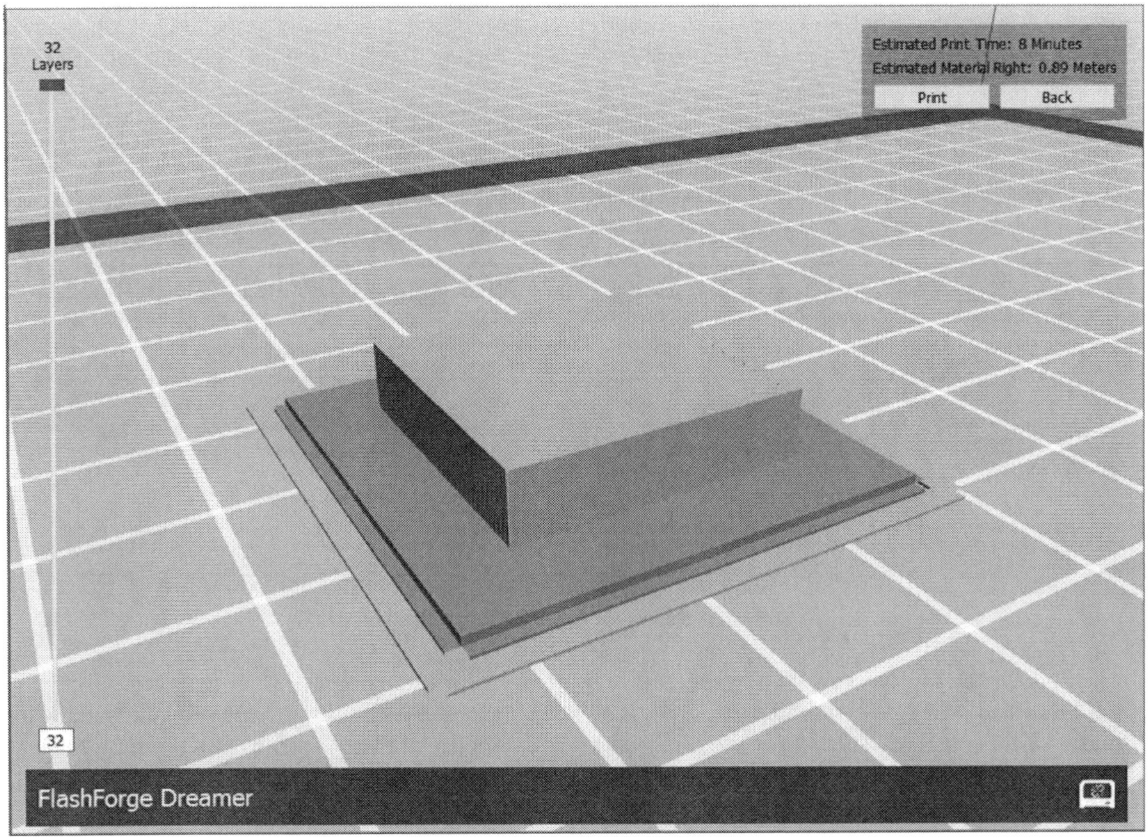

Figure 5. Slicing process of STL file using flashprint software

Based on Figure 5, the STL data was sliced into 32 series of layer. In general, it takes approximately 8 minutes to complete the printing process. Total amount of material used in this fabrication process is 0.80 meters.

3.2. Surface Coating

In this phase of experimentation, 2 out of 3 tested parts will undergo surface coating process. Chemical coating that has been chosen is XTC-3D from Smooth-On, USA. According to the manufacturer, this material is low cost, and easy to work on any 3-D printing surfaces such as FDM, SLS and SLA. It cures to hard, and impact resistant coating that can be sanded, primed and painted. All the three parts will be undergoing different layer of coating number as stated in Table 2. Brush method was used to apply the coating on the FDM parts surface.

Table 2. Three-dimensional samples with different coating layer numbers

Sample	A	B	C
XTC 3D coating Layer	0	1	2

For the first phase of the coating process, sample B and C were undergoing one layer of coating. After all samples were fully cured, second layer of coating was applied to sample C only. For each coating

IOP Conf. Series: Materials Science and Engineering **318** (2018) 012065 doi:10.1088/1757-899X/318/1/012065

process, all parts were left to be cured at room temperature (27°C). Figure 6 shows the main process flow of experimental work conducted in this study. The coating process is stopped at the second layer to maintain sample dimensional accuracy.

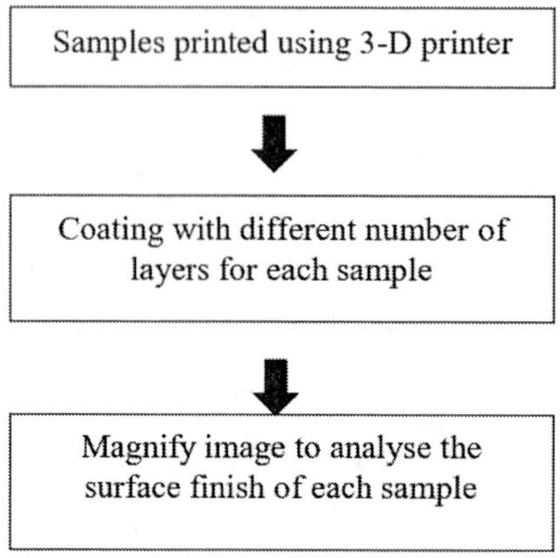

Figure 6. Process flow of experimental work

In this paper, the top layer of the sample will only be used for the comparison. A magnified image of each sample at 45X was captured using XOPTRON XST6 Stereo Microscope and were analysed using Solution Lite software. To capture and analyse the cross-sectional image of the tested parts, precision cutter was used to cut the part. The rationale of this process was to ensure the authors can observe and analyse the printed part and evaluate the coating layers.

4.Result and Discussion

As far as the effect of coating method on the surface finishing was concerned, the results of study exhibited that sample A has series of layer visible on the surface as illustrated in Figure 7. The reverse was seen for sample B and C in Figure 8 and Figure 9, respectively, which showed that the application of XTC-3D on the sample surface has fill the gap between the layer caused the better surface finish.

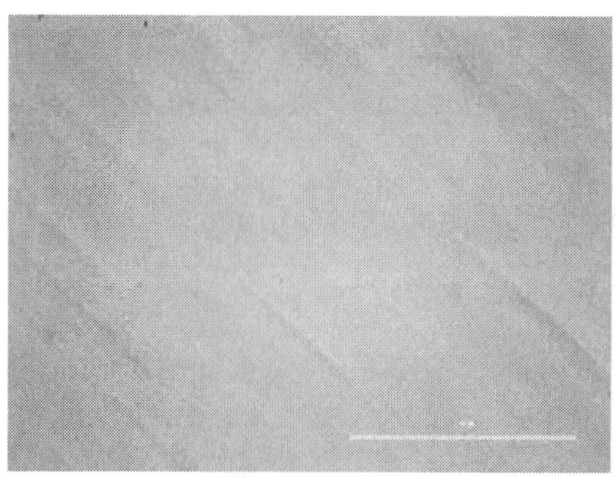

Figure 7. 45X magnifier images of 3-D print sample A

Figure 8. 45X magnifier images of 3-D print sample B

Figure 9. 45X magnifier images of 3-D print sample C

Each sample shows a significant discrepancy of surface finish quality between one another. Sample A promoted some uneven surfaces which may possibly cause by the "stairstepping" effect from the layer-by-layer process. Sample B, however, depicted a contradict finding where the surface finishing was considerably improved by the application of one layer XTC-3D coating. A similar situation was observed in Sample C where it exhibits a better surface finishing performance with two layers of XCT-3D application. Figure 10 shows the cross section of Sample C.

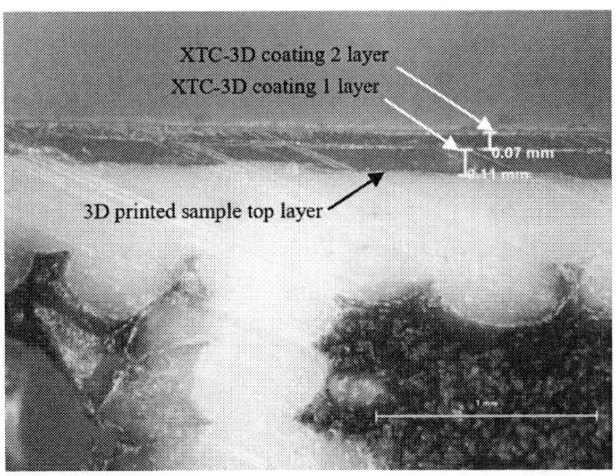

Figure 10. The cross section of Sample C

From the cross-sectional image, it shows that by applying the XTC-3D, the "stairstepping" effect was reduced to obtain a better surface finish. However, several more additional layers might be advantageous to have a better quality of surface finishing. Albeit the surface finishing could be improved, it is however noteworthy that the accuracy of part dimensions may indirectly affected and this could possibly place a major concern of issue.

5.Conclusion

"Stairstepping" effect is a common defect found in 3-D printing parts especially in FDM machine. This study shows that by applying XTC-3D coating, the surface finish can significantly be improved. However, further studies shall be conducted using different parameters of 3-D printer setting such as layer and printing resolutions, and others. The results of present study may contribute to a more economical application of XTC-3D in terms of surface finish performance.

Acknowledgments

Appreciation is given to Faculty of Engineering Technology, Universiti Malaysia Perlis. The study has been carried out with the support of Short-Term Grant (Project Number: 9001-00512) received from Research, Management and Innovation Centre, Universiti Malaysia Perlis. The authors reported no conflicts of interest related to this study.

References

[1] P. Vijay, P. Danaiah, K. V. D. Rajesh, Critical Parameters Effecting the Rapid Prototyping Surface Finish, Journal of Mechanical Engineering and Automation, vol. 1, pp. 17-20 (2011)
[2] B. Hallgrimsson, Prototyping and Model Making for Product Design, London: Laurence King Publisher, (2012)
[3] Van Gajdoš, Emil Spišák2, Ľuboš Kaščák, Volodymyr Krasinskyi,Surface finish techniques for FDM parts, Materials Science Forum, pp. 45-48 (2015)
[4] Caliskan, O.U. and Durgun, I, Silicon moulding production technology applications in automotive productdevelopment process, 12th Automotive and ProductionTechnologies Symposium, Bursa (2011)
[5] R. Noorani, Rapid Prototyping and Applications. New Jersey: John Wiley & Sons Inc., New Jersey: (2006)
[6] Gurpal Singh Bual, Parlad Kumar, Methods to Improve Surface Finish of Parts Produced by Fused Deposition Modeling, Manufacturing Science and Technology , vol. 2, no. 3, pp. 51-55 (2014)
[7] Ishak M I, Abdul Kadir M R, Sulaiman E and Abu Kasim N H 2010 Effects of different zygomatic implant body surface roughness and implant length on stress distribution *IEEE EMBS Conference on Biomedical Engineering and Sciences (IECBES) (Kuala Lumpur, Malaysia, 30 November – 2 December 2010)* p 210-15
[8] Ognjan Lužanin, Dejan Movrin, Miroslav Plančak, Experimental Investigation Of Extrusion Speed And Temperature Effects On Arithmetic Mean Surface Roughness In Fdm Built Specimens, Journal for Technology of Plasticity, vol. 38, no. 2 (2013)
[9] Bharath Vasudevarao, Dharma Prakash Natarajan, Mark Henderson, Sensitivity Of RP Surface Finish To Process Parameter Variation, Annual Internation Solid Freeform Fabrication Symposium -An Additive Manufacturing Conference,(2000)

Sugeno-Fuzzy Expert System Modeling for Quality Prediction of Non-Contact Machining Process

Sivaraos[1],*, A.Z Khalim[1], M. S Salleh[1], D. Sivakumar[2], K. Kadirgama[3]
[1]Faculty of Manufacturing Engineering, University Technical Malaysia Melaka
[2]Faculty of Mechanical Engineering, University Technical Malaysia Melaka
[3]Faculty of Mechanical Engineering, University Malaysia Pahang

E-mail: sivarao@utem.edu.my

Abstract. Modeling can be categorised into four main domains: prediction, optimisation, estimation and calibration. In this paper, the Takagi-Sugeno-Kang (TSK) fuzzy logic method is examined as a prediction modelling method to investigate the taper quality of laser lathing, which seeks to replace traditional lathe machines with 3D laser lathing in order to achieve the desired cylindrical shape of stock materials. Three design parameters were selected: feed rate, cutting speed and depth of cut. A total of twenty-four experiments were conducted with eight sequential runs and replicated three times. The results were found to be 99% of accuracy rate of the TSK fuzzy predictive model, which suggests that the model is a suitable and practical method for non-linear laser lathing process.

1. Introduction

Modelling can be divided into traditional and modern techniques. Traditional techniques include full factorial, response surface methodology, and Taguchi, while modern techniques (also known as artificial intelligence (AI)) include neural network (NN), fuzzy logic (FL) and genetic algorithm (GA). Other predictive modelling methods include tree-based, rule-based, nearest neighbour, logistic regression, graphical method and support vector machine, which were designed to solve two types of predictive modelling [1].

There are two types of fuzzy logic: Mamdani and Sugeno, also known as Takagi–Sugeno-Kang (TSK). The TSK model is one of the most powerful engineering tools for modelling and controlling complex systems [2] and was developed in an effort to create a systematic approach to generating fuzzy rules from a given input–output data set [3, 4]. It was shown that the TSK model yielded a more effective result than other AI method [5] when comparing the condition of tool wear for turning processes. Similarly, excellent results were also discovered in trade-off and practical implementation when using TSK type modelling for surface roughness and cutting force in milling operations [6]. In addition, it was shown that using TSK modelling for the prediction of surface roughness in deep drilling revealed a good relationship between the sets of input variables speed and force [7].

There has been increasing development in the engineering technology available for mechanised industries and a shift from traditional to more modern, non-traditional methods. Laser technology is a growing non-traditional method, capable of application to all types of metal and non-metal material. Laser cutting is a non-contact method using a very narrow heat affected zone and is capable of processing most engineering material with a high degree of precision and accuracy [8, 9]. The most commonly used lasers in industries are Carbon Dioxide (CO2) and Neodymium-Doped Yttrium Aluminium Garnet (Nd:Y3Al5O12).

The accuracy and precision of laser cutting has solved many conventional lathe problems, such as the deflection of the work piece whilst turning cylindrical parts with high length to diameter ratio, which causes undesirable taper on the end product. It has been proven that laser cutting has much better control of the taper on a cylindrical part, making it a more suitable process, as there is no physical contact between the laser beam (virtual tool) and the spinning work.

TSK modelling aims to understand the behaviour and scientific reasoning behind the processing phenomenon, and is therefore viewed as an optimal method in modelling the environment to best predict the response.

2. Experiment details

As to validate the performance of 3D laser lathe using modified 2D flat cutting machine, with a tangential insertion of laser beam and spinning workpiece has been precisely setup on the sacrificial table. The details of CO2 laser machine specification are shown in Table 1, while the experimental setup is shown in Figure 1. Throughout these experimentations, constant and variable processing parameters were identified.

Table 1. CO2 Laser Machine Specification

Machine	**Specification**
Manufacturer	LVD Company N.V, Belgium
Model	Helius-2513
Brand	LVD Helius
Envelope	2500x1250 mm
Maximum speed	250 mm/s
Maximum laser power	3 kW

Figure 1. Laser lathing of SS400 steel

Table 2 shows the design parameters namely laser cutting speed (V), work spinning speed (N) and depth of cut (d) which were varied through-out the experiments while Table 3 presents the laser process parameters which were kept constant. Each variable was set into two levels, low and high. The observed responses are tabulated in Table 4.

Table 2. Design parameters

Factors	Levels		
	Unit	**Low**	**High**
Laser cutting Speed (*V*)	mm/min	510	680
Work spinning speed (*N*)	rpm	1000	1500
Depth of Cut (*d*)	mm	1	1.5

The performance of laser machining plays an important role in producing reliable quality measures. Taper is the utmost critical quality to be observed when it comes to the machining of cylindrical parts especially when it is long but small in diameter, which is often called high length to diameter ratio. The taper quality observation of every processed part was made using CNC Formtester MMQ44.

Table 3. Laser process parameters (constant)

Laser Processing	**Value**
Power	1800 Watt
Frequency	1800 Hz
Duty cycle	85%
Gas pressure	0.5 bar
Laser mode	Continuous wave (CW)
Stand-off distance	1 mm
Nozzle type	Cylindrical
Beam diameter	0.5 mm
Gas jet selection	O2
Focus lens type	Cylindrical
Focal distance	0
Nozzle diameter	1.2 mm

2.1 Sugeno - Fuzzy modelling

Modeling helps to understand the information requirements, minimize time and eliminate cost before starting a real experiment. In this experiment, fuzzy logic was used to model the taper condition of non-linear laser machining environment. Fuzzy logic has become popular over the last decade because it can deal with imprecise inputs, does not need an accurate mathematical model and can handle nonlinearity [10]. Further, fuzzy logic can improve such classifications and decision support models by using fuzzy sets to define overlapping class definitions [11]. By using commercially available Matlab package, Sugeno-Fuzzy inference system was chosen for this study because it's ideal for acting as an interpolating supervisor of multiple linear controllers that are to be applied, respectively, to different operating conditions of a dynamic nonlinear system [7]. The final output of the system is the weighted average of all rule outputs, and the Sugeno rule operates are shown in Figure 2.

Table 4. Observed responses

No. Exp.	Machining Parameters			Response
	V (mm/min)	N (rpm)	d (mm)	Taper (mm)
1	510	1000	1	0.040
2			1.5	0.011
3		1500	1	0.011
4			1.5	0.006
5	680	1000	1	0.006
6			1.5	0.005
7		1500	1	0.012
8			1.5	0.005

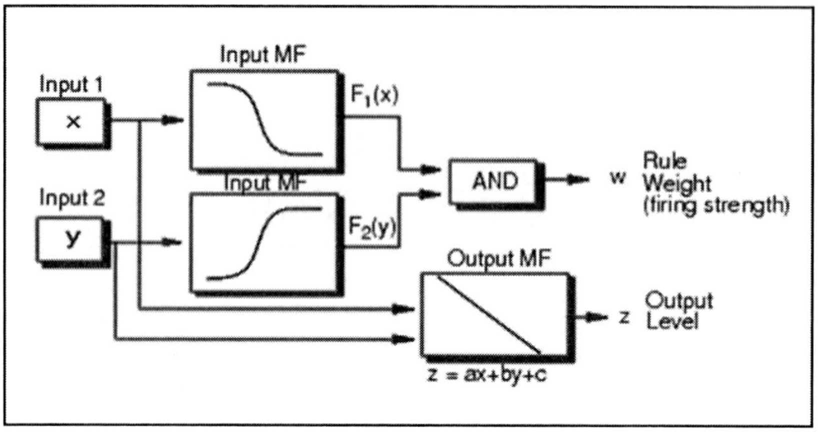

Figure 2. Sugeno final output rule

The main different between Mamdani and Sugeno is that the Sugeno output membership functions are either linear or constant but can be excellently suited for modeling nonlinear systems by interpolating between multiple linear models [7]. In this experiment, linear membership function has been adopted. Figure 3 shows the output membership function of Sugeno model. The linguistic variables, low and high was used for feed rate, cutting speed and depth of cut to represent the input numerical values. The output numerical values also represent in a similar way. Table 5 shows the fuzzy expressions of input and output parameters in numerical values.

3. Result and discussion

Based on Figure 3, eight rows of rule viewer represent the rules and three columns represent the variable parameters of laser processing. The four plots across the top of Figure 4 represent the antecedent and consequent of the first rule. The defuzzified output represent the aggregate weighted decision for the given inference system. However the decision depends on input values for the system. From the rule viewer, it was witnessed that the optimized values recommended by the Sugeno-Fuzzy are 595 mm/min for feed rate, 1250 rpm for cutting speed, and 1.25 mm for depth of cut.

Table 5. Fuzzy expression of input and output parameters

Rules	IF					THEN
	V	Connection	N	Connection	d	Taper
1	L	and	L	and	L	ML
2	L	and	L	and	H	L
3	L	and	H	and	L	L
4	L	and	H	and	H	ML
5	H	and	L	and	L	ML
6	H	and	L	and	H	ML
7	H	and	H	and	L	L
8	H	and	H	and	H	ML

*L=Low, H=High, ML=Most Lowest, MH=Most Highest

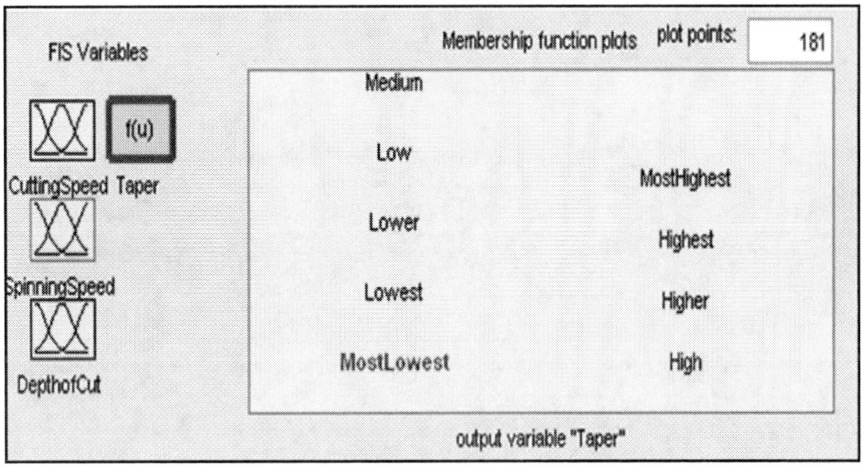

Figure 3. Sugeno output membership function.

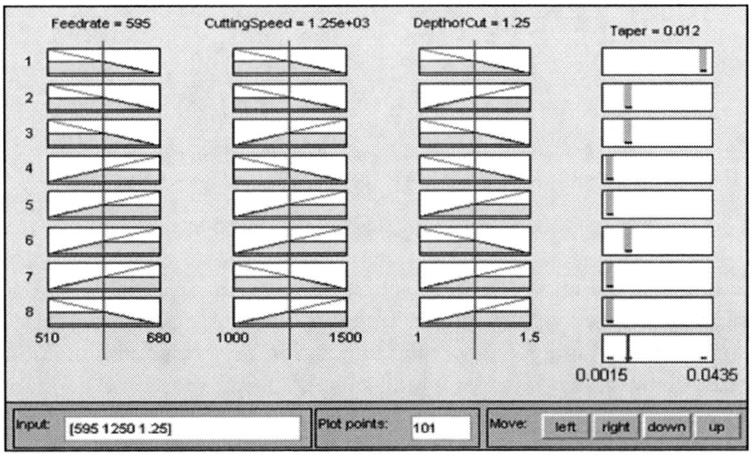

Figure 4. Rule viewer of developed sugeno model.

On the other hand, the model's estimated response of taper was 0.012 mm. The relationship between prediction and experimental values is shown in Figure 5 which shows excellent correlation among experimental and predicted values where calculated average of percentage error indicates 6.44 percent means it is in the range of acceptable value. Table 6 shows the taper values between experimental and prediction.

Table 6. Taper values experimental vs. prediction

No. Exp.	Exp.	Prediction
1	0.04	0.04
2	0.011	0.012
3	0.011	0.012
4	0.006	0.005
5	0.005	0.005
6	0.012	0.012
7	0.005	0.005
8	0.006	0.005

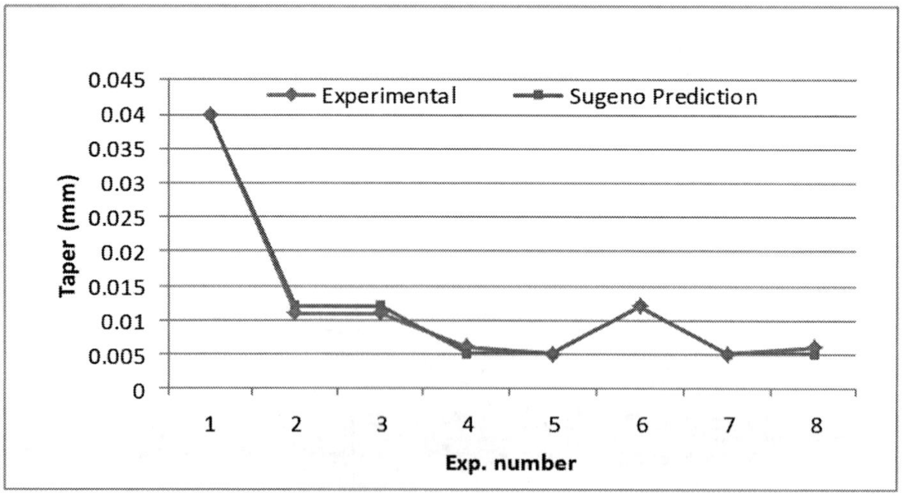

Figure 5. Comparison of prediction and experimental values

With a three-dimensional curve, the interaction between input variable and output response can be seen upon opening the surface viewer. Figure 6 shows the interaction of depth of cut and cutting speed out of taper response while Figure 7 shows the interaction between depth of cut and cutting speed. Via this surface model, it shows that lower spinning speed, cutting speed and depth of cut will increase the taper.

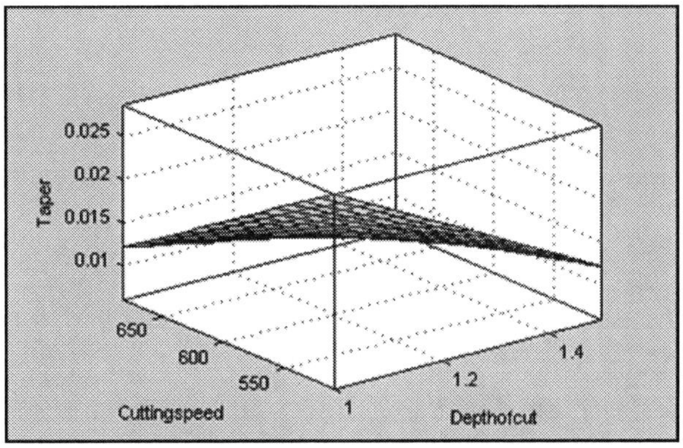

Figure 6. Interaction between depth of cut and cutting speed over taper

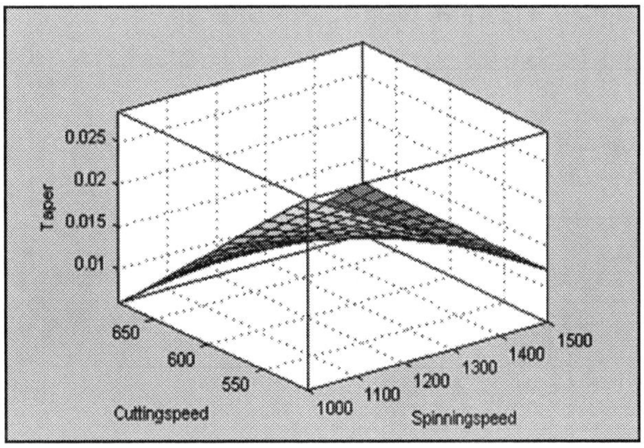

Figure 7. Interaction between depth of cut and spinning speed over taper

4. Conclusion

Prediction of taper has been done successfully and based on the model developed using Fuzzy Logic Tool Box MATLAB (R2013a), taper on the straight turning can be predicted. The results of taper prediction and experimental has been compared and it shows a very good relationship. Sugeno fuzzy model has proved their theory where based on a given input and output data, it shows significant result.

Acknowledgments

First and foremost, the authors would like to thank to Ministry of Higher Education Malaysia for awarding a research grant (Grant #: FRGS/1/2013/TK01/UTEM/03/1) to complete this research. The authors are also in debt to specially thank the top level management of Universiti Teknikal Malaysia Melaka, Centre for Research and Innovation Management, and Faculty of Manufacturing Engineering administrators for their continuous courage and support which brought this research work to a successful finding benefiting industries and mankind. Not forgetting, the LVD (Malaysia) laser technical support namely Mr. Vegiayan Ramiah for his expertise and advice in pioneering this 2D to 3D laser lathing technology.

References

[1] Mirjana Maksimović, Vladimir Vujović and Vladimir Milošević 2014 *J. of Eng. Sci. and Tech. Rev.* 7 (4) pp 89- 96
[2] Plamen P, Angelov and Dimitar P 2004 *Filev, IEEE Trans. On Sys. Man, And Cybernetics—Part B: Cybernetics,* vol. **34**, (1) pp 484-498
[3] Takagi T, and Sugeno M 1985 *IEEE Trans. On Sys. Man and Cybernetics*, vol. **15**, (1) pp 116–132
[4] Sugeno M, and Kang G. T, 1988 *Fuzzy Sets and Sys.* vol. **28**, (1) pp 15–33
[5] Qun Ren, Marek Balazinski, Luc Baron, Krzysztof Jemielniak 2011 *Eng. Appl. of Artificial Intelligence,* **24** pp 260–265
[6] Arup Kumar Nandi, 2012 *J. of the Brazilian Soc. of Mech. Sci.* vol. **XXXIV**, (1) pp 49-61
[7] Sivarao, Castilo, Tajul 2009 *Int. J. of Mech. & Mech. Eng.* vol. **9**, (9) pp 331-335
[8] Miroslav Radovanovic and Milos Madic 2011 *Nonconventional Tech.* Rev. No. 4
[9] Sivaraos, Milkey KR, Samsudin AR, Dubey AK, Kidd P 2014 *J. J. of Mech.and Ind. Eng.* **8**(1), pp 35-42
[10] Ismail MM, Ahmed F, Bendary, Alexandria 2016 *Eng. J.* **55** pp 941–949
[11] Mahalakshmi P and Ganesan K 2015 *Indian J. Fish.* **62** (1) pp 110-115

Mamdani-Fuzzy Modeling Approach for Quality Prediction of Non-Linear Laser Lathing Process

Sivaraos[1],*, A.Z Khalim[1], M. S Salleh[1], D. Sivakumar[2], K. Kadirgama[3]

[1]Faculty of Manufacturing Engineering, Universiti Teknikal Malaysia Melaka
[2]Faculty of Mechanical Engineering, Universiti Teknikal Malaysia Melaka
[3]Faculty of Mechanical Engineering, Universiti Malaysia Pahang

E-mail: sivarao@utem.edu.my

Abstract. Lathing is a process to fashioning stock materials into desired cylindrical shapes which usually performed by traditional lathe machine. But, the recent rapid advancements in engineering materials and precision demand gives a great challenge to the traditional method. The main drawback of conventional lathe is its mechanical contact which brings to the undesirable tool wear, heat affected zone, finishing, and dimensional accuracy especially taper quality in machining of stock with high length to diameter ratio. Therefore, a novel approach has been devised to investigate in transforming a 2D flatbed CO2 laser cutting machine into 3D laser lathing capability as an alternative solution. Three significant design parameters were selected for this experiment, namely cutting speed, spinning speed, and depth of cut. Total of 24 experiments were performed with eight (8) sequential runs where they were then replicated three (3) times. The experimental results were then used to establish Mamdani - Fuzzy predictive model where it yields the accuracy of more than 95%. Thus, the proposed Mamdani - Fuzzy modeling approach is found very much suitable and practical for quality prediction of non-linear laser lathing process for cylindrical stocks of 10mm diameter.

1. Introduction

Among others, the most commonly used lasers in industries are Carbon Dioxide (CO2) and Neodymium-Doped Yttrium Aluminum Garnet (Nd:Y3Al5O12). The CO2 laser is the most versatile type and widely used in materials processing. Laser cutting, being the non-contact with very narrow heat affected zone is capable of processing most engineering materials with a high degree of precision and accuracy [1].

The ability of laser in processing intricate profiles, quick setup and low cost of production makes it most preferable in sheet metal cutting industry. Over 40% medium scale manufacturers use laser as processing process mainly in cutting industries such as producing one dimensional features turbine blade, combustion chambers and aerosol nozzles [2,3]. With great advantages of laser cutting machine, the common problem faced by conventional lathe can be overcome especially deflection on workpiece while turning cylindrical parts with high length to diameter ratio which would cause undesirable taper on end product.

Thus, in order to have a better control of the taper on a cylindrical part, laser is found to be most suitable as it processes the materials without having any mechanical contact between the laser beam (virtual tool) and the spinning work. So to understand the behavior and scientific reasoning behind the actual processing phenomenon, an expert system called Mamdani-Fuzzy is seen to be suiting best in modeling the environment to predict the response. The modeling process considering expert knowledge combined with the commercially available Artificial intelligence solution [4] & modeling is one of the scientific study which gives us a clear yet brief understanding about the research atmosphere [5]. First attempt to establish a theoretical model for laser 'blind' cutting was conceptualized by using two intersecting beams which was then optimized successfully [6]. On the other hand, the performance of material removal rate (MRR), tool wear rate (TWR), overcut and taper of micro-hole during machining of D3 die steel was model using fuzzy logic [7]. In another experiment of laser cutting, a model of fuzzy logic was develop to minimize the kerf width and deviations at top and bottom sides of duralumin [8].

Fuzzy rule based modeling has also been implement for prediction of flank wear, surface roughness and specific cutting pressure in turning of titanium alloy where results showed that the model is very effective [9]. Besides that [19] proposed neural network approach to predict the behavior of the material removal process during laser machining. They also investigate and analyze the effect of laser power and grooving speed on groove shape using response surface methodology [20]. While in the other hand, the micro-grooving operation on Al2TiO5 also presents using artificial neural network (ANN) and response surface methodology (RSM) [21].

2. Experiment details

As to validate the performance of 3D laser lathe using modified 2D flat cutting machine, with a tangential insertion of laser beam and spinning workpiece has been precisely setup on the sacrificial table. The details of CO2 laser machine specification are shown in Table 1, while the experimental setup is shown in Figure 1. Throughout these experimentations, constant and variable processing parameters were identified.

Table 1. CO2 Laser Machine Specification

Machine	Specification
Manufacturer	LVD, Belgium
Model	Helius-2513
Brand	LVD Helius
Envelope	2500x1250 mm
Maximum speed	250 mm/s
Maximum laser power	3 kW

Figure 1. Laser lathing of SS400 steel

Table 2 shows the design parameters namely laser cutting speed (V), work spinning speed (N) and depth of cut (d) which were varied through-out the experiments while Table 3 presents the laser process parameters which were kept constant. Each variable was set into two levels, low and high. The observed responses are tabulated in Table 4.

Table 2. Design parameters

Factors		Levels	
	Unit	**Low**	**High**
Laser cutting Speed (*V*)	mm/min	510	680
Work spinning speed (*N*)	rpm	1000	1500
Depth of Cut (*d*)	mm	1	1.5

The performance of laser machining plays an important role in producing reliable quality measures. Taper is the utmost critical quality to be observed when it comes to the machining of cylindrical parts especially when it is long but small in diameter, which is often called high length to diameter ratio. The taper quality observation of every processed part was measured using CNC Formtester MMQ44.

Table 3. Laser process parameters (constant)

Laser Processing	Value
Power	1800 Watt
Frequency	1800 Hz
Duty cycle	85%
Gas pressure	0.5 bar
Laser mode	Continuous wave (CW)
Stand-off distance	1 mm
Nozzle type	Cylindrical
Beam diameter	0.5 mm
Gas jet selection	O2
Focus lens type	Cylindrical
Focal distance	0
Nozzle diameter	1.2 mm

2.1 Mamdani - Fuzzy modeling

Modeling helps to understand the information requirements, minimize time and eliminate cost before starting a real experiment. Further, fuzzy logic can improve such classifications and decision support models by using fuzzy sets to define overlapping class definitions [18]. In this experiment, fuzzy logic was used to model the taper condition of non-linear laser machining environment. Fuzzy logic has become popular over the last decade because it can deal with imprecise inputs, does not need an accurate mathematical model and can handle nonlinearity [10].

Table 4. Observed responses

No. Exp.	Machining Parameters			Response
	V (mm/min)	N (rpm)	d (mm)	Taper (mm)
1	510	1000	1	0.040
2			1.5	0.011
3		1500	1	0.011
4			1.5	0.006
5	680	1000	1	0.006
6			1.5	0.005
7		1500	1	0.012
8			1.5	0.005

By using commercially available Matlab package, Mamdani-Fuzzy inference system was chosen for this study because of its simplicity and non-complex algorithms often used in solving many real world problems [5]. The basic definition of Mamdani system as shown in equation (1) is the training set contains n data pairs. Each pair is made of a $p-$ dimensional input vector x and a $q-$ dimensional output vector y. The number of rules in the FIS rule base is r. Therefore, the expression of Mamdani rule i in this system expressed by [11] is written as follow.

$$\begin{aligned}&\text{If } x_1 \text{ is } A_1^i \text{ and } x_2 \text{ is } A_2^i \ldots \text{ and} x_p \text{ is } A_p^i\\&\text{Then } y_1 \text{ is } c_1^i \ldots \text{and } y_q \text{ is } c_q^i\end{aligned} \qquad (1)$$

Where, A_j^i and c_j^i are fuzzy sets that define an input and output. The first step in developing fuzzy model is to determine the input and output variables. According to [17], the process of mapping from a given set of input to an output is performed by using a fuzzy inference mechanism. There are three set of inputs namely, cutting speed, spinning speed and depth of cut, where taper is the response/output. The next step is determining the range of input and output variable. The range of input and output variables are based on low and high values of variable parameters and there after is to determine the types of membership function (MF) or known as fuzzification for input and output variables. Membership function is defines as input and output space value between 0 and 1 which mapped to a membership.

As mentioned by [12], Triangular membership function is used for the input and output as it possesses simplicity and precision in determining the value of the input and output parameters, where [13] has expressed the triangular membership function as shown in equation (2).

Triangle

$$(x\ :a,b,c)=\begin{cases} 0 & x>a \\ (x-a)/(b-a) & a\leq x\leq b \\ (c-x)/(c-b) & b\leq x\leq c \\ 0 & x>c \end{cases} \qquad (2)$$

Where, the parameters {a, *b*, c} with $a < b <$ c are the parameters of the linguistic values which determine the x coordinates of the three corners of the underlying triangular membership function and x is the range of the input parameters. The membership function of cutting speed (V), spinning speed (N), depth of cut (d) and taper are shown in Figure 2. Then after, membership function is determined, the important step is developing inference rules or fuzzy rules. This rule is based on experimental result and expertise [9].

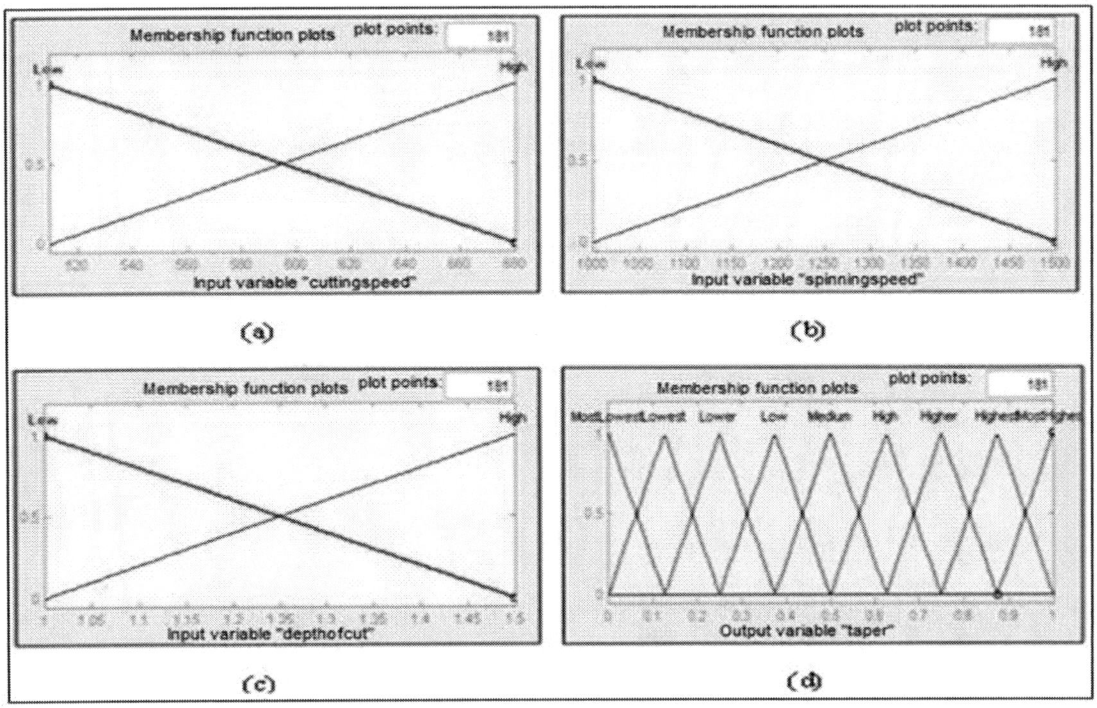

Figure 2. Input and output of fuzzification (a) Fuzzification of cutting speed, (b) Fuzzification of spinning speed, (c) Fuzzification of depth of cut, (d) Fuzzification of taper

Basically the rule is, If X is A then Y is B where, A and B are the linguistic values represented by the fuzzy sets [5]. A fuzzy logic implication statement of this form is called Fuzzy IF-THEN rules, where Fuzzy logic AND operation have been implemented in this work. As elaborated by [14] as shown in equation (3), let $A_1,\ldots,A_n$, and B be fuzzy subsets with membership functions $\mu_{A1},\ldots,\mu_{An}$, and μ_B, respectively and based on the general form fuzzy IF-THEN rule "IF a_1 is A_1 AND ... AND an is A_n THEN b is B.

$$\mu_{A_1}(a_1) \wedge \cdots \wedge \mu_{A_n}(a_n) \Rightarrow \mu_B(b) \quad (3)$$

Where,

$$\mu_{A_j}(a_j) \wedge \mu_{A_j}(a_j) = \min\{\mu_{A_j}(a_j), \mu_{A_j}(a_j)\},$$
$$1 \le i, j \le n, \text{and , therefore,}$$
$$\mu_{A_1}(a_1) \wedge \cdots \wedge \mu_{A_n}(a_n) = \min\{\mu_{A_1}(a_1), \ldots, \mu_{A_n}(a_n)\}$$

The linguistic variables, low and high was used for cutting speed, spinning speed and depth of cut to represent the input numerical values. The output numerical values also represent in a similar way. Table 5 shows the fuzzy expressions of input and output parameters in numerical values. The end of developing fuzzy model, rule and surface viewer are automatically generated for analyzing.

3. Result and discussion

The rule viewer represents a roadmap of the whole fuzzy inference modeled for the non-linear laser process, where it shows crisp behavior in great detail [15]. Based on Figure 3, eight rows of rule viewer represent the rules and three columns represent the variable parameters of laser processing.
The four plots across the top represent the antecedent and consequent of the first rule. The defuzzified output represent the aggregate weighted decision for the given inference system.

Table 5. Fuzzy expression of input and output parameters

Rules	IF					THEN
	V	Connection	N	Connection	d	Taper
1	L	and	L	and	L	ML
2	L	and	L	and	H	L
3	L	and	H	and	L	L
4	L	and	H	and	H	ML
5	H	and	L	and	L	ML
6	H	and	L	and	H	ML
7	H	and	H	and	L	L
8	H	and	H	and	H	ML

*L=Low, H=High, ML=Most Lowest, MH=Most Highest

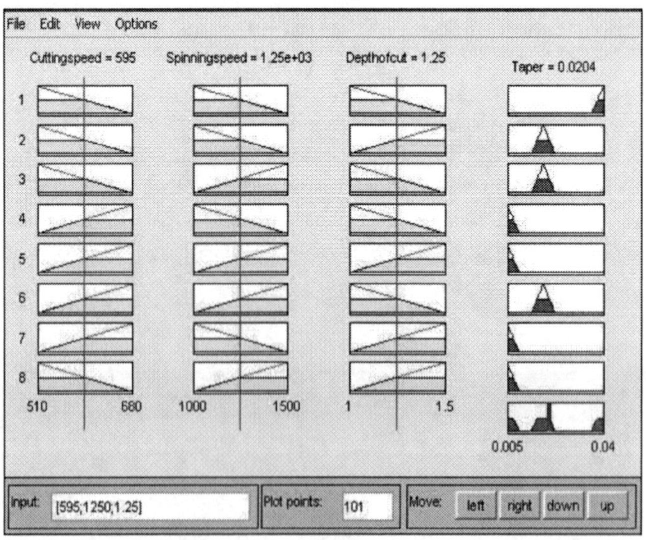

Figure 3. Rule viewer of developed fuzzy model

However the decision depends on input values for the system [12]. From the rule viewer, it was witnessed that the optimized values recommended by the Mamdani-Fuzzy are 595 mm/min for cutting speed, 1250 rpm for spinning speed, and 1.25 mm for depth of cut. On the other hand, the model's estimated response of taper was 0.0204 mm. In this model, centroid of area defuzzification method was used due to its wide acceptance and capability in giving more accurate results as compared to other methods [16]. The relationship between prediction and experimental values is shown in Figure 4 which shows excellent correlation among experimental and predicted values with minimum error of below 3 percent. Table 6 presents the comparative values of experimental and model prediction for the entire. From the table, experiment number 4 and 8 shows zero deviation which indicate that experimental and model prediction are very close.

Table 6. Taper values of experimental vs. prediction

Expn. No.	Experimental	Prediction
1	0.04	0.039
2	0.011	0.018
3	0.011	0.018
4	0.006	0.006
5	0.005	0.006
6	0.012	0.018
7	0.005	0.006
8	0.006	0.006

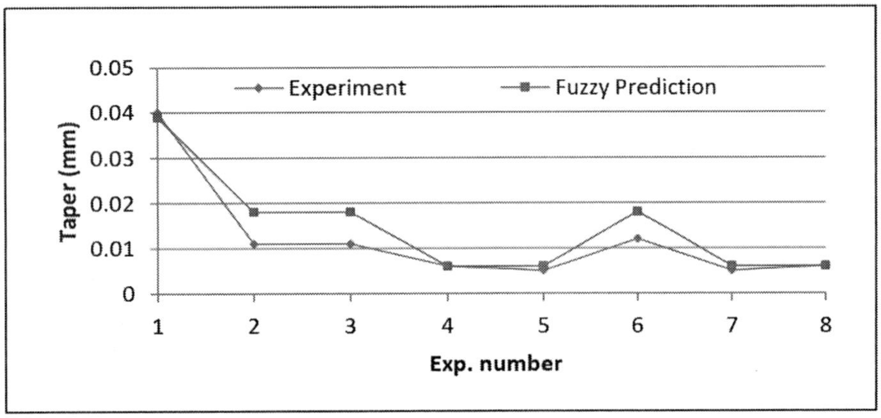

Figure 4. Comparison of prediction and experimental values

With a three-dimensional curve, the interaction between input variable and output response can be seen upon opening the surface viewer. Figure 5 shows the interaction of spinning speed and cutting speed out of taper response while Figure 6 shows the interaction between depth of cut and cutting speed. Based on surface model, the interaction between cutting speed and spinning speed shows that low cutting speed and low spinning speed will increase taper. This is because low cutting speed increase material removal rate and at the same time taper will affect. In another hand, low cutting speed and low depth of cut will also increase the taper. Again, low cutting speed increase taper which is shows that it's gives a significant effect on taper. However, moderate control of cutting speed shows taper values is quiet lower.

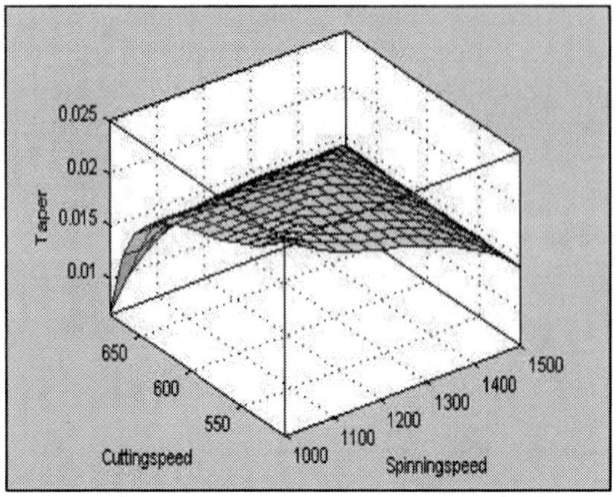

Figure 5. Interaction between cutting speed and spinning speed over taper

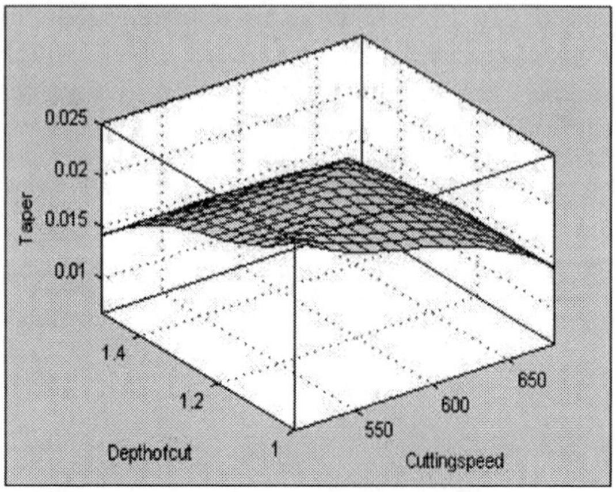

Figure 6. Interaction between cutting speed and depth of cut over taper

4. Conclusion

Prediction of taper has been done successfully and based on the model developed using Fuzzy Logic Tool Box MATLAB (R2013a), taper on the straight turning can be predicted. The results of taper prediction and experimental has been compared and it shows quite a good relationship. Therefore, fuzzy model can be used as a benchmark for selecting the input parameter before starting the real experiment.

Acknowledgment

First and foremost, the authors would like to thank to Ministry of Higher Education Malaysia for awarding a research grant to complete this research. The authors are also in debt to specially thank the top level management of Universiti Teknikal Malaysia Melaka, Centre for Research and Innovation Management, and Faculty of Manufacturing Engineering administrators for their continuous courage and support which brought this research work to a successful finding benefiting industries and mankind. Not forgetting, the LVD (Malaysia) laser technical support namely Mr. Vegiayan Ramiah for his expertise and advice in pioneering this 2D to 3D laser lathing technology.

References

[1] Sivaraos, Milkey KR, Samsudin AR, Dubey AK, Kidd P 2014, *J. of Mech. and Ind. Eng.* **8**(1), pp 35-42

[2] Caristan CL 2004 *Soc. of Mfg. Eng.*

[3] Chryssolouris G, Stavropoulos P, Salonitis K, 2014, *Handbook of Manufacturing Eng. and Technology*, 1-25

[4] Sivarao S, Brevern P, El-Tayeb NSM, Vengkatesh VC 2009 *Int. J. of Elec. & Com. Sci. IJECS IJEN* **9**(9), 37-43

[5] Chong Zhian Syn, Mohzani Mokhtar, Chin Jeng Feng, Yupiter, Manurung HP 2011 *Exp.Sys.and Appl.* **38**, 7558-7568

[6] Chryssolouris G, Bredt J, Kordas S, Wilson E 1988 *J. of eng.for ind.* **110**(1) pp 65-70

[7] Shivakoti I, Kibria G, Pradhan BB 2014 *Design and Research Conference* AIMTDR

[8] Pandey AK and Dubey AK 2012 *Optics and Lasers in Eng.* **50** pp 328-335

[9] Ramesh S, Karunamoothy L, Palanikumar K 2008 *Materials and Mfg. Processes*, **23**, pp 439-

447
[10] Ismail MM and Bendary AF 2016 *Alexandria Eng.* J. **55** pp 941–949
[11] Serge Guillaume 2001 *IEEE Trans.on Fuzzy Sys.* **9**(3) pp 426-443
[12] Sivarao S, Brevern P, El-Tayeb NSM, Vengkatesh VC 2009 *Int. J. of Mech. & Mech. Eng. IJMME-IJENS,* **9**(9) pp 1-9
[13] Imriyas K 2014 *http://scholarbank.nus.edu.sg/ bitstream/handle/10635/17612/5%20-%20Chapter% 204.pdf? sequence=5* (December 2014)
[14] Guanrong Chen and Trung Tat Pham 2014 *http://site .iugaza .edu.ps /akhateb/files/2010/05/ CHEN-PHAM-Introduction-to-Fuzzy-sets-Fuzzy-logic-and-Fuzzy-control-systems-Page-160.pdf,* (December 2014).
[15] Anonymous 2014 *http://www.mathworks.com /help/fuzzy/ruleview.html* (December 2014).
[16] Ahmed AD, Sarhan M. Sayuti, and Hamdi MA 2012 *World Academy of Sci., Eng.and Tech.,* **6**, pp 10-26
[17] Rezaei Mohammad, Asadizadeh Mostafa, Majdi Abbas, Hossaini Mohammad Farouq 2015 *Int. J. of Mining Sci. and Tech.,* **25**(1) pp 23–30
[18] Mahalakshmi P. and Ganesan K. 2015 *Indian J. Fish.* **62** (1) pp 110-115
[19] Basem F, Yousef, George K, Knopf Evgueni Bordatchev V, Suwas K, Nikumb 2003 *Int J Adv Manuf Technol* **22**: pp 41–53
[20] Klobcar D, Muhic M, Mozina J, Tusek J 2011 *Metalurgija* **50** 4 pp 235-238
[21] Dhupal D, Doloi B, Bhattacharyya B, 2007 *Proc. IMechE* Vol. **221** Part B: J. Engineering Manufacture

Sustainability of Welding Process through Bobbin Friction Stir Welding

M K Sued[1], S S M Samsuri[2], M K A M Kassim[2] and S N N M Nasir[2]

[1] Advanced Manufacturing Centre, Faculty of Manufacturing Engineering, Universiti Teknikal Malaysia Melaka, 76100, Durian Tunggal, Melaka, Malaysia
[2] Faculty of Manufacturing Engineering, Universiti Teknikal Malaysia Melaka, 76100, Durian Tunggal, Melaka, Malaysia

E-mail : kamil@utem.edu.my

Abstract. Welding process is in high demand, which required a competitive technology to be adopted. This is important for sustaining the needs of the joining industries without ignoring the impact of the process to the environment. Friction stir welding (FSW) is stated to be benefitting the environment through low energy consumption, which cannot be achieved through traditional arc welding. However, this is not well documented, especially for bobbin friction stir welding (BFSW). Therefore, an investigation is conducted by measuring current consumption of the machine during the BFSW process. From the measurement, different phases of BFSW welding process and its electrical demand are presented. It is found that in general total energy in BFSW is about 130kW inclusive of all identified process phases. The phase that utilise for joint formation is in weld phase that used the highest total energy of 120kWs. The recorded total energy is still far below the traditional welding technology and the conventional friction stir welding (CFSW) energy demand. This indicates that BFSW technology with its vast benefit able to sustain the joining technology in near future.

1. Introduction

In general, welding process and jobs can be indicated as ready for growth. This is founded based on the forecast of the increase in the manufacturing and construction employment that happening around the world [1-3]. Based on Malaysia manufacturing index expansion, 5.9% increment is shown in 2015 compared to 2014 in the same period [4]. For energy consumption the industrial sector in Malaysia used about 46.6% of the total electrical used [5]. With the excitement for progressiveness that provides stable economy, this in some way causes environmental degradation if no attention taken towards sustainability of the environment. For example, according to International Energy Agency (IEA) the CO2 emission increase to 2.2% in 2013 compared to 2012 [6] which largely contributed from the industrial sector. For that reason, manufacturing sectors are now focusing on green manufacturing approaches to conserve energy and to reduce greenhouse gas emission [7, 8].

In welding, FSW technology has been introduced by The Welding Institute (TWI), UK in 1991. It is a solid state welding process that creates the joint behind the tool as is travel forward. It used a non-consumable tool that produce frictional heat and pressure for the solid bonding formation. In general FSW process is stated to provide metallurgical, environmental and energy benefits. It has been stated that this process only consume 2.5% of the energy needed for laser weld [9]. There is no harmful gases produces and no post processing cleaning involves. The beneficial of this process is huge thus it can be suggested to provide a sustainable process in the welding fielding. This is founded on a comparison study conducted by Shrivastava, Krones and Pfefferkorn [10] whereby they found that FSW only use total energy of 174 kWs compared to gas metal arc welding (GMAW) that used 303 kWs of total energy. At high energy consumed in the GMAW process, the weld strength is still lower than FSW. Referring to Buffa, Campanella, Di Lorenzo, Fratini and Ingarao [11] the carbon footprint is not well documented. Only limited studies have been recorded its energy demands. Mostly in machining and sheet metal work. Therefore, there is a need to start publishing the finding of energy demand for realistically supporting green manufacturing. Accurate strategies can be laid by the planners based on the actual measurement.

The FSW technology can be divided into two types that are conventional friction stir welding (CFSW) and BFSW. These processes are differentiated based on the tool design as illustrated in Figure 1. Although the process is similar, however the way of the join formed is effected by the different approach [12]. The reason is because of the additional shoulder in BFSW. This cause no plunge force and tilting angle required. Additionally, the challenges on fixturing and clamping in BFSW provide more mechanical compliances that potentially alter the material flow [13, 14]. The differences are so significant; hence dedicated studies need to be conducted for the BFSW process. The finding is useful for process selection and advancement that support the reduction of greenhouse effect.

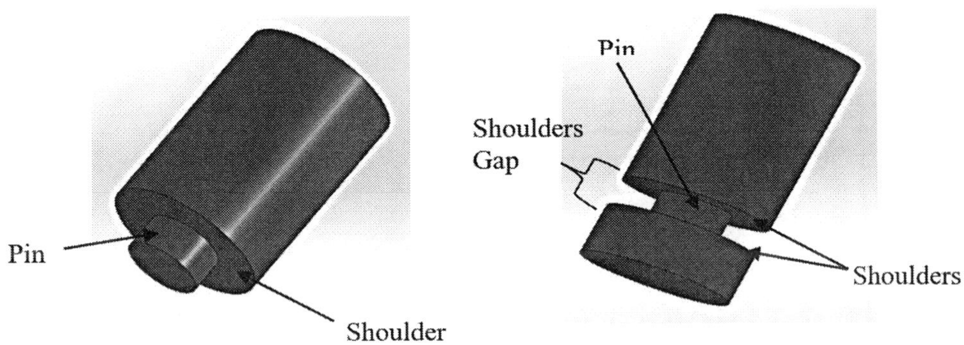

Figure 1. Types of FSW. (a) CFSW (b) BFSW

Based on the authors' findings, there are limited publication focusing on BFSW process and its characteristics especially for energy consumption. Study conducted by Buffa, Campanella, Di Lorenzo, Fratini and Ingarao [11] found that travel speed influence small changes on energy demand. The welding speed cause faster completion, hence short period of power is required. For the work conducted in [15], the investigation is about the impact of different backing plate materials on average weld energy consumption. They found that Titanium backing plate required less weld energy compared to steel and copper. The reason is of low thermal diffusivity of the material. Focusing on energy demands in FSW as a general, works that are stated above are focusing on CFSW. Therefore, in this study, electrical energy demand of BFSW and its characteristics when welding Aluminium alloy is presented.

2. Methodology

The approach of the welding is carried out by welding AA3000 series with dimension of 140 mm (length) x 140 mm (width) x 6 mm (thickness) in a butt joint configuration. Fixed gap bobbin is used having features of cylindrical shoulder and pin. The shoulder and pin have a diameter of 25 mm and

8.5 mm respectively. Tool material is H13 that has been heat treated for hardening which follow ASTM A681.

The empirical setup of the weld is demonstrated in Figure 2 and the welding parameters are shown in Table 1. Machine used is Haas 3 axis CNC milling machine as the FSW machine with 20hp capability. As the tool rotated and touch the material to be welded, current is measured using digital current clamp meter UNI-T UT232 with USB connection. The clamp is attached at the input current of the CNC machine and automatically measured and recorded the electrical demand for every second.

Figure 2. Empirical approach (a) Welding setup (b) Current measurement

The tool rotation (spindle speed) is varied while the travel speed is constant. The tool entry is from the edge and 15 secs of dwell time is applied when half of the tool body is located inside the plate. The movement of the tool towards half body tool entry is at 30 mm/min. Then full travel speed is applied.

Table 1. Welding parameter

Parameters	Setting
Tool rotation (rpm)	700, 800, 900
Travel speed (mm/min)	155
Dwell time (sec)	15
Location of tool entry	Edge

The welded plate is inspected and then being cut at the middle of the plate using CNC wire cut following ASTM E8 for the tensile test specimen. The reason for mechanical testing is to for a quality reference to the identified energy demand. The measured current by the current clamp meter is then converted to power (Watt) and plotted against time. Characterization is conducted based on the pattern. Time study is conducted to reveal the energy demand at each stage that is identified.

3. Results and discussion

Figure 3 shows one of the welded plate using bobbin tool. The plates are free from external defects such as flash, open tunnel and holes, thus categories it as an acceptable weld. These defects may present if the process parameters and tool are not suitable. Comparing the weld formation of BFSW to the arc welding it is clearly shown that the weld is clean and there is no need of post processing. Common defect in arc welding such as spatter, slag inclusion, under filling and etc. are difficult to be control and required intensive monitoring.

Figure 3. Welded plate

Referring to Figure 4, the highest tensile strength is produced at tool rotational speed of 800 rpm that is 82.72 Mpa. It is a reduction of 27.28 Mpa from the parent material. The declination is believed because of excessive heat produced by the tool shoulder and material availability around the tool [16, 17].

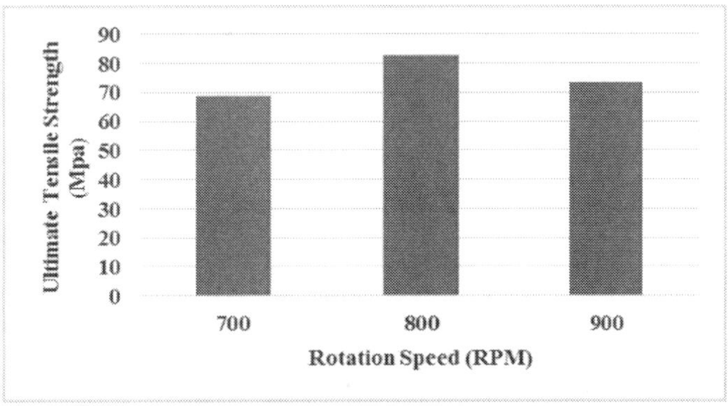

Figure 4. Maximum tensile strength for different tool rotation speed

The difference phases of energy consumption in BFSW welding are identified in Figure 5. These phases are milling power (machine ready for operation), tool entry, dwell time, weld and tool exit. In milling power phase, the tool is rotated and travel towards the material to be welded. In this phase average of 580 Watts of energy is consume. When the tool touches the solid metal, energy demand is suddenly increased to 2.4kWs with a total energy of 5.7kWs. The dwell time energy demand is as the milling power with slightly different energy demand. The reason is believed because of the work required to be performed by the tool to stir small amount of material around the tool. The utilization of energy is then increase again at the weld phase. The recorded peak energy in this phase is about

3.2kWs with a total energy of 120kWs. The tool is extensively stirring the weld and pairing the force created from shoulder compression and tool travel joint is formed behind the tool as the tool travel forward. When entering tool exit phase, energy demand is decreasing with a potential of sudden energy increase, due to releasing the tool away from the material before entering the milling phase again.

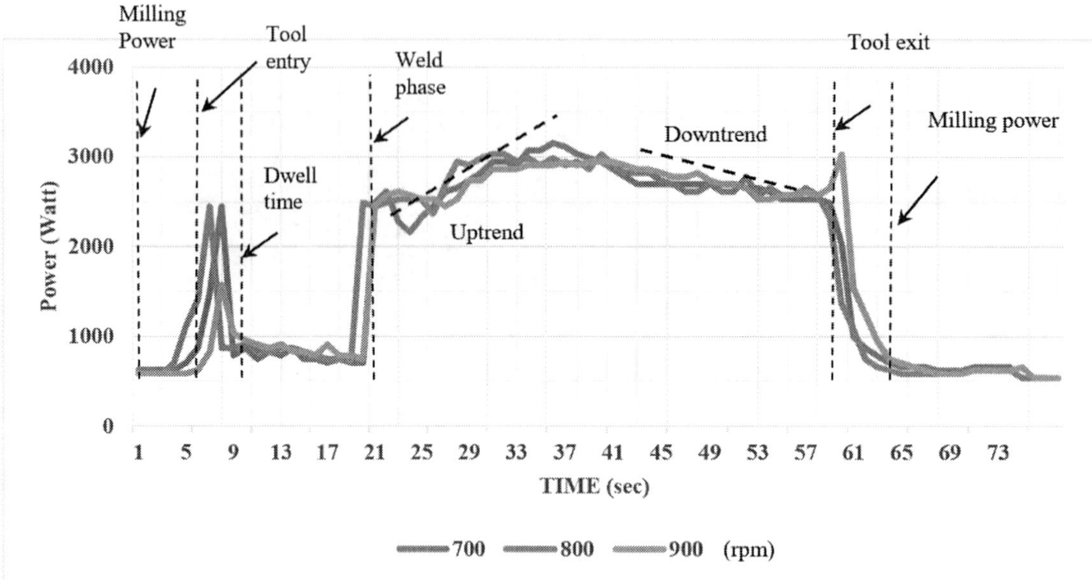

Figure 5. Power consumption during the BFSW process for different tool speed rotation.

Figure 6 shows the time studies of each phase. It can be clearly identified that most of the time is consume for weld formation that can be categories as productive time. The most non-productive time is from the dwell time. It is believed that this period can be eliminate through proper tool design such as tool features and shoulder gap [18]. Then, a faster feed rate with lower spindle speed can be introduced [16]. Furthermore, BFSW technology such as adaptive tool and floating shoulder can be implemented.

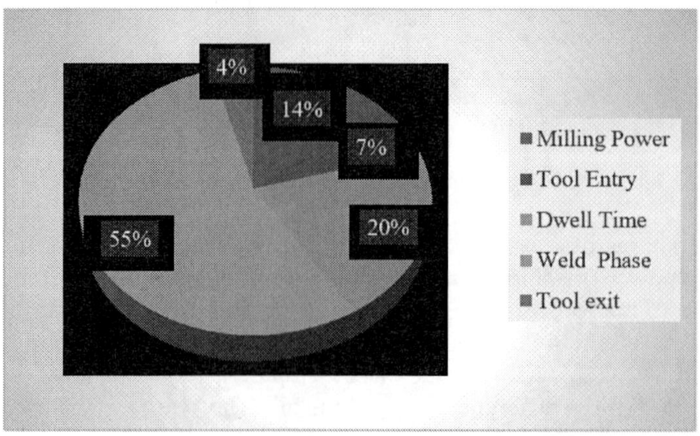

Figure 6. Time study of the BFSW process

Comparing the studies to CFSW and GMAW, the BFSW only consume about 130kWs of total energy compared to the one recorded for CFSW and GMAW. Although, the materials are slightly different in type and size, but they are welding a similar group of material that is Aluminium alloy with nearly the similar size. Additionally, in BFSW there is no plunge energy required, hence 8-13% of energy can be conserved

4. Conclusion

In this investigation, the energy consumption of BFSW has been presented. In general, FSW has the capability of welding different joint configurations which comparative to traditional welding such as GMAW. With a better energy saving of BFSW and other advantages, the welding proses that commonly labelled as non-environmentally friendly and hazardous to human through traditional welding can be eliminated. The welding process now can be better sustained and more competitive. About 50% of energy is saved between CFSW and GWAW. An additional of 20-25% energy can be further reduced when applying the BFSW technology.

Acknowledgement

The authors would like to acknowledge Universiti Teknikal Malaysia Melaka for the funding of this investigation through the funding from the research grant of PJP/2016/PROTON/FKP-AMC/S01503.

References

[1] Baker H 2011 Are Welding Jobs in Demand? : Bakers Gas & Welding Supplies)
[2] Shook R 2009 Welding Trends for 2010 and Beyond. ed G a W D A (GAWDA): Data Key Communications)
[3] Fast M 2008 Welding is One of the Main Driving Forces of the Economy. ETI School of Skilled Trades)
[4] Wan H L 2015 *Organisational Justice and Citizenship Behaviour in Malaysia,* (Singapore: Spinger) pp 21-36
[5] Government M 2016 Malaysia Energy Statistics Handbook 2016. ed E Commission (Putrajaya, Malaysia: Suruhanjaya Tenaga) p 84
[6] Ingarao G 2017 Manufacturing strategies for efficiency in energy and resources use: The role of metal shaping processes *Journal of Cleaner Production* **142, Part 4** 2872-86
[7] Maruthi G D and Rashmi R 2015 Green Manufacturing: It's Tools and Techniques that can be implemented in Manufacturing Sectors *Materials Today: Proceedings* **2** 3350-5
[8] Paul I D, Bhole G P and Chaudhari J R 2014 A Review on Green Manufacturing: It's Important, Methodology and its Application *Procedia Materials Science* **6** 1644-9
[9] Mishra R S and Ma Z Y 2005 Friction stir welding and processing *Mat Sci Eng R* **50** 1-78
[10] Shrivastava A, Krones M and Pfefferkorn F E 2015 Comparison of energy consumption and environmental impact of friction stir welding and gas metal arc welding for aluminum *CIRP Journal of Manufacturing Science and Technology* **9** 159-68
[11] Buffa G, Campanella D, Di Lorenzo R, Fratini L and Ingarao G 2017 Analysis of Electrical Energy Demands in Friction Stir Welding of Aluminum Alloys *Procedia Engineering* **183** 206-12
[12] Sued M K, Pons D, Lavroff J and Wong E H 2014 Design features for bobbin friction stir welding tools: Development of a conceptual model linking the underlying physics to the production process *Materials & Design* **54** 632-43
[13] Sued M K, Mohamad W M H W and Mohd Kassim M K A 2017 Rigidity issue in friction stir welding. In: *International symposium on research innovation and sustainability* (Melaka, Malaysia

[14] Sued M K and Pons D J 2016 Dynamic Interaction between Machine, Tool, and Substrate in Bobbin Friction Stir Welding *International Journal of Manufacturing Engineering* 14

[15] Choi W J, Morrow J D, Pfefferkorn F E and Zinn M R 2017 The Effects of Welding Parameters and Backing Plate Diffusivity on Energy Consumption in Friction Stir Welding *Procedia Manufacturing* **10** 382-91

[16] Colligan K J, O'Donnell A K, Shevock J W and Smitherman M T 2012 Friction Stir Welding of Thin Aluminium Using Fixed Gap Bobbin Tools. In: *9th International Symposium of Friction Stir Welding,* (Von Braun Center, Huntsville, Alabama, USA: TWI Limited)

[17] Sued M K, Tamadon A and Pons D 2017 Material flow visualization in bobbin friction stir welding by analogue model. In: *Mechanical Engineering Research Day,* (Universiti Teknikal Malaysia Melaka

[18] Sued M K, Pons D and Lavroff J 2014 Compression Ratio Effects in Bobbin Friction Stir Welding. In: *10th International Friction Stir Welding Symposium,* (Beijing, China: TWI Limited)

IOP Conf. Series: Materials Science and Engineering **318** (2018) 012069 doi:10.1088/1757-899X/318/1/012069

Lightweight Vehicle and Driver's Whole-Body Models for Vibration Analysis

Jamali MdSah[1], Zahari Taha[2], and Khairul Azwan Ismail[1]
[1]School of Manufacturing Engineering, UniMAP, Arau 02600 Perlis, Malaysia
[2]Faculty of Manufacturing Engineering, UMP, Pekan 26600 Pahang, Malaysia

E-mail: jamali@unimap.edu.my

Abstract. Vehicle vibration is a main factor for driving fatigue, discomfort and health problems. The ability to simulate the vibration characteristics in the vehicle and its effects on driver's whole-body vibration will give significant advantages to designers especially on the vehicle development time and cost. However, it is difficult to achieve optimal condition of ride comfort and handling when using passive suspension system. This paper presents mathematical equations that can be used to describe the vibration characteristics of a lightweight electric vehicle that had been developed. The vehicle's model was combined with the lumped-parameter model of driver to determine the whole-body vibration level when the vehicle is passing over a road hump using Matlab Simulink. The models were simulated at a constant speed and the results were compared with the experimental data. The simulated vibration level at the vehicle floor and seat were almost similar to the experimental vibration results. The suspension systems that are being used for the solar vehicle are able to reduce the vibration level due to the road hump. The models can be used to simulate and choose the optimal parameters for the suspensions.

1. Introduction

Vehicle vibration is one of the major factors that results to driving discomfort, fatigue and health problems to drivers. Improvement on wheel and seat suspension systems, and in-vehicle ergonomics consideration for drivers to minimise the effect of the vibration had long been carried out [1-3]. Whole body vibration (WBV) exposure is a common phenomenon in vehicles where vibration from road surface and structures are transmitted through the legs, buttocks and back of seated individuals. Prolonged WBV exposure may lead to adverse affects on health, activities and comfort, and may also cause motion sickness. Over-exposure may result in lumbar disc disease and tissue failure or metabolic interference. The lumbar disc could also be damaged if sitting in incorrect posture in over-exposure vibration situation. The most common injury reported due to WBV is low back pain (LBP) even though there is non-exact dose-response relationship between them [4-5].

Most of the numerical analysis studies related to the vehicle vibration were based on mass-spring-damper systems since both the spring and the damper are the two main components that form the basic system of the vehicle suspension. From the mechanical models of the systems, mathematical models were developed to analyse the characteristics of the vibration and results were compared with the allowable international standard [6] of vibration exposure to the driver. Generally, the studies were done by adopting parameters of ready made systems and necessary modification to existing systems to overcome those problems would be very limited because of the complexity of the system. However, the used of simulation software to analyse vibration characteristics for structures has enabled designers to analyse and determine the components and suitable parameters that are able to overcome the vibration problems.

2. Method

The integration of ergonomics at the early stage of design process will reduce product development time and cost [7-8]. However, the process requires tools which are able to compute the numerical analysis with respect to various human requirements and expectations. Therefore, the need for cost-effective computing tools which can incorporate ergonomics consideration during the product design stage is crucial [9].

In this study the dynamic behaviour of the vehicle and the driver is investigated using Matlab Simulink software. The combination of both vehicle and driver models will give early indication on how the vehicle suspension system will affect the driver especially on the driving comfort levels. The vehicle on suspensions can be modelled as a lumped parameters system with a rigid body connected to the wheels by the springs and dampers. The suspensions must be a linear or nonlinear elastic system and incorporate the damping function to avoid the onset of resonant oscillations [10]. A six degrees of freedom (DOFs) full vehicle model was developed by combining three quarter-vehicle suspension models [11] presenting the vertical motion and angular pitch and roll motions. The seated driver can be represented by three DOFs system consists of three body components and a seat which were coupled by spring and damping elements. The input signal to the seat is the vertical displacement of the vehicle floor assuming that the seat was placed on the centre of gravity (CG) of the vehicle. The seat is also using a passive suspension system consists of the spring and damping components. In this study, the simulation parameters used were almost similar to the components being used for the electric vehicle. The input signal or the road profile for the simulation is a road hump.

2.1. Lightweight vehicle

An electric car which uses solar energy to charge its battery has been designed and fabricated. The design has considered of using lightweight material and component because weight is an important factor for the vehicle in terms of electrical energy saving. The vehicle main body structure was fabricated using 38 mm aluminium hollow pipe of 3.14 mm thickness. Other components used such as the wheels, spring suspensions, steering and seat are common parts available off-the-shelf.

The car uses a 48V direct current (DC) permanent magnet motor of 3000 maximum r.p.m. The swing arm of the rear wheel is designed so that it can move freely in vertical direction. Four units of motorcycle spring damper suspensions are being used at the front wheels and the rear swing arm. The rear wheel consist of two 70 inches tyres which was coupled together to support the weight of the DC motor on the swing arm. The front left wheel and right front wheel consists of 21 inches tyre with aluminium sport rim. Figure 1 shows the location of the suspension systems and seat.

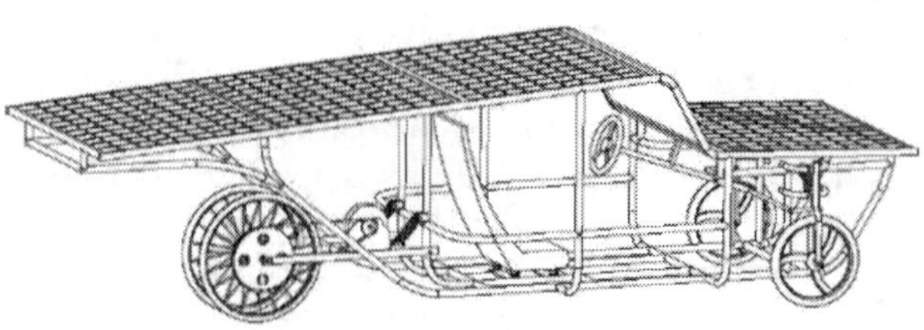

Figure 1. Lightweight electric car.

2.2. Simulation

Simulation was done using Matlab/Simulink. The mathematical model of the electric vehicle suspension system and the lumped-parameter model of a driver were developed and transferred into the Simulink block command. This analysis will help in analysing the vibration level that could be transmitted to the driver's body resulted from the road hump, through the vehicle suspension, the vehicle floor and the seat.

The simulation uses an input signal of the road hump's profile with a maximum height and width of 0.09 metres and 3 metres respectively. The velocity of the vehicle is assumed to be constant at 15 km/h, a preferred velocity when driving over the hump [12-14].

The equations of motion for the 6 DOFs full-vehicle and 3 DOFs seated driver models were programmed accordingly using blocks in Simulink. The parameters for the models are given in table 1 and table 2. There were three inputs for the simulation at both front wheels and rear wheel which represented by Z_{rfl}, Z_{rfr} and Z_{rr}. As mentioned previously, the input was a road hump signal where Z_{rfl} and Z_{rfr} were having the same signal assuming that both front wheels reached the road hump simultaneously. The signal for Z_{rr} was delayed which depends on the velocity of the vehicle, v. For the simulation, the velocity, v was set to 5.5 m/s. The time taken for the vehicle (or the rear wheel) to pass the hump is approximately 1 sec.

Table 1. The parameters for full-vehicle model.

Description	Symbol	Value
Sprung mass	M_s	310 kg
Left front - unsprung mass	M_{ufl}	15 kg
Right front - unsprung mass	M_{ufr}	15 kg
Rear - unsprung mass	M_{ur}	30 kg
Roll - moment of inertia	I_{xx}	130 kg/m^2
Pitch - moment of inertia	I_{yy}	340 kg/m^2
Left front - susp. stiffness	K_{sfl}	92 kN/m
Right front - susp. stiffness	K_{sfr}	92 kN/m
Rear - suspension stiffness	K_{sr}	184 kN/m
Left front - susp. damping	C_{sfl}	500 Ns/m
Right front - susp. damping	C_{sfr}	500 Ns/m
Rear - suspension damping	C_{sr}	1000 Ns/m
Left front tyre stiffness	K_{tfl}	38 kN/m
Right front tyre stiffness	K_{tfr}	38 kN/m
Rear tyre stiffness	K_{tr}	35 kN/m
Left front tyre damping	C_{tfl}	150 Ns/m
Right front tyre damping	C_{tfr}	150 Ns/m
Rear tyre damping	C_{tr}	100 Ns/m
Length - front wheel to C.G	a	1.4 m
length - rear wheel to C.G	b	1.5 m
Width of sprung mass	w	1.4 m
Seat mass	M_{se}	5 kg

IOP Conf. Series: Materials Science and Engineering **318** (2018) 012069 doi:10.1088/1757-899X/318/1/012069

Table 2. The parameters for full-vehicle model.

Description	Symbol	Value
Buttock and legs mass	M_{bl}	20 kg
Body mass	M_{bo}	40 kg
Head mass	M_{he}	5 kg
Seat suspension stiffness	K_{se}	7 kN/m
Seat suspension damping	C_{se}	900 Ns/m
Buttock and legs stiffness	K_{bl}	40 kN/m
Buttock and legs damping	C_{bl}	2500 Ns/m
Body stiffness	K_{bo}	35 kN/m
Body damping	C_{bo}	750 Ns/m
Head stiffness	K_{he}	130 kN/m
Head damping	C_{he}	250 Ns/m

2.3. Experimental set-up

An experiment was done to measure the vibration at the vehicle floor and seat when the vehicle passing over the speed hump. Two Kistler (Type: 8690C) piezobeam accelerometers were used and placed on the vehicle floor (under the seat) and also on the seat. Both accelerometers were connected to National Instrument NI9233 four-channel, ± 5V, 50 kS/s per channel, 24-Bit IEPE dynamic signal acquisition module and NI CompactDAQ. The analogue input was analysed using NI Sound and Vibration software using LabVIEW SignalExpress interface. The measurement rate and sample size were set to 50k Hz and 500k respectively using continuous samples acquisition mode. The data was taken for a period of 10 second for a distance of approximately 40 metres between the start and stop points. The signal was processed using Butterworth lowpass filter with 20 Hz cut-off frequency. During the experiment, the speed of the vehicle was kept constant when passing the hump.

3. Full-vehicle model

A full-vehicle suspension model can be represented as a linear six DOFs system consisting of the front and rear wheels, unsprung masses, sprung mass and suspension components as shown in figure 2.

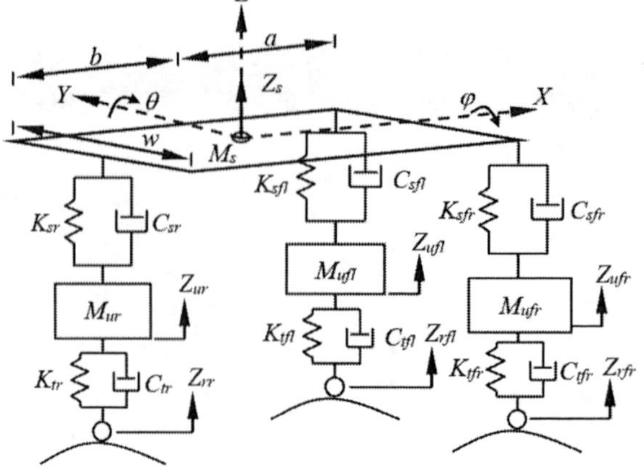

Figure 2. Six DOFs full-vehicle model.

In the model, wheels are represented by tyres having spring characteristics. The unsprung mass is made up of the weights of the wheels, axle and the components located geometrically below the suspension system while the sprung mass represents the body or chassis of the car. This six DOFs model was adapted from Ikinega [15] and the pitch and roll angles were assumed to be small. Using the Newton's second law, the equations of motion for the six DOFs model of the solar vehicle are given by (1) to (6).

(1), (2) and (3) show the equations of motion in the vertical direction for the unsprung masses M_{ufl}, M_{ufr} and M_{ur} in terms of stiffness, K, damping coefficient, C, pitch, θ and roll, φ. The vertical displacement, velocity and acceleration are given by Z , $\dot{Z}$ and $\ddot{Z}$ respectively. Similarly for pitch and roll motions, the 'dot' and 'double dot' above the symbols represent the velocity and acceleration respectively. Details of the parameters used for the equations are described in table 1.

$$\begin{aligned} M_{ufl}\ddot{Z}_{ufl} &= K_{sfl}Z_s + C_{sfl}\dot{Z}_s - aK_{sfl}\theta - aC_{sfl}\dot{\theta} + 0.5wK_{sfl}\varphi + 0.5wC_{sfl}\dot{\varphi} - K_{sfl}Z_{ufl} - C_{sfl}\dot{Z}_{ufl} \\ &- K_{tfl}Z_{ufl} + K_{tfl}Z_{rfl} - C_{tfl}\dot{Z}_{ufl} + C_{tfl}\dot{Z}_{rfl} \end{aligned} \tag{1}$$

$$\begin{aligned} M_{ufr}\ddot{Z}_{ufr} &= K_{sfr}Z_s + C_{sfr}\dot{Z}_s - aK_{sfr}\theta - aC_{sfr}\dot{\theta} - 0.5wK_{sfr}\varphi - 0.5wC_{sfr}\dot{\varphi} - K_{sfr}Z_{ufr} - C_{sfr}\dot{Z}_{ufr} \\ &- K_{tfr}Z_{ufr} + K_{tfr}Z_{rfr} - C_{tfr}\dot{Z}_{ufr} + C_{tfr}\dot{Z}_{rfr} \end{aligned} \tag{2}$$

$$M_{ur}\ddot{Z}_{ur} = K_{sr}Z_s + C_{sr}\dot{Z}_s + bK_{sr}\theta + bC_{sfr}\dot{\theta} - K_{sr}Z_{ur} - C_{sr}\dot{Z}_{ur} - K_{tr}Z_{ur} + K_{tr}Z_{rr} - C_{tr}\dot{Z}_{ur} + C_{tr}\dot{Z}_{rr} \tag{3}$$

(4), (5) and (6) show the equations of motion for the sprung mass M_s, in vertical, pitch and roll motions respectively.

$$\begin{aligned} M_s\ddot{Z}_s &= -K_{sfl}Z_s - K_{sfr}Z_s - K_{sr}Z_s - C_{sfl}\dot{Z}_s - C_{sfr}\dot{Z}_s - C_{sr}\dot{Z}_s + aK_{sfl}\theta + aK_{sfr}\theta - bK_{sr}\theta + aC_{sfl}\dot{\theta} \\ &+ aC_{sfr}\dot{\theta} - bC_{sr}\dot{\theta} + K_{sfl}Z_{ufl} + C_{sfl}\dot{Z}_{ufl} + K_{sfr}Z_{ufr} + C_{sfr}\dot{Z}_{ufr} + K_{sr}Z_{ur} + C_{sr}\dot{Z}_{ur} \end{aligned} \tag{4}$$

$$\begin{aligned} I_{yy}\ddot{\theta} &= aK_{sfl}Z_s + aK_{sfr}Z_s - bK_{sr}Z_s - aC_{sfl}\dot{Z}_s + aC_{sfr}\dot{Z}_s - bC_{sr}\dot{Z}_s - a^2K_{sfl}\theta - a^2K_{sfr}\theta - b^2K_{sr}\theta \\ &- a^2C_{sfl}\dot{\theta} - a^2C_{sfr}\dot{\theta} - b^2C_{sr}\dot{\theta} - aK_{sfl}Z_{ufl} - aC_{sfl}\dot{Z}_{ufl} - aK_{sfr}Z_{ufr} - aC_{sfr}\dot{Z}_{ufr} + bK_{sr}Z_{ur} + bC_{sr}\dot{Z}_{ur} \end{aligned} \tag{5}$$

$$\begin{aligned} I_{xx}\ddot{\varphi} &= -0.25w^2K_{sfl}\varphi - 0.25w^2K_{sfr}\varphi - 0.25w^2C_{sfl}\dot{\varphi} - 0.25w^2C_{sfr}\dot{\varphi} + 0.5wK_{sfl}Z_{ufl} + 0.5wC_{sfl}\dot{Z}_{ufl} \\ &- 0.5wK_{sfr}Z_{ufr} - 0.5C_{sfr}\dot{Z}_{ufr} \end{aligned} \tag{6}$$

From the above equations, the Simulink model was constructed. Figure 3 shows the model for the unsprung mass M_{ufl} which was given an input signal of the speed hump as shown in figure 4.

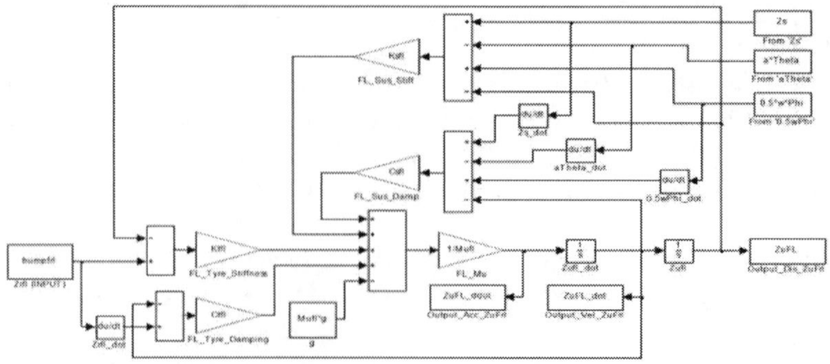

Figure 3. Simulink for the unsprung mass, M_{ufl}.

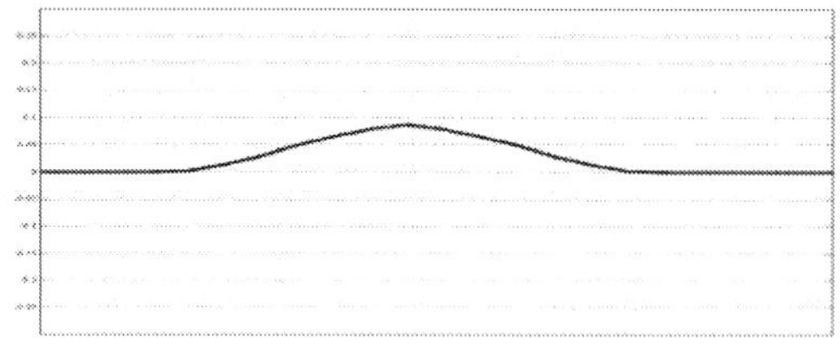

Figure 4. Profile of the road hump.

4. Driver's model

Behaviour of human body seated in a car can be modelled as a 3 DOFs system as shown in figure 5. The model was adopted from Liang and Chiang [16] and comprised of three masses which are coupled with spring and damping elements. It was assumed that the model is linear in which the driving point mechanical impedance did not vary significantly over the range of excitation type and levels in the frequency range. The three masses are comprised of the head and neck, M_{he}, the body, M_{bo} and the buttock and legs, M_{bl} in contact with the vehicle seat, M_{se}. The masses of the lower legs and feet, and the hand and arm are not included based on the assumption of their negligible contributions to the biodynamic response of the seated body.

The addition of the driver and seat models to the vehicle has changed the equation of motion for the sprung mass M_s (4) to that given by (7). The parameters used for the equations are described in table 1.

$$\begin{aligned} M_s\ddot{Z}_s &= K_{sfl}(-Z_s + a\theta + Z_{ufl}) + K_{sfr}(-Z_s + a\theta + Z_{ufr}) + K_{sr}(-Z_s - b\theta + Z_{ur}) + K_{se}(Z_{se} - Z_s) \\ &+ C_{sfl}(-\dot{Z}_s + a\dot{\theta} + \dot{Z}_{ufl}) + C_{sfr}(-\dot{Z}_s + a\dot{\theta} + \dot{Z}_{ufr}) + C_{sr}(-\dot{Z}_s - b\dot{\theta} + \dot{Z}_{ur}) + C_{se}(\dot{Z}_{se} - \dot{Z}_s) \end{aligned} \quad (7)$$

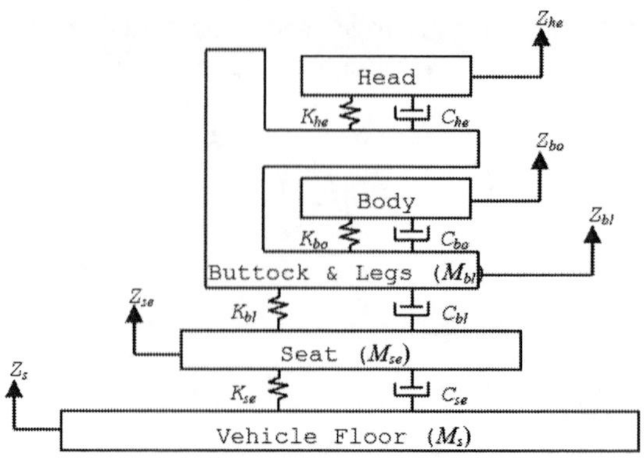

Figure 5. Three DOFs driver model.

The vertical equation of motion for the vehicle seat and the lumped-parameter model of seated driver are given by (8) to (11). The parameters used for the equations are described in table 2.

$$M_{se}\ddot{Z}_{se} = K_{se}(Z_s - Z_{se}) + K_{bl}(Z_{bl} - Z_{se}) + C_{se}(\dot{Z}_s - \dot{Z}_{se}) + C_{bl}(\dot{Z}_{bl} - \dot{Z}_{se}) \quad (8)$$

$$M_{bl}\ddot{Z}_{bl} = K_{bl}(Z_{se} - Z_{bl}) + K_{bo}(Z_{bo} - Z_{bl}) + C_{bl}(\dot{Z}_{se} - \dot{Z}_{bl}) + C_{bo}(\dot{Z}_{bo} - \dot{Z}_{bl}) \quad (9)$$

$$M_{bo}\ddot{Z}_{bo} = K_{bo}(Z_{bl} - Z_{bo}) + C_{bo}(\dot{Z}_{bl} - \dot{Z}_{bo}) \quad (10)$$

$$M_{he}\ddot{Z}_{he} = K_{he}(Z_{bl} - Z_{he}) + C_{he}(\dot{Z}_{bl} - \dot{Z}_{he}) \quad (11)$$

Figure 6 shows the Simulink model for the seat and the buttock and leg masses of the driver. It can be seen that the inputs to the seat was given by the Z_s and $\dot{Z}_s$.

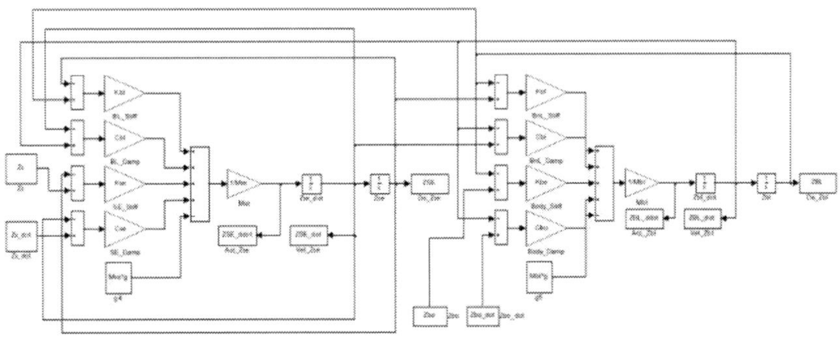

Figure 6. Simulink for the seat and buttock-legs masses.

5. Results and discussion

The input signals for left front wheel Z_{rfl} and rear wheel Z_{rr} were represented by the solid line and dashed line respectively as shown in figure 7. The input signal for right front wheel Z_{rfr} was similar to Z_{rfl}.

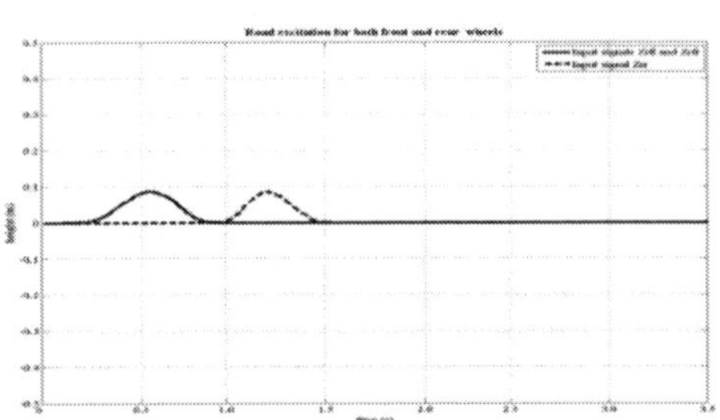

Figure 7. Input signals for the simulation.

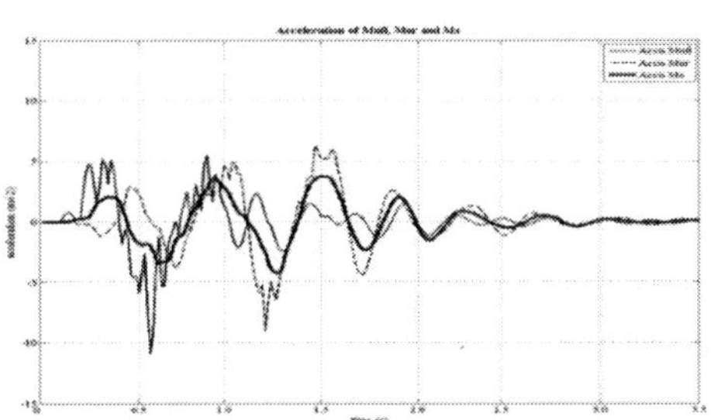

Figure 8. M_{ufl}, M_{ur} and M_s vertical accelerations.

The simulation results for vertical acceleration of the front wheel unsprung mass M_{ufl}, rear wheel unsprung mass M_{ur} and sprung mass M_s are shown in figure 8, while the frequency power spectrum of M_s is shown in figure 9. When the vehicle was driven over the hump, the accelerations of M_{ufl} and M_{ur} were between 7 ms^{-2} and -10 ms^{-2} and decaying after passing the hump. However, the acceleration of M_s was between 4 ms^{-2} and -4 ms^{-2}. Based on the simulation, the passive suspension system implemented has attenuated the vibration acceleration from the wheels to the vehicle floor quite reasonably and the significant frequency of M_s falls below 5 Hz.

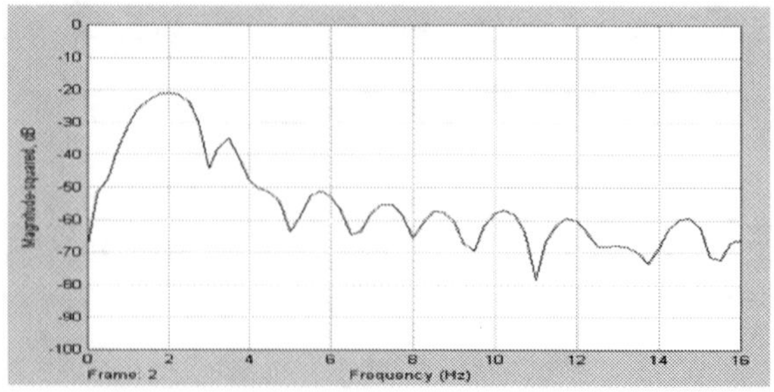

Figure 9. Frequency spectrum of M_s.

The M_{se} acceleration was lower compared to M_s acceleration as in figure 10. However, the vibration acceleration of driver's masses M_{bl}, M_{bo} and M_{he} were between 3.5 ms^{-2} and -4.8 ms^{-2}. These accelerations were slightly higher compared to the M_{se} acceleration because of the stiffness characteristics given to the masses. As can be seen in figure 11, the significant frequency of the seat was also below 5 Hz.

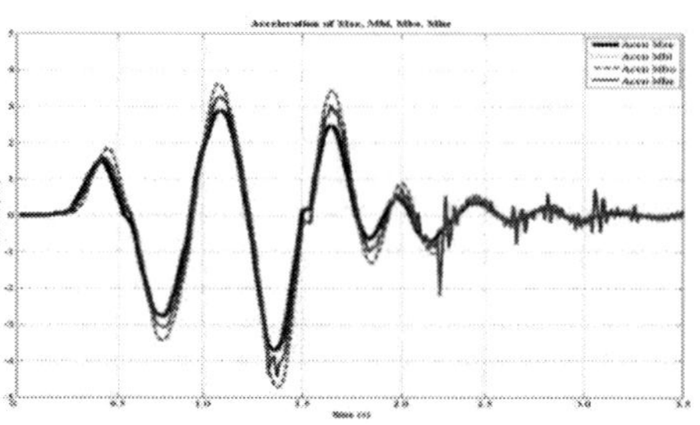

Figure 10. M_{se}, M_{bl}, M_{bo}, M_{he} vertical accelerations.

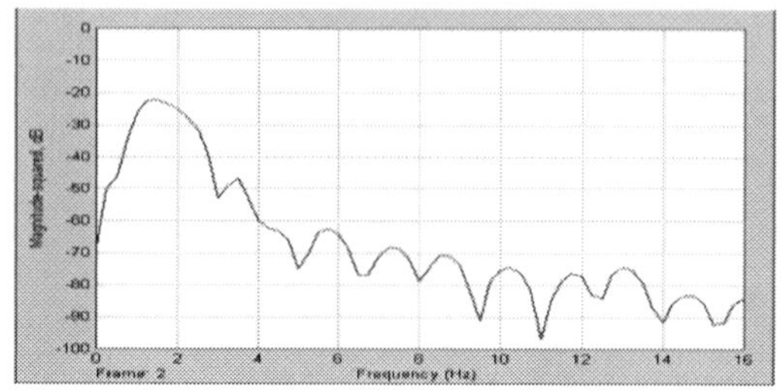

Figure 11. Power spectrum of M_{se}.

Displacements of M_{ufl}, M_{ur} and M_s are shown in figure 12. The maximum displacement of M_{ufl}, M_{ur} and M_s was 0.09 m, 0.12 m and 0.06 m respectively.

An experiment was done to measure the vertical vibration at the vehicle floor and seat. The experimental acceleration results are shown in figure 13 and figure 14. Comparison with the vertical acceleration of M_s and M_{se} shows close correlation between the simulation and experimental results.

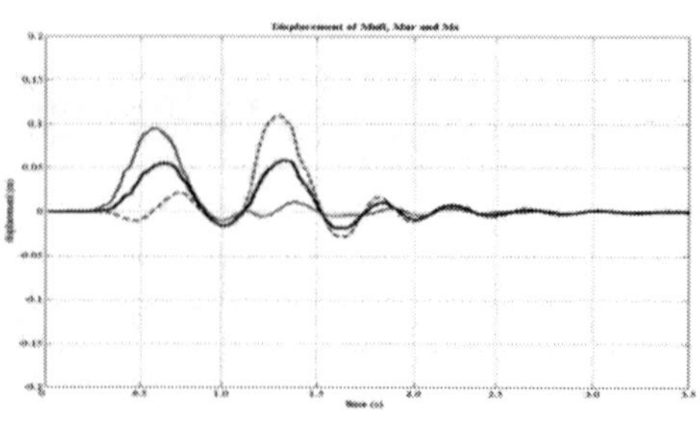

Figure 12. Vertical displacement of M_{ufl}, M_{ur} and M_s.

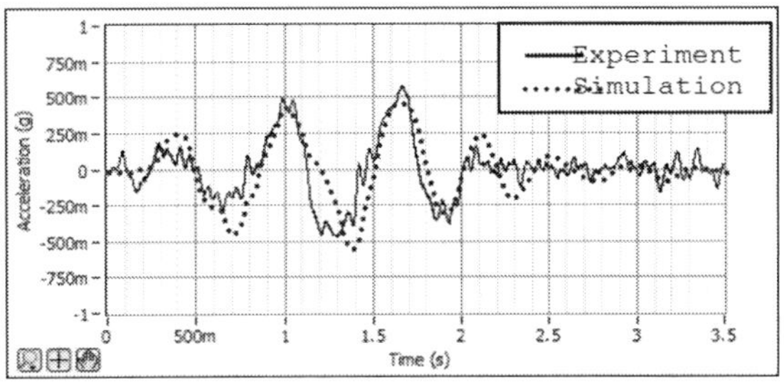

Figure 13. Vibration experimental and simulation results on the vehicle floor.

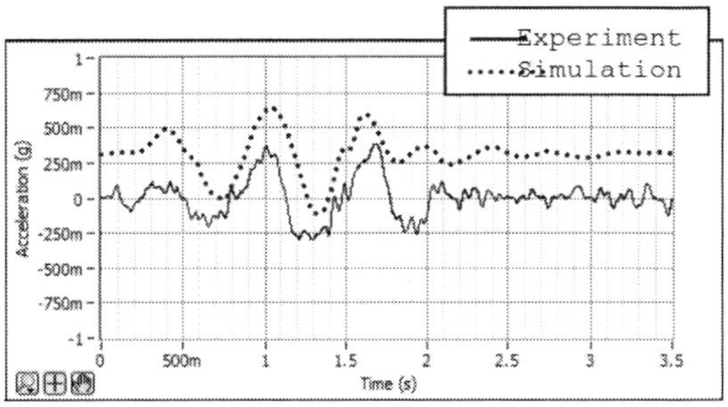

Figure 14. Vibration experimental and simulation data on the vehicle seat.

One of the possibilities to improve the comfort level for driving the vehicle is to change the damping coefficient of the suspension. Reducing the mass of the vehicle could also improve the comfort level but it is normally not possible as it has been considered at the design stage. Reducing the suspension stiffness may result to poor vehicle handling capability and also to cater for the mass of the vehicle.

6. Conclusion

Full-vehicle model of a three-wheeled lightweight electric vehicle and driver's 3 DOFs model were developed and combined to investigate the performance of the passive suspension system to attenuate the vibration when passing a road hump. The results from the simulation show that the suspensions were able to reduce the vibration level reasonably well. The experimental results were also in close agreement with the simulation. Therefore, the models can be used further for choosing optimum parameters for the suspension systems when the vehicle is driven on different type of road conditions. A simulation analysis gives great advantages to designer for selecting and determining the most effective system and parameters of a vehicle before the implementation is done.

References

[1] Donati P 2002 *J. Sound and Vibration* **253** 169-183
[2] Deboli R, Calvo A, Preti C 2017 *Int. J. Ind. Ergonomics* **58**, 69-78
[3] Hostens I, Deprez K A, Ramon H 2004 *J. Sound and Vibration* **276**, 141-156
[4] Adam S A, Jalil N A A 2017 *Proc. Eng. EPIC 2016* **170**, 435-442
[5] Makhsous M, Hendrix R, Crowther Z, Nam E, Lin F 2005 *Ergonomics*, **48**, 9, 1183-1199
[6] International Organization for Standardization 1997 Mechanical vibration and shock - evaluation of human exposure - to whole-body vibration. Part 1: General requirements. International Standard, ISO 2631-1, Second edition 1997-05-01, Corrected and reprinted 1997-07-15
[7] Sagot J C, Gouin V, Gomes S 2003 *Safety Science* **41**, 137-154
[8] Marsot J 2005 *Applied Ergonomics* **36**, 185-192
[9] Kang S J, Chun B K 2000 *Thin-Walled Struct.* **37**, 113-125
[10] Genta G 1997 *Motor Vehicle Dynamics - Modeling and Simulation*. World Scientific Publishing Co. Pte. Ltd.
[11] Kruczek A, Stribrsky A 2004 *Proc. IEEE Inter. Conf. on Mechatronics*
[12] Griffiths K R 2012 *An improved method for simulation of vehicle vibration using a journey database and wavelet analysis for the pre-distribution testing of packaging* (Doctoral dissertation) retrieved from opus.bath.ac.uk/38300/1/univBath_PhD_2012_K_Griffiths.pdf
[13] Phalke T P, Mitra A C 2017 *Materials Today: Proc.* **4** 2425-2430
[14] Talukdar R P, Talukdar S 2016 *Procedia Eng. 12th ICOVP* 2015 **144**, 1094-1101
[15] Ikenaga S, Lewis F L, Campos J, Davis L 2000 *Proc. American Control Conf. Chicago, Illinois*
[16] Liang C C, Chiang C F 2006 *Int. J. Industrial Ergonomics* **36**, 869-890

IOP Conf. Series: Materials Science and Engineering **318** (2018) 012070 doi:10.1088/1757-899X/318/1/012070

Experimental Investigation on Effect of Fin Shape on the Thermal-Hydraulic Performance of Compact Fin-and-Tube Heat Exchangers

P Moorthy[1], A N Oumer[1*], and M Ishak[1]
[1]Faculty of Mechanical Engineering, Universiti Malaysia Pahang (UMP), Malaysia

Email: ahmnur99@gmail.com

Abstract. The aim of this paper is to investigate the effect of fin shapes on the performance of compact finned flat tube heat exchangers. Three types of fin shapes namely plain, wavy, and rectangular grooved fins attached to three by three arrays of flat tube banks were considered. Moreover, the tubes were deployed in in-line and staggered arrangements. In addition to the fin shapes, the air velocity and the tube inclination angles were varied and the thermal-hydraulic performance was analysed. On the other hand, the temperatures at the tube surfaces were kept constant to produce constant heat flux throughout the study. The results showed that as flowrate increases, the heat transfer increases, however, the friction factor decreases. Staggered arrangement produces higher heat transfer and friction factor than inline fin. Moreover, the rectangular fin is the best in terms of high heat transfer however the drawback of high friction factor leads the fin to have the least efficiency of all. On the other hand, plain fin had the least heat transfer performance however the highest efficiency was achieved. Therefore, plain fin should be used when efficiency is prioritized and rectangular fin when high heat transfer is desired.

1. Introduction

Compact fin-and-tube heat exchangers are used in many engineering applications such as power plants, transport, heating systems and air conditioning [1, 2]. The performance and efficiency of such types of heat exchangers are represented through the ratio of heat transfer over pressure drop (friction factor). Most process and power industries commonly use plate fin-and-tube heat exchangers for a wide temperature ranges. This is due to their compactness, close temperature approach. Moreover, they are also easy to inspect and clean [3]. There are different shapes of fins attached to the tube banks. For instance, in some applications metal plates are clamped as one used in conventional plate heat exchangers. The main types of fins used in compact heat exchangers include plane, wavy and rectangular fins.

The wavy fin and tube arrangement is famous plate fin and tube heat exchanger where it is used in applications that are constricted by high resistance to thermal on the air side of the heat exchanger [4]. The wavy fin surface is popular because it can increase the length of the airflow in the heat exchanger by improving or enhancing the air flow mixing and increasing heat transfer performance [5]. Furthermore, wavy fin and tube heat exchangers are cheap, reliable and easier to install or conduct maintenance [6]. The waviness of the fin enhances heat transfer because wavy fin has more surface area due to corrugations and it also promotes turbulence at bigger flows [7]. Similarly, the herringbone wavy fin corrugation also increases the contact area and promotes air flow mixing [4].

Most heat exchanger designs involve forced airflow through the exchanger, and the designs vary depending on specific requirements. These heat exchangers typically use fin-and-tube geometries and modified fin patterns to attain an enhancement of the air-side heat transfer, because the major limitation on the performance arises from the air-side heat transfer resistance. While conventional fin-and-tube heat exchangers have planar fins and round tubes, many contemporary designs have been developed with serpentine fins and flat tubes. To authors best knowledge, the performance differences between flat-tube heat exchangers with different fin shapes are not fully understood, because there have been much fewer investigations of the flat-tube geometries.

Thus, the main objective of the current study is to investigate the flow and heat transfer behaviours of flat tube heat exchanger with various fin shapes.

2. Methodology

2.1. Material Preparation

The specification of the test section available at the Faculty of Mechanical Engineering, Universiti Malaysia Pahang had to be taken in to consideration during designing of the fins. For the wavy fins, the parameters of the wavy angle and height were obtained from the works of Jang and Chen [8] where their study shows that the wavy angle of 162.1° and height of 1.5 mm are suitable for enhanced heat transfer. As for the rectangular grooved fin, there was no previous study made on the shape. Therefore, the parameters were set to be 1cm x 1cm x 1cm. After finalising on the fin parameters selection, the selected parameters were converted to Solidworks for ease of manufacturing. For each fin type, a total of 10 aluminum fins were used with constant fin spacing.

2.2. Experimental Setup and Procedure

Figure 1 shows the complete experimental setup a typical example of the location of temperature sensors for inline plain fin. The working fluid was air and the experiments were conducted in an open wind tunnel to obtain the thermal-hydraulic performance of the fin and flat tube heat exchanger. The system consists of inlet section, a flow straightener, a test section, and measuring devices for air velocity, air temperature and air pressure difference, temperature controller, a suction fan, and frequency inverter. The system operates as the suction fan allows the air to be sucked into the test rig where it flows over the fin and tube heat exchanger. The tubes were heated using electrical heating rod. The test section of the duct was designed to ensure that test sample could be changed easily. This arrangement operated in a cross-flow where the coolant air passes over the heat exchanger and the tube was heated by a hot electrical rod.

The wind tunnel is a square duct of 26 cm each side. A straightener was put at inlet to laminarize the flow to evade any distortion of air velocity. The distance between the straightener and the test setions inlet was 1.2m where the distance enables the air to be fully developedbefore the air enterd the test section. The wind tunnel walls were holed at the bottom side (before and after test section) to install pitot tubes which will be connected to digital manometer. The top side was holed to install digital thermometer which measures the temperature of air flow before it reaches the test section. Then, the air leaves the extension section and flows through the fin-and-tube HEs and then to the suction fan where it is discharged to the surrounding. The wind tunnel is fixed on top the stand made of steel which is 50 cm height to minimize the noise and vibration during the operation.

Axial fan model EPM is used as the suction fan driven by a 50 W AC motor. They fan speed was varied using frequency inverter however the air velocity could not be measured. Therefore, the air velocity was measured using Digital Anemometer where frequency inverter was set when the Anemometer measures the air velocity desired. Digital Anemometer also has the function of measuring the inlet temperature of air. Mainly, Testo 110 Digital Thermometer was used to measure the inlet air temperature. REX C-100 temperature controllers were used to maintain the surface temperature of the flat tubes where the heating rods deliver constant heat. K-type thermocouples were installed to every temperature controllers to closely monitor the flat tube surface temperature. The air pressure at the inlet and the outlet of the test section were measured using the digital differential pressure manometer of model TESTO 510. Data logger was used to collect the temperature distribution shown across the fins (Figure 1b). T- type sensors were used and placed at 8 points where 6 of them measure the temperature distribution in between the flat tubes and the other 2 sensors will read the oulet temperature. The position of sensors in the test section is as shown in Figure 1b and they were fixed using high-temperature resistant plaster. The black spot marks the position of the sensors placement. Other geometric parameter of the flat tube such as tube wall thickness, tube spacing, fin spacing, etc used in the experiment are presented in Table 1.

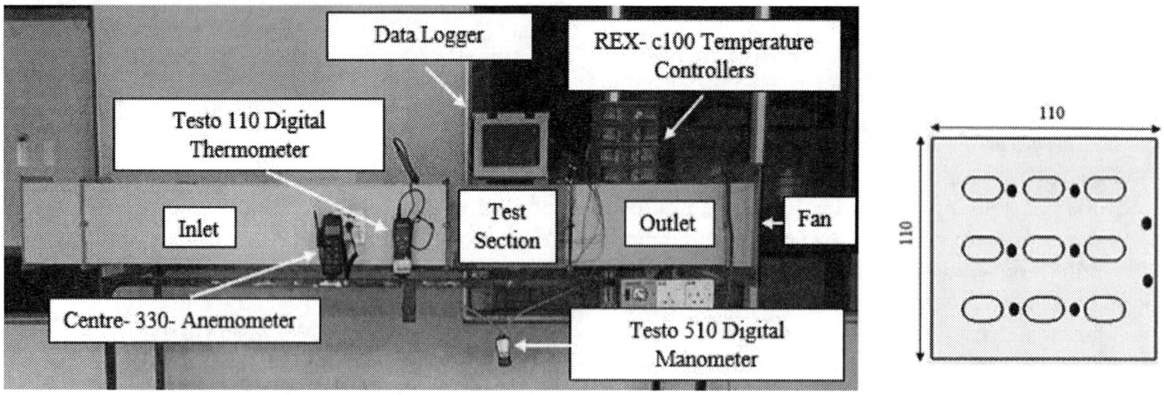

(a) **(b)**

Figure 1: (a) Complete experimental setup with measuring devices and (b) The location of temperature sensors in inline plain fin

Table 1: Geometric details of the test samples

Geometric parameter	Unit	Dimension (mm)
Tube wall thickness	δ	1
Transverse tube pitch	P_t	35
Longitudinal tube pitch	P_l	35
Tube length	L	220
Tube high outside diameter	D_{ho}	11.59
Tube wide outside diameter	D_{wo}	19.50
Fin thickness	t	0.6

2.3. Data Reduction

It is difficult to measure the wall temperature of the heat exchanger directly to acquire the convection heat transfer coefficient of the air side. Therefore, the experimental data was reduced and the thermal properties of the air was determined by the average values of the air at the inlet and outlet of the test section [9]. The main purpose of data reduction is to determine the heat transfer coefficient of the air side and pressure drop through the heat exchanger by calculating the nusselt number Nu, friction factor f and Colburn factor j of the flat tube heat exchanger from the experimental temperature data results which were recorded during each test run.

The heat transfer rate (Q_a) of the air side which used to determine the heat transfer coefficient is computed by the following equation [10]:

$$Q_a = Q_{conv} = \dot{m}\, C_p(\Delta T) \tag{1}$$

where $\dot{m}$ is the mass flow rate; ΔT is the temperature difference between the outlet and the inlet air temperature and Cp is the specific heat capacity. Due to the small thermal conductivity of air which is lower than 0.23 (w/m.°c) and the experiments were conducted at the room temperature, the heat transfer from the tube wall by conduction and radiation were neglected. Therefore, the energy balance can be expressed as following:

$$Q_{elec} = Q_{conv} = v \times I \tag{2}$$

where v and I are voltage and current. The forced convection heat transfer coefficient (h) expressed in terms of the heat transfer rate, total surface area and log mean temperature difference.

$$h = \frac{Q_{conv}}{A_t\, LMTD} \tag{3}$$

where A_t is the total heat transfer area and it is equal to the sum of the tubes and the fins surface area and LMTD is the log mean temperature difference. Thereafter, to evaluate the level of the heat transfer by convection in the air side Nu number is adopted in this study and it can be written as follows:

$$Nu = \frac{hD_h}{k} \tag{4}$$

The Colburn factor (j) and friction factor (f) of finned and unfinned flat tube have been calculated according to the following equations[11]:

$$j = \frac{h}{\rho v_{max} C_p} Pr^{\frac{2}{3}} \tag{5}$$

$$f = \frac{D}{L} \frac{2\Delta P}{\rho\, V_{max}{}^2} \tag{6}$$

where Δp is the pressure difference between the inlet and the outlet. The dimensionless number such as Nu number gives partial indication of the overall performance because sometimes the benefit gained from the heat transfer enhancement are not great enough to overcome the increase in friction losses. Therefore, thermal performance coefficient JF was calculated to compare the air side performance. Therefore, the performance evaluation criteria of the finned and flat tube heat exchanger is defined as [12]:

$$PEC = JF = \frac{j}{f} \tag{7}$$

3. Results and Discussion

3.1. Heat Transfer Performance

The heat transfer performance of the heat exchanger was evaluated in terms of dimensionless parameter Nusselt number. The main purpose was to test the effects of geometric (tube arrangement, fin shape and tube inclination angle) and flow parameters on the heat transfer performance of compact flat tube-and-fin heat exchanger.

3.1.1. Effects of flowrate.

Figure 2 shows variation of Nusselt number for plain fins against Reynolds number for inline configuration. The presented result shows that as the flow rate increases so does the Nusselt number. Moreover, it shows that the Nusselt number is directly proportional to the flowrate. This means the increase in the air velocity yields increment in the convection heat transfer of the heat exchanger. Results with similar trend has been seen from the simulation work by Kumar et al.[13]. The increasing trend is seen due to the better mixing of flow.

3.1.2. Effect of Tube Arrangement.

The comparison between inline and staggered configuration with different fin shapes and Reynolds number range was also analysed. This is to understand the effect of inline and staggered configurations on the heat transfer performance of the heat exchanger. Figure 3 shows the difference between Nusselt number variation of inline and staggered configurations of rectangular fin heat exchanger. The figure shows that the heat transfer performance of staggered configuration is higher than the inline configuration by about 7.5%. Similar trend was obtained for wavy and plain finned heat exchangers. This is due to the reason that there exist distortion to the velocity and the temperature profiles of the air flow which enhances the heat transfer properties. Another way of describing the reason for enhancement in staggered is due to the increase in the intensity in turbulence.

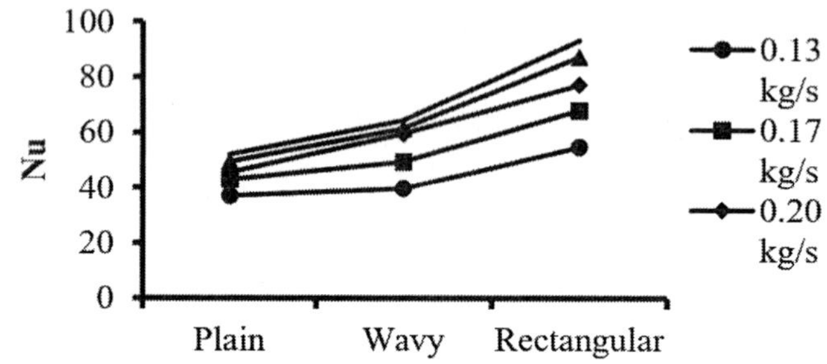

Figure 2: Variation of Nusselt number against different fin at different flowrate

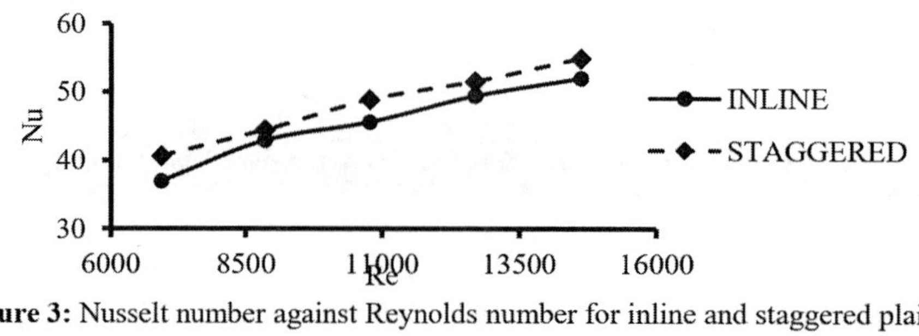

Figure 3: Nusselt number against Reynolds number for inline and staggered plain fin

3.1.3. Effect of Different Fin Shapes.

The performance of heat transfer affected by different fins shapes are presented in Figures 4a and 4b for inline and staggered arrangements respectively. The results show that the performance increases in plain, wavy, and rectangular order. Considering plain fin as reference, wavy fin shows 30% of increased performance and rectangular fin produces 69% higher performance than plain fin for inline. Rectangular fin produces the highest heat transfer performances due to the interruption done by the staggered surfaces to the flow and temperature boundary layers along the flow orientation. Similar arguments were observed in the works of Sanaye et al. [14].

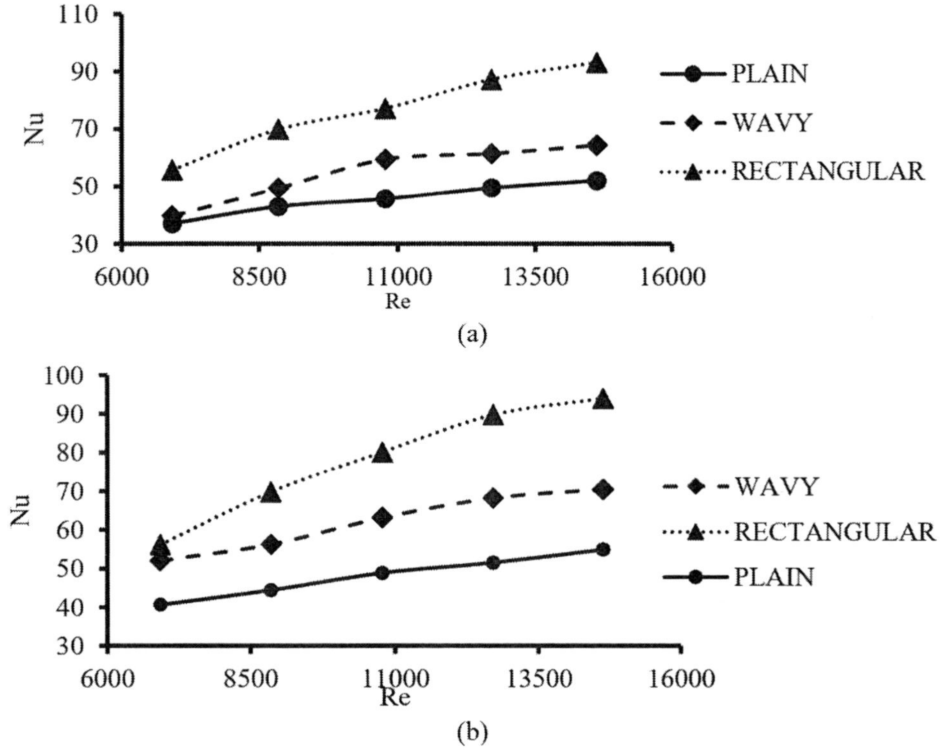

Figure 4: Variation of Nu against Reynolds number for plain, wavy, and rectangular fins for (a) inline and (b) tube arrangements

3.2. *Hydraulic Performance*
The hydraulic performance is presented in terms of friction factor, f. The main purpose is to test the effects of geometric (tube arrangement, fin shape and tube inclination angle) and flow (Reynolds number) parameters on the hydraulic performance of compact flat tube-and-fin heat exchanger.

3.2.1. Effects of air flow velocity.
Figure 5 shows variation of friction factor against Reynolds number for inline wavy fin. As the air flow velocity increases the friction factor decreases. This relationship is proved through equations where friction factor is inversely proportional to Reynolds number. As the air velocity increases the boundary layer thickness of air flow over the fin decreases and this produces decreasing trend of viscous force where viscous force is proportional to the friction factor.

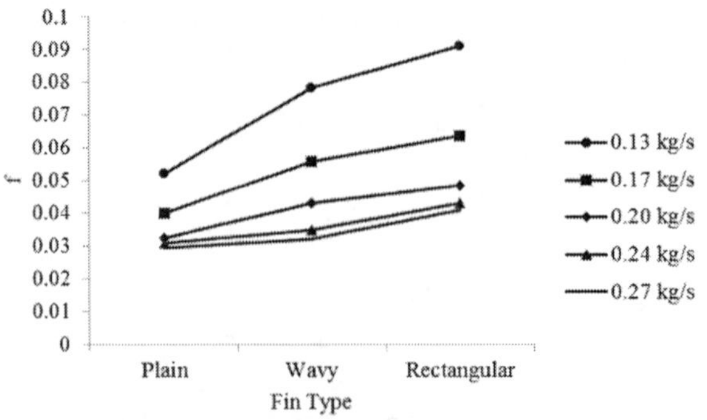

Figure 5: Friction factor against different fin at different flowrate

3.2.2. Effect of Tube Arrangement.

Figure 6 shows the comparison between inline and staggered tube arrangement against friction factor. The result shows that staggered arrangement has higher friction factor about 16.7% compared to inline tube arrangement. The results obtained for wavy and rectangular fins shows the same trend. This is due to the higher intensity in turbulence where the temperature and velocity profiles are being distorted significantly for staggered arrangement.

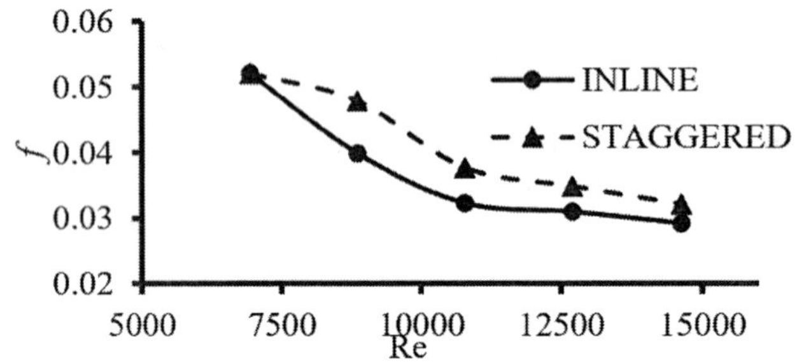

Figure 6: Comparison of friction factor between inline and staggered arrangement for plain fin

3.2.3. Effect of Different Fin Shapes

Figure 7 shows variation of friction factor against Reynolds number for different fins. The result shows that rectangular fin has the highest friction factor about 100% higher than plain fin meanwhile friction factor of wavy fin is higher by about 49.8% than plain fin. This is due the interruption that wavy and rectangular fin geometry does to the flow of air. Higher turbulence intensity leads to high friction factor.

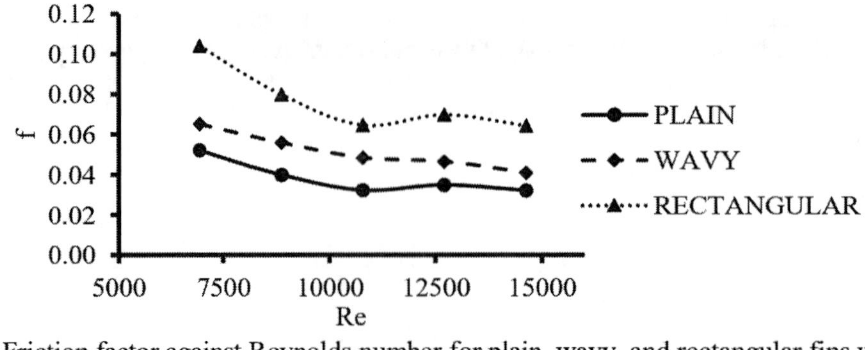

Figure 7: Friction factor against Reynolds number for plain, wavy, and rectangular fins with inline tube arrangement

3.3. Performance Evaluation Criteria

Figure 8 shows that plain fin has the highest efficiency of performance where even though plain fin has low heat transfer performance, it also has the lowest friction factor, therefore higher efficiency is obtained. On the other hand, wavy and rectangular fins have higher heat transfer performance; however, the drawback of high friction factor affects the efficiency of the fins. The usage of fin type differs according to the application where some prefer efficiency which plain fin excels. Some other applications prefer high heat transfer performance where wavy and rectangular fin excels.

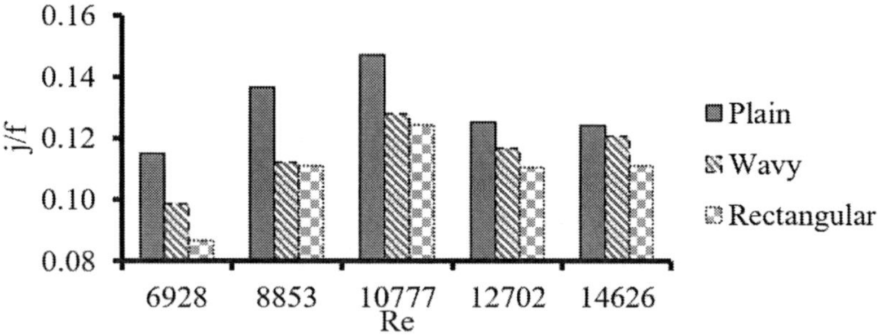

Figure 8: Efficiency of different fins against Re

4. Conclusion

From this study, it was found that the rectangular fin has the highest heat transfer performance compared to wavy and plain fin where wavy fin is higher than plain fin. Rectangular fin produces the highest heat transfer performances due to the interruption done by the staggered surfaces to the flow and temperature boundary layers along the flow orientation. The hydraulic performance has the similar trend where rectangular has the highest pressure drop compared to wavy and plain fin. This is due the interruption that wavy and rectangular fin geometry does to the flow of air. The pressure drop of plain fin is low compared to complex design of wavy and rectangular fin. Wavy and rectangular fin despite having higher heat transfer performance, they have a greater drawback in higher pressure drop. On the other hand, rectangular fin is suited for any application which prioritises thermal performance over hydraulic performance or efficiency.

Acknowledgement

The authors would like to thank Universiti Malaysia Pahang for providing necessary assistance for providing financial support under project number RDU150375.

References

[1] Tahseen T A, Ishak M, and Rahman M, 2012 *Int. J. of Auto. and Mec.Eng.* **6** 753.
[2] Tahseen T A, Ishak M, and Rahman M, 2015 *Renew. and Sust. Ener. Rev.***43** 363.
[3] Kanaris A, Mouza A, and Paras S 2009 *Int. J. of Ther. Sci.* **48** 1184.
[4] Wongwises S. and Chokeman Y 2005 *Ener. Conv. and Manag.* **46** 2216.
[5] Wang C C, Hwang Y M, and Lin Y T 2002 *Int. J. of Refr.* **25** 673.
[6] Wang C C, and Liaw J S 2012 *Int.J. of H.and M. Tra.*, **55**, 3054.
[7] Awad A and Muzychka Y S 2011 *J.of Enh. Heat Tra.***18**, 2011.
[8] Jang J Y and Chen Y K 1997 *Int. J. of H.and M. Tra.* **40**, 3981.
[9] Du X, Zeng M, Dong Z, and Wang Q 2014 *Exp. Ther.and Fl. Sci.* **52**, 146.
[10] Wang C, Fu W, and Chang C 1997 *Exp.Ther.And fl. Sci.***14** 174.
[11] He Y, Han H, Tao W, and Zhang Y 2012 *Int. J. of H.and M. Tra.* **55** 5449.
[12] He Y L, Chu P, Tao W Q, Zhang Y W, and Xie T 2013 *Appl. Ther.Eng.* **61** 770.
[13] Kumar A, Joshi J B, and Nayak A K 2017 *Int. J. H. and M. Tra.* **109**336.
[14] Sanaye S and Hajabdollahi H 2010 *Appl. Ene.* **87** 1893.

IOP Conf. Series: Materials Science and Engineering **318** (2018) 012071 doi:10.1088/1757-899X/318/1/012071

Effect of Alcohol on Diesel Engine Combustion Operating with Biodiesel-Diesel Blend at Idling Conditions

H.M. Mahmudul[1,] Ftwi.Y. Hagos[1,2*], M. Mukhtar N. A[2], Rizalman Mamat [1,2] and A. Adam Abdullah[1,2]

[1]Automotive Engineering Research Group (AERG), Faculty of Mechanical Engineering, Universiti Malaysia Pahang, 26600 Pekan, Pahang, Malaysia
[2]Auomotive Engineering Center, Universiti Malaysia Pahang,2660, Pekan, Pahang, Malaysia

E-mail: ftwi@unimap.edu.my

Abstract. Biodiesel is a promising alternative fuel to run the automotive engine. However, its blends have not been properly investigated during idling as it is the main problem to run the vehicles in a big city. The purpose of this study is to evaluate the impact of alcohol additives such as butanol and ethanol on combustion parameters under idling conditions when a single cylinder diesel engine operates with diesel, diesel-biodiesel blends, and diesel biodiesel-alcohol blends. The engine combustion parameters such as peak pressure, heat release rate and ignition delay were computed. This investigation has revealed that alcohol blends with diesel and biodiesel, BU20 blend yield higher maximum peak cylinder pressure than diesel. B5 blend was found with the lowest energy release among all. B20 was slightly lower than diesel. BU20 blend was seen with the highest peak energy release where E20 blend was found advance than diesel. Among all, the blends alcohol component revealed shorter ignition delay. B5 and B20 blends were influenced by biodiesel interference and the burning fraction were found slightly slower than conventional diesel where BU20 and E20 blends was found slightly faster than diesel So, based on the result, it can be said that among the alcohol blends butanol and ethanol can be promising alternative at idling conditions and can be used without any engine modifications.

1. Introduction

Transportation system plays a significant role in developing the economy of any country in the world. Presently, the demand of transportation sectors is fulfilled by fossil fuels either gasoline or diesel. As a result, the motor industry is developing due to high demand and as well as the average consumption of energy in the transport sector is increased by 1.1% per year. It has been reported that from the year of 2010– 2040, the transportation sector has 63% share in the step up of total global liquid fuel consumption [1, 2]. Hence, day by day the fossil fuel stocks decreasing as well as the price also increasing due to the demand and limitation. Researchers trying to find out the alternative sources to replace the natural resources. Among the sources, biodiesel is considered more viable cleaner, environment-friendly and it can be replaced as a diesel fuel replacement [1, 3]. The main advantage of using biodiesel is that it can be applied in any proportion with diesel as a biodiesel-diesel blend due to its favourable properties like petro-diesel [4]. Many researchers had conducted experiments on combustion or performances using different types of biodiesel in CI engine [5]. However, limited research was found conducted on engine combustion and performance under idling condition. Recently, idling conditions is a perilous problem for the transport sectors. Idling condition means a condition when the engine run at low-rated speed with low load. Researchers found that during the idling time, the vehicles consume higher fuel consumption and release emissions than running on the road by a factor of 1.5 [6]. Further elaborated, during idling condition time, the engine did not work at peak operating temperature and for this reason, it leads to the incomplete combustion and emissions level increase with more fuel deposits in the exhaust detected. Moreover, due to some disadvantages of the biodiesel like high viscosity, lower volatility and sometimes high emissions, researchers are in search of some additives such as alcohols added to the biodiesel fuel to improve these fuel properties [7, 8]. Recently alcohols like butanol and ethanol with the blending of diesel and biodiesel play an important role as an alternative fuel. It was found that improvement of combustion efficiencies of diesel fuel can be achieved by the adding oxygenated fuels such as ethanol, butanol, biodiesel, and vegetable oils, due to having a complete combustion. Many studies have been conducted to evaluate engine performance and emission using palm biodiesel blends in a diesel engine at idling condition [5, 9] but very few have been conducted using a high portion of alcohol as an additive with palm biodiesel-diesel blend at idling condition.

In this study, the tri-fuel composition was used consist of diesel, palm biodiesel and butanol. Another composition of tri-fuel consists of diesel, palm oil biodiesel and ethanol blends. The aim of this research was to study the combustion analysis with the new composition fuel involving pressure, energy release rate, mass burn fractions and ignition delay under idling condition.

2. Methodology

The test was done Yanmar TF120M (2012) single cylinder CI engine which was coupled with a positive displacement gear pump (model HGP-3A-F23) dynamometer. The engine was manufactured from the year of 2012.The idling condition (1200 RPM at low load was considered in this investigation. 5% and 20% (B5 and B20) blends of palm biodiesel with the diesel and 20% of alcohols of butanol and ethanol (BU20 and E20) along with palm and diesel fuel have been used to run the engine. Also, to present the effect of idling, the engine was run at no load at low engine speed and measured data were compared with the diesel at idling condition. Figure 1 presents the labelling and arrangement of the test bed component in a schematic diagram.

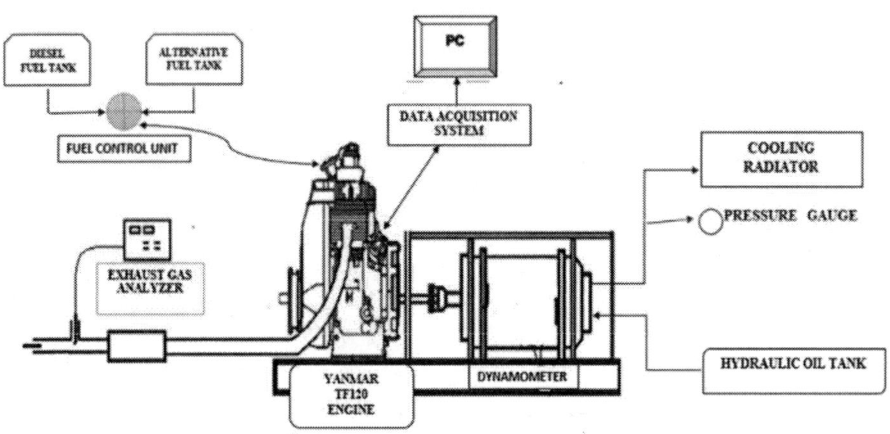

Figure 1. Schematic view of the experimental setup of engine.

Engine specification was shown in Table 1. A pressure transducer and crank angle encoder were used for combustion analysis. The relationship of the combustion pressure with that of the crank angle degree of the piston movement (pressure-crank angle degree) is provided from these two sensors. Digital data have been recorded in a computer by the using a software name TFX engineering DAQ Combustion Analyzer. The alternative fuel like palm biodiesel and the alcohol were bought from the local a market. For the blending process of Alcohol (Butanol and Ethanol) with palm and diesel, the magnetic string process was used. The basic fuel properties of diesel, biodiesel, butanol, and ethanol are shown in Table 2. The engine was run mostly at the room temperature with diesel fuel and to maintain the steady conditions at and also run for the certain time to consume the alternative fuel from the remaining experiment which was conducted carefully and repeated for 3 times. The experiment was conducted at a room temperature.

Table 1. Specifications for single cylinder engine.

Description	Specification
Engine model	TF120
Engine type	Horizontal, diesel 4 stroke
Combustion system	Direct injection
Number of cylinders	1
Bore x Stroke(mm)	92 x 96
Displacement (L)	0.638
Compression ratio	17.7
Continuous output (HP)	10.5 HP at 2400 RPM
Rated output (HP)	12 HP at 2400 RPM
Cooling system	Water cooled (radiator type)

Table 2. Basic fuel properties for diesel, biodiesel and butanol and ethanol [7, 10-12].

Fuel	Lower heating value (MJ/kg)	Density @20°C (kg/m^3)	Viscosity @40°C (MPa s)	Flash point (°C)	Cetane number	Latent heat at 25°C (kJ/kg)
Diesel	45.28	853.8	2.6	93	54.6	-
Biodiesel (PME)	41.3	867	4.53	165	67	-
Butanol	33.1	808	2.63	35	25	582
Ethanol	26.8	788	2.6	-	5-8	904

3. Result and Discussion

3.1 The in-cylinder pressure

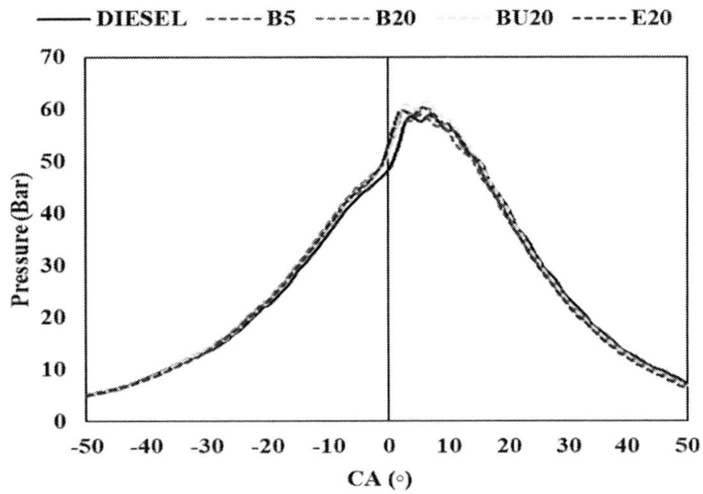

Figure 2. In-cylinder pressure vs. crank angle variation of biodiesel, biodiesel-alcohol blends compared to diesel at idling condition.

Figure 2 shows combustion pressure. Combustion peak pressure for bi-fuel category possessed almost similar reading, 59.04 for diesel, 58.08 for B5 and 60.09 for B20. Slight increment reading on peak combustion can be seen in the case of all tri-fuel blend categories, 59.80, 61.36 for E20, and BU20 respectively. Pressure graph indicated distorted fluctuation pattern at the peak which indicated slight minor engine knocking experience for all types of fuels. From Figure 2, it shows that all the biodiesel blends had slightly higher pressure than diesel. This might be happening during lower load condition as the residual gas temperature and wall temperature become low. In addition, while the injection charge becomes low, it increases the delay period and the combustion starts later. So due to the delay, the peak pressure became lower than the biodiesel blends [4]. When the engine was running at low load, residual gas temperature and wall temperature was also low. Therefore, injection charge temperatures were also significantly low, which in turns increases injection delay period. This explained why diesel fuel combustion starts later compared to biodiesel and its blends yield lower peak pressure compared to biodiesel blends. It was also found that the ethanol blends E20 had slightly lower pressure than butanol blends BU20. This might be because butanol has higher calorific value, higher density and lower latent heat of vaporization as compared to ethanol leading to a lower cooling effect that shown in table 2. This finding disagrees with the similar previous research by Lie et al. with ethanol influence in the blend yield higher peak pressure compared to the blend with butanol [12].

3.2 Heat Release Rate

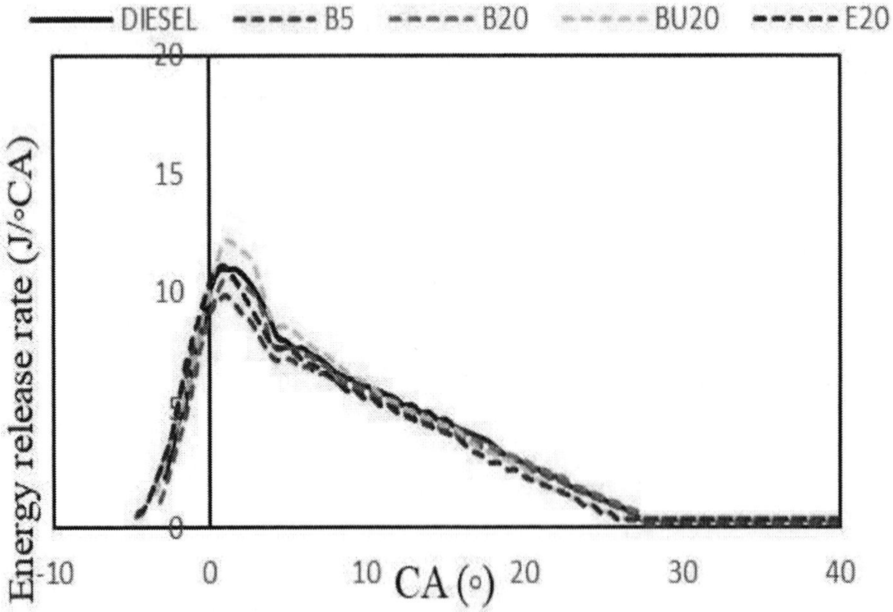

Figure 3. Energy release rate vs. crank angle variation of biodiesel, biodiesel-alcohol blends compared to diesel at idling condition.

Figure 3 illustrates the comparison of energy release rate against CA degree to see how much heat is released during the combustion and the speed of the burning with different fuel composition. The curve of energy release of all fuels maintaining the same position from top dead centre (TDC). The variation is on the peak of each fuels blend. Higher peak is an indicator for higher power thermal output [13]. Hence, higher energy release reading is more desirable. B5 blend was found with the lowest energy release among all. Slightly lower than diesel was B20. BU20 was seen with the highest peak energy release. This is an indicator of higher combustion temperature. E20 blend advanced slightly from diesel. Overall, alcohol in the blend with the lowest heating value come in second before diesel [14].

It can be seen, more energy release by B20 blend than B5 blend but conventional diesel reading still higher than both blends. With lesser diesel content, maintaining 5% biodiesel plus 20% alcohol fix the cause and boosting more energy. Alcohol influence was seen increasing the peak of energy release more than 100% diesel. To compare between the two types of alcohol, in this case, BU20 blend with Butanol beats E20 blend with ethanol. Considering the case similar to combustion pressure where butanol influence compared to ethanol influence raised up the question of why this is happening. By right, remark on oxygen content attribute should have come in handy and help the burning efficiency. But apparently, the oxygen content of ethanol which is higher than butanol did not reflect the energy release reading of BU20. This can be explained by considering the amount of energy release is subjected to the carbon oxidation state in the hydrocarbon. So, based on the ratio of hydrogen/carbon, butanol with greater H/C ratio compared to ethanol explained why more energy release of BU20 more than E20.

3.3 Ignition delay time

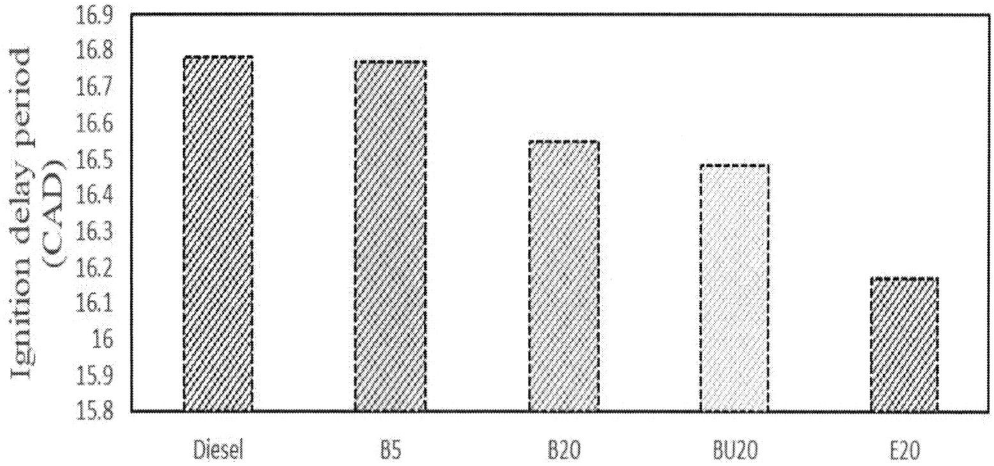

Figure 4. Ignition delay period of biodiesel, biodiesel-alcohol blends compared to diesel at idling condition.

Figure 4 illustrates the comparison of Ignition delay period of the different types of fuels under idling condition. By definition, ignition delay period is cumulated from the moment injection starts till the moment combustion actually begin to initiate [15]. It is widely accepted that volatile compound responds much faster than the rest during combustion and as a result, shorter ignition delay should be expected [16]. The result of the experiment is against the previous study by D.C. Rakopoulos et al. and M. L. Randazzo & J. R. Sodré on the statement with regards to ethanol with lower cetane number could have caused the increase of ignition delay [17] [18]. In another word, alcohol influence the blend with low cetane number would postpone the ignition more while blend under the influence of alcohol with higher cetane number eases the ignition hence shorten the duration. In this case, BU20 with butanol under higher cetane number category as compared to ethanol in the E20 blend. Furthermore, combustion pressure of E20 slightly ahead of BU20. This is an indicator that Ignition delay of E20 slightly shorter than BU20. E20 with the shortest delay period as compared to others. Potentially, shorter range ignition delay is desirable to reduce peak combustion temperature and simultaneously reduce the formation of NOx and total hydrocarbon emissions.

Agreed with Qi et al. alcohol is known as high latent heat of evaporation attribute and low cetane number which should by right increase the ignition delay period. From the increase of the ignition delay, cylinder peak pressure rise is expected [19]. M. Zöldy emphasized alcohol in tri-fuel should shorten ignition delay period prolong longer burning [20]. This assumption disagreed with M.L.Randazzo and J.R.Sodré on due to alcohol addition, ignition delay increased as a result of slow rate evaporation characteristics of alcohol [18]. Su Han Park et al. and quoted Lu et al work on ethanol blended with diesel remark; as the ratio of ethanol increased, combustion duration could be shortened but ignition delay may increase [21]. But in this experiment, with biodiesel as an additional component in the blend, yield surprising outcome on ignition delay period. All three-blended composition with alcohol component revealed shorter ignition delay as compared to blended without alcohol influence. This can be explained considering biodiesel attribute as a surfactant which may unlock alcohol hidden potential [22]. Fundamentally, the shorter the better and the peak pressure expected should be not too high. The ignition delay period is shown below in Table 3.

Table 3. The ignition delay period.

Fuel	Low load	start of injection	Start of combustion	Ignition delay(CA)
Diesel	1200	-17	-0.22	16.78
B5	1200	-17	-0.23	16.77
B20	1200	-17	-0.45	16.55
BU20	1200	-17	-0.49	16.51
E20	1200	-17	-0.78	16.22

3.4 Mass fraction Burn

Figure 5 is the amount of fuel burned during the combustion. The pattern is aligned with energy heat release pattern. The mass fraction burned of BU20 and E20 side by side has been found slightly faster than diesel. Both blends without alcohol influence burned slightly later than diesel. In one glance, the difference is not that significant. But fundamentally, MBF pattern of BU20 and E20 advance slightly is the consequences of shorter ignition delay as seen in Figure 4. In combustion cycle, this is an indicator for good energy conversion [23]. Fundamentally, MFB curve by a crank angle is to define cumulative combustion. It is good to see that there is no negative dip like typical MFB graph. theoretically, Ignition delay and combustion duration came from MBF reading. B5 and B20 influenced by biodiesel interference and the burning fraction slightly slower than conventional diesel. In another word, biodiesel slows down the burning rate while alcohol speed up the burning process.

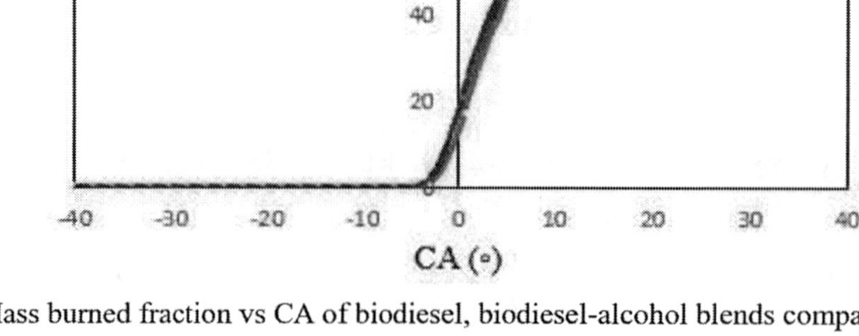

Figure 5. Mass burned fraction vs CA of biodiesel, biodiesel-alcohol blends compared to diesel at idling condition.

4. Conclusions

An experimental investigation was carried out to observe the combustion analysis of biodiesel and the alcohol blending's with diesel and biodiesel at different blending's at idling conditions. The B5 (95vol % diesel+5vol % palm biodiesel), B20 (80vol% diesel + 20 vol% palm biodiesel), BU20 (75vol% diesel + 5vol% palm biodiesel + 20vol% butanol) and, E20 (75 vol% diesel + 5vol% palm biodiesel +

20vol% ethanol) fuels were used as the test fuel at low load with 1200 rpm speeds. The main conclusions are as following below:

- The In-cylinder pressure during idling conditions for biodiesel blends and alcohol blends are a little bit higher than diesel due to lower load condition and lower the residual gas or wall temperature.
- B5 with the lowest energy release among all. Slightly lower than diesel is B20. BU20 is seen with the highest peak energy release. This is an indicator of higher combustion temperature. E20 advance slightly from diesel. Overall, alcohol in the blend with the lowest heating value come in second before diesel.
- All three-blended composition with alcohol component revealed shorter ignition delay as compared to blended without alcohol influence. This can be explained considering biodiesel attribute as a surfactant which may unlock alcohol hidden potential.
- Fundamentally, MFB curve by a crank angle is to define cumulative combustion. Theoretically, Ignition delay and combustion duration came from MBF reading. B5 and B20 influenced by biodiesel interference and the burning fraction slightly slower than conventional diesel. The mass fraction burned of BU20 and E20 side by side has been found slightly faster than diesel.

Acknowledgement

Authors would like to extend their acknowledgement to Universiti Malaysia Pahang for the facility provision and financial support through the project RDU150124 and PGRS160304.

References

1. Mahmudul, H., et al., *Production, characterization and performance of biodiesel as an alternative fuel in diesel engines–A review.* Renewable and Sustainable Energy Reviews, 2017. **72**: p. 497-509.
2. Hagos, F.Y., et al., *Effect of emulsification and blending on the oxygenation and substitution of diesel fuel for compression ignition engine.* Renewable and Sustainable Energy Reviews, 2017. **75**: p. 1281-1294.
3. Silitonga, A., et al., *A review on prospect of Jatropha curcas for biodiesel in Indonesia.* Renewable and Sustainable Energy Reviews, 2011. **15**(8): p. 3733-3756.
4. Rahman, S.A., et al., *Effect of idling on fuel consumption and emissions of a diesel engine fueled by Jatropha biodiesel blends.* Journal of cleaner production, 2014. **69**: p. 208-215.
5. Rahman, S.A., et al., *Assessing idling effects on a compression ignition engine fueled with Jatropha and Palm biodiesel blends.* Renewable Energy, 2014. **68**: p. 644-650.
6. Lambert, D.C., M. Vojtisek-Lom, and P.J. Wilson, *Roadside emissions study: Preliminary results for stationary and on-road testing of diesel trucks in tulare.* 2002.
7. Mahmudul, H., et al. *Impact of oxygenated additives to diesel-biodiesel blends in the context of performance and emissions characteristics of a CI engine.* in *IOP Conference Series: Materials Science and Engineering*. 2016. IOP Publishing.
8. Mahmudul, H., et al. *Experimental investigation of the impact of using alcohol-biodiesel-diesel blending fuel on combustion of single cylinder CI engine.* in *IOP Conference Series: Materials Science and Engineering*. 2016. IOP Publishing.
9. Rahman, S.A., et al., *Production of palm and Calophyllum inophyllum based biodiesel and investigation of blend performance and exhaust emission in an unmodified diesel engine at high idling conditions.* Energy Conversion and Management, 2013. **76**: p. 362-367.
10. IM, Y., et al., *The combustion of n-butanol-diesel fuel blends and its cycle to cycle variability in a modern common-rail diesel engine.* ARPN Journal of Engineering and Applied Sciences, 2016. **11**(4): p. 2297-2301.
11. Hafizil, M., et al., *Influence of palm methyl ester (PME) as an alternative fuel in multicylinder diesel engine.* Journal of Mechanical Engineering and Sciences (JMES), 2012. **3**: p. 331-339.

12. Liu, H., et al., *Comparison of ethanol and butanol as additives in soybean biodiesel using a constant volume combustion chamber.* Energy & fuels, 2011. **25**(4): p. 1837-1846.
13. Hulwan, D.B. and S.V. Joshi, *Performance, emission and combustion characteristic of a multicylinder DI diesel engine running on diesel–ethanol–biodiesel blends of high ethanol content.* Applied Energy, 2011. **88**(12): p. 5042-5055.
14. Armas, O., S. Martínez-Martínez, and C. Mata, *Effect of an ethanol–biodiesel–diesel blend on a common rail injection system.* Fuel Processing Technology, 2011. **92**(11): p. 2145-2153.
15. Udayakumar, R. and C.B. Kasera, *Combustion analysis of a diesel engine operating with performance improvement additives.* Combustion, 2012. **1**(2): p. 11-16.
16. Sahoo, P. and L. Das, *Combustion analysis of Jatropha, Karanja and Polanga based biodiesel as fuel in a diesel engine.* Fuel, 2009. **88**(6): p. 994-999.
17. Rakopoulos, D., et al., *Combustion heat release analysis of ethanol or n-butanol diesel fuel blends in heavy-duty DI diesel engine.* Fuel, 2011. **90**(5): p. 1855-1867.
18. Randazzo, M.L. and J.R. Sodré, *Exhaust emissions from a diesel powered vehicle fuelled by soybean biodiesel blends (B3–B20) with ethanol as an additive (B20E2–B20E5).* Fuel, 2011. **90**(1): p. 98-103.
19. Qi, D., et al., *Performance and combustion characteristics of biodiesel–diesel–methanol blend fuelled engine.* Applied Energy, 2010. **87**(5): p. 1679-1686.
20. Zöldy, M., *Ethanol–biodiesel–diesel blends as a diesel extender option on compression ignition engines.* Transport, 2011. **26**(3): p. 303-309.
21. Park, S.H., S.H. Kim, and C.S. Lee, *Mixing stability and spray behavior characteristics of diesel− ethanol− methyl ester blended fuels in a common-rail diesel injection system.* Energy & Fuels, 2009. **23**(10): p. 5228-5235.
22. Lu, X., et al., *Combustion visualization and emissions of a direct injection compression ignition engine fueled with bio-diesohol.* International journal of automotive technology, 2005. **6**(1): p. 15-21.
23. Mendera, K.Z., A. Spyra, and M. Smereka, *Mass fraction burned analysis.* Journal of KONES Internal Combustion Engines, 2002(3-4): p. p193-201.

IOP Conf. Series: Materials Science and Engineering **318** (2018) 012072 doi:10.1088/1757-899X/318/1/012072

Finite Element Simulation of Aluminium/GFRP Fibre Metal Laminate under Tensile Loading

M N M Merzuki[1], M R M Rejab[1], N K Romli [1*], D Bachtiar[1], J Siregar[1], M F Rani[2], Salwani Mohd Salleh[1]

[1]Structural Materials & Degradation Focus Group, Faculty of Mechanical Engineering, Universiti Malaysia Pahang, 26600 Pekan, Pahang, Malaysia
[2]Faculty of Engineering, DRB-Hicom University of Automotive Malaysia, 26607 Pekan, Pahang, Malaysia

Email : khaleedaromli@gmail.com

Abstract. The response of a fibre metal laminate (FML) model to the tensile loading is predicted through a computational approach. The FML consisted with layers of aluminum alloy and embedded with one layer of composite material, Glass fibre Reinforced Plastic (GFRP). The glass fibre and aluminium alloy 2024-0 was laminated by using thermoset epoxy. A compression moulding technique was used in the process of a FML fabrication. The aluminium has been roughen by a metal sanding method which to improve the bonding between the fibre and metal layer. The main objective of this paper is to determine the failure behaviour of the FML under the tensile loading. The responses on the FML under the tensile loading were numerically performed. The FML was modelled and analysed by using Abaqus/CAE 6.13 version. Based on the experimental and FE data of the tensile, the ultimate tensile stress is 120 MPa where delamination and fibre breakage happened. A numerical model was developed and agreed well with the experimental results. The laminate has an inelastic respond to increase the tensile loads which due to the plasticity of the aluminium layers.

1. Introduction

A fibre metal laminate (FML) or also known as a hybrid laminate contains metal layers and composite materials [1]. The combination of the metal layers and composite fibres produce superior characteristics such as fatigue and fracture. Besides that, the substitutions of both materials are to reduce weight of the aircraft structures and saving in oil consumption [2]. The Glare and Arall are members of fibre metal laminates. The fibre metal laminate started and applied in military application after the Second World War [3]. From the positive feedbacks of the application, the fibre metal laminate began the application widely in aircraft structures such as fuselage and wings [4]. The example of the application is in the Airbus A380 due to the improvement in fatigue, resistance to corrosion and impact and weight reduced nearly 794 kg [5].

An autoclave commonly used in manufacturing of fibre metal laminates or curing process at a high pressure. The curing temperature should be maintained as well [6]. The FMLs need to be processed up to 120°C as to avoid damage on the aluminium alloy 2024-T3. The temperature will allow excess of resin flow out and reduced its viscosity. The pressure will press and consolidate the plies then suppress the voids [7]. The shear thickening fluid is a combination of hard metal oxide particles which suspended in the liquid polymer. The STF can lead the laminates to improve resistance from ballistic impact damage [8].

The debonding process between a sheet metal layer and composite fibre layer, delamination of the composite layers, fracture due to a plasticity behaviour of a metal are commonly happened to a fibre metal laminate structure which because of the impact of a load and depended on the layup configurations [9]. The process of curing by using an autoclave will take long hours of processing and required intensive of labour cost for the whole production [10].

The purpose of this paper to investigate the failure behaviour of the fibre metal laminate modeled by using the Abaqus/CAE version 6.13 under uniaxial tensile stress. In this analysis will be more focused on the method of the analysis.

2. Experimental work

The tensile experiment was conducted by using an Instron 3369 series Universal Testing machine. The FML specimens were prepared with sample. The experiment was conducting at a displacement rate which 5 mm/min. The FML specimens were prepared through a compression moulding technique. The configuration layup of the FML was 2/1. A thermoset epoxy was mixed with hardener in a ratio 2:1 for laminating process. A moderated load was placed on the laminate for a consistent pressure application. The specimen was cured at a room temperature for 24 hours. The aluminium surface was treated with a metal sanding method to rough the metal surface and increase the bonding between the fibre and sheet metal layer. A type of sand paper, 80 grit has been used for the metal sanding.

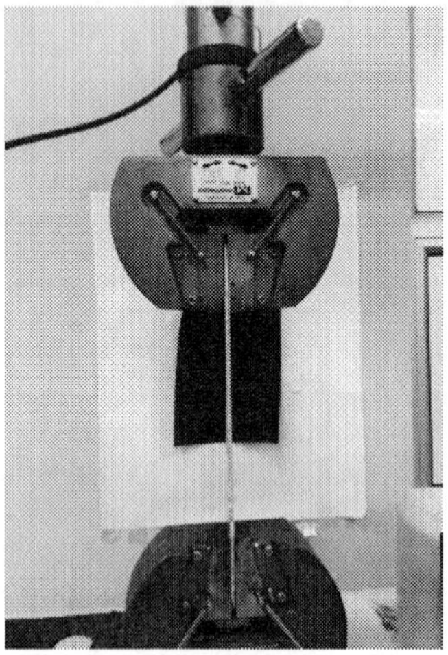

Figure. 1. Setup for the FML in tensile test.

3. Computational modelling

The alternative way to predict experimental result is by using numerical analysis. Study of numerical analysis on fibre metal laminate was become crucial as to know impact behaviour of it. The process of simulation could be done by adding the properties material used such as the Young's modulus, Poisson's ratio, density of the materials and so on. The simulation process required reasonable numbers of structural tests for a validation process. The previous researchers were combined two theories of failure criteria and plasticity as to simulate the degradation behaviour of plate and study the behaviour of nonlinear numerical model. The damage of an impact was controlled by degradation of the stiffness and fracture was because of energy releases which aimed to delamination growth. The simulation of deformation process of 3D FE model involved sensitivity of strain-rate and large changes in transverse compression. The damaged of composite plies cannot be used directly in Abaqus [11].

The Abaqus/Explicit was used to develop a numerical simulation of the FML under the quasi-static loading. The FML consisted with two different of constituent materials; 2024-0 and woven GFRP. The failure of the FML depended on the ductility of the materials.

3.1 Modelling of the 2024-0 by using Isotropic Material

The aluminium alloy 2024-0 was modelled by using an isotropic elasto-plasticity to predict elastic and plastic behavior either as a rate-dependent or rate dependent model and it has a simple form.

3.1.1 Elasticity.

A model of isotropic linear elasticity was generated for the elastic respond for a material in a numerical analysis. The material that performed linear elastic behaviour, the total stress defined through this an equation:

$$\sigma = D^{el} \varepsilon^{el} \tag{1}$$

Where σ the total stress, D^{el} is the fourth order of an elasticity tensor and ε^{el} is the total elastic strain.
From the Equation (1) the stress-strain relationship of isotropic linear elasticity strain

$$\begin{bmatrix}\sigma_{11}\\\sigma_{22}\\\sigma_{33}\\\sigma_{12}\\\sigma_{13}\\\sigma_{23}\end{bmatrix}=\frac{E}{(1+\upsilon)(1-2\upsilon)}\begin{bmatrix}1-\upsilon & \upsilon & \upsilon & 0 & 0 & 0\\ \upsilon & 1-\upsilon & \upsilon & 0 & 0 & 0\\ \upsilon & \upsilon & 1-\upsilon & 0 & 0 & 0\\ 0 & 0 & 0 & \frac{(1-2\upsilon)}{2} & 0 & 0\\ 0 & 0 & 0 & 0 & \frac{(1-2\upsilon)}{2} & 0\\ 0 & 0 & 0 & 0 & 0 & \frac{(1-2\upsilon)}{2}\end{bmatrix}\begin{bmatrix}\varepsilon_{11}\\\varepsilon_{22}\\\varepsilon_{33}\\\varepsilon_{12}\\\varepsilon_{13}\\\varepsilon_{23}\end{bmatrix} \tag{2}$$

The symbol of E is a Young's modulus and υ is a Poisson's ration. These have been defined in the equation above. The inverse relationship defined as an equation below:

$$\begin{bmatrix}\varepsilon_{11}\\\varepsilon_{22}\\\varepsilon_{33}\\\gamma_{12}\\\gamma_{13}\\\gamma_{23}\end{bmatrix}=\begin{bmatrix}\frac{1}{E} & \frac{-\upsilon}{E} & \frac{-\upsilon}{E} & 0 & 0 & 0\\ \frac{-\upsilon}{E} & \frac{1}{E} & \frac{-\upsilon}{E} & 0 & 0 & 0\\ \frac{-\upsilon}{E} & \frac{-\upsilon}{E} & \frac{1}{E} & 0 & 0 & 0\\ 0 & 0 & 0 & \frac{1}{G} & 0 & 0\\ 0 & 0 & 0 & 0 & \frac{1}{G} & 0\\ 0 & 0 & 0 & 0 & 0 & \frac{1}{G}\end{bmatrix}\begin{bmatrix}\sigma_{11}\\\sigma_{22}\\\sigma_{33}\\\sigma_{12}\\\sigma_{13}\\\sigma_{23}\end{bmatrix} \tag{3}$$

G is a shear modulus; $G=\frac{E}{2(1+\upsilon)}$

The elastic material in the properties of aluminium 2024-0 was used in the numerical modelling. The properties were taken from experimental results. The value of Young's modulus, $E=70.6$ GPa and the Poisson's ratio is $\upsilon=0.3$.

3.1.2 Yielding.

An isotropic yielding is applied to a von Mises yield surface. The surface is assumed that yielding of a metal is independent where the observation was confirmed by an experimental works. The metal was applied under a positive pressure stress. When the metal applied in that kind of situation which high pressure stress, it may be inaccurate because voids can nucleate and grow in the metal.

3.1 3 Plasticity.

Beyond the yield point is a permanent deformation or called as plastic behaviour of the aluminium. The plastic behaviour is defined in an isotropic hardening. To describe the isotropic hardening, the yield stress, σ^0 which given in a plastic strain tabular function. The interpolation of the yield stress at any plastic strain rate can be done from the table of data. The data will remain constant when it reached to the last value that have been given in the table. Decomposition of the total increment of strain defined as below:

$$d\varepsilon = d\varepsilon^{el} + de^{pl} \tag{4}$$

For rate-dependent of material the relationship of the equivalent plastic strain rate, $\dot{\bar{\varepsilon}}^{pl}$ followed by the uniaxial flow rate definition as:

$$\dot{\bar{\varepsilon}}^{pl} = \varphi\left(q, \bar{\varepsilon}^{pl}, \theta_r\right) \tag{5}$$

φ known as function, q is the von Mises equivalent stress, $\bar{\varepsilon}^{pl}$ is an equivalent of the plastic strain and θ_r is the temperature.

3.2 Failure Criteria

The development of damage and failure of ductile material are used in Abaqus/Explicit to model the damage behaviour of the aluminium. *SHEAR FAILURE and *TENSILE FAILURE or a combination of both are two material failure models will offer in the Abaqus to account the damage and failure in a ductile metal. To calculate the failure, shear and tensile failure models can use an equivalent of the plastic strain and hydrostatic cut-off stress respectively. If the failed meshes removed from the meshes it may for both failure models.

3.2.1 *Damage for Ductile Metals.*

There are two mechanisms for the ductile metals will be used in this study; ductile damage and shear damage models. The ductile fracture or ductile damage are initial criterion for predicting the onset damage which due to nucleation, growth and the voids. This model will active when the following condition is satisfied:

$$\omega_D = \int \frac{d\bar{\varepsilon}^{pl}}{\bar{\varepsilon}_D^{pl}\left(\eta, \dot{\bar{\varepsilon}}^{pl}\right)} = 1 \tag{6}$$

ω_D is a state variable that increase monotonically with plastic deformation. The model is assumed that the equivalent plastic strain at the onset of damage, $\bar{\varepsilon}_D{}^{pl}$ is a function of triaxiality, $\eta = -p/q$, and the equivalent of plastic strain, $\dot{\bar{\varepsilon}}^{pl}$. Note that p is the pressure stress and q is the Mises equivalent stress. The shear damage model is a fracture which due to shear band localisation in the ductile metals. The model will active when the following condition is satisfied:

$$\omega_s = \int \frac{d\bar{\varepsilon}^{pl}}{\bar{\varepsilon}_s^{pl}\left(\theta_s, \dot{\bar{\varepsilon}}^{pl}\right)} = 1 \tag{7}$$

ω_s is a state variable that increases monotonically with plastic deformation and is relative to the incremental change in equivalent plastic strain, $\bar{\varepsilon}^{pl}$. The model is assumed that the equivalent plastic strain at the onset of damage, $\bar{\varepsilon}_s{}^{pl}$ is a function of the shear stress ratio, $\theta_s = (q + k_s p)/\tau_{\max}$ and the equivalent plastic strain rate, $\dot{\bar{\varepsilon}}^{pl}$. Note that $\tau_{\max}$ is the maximum shear stress k_s is a material parameter ($k_s = 0.3$ for the aluminium alloy) [12].

3.3 *Modelling of composite by Hashin damage criterion*

The Hashin damage criterion is based on works of Hashin and Rotem and Hashin [13, 14]. This criterion is unlike other criteria like Tsai-Hill and Tsai-Wu [15]. Those criteria purposed an equation to predict damage initiation. The Hashin damage presents four failure modes with four corresponding indexes (a) breakage of fibre from the tension$(F_f{}^t)$, (b) buckle of fibre from the compression$(F_f{}^c)$, (c) crack of matrix from the tension $(F_m{}^t)$ and (d) crush of matrix from the compression $(F_m{}^c)$. There are four equation that applied to those failure modes respectively:

$$\left(F_f{}^t\right) = \left(\frac{\sigma_{11}}{S_{t1}}\right)^2 + \alpha\left(\frac{\sigma_{12}}{S_{12}}\right)^2 \leq 1.0 and \sigma_{11} \geq 0 \tag{8}$$

$$\left(F_f{}^c\right) = \left(\frac{\sigma_{11}}{S_{c1}}\right)^2 \leq 1.0 and \sigma_{11} < 0 \tag{9}$$

$$\left(F_m{}^t\right) = \left(\frac{\sigma_{22}}{2S_{s23}}\right)^2 + \alpha\left(\frac{\sigma_{c2}}{2S_{s23}}\right)^2 \leq 1.0 \text{ and } \sigma_{22} \geq 0 \tag{10}$$

$$\left(F_m{}^c\right) = \left(\frac{\sigma_{22}}{2S_{s23}}\right)^2 + \left[\left(\frac{S_{c2}}{2S_{s23}}\right)^2 - 1\right]\frac{\sigma_{22}}{S_{c2}} + \left(\frac{\sigma_{12}}{S_{12}}\right)^2 \leq 1.0 \text{ and } \sigma_{22} < 0 \tag{11}$$

σ_{11}, σ_{22} and σ_{12} are applied stresses while α is a coefficient of the shear stress σ_{12} which contributed in fibre breakage by a tension criterion. In the Abaqus, the criterion of Hashin could be implemented in a conjunction with a damage evolution law. The law based on a specification of four fracture energy values G_f. The values corresponded to the material degradation in each mode. The values constituent in an implementation of the Hashin-based and analysed with the damage evolution in the Abaqus. The damage variable for fibres d_f, matrix d_m and shear d_s were assessed by this expression [16]:

$$d = \frac{\delta^{f}{}_{eq}\left(\delta_{eq} - \delta^{0}{}_{eq}\right)}{\delta_{eq}\left(\delta^{f}{}_{eq} - \delta^{0}{}_{eq}\right)} \tag{12}$$

$\delta^{f}{}_{eq}$ is an equivalent of a displacement in a FE when the stress reached to zero in a stress-displacement while $\delta^{0}{}_{eq}$ is an equivalent of a displacement when the damage initiated. δ_{eq} is an equivalent of a displacement in FE for the given applied strain [16].

4. Results and Discussion

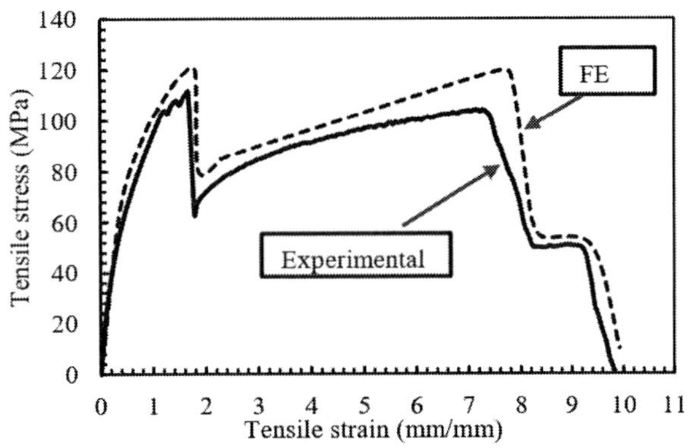

Figure 2. Graph tensile stress against tensile strain for aluminium/glass fibre laminate.

Based on the figure 2 the aluminium/glass fibre was subjected to an uniaxial tensile loading. There were a few transverse cracks as well as the delamination of the fibre and sheet metal layer. When the FML was applied and pulled by a higher loading, it was resulted in a composite structure which glass fibre degradation. This degradation was happened at the first stage. At this stage, the degradation was started by a matrix cracking which visible in the structure and continued with a delamination. Normally the matrix cracking happened at the lowest load applied. As the load increased, the delamination in the middle area was occurred as well as the delamination between metal and fibre layer. A transverse crack of the polymer composite occurred at the second stage. The matrix cracks was contained with two constituents, epoxy (matrix) and fibre. The epoxy characterised as brittle and has low resistance to a crack propagation. There are two types of matrix cracking; cracks which caused of bending and shearing. These cracks happened during high transverse shearing stress.

The 2/1 FML model based on aluminium and glass fibre was predicted through a finite element method; Abaqus/CAE. The numerical analysis was done as to evaluate the failure behaviour of the FML model under the uniaxial tensile stress. The specimen was modelled as a single layer solid which consisted with the aluminium alloy 2024-0 and glass fibre. The finite element analysis focused on a tensile behaviour. The simulation just focused on the model deformation, not the debonding and delamination. The element C3D8R was applied to the FML model. The C3D8R consisted of 8-node linear brick, reduce integration and hourglass control. The theoretical of the modelling, Young's modulus was used to determine the deformation of the FML. The mechanical properties of the FML could be predicted by the volume fraction which based on basic of the laminate theory. The model was analysed by applying Displacement/Rotation at x-axis and the other side will be fixed by a boundary condition type which called as a Encaster.

The failure mechanism of the model was recorded. The FE model became successful when at the perforation stage of the impact event. The failure area of the FML increased when the deformation have been applied increasingly. The FML exhibited and appeared with little plastic deformation at the impact location. The plastic deformation was in aluminum and there was a fracture on the composite ply. The results based on the stiffness of the material. The number of elements for the FML were with 750 elements which consisted of 250 element for each ply of 2/1 FML.

The experimental compared with finite element result which in order to verify the FE modelling approach used, FE stress and strain results. The results of both were validated and showed in the figure 3. The stress and strain relation were obtained from both, numerical analysis and experimental up to the final fracture. The failures of the FML in the numerical analysis when isotopic and Hashin damage criteria were applied to the aluminium and glass fibre, respectively. In meanwhile, the result of the FML in experimental slightly differ and lower than FE because the manufacturing process of the FML also effect to the structure.

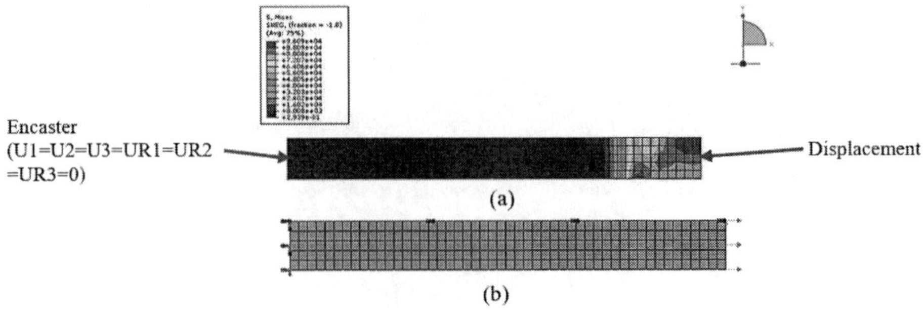

Figure 3. (a) FML analysis (b) Meshing of FML

Conclusion

The work was presented for uniaxial stress behavior and damage characterisation of a fibre metal laminate with the glass fibre. The influences of the tensile loading on the fibre metal laminate were found by the valuable information of the damaged composite structures. The damage mechanism was connected to the internal degradation of the composite material which it was a plastic behaviour of the fibre metal laminate. The failure behavior was started by a matrix cracking and followed by a delamination of the glass fibre. The delamination was a critical damage mode where found at the interface of the metal-composite in the fibre metal laminate. The plastic behavior normally happened in the metal layers. The damage of the material based on the type of the material used. The prediction of failure behaviour on the 2/1 FML was done by using the finite element analysis and were presented in details. The theoretical and calculation were considered in the analysis but adhesive strength was not involved in the analysis. The FMLs is suitable to be evaluated for the impact resistance which for automotive industry.

A numerical model was developed and agreed well with the experimental results. A glass/epoxy composite layer in the aluminium/glass laminate exhibited in a linear elasticity under a tensile loading. This show that, the laminate has an inelastic respond to increase the tensile loads which due to the plasticity of the aluminium layers.

Acknowledgement

The authors are grateful of the Ministry of Higher Education Malaysia and Universiti Malaysia Pahang (www.ump.edu.my) for funding this research RDU 1603108.

References

1. Vlot, A. and M. Krull, *Impact damage resistance of various fibre metal laminates.* Le Journal de Physique IV, 1997. 7(C3): p. C3-1045-C3-1050.
2. Zhang, Z., et al., *An experimental investigation into pin loading effects on fatigue crack growth in Fibre Metal Laminates.* Procedia Structural Integrity, 2016. **2**: p. 3361-3368.
3. Edson Cocchieri Botelho, R.A.S., Luiz Cláudio Pardini, Mirabel Cerqueira Rezende, *A Review on the Development and Properties of Continuous Fiber/epoxy/aluminum Hybrid Composites for Aircraft Structures. Materials Research*, 2006. **9**(3): p. 247-256.
4. Zenkert, D., *An introduction to sandwich structures.* Student edition, 1997.
5. Vlot, A. and J.W. Gunnink, *Fibre metal laminates: an introduction*. 2011: Springer Science & Business Media.
6. Bieniaś, J., et al., *Low-energy impact behaviour and damage characterization of carbon fibre reinforced polymer and aluminium hybrid laminates.* Archives of Civil and Mechanical Engineering, 2015. **15**(4): p. 925-932.
7. Botelho, E., et al., *Processing of the obtaining of structural reinforcement impregnated with polyamide 6/6 obtained by interfacial polymerization, using a reaction injection pultrusion system.* Brazilian Patent, PI, 2001: p. 6561-7.
8. Lee, Y.S., E.D. Wetzel, and N.J. Wagner, *The ballistic impact characteristics of Kevlar® woven fabrics impregnated with a colloidal shear thickening fluid.* Journal of materials science, 2003. **38**(13): p. 2825-2833.
9. Katnam, K., L. Da Silva, and T. Young, *Bonded repair of composite aircraft structures: A review of scientific challenges and opportunities.* Progress in Aerospace Sciences, 2013. **61**: p. 26-42.
10. Hu, Y., et al., *Application of laser peen forming to bend fibre metal laminates by high dynamic loading.* Journal of Materials Processing Technology, 2015. **226**: p. 32-39.
11. Morinière, F.D., R.C. Alderliesten, and R. Benedictus, *Modelling of impact damage and dynamics in fibre-metal laminates – A review.* International Journal of Impact Engineering, 2014. **67**: p. 27-38.
12. Liu, G., et al., *Abaqus implementation of monolithic and staggered schemes for quasi-static and dynamic fracture phase-field model.* Computational Materials Science, 2016. **121**: p. 35-47.

13. Hashin, Z., *Failure criteria for unidirectional fiber composites.* Journal of applied mechanics, 1980. **47**(2): p. 329-334.
14. Hashin, Z. and A. Rotem, *A fatigue failure criterion for fiber reinforced materials.* Journal of composite materials, 1973. **7**(4): p. 448-464.
15. Tsai, S.W. and E.M. Wu, *A general theory of strength for anisotropic materials.* Journal of composite materials, 1971. **5**(1): p. 58-80.
16. Duarte, A.P.C., A. Díaz Sáez, and N. Silvestre, *Comparative study between XFEM and Hashin damage criterion applied to failure of composites.* Thin-Walled Structures, 2017. **115**: p. 277-288.

Design and Development of Intelligent Electrodes for Future Digital Health Monitoring: A Review

A M Khairuddin[1], K N F Ku Azir[2], P Eh Kan[3]
Embedded, Networks and Advanced Computing Research Cluster (ENAC)
School of Computer and Communication Engineering, Universiti Malaysia Perlis, Kampus Pauh, 02600 Ulu Pauh, Perlis, Malaysia

E-maiil: khaiadam93@gmail.com

Abstract. Electrodes are sensors used in electrocardiography (ECG) monitoring system to diagnose heart diseases. Over the years, diverse types of electrodes have been designed and developed to improve ECG monitoring system. However, more recently, with the technological advances and capabilities from the Internet of Things (IoT), cloud computing and data analytics in personalized healthcare, researchers are attempting to design and develop more effective as well as flexible ECG devices by using intelligent electrodes. This paper reviews previous works on electrodes used in electrocardiography (ECG) monitoring devices to identify the key ftures for designing and developing intelligent electrodes in digital health monitoring devices.

1. Introduction

Traditionally, patients with cardiovascular diseases are required to monitor their heart conditions on a continuously basis. The diagnosis and measurement of the patients' vital signs need to be performed in controlled environment such as at the medical centers or hospitals. This clinic-centric approach requires patients to regularly visit the medical centers and hospitals. This traditional approach however can be very costly, tiring and time consuming for the patients. More recently, the Internet of Things (IoT) and supporting technologies such as cloud computing and big data analytics are helping to shift the clinic-centric approach towards the person-centric model or personalized healthcare [1]–[3].

The adoption of IoT, cloud computing and big data analytics in the healthcare industry is not only changing its landscape but also enables a new way of organizing the delivering of health care and health related services in the industry. For instance, with IoT, it is now possible to provide high quality personalized health to all individuals anytime, anywhere and when needed. More specifically, the IoT supported personalized healthcare allows health care and health related services to be delivered to an individual based on his or her unique biological, behavioral, cultural, and social characteristics [4]–[6].

As one of the major innovations in digital technology, the IoT is also increasingly changing the way in which the medical centers and hospitals in the health care sector operate. Since its introduction and adoption in the health care sector, this digital technology has been able to provide various medical and health care applications to the medical centers and hospitals. Among the essential medical applications provided by IoT include; remote health monitoring and delivery, chronic diseases and post-surgery care, fitness programs, elderly care, compliance with treatment as well as medication at home [7].

In addition, the IoT as interconnected technology enables medical devices such as ECG, sensors, diagnostic and imaging devices to be connected to each other as well as between doctors and their patients. These networks of IoT-based devices have helped to not only improve the detection, prevention and treatment of diseases but also reduce medical costs, increase the quality of life and enhance the experience of the patients that use these medical devices [8]–[10].

The major innovations in digital technologies have also made it possible to design and improve the capabilities of existing medical devices such as ECG devices and their essential components. More specifically, the improvement in the contemporary ECG devices has mainly resulted from the adoption of the advances and capabilities found in the IoT and supporting technologies such as cloud computing and big data analytics. However, the literature reveals not only little research but also limited information on the development of intelligent electrodes used in contemporary ECG devices. This paper reviews previous works on ECG devices to identify the key features that useful for designing and developing intelligent electrodes for digital health monitoring.

2. Limitations of Electrodes in Conventional ECG Devices

Electrocardiography (ECG) monitoring system devices are used to obtain information about the electrical activity of the human heart as well as to detect cardiovascular diseases. The electrical activity is recorded by placing the electrodes or sensors in the ECG device on the body surface of the patient.

Regarding the electrodes, the literature indicates that over the years, diverse types of electrodes have been developed and used in electrocardiography monitoring system devices. Each different type of electrode has its own advantages and disadvantages. For instance, according to [11], the traditional wet silver chloride electrode used for accumulating ECG signal from the human body requires not only certain amount of preparations but also needs the expertise of an expert to attach the electrode with wet gel [12]. The wet gel has disadvantages such as irritation, allergic reactions, inflammation, bacteria growth and undergoes dehydration that can reduce the recorded ECG signal quality [13]–[19]. Moreover, the wet electrodes cannot be used for longer period and they must also be replaced on a regular basis.

Besides the wet electrode, the dry electrode was also introduced to collect ECG signal. Like the wet electrode, the dry electrode also has disadvantages such as it often becomes stiff and it can also cause skin irritation to the patients as well. Apart from the wet and dry electrodes, the textile electrode was proposed. However, the textile electrode has a major disadvantage of requiring high pressure for good skin and electrode contact. The other disadvantage is that the textile electrode is most often embedded inside chest strap or wristband and as such it can cause not only discomfort but also constraint among the wearers [20].

In addition to the disadvantages of the conventional electrodes, the study by [21] indicated that there are also other problems linked to portable ECG monitoring devices. Among these problems included; too many cables are required to connect the electrodes with the portable device, overall cost of the ECG device has increased, patients are found to be uncomfortable during long-term monitoring as well as difficulty in acquiring high quality ECG signal because the signal is sensitive and can be affected by noises.

The more recent study by [22] reviewed past research on conventional ECG devices and also identified several issues and problems associated to these devices. Apart from recognizing the various disadvantages of traditional electrode in the conventional ECG devices, the study also verified the following issues and problems related to these devices; the present of artifacts in ECG signals, the adoption of different performance measurement criteria, the complexity of signal analysis algorithm, the impact of electrical entities on the ECG wave morphology and the difficulties in the analysis of the P-wave and the T-wave. Figure 1 and Figure 2 illustrate the conventional electrode.

Figure 1. The wet electrode for ECG by Philips [23].

Figure 2. The wet electrode for ECG by Ambu [24].

3. Intelligent Electrode and its Key Features

Given the various disadvantages, problems and issues related to the conventional electrodes, more recently, researchers have attempted to introduce more effective wearable ECG monitoring system devices that use intelligent electrode and active cable. The intelligent electrode was developed by using the advances and capabilities found in digital technologies such as IoT, cloud computing and big data analytics.

Regarding the intelligent electrode (also known as smart or intelligent sensor), the literature provides several definitions. For instance, according to [25], there are three different definitions of smart and intelligent sensor. The first definition viewed the smart sensor as the combination of a sensor, an analog interface circuit, an analog to digital converter (ADC) and a bus interface in one casing. The intelligent sensor in the second definition basically considered the sensor as having at least one or more intelligent functions. The third definition however combined the terms smart and intelligent into one single entity. In the third definition, the sensing, interfacing, signal processing, and intelligence functions of the smart sensor are placed in a single chip. The sensor is considered intelligent due to the presence of microcontroller or microprocessor although this alone may not be sufficient. The integrated smart sensor should also contain wireless communication and power management. Examples of the intelligence functions include; self-testing, self-identification, self-validation, self-checking, self-diagnosis, self-calibration, self-compensation and self-adaptation. More specifically, the previous study by [26] stated that intelligent sensor has additional functionality provided by the integration of microprocessors, microcontrollers or application specific integrated circuits (ASICS) with the sensing element itself.

Intelligent electrode can be created by embedding a system-on-chip (SoC) in each electrode to acquire high quality bio-signal [1], [21]. This is possible because the tasks of amplification, filtering, quantization, buffering and transmission of bio-signals are carried out directly on each electrode. The intelligent electrodes can communicate with each other by a serial communication link. This link which is known as the active cable connects all the intelligent electrodes together. The noise interference during ECG signal acquisition can be reduced because the analog and digital conversions happened on the electrodes rather than inside the portable device. The bio-signals are converted into the digital format and grouped in a single intelligent electrode. Then, the grouped data are transmitted wirelessly to the cloud services by using mobile devices for processing and analysis.

The research by [27] proposed a bio-signal acquisition system for both EMG and ECG that comprised intelligent electrode and data acquisition host. The system is small, light and low-cost. It specifically uses the ZigBee based communication as well as perform on-line parameter adjustment. The intelligent electrode consisted of two components that include the ZigBee module and the programmable system on chip (PSoC) chip. It also introduced additional features such as In System Serial Programming (ISSP), I2C connectivity, battery charging, and indicator LEDs. However, this system has limitations such as the software used is not user intuitive and that the prototype which is implemented on the printed circuit board (PCB) has no wearable value.

In another study, [28] developed the intelligent electrode that included low power reconfigurable application-specific integrated circuit (ASIC) and micro spike array to acquire ECG signal. The intelligent electrode developed in the study can be configured as the main electrode or sensing electrode. The main electrode gives reference signal for analog to digital conversion on the sensing electrode. It controlled and acquired ECG signal from each sensing electrodes available in the system. It also provided interface for the Bluetooth module. For the sensing electrode, it performed ECG signal conditioning and signal digitization. The active cable involved five flexible metal strips that have high conductivity and able to minimize the number of connecting cables.

The other study by [29] presented a wearable healthcare monitoring system that consisted of self-organized intelligent electrode which can be used for multiple measurements in various devices such as EEG, ECG and EMG. The intelligent electrode used in this study involved the analog front-end (AFE), successive approximation analog-to-digital converter (SARADC), active cable and digital controller. The digital controller enables the intelligent electrode to be configured to main electrode or sensing electrode.

The study by [30] specifically attempted to design intelligent electrode for personal ECG monitoring device. In this study, the intelligent electrode included configurable ASIC and micro spike array. The intelligent electrode in the study was built up with both field programmable analog array (FPAA) as well as field programmable gate array (FPGA) chips.

Importantly, a review of the previous research on the wearable ECG devices with intelligent electrode suggests several important benefits [31]. Firstly, the wearable device can sense a variety of bio-signals such as electrocardiogram (ECG), electromyogram (EMG), electrooculogram (EOG) and electroencephalogram (EEG) through the intelligent electrode. Secondly, user comfort can be improvised because the numbers of connecting cables are reduced by connecting all the electrodes with a single active cable. Thirdly, the device has serial communication protocol to set up body area network (BAN) with the group of intelligent electrodes. Finally, the intelligent electrode can provide user comfort because it is thin and flexible resulted from inkjet printing and hybrid integration technologies.

4. Design and Development of Intelligent Electrode

The design and development of intelligent electrode for future digital heath monitoring devices are vital for delivering quality health care and health related services. However, the development of these intelligent devices will not only take time and concentration but also requires the adoption of the capabilities from digital technologies such as IoT, cloud computing and big data analytics. When joined together, these technologies can integrate both the virtual and physical aspects of health monitoring and delivery. Furthermore, they are able to reduce the healthcare infrastructure cost, provide more scalability for storing vast amount of medical data as well as accessible to patients 24/7 [10].

According to [31], the ideal wearable sensing devices such as ECG devices should be ultra-thin and light, flexible, ultra-low power consumption, easy to apply and has multi sensing functions. These devices should also be energy efficient and be able to detect diverse bio-signals wirelessly. More significantly, they need to have an intelligent sensor that can determine accurately as well as independently the value of physical, biological and chemical measurements.

The research by [26], [32] proposed the development of an intelligent sensor by using a system that consists of the following three major components; hardware, software and smart sensor micro system. The following section explains briefly the three major components of the intelligent electrode and their relationships.

4.1. Hardware

The first important component of the system involves the hardware. According to the study by [26], the general hardware structure of the intelligent sensor has seven parts: (1) the sensor element that detects movement, language, light, heat, moisture, sound and transform any of them into digital data that can be mined for insights; (2) the amplifier that receives the input and helps to boost low signal levels; (3) the analogue processing has anti-aliasing filters for the conversion stage; (4) the data conversion links to the analogue-to-digital converter (ADC); (5) the digital processing software that processes sensor compensation or pattern recognition methods; (6) the data communication enables the passing, receiving data and controlling signals to the sensor bus; and (7) the control processor acts as the brain of the system and connects to the other elements. Figure 3 presents the general hardware structure of an intelligent sensor.

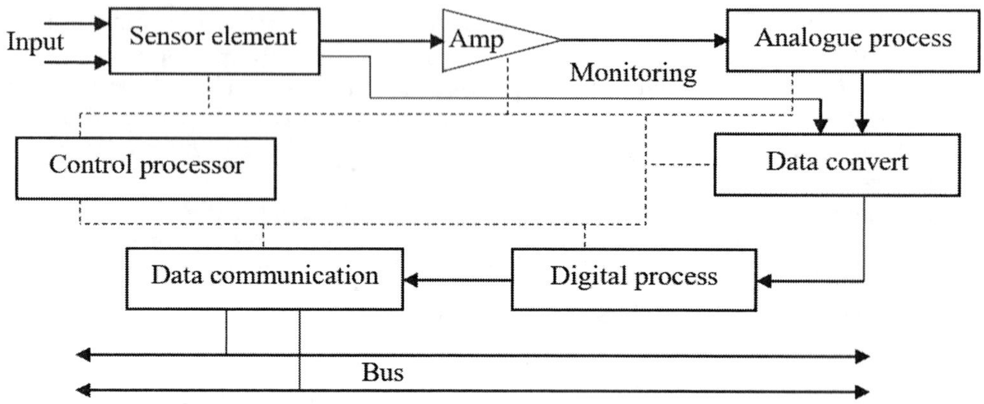

Figure 3. The general hardware structure of an intelligent sensor [26].

4.2. Software

The software is the second important component of the intelligent electrode system. The software applications enable the sensor to perform myriads tasks individually and collectively. For instance, the research by [26] indicated that the intelligent sensor should be able to evaluate the sensor signal originating from the sensing element by integrating the software with the hardware. The software applications should perform the task of evaluation that includes the ability to identify fault conditions, undertake error detection, and perform self-test or self-calibration.

4.3. Smart Sensor Microsystem

When combined, the hardware and software create a micro system for the intelligent electrode. According to [32] the smart sensor microsystem for the IoT can be realized by integrating the four essential functions which are; sensing and actuation, processing and control, power sourcing and management as well as communication. It is also important to identify and understand the application-dependent factors for each of the functions. Size, lifetime and physical interface are the application-dependent for the microsystem. For the sensing and actuation, the application-dependent are the physical parameters to sense, minimum detection limit, dynamic range, bandwidth, and sensing duty cycles. The communication function involves application-dependent such as the communication medium, distance, symmetry, protocol, data rate, and communication duty cycle. Figure 4 shows the structure of the smart sensor microsystem.

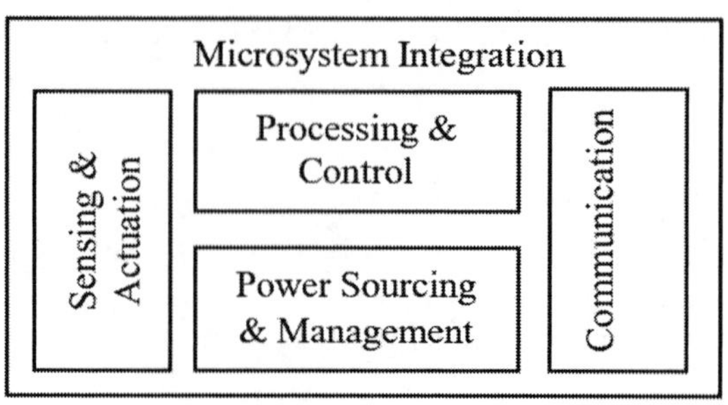

Figure 4. Smart sensor microsystem [32].

5. Conclusion

This paper reviews previous research on ECG devices and highlights how digital technologies such as IoT, cloud computing and big data analytics can help in the design and development of intelligent electrode for future digital health monitoring devices. Based on the review, the paper identified not only limitations of conventional electrodes but also emphasized on the benefits of using intelligent electrodes. In addition, the paper evaluates past studies on intelligent electrodes to determine the various key aspects of intelligent electrodes as well as to propose the system for designing and developing intelligent electrode that can be used in future digital monitoring health monitoring devices.

References

[1] L. Xie, L. Zheng, and G. Yang, "Hybrid Integration Technology for Wearable Sensor Systems," *Internet Things Adv. Appl. Healthc.*, pp. 98–137, 2017.

[2] A. Paulin, "Data Traffic Forecast in Health 4.0," *Heal. 4.0 How Virtualization Big Data are Revolutionizing Healthc.*, pp. 39–60, 2017.

[3] G. Postolache, H. Carvalho, A. Catarino, and O. A. Postolache, "Smart Clothes for Rehabilitation Context: Technical and Technological Issues," *Sensors Everyday Life*, pp. 185–219, 2017.

[4] P. Raj and A. C. Raman, *The Internet of Things: Enabling Technologies, Platforms, and Use Cases*. Taylor & Francis Group, 2017.

[5] C. Bhatt, N. Dey, and A. S. Ashour, "Internet of Things Driven Connected Healthcare," *Internet Things Big Data Technol. Next Gener. Healthc.*, pp. 3–12, 2017.

[6] O. Sangpetch and A. S. Faculty, "Security Context Framework for Distributed Healthcare IoT Platform," *Internet Things Big Data Technol. Next Gener. Healthc.*, pp. 71–76, 2016.

[7] D. Domingos, A. Respício, and R. Martinho, "Reliability of IoT- Aware BPMN Healthcare Processes," *Internet Things Adv. Appl. Healthc.*, pp. 214–248, 2017.

[8] J. Haouel;, H. Ghorbel, and H. Bargaoui, "Towards an IoT Architecture for Persons with Disabilities and Applications," *Internet Things Technol. Healthc.*, pp. 159–161, 2016.

[9] D. T. Meridou, Maria-Eleftheria Ch. Papadopoulou, A. P. Kapsalis, P. Kasnesis, A. I. D. C. Z. P. I. S. Venieris, and D. I. Kaklamani, "Improving Quality of Life with the Internet of Everything," *Beyond the Internet of Things: Everything Interconnected*, pp. 377–408, 2017.

[10] D. Malvey and D. J. Slovensky, *mHealth: Transforming Healthcare*. Springer, 2014.

[11] B. Taji, S. Shirmohammadi, V. Groza, and M. Bolic, "An ECG monitoring system using conductive fabric," *Med. Meas. Appl. Proc. (MeMeA), 2013 IEEE Int. Symp.*, pp. 309–314, 2013.

[12] L. Bor-Shyh, W. Chou, W. Hsing-Yu, H. Yan-Jun, and P. Jeng-Shyang, "Development of Novel Non-Contact Electrodes for Mobile Electrocardiogram Monitoring System," *IEEE J. Transl. Eng. Heal. Med.*, vol. **1**, no. October, pp. 1–8, 2013.

[13] D. N. Mathias, S. Il Kim, J. S. Park, and Y. H. Joung, "Real time ECG monitoring through a wearable smart T-shirt," *Trans. Electr. Electron. Mater.*, vol. **16**, no. 1, pp. 16–19, 2015.

[14] M. Abu-saude and B. I. Morshed, "Polypyrrole (PPy) Conductive Polymer Coating of Dry Patterned Vertical CNT (pvCNT) Electrode to Improve Mechanical Stability," *2016 IEEE Top. Conf. Biomed. Wirel. Technol. Networks, Sens. Syst.*, pp. 84–87, 2016.

[15] C. H. Luo *et al.*, "An ECG acquisition system prototype design with flexible PDMS dry electrodes and variable transform length DCT-IV based compression algorithm," *IEEE Sens. J.*, vol. **16**, no. 23, pp. 8244–8254, 2016.

[16] T. I. Oh *et al.*, "Nanofiber web textile dry electrodes for long-term biopotential recording," *IEEE Trans. Biomed. Circuits Syst.*, vol. **7**, no. 2, pp. 204–211, 2013.

[17] M. A. Yokus and J. S. Jur, "Fabric-based wearable dry electrodes for body surface biopotential recording," *IEEE Trans. Biomed. Eng.*, vol. **63**, no. 2, pp. 423–430, 2016.

[18] A. A. Chlaihawi, B. B. Narakathu, A. Eshkeiti, S. Emamian, S. G. R. Avuthu, and M. Z.

Atashbar, "Screen printed MWCNT/PDMS based dry electrode sensor for electrocardiogram (ECG) measurements," *IEEE Int. Conf. Electro Inf. Technol.*, vol. 2015–June, pp. 526–529, 2015.

[19] D. Pani *et al.*, "Evaluation of Novel Textile Electrodes for ECG Signals Monitoring Based on PEDOT : PSS-Treated Woven Fabrics," *Embc 2015*, pp. 3197–3200, 2015.

[20] K. Takahashi and K. Suzuki, "An ECG monitoring system through flexible clothes with elastic material," *2015 17th Int. Conf. E-Health Networking, Appl. Serv. Heal. 2015*, pp. 305–310, 2016.

[21] G. Yang, J. Chen, Y. Cao, H. Tenhunen, and L. R. Zheng, "A novel wearable ECG monitoring system based on Active-Cable and intelligent electrodes," *2008 10th IEEE Intl. Conf. e-Health Networking, Appl. Serv. Heal. 2008*, pp. 156–159, 2008.

[22] A. M. Khairuddin, K. N. F. K. Azir, and P. L. E. Kan, "Limitations and Future of Electrocardiography Devices: A Review and the Perspective from the Internet of Things," *5th Int. Conf. Res. Innov. Inf. Syst.*, 2017.

[23] B. T. Medical, "Philips Medical Systems HSG," 2015. [Online]. Available: https://www.boundtree.com/hi-tack-foam-ecg-electrodes-m2202a-pharm-11736-119.aspx. [Accessed: 19-Jul-2017].

[24] B. T. Medical, "Ambu WhiteSensor 4500M-H," 2015. [Online]. Available: https://www.boundtree.com/2741-45038-whitesensor-pharm-38470-119.aspx?search=2741-45038. [Accessed: 19-Jul-2017].

[25] S. Yurish, "Sensors: Smart vs. Intelligent," *Sensors & transducers*, vol. **114**, no. 3, pp. I–VI, 2010.

[26] N. White, "Intelligent Sensors: Systems or Components?," *Integr. Vlsi J.*, vol. 3, pp. 471–474, 2005.

[27] H. Kobayashi, "Intelligent wireless EMG / ECG electrode employing ZigBee technology," *SICE Annu. Conf. 2011*, pp. 2856–2861, 2011.

[28] G. Yang, J. Chen, F. Jonsson, and H. Tenhunen, "A 1.0 V 78 μW Reconfigurable ASIC Embedded in an Intelligent Electrode for Continuous Remote ECG Applications," *31st Annu. Int. Conf. IEEE EMBS*, pp. 2316–2319, 2009.

[29] G. Yang, J. Mao, H. Tenhunen, and L. Zheng, "Design of a Self-organized Intelligent Electrode for Synchronous Measurement of Multiple Bio-signals in a Wearable Healthcare Monitoring System," pp. 1–5, 2010.

[30] G. Yang, J. Chen, H. Tenhunen, and C. Technology, "Intelligent Electrode Design for Long-Term ECG Monitoring at Home," *2009 3rd Int. Conf. Pervasive Comput. Technol. Healthc.*, pp. 3–6, 2009.

[31] C. I. Reis and M. da S. Maximiano, *Internet of Things and Advanced Application in Healthcare*. IGI Global, 2017.

[32] M. Je, "Smart Sensor Microsystems: Application-Dependent Design and Integration Approaches," *Smart Sensors Syst. Innov. Medical, Environ. IoT Appl.*, pp. 83–107, 2017.

Effect of Drying Temperature on Rosmarinic Acid and Sinensetin Concentration in *Orthosiphon stamineus* Herbal Leaves

Sriyana Abdullah[1], Abdul Razak Shaari[1], Ibni Hajar Rukunudin[2] and Muhammad Syarhabil Ahmad[2]
[1]Faculty of Engineering Technology, University Malaysia Perlis, Kampus Uniciti, Sungai Chuchuh, Padang Besar, Perlis, Malaysia
[2]School of Bioprocess Engineering, University Malaysia Perlis, Arau, Perlis, Malaysia.

E-mail: sriyana@unimap.edu.my

Abstract. The objective of this work was to investigate the effects of drying temperature on the concentration of *Orthosiphon stamineus* biomarker compounds which were rosmarinic acid (RA) and sinensetin (SEN). The thin layer drying approach was used to dry *O. stamineus* leaves at various temperatures of 30, 40 and 50°C using a laboratory scale hot air dryer. The dried leaves were then extracted using 60% aqueous methanol prior to quantification. The RA and SEN concentrations in the dried leaves extracts were quantified by the high performance liquid chromatography. The concentration of RA for the dried leaves at 30 and 40°C were higher as compared to that of the fresh leaves. This may due to the response of the plant cells to abiotic stress. The concentration of RA also showed a significant reduction when the temperature was increased to 50°C. In contrast, the SEN concentration in *O. stamineus* dried leaf extract was lower than that of the fresh samples. The concentrations of SEN depicted insignificant effects by drying at 30 and 50°C, and the highest value was obtained in the samples dried at 40°C. Results showed that the drying process was found to affect the concentration of both compounds; therefore suitable drying conditions should be adopted to enhance the medicinal values of the plant species.

1. Introduction

Orthosiphon stamineus (*O. stamineus*) also known as misai kucing (cat whispers) or Java tea, is a popular medicinal herb in the Southeast Asia region. This herb which belongs to *Lamiaceae* family is used as a traditional medicine to cure many human illnesses and disorders especially those related to urinary tract infections and kidney stones, eruptive fever, hypertension, epilepsy, hepatitis, rheumatism, syphilis, and gonorrhea [1, 2]. A number of bioactive compounds such as rosmarinic acid (RA), sinensetin (SEN), 3'-hydroxy-5', 6,7 4'-tetramethoxyflavone, eupatorin, orthosiphol A and orthosiphol B have been identified and isolated from *O. stamineus*. These compounds define the medicinal properties of the herbal species and are classified as herbal marker compounds of the *O. stamineus* plant species [3-9].

O. staminues is always dried for extending its shelf life during long storage period prior to downstream processing. The physical properties such as color and aroma are considered as a major criterion by the consumers in determining the optimum quality of the dried herbal plant. However, the concentration of the bioactive compounds is the determinant factor that confirms the medicinal properties of the dried herbal plants. For instance, the RA that is found in abundance in *O. stamineus* and in certain plant species is a potent antioxidant due to the presence of phenolic hydroxyl group in its structure, while SEN which is commonly found in *Orthosiphon* species was reported to exhibit potential anti-cancer characteristic [10]. Many studies have examined the effects of drying on the bioactive constituents as well as the physical properties of dried medicinal and aromatic plants. For instance, the degree of change in the color and texture of the dried mint (*Mintha cordifolia*) leaves depends on several drying parameters such as drying time, drying temperature, and oxygen level [11]. Another research work reported that the microwave drying of rosemary (*Rosmarinus officinalis*) leaves resulted in the optimum retention of minerals content and color quality as compared to the oven dried leaves [12]. A similar finding was revealed in the dried thyme (*Thymus daenensis*) leaves [13]. In addition, optimum preservation of RA content in lemon balm (*Melissa officinalis*) leaves was found under drying at a low temperature [14]. To the best of our knowledge, there is no research works that have been published regarding the drying effects on *O. stamineus* marker compounds content excluded our previous work [15]. Thus, the present research emphasizes on the effect of one of the important drying parameters that is the air temperature on the RA and SEN concentrations.

2. Materials and methods

2.1. Materials and equipment

Fresh *O. stamineus* leaves aged eight to ten weeks were harvested from the School of Bioprocess Engineering, University Malaysia Perlis, Malaysia production plot. The uniformly sized fresh leaves were packed into an airtight plastic bag and capped in a plastic container to avoid dehydration. The leaves were immediately stored in the cold room at about 5°C prior to the experiment. Methanol (analytical grade), methanol and tetrahydrofuran (liquid chromatography grade) were obtained from Merck (Darmstadt, Germany). The standard marker compounds of RA and SEN were purchased from Indofine Chemical Co. (Hillborough, NJ USA). A customized laboratory scale hot air dryer was used for the drying experiments.

2.2. Thin layer drying of O. stamineus leaves

A sample size of 20 g of the fresh leaves was used for each thin layer drying experiment, which was performed twice separately. The drying experiments were performed using the hot air dryer at various air temperatures of 30, 40 and 50°C with fix air velocity of 0.8 m/s. In addition, prior to the experiments, initial moisture content of the fresh *O. stamineus* leaves was measured by oven method (105°C, 24 hr) using a laboratory drying oven (Binder). The samples were distributed evenly with a thickness of three leaves on the drying tray. The weights of the samples were taken manually during the drying process at particular time intervals, depending on the air temperature, until no weight change was recorded. In addition, the time taken to measure the weight at every interval did not exceed 30 s to avoid significant change in the surface temperature.

2.3. Extraction of bioactive compounds

The fresh and dried *O. stamineus* leaves were solvent extracted prior to the marker compounds quantification. The leaves were ground using a laboratory grinder. Then 1 g of the ground leaf was mixed with 100 ml of 60% aqueous methanol solution in a 250 ml conical flask. The conical flasks containing the samples were agitated at 150 rpm for 5 h using an orbital shaker (Model BS-1, Sartorious Certomat) at 40°C. The extracts were filtered using filter papers (Whatman), and the filtrates were kept in air tight plastic bottles in a refrigerator for further analysis.

2.4. Quantification of biomarker compounds

The quantification analysis was performed using the high performance liquid chromatography (HPLC), equipped with two pumps (Model 20A, Shimadzu), auto sampler (Model SIL 20A, Shimadzu), column oven (Model CTO, Shimadzu), and UV/Vis detector (Model SPD 20A, Shimadzu). About 1.5 ml of the previously prepared extract without dilution was filtered through a 0.45 μm syringe nylon membrane filter into an HPLC vial prior to qualification and quantification. The Merck Licrochart Purospher Start RP 18 HPLC column (250 mm, 4.6 mm i.d, 5 μm pore size) was used. The mobile phase was composed of a mixture of water: methanol: tetrahydrofuran (50: 45: 5 v/v). The mobile phase was adjusted to pH 3 by adding phosphoric acid. The samples were analyzed at the mobile phase at the flow rate of 1 ml/min, and the detector wavelength was set at 340 nm at 30°C for 40 min [2].

The standard external calibration curves were used to quantify the concentrations of the standard biomarker compounds. The biomarker compounds were prepared by dissolving the chemicals in 60% aqueous methanol solution at different concentration ranges (varying from 0 to 10 mg/l for SEN and 0 to 250 mg/l for RA). For the purpose of validation, the biomarker compounds were occasionally spiked with samples for retention time and peak confirmation.

2.5. Statistical analysis

The thin layer drying temperature that affected the marker compounds of *O. stamineus* dried leaves were compared using the Analysis of Variance (ANOVA). Unless otherwise stated, the significance level was established at probability $(P) < 0.05$. For comparison between the means, the Student's t-test was performed. Significance was established if the difference between the means for the treatments was more than the calculated least square difference (LSD).

3. Results and discussion

3.1. Effects of drying conditions on RA concentration

The concentrations of RA for the dried leaves at 30 and 40°C were higher as compared to that of the fresh leaves, as shown in Table 1. In general, the concentrations of this biomarker compound for both the temperatures were relatively high in comparison to other RA contained species [16]. However, the content of RA showed a significant reduction when the temperature was increased to 50°C. This result indicated that the RA in the leaves of *O. stamineus* was liable to a high-temperature range, as portrayed in the drastic reduction in concentration when the temperature was increased from 30 to 50°C. This trend is in accordance with our previous work which also revealed that when the leaf was dried with relatively higher temperature, the RA concentration easily destroyed [15]. Other previous studies also reported similar trends; the concentration of RA in rosemary and lemon balm leaves reduced with a progressive increase in the temperature [14, 16].

Besides, it is very interesting to observe that the dried *O. stamineus* leaves over 30 and 40°C contained higher RA than the fresh leaves (Table 1). Due to the narrow temperature range and lower maximum drying temperature, the leaves of *O. stamineus* tend to slowly lose their moisture content during drying, i.e., the leaf cells tend to be under stress. The concentration of RA in the dried leaf samples of *O. stamineus* might synthesize and accumulate as a defensive reaction. In addition, RA is classified as a defensive compound, which is produced based on the biotic or abiotic stress in the plants, such as wounding, water deficit, low temperature, and pathogen attacks [17].

As shown in Table 1, the concentrations of RA in the dried leaf sample at 30 and 40°C were significantly higher than those dried at 50°C. This pattern may cause longer drying period consumed at 30 and 40°C. This pattern is similar to the air-dried rosemary leaves samples with more than two weeks of drying period possessed the highest RA concentration in the air-dried samples that that of vacuum and oven-dried samples [16].

Table 1 Effects of drying conditions on the concentrations of biomarker compounds for *O. stamineus* leaves

T (°C)	Concentration (mg/g dw)	
	RA	**SEN**
Fresh leaf	22.34±6.71	3.71±0.68
30	$171.47 \pm 14.7_{a}$	$0.74 \pm 0.22_{ab}$
40	$98.69 \pm 25.71_{b}$	$0.88 \pm 0.12_{a}$
50	$9.10 \pm 3.58_{c}$	$0.41 \pm 0.04_{b}$

Mean ± standard deviation with different letters in a column were significantly different ($P<0.05$)

3.2. Effects of drying conditions on SEN concentration

The SEN concentration in *O. stamineus* dried leaf samples was lower than that of the fresh samples, as shown in Table 1. The concentration of SEN in samples dried at 30°C are not show a significant difference as compared with the samples dried at 40 and 50°C. However, the SEN concentration in samples dried at an air temperature of 50°C showed significant reduction as compared to the samples dried at 40°C. The pattern of SEN reduction at high temperature (50°C) may hint toward its susceptibility to heat. Apart from this, as reported in our previous work, SEN is susceptible to sunlight where it was only detected in an oven dried sample but did not appear in under shade dried and sundried samples [14]. A biomarker compound found in greater plantain (*Plantago major*), plantamojoside also reduced when the temperature increased from 30 to 50°C [18]. Another study had revealed biomarker compounds in plantain (*Plantago lanceolata*) leaves also exhibited a similar pattern [19]. The marker compounds in plantain which are iridoidglycosides catapol and aucubin, found in the dried samples at 60°C showed a decrease as compared to the fresh leaf samples. SEN is an important marker compound in *O. stamineus* and was reported to exert some medicinal benefits. For example, it exhibits antidiabetic property [20] and acts as a chemosensitizer in anticancer treatment [21]. Since, the drying process affects the concentration of SEN, appropriate drying conditions should be adopted to enhance the medicinal values of the plant species.

4. Conclusion

The concentration of RA significantly reduced with increasing temperature. The concentration of RA was found to be lower in the fresh samples as compared to the dried samples. This may be caused by the response of the plant cells to abiotic stress. The concentrations of SEN depicted insignificant effects by drying at 30 and 50°C, and the highest value was obtained in the samples dried at 40°C.

References

[1] Adam Y Somchit M N Sulaiman M R Nasaruddin A A Zuraini A Bustamam A A and Zakaria Z A 2009 Diuretic properties of *Orthosiphon stamineus* Benth *J Ethnopharmacol* **124** 154-158

[2] Akowuah G A Zhari I Norhayati I Sadikun A and Khamsah S M 2004 Sinensetin, eupatorin, 3'-hydroxy-5, 6, 7, 4'-tetramethoxyflavone and rosmarinic acid contents and antioxidative effect of *Orthosiphon stamineus* from Malaysia *Food Chem* **87** 559-566

[3] Awale S Tezuka Y Banskota A H Adnyana I K and Kadota S 2003 Highly-oxygenated isopimarane-type diterpenes from *Orthosiphon stamineus* of Indonesia and their nitric oxide inhibitory activity *Chem Phar. Bull* **51** 268-275

[4] Awale S Tezuka Y Banskota A H Adnyana I K and Kadota S 2003 Nitric oxide inhibitory isopimarane-type diterpenes from *Orthosiphon stamineus* of Indonesia *J Nat Prod* **66** 255-258

[5] Awale S Tezuka Y Banskota A H and Kadota S 2003 Inhibition of NO production by highly-oxygenated diterpenes of *Orthosiphon stamineus* and their structure-activity relationship *Biol Pharma Bull* **26** 468-473

[6] Awale S Tezuka Y Banskota A H and Kadota S 2003 Siphonols A-E: Novel nitric oxide inhibitors from *Orthosiphon stamineus* of Indonesia *Bioorg Med Chem Lett* **13** 31-35

[7] Sumaryono W Proksch P Wray V Witte L andHartmann T 1991 Qualitative and Quantitative Analysis of the Phenolic Constituents from *Orthosiphon aristatus Planta Med* **57** 176-180

[8] Pietta P G Mauri P L Gardana C and Bruno A 1991 High-performance liquid chromatography with diode-array ultraviolet detection of methoxylated flavones in Orthosiphon leaves *J Chromatogr A* **547** 439-442

[9] Nguyen M T T Awale S Tezuka Y Chien-Hsiung C and Kadota S 2004 Staminane- and Isopimarane-Type Diterpenes from *Orthosiphon stamineus* of Taiwan and Their Nitric Oxide Inhibitory Activity *J Nat Prod* **67** 654-658

[10] Dong Y J G Cao A Shi J Shi H Xie J and Wu D 2011 Effects of sinensetin on proliferation and apoptosis of human gastric cancer AGS cells. *Zhongguo Zhong Yao Za Zhi* **36** 790-794

[11] Therdthai N and Zhou W 2009 Characterization of microwave vacuum drying and hot air drying of mint leaves (*Mentha cordifolia* Opiz ex Fresen) *J Food Eng* **91** 482-489

[12] Arslan D and Ozcan M M 2008 Evaluation of drying methods with respect to drying kinetics, mineral content and colour characteristics of rosemary leaves. *Energy Convers. Manag* **49** 1258-1264

[13] Rahimmalek M and Goli S A H 2013 Evaluation of six drying treatments with respect to essential oil yield composition and color characteristics of *Thymys daenensis* subsp *daenensis* Celak leaves *Ind Crop Prod* **42** 613-619

[14] Argyropoulos D and Muller J 2014 Kinetics of change in colour and rosmarinic acid equivalents during convective drying of lemon balm (*Melissa officinalis*) *J Appl Res Med Aromat Plant* **1** 15-22

[15] Abdullah S Shaari A R Rukunudin I H Ahmad M S 2012 Effect of drying methods on metabolites composition of misai kucing (*Orthosiphon stamineus*) leaves. *Procedia APCBEE* **2** 178-182

[16] Hossain M B Barry-Ryan C Martin-Diana A B Brunton N P 2010Effect of drying method on the antioxidant capacity of six Lamiaceae herbs *Food Chem* **123** 85-91
[17] Petersen M and Simmons M S J 2003 Rosmarinic acid *Phytochem* **62** 121-125
[18] Zubair M Nybom H Lindholm C and Rumpunen K 2011 Major polyphenols in aerial organs of greater plantain (Plantago major L.) and effects of drying temperature on polyphenol contents in the leaves *Sci Hortic* **128** 523-529
[19] TamuraY and Nishibe S 2002 Changes in the concentration of bioactive compounds in plantain leaves *J Agric Food Chem* **50** 2514–2518
[20] Mohamed E A H Siddiqui M J A Lee F A Sadikun A Sue H C Soo C T Asmawi M Z and Yam M F 2012 Potent alpha-glucosidase and alpha-amylase inhibitory activities of standardized 50% ethanolic extracts and sinensetin from *Orthosiphon stamineus* Benth as anti-diabetic mechanism *BMC Complement Altern Med* **176** 1-7
[21] Choi C H Sun K H An C S Yoo J C Hahm K S Lee I H Sohng J K and Kim Y C 2002 Reversal of P-glycoprotein-mediated multidrug resistance by 5,6,7,3',4'-pentamethoxyflavone (Sinensetin) *Biochem Biophys Res Commun* **295** 832-840

IOP Conf. Series: Materials Science and Engineering **318** (2018) 012075 doi:10.1088/1757-899X/318/1/012075

Change of Bioactive Constituent in *Clinacanthus nutans* Leaves under Sun Drying

Sriyana Abdullah* **and Muhamad Faris Abdul Aziz**
Faculty of Engineering Technology, University Malaysia Perlis, Sg. Chucuh, 02000, Padang Besar, Perlis, Malaysia

E-mail: sriyana@unimap.edu.my

Abstract. *Clinacanthus nutans* (*C. nutans*) or locally known as belalai gajah is a folk medicine since ancient time. This research project was established to investigate the effects of under sun drying on the *C. nutans* bioactive constituent. The drying experiments were conducted using different drying surfaces i.e. perforated, black polythene and white polythene. The fresh and dried leaves were then extracted using a sonicator to evaluate its bioactive constituent. The total phenolic content (TPC) in the *C. nutans* extracts were determined using Follin Ciocalteu reagent method to represent the bioactive constituent. Drying over the white polythene surface showed the slowest reduction of moisture content as compared to the perforated polythene and black surfaces. Results also showed no significant effect of the drying surfaces on the TPC. However, the TPC in the dried leaves was significantly higher than in the fresh leaves. This may be due to the plant cells response to abiotic stress and the inhibition of oxidation enzymes.Therefore, drying *C. nutanc* leaves under sun light could be considered in order to preserve the concentration of phenolic compounds and for minimizing energy consumption.

1. Introduction

Clinacanthus nutans (*C. nutans*) is a popular herbal plant in Malaysia and the neighbouring countries such as Thailand and Indonesia [1]. This plant has been used since ancient time in treating skin rashes, insect and snake bits, skin lesions caused by virus, diabetis mellitus, dengue fever and has diuretics properties [2]. Through a scientific investigation this plant has proved to exert anti-herpes simplex virus [3], antiproliferative effects on cultured human cancer cell lines [4], relief of skin inflammation and incest bites and varicella-zoster lesions [5].

C. nutans is a perishable herb due to its high water content (about 80%). Perishable materials are prone to microbial degradations so that it could not be stored in its fresh form. Drying technique is commonly used to preserve them during storage by reducing moisture level to a certain safe value. This preservation method is believed to maintain beneficial medicinal properties of *C. nutans* during storage promising a premium quality herbal raw material, as for the production of high quality nutraceutical products. Several drying techniques of both natural and artificial methods have been used to dry herbal materials to preserve their bioactive constituent such as under sun, shade drying, oven, microwave, freeze drying and vacuum oven. For instance, the microwave drying of parsley did not show any significant effect on its phenolic content compared to in the fresh leaves [6]. Another research work on *C. nutans* revealed the phenolic content reduced when the leaves were dried by freeze drying as compared to the oven and air drying [7]. This result showed that the bioactive content in *C.nutans* may not susceptible to heat. Thus, our research work is conducted to investigate the effect of natural sun drying on the total phenolic content of *C. nutans* leaves for complementing the existing studies. It is important to establish the natural drying technique to our local herbs including *C. nutans* as to provide more options to the manufacturers especially in term of energy saving and maintaining bioactive constituent.

2. Material and methods

2.1. Sample preparation

The fresh *C. nutan* herbal plants were collected at the Institute of Sustainable Agrotechnology (INSAT), Universiti Malaysia Perlis (UniMAP). The uniform size leaves were sorted from the plant stems and rinsed with tap water. The leaves were kept in a fridge up to 48 hours prior to use.

2.2. Drying experiment

The fresh leaves were distributed on three different drying surfaces i.e. perforated, black polythene and white polythene. Each sample size was 30 g and the samples were prepared in three replications. The experiments were conducted under climatic condition of Campus UniCITI Alam, Universiti Malaysia Perlis, Perlis, Malaysia. The drying process was carried out during 1200 noon until 1700 afternoon. The weight of the sample was recorded every hour until it achieved saturation; weight change on the three consecutive readings was less than 0.001g. The sample weights were found to become saturated within 3 days. The samples were kept in an air-tight container when not exposed to the sun light as to maintain their moisture content.

2.3. Moisture content determination

The *C. nutans* leaves were dried in an oven at 105°C, 24 hours to determine its dry matter weight (m_{dw}). The moisture content (MC) was then calculated using an equation below:

$$MC_{db} = \frac{m_t - m_{dw}}{m_{dw}} \quad (1)$$

where,
MC_{db}= moisture content (dry basis)
m_t=weight of the sample at time, t

The graph of leaves moisture content versus drying time was plotted.

2.4. C. nutans leaves extraction

The fresh and dried *C. nutans* leaves were ground using a laboratory blander. A 1.0 g sample of ground *C. nutans* was weighed and mixed with 100 ml distilled water. Then, the samples were extracted by placing it in ultrasonic bath for 30 min at the temperature of 40°C. The extracts were filtered using Wathman No. 4 filter paper. The supernatant was used for analysis of total phenolic content (TPC).

2.5. Total phenolic compounds determination

The TPC of *C. nutans* extracts were determined using Folin Ciocalteu reagent [8]. An amount of 0.5 ml leaves extract was mixed with 0.5 ml Folin Ciocalteu reagent. Then, 1.5 ml of sodium bicarbonate was added to the mixture. The mixture was then incubated in the dark for 30 minutes. The absorbance of the mixture was measured at 765nm using a spectrophotometer. Distilled water was used as a blank solution. The concentration of phenolic compounds in the extracts were attained by comprison with a gallic acid calibration curve and expressed as mg GAE/g of extract.

2.6. Statistical analysis

The effects under sun drying using different drying surfaces on the TPC in the *C. nutans* leaves extracts were compared using the Analysis of Variance (ANOVA). Unless otherwise stated, the significance level was established at probability (P) < 0.05. For comparison between the means, the Student's t-test was performed. Significance was established if the difference between the means for the treatments was more than the calculated least square difference (LSD).

3. Results and discussion

3.1. Relationship between moisture content and drying time

Figure 1 shows the reduction of moisture content versus time of *C. nutans* leaves under sun drying using different drying surfaces. Drying on the white surface showed the slower moisture content reduction than the other surfaces especially in day 2. It is because white surface does not absorb heat, drying process was only due to the natural heat from the sunlight. Perforated and black surfaces showed similar drying occurrence. Perforated surface allows air to reach entire leaf surface thus helped the drying process made faster while, black surface is known able to absorb heat thus accelerating the drying process.

As it can be seen in Fig. 1, the trend of moisture reduction was higher during day 2 due to high temperature was recorded. Although lower temperature was recorded during day 1, the drying rate was also driven by the higher moisture content in the leaves. During day 3, the removal of moisture content for all drying surfaces were getting slower which nearly equilibrium. The equilibrium moisture content determined was 0.18 g water/g dry basis to all drying surfaces.

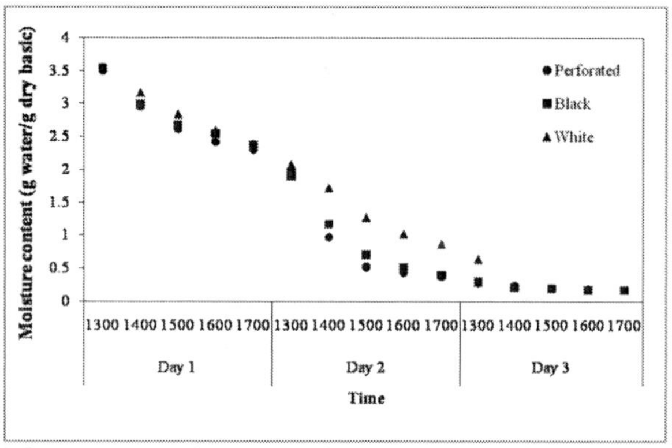

Figure 1. Variation of the *C. nutans* leaves moisture content with time during sun drying.

3.2. Effect of sun-dried on total phenolic content (TPC) in C. nutans leaves extracts

The effect of TPC in the fresh and dried *C. nutans* leaves extracts is shown in Figure 2. The highest TPC in *C. nutans* was detected in the extracts under perforated drying surface (10.28 mg GAE/g), followed by black drying surface (9.695 mg GAE/g) and the white drying surface (9.380 mg GAE/g extract). These results showed that the effect of the drying treatments on TPC was not significantly different.

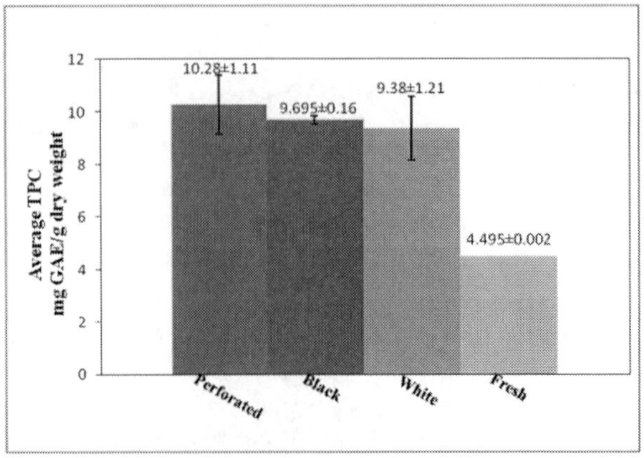

Figure 2. Effect of the drying surfaces on TPC in *C. nutans*.

The sun drying process significantly increased the TPC in the dried leaves extract as compared to that in the fresh leaves extract. This higher production of phenolic compound may be assosiated with a stress-induces responce of the plant cells caused by the moisture loss during drying [9]. Drying also inhibits oxidative enzymes, which allowing better retention of phytochemicals in the dried leaves extracts [10]. In addition, the TPC in this sun-dried *C. nutans* leaves extracts is found higher than the leaves dried by oven, air and freeze drying [7]. Another research work showed that the TPC in the *C. nutans* also higher than in the sun dried misai kucing (*Orthosiphon staminus*) leaves [11]. Regardless of its disadvantages, under sun drying technique is appropriate to dry *C. nutans* herb due to the new findings revealed in this work. Therefore, drying of *C. nutans* herb under sun light could be considered in order to preserve bioactive constituent and minimize the operating cost.

4. Conclusion

From the results obtained, it can be concluded that under sun drying on the different drying surfaces employed in this work did not show any significant difference on the TPC. However it significantly increases the content of total phenolic compounds in the *C. nutans* leaves as comparison to in the fresh leaves. Thus, the under sun drying method could be applied in order to improve the TPC content and as an option for energy saving processing method particularly on the *C. nutans* herb.

References

[1] Ariful A Sahena F Ghafoor K Hakim Juraimi A S Khatib A Zaidul I S 2016 A review of the medical pharmacologyphytochemistry, *Asian Pac J Trop Med* **9** 402–409

[2] Goonasakaran S A 2013 Preliminary Antimicrobial and Phytochemical Analysis of *Clinacanthus nutans* and Azadiracht indica, Master Thesis Technological University of Malaysia, Johor, Malaysia

[3] Pongmuangmul S Phumiamorn S Sanguansermsri P Wongkattiya N Hamilton Fraser I Sanguansermsri D 2016 Anti-herpes simplex virus activities of monogalactosyl diglyceride and digalactosyldiglyceride from *Clinacanthus nutans*, a traditional Thai herbal medicine, *Asian Pac J Trop Biomed* **6** 192–197

[4] Yong Y K Tan J J The S S Hui Mah S Chiong H S and Ahmad Z 2013 *Clinacanthus nutans* Extracts Are Antioxidant with Antiproliferative Effect on Human Cancer Cell Lines, *Evidence Based Complementary and Alternative Med* **1** 1-8

[5] Charuwichitratana S Wongrattanapasson N Timpatanapong P Bunjob M 1996 Herpes zoster: treatment with *Clinacanthus nutans* cream, Int J Dermatol **35** 665-666

[6] Magdalena S´ledz´ Małgorzata Nowacka Artur Wiktor Dorota Witrowa-Rajchert 2013 Selected chemical and physico-chemical properties of microwave- convective dried herbs, *Food and Bioprod Proc* **91** 421– 428

[7] Leng W K Ahmed Mediani Z Nur Khaleeda Zulaikha Sze W L Intan Safinar I Alfi Khatib Khozirah S Faridah A 2015 Phytochemical diversity of *Clinacanthus nutans* extracts and their bioactivity correlations elucidated by NMR based metabolomics *Phytochem Letters* **14** 123– 133

[8] Chong C H Law C L Luqman Chuah A Wan Ramli W D 2009 Drying Models and Quality Analysis of Sun- Dried Ciku *Drying Tech* **27** 985- 992

[9] Hossain M Barry-Ryan C Martin-Diana A B Brunton N 2010 Effect of drying method on the antioxidant capacity of six Lamiaceae herbs *Food Chem* **123**, 85–91

[10] Hong-fang J Ai-lin D Ling-wen Z Chun-yang X Ming-duo Y Fang-fang L 2012 Effects of drying methods on antioxidant properties in Robinia pseudoacacia L. flowers, *J Med Plants Res* **6**, 3233–3239

[11] Abdullah S Shaari A R Rukunudin I H Ahmad M S 2012 Effect of drying methods on metabolites composition of misai kucing (*Orthosiphon stamineus*) leaves *Procedia APCBEE* 178-182

IOP Conf. Series: Materials Science and Engineering **318** (2018) 012076 doi:10.1088/1757-899X/318/1/012076

Effect of Voltage and Flow Rate Electrospinning Parameters on Polyacrylonitrile Electrospun Fibers

S S S Bakar[1], K C Fong[1], A Eleyas[1] and M F M Nazeri
[1]CEGeoGTech, School of Materials Engineering, Kompleks Pengajian Jejawi 2, Taman Muhibbah, Arau Perlis

E-mail: shahnaz@unimap.edu.my

Abstract. Currently, electrospinning is a very famous technique and widely used for forming polymer nanofibers. In this paper, the Polyacrylonitrile (PAN) nanofibers were prepared in concentration of 10wt% with varied processing parameters that can affect the properties of PAN fiber in term of fiber diameter and electrical conductivity was presented. Voltage of 10, 15 and 20 kV with PAN flow rate of 10, 15 and 20 □m were applied during processing. The electrospun PAN fibers were then undergo pyrolysis at 800°C for 30 minutes. The resultant PAN nanofibers were then analysed by SEM, XRD and four point probe test after pyrolysis process. SEM image show continuos uniform and smooth surface fibrous structure of electrospun PAN fibers with average diameter of 1.81 μm. The fiber morphology is controlled by manipulating the processing parameters of electrospinning process. The results showed that the resistance of electrospun PAN fibers decreases as the processing parameter changes by increasing the applied voltage and flow rate of electrospinning.

1. Introduction

Typically, nanofiber is processed by electrospinning into long, small diameters, and high surface area per unit volume [1]. Polyacrylonitrile fiber is also known as PAN fiber has been used in both industrial and medical applications such as precursors for the production of carbon fibers, drug deliver, wound dressing, sensor materials, and composite reinforcement [2,3]. One of the main reasons for the versatility of these fibers is due to the presence of high thermal stability at the range of 200-300°C [1]. PAN fiber has a low electrical resistivity (~1,500 μΩcm) measured along the fiber direction. Normally the electrospun PAN fiber layers will have large surface area per volume ratio with large amount of small pore sizes porosity, superior mechanical properties, and flexibility, so they are most extensively used in the electrospinning process because of their excellent properties [1]. Electrospinning which also called electrostatic fiber spinning is one of the techniques that use electrical forces to form continuous nanometer diameter fibers for natural and synthetic polymers with the help of electrostatic forces [4]. Electrospinning has been known as the most effective technique for polymer nanofibers fabrication. Various polymers especially PAN fiber have been successfully electrospun into ultrafine PAN fibers in recent years [5]. During the electrospinning process, electric forces are used for fibers formation from electrospinning solutions which are mostly in solvent solution while some were in melts form [6]. Physical and chemical properties of electrospun nanofiber can be actually controlled by manipulating electrospinning parameters either in solution (e.g., concentration, viscosity) and process (e.g., applied voltage, solution feed rate) in order to meet the requirements of a specific application [7]. The common raw material used for the synthesis of electrospun PAN fibers were Polyacrylonitrile (PAN) and N-Dimethylformamide (DMF).

2. Experiments

2.1 Materials

The solutions used in the elctrospinning experiments were prepared by using Polyacrylonitrile (PAN) obtained from Sigma-Aldrich with 150000 average molecular weight without further purification. The 99.8 % N-Dimethylformamide (DMF) used was purchased from Avantor Performance Materials.The first paragraph after a heading is not indented (Bodytext style).

2.2 Methodology

This experiment has the following sections: (i) preparation of PAN polymer solution, (ii) electrospinning process of PAN polymer, (iii) measurements and characterizatiosn, (iv) X-ray diffraction, (v) pyrolysis process of PAN polymer. The studied electrospinning parameters for this research were shown in Table 1.

2.2.1 Preparation of PAN polymer solution

PAN solutions in DMF were prepared by mixing the weighed 5g PAN powder into 45g DMF solution. The concentration of 10 wt% PAN solutions was prepared by stirring the mixture at room temperature for 24 hours prior to electrospinning.

Table 1: The electrospinning parameters applied in this experiment

Voltage (kV)	Flow Rate (μL)	Tip to collector distance (cm)	Concentration (wt%)
	10		
10	15	10	10
	20		
	10		
15	15	10	10
	20		
	10		
20	15	10	10
	20		

2.2.2 Electrospinning process of PAN polymer

The syringe was placed on the syringe pump which were filled with PAN solution. A medical-used needle was used as the nozzle tip for the syringe. The tip-to-collector distance was fixed at 10cm. The studied electrospinning parameters for this study are as shown in Table 1.

2.2.3 Measurements and Characterizations

Scanning electron microscope (SEM) were performed with JEOL JSM-6460LA. X-ray diffraction (XRD) spectra were acquired with diffractometer using Cu Kα radiation in the 2θ range 15°- 60°. The resistivity were measured via four point probe test at room temperature. The formula, $Rs=KVI$ was used to calculate resistance of the electrospun PAN fibers where Rs, K, V, and I were the sheet resistance of electrospun PAN fiber layer, factor for shape of sample, voltage measured across the two voltage probe and current measured across the two current probe, respectively.

2.2.4 Pyrolysis process of PAN polymer

The electrospun PAN fibers were then undergo pyrolysis heat treatment. The electrospun PAN fibers were heated at 800 °C for 30 min under the rate of nitrogen flow at 0.1L/min. The heating rate for pyrolysis heat treatment was set at 5 °C/min. The electrospun PAN fibers undergo oxidation process when the temperature is heated to the range of 200-300°C. The pyrolyzed electrospun PAN fibers will then analysed with SEM. The pyrolysis parameters studied are as showed in Table 2.

Table 2: The pyrolysis parameters applied in this experiment

Heating time (min)	Heating temperature (°C)	Heating rate (°C/min)	Nitrogen flow rate (L/min)
30	800	5	0.1

3. Results and Discussion

The pure electrospun PAN fibers were white in colour. Electrospun PAN fibers show a continuos uniform and smooth surface fibrous structure when analyse under SEM. Figure 1 (a) shows the SEM image of electrospun PAN fibers at 10 kV, 15 μL/min before pyrolysis. The SEM image shows nonwoven thread-like structure with fiber diameter in range of 1 μm to 3 μm and an average diameter of 1.81 μm. It is a normal phenomenon obtained non-aligned nanofibers as there is no rotating drum collector used in this study. In fact, Boland et al. [8] reported that the nanofibers obtained from electrospinning can be actually aligned by using rotating drum collector with high speed. Figure 1 (b) shows SEM image of electrospun PAN fibers with 10kV, 15 μL/min after 30 min of 800°C pyrolysis.

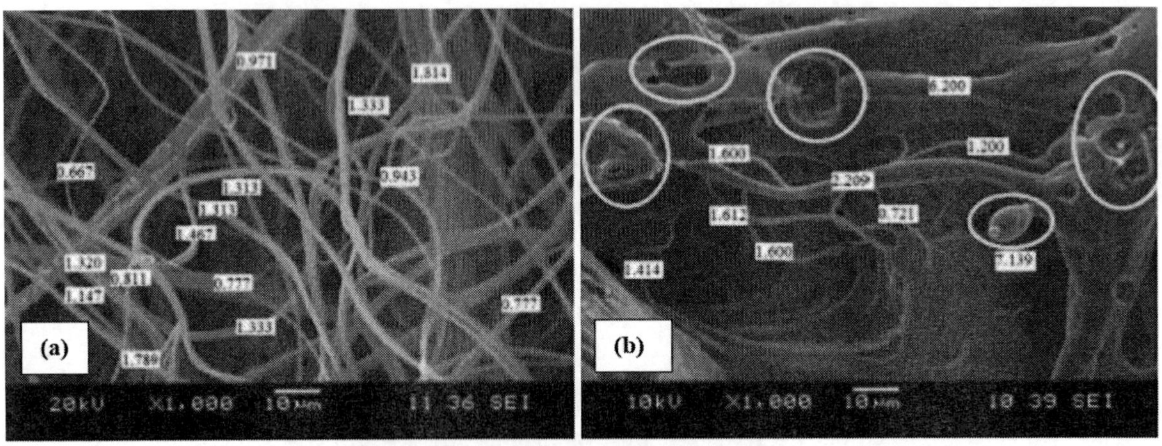

Figure 1. SEM micrographs of electrospun PAN fibers yield by 10kV at 15□L/min (a) before pyrolysis and (b) after 800°C pyrolysis for 30min

The average diameter for electrospun PAN fibers with flow rate obtained was 3.17 μm. From the SEM image, the electrospun PAN fibers undergo slight structural changes, but overall, the morphology is remain almost the same. This also in agreement with the finding by Voepel et al [86], as they reported in their study that the electrospun PAN fibers undergo slight structural changes. The pyrolysis had fused some of the elecrospun fibers which resulted in the larger fiber diameters. The overall average diameter measured of the electrospun PAN fibers before and after pyrolysis are presented in Table 3.

Figure 2 shows the XRD patterns of electrospun PAN fibers with different processing parameters. The XRD pattern of electrospun PAN fibers show the strongest diffraction peak at around $2\theta = 16.72°$ which corresponds to a planar spacing d = 5.2975Å. The electrospun PAN fibers is more on crystalline phase rather than amorphous. The crystalline peak shown was corresponding to carbon. By calculating the d-spacing for both the peaks, it is found that the d-spacing is closely to the ratio of$\sqrt{3}$:1. Hence, this determined that the PAN chains was actually in hexagonal packing. Besides, there is a weak hill at around 2θ = 29.26° which determine the amorphous structure of PAN fibers. The intensity of the peak is lower compared to the carbon peak at 2θ = 16.72.

Table 3: The average electrospun PAN fibers diameters before and after pyrolysis

Processing parameters		Average diameter	
Voltage (kV)	Flow rate (µL/min)	Before pyrolysis (µm)	After pyrolysis (µm)
10	10	1.81	3.17
	15	1.18	3.72
	20	1.27	3.84
15	10	1.33	3.72
	15	1.29	2.65
	20	1.21	2.36
20	10	2.35	2.59
	15	1.50	2.57
	20	1.40	2.92

Figure 3 shows the XRD patterns of electrospun PAN fibers with processing parameters of 15 kV and 15 µL/min that undergoes pyrolysis at 800°C for 30 mins. The XRD pattern of electrospun PAN fibers after pyrolysis show the strongest diffraction peak at around 2θ = 25.68° which corresponds to a planar spacing d = 3.466 Å. Besides that, there is a weak hill at around 2θ = 42.04° which determine the amorphous structure of PAN fibers. The intensity of the peak is lower compared to first carbon peak at 2θ = 25.68°.

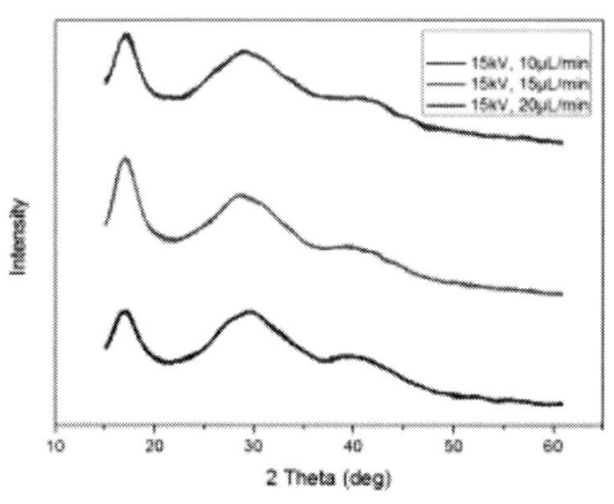

Figure 2. XRD patterns of electrospun PAN fibers that synthesized at 15kV at various flow rate

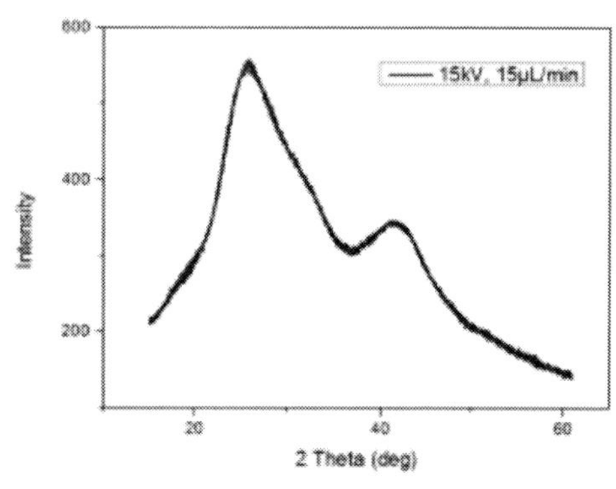

Figure 3. XRD patterns of electrospun PAN fibers at 15kV after pyrolysis at 800°C for 30 mins.

Table 4 shows that the resistance of electrospun PAN fibers before pyrolysis decreases as the processing parameter changes by increasing the applied voltage and flow rate of electrospinning. For the first sample which is electrospun PAN fibers produced with 10 kV at 10 µL/min shows the highest resistance to electrical conductivity with 512.6 Ω. The electrospun PAN fibers were then undergo a decrease in resistance when the processing parameter changes with increasing the applied voltage. The resistance of electrospun PAN fibers did not varies much in resistance. In other word, the conductivity of the samples are having almost the same electrical conductivity.

Table 4: The resistance of electrospun PAN fibers before and after pyrolysis at various processing parameters

Processing parameters		Resistance (Ω)	
Voltage (kV)	Flow rate (μL/min)	Before pyrolysis (μm)	After pyrolysis (μm)
10	10	512.60	468.80
	15	444.84	464.50
	20	403.23	459.19
15	10	414.15	465.90
	15	412.28	462.90
	20	407.44	454.67
20	10	419.32	450.80
	15	412.01	448.40
	20	410.11	445.57

The electrospun PAN fibers have exhibited sharp increase in resistance after pyrolysis process. The decreasing value trend of resistance when electrospinning voltage and flow rate were increased were similar to the results before fibers undergoes pyrolysis. Overall the conductivity of pyrolysed electrospun PAN fibers still consider as non-conductive fiber due to too high resistance to electrical conductivity [9].

4. Conclusion

In this work, the effects of processing parameters on the electrospun PAN fibers towards the morphology of PAN fibers, electrical conductivity properties and also the effect of pyrolysis on the shrinkage of fiber diameter has been explored. It was found that the properties of electrospun PAN fibers produced is strongly depending on the electrospinning processing parameters such as voltage applied and flow rate, and also the pyrolysis heating temperature. With increasing voltage applied the fiber diameter of electrospun PAN fibers increased, while increasing flow rate has resulted in decreasing fiber diameters. The pyrolysed PAN fibers show increased in fiber diameter due to the fiber diffusity and also decreased the electrical conductivity of the fibers.

5. Acknowledgements

This work was financially supported by FRGS project under grant 9003-00400 Derivation of Carbon Nanowires from Electrospinning of Photoresist Polymer.

References

[1] S. S. Qavamnia and K. Nasouri. Polymer Science Series A, Vol. 57, pp. 343-349. (2015)
[2] Z. Li and C. Wang. One-Dimensional nanostructures: Electrospinning Technique and Unique Nanofibers. Springer-Verlag Berlin Heidelberg (2013).
[3] C. J. Thompson, G. G. Chase, A. L. Yarin, and D. H. Reneker. Polymer, Vol. 48, pp. 6913-6922. (2007)
[4] S. Mohammadzadehmoghadam, Y. Dong, S. Barbhuiya, L. Guo, D. Liu, R. Umer, X. Qi, and Y. Tang, "Electrospinning: Current Status and Future Trends," in Nano-size Polymers: Preparation, Properties, Applications, S. Fakirov, Ed., ed Cham: Springer International Publishing, pp. 89-154. (2016)
[5] H. R. Darrell and C. Iksoo. Nanotechnology, Vol. 7, p. 216. (1996)
[6] O. S. Yördem, M. Papila, and Y. Z. Menceloğlu. Materials & Design, Vol. 29, pp. 34-44. (2008)

[7] S. G. Kumbar, R. James, S. P. Nukavarapu, and C. T. Laurencin. Biomedical Materials, Vol. 3, pp. 034002. (2008)
[8] G. E. W. Eugene D. Boland, David G. Simpson, and G. L. B. Kristin J. Pawlowski. Journal of Macromolecular Science, Part A. Vol. 38, pp. 1231-1243. (2001)
[9] M. Panapoy, A. Dankeaw, and B. Ksapabutr, Thammasat Int J Sc Tech, Vol. 13, pp. 11-17. (2008)

IOP Conf. Series: Materials Science and Engineering **318** (2018) 012077 doi:10.1088/1757-899X/318/1/012077

Evaluation of FSO System Availability in Haze Condition

A. A. Anis[1], C. B. M. Rashidi[1*], S. A. Aljunid[1] and A. K. Rahman[2]
[1] Advance Communication Engineering, Centre of Excellence School of Computer & Communication Eng. Universiti Malaysia Perlis, (UniMAP) Kangar, 01000, Kangar Perlis
[2] Departments of Electrical and Electronic Eng. Universiti Malaysia Sarawak, (UNIMAS), Sarawak

E-mail: rashidibeson@unimap.edu.my

Abstract. In this paper, we proposed the evaluation of FSO system availability in haze condition. The atmospheric attenuation by weather conditions in the atmosphere as the most challenging problem of FSO system as the system performance is severely degraded and causing the signal optic to be transmitted poorly. The effects of haze condition on the performance of FSO system is stressed out and focused in this paper. From the evaluation of the analysis, designs of FSO system are proposed to obtain a system with improved link performance in haze conditions. The scattering coefficient and the atmospheric attenuation are determined using Beer's Lambert equation. From the research, the link performance of the system is greatly improved using Design 2 with minimum BER of 10^{-127} and maximum Q Factor of 23.98. The FSO system using Design 2 has better performance compared to Design 1 in haze condition as the optical signals could penetrate the dense haze better without losing much optical power during the transmission to the scattering.

1. Introduction

Free space optic (FSO) communication is the transmission of modulated near-infrared (NIR) beams. They are transmitted via the atmosphere to obtain optical communication [1]. There are several important features of FSO system such as that it is unaffected by electromagnetic interference and radio frequency interference, high bandwidth and requires no licensing. Due to these interesting advantages, FSO system has been use in many applications. They are metro network extension, fiber backup, backhaul, enterprise connectivity, and military applications [2].

Despite its multiples advantages, FSO systems have its weaknesses. One of the primary drawbacks which could effectively degrade the Free Space Optics (FSO) link performance is none other than the weather condition. Several parameters such as the link range, the bit error rate (BER), quality factor, and optical power will experience extreme attenuation in the weather environment. Atmospheric attenuation due to this problem happens to prevent the link transmission in the weather conditions effectively thus achieving a high performance FSO system is impossible [3]. It can affect the system by means of absorption and scattering of laser beam by aerosols and particles in mid-air [4].

One of the local weather conditions which severely affect the optical transmission of FSO system is haze. The signal transmission during hazy days is really bad especially in low visibility haze condition. A laser beam experiences power loss when it travels through the air. The selection of distance between transmitter and receiver, bit rate, attenuation, and diameter aperture can be adjusted to greatly reduce the attenuation effect on FSO system. The scattering coefficient and atmospheric attenuation could be obtained by theoretical analysis of the local weather studies to determine the severity of weather conditions. The information obtain could help in expecting potential impact on signal transmission quality and transmission interference system operation. Furthermore, FSO system design could be enhanced to overcome possible interference and thus improve the system operation.

The common wavelengths used for FSO communication are 785 nm, 850 nm, and 1550 nm. The choice of wavelength for FSO system design happens to be one of the most decisive. Nonetheless, the FSO transmission link power is carefully exploited as it is potentially harm the eyes [5-6].

2. Free Space Optics (FSO) Link Budget

The scattering coefficient is used to evaluate the haze on FSO systems and it is affected by the visibility conditions. In addition, it is used to calculate the atmospheric attenuation. In low visibility of haze conditions, the particles size and concentration are higher compared to average visibility. So, scattering and attenuation is easily triggered in low visibility conditions. There are two factors affecting the attenuation coefficient. They are scattering and absorption [7]. These absorption and scattering are happening due to aerosols and gaseous particle exists in the atmosphere. Typically, the laser bit wavelengths used (785 nm, 850 nm and 1550 nm) were chosen for the system. These wavelengths fall inside the transmission windows in the atmospheric absorption spectra. Thus, the contributions of absorption to the total attenuation coefficient were found to be very small. So, the total attenuation coefficient is influenced by the effects of scattering. The size of the atmospheric particle with respect to the transmission laser wavelength determines the type of scattering. There are three types of scattering which are Mie scattering, Rayleigh scattering, and non- selective scattering.

The scattering coefficient can be expressed as a function of the visibility and wavelength. Hypothetically, low scattering coefficient caused high visibility caused by higher concentration and bigger particle sizes of haze along the transmission path of optical signal. Equation (1) is used to determine the scattering coefficient in hazy days.

$$\beta = \frac{3.91}{V}\left(\frac{\lambda}{550\text{nm}}\right)^{-q} \tag{1}$$

where
V=visibility in kilometres
λ = wavelength in nanometers
q = the distribution size of the scattering particles

$$q = \begin{cases} 1.6 & \text{for} \quad V>50\text{Km} \\ 1.3 & \text{for} \quad 6\text{Km}<V<50\text{Km} \\ 0.585V^{1/3} & \text{for} \quad V<6\text{Km} \end{cases}$$

There are three parameter which affected the visibility which are the degree of sources coherence, the different in length between the paths and the detector location with based on the source. The performance of FSO systems degrades when the visibility is low when the concentration and size of the particles are higher compared to average visibility. The atmospheric attenuation can be calculated using the following Beer's Law Equation [8].

$$\tau(R) = -\frac{P(R)}{P(0)} = e^{-(\beta)R} \tag{2}$$

where
$\tau(R)$ = transmittance at distance R,
$P(R)$ = laser power of R,
$P(0)$ = laser power at the source
β = attenuation or total extinction coefficient (per unit length)

If Beer;s Lambert equation can be derived into equation (2)

$$= -10 \log \frac{P(R)}{P(O)} \quad (3)$$

$$= -10 \log e^{-\beta R}$$

$$= 10 \log e^{\beta R} \quad (4)$$

$$\tau(R) = \log e^{[(\frac{3.91}{V}) * (\frac{\lambda}{550nm})^{-q} * R]} dB \quad (5)$$

Equation (1) is substituted into equation (3) to form equation (4) which is used to determine the atmospheric attenuation of the FSO system in haze conditions.

$$P_{received} = P_{transmit} \frac{A_{receiver}}{(\theta R)^2} e^{-\alpha R} \quad (6)$$

where
$P_{transmit}$ = transmitted power (Watt)
$A_{receiver}$ = area of the receiver (m^2)
θ = beam divergence in radian (rad)
R = distance between transmitter and receiver (m)
α = average atmospheric attenuation coefficient (dB/km)

3. Simulations and Setup Analysis

Two designs of FSO systems are proposed in this paper to study the attenuation effects of haze conditions on the FSO system and the availability of the system in low visibility condition. Figure 1 shows the block diagram of Design 1. It is the basic design of FSO system with a CW laser, a Mach-Zehnder Modulator, an FSO channel, and a PIN photodetector installed. A BER analyzer is installed at the receiver of the system to analyse and determine the BER of the system. Figure 2 shows the block diagram of Design 2. It is the new and improved design of FSO system based on the performance evaluation of Design 1 in the haze conditions. Similar to Design 1, this design has a single transmitter and a single receiver installed. However, five optical channels and an optical amplifier at each end of the channel are included in the design. Both designs are made and simulated under haze conditions to observe and analyse their performance using software called *OptiSys*^{TM} [9].

Design 1

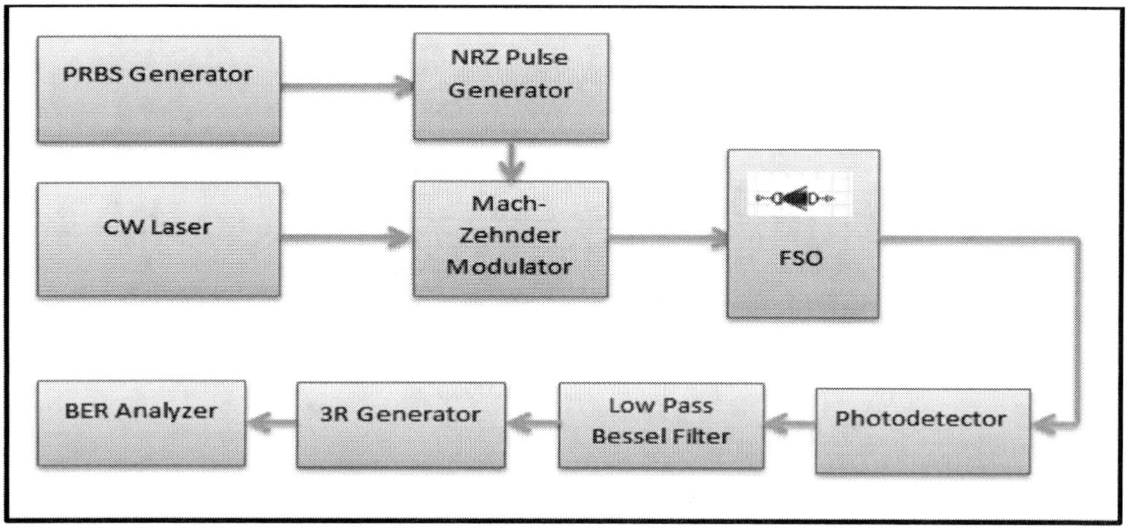

Figure 1. The block diagram of Design 1.

Design 2

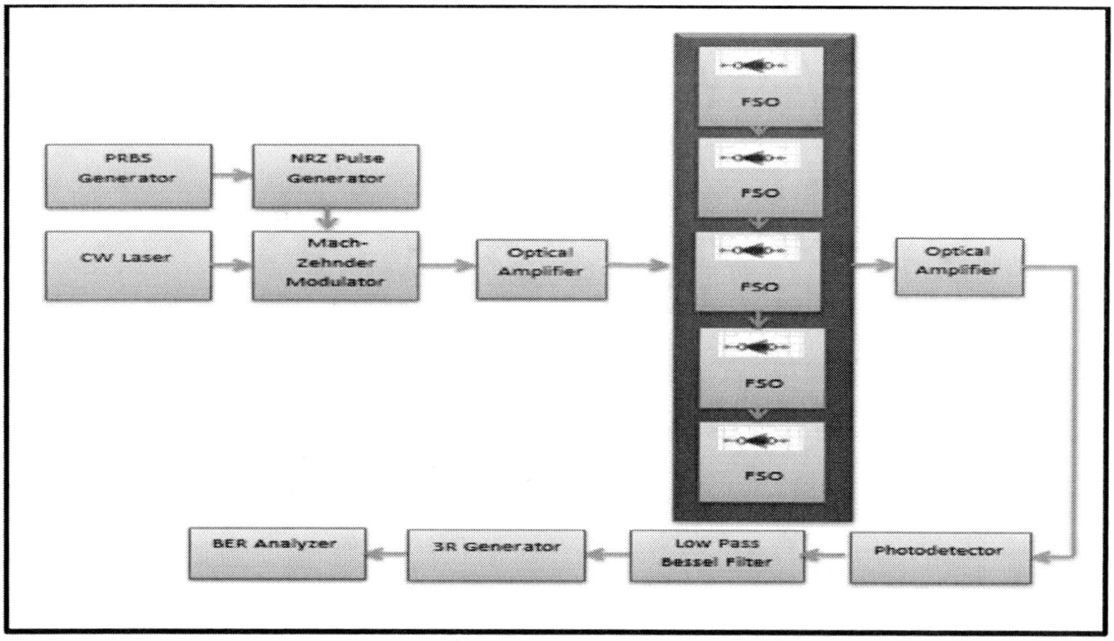

Figure 2. The block diagram of Design 2.

4. Simulation and Setup Analysis

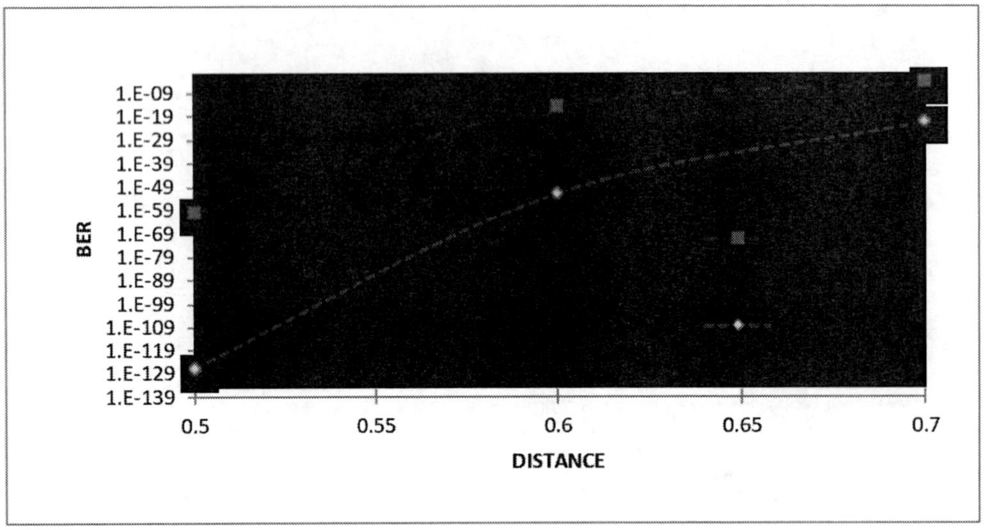

Figure 3. The graph of bit error rate versus the distance.

Figure 3 shows the graph of bit error rate (BER) versus link distance in haze condition. The distance investigated in this research ranges from 0.5 km to 0.7 km. At 0.5 km distance, the BER value of both designs achieved their minimum point which is 10^{-60} for Design 1, and 10^{-127} for Design 2. As the distance increases, the BER value for each design increases too. It is due to the haze attenuation along the path. The longer the distance between the transmitter and the receiver, the greater the attenuation caused by haze particles, and thus the higher the BER. Therefore, the performance of FSO system becomes poorer.

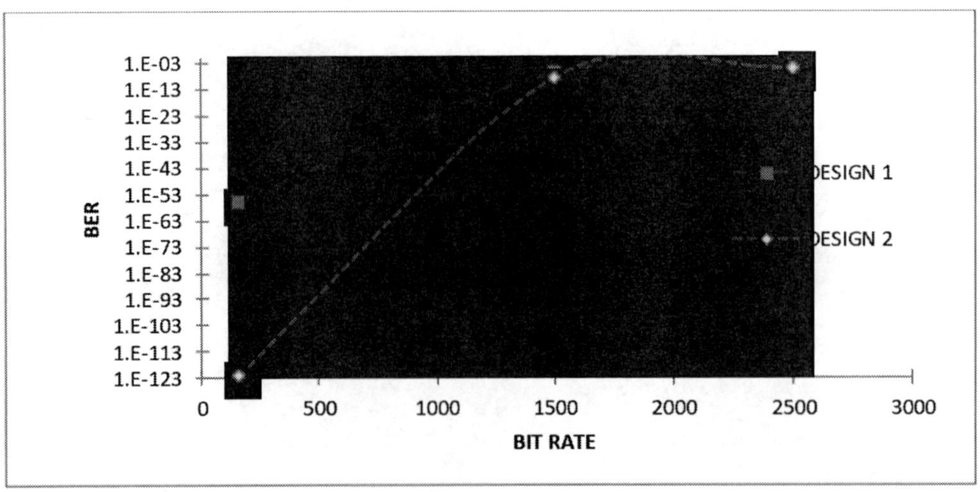

Figure 4. The graph of bit error rate versus the bit rate.

Figure 4 shows the bit error rate (BER) versus bit rate in haze condition. From the figure, it can be observed that Design 2 shows promising capabilities in haze condition. As the bit rate used is increased, more bits are transmitted through the transmission channel. This caused the increment of error bits. From the graph, it is clearly shown that Design 2 has the lower BER value. At bit rate 155 Mbps, the BER of Design 2 reaches until 10^{-123} compared to BER of Design 1 which is 10^{-56} respectively. Obviously, Design 2 is the best design for haze condition as it could maintain better BER value compared to Design 1.

5. Conclusions

Optical communication system especially FSO system is the first choice in transmitting data at high speed since it uses light for signal transmission. However, the high atmospheric attenuation caused on the FSO link by the weather conditions makes the system performance to be poor. In this research, Design 2 is proposed for the system to efficiently transmit the optical signal eventhough in dense haze conditions. The attenuation effects on the link performance is greatly reduced as the minimum BER achieved is 10^{-127}.

References

[1] Alkholidi, A. G., & Altowij, K. S. (2014). Free space optical communications-Theory and practices. InTech

[2] Zhu, X., & Kahn, J. M. (2002). Free-space optical communication through atmospheric turbulence channels. IEEE Transactions on communications, 50(8), 1293-1300.

[3] Kim, I. I., McArthur, B., & Korevaar, E. J. (2001, February). Comparison of laser beam propagation at 785 nm and 1550 nm in fog and haze for optical wireless communications. In Information Technologies 2000 (pp. 26-37). International Society for Optics and Photonics.

[4] Alnajjar, S. H., Noori, A. A., & Moosa, A. A. (2017). Enhancement of FSO communications links under complex environment. Photonic Sensors, 7(2), 113-122.

[5] Anis, A. A., Rahman, A. K., Rashidi, C. B. M., & Aljunid, S. A. (2016, August). Investigation of haze effects via utilizing different wavelength on free space optical communication. In *Electronic Design (ICED), 2016 3rd International Conference on* (pp. 358-360). IEEE.

[6] Anis, A. A., Rahman, A. K., Rashidi, C. B. M., & Aljunid, S. A. (2016, August). Link budget analysis for free space optical (FSO) communication under haze condition with adverse wavelength. *International Conference on Electronic Design* (pp. 354-357). IEEE.

[7] Rashidi, C.B.M., Anuar, M.S., Aljunid, S.A.(2010). Study of direct detection technique for zero cross correlation code in OCDMA, International Conference on Computer and Communication Engineering. 10.1109/ICCCE.2010.5556830, IEEE.

[8] Killinger, D. (2002). Free space optics for laser communication through the air. Optics and Photonics News, 13(10), 36-42.

[9] Rashidi, C.B.M., Aljunid, S.A. Ghani. F., Anuar, M.S, Fadhil, H. A. (2012). Code length optimization using Flexible Cross Correlation (FCC) code in OCDMA networks, 3rd International Conference on Photonics. 10.1109/ICP.2012.6379828. Pages 355-359.